GEOMETRIC REGULAR POLYTOPES

Regular polytopes and their symmetry have a long history stretching back two and a half millennia, to the classical regular polygons and polyhedra. Much of modern research focuses on abstract regular polytopes, but significant recent developments have been made on the geometric side, including the exploration of new topics such as realizations and rigidity, which offer a different way of understanding the geometric and combinatorial symmetry of polytopes.

This is the first comprehensive account of the modern geometric theory, and includes a wide range of applications, along with new techniques. While the author explores the subject in depth, his elementary approach to traditional areas such as finite reflexion groups makes this book suitable for beginning graduate students as well as more experienced researchers.

Encyclopedia of Mathematics and Its Applications

This series is devoted to significant topics or themes that have wide application in mathematics or mathematical science and for which a detailed development of the abstract theory is less important than a thorough and concrete exploration of the implications and applications.

Books in the **Encyclopedia of Mathematics and Its Applications** cover their subjects comprehensively. Less important results may be summarized as exercises at the ends of chapters. For technicalities, readers can be referred to the bibliography, which is expected to be comprehensive. As a result, volumes are encyclopedic references or manageable guides to major subjects.

All the titles listed below can be obtained from good booksellers or from Cambridge University Press. For a complete series listing visit
www.cambridge.org/mathematics.

124 F. W. King *Hilbert Transforms I*
125 F. W. King *Hilbert Transforms II*
126 O. Calin and D.-C. Chang *Sub-Riemannian Geometry*
127 M. Grabisch *et al. Aggregation Functions*
128 L. W. Beineke and R. J. Wilson (eds.) with J. L. Gross and T. W. Tucker *Topics in Topological Graph Theory*
129 J. Berstel, D. Perrin and C. Reutenauer *Codes and Automata*
130 T. G. Faticoni *Modules over Endomorphism Rings*
131 H. Morimoto *Stochastic Control and Mathematical Modeling*
132 G. Schmidt *Relational Mathematics*
133 P. Kornerup and D. W. Matula *Finite Precision Number Systems and Arithmetic*
134 Y. Crama and P. L. Hammer (eds.) *Boolean Models and Methods in Mathematics, Computer Science, and Engineering*
135 V. Berthé and M. Rigo (eds.) *Combinatorics, Automata and Number Theory*
136 A. Kristály, V. D. Rădulescu and C. Varga *Variational Principles in Mathematical Physics, Geometry, and Economics*
137 J. Berstel and C. Reutenauer *Noncommutative Rational Series with Applications*
138 B. Courcelle and J. Engelfriet *Graph Structure and Monadic Second-Order Logic*
139 M. Fiedler *Matrices and Graphs in Geometry*
140 N. Vakil *Real Analysis through Modern Infinitesimals*
141 R. B. Paris *Hadamard Expansions and Hyperasymptotic Evaluation*
142 Y. Crama and P. L. Hammer *Boolean Functions*
143 A. Arapostathis, V. S. Borkar and M. K. Ghosh *Ergodic Control of Diffusion Processes*
144 N. Caspard, B. Leclerc and B. Monjardet *Finite Ordered Sets*
145 D. Z. Arov and H. Dym *Bitangential Direct and Inverse Problems for Systems of Integral and Differential Equations*
146 G. Dassios *Ellipsoidal Harmonics*
147 L. W. Beineke and R. J. Wilson (eds.) with O. R. Oellermann *Topics in Structural Graph Theory*
148 L. Berlyand, A. G. Kolpakov and A. Novikov *Introduction to the Network Approximation Method for Materials Modeling*
149 M. Baake and U. Grimm *Aperiodic Order I: A Mathematical Invitation*
150 J. Borwein *et al. Lattice Sums Then and Now*
151 R. Schneider *Convex Bodies: The Brunn–Minkowski Theory (Second Edition)*
152 G. Da Prato and J. Zabczyk *Stochastic Equations in Infinite Dimensions (Second Edition)*
153 D. Hofmann, G. J. Seal and W. Tholen (eds.) *Monoidal Topology*
154 M. Cabrera García and Á. Rodríguez Palacios *Non-Associative Normed Algebras I: The Vidav–Palmer and Gelfand–Naimark Theorems*
155 C. F. Dunkl and Y. Xu *Orthogonal Polynomials of Several Variables (Second Edition)*
156 L. W. Beineke and R. J. Wilson (eds.) with B. Toft *Topics in Chromatic Graph Theory*
157 T. Mora *Solving Polynomial Equation Systems III: Algebraic Solving*
158 T. Mora *Solving Polynomial Equation Systems IV: Buchberger Theory and Beyond*
159 V. Berthé and M. Rigo (eds.) *Combinatorics, Words and Symbolic Dynamics*
160. B. Rubin *Introduction to Radon Transforms: With Elements of Fractional Calculus and Harmonic Analysis*
161 M. Ghergu and S. D. Taliaferro *Isolated Singularities in Partial Differential Inequalities*
162 G. Molica Bisci, V. D. Radulescu and R. Servadei *Variational Methods for Nonlocal Fractional Problems*
163 S. Wagon *The Banach–Tarski Paradox (Second Edition)*
164 K. Broughan *Equivalents of the Riemann Hypothesis I: Arithmetic Equivalents*
165 K. Broughan *Equivalents of the Riemann Hypothesis II: Analytic Equivalents*
166 M. Baake and U. Grimm (eds.) *Aperiodic Order II: Crystallography and Almost Periodicity*
167 M. Cabrera García and Á. Rodríguez Palacios *Non-Associative Normed Algebras II: Representation Theory and the Zel'manov Approach*
168 A. Yu. Khrennikov, S. V. Kozyrev and W. A. Zúñiga-Galindo *Ultrametric Pseudodifferential Equations and Applications*
169 S. R. Finch *Mathematical Constants II*
170 J. Krajíček *Proof Complexity*
171 D. Bulacu, S. Caenepeel, F. Panaite and F. Van Oystaeyen *Quasi-Hopf Algebras*
172 P. McMullen *Geometric Regular Polytopes*
173 M. Aguiar and S. Mahajan *Bimonoids for Hyperplane Arrangements*

Encyclopedia of Mathematics and its Applications

Geometric Regular Polytopes

PETER MCMULLEN
University College London

CAMBRIDGE
UNIVERSITY PRESS

CAMBRIDGE
UNIVERSITY PRESS

University Printing House, Cambridge CB2 8BS, United Kingdom

One Liberty Plaza, 20th Floor, New York, NY 10006, USA

477 Williamstown Road, Port Melbourne, VIC 3207, Australia

314–321, 3rd Floor, Plot 3, Splendor Forum, Jasola District Centre, New Delhi – 110025, India

79 Anson Road, #06–04/06, Singapore 079906

Cambridge University Press is part of the University of Cambridge.

It furthers the University's mission by disseminating knowledge in the pursuit of education, learning, and research at the highest international levels of excellence.

www.cambridge.org
Information on this title: www.cambridge.org/9781108489584
DOI: 10.1017/9781108778992

First published 2020

Printed in the United Kingdom by TJ International Ltd. Padstow Cornwall

A catalogue record for this publication is available from the British Library.

ISBN 978-1-108-48958-4 Hardback

Contents

Foreword

As might have been hoped, if not actually expected, since the publication of the monograph *Abstract Regular Polytopes* [99] by McMullen and Schulte there have been considerable advances in the subject. However, despite the obvious appeal of the geometric side of the theory, rather little space was devoted to it there. Indeed, the only systematic classification problems addressed in the book were those of the classical regular polytopes and honeycombs of Coxeter's seminal work *Regular Polytopes* [27], and what occurred in at most three dimensions. Otherwise, only sporadic examples were considered, such as ones illustrating aspects of realization theory or of regular polytopes whose universality is based on geometric constructions. In a sense, then, this book attempts to redress what might be perceived as an imbalance.

As we have just said, up to the publication of [99], the only systematic dimension-by-dimension investigation of regular polytopes had concentrated on small dimensions (at most three). In a sequence of subsequent papers [82, 83, 84, 86], McMullen has extended such classifications in several ways, by restricting attention to regular polytopes of full or nearly full rank (terms which will be defined in the text – the paper [98] by McMullen and Schulte can be regarded as the first of the sequence). To give examples, the subjects of [27] are the classical regular polytopes of full rank, though not all the regular polytopes and apeirotopes (that is, infinite polytopes) of full rank are classical.

It has therefore seemed appropriate to consolidate this line of research, and – with some rethinking of the basics – attempt to present it in a coherent way. Our main agenda are simple: to classify the regular polytopes and apeirotopes of full or nearly full rank. However, we shall also look at other interesting families that arose out of our investigations.

Unlike all Gaul, the book will be divided into four parts. Part I will cover those aspects of the abstract theory which we need subsequently, while Parts II and III will treat the cases of full and nearly full rank, respectively. Part IV contains somewhat of a miscellany. As we have indicated, it is our intention to recast the previously published material where we feel this to be necessary; we also take the opportunity to correct a number of mistakes in earlier treatments and repair some omissions.

In more detail, in Part I we describe the background to the theory of regular polytopes, both abstract and geometric. Chapter 1 sets the scene, by looking at some geometric foundations; in particular, it treats the discrete hyperplane reflexion groups, which play a central rôle in the whole book. In Chapter 2, we outline the abstract theory; this is a brief epitome of those parts of [99, Chapters 2–4] which have not been treated in Chapter 1. The geometric aspect is introduced in Chapters 3 and 4 through the theory of realizations; this material has been completely reworked from the earlier account in [99], and incorporates that from the recent new [85, 88], as well as some corrections of the theory by Ladisch [67]. Indeed, some concepts which were used in the earlier accounts have now disappeared, since they have proved redundant; conversely, more recent ideas – including some not previously in print – have resulted in a theory with considerable power. In Chapter 5 we consider various abstract and geometric operations and constructions on regular polytopes and apeirotopes. Finally, Chapter 6 introduces from [89] another recent concept, that of rigidity of regular polytopes; the basic question here asks to what extent the geometry of a regular polytope is determined by restricted geometric data.

There are two chapters in Part II. Chapter 7 successively covers the classical regular polytopes – including the star-polytopes – and apeirotopes, including an approach to the enumeration which is subsequent to that by Coxeter in [27]. One feature is the complete description in Section 7K of the realization domain of the 120-cell, which expands on the treatment in [92]. To a considerable extent (for the 4-dimensional polytopes) we follow the treatment of Du Val in the use of quaternions. The non-classical examples of full rank are covered in Chapter 8; a core feature of the classification is a restriction on the dimensions of the mirrors of their generating reflexions established in Section 4B.

Part III on the cases of nearly full rank is the longest, with five chapters. Attention must be drawn to the fact that these contain many corrections and additions to the papers on which they are based. Chapter 9 begins by treating the cases where blending is involved, and then classifying the various families which occur in each dimension. The remaining chapters deal with the pure polytopes and apeirotopes. Chapter 10 looks at the 3-dimensional apeirohedra; the material is extracted from [98], which was reproduced in [99, Section 7F], but we have augmented it by a discussion of the rigidity of the apeirohedra. (Indeed, wherever appropriate we say whether polytopes under consideration are rigid.) Chapter 11 then deals with the 4-dimensional polyhedra, expanding [83] a little, and with an excursion into an interesting family related to quasi-regular 3-dimensional polyhedra. The 4-dimensional apeirotopes of rank 4 are described in Chapter 12, which has been the most substantially reworked from the original paper [84]. Finally, following [86], Chapter 13 treats the regular polytopes and apeirotopes of nearly full rank in all higher dimensions.

The material in Part IV is hitherto unpublished. It illustrates further aspects of the foregoing theory, particularly those of realizations. In Chapter 14, we look more closely at the Gosset-Elte polytopes which are intimately related to several families of polytopes of nearly full rank; this provides a range of ways of applying realization theory, as well as being (we think) of intrinsic interest. The next Chapter 15 describes the realization domains of some of the locally toroidal polytopes that featured prominently in [99]; it also introduces a new class of universal such polytopes. Chapter 16 treats a family of 4-polytopes that displays some remarkable parallels with the pentagonal 4-polytopes of Chapter 7, while Chapter 17 deals with a family of 5-polytopes that combines both families of 4-polytopes as facets or vertex-figures. In part, as well, these two chapters provide further examples of the techniques used to determine realizations.

One feature of our treatment deserves special mention. It has always been our philosophy that one should try to approach any aspect of the subject using techniques which are as elementary as possible. This shows itself in various ways; for instance, in Section 1E we classify the finite and euclidean reflexion groups without appealing in any serious way to the theory of quadratic forms, as well as reproducing from [99, Section 3E] the earlier calculations of the finite orders using a convexity argument. Similarly, in Section 7E we find the Petrie polygons of the regular 600-cell and its relatives without solving trigonometric equations, while in Section 7G we determine the abstract groups of the 4-dimensional regular star-polytopes without going through the laborious changes of generators of [81] or [99, Theorem 7D16]. Even more striking, perhaps, though there are clear parallels between representations of finite groups and realizations of regular polytopes, particularly in orthogonality relations, we make almost no appeal to representation theory in our treatment of realizations.

It is inevitable that there is some overlap with [99]. For completeness of the exposition we need to cover a certain part of the introductory material, and the two accounts touch at various other places. However, even when we are not looking at things in a different way, for instance by adopting a more intuitive (but equivalent) definition of abstract polytopes, as we have indicated we have often reworked what we have written in various papers, as well as making some necessary corrections of them. Moreover, we have occasionally added background material that was omitted in [99]; for example, in Section 1G we give proofs of the Brianchon–Gram and Somerville theorems that lie behind the calculations of the orders of the finite Coxeter groups in Section 1H.

Several items in the bibilography are included for further reading, and are not cited in the text.

We have been encouraged to produce this book by friends and colleagues, of whom particular mention should be made to Asia Ivić Weiss, Barry Monson, Egon Schulte and Marston Conder. Apologies are also due to them for the time the book has taken to appear. Thanks are also due to the staff of Cambridge University Press, for their help and forbearance in getting this book into print.

I

Regular Polytopes

1

Euclidean Space

The main purpose of this chapter is to discuss groups generated by reflexions, concentrating here on the finite and discrete ones in euclidean spaces. There are several reasons for this. One rather important one is that this topic does not depend on anything that follows; indeed, to the contrary, we shall constantly appeal to reflexion groups for examples to illustrate the subsequent theory. In fact, until Part IV all but a single family of the regular polytopes described in this monograph have symmetry groups which are closely related to reflexion groups (if they are not reflexion groups themselves, then they are subgroups of them, or are obtained by twisting them with automorphisms). Moreover, it turns out that, except for one family, all finite or discrete affine reflexion groups are the symmetry groups of some regular polytopes or apeirotopes, even those which do not have linear diagrams.

The chapter contains ten sections. There are four preliminary ones, mainly to establish notation and conventions. Section 1A surveys the algebraic properties of euclidean spaces, while Sections 1C and 1D look at their metrical properties; in between, Section 1B covers the main features of convex sets that we shall need to appeal to. In the core Section 1E we classify the finite and discrete infinite reflexion groups in euclidean spaces; the initial part of our treatment is novel. In the next Section 1F we briefly comment on subgroup relationships among these groups. We also need to know the orders of the finite Coxeter groups; we find these by purely elementary geometric methods in Section 1H using angle-sum relations established in the previous Section 1G. The lower-dimensional spaces are somewhat special. For the following section, we need to know what the finite rotation groups in $\mathbb{E}^3$ are; this problem is solved in Section 1J. In 4-dimensional space $\mathbb{E}^4$, quaternions provide an alternative approach to finite orthogonal groups, and are actually needed to describe certain regular polyhedra in that space; what we want is covered in Section 1K.

We should emphasize that, by and large, we will only prove assertions made in this chapter if we need to employ them subsequently. Thus we shall include certain peripheral material as background, but not go into it in any great detail. And, of course, we shall try not to insult the reader by proving too many standard results in algebra and analysis.

1A Algebraic Properties

In this section, we are mainly interested in $\mathbb{E}^d$ as a linear (vector) or affine space; the extra properties induced by the inner product and norm will be discussed in Section 1C. As we said in the preamble to the chapter, a main purpose of this and the next section is to establish notation and conventions.

Linear Spaces

For the moment, therefore, we just consider finite dimensional real linear (or vector) spaces. Indeed, in this section, only the fact that the real numbers $\mathbb{R}$ form a field is material; at this stage, it is not important that $\mathbb{R}$ is ordered. Thus, $\mathbb{X}$, $\mathbb{Y}$, and so on, will be finite dimensional linear spaces over $\mathbb{R}$.

We assume that the reader is familiar with the fundamental algebraic ideas of groups, rings, fields and linear spaces. In particular, in the latter context, the notions of linear combination, linear dependence and independence, linear subspace and (linear) basis will be taken for granted. The only point we wish to make here is notational: the *linear hull* $\operatorname{lin} X$ of $X \subseteq \mathbb{X}$ is

- the set of linear combinations of vectors in X,
- the intersection of the linear subspaces of $\mathbb{X}$ which contain X.

By definition, the zero vector (or origin) $o \in \operatorname{lin} X$ always.

The basic operations of a linear space extend to subsets. Thus, for $X, Y \subseteq \mathbb{X}$ and $\lambda \in \mathbb{R}$, we define the *sum* $X + Y$ and *scalar multiple* λX by

1A1 $$X + Y := \{x + y \mid x \in X,\ y \in Y\},$$

1A2 $$\lambda X := \{\lambda x \mid x \in X\}.$$

In particular, we write $X + t := X + \{t\}$ for the *translate* of X by $t \in \mathbb{X}$; then t is called the corresponding *translation vector*.

Affine Properties

In some contexts, though, it is inconvenient to have the zero vector o playing a special rôle, and so it is preferable to regard $\mathbb{X}$ as an affine space. The *line* xy through $x, y \in \mathbb{X}$ is

1A3 $$xy := \{(1 - \lambda)x + \lambda y \mid \lambda \in \mathbb{R}\} \subseteq A;$$

an *affine subspace* A in $\mathbb{X}$ is determined by the fact that, if $x, y \in A$ then $xy \subseteq A$ (see the notes at the end of the section). Actually, the definition allows the empty set $\emptyset$ and point-sets to be affine subspaces as well (note that $xx = \{x\}$); contrast the former with the fact that *linear* subspaces always contain o, and so are non-empty. An easy exercise shows the following. Define the *affine hull* $\operatorname{aff} X$ of a subset $X \subseteq \mathbb{X}$ by

1A4 $$\operatorname{aff} X := \bigcap\{A \subseteq \mathbb{X} \mid A \text{ an affine subspace, and } X \subseteq A\}.$$

Then aff X consists of all *affine* combinations

1A5 $$\lambda_0 x_0 + \cdots + \lambda_k x_k, \qquad \lambda_0 + \cdots + \lambda_k = 1,$$

of points $x_0, \ldots, x_k \in X$. We also say that A *spans* aff A *affinely*. Moreover, we have

1A6 Proposition *A non-empty affine subspace A is a translate $A = L + t$ of a (unique) linear subspace L.*

Proof. Indeed, if $t \in A$ is any point, then it is straightforward to show that $L := A - t$ is a linear subspace. Observe that, if $t' \in A$ also, then $t' - t \in L$, so that

$$A - t' = (A - t) - (t' - t) = L - (t' - t) = L;$$

the uniqueness of L is a consequence. □

We say that two affine subspaces A_1, A_2 are *parallel* if A_2 is a translate of A_1; hence, parallel affine subspaces are translates of the same linear subspace.

The concepts of affine dependence, independence and basis are the natural extensions of the linear notions. Thus an *affinely independent* set B is such that no one of its points is an affine combination of the others. Equivalently, $B = \{b_0, \ldots, b_k\}$ is affinely independent if $\xi_0 b_0 + \cdots + \xi_k b_k = o$ for $\xi_0, \ldots, \xi_k \in \mathbb{R}$ such that $\xi_0 + \cdots + \xi_k = 0$ implies that $\xi_0 = \cdots = \xi_k = 0$. An *affine basis* of $\mathbb{X}$ is an affinely independent set $B \subseteq \mathbb{X}$ which spans $\mathbb{X}$ affinely.

1A7 Proposition *An affine basis of a d-dimensional space $\mathbb{X}$ consists of $d+1$ points.*

Proof. It is easily shown that $\{b_0, \ldots, b_d\}$ is affinely independent if and only if $\{b_1 - b_0, \ldots, b_d - b_0\}$ is linearly independent; the claim then follows. □

The obvious definition of the *dimension* dim A of an affine subspace A is $\dim A := \dim L$, if $A = L + t$ for some linear subspace L. Thus an affine basis of A has $\dim A + 1$ points; compare Proposition 1A7. For the empty set, the natural definition is thus $\dim \emptyset := -1$. Often also useful is the notion of *codimension* $\operatorname{codim} A := \dim \mathbb{X} - \dim A$. In particular, an affine subspace of codimension 1 is called a *hyperplane*.

1A8 Remark In some contexts, like those of realizations (see, for example, Section 3L) we find it useful to work in linear spaces over ordered fields other than $\mathbb{R}$. Of particular interest are the *rational numbers* $\mathbb{Q}$ and (real) *algebraic numbers* $\mathbb{A}$.

Mappings

We next look at mappings. Again, we assume that the reader is familiar with linear mappings; however, we wish to recall some terminology and introduce some notation.

If $\Phi\colon \mathbb{X} \to \mathbb{Y}$ is a linear mapping, then we denote by $\operatorname{im}\Phi$ its *image space* and by $\ker\Phi$ its *kernel*; their dimensions are the *rank* $\operatorname{rank}\Phi$ and *nullity* $\operatorname{null}\Phi$, respectively. Recall that the latter are related by

$$\operatorname{rank}\Phi + \operatorname{null}\Phi = \dim\mathbb{X}.$$

For fixed $\mathbb{X}$ and $\mathbb{Y}$, the family of linear mappings $\Phi\colon \mathbb{X} \to \mathbb{Y}$ forms, in a natural way, a linear space $\operatorname{Hom}(\mathbb{X}, \mathbb{Y})$ of dimension $\dim\mathbb{X}\dim\mathbb{Y}$.

A linear mapping $u\colon \mathbb{X} \to \mathbb{R}$ is called a *linear functional*; in this special case, we have the *dual space* $\mathbb{X}^* := \operatorname{Hom}(\mathbb{X}, \mathbb{R})$. There is a natural pairing $\langle\cdot,\cdot\rangle$ on $\mathbb{X} \times \mathbb{X}^*$, so that we write the image of $x \in \mathbb{X}$ under $u \in \mathbb{X}^*$ as $\langle x, u\rangle = \langle u, x\rangle$, thus emphasizing the underlying symmetry. Corresponding to a basis $E = (e_1, \ldots, e_d)$ of $\mathbb{X}$ is a *dual basis* $E^* = (e_1^*, \ldots, e_d^*)$ of $\mathbb{X}^*$, which satisfies

$$\langle e_j, e_k^*\rangle = \delta_{jk} := \begin{cases} 1, & \text{if } j = k, \\ 0, & \text{if } j \neq k, \end{cases}$$

the Kronecker delta.

If $x \in \mathbb{X}$, then writing $x = \sum_{j=1}^{d} \xi_j e_j$ and applying e_k^* shows that $\xi_k = \langle x, e_k^*\rangle$. In other words, if $x \in \mathbb{X}$, then

1A9
$$x = \sum_{j=1}^{d} \langle x, e_j^*\rangle e_j.$$

Another familiar fact is

1A10 Proposition *A hyperplane H in $\mathbb{X}$ can be represented as $H(u, \beta) := \{x \in \mathbb{X} \mid \langle x, u\rangle = \beta\}$, for some non-zero $u \in \mathbb{X}^*$ and $\beta \in \mathbb{R}$.*

Proof. By Proposition 1A6, H is a translate $H = H_0 + t$ of a linear hyperplane H_0 of $\mathbb{X}$. If $\dim\mathbb{X} = d$, choose any (linear) basis $\{a_1, \ldots, a_{d-1}\}$ of H_0, and extend to a basis $\{a_1, \ldots, a_d\}$ of $\mathbb{X}$. If $\{a_1^*, \ldots, a_d^*\}$ is the dual basis of $\mathbb{X}^*$, then we define $u := a_d^*$, so that $H_0 = \{x \in \mathbb{X} \mid \langle x, u\rangle = 0\}$. It follows at once that $H = H(u, \beta)$, with $\beta := \langle t, u\rangle$, which is as asserted. □

For now, we only need the idea of an *affine mapping* $\Phi\colon \mathbb{X} \to \mathbb{Y}$, with $\mathbb{X}, \mathbb{Y}$ finite dimensional real linear spaces: this is such that

1A11
$$((1-\lambda)x + \lambda y)\Phi = (1-\lambda)x\Phi + \lambda y\Phi$$

for all $x, y \in \mathbb{X}$ and $\lambda \in \mathbb{R}$. A straightforward inductive argument shows that affine mappings preserve arbitrary affine combinations. Moreover, we actually have

1A12 Proposition *If $\Phi\colon \mathbb{X} \to \mathbb{Y}$ is an affine mapping, then there is a $t \in \mathbb{Y}$ and a linear mapping $\Psi\colon \mathbb{X} \to \mathbb{Y}$ such that $x\Phi = x\Psi + t$ for all $x \in \mathbb{X}$.*

Proof. Define $t := o\Phi$ and Ψ by $x\Psi := x\Phi - t$. We leave to the reader the easy exercise of completing the proof (that is, showing that Ψ is linear). □

An invertible affine mapping $\Phi\colon \mathbb{X} \to \mathbb{X}$ is also called an *affinity*; clearly Φ is invertible just when the associated linear mapping Ψ of Proposition 1A12 is invertible. As a particular affinity, we have the *translation* $x \mapsto x+t$, with $t \in \mathbb{X}$ as before the corresponding translation vector.

Matrices

As is well known, a linear mapping $\Phi\colon \mathbb{X} \to \mathbb{Y}$ can be represented by a matrix with respect to a choice of bases of $\mathbb{X}$ and $\mathbb{Y}$; this matrix will be invertible just when Φ is invertible.

We shall often associate an ordered set of vectors $(a_1, \ldots, a_k)$ in $\mathbb{R}^m$ with the $k \times m$ matrix A whose rows are the a_i. If B is an $m \times n$ matrix, then it is sometimes useful to think of the entries of the product AB as $\langle a_i, b_j \rangle$, where $b_1, \ldots, b_m$ are now the columns of B, regarded as vectors in the dual space $(\mathbb{R}^m)^*$.

Recall that the *trace* $\operatorname{tr} A$ of an $m \times n$ matrix $A = (\alpha_{ij})$ is

1A13
$$\operatorname{tr} A := \sum_j \alpha_{jj},$$

the range of summation being $1 \leqslant j \leqslant \min\{m, n\}$. Thus $\operatorname{tr} A = \operatorname{tr} A^\mathsf{T}$, with $A^\mathsf{T} = (\beta_{ij})$ the *transpose* of A, so that $\beta_{ij} = \alpha_{ji}$ for each i, j. Moreover, if B is an $n \times m$ matrix, then we have

1A14
$$\operatorname{tr}(AB) = \operatorname{tr}(BA),$$

as is easy to see.

If the linear mapping $\Phi\colon \mathbb{X} \to \mathbb{Y}$ is represented by the matrix A with respect to given bases of $\mathbb{X}$ and $\mathbb{Y}$, then the dual mapping $\Phi^*\colon \mathbb{Y}^* \to \mathbb{X}^*$ is represented by the transpose matrix A^T with respect to the dual bases of $\mathbb{Y}^*$ and $\mathbb{X}^*$.

Groups

A finite group $\mathbf{G}$ of affinities on $\mathbb{X}$ with order $|\mathbf{G}| := \operatorname{card} \mathbf{G}$ has a fixed point, namely, the centroid

$$c := \frac{1}{|\mathbf{G}|} \sum_{\Phi \in \mathbf{G}} x\Phi$$

of the images of an arbitrary point $x \in \mathbb{X}$ under $\mathbf{G}$ (there may be more than one such point c). Conjugating $\mathbf{G}$ by a translation which takes c to the origin shows that we lose no generality in assuming that $\mathbf{G}$ is a subgroup of the *general linear group* $\mathbf{GL}(\mathbb{X}) := \operatorname{Hom}(\mathbb{X}, \mathbb{X})$ of invertible *linear* mappings on $\mathbb{X}$.

We say that $\mathbb{L} \leqslant \mathbb{X}$ is an *invariant subspace* of $\mathbf{G}$ if $x\Phi \in \mathbb{L}$ for all $x \in \mathbb{L}$ and $\Phi \in \mathbf{G}$. We call $\mathbf{G}$ *irreducible* if its only invariant subspaces are $\{o\}$ and $\mathbb{X}$ itself.

We call two subgroups $\mathbf{G}, \mathbf{H} \leqslant \mathbf{GL}(\mathbb{X})$ of linear mappings *linearly equivalent* if they are conjugate under some $\Theta \in \mathbf{GL}(\mathbb{X})$, so that $\mathbf{H} = \Theta^{-1}\mathbf{G}\Theta$. Observe that, if $\mathbf{G}$ is irreducible, then so is $\mathbf{H}$.

Direct Sums

Linear spaces $\mathbb{X}$ and $\mathbb{Y}$ can be combined in two ways (for our purposes). First, we have the ordinary *direct sum* or *cartesian product* $\mathbb{X} \oplus \mathbb{Y}$. The usual way of expressing a vector $z \in \mathbb{X} \oplus \mathbb{Y}$ is as $z = (x, y)$; this is particularly appropriate when x and y are coordinate vectors (with respect to chosen bases of the two spaces). We thus have $\dim(\mathbb{X} \oplus \mathbb{Y}) = \dim \mathbb{X} + \dim \mathbb{Y}$.

1A15 Remark It is frequently the case that we express a given linear space as an *internal* direct sum of subspaces $\mathbb{L}_1, \mathbb{L}_2, \ldots$, and so write

$$\mathbb{X} = \mathbb{L}_1 \oplus \mathbb{L}_2 \oplus \cdots .$$

Rings and Algebras

When we come to realizations in Chapter 3, we shall want to phrase things in terms of direct sums of rings or, rather, of algebras. An *algebra* $\mathbb{K}$ over $\mathbb{R}$ (this is all we need) is a real vector space for addition and scalar multiplication; further, it has an associative multiplication which distributes over addition and scalar multiplication, and so satisfies

$$\begin{aligned} a(b+c) &= ab + ac, \\ (a+b)c &= ac + bc, \\ (\lambda a)b = a(\lambda b) &= \lambda(ab), \end{aligned}$$

for all $a, b, c \in \mathbb{K}$ and $\lambda \in \mathbb{R}$. The *direct sum* $\mathbb{K}_1 \oplus \mathbb{K}_2$ of two algebras $\mathbb{K}_1$ and $\mathbb{K}_2$ then just satisfies the algebra properties in each coordinate separately, so that the basic operations are

$$\begin{aligned} (a_1, a_2) + (b_1, b_2) &= (a_1 + b_1, a_2 + b_2), \\ \lambda(a_1, a_2) &= (\lambda a_1, \lambda a_2), \\ (a_1, a_2)(b_1, b_2) &= (a_1 b_1, a_2 b_2). \end{aligned}$$

For further properties of direct sums of rings, see for example [4, Chapter 13] or [64, Chapter 1].

Tensor Products

The other combination is the *tensor product* $\mathbb{X} \otimes \mathbb{Y}$, an element of which is called a *tensor*. Formally, this is the *universal* linear space $\mathbb{W}$ for mappings Φ on $\mathbb{X} \times \mathbb{Y}$ that are *bilinear*, in that

1A16 $$(\lambda x + \mu z, y)\Phi = \lambda(x, y)\Phi + \mu(z, y)\Phi,$$

with symmetric expressions for the second term, and which is such that any bilinear mapping Φ on $\mathbb{X} \times \mathbb{Y}$ induces a linear mapping on $\mathbb{W}$. We need to know two things about the tensor product. First, $\dim(\mathbb{X} \otimes \mathbb{Y}) = \dim \mathbb{X} \dim \mathbb{Y}$: if

$\{f_1, \ldots, f_d\}$ is a (linear) basis of $\mathbb{X}$ and $\{g_1, \ldots, g_c\}$ is one of $\mathbb{Y}$, then $e_{ij} := f_i \otimes g_j$ gives a basis of $\mathbb{X} \otimes \mathbb{Y}$. Second, linear mappings Φ on $\mathbb{X}$ and Ψ on $\mathbb{Y}$ induce a linear mapping $\Phi \otimes \Psi$ on $\mathbb{X} \otimes \mathbb{Y}$ by

1A17 $$(x \otimes y)(\Phi \otimes \Psi) := (x\Phi) \otimes (y\Psi).$$

In our treatment, we never have to deal with other than a *simple* tensor $x \otimes y$, with $x \in \mathbb{X}$ and $y \in \mathbb{Y}$.

1A18 Remark There is a natural isomorphism $\operatorname{Hom}(\mathbb{X}, \mathbb{Y}) \cong \mathbb{X}^* \otimes \mathbb{Y}$. Observe that, *up to isomorphism*, the tensor product is associative and commutative.

An element φ of the dual space $\mathbb{X}^* \otimes \mathbb{Y}^*$ of $\mathbb{X} \otimes \mathbb{Y}$ is also called a *bilinear form*. If $\mathbb{X} = \mathbb{Y}$ and $\varphi(x, y) = \varphi(y, x)$ for all $x, y \in \mathbb{X}$, then we call the form *symmetric*. The case $x = y$ gives a *quadratic form*. We say that φ is *positive semi-definite* if $\varphi(x, x) \geqslant 0$ for all $x \in \mathbb{E}^d$, and *positive definite* if $\varphi(x, x) > 0$ whenever $x \neq o$. If $\dim \mathbb{X} = d$, then we can write a quadratic form as $\varphi(x, x) = xAx^{\mathsf{T}}$, with now x regarded as a coordinate vector with respect to a chosen basis of $\mathbb{X}$ and $A = (\alpha_{jk})$ a $d \times d$ *symmetric* matrix, meaning that $\alpha_{jk} = \alpha_{kj}$ for all $j, k = 1, \ldots, d$.

When $\mathbb{X} = \mathbb{Y}$, we have refinements of the tensor product. In the space of r-fold tensors $x_1 \otimes \cdots \otimes x_r$, we can make two kinds of identification. First, taking $x_j \otimes x_{j+1} = x_{j+1} \otimes x_j$ (for $j = 1, \ldots, r-1$) gives a *symmetric* tensor. In this case, we can write $x_j \otimes x_{j+1} =: x_j x_{j+1}$ as an ordinary product. Second, setting $x_j \otimes x_{j+1} + x_{j+1} \otimes x_j = o$ gives an *alternating* tensor; here, it is important that we impose identifications on *adjacent* terms in an alternating tensor product $x_1 \wedge \cdots \wedge x_k$, so that $x_{j+1} \wedge x_j = -(x_j \wedge x_{j+1})$. Both notions occur in the theory of realizations, but we shall see that alternating tensors only occasionally play a useful rôle.

A linear mapping Φ on $\mathbb{X}$ induces a linear mapping $\wedge^r \Phi$ on $\wedge^r \mathbb{X}$. If $\dim \mathbb{X} = d$, then $\wedge^d \mathbb{X} \cong \mathbb{R}$, so that, if $\Phi \colon \mathbb{X} \to \mathbb{X}$, then $\wedge^d \Phi \colon \mathbb{R} \to \mathbb{R}$ is just multiplication by a scalar, which is denoted $\det \Phi$ and called the *determinant* of Φ.

In a similar way, a $d \times d$ matrix A induces a linear mapping on $\mathbb{X}$ with respect to a chosen basis $(e_1, \ldots, e_d)$, say. Then $\det A$ is independent of the basis, and is naturally called the determinant of A. We can identify $a_j = e_j A$ with the jth row of A. The determinant shows that there is a natural identification of $(d-1)$-fold alternating tensors with linear functionals on $\mathbb{X}$, namely,

$$\langle x, a_2 \wedge \cdots \wedge a_d \rangle = \det(x, a_2, \ldots, a_d).$$

With the single exception $d = 3$, this is of little general interest here; the exception involves the definition of quaternions in Section 1K.

Notes to Section 1A

1. Our convention, in which we follow Coxeter [27], is to write mappings *after* their arguments. We rarely have compositions of mappings except in the context of

groups; here, it seems to us more natural to compose mappings in the order in which they are applied. Moreover – and this is useful in another way – we can then think of vectors as *rows* rather than columns, which obviates the need for any special conventions in text.
2. Affine subspaces are also known as *flats*. We do not use the term in this context, because later on we encounter a quite different concept of flatness.
3. A hint to Proposition 1A12 is to note that $\lambda x + \mu y = \lambda x + \mu y + (1 - \lambda - \mu)o$, as an affine combination.
4. Observe that, for dual spaces, we have $(\mathbb{X} \otimes \mathbb{Y})^* = \mathbb{X}^* \otimes \mathbb{Y}^*$.
5. Similarly, a linear mapping $\Phi \colon \mathbb{X} \to \mathbb{Y}$ induces a dual linear mapping $\Phi^* : \mathbb{Y}^* \to \mathbb{X}^*$, defined by $\langle x, v\Phi^* \rangle := \langle x\Phi, v \rangle$ for all $x \in \mathbb{X}$ and $v \in \mathbb{Y}^*$. But we make no future use of this or the previous concept.

1B Convexity

From now on, we use the fact that $\mathbb{R}$ is an *ordered* field, and so it should not be surprising that concepts such as positivity come into play. We shall make extensive use of this and the closely related concept of convexity in various places in the book.

Positive Combinations

For the analogue of 'linear', we call $C \subseteq \mathbb{X}$ a (*convex*) *cone* if $\lambda x + \mu y \in C$ whenever $x, y \in C$ and $\lambda, \mu \geqslant 0$. Thus the *positive hull* of $X \subseteq \mathbb{X}$ is

1B1 $$\operatorname{pos} X := \bigcap \{C \subseteq \mathbb{X} \mid C \text{ a cone, and } X \subseteq C\}.$$

Then $\operatorname{pos} X$ consists of all *positive combinations*

1B2 $$\lambda_1 x_1 + \cdots + \lambda_k x_k, \qquad \lambda_1, \ldots, \lambda_k \geqslant 0,$$

of points $x_1, \ldots, x_k \in X$; we also say that $\operatorname{pos} X$ *spans* X *positively*.

As a special case, if X is linearly independent, then we call $K := \operatorname{pos} X$ a *simple* cone. If C is a cone of dimension $\dim C := \dim \operatorname{lin} C = d$, then we refer to C as a *d-cone*.

In a couple of places we shall need an important result from convexity. We state the theorem first as it applies to cones. A *ray* $[oa$ (or *half-line*) is a set of the form

1B3 $$[oa := \{\lambda a \mid \lambda \geqslant 0\},$$

for some $a \in \mathbb{E}^d$. If C is a closed convex cone, then a ray $E \subseteq C$ is called *extreme* if, whenever $a \in E$ and $x, y \in E$ satisfy $a = x + y$, then x and y are scalar multiples of a.

We then have *Carathéodory's Theorem* (see the notes at the end of the section); we refer to (for example) [132, Proposition 1.15] for more details.

1B4 Theorem *If C is a closed convex cone of dimension d, then each point of C is a sum of at most d points on extreme rays of C.*

Sketch of proof. First observe that, by definition of extreme, a general point of C is a positive combination (and hence a sum) of points on extreme rays of C. So, express $a \in C$ as a minimal positive combination, say

$$a = \sum_{i=1}^{k} \lambda_i b_i,$$

with $b_j \in C$ and $\lambda_j > 0$ for $j = 1, \ldots, k$. If $\{b_1, \ldots, b_k\}$ is linearly dependent, then

$$o = \sum_{i=1}^{k} \mu_i b_i$$

for some $\mu_1, \ldots, \mu_k$, not all zero; without loss of generality, we may assume that some $\mu_j > 0$. Defining

$$\nu := \max\{\lambda_j/\mu_j \mid \mu_j > 0\},$$

we then express

$$a = a - \nu o = \sum_{i=1}^{k} (\lambda_i - \nu\mu_i) b_i$$

as a positive combination of fewer points of C. □

It hardly needs saying that, if X positively spans a linear subspace $\mathbb{L}$ of $\mathbb{X}$, then $\mathbb{L} = \operatorname{lin} X$ also. There is an analogous concept of *positive basis*, that is, a set $B \subseteq \mathbb{X}$ which positively spans $\mathbb{X}$, but no proper subset of which also positively spans. The theory of positive bases is much more elaborate than that of linear or affine bases. Indeed, if $\dim \mathbb{X} = d$, then a positive basis B of $\mathbb{X}$ satisfies $d + 1 \leqslant \operatorname{card} B \leqslant 2d$, and the whole range is attainable; we refer the interested reader to [73, 120] for further details. If the positive basis B consists of $d + 1$ vectors, then we call it *minimal*. We make the following observation.

1B5 Proposition *Every subset of d vectors in a minimal positive basis of a d-dimensional real vector space $\mathbb{X}$ is linearly independent.*

Proof. Let $B = \{b_0, \ldots, b_d\}$ be the given minimal positive basis and suppose that $\{b_1, \ldots, b_d\}$, say, is linearly dependent. Now $-b_0 \in \mathbb{X}$, and so we can find $\lambda_0, \ldots, \lambda_d \geqslant 0$ such that

$$-b_0 = \lambda_0 b_0 + \cdots + \lambda_d b_d \implies (1 + \lambda_0) b_0 = -(\lambda_1 b_1 + \cdots + \lambda_d b_d).$$

However, this implies that b_0 lies in the (at most) $(d - 1)$-dimensional subspace spanned by $b_1, \ldots, b_d$, so that $\mathbb{X}$ itself is at most $(d - 1)$-dimensional, contrary to what we were told initially. □

1B6 Remark We can say rather more about the situation of Proposition 1B5. Since a minimal positive basis $B = \{b_0, \ldots, b_d\}$ consists of $d+1$ vectors, it must be linearly dependent. Of course, by what Proposition 1B5 tells us the linear dependence must be unique up to scaling, and must involve all $d+1$ vectors in B. But now we can employ exactly the same argument as above to write

$$(1+\lambda_0)b_0 + \lambda_1 b_1 + \cdots + \lambda_d b_d = o,$$

showing that all the coefficients in the linear dependence can be taken to be non-negative, and therefore actually positive.

In fact, we have a converse result.

1B7 Theorem *Let $B = \{b_0, \ldots, b_k\} \subseteq \mathbb{X}$ be minimally linearly dependent, in that none of its proper subsets is linearly dependent. If $\lambda_0 b_0 + \cdots + \lambda_k b_k = o$ with all λ_j of the same sign, then* $\operatorname{pos} B = \operatorname{lin} B$.

Proof. We can clearly take each $\lambda_j > 0$. If $x \in \operatorname{lin} B$, say $x = \mu_0 b_0 + \cdots + \mu_k b_k$ for some $\mu_0, \ldots, \mu_k \in \mathbb{R}$, then we can find $\alpha \geqslant 0$ such that $\alpha\lambda_j + \mu_j \geqslant 0$ for each $j = 0, \ldots, k$. But now we have

$$x = \alpha o + x = \sum_{i=0}^{k} (\alpha\lambda_i + \mu_i) b_i \in \operatorname{pos} B,$$

as claimed. □

Convex Sets

The other concept along these lines is that of convexity. We call $K \subseteq \mathbb{X}$ *convex* if the (*line*) *segment* $[xy]$

1B8 $$[xy] := \{(1-\lambda)x + \lambda y \mid 0 \leqslant \lambda \leqslant 1\}$$

is contained in K for all $x, y \in K$. The *convex hull* of $X \subseteq \mathbb{X}$ is

1B9 $$\operatorname{conv} X := \bigcap\{K \subseteq \mathbb{X} \mid K \text{ convex, and } X \subseteq K\}.$$

Thus, in exact analogy to what we have just seen, $\operatorname{conv} X$ consists of each *convex combination*

1B10 $$\lambda_0 x_0 + \cdots + \lambda_k x_k, \qquad \lambda_0 + \cdots + \lambda_k = 1, \quad \lambda_0, \ldots, \lambda_k \geqslant 0,$$

of points $x_0, \ldots, x_k \in X$.

1B11 Remark Observe that, according to our definition, a cone is convex.

The *dimension* of a convex set K is that of its affine hull: $\dim K := \dim \operatorname{aff} K$. Recall that the dimension of an affine subspace $A = L + t$, with L a linear subspace and t a translation vector, is $\dim A = \dim L$. The *interior* of K is denoted $\operatorname{int} K$; it is an open convex set. The *relative interior* $\operatorname{relint} K$ is the interior of K relative to $\operatorname{aff} K$, its affine hull. An important fact is that, if $K \neq \emptyset$, then $\operatorname{relint} K \neq \emptyset$.

An *extreme point* a of a closed convex set K is such that, if $a = (1-\lambda)x + \lambda y$ for some $x, y \in K$ and $0 < \lambda < 1$, then $x = a = y$. We next have Carathéodory's Theorem for convex sets; this can be deduced from Theorem 1B4, or refer to [114] or [132] again.

1B12 Theorem *If K is a compact convex set of dimension d, then each point of K is a convex combination of at most $d+1$ extreme points of K.*

Polyhedral Sets and Polytopes

An important – indeed, core – example we need later is a *convex polytope*, namely, the convex hull of a *finite* set. For now, we just define a *k-simplex* to be the convex hull of an affinely independent set of $k+1$ points, but we shall return to convex polytopes in a moment.

The intersection of finitely many closed half-spaces of $\mathbb{X}$ is called a *polyhedral set*. It is clear that polyhedral sets are convex. Indeed, we have

1B13 Proposition *A convex polytope in $\mathbb{X}$ is a polyhedral set. Conversely, a bounded polyhedral set is a convex polytope.*

In saying that a set $X \subseteq \mathbb{X}$ is *bounded* we mean that X is a subset of some simplex. Alternatively, we can employ the order topology, induced by intersections with lines (indeed, this is all the topology we need). We refer the reader to, for example, [56, 132] for proofs and more details about convex polytopes and polyhedral sets.

Associated with a polyhedral set K in $\mathbb{X}$ is its *recession* cone $\operatorname{rec} K$, defined by

1B14 $$\operatorname{rec} K := \{y \in \mathbb{X} \mid x + \lambda y \in K \text{ for all } x \in K \text{ and } \lambda \geqslant 0\}.$$

If $K = \{x \in \mathbb{X} \mid \langle x, u_j \rangle \leqslant \beta_j \text{ for } j = 1, \ldots, n\}$ is non-empty, then

$$\operatorname{rec} K = \{y \in \mathbb{X} \mid \langle y, u_j \rangle \leqslant 0 \text{ for } j = 1, \ldots, n\}.$$

Support and Separation

A (*closed*) *half-space* of $\mathbb{X}$ is a set of the form $H^-(u, \beta) := \{x \in \mathbb{X} \mid \langle x, u \rangle \leqslant \beta\}$, where $u \in \mathbb{X}^* \setminus \{o\}$ and $\beta \in \mathbb{R}$. The vector u is called an *outer normal* to the half-space $H^-(u, \beta)$. Of course, the boundary of $H^-(u, \beta)$ is the hyperplane $H(u, \beta)$.

If the closed convex set K is such that $K \cap H(u, \beta) \neq \emptyset$ and $K \subseteq H^-(u, \beta)$, then we say that $H(u, \beta)$ is a *support* hyperplane to K at each point $a \in K \cap H(u, \beta)$; moreover, $K \cap H(u, \beta)$ is called a *face* of K; see the notes at the end of the section.

While there are several different types of separation in convexity, we shall only need one, namely, *strict* separation. Two subsets A, B of a euclidean space $\mathbb{E}$ are *strictly separated* by a hyperplane H in $\mathbb{E}$ if A is contained in one open half-space bounded by H and B is contained in the other. The only result we require is

1B15 Theorem *If P is a polyhedral set in $\mathbb{E}$ and $x \in \mathbb{E} \setminus P$, then x is strictly separable from P.*

Proof. We can write $P = \bigcap\{H^-(u_j, \beta_j) \mid j = 1, \dots, m\}$ for some closed half-spaces $H^-(u_j, \beta_j)$ of $\mathbb{E}$. Since $x \notin P$, there is at least one k such that $\langle x, u_k \rangle > \beta_k$. If $\beta_k < \gamma < \langle x, u_k \rangle$, then $H := H(u_k, \gamma)$ strictly separates x from P. □

1B16 Remark A more algebraic way of describing separation here is by a linear functional u. This particularly applies to point-sets: $u \in \mathbb{X}^*$ separates $x, y \in \mathbb{X}$ if $\langle x, u \rangle \neq \langle y, u \rangle$.

Polarity

Two further ideas are important. For any set $X \subseteq \mathbb{X}$, define

1B17 $$X^* := \{y \in \mathbb{X}^* \mid \langle x, y \rangle \leqslant 1 \text{ for all } x \in X\}.$$

Then X^* is a closed convex set with $o \in X^*$; moreover, $X^{**}(= (X^*)^*) = \operatorname{cl}\operatorname{conv} X$, the closure of the convex hull of X. There are two special cases.

First, if P is a convex polytope in $\mathbb{X}$ with $o \in \operatorname{int} P$, then P^* is its *polar*. It can be shown (using notions like those of Proposition 1B13, for example) that P^* is another convex polytope, whose combinatorics are closely related to those of P. More specifically, faces $F \leqslant P$ and $\widetilde{F} \leqslant P^*$ are paired up by the polarity, with $\dim F + \dim \widetilde{F} = \dim \mathbb{X} - 1$. If $a \in \operatorname{relint} F$ is any point, then

$$\widetilde{F} = \{y \in P^* \mid \langle a, y \rangle = 1\}.$$

Second, if K is a cone, then it should be clear that the definition reduces to

$$K^* = \{y \in \mathbb{X}^* \mid \langle x, y \rangle \leqslant 0 \text{ for all } x \in K\}.$$

There are similar relationships between faces of K and K^* to those between faces of polar polytopes, but we do not need to appeal to them.

1B18 Remark It can also be seen that the positive hull of a finite set of vectors is a polyhedral cone; if it is full-dimensional, then, as a polyhedral set, it is a simple cone when the outer normal vectors determining its half-spaces are linearly independent. A polyhedral cone K contains a maximal linear subspace, its *face of apices*.

Euler Characteristic

In order to make sense of the definition of the Euler characteristic, which we shall need to appeal to in Section 1G, we must first prove *Euler's Theorem*.

1B19 Theorem *If P is a convex polytope, then*

$$\sum_{F \leqslant P} (-1)^{\dim F} = 1,$$

where the sum extends over all non-empty faces F of P, including P itself.

Sketch proof. Our proof is an adaptation of that of [56]. That proof concealed a Morse-theory like approach; ours makes this approach more overt (see also, for instance, [132]). We write

1B20 $$f(P;\mathsf{t}) := \sum_{F \leqslant P} \mathsf{t}^{\dim F},$$

where t is an indeterminate, and the latter sum extends, as above, over all the non-empty faces of P. Then

1B21 $$\chi(P) := f(P;-1),$$

which is the sum occurring in the statement of Euler's theorem, is called the *Euler characteristic* of P. (Strictly speaking, $\chi(P)$ is the Euler characteristic of the face-complex of P, but we do not need to know this.)

Our argument proceeds by induction on $\dim P$; the result is trivial if $\dim P = 0$ or 1 (or even 2).

Let v be a general vector for P, by which we mean that $\langle x, v\rangle \neq \langle y, v\rangle$ for all distinct $x, y \in \operatorname{vert} P$. We then *sweep* the variable half-space $H^- = H^-(v, \beta)$ through P; that is, we increase β from $-\infty$ to ∞. We write

$$\chi(P, v, \beta) := \sum \{(-1)^{\dim F} \mid F \leqslant P \text{ and } F \subseteq H^-(v,\beta)\},$$

again with F ranging over all non-empty faces of P. Further, we write

$$f(P, v, \beta; \mathsf{t}) := \sum \{\mathsf{t}^{\dim F} \mid F \leqslant P \text{ and } F \subseteq H^-(v,\beta)\}.$$

We study how these two functions change as β increases; note that $\chi(P, v, \beta) = f(P, v, \beta; -1)$. In fact, we make the following inductive assumption: *if Q is a $(d-1)$-polytope and $H^-(v,\beta) \cap Q \neq \emptyset$, then $\chi(Q, v, \beta) = 1$.*

Now $f(P, v, \beta; \mathsf{t})$ only changes when the sweeping half-space H^- encounters a vertex x of P (x is unique, because of the choice of v). We write $\Delta\chi$ and Δf for the changes in χ and f at x or, more exactly, at $\beta := \langle x, v\rangle$. At $\beta = -\infty$, both χ and f start at 0. If x is the first vertex of P to be met by H^-, then

$$\Delta\chi = \Delta f = 1.$$

Thereafter, there is a hyperplane J in $\mathbb{X}$ which strictly separates x from the remaining vertices of P, that is, from $\operatorname{conv}(\operatorname{vert} P \setminus \{x\})$. We can perturb J if necessary, so that v is not a normal vector to J. Then $Q := J \cap P$ is a polytope of dimension $\dim P - 1$, and to each j-face F of P which contains x corresponds a $(j-1)$-face $G := J \cap F$ of Q, and conversely. Moreover, $F \subseteq H^-$ if and only if $G \subseteq H^-$.

Taking account of x itself, it follows at once that

$$\Delta f(P, v, \beta; \mathsf{t}) = 1 + \mathsf{t} f(Q, v, \beta; \mathsf{t}).$$

Put $\mathsf{t} = -1$, note that $H^- \cap Q \neq \emptyset$, and use the inductive assumption above, to conclude that $\Delta\chi(P, v, \beta) = 0$. This establishes the inductive assumption, and hence, when H^- contains P, the theorem itself. □

1B22 Remark We can avoid introducing the indeterminate t in the proof of Theorem 1B19, but it clarifies how the induction works.

In fact, we need a variant of Theorem 1B19 in the proof of Theorems 1G1 and 1G2. However, we do not wish to be distracted by too much detail. In extending the Euler characteristic to relatively open convex sets (that is, sets that are open in their affine hulls), we shall demand that it be a simply additive topological invariant, so that $\chi(A \cup B) = \chi(A) + \chi(B)$ if $A \cap B = \emptyset$. Since a convex polytope is the disjoint union of the relative interiors of its faces, this then implies

1B23 Theorem *If C is a relatively open convex set, then* $\chi(C) = (-1)^{\dim C}$.

Notes to Section 1B

1. We should, strictly, talk about *non-negative* rather than positive combinations. However, the term used here is, by now, too firmly embedded in the language to be changed very easily.
2. Carathéodory's Theorem is more usually stated for convex sets; Ziegler [132] gives both versions.
3. There are interesting connexions of Carath'eodory's theorem with two other famous theorems in convexity, namely, those of Radon and Helly. See the seminal article [37] by Danzer, Grünbaum and Klee, or [114, Section 1.1].
4. Strictly speaking, what we have called a face is an *exposed* face, a particular case of which is an exposed point. An *extreme* face of C is a subset F such that, if $a, b \in C$ and $[ab]$ meets relint F, then $a, b \in F$. For polyhedral sets, the two notions coincide.
5. More generally, an extreme point is the limit of exposed points; se the references cited next.
6. An excellent general reference to convexity theory is Schneider [114]. For convex polytopes in particular, see Grünbaum [56] or Ziegler [132].

1C Euclidean Structure

In the next two sections, we give a brief survey of euclidean spaces and their isometry groups. We also say a few words about unitary spaces.

Inner Product and Norm

We can look at euclidean spaces in several ways. Following on from our algebraic treatment in Section 1A, we define a *euclidean space* to be a finite dimensional real linear space $\mathbb{E}$ identified with its dual space $\mathbb{E}^*$ by identifying a basis $E = (e_1, \ldots, e_d)$ (say) of $\mathbb{E}$ with the dual basis $E^* = (e_1^*, \ldots, e_d^*)$ of $\mathbb{E}^*$. The resulting *scalar* or *inner product* $\langle \cdot, \cdot \rangle$ then has the following properties:

- $\langle y, x \rangle = \langle x, y \rangle$ for all $x, y \in \mathbb{E}$;
- $\langle \lambda x + \mu y, \cdot \rangle = \lambda \langle x, \cdot \rangle + \mu \langle y, \cdot \rangle$ for all $x, y \in \mathbb{E}$ and $\lambda, \mu \in \mathbb{R}$;

- $\langle\cdot,\cdot\rangle$ is *positive definite*, in that $\langle x, x\rangle \geqslant 0$, with equality if and only if $x = o$;
- if the *euclidean norm* $\|\cdot\|$ is defined by $\|x\| := \sqrt{\langle x, x\rangle}$, then $\langle x, y\rangle \leqslant \|x\|\|y\|$, with equality if and only if one of x and y is a non-negative multiple of the other;
- $\|\lambda x\| = |\lambda|\|x\|$ for all $x \in \mathbb{E}$ and $\lambda \in \mathbb{R}$;
- the norm satisfies the *triangle inequality* $\|x+y\| \leqslant \|x\|+\|y\|$, with equality just as previously.

We say that $x \in \mathbb{E}$ is a *unit vector* if $\|x\| = 1$.

The norm enables us to define the *euclidean distance* $\|x - y\|$ between x and y (we do not introduce a special notation for the distance); the triangle inequality now takes the form

$$\|x - z\| \leqslant \|x - y\| + \|y - z\|$$

for $x, y, z \in \mathbb{E}$.

Recall that two vectors $x, y \in \mathbb{E}$ are *orthogonal* if $\langle x, y\rangle = 0$. More generally, the *orthogonal complement* of a subset $V \subseteq \mathbb{E}$ is

1C1 $$V^{\perp} := (\operatorname{lin} V)^{\perp} := \{y \in \mathbb{E} \mid \langle x, y\rangle = 0 \text{ for all } x \in V\}.$$

Observe that $V^{\perp\perp} := (V^{\perp})^{\perp} = \operatorname{lin} V$. Further, we call a set $\{u_1, \ldots, u_k\} \subset \mathbb{E}$ *orthonormal* if each u_j is a unit vector and u_i and u_j are orthogonal if $i \neq j$. There are fundamental properties of a euclidean space $\mathbb{E}$:

- an orthonormal set is linearly independent;
- each linear subspace of $\mathbb{E}$ has an orthonormal basis;
- any orthonormal set in $\mathbb{E}$ can be extended to an orthonormal basis of $\mathbb{E}$.

As a special case of (1A9), and then taking the inner product of x with itself, we have

1C2 Lemma *If $\{u_1, \ldots, u_d\}$ is an orthonormal basis of the euclidean space $\mathbb{E}$ and $x \in \mathbb{E}$, then*

$$x = \sum_{j=1}^{d} \langle x, u_j\rangle u_j, \qquad \|x\|^2 = \sum_{j=1}^{d} \langle x, u_j\rangle^2.$$

Observe that the *Gram–Schmidt orthogonalization process* guarantees the properties of $\mathbb{E}$ listed above; this turns an arbitrary basis into an orthogonal (or even orthonormal) one with respect to a given positive definite inner product. However, while we shall freely choose orthonormal bases in a euclidean space $\mathbb{E}$, in the other context where orthogonality plays a central rôle – for realizations in Sections 3F and 3G – a different obvious candidate for an orthogonal basis presents itself.

1C3 Remark Recall that the transpose of a matrix $A = (\alpha_{jk})$ is the matrix $A^{\mathsf{T}} = (\beta_{jk})$ with $\beta_{jk} = \alpha_{kj}$ for each j and k. We call an $d \times d$ matrix U *orthogonal* if $U^{-1} = U^{\mathsf{T}}$. If the rows $u_1, \ldots, u_d$ of a $d \times d$ matrix U form an orthonormal basis of $\mathbb{E}^d$, then $UU^{\mathsf{T}} = I_d$, the $d \times d$ identity matrix. This is equivalent to $U^{-1} = U^{\mathsf{T}}$, so that U is orthogonal and $U^{\mathsf{T}}U = I_d$ also; hence the columns of U form an orthonormal basis as well.

The euclidean and affine structures of $\mathbb{E}$ are strongly connected.

1C4 Proposition *The euclidean distances of a point $x \in \mathbb{E}$ from the points of an affine basis A determine the position of x.*

Sketch of proof. Let $A = \{a_0, \ldots, a_d\}$, and write $\delta_j := \|x - a_j\|$ for $j = 0, \ldots, d$. For $0 \leqslant j < k \leqslant d$, we have

$$\|x - a_j\|^2 - \|x - a_k\|^2 = \langle 2x - a_j - a_k, a_k - a_j \rangle,$$

which yields

$$\langle x, a_k - a_j \rangle = \tfrac{1}{2}\big(\|a_k\|^2 - \delta_k^2 - \|a_j\|^2 + \delta_j^2\big).$$

This reduces the problem to that of solving a system of *linear* equations; since the vectors $a_1 - a_0, \ldots, a_d - a_0$ are linearly independent, this has a unique solution (it is easy to check that the equations are consistent). $\square$

Eutaxy

Recall that, if $\mathbb{L} \leqslant \mathbb{E}$ is a linear subspace, then *orthogonal projection* on $\mathbb{L}$ is the mapping $\Pi = \Pi_{\mathbb{L}} \colon \mathbb{E} \to \mathbb{L}$ given by

1C5
$$x\Pi := \sum_{i=1}^{r} \langle x, u_i \rangle u_i,$$

where $\{u_1, \ldots, u_r\}$ is any orthonormal basis of $\mathbb{L}$. This follows from the fact that we can extend $\{u_1, \ldots, u_r\}$ to an orthonormal basis $\{u_1, \ldots, u_d\}$ of $\mathbb{E}$, and then $\{u_{r+1}, \ldots, u_d\}$ is an orthonormal basis of the orthogonal complement $\mathbb{L}^{\perp}$ of $\mathbb{L}$.

There is an important concept that will be needed when we come to discuss realizations in Chapter 3. When $d < n$, it will be convenient to identify $\mathbb{E}^d$ with the subspace of vectors of the form $(\xi_1, \ldots, \xi_d, 0, \ldots, 0) \in \mathbb{E}^n$; then its orthogonal complement is naturally to be identified with $\mathbb{E}^{n-d}$. We call $(a_1, \ldots, a_n) \subseteq \mathbb{E}^d$ a *eutactic star* if it is the image of some orthonormal basis $(e_1, \ldots, e_n)$ of $\mathbb{E}^n$ under orthogonal projection on $\mathbb{E}^d$ (see the notes at the end of the section). Then we have

1C6 Theorem *The rows $a_1, \ldots, a_n$ of an $n \times d$ matrix A form a eutactic star if and only if $A^{\mathsf{T}}A = I_d$.*

Proof. First, suppose that $(a_1, \dots, a_n)$ is a eutactic star. If $(e_1, \dots, e_n)$ is the corresponding orthonormal basis of $\mathbb{E}^n$, then the matrix E with rows $e_1, \dots, e_n$ is orthogonal. From Remark 1C3, the columns of E also form an orthonormal basis of $\mathbb{E}^n$. Since the first d columns of E are those of A, it follows that $A^{\mathsf{T}}A = I_d$, as claimed.

For the converse, if $A^{\mathsf{T}}A = I_d$, then the columns of A form an orthonormal set. Thus there is an $n \times (n-d)$ matrix B such that the columns of A and B together form an orthonormal basis of $\mathbb{E}^n$. Writing $E := [A\ B]$, it follows from Remark 1C3 again that the rows e_j of E also form an orthonormal basis of $\mathbb{E}^n$. If $e_j := (a_j, b_j)$ with b_j the jth row of B for $j = 1, \dots, n$, then clearly a_j is the image of e_j under orthogonal projection on $\mathbb{E}^d$, which completes the proof. □

1C7 Theorem *If $\mathbb{L}$ is a subspace of $\mathbb{E}^n$ and $(a_1, \dots, a_n)$ is the image under orthogonal projection on $\mathbb{L}$ of some orthonormal basis of $\mathbb{E}^n$, then*

$$\sum_{k=1}^{n} \|a_k\|^2 = \dim \mathbb{L}.$$

Proof. We lose no generality in taking $\mathbb{L} = \mathbb{E}^d$, the coordinate subspace, as before; thus $\dim \mathbb{L} = d$. With A the matrix whose rows are $a_1, \dots, a_n$, we then have

$$\sum_{k=1}^{n} \|a_k\|^2 = \operatorname{tr}(AA^{\mathsf{T}}) = \operatorname{tr}(A^{\mathsf{T}}A) = \operatorname{tr} I_d = d,$$

by an appeal to (1A14) and Theorem 1C6. □

Where realizations are concerned, we need a special case of Theorem 1C7.

1C8 Corollary *If, under the conditions of Theorem 1C7, the vectors a_j all have the same norm $\|a_j\| = \rho$, then $\rho^2 = \dim \mathbb{L}/n$.*

Inherited Euclidean Structures

The euclidean structure is induced in the usual way on the direct sum of two euclidean spaces $\mathbb{E}_1$ and $\mathbb{E}_2$ by

1C9 $$\langle (x_1, x_2), (y_1, y_2) \rangle := \langle x_1, y_1 \rangle + \langle x_2, y_2 \rangle,$$

with $x_j, y_j \in \mathbb{E}_j$ for $j = 1, 2$.

Matrices of the same shape inherit a euclidean structure. We introduced the trace of a matrix in (1A13), and noted that $\operatorname{tr} A = \operatorname{tr}(A^{\mathsf{T}})$. We then define the *inner product* of the $k \times m$ matrices $A = (\alpha_{ij})$ and $B = (\beta_{ij})$ by

1C10 $$\langle A, B \rangle := \operatorname{tr}(AB^{\mathsf{T}}).$$

From this definition, we have

$$\langle A, B \rangle = \sum_{i=1}^{k} \sum_{j=1}^{m} \alpha_{ij}\beta_{ij},$$

and so it easily follows that $\langle\cdot,\cdot\rangle$ is positive definite and symmetric.

We can now apply this notion to tensors. If $a \in \mathbb{E}^k$ and $b \in \mathbb{E}^m$ (as coordinate row vectors with respect to chosen bases), then we can identify $a \otimes b$ with the $k \times m$ matrix $a^{\mathsf{T}} b$. As a consequence,

1C11 $$\langle a_1 \otimes b_1, a_2 \otimes b_2\rangle = \langle a_1, a_2\rangle\langle b_1, b_2\rangle.$$

Indeed, we could use (1C11) as a definition of the inner product of tensors.

When $k = m$, we can almost identify the alternating tensor $a \wedge b$ with the anti-symmetric matrix $a^{\mathsf{T}} b - b^{\mathsf{T}} a$. Strictly speaking, since each coordinate of $a \wedge b$ occurs twice in the matrix (with opposite signs), we should divide the matrix by $\sqrt{2}$ to preserve the norm. With this in mind, applying (1C11) then yields

1C12 $$\langle a_1 \wedge b_1, a_2 \wedge b_2\rangle = \langle a_1, a_2\rangle\langle b_1, b_2\rangle - \langle a_1, b_2\rangle\langle a_2, b_1\rangle.$$

Note that, as it must, interchanging a_1 and b_1 changes the sign of the inner product.

Determinants

The main thing we need to recall about determinants is that they measure (signed) *volume*. In fact, a useful concept is the following. If $A = \{a_1, \ldots, a_k\} \subseteq \mathbb{E}^d$, then the *absolute determinant* $\operatorname{Det} A$ is defined to be

1C13 $$\operatorname{Det} A = \sqrt{\det\bigl(\langle a_i, a_j\rangle \mid 1 \leqslant i, j \leqslant k\bigr)};$$

this measures the k-dimensional volume of the box

$$B(a_1, \ldots, a_k) := \{\xi_1 a_1 + \cdots + \xi_k a_k \mid 0 \leqslant \xi_j \leqslant 1 \text{ for } j = 1, \ldots, k\};$$

naturally, $\operatorname{Det} A = 0$ if the vectors $a_1, \ldots, a_k$ are linearly dependent. This idea lies behind certain metrical relationships between classical regular polytopes and Schläfli determinants, which we shall meet in Sections 7A and 7G.

If $\Phi\colon \mathbb{E} \to \mathbb{E}$ is a linear mapping, then its *determinant* $\det \Phi$ is the determinant of the matrix of Φ with respect to any basis of $\mathbb{E}$ (it is independent of the choice of basis). In particular, the condition for $\Phi \in \mathbf{GL}(\mathbb{E})$ is just $\det \Phi \neq 0$. The *orthogonal group* $\mathbf{O}$ is defined by

1C14 $$\mathbf{O} = \mathbf{O}_d = \mathbf{O}(\mathbb{E}^d) := \{\Phi\colon \mathbb{E} \to \mathbb{E} \mid \langle x\Phi, y\Phi\rangle = \langle x, y\rangle \text{ for all } x, y \in \mathbb{E}\}$$

(we usually drop mention of the dimension d if it is clear from the context). Observe that $\Psi \in \mathbf{O}$ if and only if its matrix U with respect to any orthonormal basis of $\mathbb{E}^d$ is orthogonal. If $\Phi \in \mathbf{O}$, then $\det \Phi = \pm 1$; the subgroup

1C15 $$\mathbf{SO}_d = \mathbf{SO}(\mathbb{E}^d) := \{\Phi \in \mathbf{O}_d \mid \det \Phi = 1\}$$

is the *special orthogonal group* $\mathbf{SO}_d$. As before, we write $\mathbf{SO}$ if there is no need to specify the dimension d.

Unitary Space

While we generally work in euclidean spaces, it is occasionally necessary that the setting be unitary space $\mathbb{C}^n$. This is similarly equipped with an inner product

1C16 $$\langle x, y \rangle := \sum_{j=1}^{n} \xi_j \overline{\eta}_j$$

with coordinates relative to some chosen basis $U = \{u_1, \ldots, u_n\}$ of $\mathbb{C}^n$, which is then (by definition) orthonormal. A *unitary* mapping $\Phi\colon \mathbb{C}^n \to \mathbb{C}^n$ preserves the inner product, so that $\langle x\Phi, y\Phi \rangle = \langle x, y \rangle$ for all $x, y \in \mathbb{C}^n$; the group of such unitary mappings is denoted $\mathbf{U} = \mathbf{U}_n$. An $n \times n$ *unitary* matrix A is such that

1C17 $$A^{-1} = A^* := \overline{A^{\mathsf{T}}} = \overline{A}^{\mathsf{T}},$$

the complex conjugate of the transpose. Then the linear mapping $\Phi\colon \mathbb{C}^n \to \mathbb{C}^n$ is unitary if and only if its matrix with respect to any orthonormal basis U of $\mathbb{C}^n$ is unitary.

1C18 Remark There is a natural identification $\mathbb{C}^n \cong \mathbb{E}^{2n}$, when we regard the former as a *real* linear space. In this context, it may be seen that the real inner product of $x, y \in \mathbb{C}^n$ is $\Re\langle x, y \rangle = \frac{1}{2}(\langle x, y \rangle + \langle y, x \rangle)$, with $\Re$ denoting the real part. Observe that the real inner product and norm are preserved by unitary mappings, as they must be.

Notes to Section 1C

1. It is easy to see that the linear equations arising in the proof of Proposition 1C4 are consistent.
2. Orthogonal projection (1C5) is a special case of *metric projection*, which, with C a closed convex set in $\mathbb{E}$, sends $x \in \mathbb{E}$ to the (unique) nearest point of C to x.
3. In [27, Section 13.7], Coxeter has a slightly different definition of eutactic star, adjoining the negatives of its vectors. Following Hadwiger [59], Coxeter makes rather heavy work of the proof of Theorem 1C6 (and only in dimension 3).

1D Isometries

We now move on to groups acting on $\mathbb{E}$ which respect the euclidean structure. An affine mapping $\Phi\colon \mathbb{E} \to \mathbb{E}$ is an *isometry* if $\|x\Phi - y\Phi\| = \|x - y\|$ for all $x, y \in \mathbb{E}$. It is clear that the isometries of $\mathbb{E}^d$ form a group, called the *isometry group*, which we designate by $\mathbf{M} = \mathbf{M}_d = \mathbf{M}(\mathbb{E}^d)$. The isometries fixing the origin o form the orthogonal subgroup $\mathbf{O}$. If two subsets $X, Y \subseteq \mathbb{E}$ are such that $Y = X\Phi$ for some isometry $\Phi \in \mathbf{M}$, then we say that X and Y are *congruent*.

1D1 Remark It should be clear that a general isometry $\Phi \in \mathbf{M}$ is of the form $x\Phi = x\Psi + t$, with $\Psi \in \mathbf{O}$ and $t \in \mathbb{E}$. Moreover, if $\mathbf{T} \lhd \mathbf{M}$ denotes the (normal) subgroup of translations $x \mapsto x + t$, then $\mathbf{M}/\mathbf{T} \cong \mathbf{O}$, and

$$\mathbf{M} = \mathbf{T} \rtimes \mathbf{O}$$

is a semidirect product. Recall that the notation means that $\mathbf{O}$ acts as a group of automorphisms of $\mathbf{T}$ (in the natural way on the right).

If $\mathbf{G} \leqslant \mathbf{M}$ is any subgroup, then its *point-group* $\mathbf{SG}$ is its image under the quotient mapping of Remark 1D1, namely,

1D2 $$\mathbf{SG} := \mathbf{G}/(\mathbf{G} \cap \mathbf{T}) \cong \mathbf{GT}/\mathbf{T}$$

(see the notes at the end of the section).

1D3 Remark If A and B are matrices of the same shape and $A' = UA$ and $B' = UB$ for some orthogonal matrix U, then $\langle A', B' \rangle = \langle A, B \rangle$, with the same result for multiplication on the right, so that the inner product of matrices respects the euclidean structure. Naturally, this carries over to the various tensor products.

Now recall that λ is an *eigenvalue* of the $d \times d$ (symmetric) matrix A if $xA = \lambda x$ for some vector $x \neq o$; then x is an *eigenvector* corresponding to λ. The quadratic form $\varphi(x, x) = xAx^{\mathsf{T}}$ is positive (semi-)definite if and only if every eigenvalue of A is positive (non-negative, respectively). In calling a matrix $L = (\lambda_{jk})$ *diagonal*, we mean that $\lambda_{jk} = \lambda_j \delta_{jk}$ for some λ_j with $j = 1, \ldots, d$; we then write $L = \operatorname{diag}(\lambda_1, \ldots, \lambda_d)$. Bear in mind that a symmetric matrix A can always be diagonalized by some orthogonal matrix U, so that $A = U^{\mathsf{T}}LU$ with L diagonal; here, each row u_j of U is a unit eigenvector of A with eigenvalue λ_j.

A more general mapping of $\mathbb{E}$ than an isometry is a *similarity*, which is a mapping of the form $x \mapsto \alpha(x\Phi)$, with $\alpha \neq 0$ fixed and $\Phi \in \mathbf{M}$; we call two sets X, Y *similar* if there is a similarity mapping X to Y. We briefly refer to the *similarity class* of a set X (that is, the family of sets similar to X) as its *shape*. If two subsets of $\mathbb{E}$ are similar, then we also say that they have the *same shape*.

We have the following characterization of similarities, whose easy proof we leave to the reader (see the notes at the end of the section). Recall that the *sphere* with *centre* c and *radius* $\rho > 0$ is $S(c, \rho) := \{x \in \mathbb{E} \mid \|x - c\| = \rho\}$. In particular, $\mathbb{S} := S(o, 1)$ is the *unit* sphere

1D4 Proposition *An affine mapping Φ on $\mathbb{E}$ is a similarity if and only if $S\Phi$ is a sphere for some sphere S in $\mathbb{E}$.*

1D5 Remark More generally, the image of a sphere under an affinity will be an *ellipsoid* with centre c (say), of the form

$$\{x \in \mathbb{E} \mid (x - c)A(x - c)^{\mathsf{T}} = 1\}$$

for some positive definite symmetric matrix A. If $UAU^\top = L$ is a diagonal matrix as above and $c = o$, then $y = \eta_1 u_1 + \cdots + \eta_d u_d$ is a general point of the ellipsoid just when

$$\frac{\eta_1^2}{\alpha_1^2} + \cdots + \frac{\eta_d^2}{\alpha_d^2} = 1,$$

with $\alpha_j^2 = 1/\lambda_j$ for $j = 1, \ldots, d$. We call α_j (> 0) the *semi-axis* corresponding to u_j; note that semi-axes need not be distinct (and then U is not unique).

Reflexions

A *reflexion* in $\mathbb{E}$ is an involutory isometry in $\mathbf{M}$. A particularly important kind of reflexion is that in a *hyperplane* $H(u, \beta)$, which is denoted $R(u, \beta)$, so that

1D6
$$xR(u, \beta) := x - 2(\langle x, u\rangle - \beta)u,$$

with u a unit vector and $\beta \in \mathbb{R}$. To verify that $R = R(u, \beta)$ is an isometry, observe that, if $x, y \in \mathbb{E}$, then

$$\begin{aligned}\|xR - yR\|^2 &= \langle x - y - 2\langle x - y, u\rangle u, x - y - 2\langle x - y, u\rangle u\rangle \\ &= \langle x - y, x - y\rangle - 4\langle x - y, \langle x - y, u\rangle u\rangle + 4\langle x - y, u\rangle^2 \|u\|^2 \\ &= \|x - y\|^2,\end{aligned}$$

as required.

In the same spirit as Proposition 1C4, we have

1D7 Proposition *If $A = (a_0, \ldots, a_d)$ and $B = (b_0, \ldots, b_d)$ are two ordered affine bases of $\mathbb{E}$ such that $\|a_j - a_k\| = \|b_j - b_k\|$ for each $0 \leqslant j < k \leqslant d$, then A and B are congruent.*

Sketch of proof. In fact, we will show more. We make the inductive assumption that, for some k, we have already found an isometry Φ_k such that $a_j \Phi_k = b_j$ for $j < k$ (with $\Phi_0 = I$). Relabel, so that we now write a_j instead of $a_j \Phi_k$ for each j. If $b_k \neq a_k$, then, with

$$\begin{aligned}u &:= \|b_k - a_k\|^{-1}(b_k - a_k), \\ \beta &:= \langle \|b_k - a_k\|^{-1}(b_k - a_k), \tfrac{1}{2}(b_k + a_k)\rangle = \tfrac{1}{2}(\|b_k\|^2 - \|a_k\|^2)/\|b_k - a_k\|,\end{aligned}$$

we have $a_j = b_j \in H(u, \beta)$ for $j < k$, so that

$$a_j R(u, \beta) = b_j$$

for $j \leqslant k$. Thus taking $\Phi_{k+1} := \Phi_k R(u, \beta)$ yields the inductive step.

To complete the proof, we merely appeal to Proposition 1C4 and the definition of isometry. □

Thus, in fact, what we have shown about $\mathbf{M}$ is the following.

1D8 Corollary *The isometry group* $\mathbf{M}$ *is generated by hyperplane reflexions.*

In preparation for Lemma 3G6, it is useful to have a rather stronger result than Corollary 1D8 about the orthogonal group $\mathbf{O}$. Let $(e_1, \ldots, e_d)$ be a fixed orthonormal basis of $\mathbb{E}$. For $s \neq t$ and $\mu, \nu \in \mathbb{R}$ such that $\mu^2 + \nu^2 = 1$, define the hyperplane reflexion $R_{st}(\mu, \nu)$ by

1D9
$$R_{st}(\mu, \nu) \colon e_j \mapsto \begin{cases} \mu e_s + \nu e_t, & \text{if } j = s, \\ \nu e_s - \mu e_t, & \text{if } j = t, \\ e_j, & \text{otherwise.} \end{cases}$$

Then we have

1D10 Theorem *The hyperplane reflexions* $R_{st}(\mu, \nu)$ *generate the orthogonal group* $\mathbf{O}$.

Proof. Note first that $(\mu, \nu) = (-1, 0)$ changes the sign of e_s, while $(\mu, \nu) = (0, 1)$ interchanges e_s and e_t. In general, suppose that $A \in \mathbf{O}_d$ has matrix (α_{ij}) with respect to $(e_1, \ldots, e_d)$. Write $a := a_1 = \alpha_1 e_1 + \cdots + \alpha_d e_d$ for the first row of A (this is, of course, a unit vector). Permuting the e_j and changing signs, if necessary, we may assume that $\alpha_1 > 0$. Successively, for each $t = 2, \ldots, d$ such that $\alpha_t \neq 0$, apply the reflexion $R_{1t}(\mu, \nu)$ with

$$(\mu, \nu) = \frac{1}{\sqrt{\alpha_1^2 + \alpha_t^2}} (\alpha_1, \alpha_t);$$

retaining the same notation as we go along, this produces a new a for which $\alpha_t = 0$. Restoring $a_1 = a$, and noting that this process can only terminate at e_1, we conclude that we can find an orthogonal matrix U_1, which is a product of suitable reflexions $R_{st}(\mu, \nu)$, for which $a_1 U_1 = e_1$. The rows $a_2, \ldots, a_d$ of the new matrix A are orthogonal to e_1, and so have first coordinate 0. We can now proceed by induction on d, ending with a product U of reflexions $R_{st}(\mu, \nu)$ such that $AU = I$. It follows that $A = U^{-1}$ is a product of reflexions $R_{st}(\mu, \nu)$ (in the reverse order), as claimed. □

The *axis* of an isometry Φ is

1D11
$$\operatorname{axis} \Phi := \{x \in \mathbb{E} \mid x\Phi = x\}.$$

An extremely useful convention for a reflexion R is to identify it with its axis or *mirror* $\operatorname{axis} R$; we thus often employ the same symbol for both. In a similar way, the *axis* of a subgroup $\mathbf{G}$ of isometries is

1D12
$$\operatorname{axis} \mathbf{G} := \{x \in \mathbb{E} \mid x\Phi = x \text{ for each } \Phi \in \mathbf{G}\}.$$

We next have a very general result about products of reflexions. This is most succinctly phrased in terms of an ambient space, which may be spherical, euclidean or hyperbolic, though the last will be of much less interest in this book.

1D13 Theorem *Let $\mathbb{U}$ be d-dimensional spherical, euclidean or hyperbolic space, and let T be a d-dimensional simplex in $\mathbb{U}$ bounded by the hyperplanes $R_0, \ldots, R_d$. Then, regarded as reflexions, the product $R_0 \cdots R_d$ has no fixed points in $\mathbb{U}$.*

Proof. We appeal to induction on d, with the result trivial if $d = 1$. For larger d, suppose that $v \in \mathbb{U}$ is fixed by the product. First consider the sub-product $R_0 \cdots R_{d-1}$. We cannot have $v \in R_0 \cap \cdots \cap R_{d-1}$, because then v is fixed by $R_0 \cdots R_{d-1}$, but is not fixed by R_d since $v \notin R_d$. Otherwise, let $\mathbb{S}$ be the sphere centred at $R_0 \cap \cdots \cap R_{d-1}$ containing v. The inductive assumption says that $v \neq vR_0 \cdots R_{d-1} \in \mathbb{S}$ also. We now apply R_d, deducing that $v = vR_0 \cdots R_{d-1}R_d \in \mathbb{S}$. However, because the centre of $\mathbb{S}$ does not lie on R_d, the only points of $\mathbb{S}$ which R_d can take into points of $\mathbb{S}$ are those of the intersection $R_d \cap \mathbb{S}$. But then this would imply that $vR_d = v$, and hence that $v = vR_0 \cdots R_{d-1}$ after all; we have now achieved the required contradiction. □

1D14 Remark If $\mathbb{U}$ is a spherical space, hyperplane reflexions are reflexions in great spheres of $\mathbb{U}$, and correspond to reflexions in hyperplanes through the centre of $\mathbb{U}$, if we think of $\mathbb{U}$ as sitting in euclidean space of dimension $d + 1$.

In preparation for the following sections, it is appropriate to go further into the geometry of reflexions, particularly their commutativity properties. For convenience in the ensuing discussion, all reflexions will be assumed to be linear; the generalization to arbitrary reflexions will be obvious.

We begin with the commutativity criterion itself, the motive being that we need one which is not symmetric between the reflexions.

1D15 Theorem *If R and S are linear reflexions in the euclidean space $\mathbb{E}$, then the following conditions are equivalent:*

(a) *$R \rightleftharpoons S$, meaning that R and S commute, so that $RS = SR$;*

(b) *$S = (R \cap S) \oplus (R^{\perp} \cap S)$, with both R and S taken as mirrors;*

(c) *$SR = S$, where S only is taken to be a mirror.*

Proof. First, suppose that $R \rightleftharpoons S$. Let $x \in S$ be arbitrary, and write $x = y + z$ with $y \in R$ and $z \in R^{\perp}$ (regarded as mirrors, of course). Thus $xR = (y+z)R = y - z$ and $xS = x$, so that $xRS = xSR = xR$ implies that $xR \in S$. Hence $y + z, y - z \in S$, implying that $y, z \in S$ and therefore $y \in R \cap S$ and $z \in R^{\perp} \cap S$. This shows that $S = (R \cap S) \oplus (R^{\perp} \cap S)$ (the sum is clearly direct), so that (a) implies (b).

The decomposition $S = (R \cap S) \oplus (R^{\perp} \cap S)$ immediately implies that $SR = S$, and therefore (b) implies (c).

The argument is reversible. If $SR = S$, then express $x \in S$ uniquely as $x = y + z$ with $y \in R$ and $z \in R^{\perp}$. Hence $y - z = yR + zR = xR \in S$, from which we deduce that $y, z \in S$. Then $(R \cap S) \oplus (R^{\perp} \cap S) = S$ follows at once, so that (c) implies (b).

Finally, suppose that $S = (R \cap S) \oplus (R^{\perp} \cap S)$. We choose orthonormal bases U_{++} of $R \cap S$ and U_{-+} of $R^{\perp} \cap S$; consequently, $U_{++} \cup U_{-+}$ is an orthonormal basis of S. We now extend U_{++} by U_{+-} to an orthonormal basis $U_+ = U_{++} \cup U_{+-}$ of R, and U_{-+} by U_{--} to an orthonormal basis $U_- = U_{-+} \cup U_{--}$ of $R^\perp$. Since $U = U_+ \cup U_-$ is a basis of $\mathbb{E}$, it follows at once that $U_{+-} \cup U_{--}$ is a basis of $S^\perp$. This gives complete symmetry between R and S, and clearly we now have $R \rightleftharpoons S$. Thus (b) implies (a), which completes the proof. □

There is an important consequence of Theorem 1D15.

1D16 Corollary *Let R, S_1, ..., S_k be linear reflexions in the euclidean space $\mathbb{E}$, such that $R \rightleftharpoons S_j$ for $j = 1, \ldots, k$. Then $R \rightleftharpoons S_1 \cap \cdots \cap S_k$.*

Proof. It is enough to establish the case $k = 2$, so we suppose that $R \rightleftharpoons S, T$ (with S, T reflexions), and show that $R \rightleftharpoons S \cap T$. We now appeal to Theorem 1D15. Since $(R \cap S) \oplus (R^\perp \cap S) = S$ and $(R \cap T) \oplus (R^\perp \cap T) = T$, it readily follows that

$$(R \cap (S \cap T)) \oplus (R^\perp \cap (S \cap T)) = S \cap T,$$

and hence $R \rightleftharpoons S \cap T$, as required. □

We end with a general remark. If S and T are linear reflexions, then, since $ST = (-S)(-T) = S^\perp T^\perp$ (thus identifying $-S$ with its mirror $S^\perp$, and so on), $S \cap T$ and $S^\perp \cap T^\perp$ are both pointwise fixed by the product. That is, the axis of ST is

1D17 $$\operatorname{axis}(ST) = (S \cap T) \oplus (S^\perp \cap T^\perp) = (S \cap T) \oplus (S + T)^\perp.$$

In particular, if S and T commute, then (1D17) is the mirror of their product $ST = TS$, which is again a reflexion. There is an obvious generalization to any two reflexions whose mirrors meet.

Discrete Groups

Our almost exclusive interest in subgroups $\mathbf{G}$ of isometries acting on $\mathbb{E}$ is in those which are *discrete*, in that, for each $x \in \mathbb{E}$, there is a minimal distance $\delta = \delta(x) > 0$ such that $\|xG - x\| \geqslant \delta$ whenever $G \in \mathbf{G}$ is such that $xG \neq x$. Note that *finite* subgroups of $\mathbf{M}$ must be discrete.

1D18 Remark The point-group of a discrete group need not be finite. As a simple example, let ϑ be an irrational multiple of π, and consider the group $\mathbf{G}$ consisting of the isometries Φ_k of $\mathbb{E}^3$ defined by

$$(\xi_1, \xi_2, \xi_3)\Phi := (\xi_1 + k, \cos(k\vartheta)\xi_2 - \sin(k\vartheta)\xi_3, \sin(k\vartheta)\xi_2 + \cos(k\vartheta)\xi_3).$$

The $\mathbf{SG}$ is the group generated by rotation through the angle ϑ, which is far from finite.

Let $\mathbf{G}$ be a discrete group of isometries acting on $\mathbb{E}$. We say that $K \subseteq \mathbb{E}$ is a *fundamental region* for $\mathbf{G}$ if the following hold:

- K is the closure of its interior,

- $K\mathbf{G} = \mathbb{E}$,
- if $\Phi \in \mathbf{G} \setminus \{I\}$, then $\operatorname{int} K \cap \operatorname{int}(K\Phi) = \emptyset$.

What the second condition says is that the images of K under $\mathbf{G}$ cover $\mathbb{E}$. From the third we deduce that $\operatorname{int}(K\Phi) \cap \operatorname{int}(K\Psi) = \emptyset$ whenever $\Phi, \Psi \in \mathbf{G}$ with $\Phi \neq \Psi$, so that the interiors of distinct copies of K do not overlap.

Since $\mathbf{G}$ is discrete, it is clear that there exists $c \in \mathbb{E}$ such that $c \notin \operatorname{axis} \Phi$ for any $\Phi \in \mathbf{G} \setminus \{I\}$. The *Voronoĭ region* of $\mathbf{G}$ with *centre* c is

1D19 $$\operatorname{vor}(\mathbf{G}, c) := \{x \in \mathbf{G} \mid \|x - c\| \leqslant \|x - c\Phi\| \text{ for each } \Phi \in \mathbf{G}\}.$$

If we rewrite the defining inequality as

$$\|x - c\|^2 \leqslant \|x - c\Phi\|^2 \implies \langle x, c\Phi - c\rangle \leqslant \tfrac{1}{2}\big(\|c\Phi\|^2 - \|c\|^2\big),$$

it follows that $\operatorname{vor} \mathbf{G}, c)$ is an intersection of closed half-spaces, and so is a polyhedral set. Indeed, we readily see that

1D20 Proposition *The Voronoĭ region* $\operatorname{vor}(\mathbf{G}, c)$ *is a fundamental region for* $\mathbf{G}$ *in* $\mathbb{E}$.

The translations in a group $\mathbf{G}$ of isometries form an abelian subgroup $\mathbf{T}$, say. If $\mathbf{G}$ is discrete, then so is $\mathbf{T}$, and hence consists of the translations by vectors of the form $\nu_1 b_1 + \cdots + \nu_r b_r$ for some $r \leqslant d$. Such a set of vectors is called a *lattice*, and r is called its *rank*. We quote here a core result due to Bieberbach [3] (see also [110, §7.4]).

1D21 Theorem *If* $\mathbf{G}$ *is a discrete group of isometries of* $\mathbb{E}^d$ *with* $\operatorname{conv}(v\mathbf{G}) = \mathbb{E}^d$ *for some* $v \in \mathbb{E}^d$, *then* $\mathbf{G}$ *contains a subgroup* $\mathbf{T}$ *of translations of* $\mathbb{E}^d$ *forming a lattice of full rank* d.

The following is a useful observation.

1D22 Proposition *If* $\mathbf{G}$ *is a discrete group of isometries of* $\mathbb{E}^d$ *such that* $\operatorname{conv}(v\mathbf{G}) = \mathbb{E}^d$ *for some* $v \in \mathbb{E}^d$, *then* $\mathbf{G}$ *contains no rotation about an axis of codimension* 2 *of period other than* 2, 3, 4 *or* 6.

Proof. If A is a $(d-2)$-dimensional axis of rotation in $\mathbf{G}$, then Theorem 1D21 implies that there are translates of A under $\mathbf{G}$ distinct from A; let B be one of them at minimal distance from A. If the corresponding rotation Φ about B has period $q > 7$, then we can take a suitable power, and assume that Φ is a rotation through $2\pi/q$. But then $A\Phi$ is nearer to A than B, giving a contradiction. The case $q = 5$ is only a little more complicated. Here, suitable rotations about A and B produce two rotation images of them closer together than A and B; we leave the details to the reader. □

Reducibility

In the discussion of realizations in Section 3G we shall need some important facts about irreducibility; these mainly concern finite groups, but the ideas cover the general case. To begin with, we have

1D23 Theorem *A subgroup* $\mathbf{G} \leqslant \mathbf{O}$ *is irreducible if and only if every ellipsoid invariant under* $\mathbf{G}$ *is a sphere.*

Proof. Since spheres are invariant under $\mathbf{G}$, we need only show the converse. If some non-spherical ellipsoid is invariant under $\mathbf{G}$, let $\mathbb{L}$ be the proper subspace of $\mathbb{E}$ spanned by those u_j of Remark 1D5 with maximal semi-axes α_j. Then $\mathbb{L}$ is invariant under $\mathbf{G}$, contrary to assumption. □

1D24 Theorem *If* $\mathbf{G}, \mathbf{H} \leqslant \mathbf{O}$ *are equivalent with* $\mathbf{G}$ *irreducible, then they are equivalent under some similarity.*

Proof. We have already remarked that $\mathbf{H}$ is also irreducible. Let $\Theta \in \mathbf{GL}(\mathbb{E})$ be the equivalence, so that $\mathbf{H} = \Theta^{-1}\mathbf{G}\Theta$. If $\mathbb{S}$ is the unit sphere in $\mathbb{E}$, then the ellipsoid $\mathbb{S}\Theta$ is invariant under $\mathbf{H}$. But Theorem 1D23 now says that $\mathbb{S}\Theta$ is a sphere, so that Θ is a similarity by Proposition 1D4. □

1D25 Remark Since $\Theta = \lambda\Phi$ for some $\Phi \in \mathbf{O}$, we can replace Θ by Φ in Theorem 1D24, so that $\mathbf{G}$ and $\mathbf{H}$ are actually orthogonally equivalent.

Signed Permutations

We shall employ an abbreviated notation for certain kinds of isometry. We regard $\mathbb{E}^d$ as the usual coordinate vector space with respect to the standard orthonormal basis $\{e_1, \ldots, e_d\}$. A permutation $\rho \in \boldsymbol{S}_d$ then acts in the natural way on this basis: $e_j\rho := e_{j\rho}$. In particular, the *transposition* $(j\ k)$ interchanges e_j and e_k, while fixing e_i if $i \neq j, k$. We use $\overline{k}$ to mean the change of sign of e_k (alone). We can combine these to form transpositions such as $(j\ \overline{k})$, which interchanges e_j and $-e_k$ (and $-e_j$ and e_k, of course).

We can extend such notation to more general permutations, but a little care is needed. Thus we must write $(1\ \overline{2}\ \overline{1}\ 2)$ for the isometry of period 4 that permutes $e_1, -e_2, -e_1, e_2$ cyclically, and observe that this is different from $(1\ \overline{2})$. Finally, we can additionally change all signs, for example obtaining involutions such as $-(j\ \overline{k})$. We refer to an isometry expressed in such a way as a *signed permutation*.

We can further write, for example, $x \mapsto x\rho + t$ for an affine isometry of this kind, with ρ now a signed permutation and $t \in \mathbb{E}^d$ a translation vector.

Notes to Section 1D

1. The term *special group* is employed by Coxeter [27] and others for what we call the point-group. The latter term is that preferred in crystallography, and avoids using 'special' in a technical sense.
2. The Voronoĭ region has been given many other names, the most familiar of which is probably *Dirichlet cell*.
3. For the proof of Proposition 1D4, it is helpful to remember that $\Phi \in \mathbf{O}$ just when $\mathbb{S}\Phi = \mathbb{S}$.

1E Reflexion Groups

In contrast to the much wider coverage of [99, Sections 3A, 3B], we shall confine our attention here strictly to the discrete groups in a euclidean space $\mathbb{E}$ which are generated by hyperplane reflexions (little of what we do here actually requires us to specify the dimension of our space). Recall that, if $u \in \mathbb{E}$ is a unit vector and $\beta \in \mathbb{R}$, then the reflexion R in the hyperplane $H(u, \beta) := \{x \in \mathbb{E} \mid \langle x, u \rangle = \beta\}$ is the affinity (invertible affine mapping) given by

$$xR = x - 2(\langle x, u \rangle - \beta)u.$$

Observe that R is involutory ($R^2 = I$, the identity) and that each point of $H(u, \beta)$ is fixed by R. As in Section 1C we refer to $H(u, \beta)$ as the mirror of R, which we often identify with R, and so use the same symbol for both.

We begin by looking at subgroups generated by two hyperplane reflexions S and T, say. There are two possibilities. First, as hyperplane mirrors, S and T are parallel, and so have the same unit normal vector u. If $S = H(u, \beta)$ and $T = H(u, \gamma)$, then ST is the translation $x \mapsto x + a$ with $a = 2(\gamma - \beta)u$. Second, S is as before, but $T = H(v, \gamma)$ with $v \neq \pm u$ another unit vector. In this case, define the angle ϑ with $0 < \vartheta \leqslant \pi/2$ by $\cos \vartheta = \pm\langle u, v \rangle$. Then ST is a rotation about the axis $S \cap T$ through angle 2ϑ; in particular, ST has finite period if and only if ϑ is *rational*, by which we mean a rational multiple of π.

We state the following very familiar result purely for reference purposes.

1E1 Theorem *If the hyperplane reflexions S and T generate a finite group $\mathbf{D} := \langle S, T \rangle$, then this group $\mathbf{D}$ is dihedral of order $2p$, where the angle between S and T is $k\pi/p$ for some $1 \leqslant k < \frac{1}{2}p$ with greatest common divisor $(k, p) = 1$. Moreover, $\mathbf{D}$ has p hyperplane reflexions, with mirrors equally spaced at angles π/p.*

We call a subset $U \subseteq \mathbb{E}$ of non-zero vectors a *nip-set* if $\langle u, v \rangle \leqslant 0$ whenever $u, v \in U$ with $u \neq v$ (we make no initial assumption that U is finite). Here, 'nip' stands for 'non-positive inner product'. Our subsequent treatment depends on specifically describing nip-sets. We need some temporary notation; we define

1E2 $$|x| := (|\xi_1|, \ldots, |\xi_k|)$$

for $x = (\xi_1, \ldots, \xi_k) \in \mathbb{E}^k$, where $|\alpha| := \max\{\alpha, -\alpha\}$ is the usual absolute value of $\alpha \in \mathbb{R}$.

1E3 Lemma *Let $V = \{v_1, \ldots, v_k\}$ be a nip-set, and for $x = (\xi_1, \ldots, \xi_k) \in \mathbb{E}^k$ define $\varphi(V, x) := \|\xi_1 v_1 + \cdots + \xi_k v_k\|^2$. Then*

$$\varphi(V, |x|) \leqslant \varphi(V, x).$$

Proof. This is straightforward. Indeed,

$$\varphi(V, x) = \sum_{i,j=1}^{k} \xi_i \xi_j \langle v_i, v_j \rangle,$$

and $|\xi_i||\xi_j|\langle v_i, v_j \rangle \leqslant \xi_i \xi_j \langle v_i, v_j \rangle$ whenever $i \neq j$, because $\langle v_i, v_j \rangle \leqslant 0$. □

The required description of nip-sets now follows.

1E4 Theorem *A nip-set in $\mathbb{E}$ admits a decomposition into mutually orthogonal subsets, each of which is linearly independent or minimally positively dependent. In particular, a nip-set is finite.*

Proof. Let $U \subseteq \mathbb{E}$ be a nip-set, and let $V = \{v_1, \ldots, v_k\} \subseteq U$ be a minimal linearly dependent subset (if one exists), so that there are non-zero $\xi_1, \ldots, \xi_k$ such that $\xi_1 v_1 + \cdots + \xi_k v_k = o$. From the definition of $\varphi(V, \cdot)$ and Lemma 1E3 we have

$$0 \leqslant \varphi(V, |x|) \leqslant \varphi(V, x) = 0,$$

and so we also have $\zeta_1 v_1 + \cdots + \zeta_k v_k = o$, where $z = (\zeta_1, \ldots, \zeta_k) := |x|$. The minimality then forces V to be positively dependent (and clearly minimally so).

If this $V \subset U$ (proper subset), then pick any $u \in U \setminus V$. If $\langle v_i, u \rangle \neq 0$ for some $i = 1, \ldots, k$, then $\langle x, u \rangle > 0$ for some $x \in \operatorname{lin} V = \operatorname{pos} V$; here we appeal to Theorem 1B7. That is, there are $\xi_1, \ldots, \xi_k \geqslant 0$, not all zero, such that $x = \xi_1 v_1 + \cdots + \xi_k v_k$, so that

$$\sum_{i=1}^{k} \xi_i \langle v_i, u \rangle = \langle x, u \rangle > 0.$$

It follows that $\langle v_i, u \rangle > 0$ for some $i = 1, \ldots, k$, contrary to the definition of a nip-set. In other words, the complement $U \setminus V$ of V in U lies in the orthogonal complement $V^{\perp}$ of V in $\mathbb{E}$.

We now proceed in the same manner with $U \setminus V$, until we have exhausted U; this takes finitely many steps (since we decrease the dimension of the complement each time), so that U itself must be finite. Note that, if we are left with a set to which this procedure can no longer be applied, then it is linearly independent. □

Now suppose that $\mathbf{G}$ is a group of isometries acting discretely on $\mathbb{E}$ which is generated by hyperplane reflexions. If $\mathcal{R}$ is the family of all the mirrors of the reflexions in $\mathbf{G}$, then $\mathcal{R}$ partitions $\mathbb{E}$ into congruent convex sets, which are permuted by the elements of $\mathbf{G}$. We first claim

1E5 Lemma *Each set in the partition of $\mathbb{E}$ by $\mathcal{R}$ is a fundamental region for $\mathbf{G}$.*

Proof. Choose K to be any one of these sets. If K is not a fundamental region for $\mathbf{G}$, then some image $K\Phi$ of K, with $\Phi \in \mathbf{G}$, is such that $K \neq K\Phi$ but $\operatorname{int} K \cap \operatorname{int}(K\Phi) \neq \emptyset$. This forces some bounding hyperplane H of $K\Phi$ to cut K; however, since $H \in \mathcal{R}$, this contradicts the definition of K. □

Now let $U(K)$ be the set of unit outer normal vectors to the mirrors in $\mathcal{R}$ which are spanned by facet hyperplanes of any given fundamental region K. We next have

1E6 Lemma *The set $U = U(K)$ is a nip-set.*

Proof. Let $S, T \in \mathcal{R}$ be facet hyperplanes of K, whose outer unit normals are $v, w \in U$. If $\langle v, w \rangle > 0$, then the corresponding dihedral angle subtended by S and T is greater than $\pi/2$. Now the reflexions in S and T generate a dihedral subgroup $\mathbf{D}$ and, as we observed in Theorem 1E1, $\mathbf{D}$ contains another mirror R which splits this dihedral angle. But then R similarly splits K, so that K could not, after all, have been a fundamental region for $\mathbf{G}$. □

1E7 Remark Observe particularly in this proof that it is not necessary that $S \cap T$ meet K at all.

We can say a little more at this stage. In fact, appealing to Theorem 1E1 again, the proof of Lemma 1E6 shows that the dihedral angle between S and T must be π/q for some $q \in \mathbb{N}$ (the case $q = 1$ accounts for two parallel hyperplanes bounding a slab which contains K). The previous discussion then yields

1E8 Theorem *A fundamental region for a discrete group on a euclidean space* $\mathbb{E}$ *generated by hyperplane reflexions is an orthogonal direct sum of simplices, simple polyhedral cones and linear subspaces. Moreover, each of its dihedral angles is a submultiple of* π.

We are now in a position to introduce Coxeter diagrams (see the notes at the end of the section). Let K be a fundamental region for a discrete group $\mathbf{G}$ generated by hyperplane reflexions acting on $\mathbb{E}$. The *Coxeter diagram* (or *diagram* for short) for $\mathbf{G}$ is a graph; however, to distinguish such graphs from (for example) edge-graphs of polytopes, we use the terms *node* and *branch* instead of vertex and edge. With each mirror bounding K along a facet we associate a node; two nodes are then joined by a branch marked p if the dihedral angle of K between the mirrors is π/p, with the following exceptions:

- if the mark would be 2, then the branch is omitted,
- a mark 3 on a branch is omitted.

Note that parallel mirrors lead to a mark ∞ on the corresponding branch (their normals are opposite to each other). The omission of marks 3 is due to their frequency. As for omission of branches carrying a mark 2, we have

1E9 Proposition *The connected components of a Coxeter diagram correspond to the irreducible components of the corresponding reflexion group.*

Proof. This is straightforward: two hyperplane reflexions commute just when the angle between their mirrors is $\pi/2$. □

Henceforth, therefore, we shall largely concentrate on connected diagrams, corresponding to irreducible reflexion groups.

1E10 Remark If we think of the nodes of a diagram as representing instead the outer normal vectors to the fundamental region, then a mark p corresponds to inner product $-\cos\frac{\pi}{p}$ between the normals. We shall later put non-integer marks on diagrams, and the actual normal vectors will then take on an added importance.

The following well-known classification results for discrete reflexion groups were obtained by Coxeter [21] (see also [18, 22]).

1E11 Theorem *The irreducible finite reflexion groups are precisely those with a diagram listed in Table 1E12.*

1E12

Notation	Diagram	Order
$\mathbf{A}_d$		$(d+1)!$
$\mathbf{B}_d$	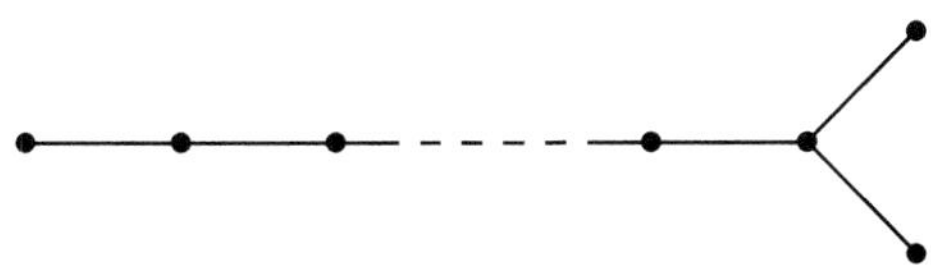	$2^{d-1}d!$
$\mathbf{C}_d$	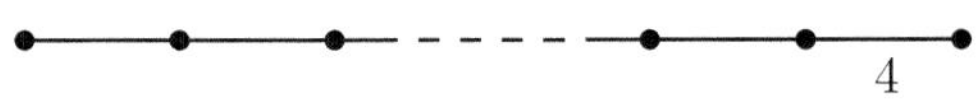	$2^d d!$
$\mathbf{D}_2^p$	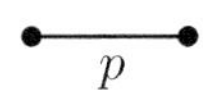	$2p$
$\mathbf{E}_6$	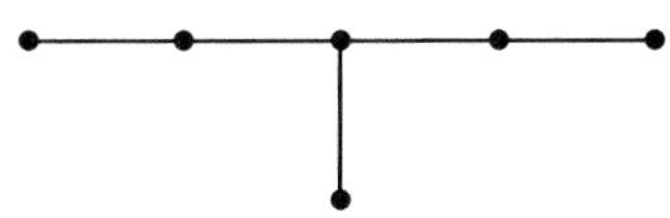	$72 \cdot 6!$
$\mathbf{E}_7$	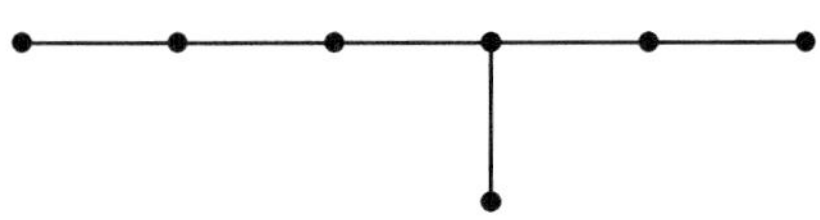	$8 \cdot 9!$
$\mathbf{E}_8$		$192 \cdot 10!$
$\mathbf{F}_4$	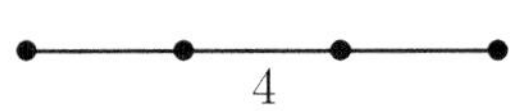	1152
$\mathbf{G}_3$		120
$\mathbf{G}_4$		14400

The finite reflexion groups

1E13 Theorem *The irreducible discrete affine reflexion groups are precisely those with a diagram listed in Table 1E14.*

1E14

Notation	Diagram
$\mathbf{P}_{d+1}$	
$\mathbf{Q}_{d+1}$	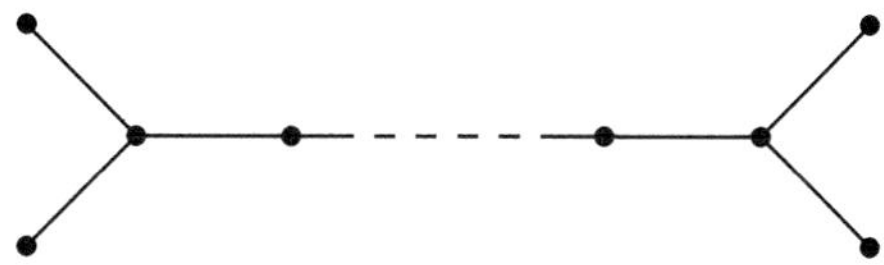
$\mathbf{R}_{d+1}$	
$\mathbf{S}_{d+1}$	
$\mathbf{T}_7$	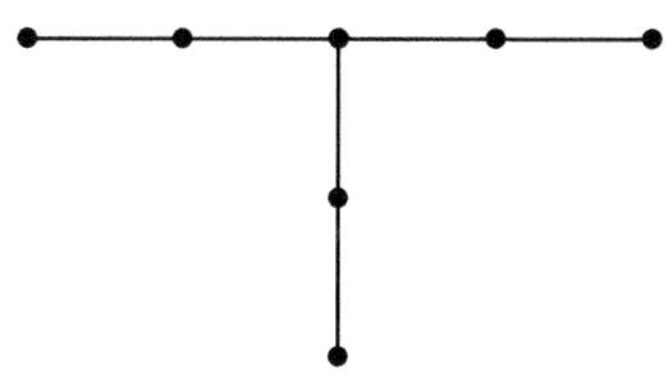
$\mathbf{T}_8$	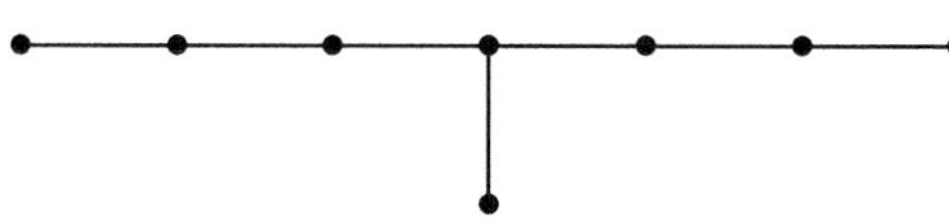
$\mathbf{T}_9$	
$\mathbf{U}_5$	
$\mathbf{V}_3$	
$\mathbf{W}_2$	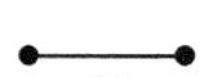

The discrete affine reflexion groups

Throughout, we follow the original notation of Coxeter [27, Sections 11.4, 11.5 and Table IV] for the diagrams listed in Tables 1E12 and 1E14; thus the subscript at a diagram name is its number of nodes. We use $d+1$ rather than d as an index for the affine groups, because we prefer to think of them as acting on $\mathbb{E}^d$. For a geometric method of calculating the orders of the finite reflexion groups (given in Table 1E12), see Section 1H.

Proof of Theorems 1E11 and 1E13. For the proof, we shall list the unit outer normals for the diagrams of the infinite groups; the method of proof requires us to introduce two more diagrams in (1E16), which are not actually those of affine reflexion groups, but act as ones with which we can compare putative diagrams that are ultimately to be rejected.

It remains to list the appropriate unit normals and positive dependences. In what follows, the e_j are the vectors of the appropriate standard orthonormal basis.

For $\mathbf{P}_{d+1}$, as with other groups which are closely related to the *symmetric group* $\mathbf{A}_d$, we often work in the *symmetric hyperplane*

1E15 $$\mathbb{L}^d := \{x = (\xi_0, \ldots, \xi_d) \in \mathbb{E}^{d+1} \mid \xi_0 + \cdots + \xi_d = 0\};$$

this notation (with different dimensions d) will retain the same meaning from now on. We can then take the outer unit normals $u_0, \ldots, u_d$ to be given by

$$\sqrt{2}u_j := e_j - e_{j-1}$$

for $j = 0, \ldots, d$, where the indices are to be taken modulo $d+1$; correspondingly, we have $z = (1, \ldots, 1)$.

For $\mathbf{Q}_{d+1}$, $\mathbf{R}_{d+1}$ and $\mathbf{S}_{d+1}$, the normals $u_1, \ldots, u_{d-1}$ are common to all three groups, and are given by

$$\sqrt{2}u_j := e_{j+1} - e_j.$$

Then u_0, u_d and z are given by

$$\begin{array}{llllll}
\mathbf{Q}_{d+1}: & \sqrt{2}u_0 = e_1 + e_2, & \sqrt{2}u_d = -(e_{d-1} + e_d), & z = (1,1,2,\ldots,2,1,1), \\
\mathbf{R}_{d+1}: & u_0 = e_1, & u_d = -e_d, & z = (1,\sqrt{2},\ldots,\sqrt{2},1), \\
\mathbf{S}_{d+1}: & \sqrt{2}u_0 = e_1 + e_2, & u_d = -e_d, & z = (1,1,2,\ldots,2,\sqrt{2}).
\end{array}$$

For $\mathbf{U}_5$, we have

$$\begin{array}{rclrcl}
\sqrt{2}u_0 & = & e_1 + e_2, & \sqrt{2}u_j & = & e_{j+2} - e_{j+1} \quad (j = 1, 2), \\
u_3 & = & -e_4, & u_4 & = & \tfrac{1}{2}(-1, 1, 1, 1),
\end{array}$$

with $z = (1, 2, 3, 2\sqrt{2}, \sqrt{2})$.

For $\mathbf{V}_3$,

$$u_0 = e_1, \quad u_1 = \tfrac{1}{2}(\sqrt{3}e_2 - e_1), \quad u_2 = -e_2, \qquad z = (1, 2, \sqrt{3}).$$

Finally, for $\mathbf{W}_2$,

$$u_0 = e_1, \quad u_1 = -e_1, \qquad z = (1, 1).$$

The auxiliary diagrams are

1E16

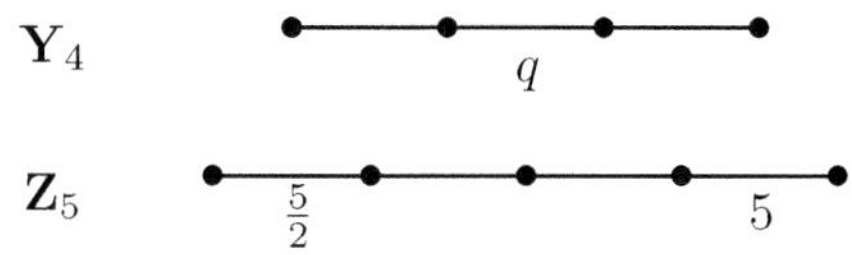

Here, $\pi/q = \arccos \frac{3}{4}$. For $\mathbf{Y}_4$, we have

$$u_0 = e_1, \quad u_1 = \tfrac{1}{2}(\sqrt{3}e_2 - e_1), \quad u_2 = \tfrac{1}{2}(e_3 - \sqrt{3}e_2), \quad u_3 = -e_3,$$

with $z = (1, 2, 2, 1)$. For $\mathbf{Z}_5$,

$$u_0 = e_1,\ u_1 = \tfrac{1}{2}(-\tau^{-1}, \tau, 1, 0),\ u_2 = -e_3,\ u_3 = \tfrac{1}{2}(0, -\tau^{-1}, 1, \tau),\ u_4 = -e_4,$$

with $z = (\tau^{-2}, 2\tau^{-1}, \sqrt{5}, 2\tau, \tau^2)$ and $\tau := \frac{1}{2}(1 + \sqrt{5})$, the golden section.

We treat the groups $\mathbf{T}_7$, $\mathbf{T}_8$ and $\mathbf{T}_9$ in a slightly different way; for these, we consider the diagram of (1E17):

1E17

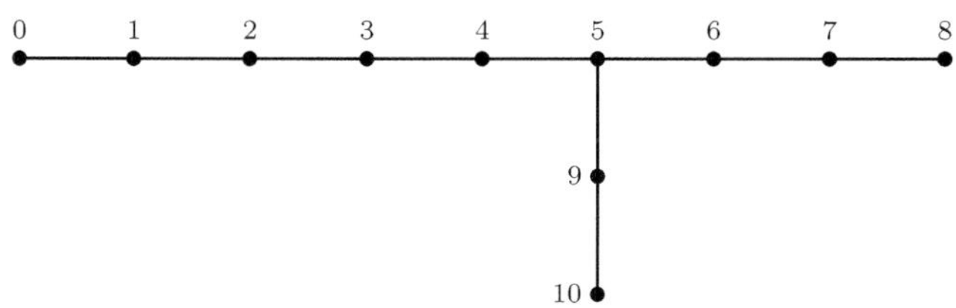

It should be emphasized that we have omitted two branches (see the notes at the end of the section) – we have only drawn what will actually be used. A mark j at a node indicates the normal vector u_j, given as follows:

$$\begin{aligned}
\sqrt{8}u_0 &= (1, 1, 1, 1, 1, 1, 1, 1),\\
\sqrt{2}u_1 &= -(e_1 + e_8),\\
\sqrt{2}u_j &= e_{j-1} - e_j, \quad \text{for } j = 2, \ldots, 8,\\
\sqrt{8}u_9 &= (-1, -1, -1, -1, 1, 1, 1, 1),\\
\sqrt{2}u_{10} &= e_1 - e_8.
\end{aligned}$$

Then the appropriate normals are as follows.

$$\begin{aligned}
&\mathbf{T}_7\colon && u_3,\ \ldots,\ u_7,\ u_9,\ u_{10}, && z = (1, 2, 3, 2, 1, 2, 1);\\
&\mathbf{T}_8\colon && u_2,\ \ldots,\ u_8,\ u_9, && z = (1, 2, 3, 4, 3, 2, 1, 2);\\
&\mathbf{T}_9\colon && u_0,\ \ldots,\ u_7,\ u_9, && z = (1, 2, 3, 4, 5, 6, 4, 2, 3).
\end{aligned}$$

Observe that this description has $\mathbf{T}_6$ and $\mathbf{T}_7$ acting (effectively) on subspaces of $\mathbb{E}^8$.

The proof that there can be no more diagrams for affine reflexion groups than those listed is completed by two crucial observations:

- no mark on any branch can be increased;
- no marked branch can be added.

For the first, let the nip-set of unit outer normals to the fundamental simplex be $U = \{u_0, \ldots, u_d\}$, with corresponding positive dependence $z(U) = (\zeta_0, \ldots, \zeta_d) \in \mathbb{E}$. Suppose, if possible, that we have unit normals $V = \{v_0, \ldots, v_d\}$ corresponding to a diagram with at least one mark strictly increased from that of U (and the remainder no smaller). Because $\langle v_i, v_k \rangle < \langle u_i, u_j \rangle$ for at least one pair $\{i, j\}$ with $i \neq j$, we should then have

$$0 \leqslant \varphi(V, z(U)) < \varphi(U, z(U)) = 0,$$

a contradiction. The second claim follows directly from the description of nip-sets in Theorem 1E4.

We can now repeat the analysis of [27, Section 11.5], to classify the diagrams of the finite as well as affine reflexion groups. Consider a possible (new) diagram $\mathbf{X}$; we first suppose that $\mathbf{X}$ is not a string diagram.

- Since each $\mathbf{P}_{d+1}$ is euclidean, $\mathbf{X}$ must be a tree (that is, it has no circuits).
- Since $\mathbf{Q}_5$ is euclidean, no node of $\mathbf{X}$ belongs to more than three branches.
- Since each $\mathbf{Q}_{d+1}$ is euclidean for $d > 4$, no more than one node of $\mathbf{X}$ can belong to three branches.
- Since each $\mathbf{S}_{d+1}$ is euclidean, if $\mathbf{X}$ has three branches at some node, then no branch can be marked.
- If $\mathbf{X}$ has three branches with r, s, t (extra) nodes, where $r \geqslant s \geqslant t \geqslant 1$, then $s \geqslant 1$ since $\mathbf{X} \neq \mathbf{B}_d$ for any d. Since $\mathbf{T}_7$ is euclidean, $t = 1$. Since $\mathbf{T}_8$ is euclidean, $s = 2$. Since $\mathbf{T}_9$ is euclidean, $r \leqslant 5$.

We are now reduced to a string diagram $\mathbf{X}$, again assumed not to be in our list already.

- Since $\mathbf{R}_{d+1}$ is euclidean, only one branch of $\mathbf{X}$ can be marked.
- Since $\mathbf{V}_3$ is euclidean, no mark can exceed 5.
- Since $\mathbf{Z}_4$ is euclidean and $q < 5$, no mark on a middle branch can exceed 4.
- Since $\mathbf{U}_5$ is euclidean, a mark 4 or 5 can only occur on an end branch.
- Since $\mathbf{Y}_5$ is euclidean and $\frac{5}{2} < 3$, $\mathbf{X}$ cannot have any branch with mark 5.

But this has now excluded any new diagrams, and hence the enumeration is complete.

All that remains is to show that each admissible diagram indeed corresponds to a finite or affine reflexion group (as appropriate). For this, we employ a versatile argument, which appears in several contexts; for regularity, it was used in [70]. Let K be a region determined by the diagram; that is, K is a simple cone or simplex whose unit outer normal vectors correspond to the diagram. Note that we have specified such vectors in the earlier part of the proof. Let $\mathbf{G}$ be generated by the reflexions in the bounding hyperplanes of K. Observing that no problems arise with one or two generators, we appeal to an inductive assumption that the copies of K under $\mathbf{G}$ fit together nicely around each edge (for K a cone) or vertex (for K a simplex). In other words, the copies completely surround the edge or vertex without their interiors overlapping. We then deduce that all the copies of K under $\mathbf{G}$ cover $\mathbb{E}$. If we had an overlap somewhere, say with a chain $K_0, \ldots, K_r$ of copies with K_{j-1} and K_j meeting on a common facet for each j and $\operatorname{int} K_0 \cap \operatorname{int} K_r \neq \emptyset$, then we could construct a polygonal circuit $a_0, \ldots, a_r = a_0$, with $a_j \in \operatorname{int} K_j$ which passes from K_{j-1} to K_j across their common facet. Now contract this circuit to a_0, avoiding the common apex of the cones in the former case. Since at each stage of the contraction everything fits together nicely, we see that no contradiction can arise, and thus the copies of K under $\mathbf{G}$ form fundamental regions for $\mathbf{G}$. □

1E18 Remark There is an alternative notation for the groups $\mathbf{B}_d$ ($d \geqslant 4$), $\mathbf{Q}_5$, $\mathbf{E}_d$ ($d = 6, 7, 8$) and $\mathbf{T}_d$ ($d = 7, 8, 9$). If the Coxeter diagram is a tree, with unmarked branches of length $r, s, t, \ldots$ coming out of a common node, then we write $[3^{r,s,t,\ldots}]$ for the corresponding group. Hence there are $r + s + t + \cdots + 1$ nodes in all. Thus, for example, $\mathbf{T}_8 = [3^{3,3,1}]$. Note that we could also write $[3^{r,s}]$ (or even $[3^{r,s,0}]$) for $\mathbf{A}_{r+s+1}$, with any r and s.

Coxeter Groups

So far, we have just talked about reflexion groups. However, there is an important generalization of reflexion group. A *Coxeter group* is an abstract group of the form

1E19 $$\boldsymbol{G} = \langle \boldsymbol{r}_1, \ldots, \boldsymbol{r}_d \mid (\boldsymbol{r}_j \boldsymbol{r}_k)^{p_{jk}} = \boldsymbol{e} \rangle,$$

where $p_{jj} = 1$ and $p_{jk} = p_{kj} \geqslant 2$ for $j \neq k$ (possibly $p_{jk} = \infty$, giving no relation). The argument of the last part of the proof of Theorems 1E11 and 1E13 needs very little added to show

1E20 Proposition *A finite or discrete affine reflexion group is isomorphic to the corresponding Coxeter group.*

Proof. The abstract relations of the reflexion group $\mathbf{G}$, with generators (R_0, if present), $R_1, \ldots, R_d$ corresponding to the nodes of the diagram, are just those given by the marks on the branches of the diagram. To see this, note that a relation on $\mathbf{G}$ in terms of its generators is induced by a circuit though copies of the fundamental region K for $\mathbf{G}$; the contraction argument in the proof of Theorems 1E11 and 1E13 shows that only the relations arising from the circuit passing across faces of codimension 2 are needed, and these come from the diagram. □

If the Coxeter group $\boldsymbol{G}$ of (1E19) additionally satisfies $p_{jk} = 2$ whenever $|j - k| \geqslant 2$, then we call $\boldsymbol{G}$ a *string* Coxeter group. We then write $p_j := p_{j,j+1}$ for each $j = 1, \ldots, j - 1$, and denote $\boldsymbol{G}$ by

1E21 $$\boldsymbol{G} =: [p_1, \ldots, p_{d-1}].$$

Thus Proposition 1E20 says that $[p_1, \ldots, p_{d-1}]$ is isomorphic to the hyperplane reflexion group with diagram

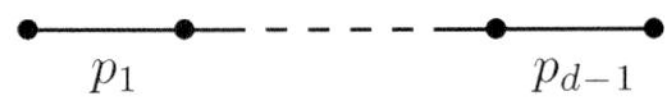

when the reflexion group is finite or affine. In this spirit, we may associate with every Coxeter group $\boldsymbol{G}$ of (1E19) a corresponding Coxeter diagram: for each generator $\boldsymbol{r}_j$ we have a node j, with two nodes j, k joined by a branch marked p_{jk}. As with our previous diagrams, branches marked 2 are omitted, and a mark 3 on a branch is omitted. Naturally, associated with a string Coxeter group is a *string* diagram.

Two Coxeter groups, and their rotation subgroups, occur frequently enough in abstract contexts to deserve special notation. Thus we have the following.

- The *symmetric group* S_n is the group of all permutations on n elements, say $\{1, \ldots, n\}$. In this notation, which we shall often use, α^k is shorthand for a string $\alpha, \ldots, \alpha$ of length k. We can identify S_n with the permutations of the standard basis of $\mathbb{E}^n$. In a natural way, $\mathbf{A}_d \cong \mathrm{S}_{d+1}$.
- The subgroup of S_n consisting of the even permutations on $\{1, \ldots, n\}$ is the *alternating group* A_n.
- The *dihedral group* $\mathrm{D}_n = [n]$ is just the abstract group corresponding to $\mathbf{D}_2^n$; its even subgroup is the *cyclic group* C_n.

As in Section 1D, we can extend the notation to signed permutations.

Unitary Reflexion Groups

Unitary reflexion groups do not play a central rôle in this text. However, we shall appeal to some of them to provide examples of realization domains in, for example, Sections 4G and 15F. Moreover, we also have an excuse to introduce some notation which we need later on.

The basic case we have is that of a circuit of m branches, one marked q and the others unmarked, with a mark $p \geqslant 2$ inside the circuit. If we label the corresponding involutory generators $\boldsymbol{r}_1, \ldots, \boldsymbol{r}_m$ in cyclic order, then the central mark imposes an additional relation

1E22 $$(\boldsymbol{r}_1\boldsymbol{r}_2 \cdots \boldsymbol{r}_m\boldsymbol{r}_{m-1} \cdots \boldsymbol{r}_2)^p = \boldsymbol{e}$$

on the corresponding infinite group (which is $\mathbf{P}_m$ if $q = 3$ also – it may be seen that the starting point in the labelling is irrelevant so far as the additional relation is concerned). In case $q = 3$, as a geometric group it acts on $\mathbb{C}^m$, and is generated by all involutory reflexions in the hyperplanes with normals $e_j - \vartheta e_k$, with $1 \leqslant j < k \leqslant m$ and any pth root ϑ of 1. If we take ϑ to be a primitive pth root, then the group consists of all mappings of the form

$$(\zeta_1, \ldots, \zeta_m) \mapsto (\vartheta^{k_1}\zeta_{1\sigma}, \ldots, \vartheta^{k_m}\zeta_{m\sigma}),$$

with σ a permutation of $\{1, \ldots, m\}$ and $k_1 + \cdots + k_m \equiv 0 \pmod{p}$.

The case $p = 2$ and $q = 3$ occurs quite often as a subdiagram (for some m); we then have the real group $\mathbf{B}_m$, but with a different way of representing it by a diagram. Since we will need to identify the whole group, we show how to change generators appropriately. We begin with the diagram of Figure 1E23; the labels on the nodes indicate the initial generators $\boldsymbol{s}_1, \ldots, \boldsymbol{s}_m$.

1E23

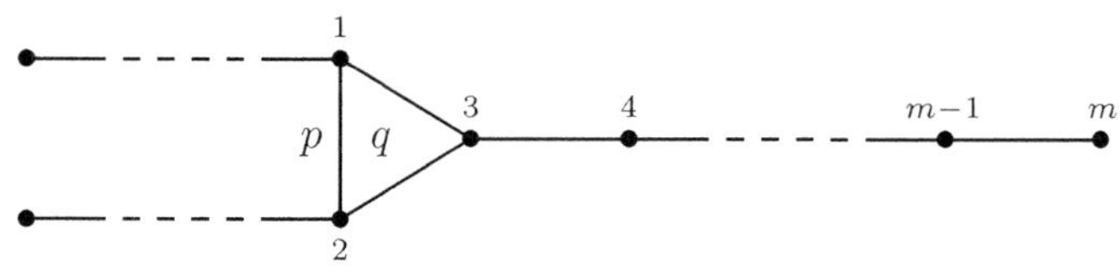

No part of the diagram to the left of the triangle plays any rôle, so long as there are no circuits involving the branch marked p.

We now define an invertible mixing operation by

1E24

$$\begin{aligned} \boldsymbol{r}_1 &:= \boldsymbol{s}_m \boldsymbol{s}_{m-1} \cdots \boldsymbol{s}_4 \boldsymbol{s}_3 \cdot \boldsymbol{s}_1 \cdot \boldsymbol{s}_3 \boldsymbol{s}_4 \cdots \boldsymbol{s}_{m-1} \boldsymbol{s}_m, \\ \boldsymbol{r}_j &:= \boldsymbol{s}_j, \quad \text{for } j = 2, \ldots, m; \end{aligned}$$

thus we replace $\boldsymbol{s}_1$ by its conjugate $\boldsymbol{r}_1$ (as indicated in (1E24) – see the notes at the end of the section). Then the diagram corresponding to the new generators $\boldsymbol{r}_1, \ldots, \boldsymbol{r}_m$ is that of Figure 1E25; again, the labels on the nodes indicate the generators.

1E25

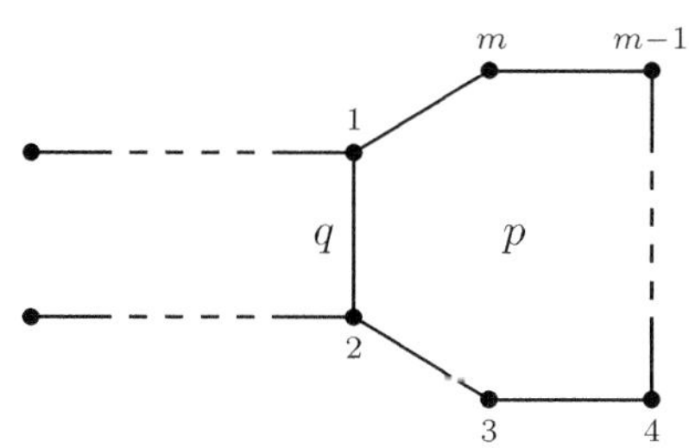

We shall not give the fairly routine proof here; instead, we refer the interested reader to [99, p. 319].

1E26 Remark When $p = 2$ and $q = 3$, the new diagram of Figure 1E25 will give an alternative way of presenting a group $\mathbf{B}_d$, $\mathbf{E}_d$ or $\mathbf{T}_{d+1}$ (for some d).

1E27 Remark In [99, Chapter 9], there is a detailed discussion of yet other ways of representing unitary reflexion groups by diagrams, and the connexions among them. We shall say no more about this topic here.

Notes to Section 1E

1. Our treatment of the classification problem is based on that of Coxeter in [27, Chapters X, XI], but with simplifications coming from our application of convexity arguments. In particular, using nip-sets enables us to avoid any discussion of quadratic forms, except in a very minor way (which does, however, correspond to a crucial part of Coxeter's argument).
2. It might be thought that, in the last part of the proof of Theorems 1E11 and 1E13, we should ensure that at no stage of the contraction does the circuit meet a face of any K_j of codimension 3 (or more). In fact, the inductive assumption covers all but vertices of K in the conical case.
3. Coxeter has described the origins of Coxeter diagrams in [31]; we give a very abbreviated account here. He introduced them in [18], in a way that already allowed for possibly fractional marks (he talked there about dots and links, while we have followed his later usage of nodes and branches); the first drawings of Coxeter diagrams are in [21, 22]. Witt [130] found something close to Coxeter diagrams around 1941; he had a branch with integer mark q represented by $q-2$ links. Though Dynkin is often quoted as an independent discoverer, in [44] (according to Coxeter's account) his diagrams had $4\cos^2(\pi/q)$ links instead; this implies that he could only draw diagrams for crystallographic groups. In view of this latter restriction, it seems perverse to credit Dynkin's so-called rediscovery; we feel that Coxeter himself was over-generous in this respect.
4. The missing branches in Figure 1E17 join node 10 to nodes 2 and 8, with branches marked $\frac{3}{2}$. Thus these branches do not contribute to our arguments.
5. The fact that the positive dependences for $\mathbf{T}_7, \mathbf{T}_8$ and $\mathbf{T}_9$ can be chosen to consist of small integers (including 1), moreover in arithmetic progressions up to the trivalent node, is easy to see, but has additional significance that we shall not go into here; see (for example) [27, 11.9].
6. A converse to Proposition 1E20 holds, in that a finite Coxeter group is isomorphic to a reflexion group. There is an extension to certain infinite Coxeter groups – precisely those corresponding to affine reflexion groups. However, to describe this would involve a discussion of quadratic forms that we have been at pains to avoid.
7. A good recent account of the conventional approach to classifying the discrete reflexion groups is to be found in Humphreys [63].
8. Mixing operations in general will be discussed in Sections 5A and 5B.

1F Subgroup Relationships

We briefly describe here the subgroup relationships among the finite and discrete affine reflexion groups. We begin with some obvious ones.

1F1 Proposition *There are the following connexions among the hyperplane reflexion groups:*

(a) $\mathbf{B}_d < \mathbf{C}_d$, *with index* 2;
(b) *if d is odd, then* $\mathbf{C}_d = \mathbf{B}_d \times \mathbf{Z}_2$, *with* $\mathbf{Z}_2 = \{\pm I\}$;
(c) $\mathbf{D}_2^3 = \mathbf{A}_2$, $\mathbf{D}_2^4 = \mathbf{C}_2$;
(d) $\mathbf{B}_3 = \mathbf{A}_3$;
(e) $\mathbf{Q}_{d+1} < \mathbf{S}_{d+1} < \mathbf{R}_{d+1}$, *with indices* 2;

(f) $\mathbf{P}_4$ *plays the rôle of* $\mathbf{Q}_4$.

Observe that we could set $\mathbf{G}_2 = \mathbf{D}_2^5$ to begin that small sequence. Since we do not need the diagram $\mathbf{Z}_5$ any longer, this frees $\mathbf{Z}_k$ to stand for a geometric cyclic group of order k (the case $k = 2$ is the usual one, as above).

We next prove a result to which we shall appeal in Section 1H as well.

1F2 Theorem *The point-group* **SG** *of a discrete affine reflexion group* **G** *is isomorphic to a subgroup of* **G**.

Proof. Bieberbach's Theorem 1D21 shows that **G** has a lattice of translations of full rank; it is not too hard to prove this directly (all we need is one translation, and then its conjugates under any finite irreducible subgroup of **G** provide a lattice of full rank). Thus **G** has just finitely many families of parallel reflexion hyperplanes; the point-group **SG** is then generated by the hyperplanes through o parallel to these families. The discreteness of **G** implies that **SG** is finite, and so (in $\mathbb{E}^d$) **SG** is generated by d of these reflexions. Now pick arbitrary reflexion hyperplanes among those of **G** that are parallel to these d. These hyperplanes meet in a point, which must be a vertex of some fundamental region of **G**. The subgroup of **G** fixing this vertex is then isomorphic to **SG**, which establishes the claim. □

What Theorem 1F2 says is that the point-group of an affine reflexion group can be identified with its largest (reflexion) subgroup. But we observe that the relationship also goes the other way: if **S** is a finite subgroup of an affine reflexion group **G**, and **S** is the point-group of an affine reflexion group **H**, then **H** is a subgroup of **G**. (Bear in mind here that, for instance, an affine reflexion group acting irreducibly on $\mathbb{E}^d$ is a subgroup of index 2^d in itself.) Thus we can add some exceptional cases to Proposition 1F1.

1F3 Proposition *The following subgroup relationships hold:*

- $\mathbf{P}_3 < \mathbf{V}_3$,
- $\mathbf{C}_4 < \mathbf{F}_4$,
- $\mathbf{A}_4, \mathbf{B}_4 < \mathbf{G}_4$,
- $\mathbf{R}_5 < \mathbf{U}_5$,
- $\mathbf{A}_7 < \mathbf{E}_7$,
- $\mathbf{P}_8 < \mathbf{T}_8$,
- $\mathbf{A}_8, \mathbf{B}_8 < \mathbf{E}_8$,
- $\mathbf{P}_9, \mathbf{Q}_9 < \mathbf{T}_9$.

We shall not prove any of these here. The first is trivial, and the next three will emerge from the constructions in Chapter 7. The last four are consequences of the preceeding discussion. We shall also meet some of these relationships in Section 1H.

1F4 Remark The cases arising from the groups $\mathbf{T}_{d+1}$ for $d = 7$ and 8 are particularly interesting. Thus we have

- $\mathbf{E}_7$ has a subgroup $\mathbf{A}_7$ of index 72,
- $\mathbf{E}_8$ has subgroups $\mathbf{A}_8$ of index 1920 and $\mathbf{B}_8$ of index 135.

The relationship $\mathbf{B}_8 < \mathbf{E}_8$ plays a central rôle in the realization domain of the (abstract) Gosset–Elte polytope $\{3^2, 3^{4,1}\}$, as will be seen in Section 14G.

1G Angle-Sum Relations

In Section 1H, we calculate the orders of the finite Coxeter groups using various angle-sum relations. We establish those relations here.

Let K be a d-dimensional polyhedral set in $\mathbb{E}^d$ and let F be a non-empty face of K. The *(inner) normalized angle* $\alpha(F, K)$ of K at F is that proportion of a sufficiently small ball centred at a relatively interior point a of F which lies in K:

$$\alpha(F, K) = \lim_{\lambda \searrow 0} \frac{\mathrm{vol}(B(a, \lambda) \cap K)}{\mathrm{vol}(B(x, \lambda))},$$

where $B(a, \lambda) := \{x \in \mathbb{E}^d \mid \|x - a\| \leqslant \lambda\}$ is the *ball* of radius $\lambda > 0$ with centre a, and vol as before denotes volume. We shall employ two *angle-sum relations*. The first is the *Brianchon–Gram Theorem* (see [8, 55]).

1G1 Theorem *If $d \geqslant 1$ and P is a convex d-polytope in $\mathbb{E}^d$, then*

$$\sum_F (-1)^{\dim F} \alpha(F, P) = 0.$$

The second is the *Sommerville Theorem* (see [123]).

1G2 Theorem *If P is a polyhedral cone in $\mathbb{E}^d$, then*

$$\sum_F (-1)^{\dim F} \alpha(F, P) = (-1)^d \alpha(A, P),$$

where A is the face of apices of P.

In both theorems, the sums extend over all non-empty faces F (including P itself).

Sketch proof of Theorems 1G1 and 1G2. Let K be a polyhedral set in $\mathbb{E}^d$ of full dimension, say

$$K = \{x \in \mathbb{E}^d \mid \langle x, u_j \rangle \leqslant \beta_j \text{ for } j = 1, \ldots, n\}.$$

Now let $y \in \mathbb{E}^d$ be a general direction, that is, $\langle y, u_j \rangle \neq 0$ for any j. If F is a proper face of K, then $y \in A(F, K)$, the *angle cone* of K at F, if $x + \lambda y \in K$ for each $x \in \operatorname{relint} F$ and all small enough $\lambda > 0$. Projecting all these sets $\operatorname{relint} F$ in direction y yields an open convex set (of dimension $d - 1$) with Euler characteristic $(-1)^{d-1}$ from Theorem 1B23. Now if there are such faces F for y, then there are facets, and so $\langle y, u_j \rangle < 0$ for at least one j. Hence, the only directions y for which there are no such faces are those for which $\langle y, u_j \rangle > 0$

for all j, in other words, those in $-\operatorname{rec} K = \operatorname{rec}(-K)$, with the recession cone as defined in (1B14).

The contribution from K itself, which is $(-1)^d$ for all directions, cancels out everything except the 0 for $y \in \operatorname{rec}(-K)$; we thus arrive at

$$\sum_F (-1)^{\dim F} \alpha(F, K) = (-1)^d \alpha(-A, \operatorname{rec}(-K)) = (-1)^d \alpha(A, \operatorname{rec} K),$$

where A is the face of apices of $\operatorname{rec} K$ as defined in Remark 1B18. This is the general result.

For the specific cases, we just note that, if K is a polytope, then $\operatorname{rec} K = \{o\}$; similarly, if K is a cone, then $\operatorname{rec} K = K$. □

This common generalization to arbitrary polyhedral sets of the Brianchon–Gram and Sommerville Theorems is the metric version of an equidissectability result proved in [74]. The point here is that the theorems belong to affine rather than euclidean geometry; for further details, see the notes at the end of the section.

Notes to Section 1G

1. The result we have called the Brianchon–Gram Theorem was formerly attributed solely to Gram [55] (perhaps first in [56, Section 14.1], wherein are to be found other historical references), who proved it in the 3-dimensional case. However, McMullen discovered that Gram had been anticipated by Brianchon [8], whence the name given here for the theorem.
2. So long as we distinguish between a cone K and its negative $-K$, any angle function α can be used instead of the euclidean angle in the common generalization of Theorems 1G1 and 1G2. For instance, we can take an arbitrary convex body C with $o \in \operatorname{int} C$, and define
$$\alpha(K) := \frac{\operatorname{vol}(C \cap K)}{\operatorname{vol} C};$$
then the theorems still hold for this new angle α.
3. While Schläfli [112] seems not to know the Sommerville theorem in full generality, he does give a reduction from odd dimension to the lower even dimensions in case of a simple cone.
4. More about the background to these results can be found in [56, Section 14.1] and [74]; the present proof of Theorem 1G1 was first given by Shephard [119].

1H Group Orders

In this section, we describe how the Brianchon–Gram Theorem for angle sums of convex polytopes can be employed to calculate the order of a finite Coxeter group in a purely elementary manner. We largely follow [99, Section 3E], which somewhat simplified the original treatment in [77]. However, we also prove the simplex dissection results that we merely appealed to in [99], in part at least because they are useful subsequently.

Let $\mathbf{G}$ be a finite group generated by reflexions in hyperplanes in euclidean space $\mathbb{E}^d$. A convention that we introduced in Section 1C identified a reflexion R (or, more generally, an involutory isometry of $\mathbb{E}^d$), with its mirror of fixed points $\{x \in \mathbb{E}^d \mid xR = x\}$. These hyperplane mirrors must contain a common point; if we take this point to be the origin o of coordinates, then $\mathbf{G}$ is an orthogonal group. Recall that these groups are listed in Table 1E12.

As the discussion in Section 1E showed, the images under $\mathbf{G}$ of the mirrors of the generating reflexions R_j dissect the space $\mathbb{E}^d$ into congruent convex cones, which are fundamental regions for $\mathbf{G}$, and whose number is obviously the order $|\mathbf{G}|$ of $\mathbf{G}$. Thus, to find $|\mathbf{G}|$, it is only necessary to count these cones or, equivalently, measure their normalized volumes or *angles*. However, until [77], if $d \geqslant 4$ no strictly elementary way of doing this was available.

Let us assume that $\mathbf{G}$ is *irreducible*, that is, acts irreducibly on $\mathbb{E}^d$ (with no non-trivial invariant subspaces). Then, again as seen in Section 1E (see also [27, Chapter 11]), $\mathbf{G}$ is generated by precisely d reflexions, whose mirrors may be chosen to bound any one of the fundamental cones in the dissection of $\mathbb{E}^d$ just described.

If these mirrors are $R_1, \ldots, R_d$, then, for $1 \leqslant j < k \leqslant d$, the dihedral angle between R_j and R_k is π/p_{jk} for some integer $p_{jk} \geqslant 2$, and $\mathbf{G}$ has the presentation

$$G = \langle R_1, \ldots, R_d \mid (R_j R_k)^{p_{jk}} = E \text{ for } 1 \leqslant j \leqslant\ k \leqslant n \rangle,$$

where $p_{jj} = 1$ for $1 \leqslant j \leqslant d$, and E is the identity. We recall that abstract groups with such presentations are known as Coxeter groups; we saw in Section 1E (see also Coxeter [22]) that all finite or affine reflexion groups are, in fact, isomorphic to Coxeter groups.

Let us briefly survey the earlier methods for calculating the order $|\mathbf{G}|$ of $\mathbf{G}$. First, we may associate with $\mathbf{G}$ a convex polytope, the numbers of whose faces of various kinds are the indices $[\mathbf{G} : \mathbf{H}]$ of certain subgroups $\mathbf{H}$ of $\mathbf{G}$ which are also generated by reflexions. If d is odd, then Euler's Theorem (see, for example, [56, Chapter 8]) and the knowledge of the orders $|\mathbf{H}|$ will yield $|\mathbf{G}|$ (see also the remarks below), so that the actual polytope need not be constructed. When d is even, so that Euler's Theorem on its own can only yield the ratios of the numbers of faces, a suitable polytope can often be constructed by synthetic methods, and again the value of $|\mathbf{G}|$ results. (From an historical point of view, of course, reflexion groups arise from polytopes, rather than the other way round.) In fact, various simplex dissection results (for example, in the case of the group $[3, 3, 5]$ below; see also [29], which contains many other useful references) enable us to avoid such arguments in all but a very few cases; unfortunately, such cases are the most interesting.

Second, for $d = 4$, the order of the symmetry group of a convex regular polytope (which excludes only one of the five cases, and this is in any event a subgroup of index 2 in one of the others) can be calculated with the aid of a solution of a trigonometrical equation; we refer the reader to [27, Section 12.8], whose details we do not reproduce. An alternative method involves evaluating certain integrals due to Schläfli; since the most relevant one of these cannot be evaluated directly, recourse must be had once more to the simplex dissection results (again, see [29]). For more background material, see the notes at the end of the section.

Finally, for other even $d \geqslant 6$ ($d = 6$ and $d = 8$ are the only important cases, but also see the notes at the end of the section), the group $\mathbf{G}$ can be associated with a honeycomb, and $|\mathbf{G}|$ can be found from the relative numbers of faces of this honeycomb. However, none of these last three methods is elementary; in particular, the last depends upon the somewhat deep result (see [27, Section 9.8]) that these relative numbers exist, and that the analogue of Euler's Theorem holds for them. A variant of this technique appears in [17], but it is used there with knowledge of the order of the group $[3, 3, 5]$ to calculate the densities of the regular 4-dimensional star-polytopes. It is described in [27, Chapter 14] as resting 'on rather flimsy foundations'; however, see the case of $[3, 3, 5]$ below.

The method which we shall describe here is, in a vague sense, related to the last of these approaches (we shall make the connexion more explicit later), but the results to which we shall appeal, namely, Theorems 1G1 and 1G2, only rely (as we saw) on the ordinary Euler Theorem (in $\mathbb{E}^{d-1}$).

It is worth making an additional remark at this stage. Our calculations will be purely geometric; in other words, though we often use the language of group theory, we do not really make use of the fact that the cones whose sizes we find are the fundamental regions of groups. What we do use is the fact that certain hyperplanes, with given angles between them, determine either simple cones or simplices. We shall need to augment our lists in Tables 1E12 and 1E14 not only by $\mathbf{Z}_5$ but also by other diagrams with marks $\frac{5}{2}$; in the case of simplices, we do not obtain discrete groups.

We could employ the angle-sum relations directly; this is what was done in [77]. However, it is better to make a simplification of the resulting formula first. In applying Theorem 1G1, we can confine our attention to a d-simplex T, since we know from Theorem 1E8 that we only need consider this case. We label the facets of T by $T_0, \ldots, T_d$, with T_0 subsequently playing a distinguished rôle. The *angle graph* $\mathcal{G}$ of T has node set $\mathsf{D} := \{0, \ldots, d\}$, and the branch-set B contains $\{j, k\}$ (for $0 \leqslant j < k \leqslant d$) precisely when the dihedral angle between T_j and T_k is *not* $\pi/2$. In the case when T is the fundamental region of some Coxeter group $\mathbf{G}$, this means that $\mathcal{G}$ is just the Coxeter diagram of $\mathbf{G}$, but with the labels left off (a branch corresponding to a label 2 is still omitted). We may then talk about connected subgraphs, components, and so on, of $\mathcal{G}$. Since T is a euclidean simplex, it is not hard to see that $\mathcal{G}$ itself is connected.

If $\mathsf{J} \subseteq \mathsf{D}$, then we write $T_{\mathsf{J}} := \bigcap\{T_i \mid i \in \mathsf{J}\}$ (thus T_j is an abbreviation for $T_{\{j\}}$). If $\mathsf{J} \subset \mathsf{D}$ (a proper subset), then Theorem 1G2 implies that

$$\sum_{\mathsf{K} \subseteq \mathsf{J}} (-1)^{\operatorname{card} \mathsf{K}} \alpha(T_{\mathsf{K}}, T) = \alpha(T_{\mathsf{J}}, T),$$

since $\dim T_{\mathsf{K}} = d - \operatorname{card} \mathsf{K}$. We can now decompose the formula in Theorem 1G1 in the following way. Let $\mathsf{M} \subset \mathsf{D}$ be a proper subset which induces a connected subgraph $\mathcal{G}(\mathsf{M})$ of $\mathcal{G}$ containing node 0; formally, we also allow $\mathsf{M} = \emptyset$ here. We write $\mathcal{M} := \mathcal{M}(\mathsf{D})$ for the set of such M. Define

$$\mathsf{M}^* := \{i \in \mathsf{D} \mid \{i,j\} \notin \mathsf{B} \text{ for any } j \in \mathsf{M}\},$$

with $\mathsf{M}^* := \mathsf{D} \setminus \{0\}$ if $\mathsf{M} = \emptyset$. Thus M^* is the set of nodes of $\mathcal{G}$ obtained by deleting M and any neighbours of M in $\mathcal{G}$, with 0 itself taken as the sole neighbour if $\mathsf{M} = \emptyset$. In other words, M^* induces the *disjoint* subgraph of $\mathcal{G}$, namely, the maximal subgraph $\mathcal{G}(\mathsf{M}^*)$ of $\mathcal{G} \setminus \{0\}$ which is disconnected from $\mathcal{G}(\mathsf{M})$.

We now collect together the contributions $(-1)^{\dim F}\alpha(F,T)$ in Theorem 1G1 from faces F of T as follows. Suppose that $F = T_{\mathsf{J}}$, with $\mathsf{J} \subset \mathsf{D}$ (note that $T_{\mathsf{D}} = \emptyset$, which is not counted; however, $T_{\emptyset} = T$ is counted). Write $\mathsf{M} := \mathsf{M}(F)$ for the node-set of the connected component of the induced subgraph $\mathcal{G}(\mathsf{J})$ of $\mathcal{G}$ which contains 0, and let $\mathsf{M}^* := \mathsf{M}^*(F)$ be the corresponding vertex-set of the disjoint subgraph.

We thus see that $F = T_{\mathsf{M} \cup \mathsf{K}}$ for some $\mathsf{K} \subseteq \mathsf{M}^*$ and, since our angles are scaled so that the total angle is 1, we have

$$\alpha(F,T) = \alpha(T_{\mathsf{M}},T)\alpha(T_{\mathsf{K}},T).$$

Adding together first the contributions for a fixed subset M yields

$$\begin{aligned}\sum_{\mathsf{M}(F)=\mathsf{M}} (-1)^{\dim F}\alpha(F,T) &= \sum_{\mathsf{K} \subseteq \mathsf{M}^*} (-1)^{d-\operatorname{card}\mathsf{M}-\operatorname{card}\mathsf{K}}\alpha(T_{\mathsf{M}\cup\mathsf{K}},T) \\ &= \sum_{\mathsf{K} \subseteq \mathsf{M}^*} (-1)^{d-\operatorname{card}\mathsf{M}-\operatorname{card}\mathsf{K}}\alpha(T_{\mathsf{M}},T)\alpha(T_{\mathsf{K}},T) \\ &= (-1)^{\operatorname{card}\mathsf{M}}\alpha(T_{\mathsf{M}},T)\alpha(T_{\mathsf{M}^*},T).\end{aligned}$$

We can summarize the discussion as

1H1 Theorem *If T is a d-simplex, then*

$$\sum_{\mathsf{M} \in \mathcal{M}} (-1)^{\operatorname{card}\mathsf{M}}\alpha(T_{\mathsf{M}},T)\alpha(T_{\mathsf{M}^*},T) = 0.$$

Exactly similar considerations yield a simplification of Theorem 1G2. The only difference in the calculations is that we now have a simple d-cone T, bounded by $T_0, \ldots, T_{d-1}$ (we have changed the notation for future convenience); we thus take $\mathsf{D} := \{0, \ldots, d-1\}$. We now allow $\mathsf{M} = \mathsf{D}$ to give a connected subgraph of $\mathcal{G}(T)$, and our conclusion is

1H2 Theorem *If T is a simple d-cone, then*

$$\sum_{\mathsf{M} \in \mathcal{M}} (-1)^{\operatorname{card}\mathsf{M}}\alpha(T_{\mathsf{M}},T)\alpha(T_{\mathsf{M}^*},T) = (-1)^d\alpha(O,T),$$

where $O := \{o\}$ is the apex of T.

We now apply Theorems 1H2 and 1H1 to finite groups $\mathbf{G}$ generated by reflexions in $\mathbb{E}^d$. Actually, while we talk about orders of subgroups, we are really only making angle-sum calculations, and so our results sit in a wider context. Suppose that $\mathbf{G} = \langle R_j \mid j \in N \rangle$ is a finitely generated reflexion group in $\mathbb{E}^d$; we may have to allow $\mathbf{G}$ to be infinite, or even non-discrete. We write $\mathbf{G}_\mathsf{J} := \langle R_j \mid j \in \mathsf{J} \rangle$ for each subset $\mathsf{J} \subseteq \mathsf{D}$, and $g_\mathsf{J} := |\mathbf{G}_\mathsf{J}|$ when this order is finite.

We begin with the easier cases. First, let $\mathbf{G}$ be a finite Coxeter group; we may suppose that $\mathbf{G}$ acts irreducibly on $\mathbb{E}^d$, so that $\mathsf{D} = \{0, \ldots, d-1\}$. Then the mirrors R_j bound a simple d-cone T with facets $T_0, \ldots, T_{d-1}$. Moreover, T is a fundamental region for $\mathbf{G}$, and T and its images form a dissection of $\mathbb{E}^d$ into congruent cones. Indeed, for each $\mathsf{J} \subseteq \mathsf{D}$, the images of T under $\mathbf{G}_\mathsf{J}$ surround the face T_J of T, so that the total angle of these cones at T_J is 1. The initial discussion and the definition of angle above yield at once

1H3 Lemma *For each* $\mathsf{J} \subset \mathsf{D}$,

$$\alpha(T_\mathsf{J}, T) = \frac{1}{g_\mathsf{J}}.$$

Note here that, for the case $\mathsf{J} = \emptyset$, we have $T_\emptyset = \{E\}$, the identity subgroup, and $g_\emptyset = 1$. We use the abbreviation $g := g_\mathsf{D} = |\mathbf{G}|$.

Now, when d is even, the term $\alpha(T_\mathsf{D}, T) = \alpha(O, T) = 1/g$ occurs on each side of the formula of Theorem 1H2 (or of Theorem 1G2), and hence the formula yields no useful information about g. However, if d is odd, then we can move the term $1/g$ from the left to the right side of the equation, and deduce

1H4 Theorem *If* $\mathbf{G} = \langle R_0, \ldots, R_{d-1} \rangle$ *is a finite Coxeter group with d odd, then*

$$\frac{2}{g} = \sum_{\mathsf{M} \in \mathcal{M} \setminus \{\mathsf{D}\}} \frac{(-1)^{\operatorname{card} \mathsf{M} + 1}}{g_\mathsf{M} g_{\mathsf{M}^*}}.$$

1H5 

The simplex $S(p, q)$

As an example of Theorem 1H4, if $d = 3$, p, q are integers, and $\mathbf{G} := [p, q]$ has Coxeter diagram as in Figure 1H5, then $g := g(p, q) = |\mathbf{G}|$ is given by

$$\frac{2}{g} = \frac{1}{2q} - \frac{1}{2 \cdot 2} + \frac{1}{2p}.$$

Hence, after simplification,

$$g(p, q) = \frac{8pq}{4 - (p-2)(q-2)},$$

a formula familiar from [27, 5.43]. We stress here that, if we interpret the left side of the first expression as an angle, we do not actually have to assume that p and q are integers (or even rationals).

Whether d is even or odd, in the particular case when $\mathbf{G} = [p_1, \ldots, p_{d-1}]$ is a string group, with the usual labelling of its generators, we shall show in Section 4A that it is the symmetry group $\mathbf{G}(P)$ of the regular convex d-polytope $P = \{p_1, \ldots, p_{d-1}\}$. The base m-face F_m of P is stabilized by $\langle R_j \mid j \neq m \rangle$, so that the number $f_m = f_m(P)$ of m-faces of P is $g/g_{\mathsf{D}\setminus\{m\}}$. Indeed, this also holds for the single d-face P itself of P. Since $g_{\mathsf{D}\setminus\{m\}} = g_{\{0,\ldots,m-1\}}g_{\{m+1,\ldots,d-1\}}$, and since each $\mathsf{M} \in \mathcal{M}$ is of the form $\mathsf{M} = \{0, \ldots, m-1\}$, with $\mathsf{M}^* = \{m+1, \ldots, d-1\}$, then multiplying the formula of Theorem 1H2 by g just yields a special case of Euler's Theorem 1B19:

1H6 Theorem *If P is a regular convex d-polytope and f_m is the number of m-faces of P for $m = 0, \ldots, d$, then*

$$\sum_{m=0}^{d} (-1)^m f_m = 1.$$

In order to treat the case when d is even, we need to embed the finite Coxeter group $\mathbf{G}$ in an infinite group. We call $\mathbf{G}$ *crystallographic* if it is a subgroup of an infinite discrete group $\mathbf{H}$ generated by reflexions (in the same euclidean space). As we saw in Section 1E, the mirrors of all the reflexions in $\mathbf{H}$ dissect $\mathbb{E}^d$ into fundamental regions, which are simplices if $\mathbf{H}$ (or even $\mathbf{G}$) is irreducible. We apply Theorem 1H1 to these simplices.

Let T be such a simplex, bounded by the mirrors $R_0, \ldots, R_d$, which we identify with the corresponding reflexions. We shall always be able to suppose that $\mathbf{G} = \langle R_1, \ldots, R_d \rangle$, and that any subgroups generated by other proper subsets of the R_j are isomorphic to subgroups of $\mathbf{G}$ (that is, that $\mathbf{G} = \mathbf{SH}$ is the point-group of $\mathbf{H}$ as defined in Section 1C); some of these subgroups may actually be isomorphic to $\mathbf{G}$ itself. In any event, for each proper subset $\mathsf{J} \subset \mathsf{D} := \{0, \ldots, d\}$, we see that $\mathbf{G}_{\mathsf{J}} := \langle R_j \mid j \in \mathsf{J} \rangle$ is a finite subgroup of $\mathbf{H}$, which leaves invariant the affine subspace $H_{\mathsf{J}} := \bigcap\{R_j \mid j \in \mathsf{J}\}$, and hence the face $T_{\mathsf{J}} := H_{\mathsf{J}} \cap T$ of T. (The discreteness of $\mathbf{H}$ ensures that all these subgroups $\mathbf{G}_{\mathsf{J}}$ are actually finite, but we emphasize once again that we shall be performing pure angle calculations, which do not depend on this discreteness.) Theorem 1H1 and Lemma 1H3 immediately imply

1H7 Theorem *Let $\mathbf{H}$ be an irreducible discrete infinite reflexion group in $\mathbb{E}^d$, with generating reflexions $R_0, \ldots, R_d$ in the bounding hyperplanes of its fundamental region T. If $\mathsf{D} := \{0, \ldots, d\}$ and $\mathcal{M} := \mathcal{M}(\mathsf{D})$ are as above, then*

$$\sum_{\mathsf{M} \in \mathcal{M}} \frac{(-1)^{\operatorname{card} \mathsf{M}}}{g_{\mathsf{M}} g_{\mathsf{M}^*}} = 0.$$

We shall give the important examples of this result; in view of Theorem 1H4, of course, we are mostly interested in the cases when d is even. The interested reader will easily determine which infinite discrete reflexion group has a given finite crystallographic reflexion group as its point-group (comparing the lists in Tables 1E12 and 1E14 or in [27, Table IV] makes this straightforward). Let us find the orders of the groups $[3,3,4]$, $[3,4,3]$, $\mathbf{E}_6$, $\mathbf{E}_7$ and $\mathbf{E}_8$. (We should remark that the order of $[3,3,4]$ can be found more simply with the aid of the generic simplex dissection results we shall discuss later; $[3,4,3]$ can be dealt with by another generic simplex dissection which we shall not need here, but $\mathbf{E}_6$ and $\mathbf{E}_8$ present problems of a deeper kind.)

1H8

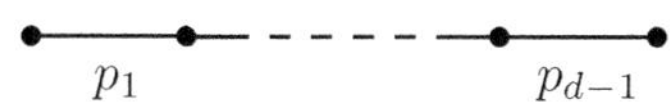

The simplex $S(p_1,\dots,p_{d-1})$

For convenience, let us denote by $S(p_1,\dots,p_{d-1})$ the simple cone or simplex corresponding to the diagram of Figure 1H8, and let $\alpha(p_1,\dots,p_{d-1})$ denote its angle, which is taken as 0 for a euclidean simplex – here, we shall only use the cases $d=4$ or 5.

1H9

The simplex $S(4,3,3,4)$

First, $[3,3,4]$ is the point-group of the group $[4,3,3,4]$. Applying Theorem 1H7 to the simplex of Figure 1H9 yields for the unknown order $g = g(3,3,4)$:

$$\frac{1}{g} - \frac{1}{2\cdot 48} + \frac{1}{8\cdot 8} - \frac{1}{48\cdot 2} + \frac{1}{g} = 0.$$

(We have substituted the (assumed known) orders of the lower-dimensional groups.) From this easily follows $g = 384$.

1H10

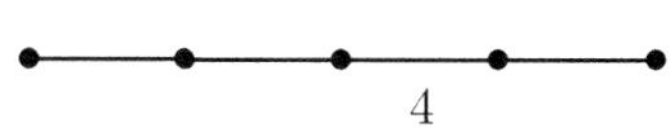

The simplex $S(3,3,4,3)$

In turn, $[3,4,3]$ is the point-group of $[3,3,4,3]$. Using the just found order $g(3,3,4) = 384$, Theorem 1H7 applied to the simplex of Figure 1H10 yields for the order g:

$$\frac{1}{g} - \frac{1}{2\cdot 48} + \frac{1}{6\cdot 6} - \frac{1}{24\cdot 2} + \frac{1}{384} = 0,$$

which implies that $g = 1152$.

We next come to $\mathbf{E}_6 = [3^{2,2,1}]$. This is the point-group of $\mathbf{T}_7 = [3^{2,2,2}]$ (see Table 1E14 for its diagram), and assuming that we have already found the orders of the lower-dimensional groups ($|\mathbf{A}_k| = (k+1)!$ and $|\mathbf{B}_k| = 2^{k-1}k!$ for $k = 4$ or 5 are the only extra orders we need; see [27, Table IV]), the order g satisfies

$$\frac{1}{g} + \frac{1}{2 \cdot 6!} + \frac{1}{3! \cdot 3! \cdot 3!} - \frac{1}{4! \cdot 2 \cdot 2} + \frac{2}{5! \cdot 2} - \frac{2}{6! \cdot 2} - \frac{1}{2^4 \cdot 5!} + \frac{2}{g} = 0,$$

which yields $g = 72 \cdot 6!$.

For $\mathbf{E}_7$, Theorem 1H4 applies, and we have

$$\frac{2}{g} = \frac{1}{72 \cdot 6!} - \frac{1}{2 \cdot 2^4 \cdot 5!} + \frac{1}{3! \cdot 5!} - \frac{1}{4! \cdot 3! \cdot 2} + \frac{1}{7!} + \frac{1}{2^5 \cdot 6!},$$

and hence $g = 8 \cdot 9!$.

Finally, we regard $\mathbf{E}_8$ as the point-group of $\mathbf{T}_9$, and Theorem 1H7 yields

$$\frac{1}{g} - \frac{1}{2 \cdot 8 \cdot 9!} + \frac{1}{3! \cdot 72 \cdot 6!} - \frac{1}{4! \cdot 2 \cdot 2^4 \cdot 5!} + \frac{1}{5! \cdot 5!} - \frac{1}{6! \cdot 3! \cdot 2} + \frac{1}{9!} + \frac{1}{2^7 \cdot 8!} = 0,$$

from which we deduce that $g = 192 \cdot 10!$.

The Group $[3,3,5]$

The non-crystallographic reflexion groups were not treated so far. However, the only group which actually escaped our treatment was $[3,3,5]$, the group of the regular 600-cell (or 120-cell) in $\mathbb{E}^4$. We shall now see that even this group is amenable to our approach; as a bonus, we shall also be able to calculate the densities of the regular star-polytopes in $\mathbb{E}^4$. We shall discuss the star-polytopes in Section 7G. (A recursive definition of *density* is given in [27, p.94]; we shall define it differently later, but the concept does not concern us overmuch. In any event, we are performing pure angle computations.)

Our basic result, which is just Theorem 1H1 with substitution for the angles of products of cones of dimension at most 3, is

1H11 Theorem *If $S(p,q,r,s)$ is a 4-simplex, then*

$$\alpha(p,q,r) + \alpha(q,r,s) = \frac{1}{8}\left(\frac{1}{p} + \frac{1}{q} + \frac{1}{r} + \frac{1}{s} - \frac{2}{ps} - 1\right).$$

The result of [17] which Coxeter held in [27, Chapter 14] to "rest on rather flimsy foundations" is basically this, interpreted in terms of orders of groups and densities.

We now apply Theorem 1H11 to various simplices arising in the dissection of $\mathbb{E}^4$ by the fundamental cones of $[3,3,5]$. We give these *generic simplex dissection* results here, but we shall also appeal to them later. They are called generic because they do not depend on our working in any particular group; indeed, the simplices involved do not have to correspond to any group, and may be spherical or even hyperbolic (in the former case, a spherical simplex arises as the intersection of a cone with the unit sphere).

1H12 Theorem *If $0 < \alpha < \pi$ is an arbitrary angle, $q := 2\pi/\alpha$ and $r = 0, \dots, m$, then*

(a) *the simplex $S(3^{m-2-r}, q, \frac{q}{2}, q, 3^{r-2})$ can be dissected into $\binom{m}{r}$ copies of $S(3^{m-2}, q)$,*

(b) *the simplex $S(3^{m-2-r}, 4, q, q, 4, 3^{r-2})$ can be dissected into $\binom{m}{r}$ copies of $S(3^{m-2}, 4, q)$.*

Sketch of proof. The conventions for extreme values of r should be obvious; just think of the blocks $q, \frac{q}{2}, q$ or $4, q, q, 4$ as migrating through a sequence of 3s. We have chosen the indices so that part (a) can be applied to part (b) without changing them. Note that we shall say more about constructions like those following in Sections 4A and 7A.

First, consider the regular $(m-1)$-simplex T with dihedral angle 2α and vertices $v_1, \dots, v_m$. Write H_{jk} for the hyperplane which bisects the dihedral angle at the $(m-3)$ face opposite the edge $v_j v_k$ (and so also bisects the edge). The hyperplanes H_{jk} (which bound the fundamental regions for the symmetry group of T) dissect T into $m!$ copies of $S(3^{m-2}, q)$.

A typical copy of $S(3^{m-2-r}, q, \frac{q}{2}, q, 3^{r-2})$ is bounded by $S_j := H_{j,j+1}$ for $j = 1, \dots, r-1$,

$$S_r := \operatorname{span}\{v_1, \dots, v_{r-1}, v_{r+1}, \dots, v_m\},$$
$$S_{r+1} := \operatorname{span}\{v_1, \dots, v_r, v_{r+2}, \dots, v_m\},$$

and $S_j := H_{j-1,j}$ for $j = r+2, \dots, m$. A little work (which we leave to the interested reader) shows that $S(3^{m-2-r}, q, \frac{q}{2}, q, 3^{r-2})$ is dissected into $r!(m-r)!$ copies of $S(3^{m-2}, q)$. Alternatively, one may think of $S(3^{m-2-r}, q, \frac{q}{2}, q, 3^{r-2})$ as the convex hull of fundamental regions for complementary $(r-1)$- and $(m-r-1)$-faces of T.

For part (b), consider the regular m-staurotope X with dihedral angles 2α, centre v_0 and vertices $v_1, \dots, v_m$ and their (unneeded) antipodes; see Section 2F. Performing the previous construction in the facet $\operatorname{conv}\{v_1, \dots, v_m\}$ and taking its convex hull with v_0 yields $S(3^{m-2-r}, 4, q, q, 4, 3^{r-2})$. Bearing in mind the convention about extreme values of r, we see that this can be dissected into $\binom{m}{r}$ copies of $S(3^{m-2}, 4, q)$, as was claimed. □

1H13 Remark When $q = 4$, Theorem 1H12 yields an alternative calculation for the order of $[3^{m-2}, 4]$.

At this stage, the special cases we use all occur in [29].

1H14 Corollary *For each $q > 2$,*

(a) $\alpha(3, q, \frac{q}{2}) = 4\,\alpha(3, 3, q)$,

(b) $\alpha(q, \frac{q}{2}, q) = 6\,\alpha(3, 3, q)$.

As a useful convention, whenever $q > 1$ we shall define q' by

$$\frac{1}{q} + \frac{1}{q'} = 1.$$

Then we have

1H15 Lemma *If C is a simple 4-cone in $\mathbb{E}^4$ with dihedral angles π/q_{ij} for $1 \leqslant j < k \leqslant 4$ and C' is the cone obtained by reqlacing q_{j4} by q'_{j4} for $j = 1, 2, 3$, then*

$$\alpha(C) + \alpha(C') = \frac{1}{4}\left\{\frac{1}{q_{12}} + \frac{1}{q_{13}} + \frac{1}{q_{23}} - 1\right\}.$$

Proof. The two cones C and C' will fit together along their common fourth face, to form the (orthogonal) product of a line with a 3-dimensional cone D whose dihedral angles are π/q_{ij} for $1 \leqslant j < k \leqslant 3$. Thus $\alpha(C) + \alpha(C')$ is the angle of this product cone, which is just that of D, and so the number given in the theorem. □

We shall need two consequences of Lemma 1H15.

1H16 Corollary *For each p, q, r,*

$$\alpha(p, q, r') = \frac{1}{4}\left\{\frac{1}{p} + \frac{1}{q} - \frac{1}{2}\right\} - \alpha(p, q, r).$$

1H17 Corollary *For each p, q, r,*

$$\alpha(p, q', r) = \alpha(p, q, r) + \frac{1}{4}\left\{\frac{1}{2} - \frac{1}{q}\right\}.$$

Proof. Corollary 1H16 is a direct application of Lemma 1H15, which also yields

$$\alpha(p, q', r) = \frac{1}{4}\left\{\frac{1}{p} + \frac{1}{2} + \frac{1}{2} - 1\right\} - \alpha(p, q, r'),$$

and Corollary 1H17 follows at once from Corollary 1H16. □

We now employ these results to calculate $g := g(3, 3, 5)$. For appropriate simplices $S(p, q, r)$, we define the *density* $d(p, q, r)$ by

1H18 $$\alpha(p, q, r) := d(p, q, r)\alpha(3, 3, 5).$$

It is worth stressing at this point that we shall make no prior assumption that $d(p, q, r)$ is even rational, let alone an integer. (Of course, the discussion of Section 7G will show that they must both be integers, because the regular star-polytopes $\{p, q, r\}$ corresponding to the simplices $S(p, q, r)$ have the same symmetry group $[3, 3, 5]$ as $\{3, 3, 5\}$; however, that result is independent of the present one. For more about what 'density' means, see the notes at the end of the section.)

For brevity, we set $d := d(3, 3, \frac{5}{2})$. From the simplex $S(\frac{5}{2}, 3, 3, 5)$ (which already we met in Section 1E) we have

$$\alpha(3, 3, \tfrac{5}{2}) + \alpha(3, 3, 5) = \frac{1}{8}\left\{\frac{2}{5} + \frac{1}{3} + \frac{1}{3} + \frac{1}{5} - \frac{4}{25} - 1\right\},$$

and hence

1H19 $$\frac{d+1}{g} = \frac{1}{75} = \frac{192}{14400}.$$

Next, for the simplex $S(3, \frac{5}{2}, 5, 3)$ we verify that we have unit normals and positive dependence

$$\tfrac{1}{2}(0, -\tau^{-1}, 1, \tau), \quad -e_3, \quad \tfrac{1}{2}(\tau, 1, \tau^{-1}, 0), \quad -e_1, \quad \tfrac{1}{2}(1, -\tau, 0, -\tau^{-1});$$
$$(\tau^{-1}, 2\tau^{-1}, 3, 2\tau, \tau).$$

We thus obtain

$$\alpha(3, \tfrac{5}{2}, 5) + \alpha(3, 5, \tfrac{5}{2}) = \frac{1}{8}\left\{\frac{1}{3} + \frac{2}{5} + \frac{1}{5} + \frac{1}{3} - \frac{2}{9} - 1\right\} = \frac{1}{180}.$$

Now,

$$\alpha(3, 5, \tfrac{5}{2}) = 4\,\alpha(3, 3, 5) = \frac{4}{g}$$

from Corollary 1H14, while

$$\begin{aligned}\alpha(3, \tfrac{5}{2}, 5) &= \frac{1}{4}\left\{\frac{1}{3} + \frac{2}{5} - \frac{1}{2}\right\} - \alpha(3, \tfrac{5}{2}, \tfrac{5}{4}) \\ &= \frac{7}{120} - 4\,\alpha(3, 3, \tfrac{5}{2}) \\ &= \frac{7}{120} - \frac{4d}{g},\end{aligned}$$

from Corollary 1H16 and the previous calculation. Substituting, we have

1H20 $$\frac{d-1}{g} = \frac{1}{4}\left\{\frac{7}{120} - \frac{1}{180}\right\} = \frac{19}{1440} = \frac{190}{14400}.$$

From (1H19) and (1H20) we deduce immediately that $g = 14400$ and $d = 191$.

We can now write

$$d(p, q, r) = g\,\alpha(p, q, r) = 14400\,\alpha(p, q, r)$$

for the density of $S(p, q, r)$ (for appropriate p, q, r). We first have the obvious values

$$\begin{aligned}d(3, 5, \tfrac{5}{2}) &= 4, \\ d(5, \tfrac{5}{2}, 5) &= 6,\end{aligned}$$

from Corollary 1H14, while that result and Corollaries 1H16 and 1H17 yield

$$\begin{aligned}d(\tfrac{5}{2}, 5, \tfrac{5}{2}) &= 14400\left\{\alpha(\tfrac{5}{2}, \tfrac{5}{4}, \tfrac{5}{2}) + \frac{1}{4}\left[\frac{1}{2} - \frac{4}{5}\right]\right\} = 6d - 1080 = 66, \\ d(3, \tfrac{5}{2}, 5) &= 14400\,\frac{7}{120} - 4d = 76,\end{aligned}$$

where in the last equation we have not repeated our previous calculations.

It remains to find $d(\frac{5}{2},3,5)$ (of the simplices which arise in Section 7G). Here we use the simplex $S(\frac{5}{2},3,5,\frac{5}{2})$, for which we can check that we have unit normals and positive dependence

$$\tfrac{1}{2}(0,\tau^{-1},-1,\tau), \quad -e_2, \quad \tfrac{1}{2}(\tau,1,\tau^{-1},0), \quad -e_1, \quad \tfrac{1}{2}(\tau^{-1},0,-\tau,-1);$$
$$(\tau^{-1},2,\tau\sqrt{5},2\tau,1).$$

From this, we see that

$$\alpha(\tfrac{5}{2},3,5)+\alpha(3,5,\tfrac{5}{2})=\frac{1}{8}\left\{\frac{2}{5}+\frac{1}{3}+\frac{1}{5}+\frac{2}{5}-\frac{8}{25}-1\right\}=\frac{1}{600},$$

and hence

$$d(\tfrac{5}{2},3,5)=14400\,\frac{1}{600}-4=20.$$

This completes our calculations. We may observe that we have not used the simplices $S(\frac{5}{2},5,\frac{5}{2},5)$ or $S(5,3,\frac{5}{2},5)$, which yield no new information.

Notes to Section 1H

1. The dissections of Theorem 1H12 seem to be part of the folklore of the subject. In full generality, they were first proved by Debrunner [40].
2. Theorem 1H12 can be used instead of some of our calculations. For example, $\alpha(3,3,4)=4\alpha(3,4,2)$ from part (a) yields $g(3,3,4)=4\cdot g(3,4)\cdot 2=384$, while $\alpha(4,3,3)=3\alpha(3,4,3)$ from part (b) tells us that $g(3,4,3)=3g(3,3,4)=3\cdot 384=1152$. Indeed, similar calculations give $g(3^{d-2},4)$ by induction on d.
3. In [28, Section 3.5], Coxeter uses a certain dissection result and evaluates a related integral to find the orders of the groups $\mathbf{E}_d$ for $d=6,7,8$. This approach is more complicated than our method.
4. The integrals due to Schläfli [112] provide an alternative approach to calculating the order of $\mathbf{G}_4=[3,3,5]$ and the densities; see [29] once more. However, the most important of these integrals cannot be evaluated directly.
5. It is a curious fact that, in odd dimension d, the simplified formula of Theorem 1H7 may actually lose the group order that was sought. For instance, if we ask for the order g of $[4,3]$ from $S(4,3,4)$, we get
$$\frac{1}{g}-\frac{1}{2\cdot 8}+\frac{1}{8\cdot 2}-\frac{1}{g}=0,$$
which tells us nothing. Fortunately, these cases are covered by Theorem 1H4 which, conversely, yields nothing when d is even.
6. As we said, Coxeter [27, Section 6.4] defines *density* recursively, in a somewhat unintuitive way geometrically; fortunately, there is an obvious picture of the cones $S(p,q,r)$ giving a $d(p,q,r)$-fold covering of a general point of $\mathbb{E}^4$. In [27, p. 283], he finds the densities by successive simplex dissections; our approach obviates these elaborate calculations.
7. Nevertheless, the simplex dissections do correspond in a straightforward way to changes of generators. Tracing these changes leads to generatrices of the regular star-polytopes in $\mathbb{E}^4$ in terms of that of the 600-cell $\{3,3,5\}$; see Section 7G.

1J Ordinary Space

The aim of this section is to describe the finite rotation groups in $\mathbb{E}^3$. In fact, we shall establish

1J1 Theorem *Each finite subgroup of the special orthogonal group* $\mathbf{SO}_3$ *is a subgroup of index* 2 *of some reflexion group.*

We begin with two lemmas, which already indicate the important rôle played by reflexions.

1J2 Lemma *If* $\Phi \in \mathbf{SO}_3$ *is a rotation with axis* $A \leqslant \mathbb{E}^3$*, then there are plane reflexions* R, S *whose mirrors contain* A *such that* $\Phi = RS$*; moreover, either* R *or* S *can be chosen arbitrarily.*

Proof. With an arbitrary choice of plane $R \geqslant A$ regarded as a reflexion, define $S := R\Phi$. Then $S \in \mathbf{O}_3 \setminus \mathbf{SO}_3$ fixes A, and so must also be a plane reflexion. Then $\Phi = RS$, as wanted. The proof with S chosen instead of R is almost identical. □

The second lemma is a trivial observation, given Lemma 1J2.

1J3 Lemma *Let* $\Phi, \Psi \in \mathbf{SO}_3$ *be rotations with distinct axes* A, B*, respectively. If* $S := \operatorname{lin}(A \cup B)$ *and further reflexions* R *and* T *are such that* $\Phi = RS$ *and* $\Psi = ST$*, then* $\Phi\Psi = RT$*.*

We need two more preliminary results. The first is a well-known property of spherical triangles, and the second is probably also fairly familiar. For completeness, we give their proofs.

1J4 Lemma *If the dihedral angles of a simple* 3*-cone* K *are* α, β, γ*, then* $\alpha + \beta + \gamma > \pi$*.*

Proof. The case $d = 3$ of Theorem 1G2 shows that the normalized angle of K is $(\alpha+\beta+\gamma-\pi)/4\pi$. Since this is positive, we have the assertion of the lemma. □

We next need the *spherical sine formula.*

1J5 Lemma *If* K *is a simple* 3*-cone whose dihedral angles are* α, β, γ *and corresponding opposite face angles are* λ, μ, ν*, then*

$$\frac{\sin\lambda}{\sin\alpha} = \frac{\sin\mu}{\sin\beta} = \frac{\sin\nu}{\sin\gamma}.$$

Proof. Take unit vectors a, b, c in the directions of the edges of K (with α corresponding to a, and so on), and let u, v, w be unit outer normals to K at its faces (with u opposite a, and so on). With the orientation induced by the ordering (a, b, c), the exterior (or vector) product of b and c is $b \wedge c = -\sin\lambda\, u$, so that the determinant is

$$\det(a, b, c) = \langle a, b \wedge c\rangle = -\langle a, u\rangle \sin\lambda.$$

Applying cyclic permutations to (a, b, c) (which preserve the determinant) thus yields

$$\langle a, u\rangle \sin\lambda = \langle b, v\rangle \sin\mu = \langle c, w\rangle \sin\nu.$$

Now consider the polar cone K^* of K, whose edges are in directions u, v, w. Then K^* has dihedral angles $\pi - \lambda, \pi - \mu, \pi - \nu$ and corresponding face angles $\pi - \alpha, \pi - \beta, \pi - \gamma$, and so the foregoing argument says that

$$\langle u, a\rangle \sin\alpha = \langle v, b\rangle \sin\beta = \langle w, c\rangle \sin\gamma.$$

The claim of the lemma now follows at once. □

Proof of Theorem 1J1. We basically follow the treatment of [43, Section 2.10], except that our ultimate claim is a little stronger. Let $\mathbf{G}$ be the given (non-trivial) finite subgroup. If $\mathbf{G}$ has only one rotation axis, then clearly it is cyclic, and we can appeal to Lemma 1J2 for the claim of the theorem. As Lemma 1J3 shows, we cannot have just two distinct rotation axes. We can have three, which are mutually perpendicular, about which there are half-turns. If there is a single rotation Φ with period $p > 2$, then the remaining rotations are half-turns, whose axes are perpendicular to the axis of Φ; then $\mathbf{G}$ is actually isomorphic to a dihedral group. The second case can be thought of as a special case of the third, and adjoining to $\mathbf{G}$ the plane reflexion R whose mirror contains the 2-fold rotation axes produces a product of a dihedral group $\mathbf{D}_2^p$ and $\langle R\rangle$ for some p; the second case is $p = 2$, giving the group generated by the reflexions in three mutually perpendicular planes.

Finally, then, we can suppose that $\mathbf{G}$ contains rotations Φ, Ψ of periods $p, q > 2$ about two distinct axes A, B, say. We choose such Φ and Ψ so that the angle between A and B is minimal. We now carry out the construction of Lemma 1J3, supposing (as we may) that $\Phi = RS$ and $\Psi = ST$ are actually rotations through $2\pi/p$ and $2\pi/q$, respectively. Then the angle between R and S is π/p, and that between S and T is π/q; replacing Ψ by its inverse if necessary, we can suppose that the planes R, S, T bound a simple cone K. Now $\Phi\Psi = RT$ is a rotation through 2α for some angle α. Choosing (outer) unit normals u, v, w to R, S, T such that $\langle u, v\rangle = -\cos(\pi/p)$ and $\langle v, w\rangle = -\cos(\pi/q)$, we have $\langle u, w\rangle = -\cos\alpha$. From Lemma 1J4, we have

$$\alpha > \pi - \frac{\pi}{p} - \frac{\pi}{q} \geqslant \frac{\pi}{3};$$

hence, if $C := R \cap T$ (as mirrors), it follows from Lemma 1J5 that the angle between A and C is less than that between A and B, and so we obtain a contradiction to our initial choice unless $\alpha = \pi/2$.

Now K is a fundamental region for the reflexion group $\langle R, S, T\rangle$, and we further note that $K \cup KS$ is a fundamental region for the group $\mathbf{H} := \langle\Phi, \Psi\rangle$. If $\mathbf{H} \neq \mathbf{G}$, then (up to conjugacy) some further rotation $\Theta \in \mathbf{G}$ has axis D (say) which meets $K \cup KS$ other than on one of its edges. In fact, either by replacing K by KS or by conjugating the whole group $\mathbf{G}$ by S, we can assume that D actually meets K non-trivially. Since the angle between A and D is less than that between A and B, by the same reasoning as before it follows that Θ must have period 2.

There are two possibilities; these become clearer if we look at how the various lines, planes and cones meet the unit sphere $\mathbb{S}$. Thus we write $A \cap \mathbb{S} =: \{a\}$, and so on. First, $D \leqslant S$, in which case D must bisect the angle between A and B, and Θ interchanges A and B. Then – see Figure 1J6 – we can write $\Theta = SU$, where U is the angle bisector of K at C, from which we easily deduce that $\Phi\Theta = RU$ is a rotation about C through $\pi/2$, contradicting the initial choice of A and B (or the fact that the rotation about C is a half-turn).

1J6

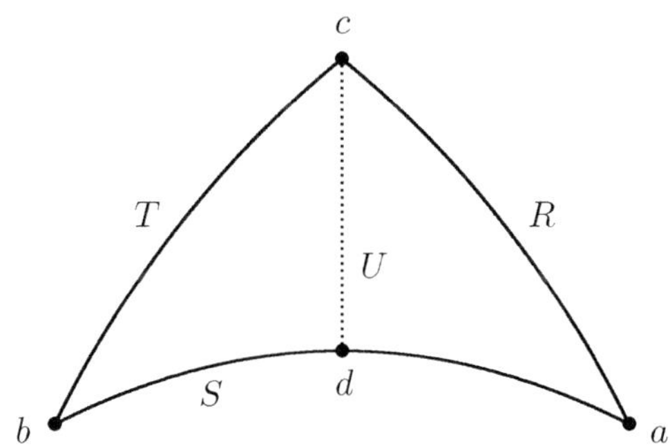

Otherwise, we have the situation depicted in Figure 1J7. Here, the angle at D between the planes spanned by D and A and B is obtuse, and consequently that between the axis $A' := A\Theta$ of the conjugate $\Theta\Psi\Theta$ of Ψ and B is less than that between A and B; this again gives a contradiction to our original choice of Φ and Ψ.

1J7

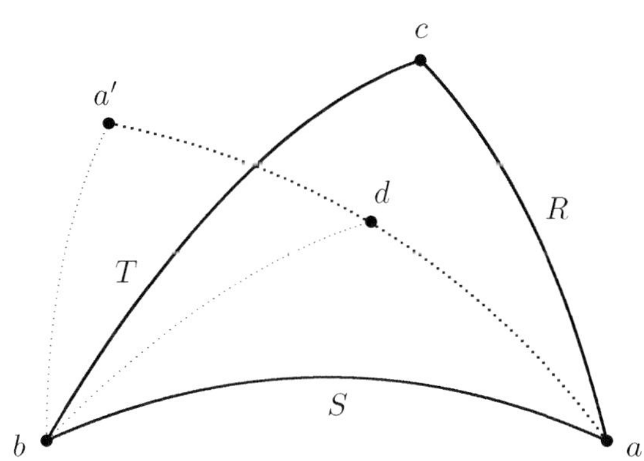

Thus $\mathbf{H} = \mathbf{G}$ in either case, as required. With the observation that $\mathbf{G}$ is a subgroup of $\langle R, S, T\rangle$ of index 2, this completes the proof. □

1J8 Remark With p, q as in the proof of Theorem 1J1, the reflexion group $\langle R, S, T\rangle$ is also denoted $[p, q]$, with its rotation subgroup written $[p, q]^+$. We can extend this to the dihedral group $[2, q]^+$ in the third case of the proof, with the added possibility that $q = 2$ also.

1K Quaternions

In Section 11E we shall encounter a family of regular polyhedra in $\mathbb{E}^4$ whose symmetry groups consist of rotations (direct isometries) only. These groups are anomalous, in that there need be no immediate relationships to reflexion groups; indeed, in some cases, no such relationship exists. Our treatment of this family will employ quaternions. We shall also see that quaternions are useful for investigating properties of regular 4-polytopes in Sections 7E and 7F. In this section, we give a brief outline of that part of the theory of quaternions which we need; for further details, consult [43].

We recall that the general *quaternion* is a linear combination of the form $\mathbf{x} = \xi_0 + \xi_1\mathbf{i} + \xi_2\mathbf{j} + \xi_3\mathbf{k}$, where $\xi_j \in \mathbb{R}$ for $j = 0, \ldots, 3$, with an associative (but not commutative) multiplication induced by

1K1 $$\mathbf{i}^2 = \mathbf{j}^2 = \mathbf{k}^2 = \mathbf{ijk} = -1.$$

It is convenient to identify $\mathbf{x}$ with the vector $(\xi_0, \xi_1, \xi_2, \xi_3) \in \mathbb{E}^4$. The *real part* of $\mathbf{x}$ is $\Re(\mathbf{x}) := \xi_0$, and its *imaginary part* is $\Im(\mathbf{x}) := \xi_1\mathbf{i} + \xi_2\mathbf{j} + \xi_3\mathbf{k}$. The *conjugate* of $\mathbf{x}$ is $\widetilde{\mathbf{x}} := \xi_0 - \xi_1\mathbf{i} - \xi_2\mathbf{j} - \xi_3\mathbf{k}$; then $\widetilde{\mathbf{x}}\mathbf{x} = \mathbf{x}\widetilde{\mathbf{x}} = \|\mathbf{x}\|^2$ (as a real vector). Thus, if $\mathbf{x} \neq 0$, then $\mathbf{x}$ is invertible. In particular, a *unit quaternion* $\mathbf{x}$ is such that $\|\mathbf{x}\| = 1$; the set $\mathbf{Q}$ of unit quaternions forms a group, with the inverse of $\mathbf{x} \in \mathbf{Q}$ being $\widetilde{\mathbf{x}}$.

As we suggested at the end of Section 1A, we can also define quaternionic multiplication using alternating tensors. If we write quaternions in the form $\mathbf{x}_j = \eta_j + \mathbf{z}_j$, where $\eta_j := \Re(\mathbf{x}_j)$ and $\mathbf{z}_j := \Im(\mathbf{x}_j)$ for $j = 1, 2$, then

1K2 $$\mathbf{x}_1\mathbf{x}_2 = (\eta_1\eta_2 - \langle \mathbf{z}_1, \mathbf{z}_2 \rangle) + \eta_1\mathbf{z}_2 + \eta_2\mathbf{z}_1 + \mathbf{z}_1 \wedge \mathbf{z}_2,$$

with $\wedge$ the vector product of 3-vectors as introduced there.

1K3 Remark Multiplication on the right by a unit quaternion results in a left screw, and conversely (see [43] for more details). This is illustrated by

$$(1, \mathbf{i}, \mathbf{j}, \mathbf{k})\mathbf{i} = (\mathbf{i}, -1, -\mathbf{k}, \mathbf{j}),$$

which is a positive quarter-turn in the $(1, \mathbf{i})$-plane, and a negative quarter-turn in the $(\mathbf{j}, \mathbf{k})$-plane. Compare the comment on conjugacy below; further, see the notes at the end of the section.

If $\mathbf{u}$ is a unit quaternion, then the linear reflexion $R(\mathbf{u})$ in $\mathbb{E}^4$ whose mirror is the hyperplane with normal $\mathbf{u}$ is

1K4 $$R(\mathbf{u})\colon \mathbf{x} \mapsto -\mathbf{u}\widetilde{\mathbf{x}}\mathbf{u}.$$

This is easily checked directly, but observe that, since $2\langle \mathbf{x}, \mathbf{u} \rangle = 2\Re(\mathbf{x}\widetilde{\mathbf{u}}) = \mathbf{x}\widetilde{\mathbf{u}} + \mathbf{u}\widetilde{\mathbf{x}}$, we have

$$\mathbf{x}R(\mathbf{u}) = \mathbf{x} - 2\langle \mathbf{x}, \mathbf{u} \rangle \mathbf{u} = \mathbf{x} - (\mathbf{x}\widetilde{\mathbf{u}} + \mathbf{u}\widetilde{\mathbf{x}})\mathbf{u} = -\mathbf{u}\widetilde{\mathbf{x}}\mathbf{u},$$

as claimed. Because an element of the orthogonal group $\mathbf{O}_4$ is the product of such reflexions (compare Corollary 1D8), it follows that it can be represented by a transformation of the form

1K5 $$\mathbf{x} \mapsto \mathbf{x}\,\mathbf{g}(\mathbf{a},\mathbf{b}) := \widetilde{\mathbf{a}}\mathbf{x}\mathbf{b} \text{ or } \widetilde{\mathbf{x}}\,\mathbf{g}(\mathbf{a},\mathbf{b}) = \widetilde{\mathbf{a}}\widetilde{\mathbf{x}}\mathbf{b}$$

where $\mathbf{a}, \mathbf{b} \in \mathbf{Q}$; the maps of the former kind belong to the rotation subgroup $\mathbf{SO}_4$. In keeping with our usual conventions, mappings are thought of as acting on the right; thus it must be the inverse of a quaternion which provides an appropriate mapping when acting on the left. The group $\mathbf{G}$ of a polyhedron of the kind mentioned at the beginning of the section is clearly a finite subgroup of $\mathbf{SO}_4$, and so can be thought of as consisting of such elements $\mathbf{g}(\mathbf{a}, \mathbf{b})$.

Before we go further, it is appropriate to comment on the rôle of conjugacy, which gives insights into the geometry of the mappings $\mathbf{g}(\cdot,\cdot)$ (compare [43] here). The left-acting quaternions $\mathbf{a}$ occurring in a finite subgroup $\mathbf{G} \leqslant \mathbf{SO}_4$ clearly form a finite *left subgroup* $\mathbf{G}_L$ of the whole group $\mathbf{Q}$ of unit quaternions. There is similarly a *right subgroup* $\mathbf{G}_R$ consisting of the right-acting quaternions $\mathbf{b}$. Then it is easy to see that conjugating the whole group $\mathbf{G}$ leads to conjugating the left and right groups $\mathbf{G}_L$ and $\mathbf{G}_R$, and conversely; this follows from

1K6 $$\widetilde{\mathbf{g}}(\widetilde{\mathbf{a}}\mathbf{x}\mathbf{b})\mathbf{h} = (\widetilde{\mathbf{g}}\,\widetilde{\mathbf{a}}\mathbf{g})(\widetilde{\mathbf{g}}\mathbf{x}\mathbf{h})(\widetilde{\mathbf{h}}\mathbf{b}\mathbf{h}) = (\widetilde{\widetilde{\mathbf{g}}\mathbf{a}\mathbf{g}})(\widetilde{\mathbf{g}}\mathbf{x}\mathbf{h})(\widetilde{\mathbf{h}}\mathbf{b}\mathbf{h}),$$

with $\mathbf{a}, \mathbf{b}, \mathbf{g}, \mathbf{h} \in \mathbf{Q}$. This freedom to take conjugates, which amounts to a free choice of suitable coordinates for the subgroups $\mathbf{G}_L$ and $\mathbf{G}_R$, will prove very useful.

In fact, $\mathbf{G}$ is a certain quotient of $\mathbf{G}_L \times \mathbf{G}_R$. If we define

$$\mathbf{N}_L := \{\mathbf{a} \in \mathbf{G}_L \mid \mathbf{g}(\mathbf{a}, 1) \in \mathbf{G}\}$$

(that is, $\mathbf{x} \mapsto \widetilde{\mathbf{a}}\mathbf{x}$ is in $\mathbf{G}$), and $\mathbf{N}_R$ similarly, then $\mathbf{N}_L \lhd \mathbf{G}_L$ and $\mathbf{N}_R \lhd \mathbf{G}_R$ are normal subgroups such that $\mathbf{G}_L/\mathbf{N}_L \cong \mathbf{G}_R/\mathbf{N}_R$. To see this, just observe that $\mathbf{N}_L\mathbf{a} \leftrightarrow \mathbf{N}_R\mathbf{b}$ gives a one-to-one correspondence. Following [43, Section 21], we denote the group $\mathbf{G}$ by

1K7 $$(\mathbf{G}_L/\mathbf{N}_L; \mathbf{G}_R/\mathbf{N}_R);$$

we only meet such groups with $\mathbf{G}_L \neq \mathbf{G}_R$ in Section 11E (however, see the notes at the end of the section).

If a subgroup $\mathbf{G}$ of $\mathbf{O}_4$ does not consist solely of rotations (that is, elements of $\mathbf{SO}_4$), then it has mappings of the form $x \mapsto \mathbf{a}\widetilde{\mathbf{x}}\mathbf{b}$, whose inverse is $\mathbf{x} \mapsto \mathbf{b}\widetilde{\mathbf{x}}\mathbf{a}$. Moreover, the composition of such a mapping and $\mathbf{x} \mapsto \mathbf{c}\widetilde{\mathbf{x}}\mathbf{d}$ is

$$\mathbf{x} \mapsto \mathbf{c}\widetilde{\mathbf{a}\widetilde{\mathbf{x}}\mathbf{b}}\mathbf{d} = (\mathbf{c}\widetilde{\mathbf{b}})\mathbf{x}(\widetilde{\mathbf{a}}\mathbf{d}).$$

What this shows is

1K8 Proposition *In a subgroup* $\mathbf{G} \leqslant \mathbf{O}_4$ *not consisting solely of rotations, the left and right elements* $\mathbf{a}$ *and* $\mathbf{b}$ *of mappings* $\mathbf{x} \mapsto \mathbf{a}\widetilde{\mathbf{x}}\mathbf{b}$ *in* $\mathbf{G}$ *form the same subgroup* $\mathbf{G}_L = \mathbf{G}_R$ *of* $\mathbf{Q}$.

With $\mathbf{H} := \mathbf{G}_L = \mathbf{G}_R$ and $\mathbf{N} = \mathbf{N}_L = \mathbf{N}_R$, we can write $\mathbf{G}$ as

1K9 $$(\mathbf{H}/\mathbf{N}; \mathbf{H}/\mathbf{N})^* := \{\mathbf{x} \mapsto \widetilde{\mathbf{a}}\mathbf{x}\mathbf{b} \text{ or } \widetilde{\mathbf{a}}\widetilde{\mathbf{x}}\mathbf{b} \mid \widetilde{\mathbf{a}}\mathbf{b} \in \mathbf{N}\},$$

because this is what the condition on the quotients amounts to.

Now it is clear that we can always write a unit quaternion $\mathbf{a}$ in the form $\mathbf{a} = \cos\vartheta + \sin\vartheta\,\mathbf{u}$ for some $0 \leqslant \vartheta < 2\pi$, where $\mathbf{u}$ is a pure imaginary unit quaternion. If $\mathbf{a} = \cos\vartheta + \sin\vartheta\mathbf{u}$ and $\mathbf{b} = \cos\varphi + \sin\varphi\mathbf{v}$, with $\mathbf{u}, \mathbf{v}$ pure imaginary unit quaternions, then $\mathbf{g}(\mathbf{a}, \mathbf{b})$ is (in general) a double rotation through angles $\pm\varphi\pm\vartheta$; thus it is a planar rotation (with a 2-dimensional axis) just when $\varphi = \pm\vartheta$. To see this, in view of the previous comment about conjugacy, it suffices to take $\mathbf{v} = \mathbf{u} = \mathbf{i}$, say. It further follows that, for $\mathbf{g}(\mathbf{a}, \mathbf{b})$ to be a reflexion (that is, involutory), other than the central inversion $\mathbf{x} \mapsto -\mathbf{x}$, both $\mathbf{a}$ and $\mathbf{b}$ must be pure imaginary.

The standard scalar product in $\mathbb{E}^4$ is given by $\langle \mathbf{x}, \mathbf{y} \rangle = \Re(\widetilde{\mathbf{x}}\mathbf{y}) = \Re(\mathbf{x}\widetilde{\mathbf{y}})$, and so we see that, if $\mathbf{a}, \mathbf{b} \in \mathbf{Q}$ are pure imaginary (so that $\mathbf{a}^2 = \mathbf{b}^2 = -1$), then $\mathbf{ab}$ is of the form $\cos\vartheta + \sin\vartheta\mathbf{u}$ for some pure imaginary $\mathbf{u} \in \mathbf{Q}$, where $\cos\vartheta = -\langle \mathbf{a}, \mathbf{b} \rangle$.

The unit quaternion $\mathbf{a} = \cos\vartheta + \sin\vartheta\mathbf{u}$ (as above) induces an element of $\mathbf{SO}_3$, given by

1K10 $$\mathbf{x} \mapsto \mathbf{x}\mathbf{g}(\mathbf{a}, \mathbf{a}) = \widetilde{\mathbf{a}}\mathbf{x}\mathbf{a},$$

where $\mathbf{x}$ is pure imaginary; thus we regard $\mathbb{E}^3$ as the subspace of pure imaginary quaternions. This element is a rotation through the angle -2ϑ about the axis $\operatorname{lin}\{\mathbf{u}\}$ (or about $\operatorname{lin}\{1, \mathbf{u}\}$ if thought of as in $\mathbb{E}^4$ – see the notes at the end of the section); from this, it follows that the kernel of the homomorphism from $\mathbf{Q}$ to $\mathbf{SO}_3$ is $\{\pm 1\}$. Thus the left subgroup $\mathbf{G}_L \leqslant \mathbf{Q}$ has a quotient $\widetilde{\mathbf{G}}_L \leqslant \mathbf{SO}_3$ of half its order, unless $-1 \notin \mathbf{G}_L$ when we have an isomorphism; similarly, $\mathbf{G}_R$ has such a quotient $\widetilde{\mathbf{G}}_R$. Two such groups $\mathbf{G}, \mathbf{G}'$ are conjugate in $\mathbf{SO}_4$ if and only if the corresponding pairs of rotation groups $\widetilde{\mathbf{G}}_L, \widetilde{\mathbf{G}}'_L$ and $\widetilde{\mathbf{G}}_R, \widetilde{\mathbf{G}}'_R$ are conjugate in $\mathbf{SO}_3$. However, the fact that each element of $\mathbf{SO}_3$ lifts to *two* elements of $\mathbf{Q}$ will cause some problems.

For future reference, we list the quaternion groups which play a rôle later; the cyclic groups need no mention. The notation is that of [43, Chapter 3]; there are slight conflicts with the notation for other geometric groups, but these should cause no problems. It must be emphasized that the coordinates chosen here are, to a certain extent, arbitrary. Their elements are quaternions $\nu_0 + \nu_1\mathbf{i} + \nu_2\mathbf{j} + \nu_3\mathbf{k}$, with the vectors $(\nu_0, \nu_1, \nu_2, \nu_3) \in \mathbb{E}^4$ taking the following values.

Binary Dihedral Group

First, we have the *binary dihedral group* $\mathbf{D}_n$ of order $4n$, with all elements

1K11 $$(\cos(k\pi/n), \sin(k\pi/n), 0, 0), \quad (0, 0, \cos(k\pi/n), \sin(k\pi/n)),$$

for $k = 0, \ldots, 2n-1$. This is the double cover of $[2, n]^+$ (which is isomorphic to the dihedral group $\mathbf{D}_2^n$). In practice, binary dihedral groups will turn up in ways different from this.

Binary Octahedral Group

Next, we have the *binary octahedral group* $\mathbf{O}$ of order 48, consisting of all permutations of

1K12 $(\pm1,0,0,0),\quad \frac{1}{2}(\pm1,\pm1,\pm1,\pm1),\quad \frac{1}{2}(\pm\sqrt{2},\pm\sqrt{2},0,0).$

These form the vertices of two dual regular 24-cells in $\mathbb{E}^4$, as we shall see in Section 7C. The group is the double cover of the octahedral group $[3,4]^+$.

Binary Icosahedral Group

Last, we have the *binary icosahedral group* $\mathbf{I}$ of order 120, consisting of all even permutations of

1K13 $(\pm1,0,0,0),\quad \frac{1}{2}(\pm1,\pm1,\pm1,\pm1),\quad \frac{1}{2}(\pm\tau,\pm1,\pm\tau^{-1},0),$

where $\tau = \frac{1}{2}(1+\sqrt{5})$ is the golden number. The quaternions of (1K13) are just those lifted from the icosahedral group $[3,5]^+$, the rotational symmetries of the icosahedron in Theorem 7D2. The odd permutations would serve as well, and for the most part we need not worry about which choice is made in any given context (see the notes at the end of the section). However, in Section 7F both choices are involved; we shall see in Section 7E that each forms the vertex-set of a regular 600-cell.

Binary Tetrahedral Group

Observe that the latter two groups $\mathbf{O}$ and $\mathbf{I}$ have a common subgroup, the *binary tetrahedral group* $\mathbf{T}$, consisting of the first 24 elements of each (which form the vertices of a 24-cell – see Section 7C); see the notes at the end of the section. This is the double cover of the tetrahedral group $[3,3]^+$.

We shall see in Section 11E that the only groups that occur as appropriate groups $\widetilde{\mathbf{G}}_L$ or $\widetilde{\mathbf{G}}_R$ are the dihedral groups of order $4n$ (generated by half-turns rather than plane reflexions), the octahedral group $[3,4]^+$ or the icosahedral group $[3,5]^+$. By contrast, the cyclic groups (rotations about a single axis), dihedral groups of order $4n+2$ and tetrahedral group $[3,3]^+$ do not contain enough half-turns (and, in particular, are not generated by them).

We have a special notation for the group $\mathbf{V} := \mathbf{D}_2 = \{\pm1,\pm\mathbf{i},\pm\mathbf{j},\pm\mathbf{k}\}$; this is the lifting of the *four-group* $[2,2]^+$, consisting of the half-turns about the coordinate axes of $\mathbb{E}^3$.

Completeness of Enumeration

For completeness, we prove

1K14 Theorem *Apart from cyclic groups, up to isomorphism the only finite subgroups of* $\mathbf{Q}$ *are* $\mathbf{D}_n$ *for some* n, $\mathbf{T}$, $\mathbf{O}$ *and* $\mathbf{I}$.

Proof. Consider the rotation group $\widetilde{\mathbf{G}} \leqslant \mathbf{SO}_3$ corresponding to the given finite subgroup $\mathbf{G} \leqslant \mathbf{Q}$. If $\widetilde{\mathbf{G}}$ is cyclic, say consisting of rotations about the axis $A := \operatorname{lin}\{\mathbf{u}\}$ with $\mathbf{u} \in \mathbb{E}^3$ a unit vector, then $\mathbf{G}$ can only consist of quaternions of the form $\cos\vartheta + \sin\vartheta\mathbf{u}$, and so must also be cyclic. Otherwise, $\widetilde{\mathbf{G}} = [p, q]^+$ for some p and q, in the notation of Section 1J. Since $\widetilde{\mathbf{G}}$ contains 2-fold rotations, we see that $\mathbf{G}$ must contain pure imaginary quaternions, and hence also -1; it follows that $\mathbf{G}$ doubly-covers $\widetilde{\mathbf{G}}$. The double covers of $[2, n]^+$, $[3, 3]^+$, $[3, 4]^+$ and $[3, 5]^+$ are just (up to conjugacy) $\mathbf{D}_n$, $\mathbf{T}$, $\mathbf{O}$ and $\mathbf{I}$, respectively. □

1K15 Remark We shall justify the expressions in (1K11), (1K12) and (1K13) of these quaternion groups when we find coordinates for the classical regular polyhedra in Sections 7B and 7D, since these will show what the finite rotation groups look like.

Notes to Section 1K

1. The Preface to [43] contains an historical account of the origins of quaternions.
2. The notation $\mathbf{V} := \mathbf{D}_2$ recalls the German name 'Viergruppe' for the four-group. Observe that it is a normal subgroup of $\mathbf{T}$ of index 3.
3. Remark 1K3 indicates that a mismatch of orientations is at the heart of some disparities; this is a consequence of the way that quaternionic multiplication is set up.
4. As we pointed out, in a right-handed system the mapping of (1K10) gives a rotation through -2ϑ. We can see this by picking $\mathbf{u} = \mathbf{i}$; then $\mathbf{j}\mathbf{g}(\mathbf{a}, \mathbf{a}) = \cos(2\vartheta)\mathbf{j} - \sin(2\vartheta)\mathbf{k}$, which is indeed a rotation through -2ϑ. The same will then hold whenever $\mathbf{u}, \mathbf{v}, \mathbf{w} \in \mathbb{E}^3$ form an orthonormal basis with the same orientation. This is at variance with our writing mappings *after* their arguments, but again follows from the definition of multiplication of quaternions.
5. It is useful to have a separate symbol $\mathbf{U} := \mathbf{O} \setminus \mathbf{T}$ for the other coset of $\mathbf{T}$ in $\mathbf{O}$. (No confusion with the saame notation for a unitary group is likely.)
6. An alternative name for the binary icosahedral group is the *icosians*.
7. If we choose the odd permutations of $\frac{1}{2}(\pm\tau, \pm 1, \pm\tau^{-1}, 0)$ instead of the even ones to form the binary icosahedral group, then we denote the result by $\mathbf{I}^{\ddagger}$. For both choices, we follow the convention of [43, §3.20].
8. There are two groups not covered by the notation introduced in this section. For these, the interaction between left and right multiplication involves the involutory isomorphism between $\mathbf{I}$ and $\mathbf{I}^{\ddagger}$ that changes the sign of $\sqrt{5}$, so that $\tau \leftrightarrow -\tau^{-1}$. When we come to discuss the regular polytopes with group $[3, 3, 5]$ in Sections 7E and 7F, we shall see that $\mathbf{I}$ and $\mathbf{I}^{\ddagger}$ interact in interesting ways.

2

Abstract Regular Polytopes

There is a very rich theory of abstract regular polytopes, which was mostly established in various papers that culminated in the monograph [99]. We only need certain basic aspects of that theory; this chapter is devoted to describing those basics. The general principle here will be that we prove what we use subsequently, referring the reader to [99] for more general results along the same lines and for further ramifications of them. Since this book will concentrate on the *geometric* side of the subject, it happens that a fair amount of the abstract theory is only of incidental interest (though we should usually bear it in mind). Thus we present here a somewhat brief, partial, account of the theory.

It is no part of our purpose to go into the general theory of abstract polytopes and their automorphism groups. However, we do need some basic facts about abstract polytopes in general, and so in Section 2A we introduce a new recursive definition, which corresponds much more closely than that of [99, Chapter 2] to one's intuitive idea of what a polytope should be. In Section 2B, we introduce regularity of abstract polytopes and the central idea of string C-groups, and show that the two concepts are equivalent. The intersection property defines a C-group; in Section 2C we establish conditions that ensure it, in particular some quotient criteria. In Section 2D we discuss presentations of the groups of regular polytopes, including the circuit criterion, and introduce some related general concepts and notation. Maps or polyhedra – the polytopes of rank 3 – form an important class of regular polytopes; Section 2E is devoted to describing some of their properties and giving some useful examples. Finally, in Section 2F we discuss certain special properties of regular polytopes, such as central symmetry, flatness and collapsibility, that will make frequent appearances.

The reader will do well to bear in mind a core motivation for the theory presented here. What we study has its ultimate origins in the description and classification of the five so-called Platonic polyhedra in Euclid's *Elements* ($\Sigma\tau o\iota\chi\varepsilon\iota\alpha$). Thus the term abstract polytope generalizes the properties of the face-lattice of a convex polytope; we refer the reader to our brief introduction in Section 1B, and for more detail to [56, 132] for the general background to this extensive theory, which we shall occasionally dip into. Further, while the idea of regularity, in its modern algebraic sense of action of a group of symmetries, was only very vaguely appreciated two millennia ago, our definition will correspond naturally with what our predecessors would have understood. In other words, what we do is pick out key properties of regular convex polytopes, and elevate them to form our definitions.

2A Abstract Polytopes

As we have said, our definition of abstract polytope will be recursive. We shall introduce the basic properties, and discuss them as we go along.

Posets

We first recall some preliminaries. Let $(\mathcal{P}, \leqslant)$ be a *poset* (partially ordered set), so that $\mathcal{P}$ is a set with a transitive antisymmetric binary relation $\leqslant$. Thus, if $\boldsymbol{A}, \boldsymbol{B}, \boldsymbol{C} \in \mathcal{P}$ with $\boldsymbol{A} \leqslant \boldsymbol{B} \leqslant \boldsymbol{C}$ then $\boldsymbol{A} \leqslant \boldsymbol{C}$, and if $\boldsymbol{A} \leqslant \boldsymbol{B} \leqslant \boldsymbol{A}$ then $\boldsymbol{A} = \boldsymbol{B}$. If $\boldsymbol{A} \leqslant \boldsymbol{B}$ or $\boldsymbol{B} \leqslant \boldsymbol{A}$, then we call $\boldsymbol{A}$ and $\boldsymbol{B}$ *incident*. A poset $\mathcal{P}$ is *connected* if, given any $\boldsymbol{A}, \boldsymbol{B} \in \mathcal{P}$, there is a sequence $\boldsymbol{A} = \boldsymbol{C}_0, \ldots, \boldsymbol{C}_k = \boldsymbol{B}$ in $\mathcal{P}$ (for some k), such that $\boldsymbol{C}_{j-1}$ and $\boldsymbol{C}_j$ are incident for $j = 1, \ldots, k$. We write $\boldsymbol{A} < \boldsymbol{B}$ if $\boldsymbol{A} \leqslant \boldsymbol{B}$ but $\boldsymbol{A} \neq \boldsymbol{B}$. If $\boldsymbol{A} < \boldsymbol{B}$, but there is no $\boldsymbol{C}$ such that $\boldsymbol{A} < \boldsymbol{C} < \boldsymbol{B}$, then $\boldsymbol{B}$ is said to *cover* $\boldsymbol{A}$.

At this stage, we do not distinguish between two *isomorphic* posets $(\mathcal{P}, \leqslant)$ and $(\mathcal{Q}, \leqslant)$, for which there is a one-to-one correspondence between $\mathcal{P}$ and $\mathcal{Q}$ which preserves the binary relation $\leqslant$.

Polytopes

We did not describe even the bare bones of the combinatorial structure of a convex polytope in Section 1B, precisely because the abstract definition that follows is specifically designed to mimic this structure; see the end of the section, particularly Remark 2A12. Thus, an *abstract polytope* is a poset $\mathcal{P}$ with the following properties.

- $\mathcal{P}$ has a unique minimal element $\boldsymbol{0}$ and a unique maximal element $\boldsymbol{1}$.

The notation $\boldsymbol{0}$ and $\boldsymbol{1}$ is temporary. The elements $\boldsymbol{0}$ and $\boldsymbol{1}$ are *improper*; all other elements are *proper*.

An element of an abstract polytope $\mathcal{P}$ that covers $\boldsymbol{0}$ (that is, an *atom* in the usual language of posets) is called a *vertex*, and one which is covered by $\boldsymbol{1}$ (that is, a *co-atom*) is a *cell*. If $\boldsymbol{V} \in \mathcal{P}$ is a vertex, then the subposet

$$\{\boldsymbol{B} \in \mathcal{P} \mid \boldsymbol{V} \leqslant \boldsymbol{B} \leqslant \boldsymbol{1}\}$$

is called the *vertex-figure* of $\mathcal{P}$ at $\boldsymbol{V}$. Similarly, if $\boldsymbol{C} \in \mathcal{P}$ is a cell, then the subposet

$$\{\boldsymbol{B} \in \mathcal{P} \mid \boldsymbol{0} \leqslant \boldsymbol{B} \leqslant \boldsymbol{C}\}$$

is the *facet* of $\mathcal{P}$ at $\boldsymbol{C}$.

Next, we need the small abstract polytopes, which are the following:

- the (-1)-polytope has $\boldsymbol{0} = \boldsymbol{1}$;
- the 0-polytope, denoted $\{1\}$ and called the *henogon*, has only the improper elements $\boldsymbol{0}$ and $\boldsymbol{1}$;
- the 1-polytope has exactly two proper elements, so that each covers $\boldsymbol{0}$ and is covered by $\boldsymbol{1}$; it is called the *digon*, and is denoted $\{2\}$.

Now suppose that $m \geqslant 2$. Finally, then, an *m-polytope* $\mathcal{P}$ additionally satisfies

- each vertex-figure and each facet of $\mathcal{P}$ is an $(m-1)$-polytope;
- the subposet $\mathcal{P}_{0,m-1}$ of $\mathcal{P}$ composed of its vertices and cells is connected.

We say that a poset $\mathcal{P}$ is *ranked* if it has a *rank function*, denoted rank, such that, if $\boldsymbol{B}$ covers $\boldsymbol{A}$, then $\operatorname{rank} \boldsymbol{B} = \operatorname{rank} \boldsymbol{A} + 1$. Then we have

2A1 Proposition *An abstract polytope $\mathcal{P}$ is a ranked poset.*

Proof. We begin by settting $\operatorname{rank} \boldsymbol{0} = -1$ for every $\mathcal{P}$. For $\boldsymbol{B} \in \mathcal{P} \setminus \{\boldsymbol{0}\}$, we set $\operatorname{rank} \boldsymbol{B} = j$, and call $\boldsymbol{B}$ a *j-element* if $\boldsymbol{B}$ covers some $\boldsymbol{A} \in \mathcal{P}$ with $\operatorname{rank} \boldsymbol{A} = j-1$. We then easily see that $\operatorname{rank} \boldsymbol{1} = m$ if $\mathcal{P}$ is an m-polytope. □

2A2 Remark Of course, in introducing the term 'm-polytope' we anticipated Proposition 2A1. We shall also use the names henogon and digon in a geometric sense, there being no call to introduce separate terms.

If $\mathsf{K} \subseteq \mathsf{M} := \{0, \ldots, m-1\}$, then we define $\mathcal{P}_{\mathsf{K}} := \{\boldsymbol{A} \in \mathcal{P} \mid \operatorname{rank} \boldsymbol{A} \in \mathsf{K}\}$. In particular, we write $\mathcal{P}_k := \mathcal{P}_{\{k\}}$ for the set of k-elements of an abstract polytope $\mathcal{P}$; similarly, $\mathcal{P}_{0,m-1} = \mathcal{P}_0 \cup \mathcal{P}_{m-1}$ is an abbreviation for $\mathcal{P}_{\{0,m-1\}}$.

Infaces and Cofaces

Our notion of abstract polytopes leads quickly to various other properties. These follow directly from obvious inductive assumptions, and so we shall omit the proofs (which are, in any case, very similar to those in [99, Section 2A]).

Throughout the remainder of the section, let $\mathcal{P}$ be a given m-polytope. If $\boldsymbol{A} \in \mathcal{P}$ is an $(r-1)$-element and $\boldsymbol{C} \in \mathcal{P}$ is a s-element with $\boldsymbol{A} \leqslant \boldsymbol{C}$, then the subposet $\{\boldsymbol{B} \mid \boldsymbol{A} \leqslant \boldsymbol{B} \leqslant \boldsymbol{C}\}$ is called an *(r,s)-inface* of $\mathcal{P}$; see the notes at the end of the section. It is sometimes the case that only the rank $k := s - r$ is important, in which case we talk about a *k-inface*. The special case $\boldsymbol{A} = \boldsymbol{0}$ gives a *k-face*, and $\boldsymbol{C} = \boldsymbol{1}$ gives a *k-coface*. What is obvious from the recursive definition is the following.

2A3 Proposition *A k-inface of a polytope is a k-polytope.*

Extending our use of geometric language, we also call a 1-face of a polytope an *edge* and an $(m-2)$-face of an m-polytope a *ridge*. We also frequently refer to $\mathcal{P}_{01} := \mathcal{P}_0 \cup \mathcal{P}_1$ as the *edge-graph* of the polytope $\mathcal{P}$.

Flags

A *chain* of $\mathcal{P}$ is a totally ordered subset of elements. A maximal chain of $\mathcal{P}$ is a *flag*. Strictly speaking, a flag of $\mathcal{P}$ contains $\boldsymbol{0}$ and $\boldsymbol{1}$, but usually we shall leave the improper elements of a flag to be understood. We write $\mathcal{F} = \mathcal{F}(\mathcal{P})$ for the family of flags of $\mathcal{P}$. The recursive definition implies

2A4 Proposition *Every chain of a polytope $\mathcal{P}$ is contained in some flag. In particular, if $m \geqslant 1$, then $\mathcal{F}(\mathcal{P}) \neq \emptyset$. Each flag of $\mathcal{P}$ contains m proper elements (of ranks $0, \ldots, m-1$).*

Two flags $\boldsymbol{\Phi}, \boldsymbol{\Psi} \in \mathcal{F}$ are called *adjacent* if they differ by one element; if this element has rank j, then we say that $\boldsymbol{\Phi}$ and $\boldsymbol{\Psi}$ are *j-adjacent*, and write $\boldsymbol{\Psi} = \boldsymbol{\Phi}^j$. Note that the definition of a digon (1-polytope) ensures that $\boldsymbol{\Phi}^j$ is unique. We further write $\boldsymbol{\Phi}^{jk} = (\boldsymbol{\Phi}^j)^k$, and so on. We next have

2A5 Proposition *For $\boldsymbol{\Phi} \in \mathcal{F}(\mathcal{P})$ and $j, k = 0, \ldots, m-1$,*

(a) $\boldsymbol{\Phi}^{jj} = \boldsymbol{\Phi}$,

(b) $\boldsymbol{\Phi}^{jk} = \boldsymbol{\Phi}^{kj}$ *if* $|k - j| \geqslant 2$.

Connectedness

We say that a ranked poset $\mathcal{P}$ is *flag-connected* if, whenever $\boldsymbol{\Phi}, \boldsymbol{\Psi} \in \mathcal{F}$, there is a sequence $\boldsymbol{\Phi} = \boldsymbol{\Omega}_0, \ldots, \boldsymbol{\Omega}_r = \boldsymbol{\Psi}$ in $\mathcal{F}$ such that $\boldsymbol{\Omega}_{s-1}$ and $\boldsymbol{\Omega}_s$ are adjacent for $s = 1, \ldots, r$. Moreover, $\mathcal{P}$ is *strongly* flag-connected if $\boldsymbol{\Omega}_s \supseteq \boldsymbol{\Phi} \cap \boldsymbol{\Psi}$ for each such s.

2A6 Proposition *An abstract polytope $\mathcal{P}$ is strongly flag-connected.*

Sketch of proof. Let $\boldsymbol{\Phi}, \boldsymbol{\Psi}$ be the given flags. If the chain $\boldsymbol{\Omega} := \boldsymbol{\Phi} \cap \boldsymbol{\Psi}$ contains proper elements, say $\boldsymbol{F}_1 < \cdots < \boldsymbol{F}_{k-1}$, then we can appeal to induction on rank for the infaces $\boldsymbol{F}_j/\boldsymbol{F}_{j-1}$ for each j (with $\boldsymbol{F}_0 = \boldsymbol{0}$ and $\boldsymbol{F}_k = \boldsymbol{1}$) to move the part of $\boldsymbol{\Phi}$ in the inface to the corresponding part of $\boldsymbol{\Psi}$ through adjacent flags. If they have no proper elements in common, take $(m-1)$-elements $\boldsymbol{F} \in \boldsymbol{\Phi}$, $\boldsymbol{G} \in \boldsymbol{\Psi}$ and use the connectedness of the subposet $\mathcal{P}_{0,m-1}$. We can now apply the first part to facets $\boldsymbol{C}_j/\boldsymbol{0}$ and vertex-figures $\boldsymbol{1}/\boldsymbol{V}_j$ of the sequence

$$\boldsymbol{F} = \boldsymbol{C}_0 > \boldsymbol{V}_1 < \boldsymbol{C}_2 > \cdots > \boldsymbol{V}_{2k-1} < \boldsymbol{C}_{2k} = \boldsymbol{G}$$

(for some k) guaranteed by this connectedness, through intermediate flags that contain successive pairs. □

If we fix a flag $\boldsymbol{\Phi}$, then the strong flag-connectedness implies that we can express a general flag $\boldsymbol{\Psi}$ in the form

2A7 $$\boldsymbol{\Psi} = \boldsymbol{\Phi}^{j(1)\cdots j(r)}.$$

The sequence $\mathsf{J} := j(1) \cdots j(r)$ of (2A7) is called a *flag-sequence* associated with $\boldsymbol{\Psi}$. We shall see the utility of this idea in Section 2D and subsequently.

The flag-connectedness of Proposition 2A6 induces further connectedness properties of the elements of $\mathcal{P}$; of course, the case $j = 0$, $k = m-1$ is part of the initial definition, which we used in the proof of Proposition 2A6.

2A8 Proposition *If $\mathcal{P}$ is an abstract m-polytope and $0 \leqslant j, k \leqslant m-1$ with $j \neq k$, then, given any two j-elements $\boldsymbol{A}$ and $\boldsymbol{B}$ of $\mathcal{P}$, there is a sequence*

$$\boldsymbol{A} = \boldsymbol{F}_0,\ \boldsymbol{F}_1,\ \ldots,\ \boldsymbol{F}_r = \boldsymbol{B}$$

of alternating j- and k-elements for some (even) r, such that $\boldsymbol{F}_{s-1}$ and $\boldsymbol{F}_s$ are incident for each $s = 1, \ldots, r$.

Proof. Choose any flags $\boldsymbol{\Phi}$ and $\boldsymbol{\Psi}$ of $\mathcal{P}$ such that $\boldsymbol{A} \in \boldsymbol{\Phi}$ and $\boldsymbol{B} \in \boldsymbol{\Psi}$. Then Proposition 2A6 implies that there is a flag-sequence $\mathsf{J} = j(1) \cdots j(r)$ such that $\boldsymbol{\Psi} = \boldsymbol{\Phi}^{\mathsf{J}}$. We only change a j-element in successive flags of the sequence at an entry $j(s) = j$; this does not change the current k-element. Two or more successive changes of the j-element while keeping the same k-element can clearly be regarded as a single change of j-element. Interchanging j and k in this argument results in the alternating sequence of j- and k-elements asked for. □

This connectedness property of $\mathcal{P}$ implies that $\mathcal{P}$ is *strongly connected*, in that the same connectedness holds for each inface of $\mathcal{P}$ of rank at least 2. In fact, we decree that infaces of rank less than 2 are automatically connected. As a special case, the edge-graph $\mathcal{E} = \mathcal{E}(\mathcal{P})$ of an m-polytope $\mathcal{P}$ with $m \geqslant 2$ is connected (see the notes at the end of the section).

Duality

If $\leqslant$ is an ordering of a poset, then the opposite ordering $\leqslant^*$ is given by $\boldsymbol{A} \leqslant^* \boldsymbol{B}$ if $\boldsymbol{B} \leqslant \boldsymbol{A}$. The *dual* $\mathcal{P}^\delta$ of an abstract polytope $\mathcal{P}$ has the same set of elements, but with the opposite ordering. Thus we have

2A9 Proposition *If $\mathcal{P}$ is an m-polytope, then so is its dual $\mathcal{P}^\delta$.*

Note that a j-element of $\mathcal{P}$ becomes an $(m-j-1)$-element of $\mathcal{P}^\delta$.

Order Complex

For $m \geqslant 2$, we can associate with an abstract m-polytope $\mathcal{P}$ its *order complex* $\mathcal{C} = \mathcal{C}(\mathcal{P})$. A j-simplex in $\mathcal{C}$ is a chain of $j+1$ proper elements; thus $\mathcal{C}$ is a pure simplicial $(m-1)$-complex, which may be countably infinite, whose maximal simplices are the flags of $\mathcal{P}$. Observe that $\mathcal{P}$ and $\mathcal{P}^\delta$ have the same order complexes. Notice that $|\mathcal{C}|$ will generally be far from having the structure of a manifold. The exception is that of rank 3, when the underlying surface will often help to explain what is happening under the application of various operations.

We call a polytope *orientable* if its order complex is orientable, meaning that its maximal simplices admit a 2-colouring.

Atomicity

By definition, an abstract polytope $\mathcal{P}$ is a poset, but not necessarily a *lattice*, in the sense that, for each $\boldsymbol{F}, \boldsymbol{G} \in \mathcal{P}$, there are a unique maximal element $\boldsymbol{F} \wedge \boldsymbol{G}$

contained in $\boldsymbol{F}$ and $\boldsymbol{G}$ and a unique minimal element $\boldsymbol{F} \vee \boldsymbol{G}$ containing $\boldsymbol{F}$ and $\boldsymbol{G}$; moreover, these satisfy

2A10
$$\begin{aligned} \boldsymbol{F}_1 \wedge (\boldsymbol{F}_2 \vee \boldsymbol{F}_3) &= (\boldsymbol{F}_1 \wedge \boldsymbol{F}_2) \vee (\boldsymbol{F}_1 \wedge \boldsymbol{F}_3), \\ \boldsymbol{F}_1 \vee (\boldsymbol{F}_2 \wedge \boldsymbol{F}_3) &= (\boldsymbol{F}_1 \vee \boldsymbol{F}_2) \wedge (\boldsymbol{F}_1 \vee \boldsymbol{F}_3), \end{aligned}$$

for all elements $\boldsymbol{F}_1, \boldsymbol{F}_2, \boldsymbol{F}_3 \in \mathcal{P}$.

A natural weaker condition in the geometric context is atomicity. We say $\mathcal{P}$ is *atomic* if, for each $k = 1, \ldots, m-1$, a k-element $\boldsymbol{F} \subseteq \mathcal{P}$ is determined by the subset $\mathcal{F} := \{\boldsymbol{V} \in \mathcal{P}_0 \mid \boldsymbol{V} \leqslant \boldsymbol{F}\} \subseteq \mathcal{P}_0$ of its vertices; that is,

2A11
$$\boldsymbol{F} = \bigvee \mathcal{F}.$$

There is a corresponding dual condition, called *co-atomicity*, but we shall not need to appeal to it.

2A12 Remark In contrast, the set $\mathcal{F}(P)$ of faces of a convex polytope P (including the improper faces $\emptyset$ and P itself) do form a lattice under $\leqslant \; = \; \subseteq$, with $F \wedge G = F \cap G$ and $F \vee G = \bigcap\{H \in \mathcal{F}(P) \mid F, G \leqslant H\}$.

Notes to Section 2A

1. We refer to [99, Section 2A] for the alternative approach to abstract polytopes. It is a straightforward exercise (which we leave to the interested reader) to show that our definition is equivalent to the earlier one, although ours is ostensibly stronger.
2. We have used the term *element* to distinguish the concept from *face*; thus 'face' and 'coface', which are both polytopes of the appropriate rank, denote dual ideas. However, later we shall be less concerned to make the distinction.
3. As a special case, it has not previously been customary to distinguish between a cell and its facet. In fact, cells are generally less important than their facets.
4. We have introduced the new term *inface*, instead of *section* as was used in [99], because we occasionally wish to use the term section with its geometric meaning as a subset of an object obtained by slicing it with (generally) a hyperplane.
5. In previous work, for the most part the dual of $\mathcal{P}$ has been denoted $\mathcal{P}^*$. However, for polyhedra (3-polytopes) it has been more useful to employ $\mathcal{P}^\delta$, and so we have chosen here to make this notation uniform throughout the book.
6. Through $|\mathcal{C}|$, the order complex $\mathcal{C}$ provides some underlying topology, although (as we said) we rarely have a manifold structure; it will play a useful rôle later. The point-set $|\mathcal{C}|$ is sometimes called the polyhedron of the complex $\mathcal{C}$, though (for obvious reasons) we shall avoid the term.

2B Regularity

As we said at the beginning of the chapter, our aim in this section is to introduce the concepts of regularity of polytopes and string C-groups, and show that they are essentially interchangeable. Throughout the section, $\mathcal{P}$ will be a fixed (abstract) m-polytope.

Automorphisms

An *automorphism* of $\mathcal{P}$ is an order-preserving permutation of its elements. It is clear that these automorphisms form a group $\boldsymbol{G} = \boldsymbol{G}(\mathcal{P})$, which is called the *automorphism group* of $\mathcal{P}$. If $\boldsymbol{G}$ is transitive on the family $\mathcal{F}$ of flags of $\mathcal{P}$, then we say that $\mathcal{P}$ is *regular*.

Let $\mathcal{P}$ be a regular polytope; we fix a flag $\boldsymbol{\Phi} \in \mathcal{F}(\mathcal{P})$, and call it the *base* flag. Then, for each $j \in \mathsf{M}$, there is an $\boldsymbol{r}_j \in \boldsymbol{G} = \boldsymbol{G}(\mathcal{P})$ such that

2B1 $$\boldsymbol{\Phi}^j = \boldsymbol{\Phi}\boldsymbol{r}_j.$$

The next two theorems describe the essential features of $\boldsymbol{G}$. A key observation, which follows from the uniqueness of the flag j-adjacent to a given flag, is

2B2 Lemma *If $\boldsymbol{\Psi}$ is any flag of $\mathcal{P}$ and $\boldsymbol{s} \in \boldsymbol{G}$, then $\boldsymbol{\Psi}^j\boldsymbol{s} = (\boldsymbol{\Psi}\boldsymbol{s})^j$ for each $j = 0, \ldots, m-1$.*

The first result is

2B3 Theorem *If $\mathcal{P}$ is a regular m-polytope, then $\boldsymbol{G}(\mathcal{P}) = \langle \boldsymbol{r}_0, \ldots, \boldsymbol{r}_{m-1} \rangle$ with $\boldsymbol{r}_j$ defined by (2B1), and is simply transitive on $\mathcal{F}(\mathcal{P})$. Moreover,*

(a) $\boldsymbol{r}_j^2 = \boldsymbol{e}$, *the identity, for each* j,

(b) $\boldsymbol{r}_j \rightleftharpoons \boldsymbol{r}_k$ *whenever* $|k - j| \geqslant 2$.

Recall that $\boldsymbol{r}_j \rightleftharpoons \boldsymbol{r}_k$ means that $\boldsymbol{r}_j$ and $\boldsymbol{r}_k$ commute.

Proof. From (2A7) and an appeal to Lemma 2B2, we can express a general flag $\boldsymbol{\Psi}$ in the form
$$\boldsymbol{\Psi} = \boldsymbol{\Phi}^{j(1)\cdots j(s)} = \boldsymbol{\Phi}\boldsymbol{r}_{j(s)} \cdots \boldsymbol{r}_{j(1)},$$
for some $j(1), \ldots, j(s) \in \mathsf{M}$. This establishes the initial claim. Next, note that an automorphism that fixes *any* flag Ψ also fixes every adjacent flag Ψ^j; the simple transitivity is then a consequence of Proposition 2A6. Finally, $\boldsymbol{r}_j^2 = \boldsymbol{e}$ follows from $\Phi^{jj} = \Phi$, and $\boldsymbol{r}_j \rightleftharpoons \boldsymbol{r}_k$ for $|k - j| \geqslant 2$ follows from $\Phi^{jk} = \Phi^{kj}$, both using Proposition 2A5. □

These $\boldsymbol{r}_j$ are called the *distinguished* generators of $\boldsymbol{G}$, relative (of course) to the choice of base flag. As a short way of talking about the ordered set $(\boldsymbol{r}_0, \ldots, \boldsymbol{r}_{m-1})$ of distinguished generators, we shall call it the *generatrix* of $\boldsymbol{G}$ and of the corresponding regular polytope $\mathcal{P}$.

If $\mathsf{J} \subseteq \mathsf{M}$, then we define $\boldsymbol{G}_\mathsf{J}$ by

2B4 $$\boldsymbol{G}_\mathsf{J} := \langle \boldsymbol{r}_i \mid i \in \mathsf{M} \setminus \mathsf{J} \rangle;$$

we call $\boldsymbol{G}_\mathsf{J}$ a *distinguished* subgroup of $\boldsymbol{G}$. The reason behind this notation will become clear later. When $\mathsf{J} = \{j\}$ is a singleton, we abbreviate $\boldsymbol{G}_{\{j\}}$ to $\boldsymbol{G}_j$; extending this shorthand, among such subgroups that we meet frequently are the $\boldsymbol{G}_{jk} := \boldsymbol{G}_{\{j,k\}}$.

2B5 Remark Theorem 2B3(b) implies that, for $j = 1, \dots, m-2$, the subgroup $\boldsymbol{G}_j$ is an (internal) direct product:

$$\boldsymbol{G}_j = \langle \boldsymbol{r}_0, \dots, \boldsymbol{r}_{j-1} \rangle \langle \boldsymbol{r}_{j+1}, \dots, \boldsymbol{r}_{m-1} \rangle = \langle \boldsymbol{r}_0, \dots, \boldsymbol{r}_{j-1} \rangle \times \langle \boldsymbol{r}_{j+1}, \dots, \boldsymbol{r}_{m-1} \rangle.$$

C-Groups

We now introduce a core concept, of which a special case pertains to regular polytopes. A group $\boldsymbol{G}$ generated by involutions $\{\boldsymbol{r}_j \mid j \in \mathsf{M}\}$ for some (finite) index set M is called a *C-group* with respect to the generatrix $(\boldsymbol{r}_j \mid j \in \mathsf{M})$ if it satisfies the *intersection property*, namely,

2B6
$$\langle \boldsymbol{r}_i \mid i \in \mathsf{J} \rangle \cap \langle \boldsymbol{r}_i \mid i \in \mathsf{K} \rangle = \langle \boldsymbol{r}_i \mid i \in \mathsf{J} \cap \mathsf{K} \rangle$$

whenever $\mathsf{J}, \mathsf{K} \subseteq \mathsf{M}$ (see the notes at the end of the section).

There is a further property that the generatrix holds; this is central to what we mean by a regular polytope.

2B7 Theorem *The distinguished subgroups of the automorphism group* $\boldsymbol{G}$ *of a regular* m*-polytope* $\mathcal{P}$ *satisfy the intersection property* (*2B6*).

Proof. This is actually equivalent to the strong flag-connectedness of $\mathcal{P}$, namely, Proposition 2A6. □

The properties of Theorems 2B3 and 2B7 define $\boldsymbol{G}$ to be a *string* C-group.

Duality

We next make an obvious observation.

2B8 Proposition *If the polytope* $\mathcal{P}$ *is regular, then so is its dual* $\mathcal{P}^{\delta}$.

Indeed, it is clear that, if $(\boldsymbol{r}_0, \dots, \boldsymbol{r}_{m-1})$ is the generatrix of the string C-group $\boldsymbol{G}$ and m-polytope $\mathcal{P}$, and we define $\boldsymbol{s}_j := \boldsymbol{r}_{m-j-1}$ for each j, then $(\boldsymbol{s}_0, \dots, \boldsymbol{s}_{m-1})$ is also the generatrix of the same string C-group, with the partial ordering $\leqslant$ reversed; this gives the dual $\mathcal{P}^{\delta}$. It goes without saying that any combinatorial property of $\mathcal{P}$ translates into a corresponding one of $\mathcal{P}^{\delta}$; we shall often use this fact with little additional comment.

If $\mathcal{P}$ is a regular polytope, then its facets are all isomorphic, as are its vertex-figures. We can then talk about the facet of $\mathcal{P}$, which we denote by $\mathcal{P}^{\boldsymbol{f}}$; similarly, the vertex-figure of $\mathcal{P}$ is denoted $\mathcal{P}^{\boldsymbol{v}}$. (The use of affixes here is to accord with that of other operations; compare the notation for the dual, and see also Section 5B.) Observe that $\mathcal{P}^{\delta\boldsymbol{f}} = \mathcal{P}^{\boldsymbol{v}\delta}$.

This notation can be extended. We define the *ridge* of $\mathcal{P}$ to be $\mathcal{P}^{\boldsymbol{r}} := \mathcal{P}^{\boldsymbol{ff}}$, and the *edge-figure* of $\mathcal{P}$ to be $\mathcal{P}^{\boldsymbol{e}} := \mathcal{P}^{\boldsymbol{vv}}$.

Combinatorial Structure

We now describe how to recover the combinatorial structure of a regular m-polytope $\mathcal{P} = \mathcal{P}(\boldsymbol{G})$ from a string C-group $\boldsymbol{G}$ with generatrix $(\boldsymbol{r}_0, \dots, \boldsymbol{r}_{m-1})$.

We define the ranked poset $\mathcal{P}$ as follows. For each $j \in \mathsf{M}$, a j-element of $\mathcal{P}$ is a right coset $\boldsymbol{G}_j\boldsymbol{s}$ of the distinguished subgroup $\boldsymbol{G}_j$; thus j is the rank of the element. We say that two elements are incident just when they intersect (as cosets), and define the binary relation $\leqslant$ on $\mathcal{P}$ by

2B9 $$\boldsymbol{G}_j\boldsymbol{s} \leqslant \boldsymbol{G}_k\boldsymbol{t} \iff \boldsymbol{G}_j\boldsymbol{s} \cap \boldsymbol{G}_k\boldsymbol{t} \neq \emptyset \text{ and } j \leqslant k.$$

Formally, we also adjoin two copies of $\boldsymbol{G}$ itself, labelled $\boldsymbol{G}_{-1}$ and $\boldsymbol{G}_m$, as the (unique) (-1)- and m-elements of $\mathcal{P}$. Thus the elements $\boldsymbol{G}_{-1}$ and $\boldsymbol{G}_m$ are improper; the other elements of $\mathcal{P}$ are proper.

The relation (2B9) can be rephrased as

2B10 Proposition *In the string C-group $\boldsymbol{G}$, if $0 \leqslant j < k \leqslant m-1$ and $\boldsymbol{s}, \boldsymbol{t} \in \boldsymbol{G}$, then $\boldsymbol{G}_j\boldsymbol{s} \leqslant \boldsymbol{G}_k\boldsymbol{t}$ if and only if*

$$\boldsymbol{s}\boldsymbol{t}^{-1} \in \langle \boldsymbol{r}_{j+1}, \ldots, \boldsymbol{r}_{m-1}\rangle\langle \boldsymbol{r}_0, \ldots, \boldsymbol{r}_{k-1}\rangle.$$

Proof. If the given elements are incident, then there are $\boldsymbol{a} \in \boldsymbol{G}_j$ and $\boldsymbol{b} \in \boldsymbol{G}_k$ such that $\boldsymbol{a}\boldsymbol{s} = \boldsymbol{b}\boldsymbol{t}$. From this, and using the way that the distinguished generators $\boldsymbol{r}_i$ commute, we see that

$$\begin{aligned} \boldsymbol{s}\boldsymbol{t}^{-1} &= \boldsymbol{a}^{-1}\boldsymbol{b} \in \boldsymbol{G}_j\boldsymbol{G}_k \\ &= \langle \boldsymbol{r}_0, \ldots, \boldsymbol{r}_{j-1}\rangle\langle \boldsymbol{r}_{j+1}, \ldots, \boldsymbol{r}_{m-1}\rangle\langle \boldsymbol{r}_0, \ldots, \boldsymbol{r}_{k-1}\rangle\langle \boldsymbol{r}_{k+1}, \ldots, \boldsymbol{r}_{m-1}\rangle \\ &= \langle \boldsymbol{r}_{j+1}, \ldots, \boldsymbol{r}_{m-1}\rangle\langle \boldsymbol{r}_0, \ldots, \boldsymbol{r}_{k-1}\rangle \end{aligned}$$

as required. The argument is reversible, and this establishes the proposition. □

As an important consequence, we have

2B11 Corollary *The relation $\leqslant$ on the poset $\mathcal{P}$ is a partial ordering.*

Proof. First, suppose that $\boldsymbol{G}_j\boldsymbol{s} \leqslant \boldsymbol{G}_k\boldsymbol{t} \leqslant \boldsymbol{G}_j\boldsymbol{s}$. Then $j = k$, and hence $\boldsymbol{G}_j\boldsymbol{s} = \boldsymbol{G}_j\boldsymbol{t}$ since *distinct* cosets of $\boldsymbol{G}_j$ are disjoint. Second, let $i < j < k$, and suppose that $\boldsymbol{a}, \boldsymbol{b}, \boldsymbol{c} \in \boldsymbol{G}$ are such that $\boldsymbol{G}_i\boldsymbol{a} \leqslant \boldsymbol{G}_j\boldsymbol{b} \leqslant \boldsymbol{G}_k\boldsymbol{c}$. Then

$$\begin{aligned} \boldsymbol{a}\boldsymbol{b}^{-1} &\in \langle \boldsymbol{r}_{i+1}, \ldots, \boldsymbol{r}_{m-1}\rangle\langle \boldsymbol{r}_0, \ldots, \boldsymbol{r}_{j-1}\rangle, \\ \boldsymbol{b}\boldsymbol{c}^{-1} &\in \langle \boldsymbol{r}_{j+1}, \ldots, \boldsymbol{r}_{m-1}\rangle\langle \boldsymbol{r}_0, \ldots, \boldsymbol{r}_{k-1}\rangle, \end{aligned}$$

from which follows

$$\begin{aligned} \boldsymbol{a}\boldsymbol{c}^{-1} &= \boldsymbol{a}\boldsymbol{b}^{-1}\boldsymbol{b}\boldsymbol{c}^{-1} \\ &\in \langle \boldsymbol{r}_{i+1}, \ldots, \boldsymbol{r}_{m-1}\rangle\langle \boldsymbol{r}_0, \ldots, \boldsymbol{r}_{j-1}\rangle\langle \boldsymbol{r}_{j+1}, \ldots, \boldsymbol{r}_{m-1}\rangle\langle \boldsymbol{r}_0, \ldots, \boldsymbol{r}_{k-1}\rangle \\ &= \langle \boldsymbol{r}_{i+1}, \ldots, \boldsymbol{r}_{m-1}\rangle\langle \boldsymbol{r}_{j+1}, \ldots, \boldsymbol{r}_{m-1}\rangle\langle \boldsymbol{r}_0, \ldots, \boldsymbol{r}_{j-1}\rangle\langle \boldsymbol{r}_0, \ldots, \boldsymbol{r}_{k-1}\rangle \\ &= \langle \boldsymbol{r}_{i+1}, \ldots, \boldsymbol{r}_{m-1}\rangle\langle \boldsymbol{r}_0, \ldots, \boldsymbol{r}_{k-1}\rangle, \end{aligned}$$

so that $\boldsymbol{G}_i\boldsymbol{a} \leqslant \boldsymbol{G}_k\boldsymbol{c}$. These two properties say that $\leqslant$ is a partial ordering, as claimed. □

The distinguished subgroups $\boldsymbol{G}_0$ and $\boldsymbol{G}_{m-1}$ are clearly themselves string C-groups. We adopt the language of Section 2A by calling a coset of $\boldsymbol{G}_0$ a vertex and a coset of $\boldsymbol{G}_{m-1}$ a cell; the obvious inductive assumption lets us call the corresponding polytopes the (initial) vertex-figure and facet of $\mathcal{P}$.

In view of the partial ordering of Corollary 2B11 we can talk about chains of $\mathcal{P}$ and, in particular, about flags; the same circle of ideas shows that every chain is contained in some flag. The initial flag of $\mathcal{P}$ is $\boldsymbol{\Phi} = (\boldsymbol{G}_0, \ldots, \boldsymbol{G}_{m-1})$ (as usual ignoring the improper elements). The intersection property (2B6) implies that

$$\boldsymbol{G}_0 \cap \cdots \cap \boldsymbol{G}_{m-1} = \{e\},$$

from which follows

2B12 Lemma *To each flag $\boldsymbol{\Psi}$ of $\mathcal{P}$ corresponds a unique $\boldsymbol{g} \in \boldsymbol{G}$ such that $\boldsymbol{\Psi} = \boldsymbol{\Phi g}$. Thus $\boldsymbol{G}$ is simply transitive on the set $\mathcal{F}$ of flags of $\mathcal{P}$.*

2B13 Remark Observe that, in fact, $\boldsymbol{G}$ is transitive on the chains of $\mathcal{P}$ of each type. A core feature is that, if $\boldsymbol{g} \in \boldsymbol{G}$ fixes *any* flag $\boldsymbol{\Psi}$ (so that $\boldsymbol{\Psi g} = \boldsymbol{\Psi}$), then $\boldsymbol{g} = \boldsymbol{e}$.

We deduce from this that the subposet $\mathcal{P}_{0,m-1}$ is connected, which is the final step. Summarizing the discussion, we have proved

2B14 Theorem *A string C-group $\boldsymbol{G}$ is the automorphism group of an abstract regular polytope $\mathcal{P} = \mathcal{P}(\boldsymbol{G})$.*

2B15 Remark The initial j-face of $\mathcal{P}$ has automorphism group $\langle \boldsymbol{r}_0, \ldots, \boldsymbol{r}_{j-1} \rangle$, while $\langle \boldsymbol{r}_{m-j}, \ldots, \boldsymbol{r}_{m-1} \rangle$ is the automorphism group of the initial j-coface. The automorphism groups of infaces are expressed similarly.

Of particular importance are the infaces of $\mathcal{P}$ of rank 2. These are regular polygons; thus, for each $j = 1, \ldots, m-1$, the $(j-2, j+1)$-inface is a polygon $\{p_j\}$ for some $p_j \geqslant 2$, including possibly $p_j = \infty$. We then call $\{p_1, \ldots, p_{m-1}\}$ the *Schläfli type* of $\mathcal{P}$.

2B16 Remark More generally, we refer to $\{p_1, \ldots, p_{m-1}\}$ as the *Schläfli symbol* of $\mathcal{P}$, particularly when it completely specifies $\mathcal{P}$. We shall often employ the useful notation that (in a vector, or the like) a string $\alpha, \ldots, \alpha$ of length k is abbreviated to α^k; it will not be employed when confusion with an ordinary power is possible. For instance, we shall see that the regular m-simplex has Schläfli symbol $\{3^{m-1}\}$.

Allomorphism

A situation occurs often enough that it is worth having a term for. If two regular polytopes $\mathcal{P}$ and $\mathcal{Q}$ are distinct, but nevertheless are isomorphic and have the same vertices, then we call them *allomorphic* (see the notes at the end of the section); we also refer to $\mathcal{Q}$ as an *allomorph* of $\mathcal{P}$. As examples, every (finite) regular polygon $\{p\}$ has allomorphs, except for $p = 3, 4, 6$ (the crystallographic polygons). Thus, for $p \leqslant 10$, we have two allomorphs for the pentagon $\{5\}$, octagon $\{8\}$ and decagon $\{10\}$, and three for the heptagon $\{7\}$ and enneagon $\{9\}$.

Notes to Section 2B

1. As was pointed out in [99], strictly speaking we should talk about regular abstract polytopes. The term we use is influenced by euphony.
2. Observe that the concepts of 'complex polytope' introduced in [116] (see [30] for a comprehensive account) or 'incidence polytope' of [38] are more general than we wish to consider here, and their regularity properties require a correspondingly deeper treatment. Indeed, regarded as geometric objects in the appropriate unitary space, regular complex polytopes have fewer symmetries than they would have regarded as incidence polytopes in the corresponding euclidean space (of twice the dimension).
3. The 'C' in the term 'C-group' stands for 'Coxeter'.
4. In common usage, an edge of a graph is a pair of its vertices; thus 'edge' is the appropriate term in 'edge-graph', rather than '1-element'.
5. We have introduced here a convention that we hope the reader will find useful. Heavy braces are used to denote *abstract* regular polytopes; we indicate *geometric* regular polytopes (whose occurrence is more frequent – they will be introduced in Chapter 4) by ordinary braces.
6. The term *allomorph* has quite a different meaning in philology, but it is unlikely that any confusion will arise. We have actually borrowed it from chemistry, where it meant chemically identical molecules which have different geometric structures.

2C Regularity Criteria

In this section, we consider various conditions which guarantee that we have a string C-group, that is, the automorphism group of a regular polytope. Several of these involve a given group and a known string C-group, between which is a quotient relation.

Pre-polytopes

If we drop the intersection property (2B6) from the definition of a string C-group, then we have what we call a *string group generated by involutions*, or *sggi* for short. An sggi $\boldsymbol{G}$ will then play the rôle of automorphism group for a regular *pre-polytope* $\mathcal{P}$. All the properties of regular polytopes generalize, but the local structure of infaces (such as connectedness) is no longer preserved.

2C1 Remark We shall see that, in the geometric context, pre-polytopes arise very naturally and, indeed, are unavoidable. It is thus appropriate to ask why we restrict the main thrust of our treatment to polytopes, that is, to those whose groups do satisfy the intersection property. There are two reasons. First, the geometric: we generalize the strong-connectness properties of convex polytopes. Second, the algebraic: the groups generalize the basic (intersection) property of

Coxeter groups. In fact, we could add a third reason: the general sggi is defined so broadly that it is not actually sufficiently interesting. We therefore do not study regular pre-polytopes for their own sake, if for no other reason that they form a much larger collection of objects than do the regular polytopes (see the notes at the end of the section for more details). Nevertheless, since they occur quite naturally in realization theory, as we shall see in Chapter 4, we must bear them in mind.

Intersection Property

The core problem addressed in this section is how to tell whether a general sggi is actually a C-group, that is, satisfies the intersection property. We shall give a brief outline of the basic results in this area, and illustrate them with an example which, while fairly obvious, seems not to have been dealt with formally so far.

There is an imbalance in our approach, in that – for geometric reasons – we usually know exactly what a putative vertex-figure should be. Thus our existence criteria generally presuppose that we are given a string C-group to play the part of the group of the vertex-figure, and we then have to determine whether adjoining some initial involution – to form an sggi – yields a new string C-group. A starting observation is that two involutions always generate a string C-group.

The core result, to which we frequently appeal, is [99, Proposition 2E16] in dual formulation.

2C2 Theorem *Let $\boldsymbol{G} = \langle \boldsymbol{r}_0, \ldots, \boldsymbol{r}_{m-1} \rangle$ be an sggi such that the distinguished subgroup $\boldsymbol{G}_0 = \langle \boldsymbol{r}_1, \ldots, \boldsymbol{r}_{m-1} \rangle$ is a C-group. Then $\boldsymbol{G}$ is a C-group if either of the following holds:*

(a) $\boldsymbol{G}_{m-1} = \langle \boldsymbol{r}_0, \ldots, \boldsymbol{r}_{m-2} \rangle$ *is also a C-group, and* $\boldsymbol{G}_{m-1} \cap \boldsymbol{G}_0 = \boldsymbol{G}_{0,m-1}$*;*

(b) $\langle \boldsymbol{r}_0, \ldots, \boldsymbol{r}_j \rangle \cap \boldsymbol{G}_0 = \langle \boldsymbol{r}_1, \ldots, \boldsymbol{r}_j \rangle$ *for each* $j = 0, \ldots, m-2$.

Proof. We first show that (b) follows from (a). Recall first that $\boldsymbol{G}_{0,m-1} = \langle \boldsymbol{r}_1, \ldots, \boldsymbol{r}_{m-2} \rangle$. Since $\boldsymbol{G}_{m-1}$ is a C-group, under the given condition we have

$$\begin{aligned}\langle \boldsymbol{r}_0, \ldots, \boldsymbol{r}_j \rangle \cap \boldsymbol{G}_0 &= \langle \boldsymbol{r}_0, \ldots, \boldsymbol{r}_j \rangle \cap (\boldsymbol{G}_{m-1} \cap \boldsymbol{G}_0) \\ &= \langle \boldsymbol{r}_0, \ldots, \boldsymbol{r}_j \rangle \cap \langle \boldsymbol{r}_1, \ldots, \boldsymbol{r}_{m-2} \rangle \\ &= \langle \boldsymbol{r}_1, \ldots, \boldsymbol{r}_j \rangle,\end{aligned}$$

as claimed.

So, let us assume (b) for each j. Writing $\boldsymbol{H} := \boldsymbol{G}_0$ and

$$\boldsymbol{H}_j := \langle \boldsymbol{r}_i \mid i = 1, \ldots, m-1,\ i \neq j \rangle$$

for each $j = 1, \ldots, m-1$, we thus have

$$\begin{aligned}\boldsymbol{H} \cap \boldsymbol{G}_j &= \boldsymbol{H} \cap \langle \boldsymbol{r}_0, \ldots, \boldsymbol{r}_{j-1} \rangle \langle \boldsymbol{r}_{j+1}, \ldots, \boldsymbol{r}_{m-1} \rangle \\ &= (\boldsymbol{H} \cap \langle \boldsymbol{r}_0, \ldots, \boldsymbol{r}_{j-1} \rangle) \langle \boldsymbol{r}_{j+1}, \ldots, \boldsymbol{r}_{m-1} \rangle \\ &= \langle \boldsymbol{r}_1, \ldots, \boldsymbol{r}_{j-1} \rangle \langle \boldsymbol{r}_{j+1}, \ldots, \boldsymbol{r}_{m-1} \rangle \\ &= \boldsymbol{H}_j.\end{aligned}$$

Now, to establish the intersection property in general, it suffices to show that, for each $\mathsf{K} \subseteq \mathsf{M} := \{0, \ldots, m-1\}$, we have

$$\boldsymbol{G}_{\mathsf{K}} := \langle \boldsymbol{r}_i \mid i \notin \mathsf{K} \rangle = \bigcap \{\boldsymbol{G}_j \mid j \in \mathsf{K}\}.$$

If $0 \in \mathsf{K}$, then

$$\begin{aligned} \bigcap \{\boldsymbol{G}_j \mid j \in \mathsf{K}\} &= \bigcap \{\boldsymbol{H} \cap \boldsymbol{G}_j \mid j \in \mathsf{K}\} \\ &= \bigcap \{\boldsymbol{H}_j \mid j \in \mathsf{K} \setminus \{0\}\} \\ &= \boldsymbol{H}_{\mathsf{K}\setminus\{0\}} \\ &= \boldsymbol{G}_{\mathsf{K}}. \end{aligned}$$

Here, we have used the fact that $\boldsymbol{G}_0$ is a C-group. On the other hand, if $0 \notin \mathsf{K}$, suppose that $k \in \mathsf{K}$ is its smallest member. Since $\langle \boldsymbol{r}_{k+1}, \ldots, \boldsymbol{r}_{m-1} \rangle \subseteq \boldsymbol{G}_0 \cap \boldsymbol{G}_k$, we see that

$$\begin{aligned} \bigcap \{\boldsymbol{G}_j \mid j \in \mathsf{K}\} &= \boldsymbol{G}_k \cap \bigcap \{\boldsymbol{G}_j \mid j \in \mathsf{K} \setminus \{k\}\} \\ &= \langle \boldsymbol{r}_0, \ldots, \boldsymbol{r}_{k-1} \rangle \Big(\bigcap \{\boldsymbol{G}_j \mid j \in \mathsf{K} \setminus \{k\}\} \cap \langle \boldsymbol{r}_{k+1}, \ldots, \boldsymbol{r}_{m-1} \rangle \Big) \\ &\subseteq \langle \boldsymbol{r}_0, \ldots, \boldsymbol{r}_{k-1} \rangle \Big(\bigcap \{\boldsymbol{G}_j \mid j \in \mathsf{K} \setminus \{k\}\} \cap \boldsymbol{G}_0 \cap \boldsymbol{G}_k \Big) \\ &= \langle \boldsymbol{r}_0, \ldots, \boldsymbol{r}_{k-1} \rangle \langle \boldsymbol{r}_j \mid j \notin \mathsf{K} \cup \{0\} \rangle \\ &= \boldsymbol{G}_{\mathsf{K}}, \end{aligned}$$

where we have used what we showed previously. Since the opposite inclusion is trivial, this completes the proof. □

2C3 Remark Of course, there is a dual result, with $\boldsymbol{G}_{m-1}$ instead of $\boldsymbol{G}_0$. We shall take for granted the fact that all of the results presented here can be phrased in dual terms, and only occasionally comment on this subsequently.

Quotients

For $1 \leqslant k \leqslant m$, let $(\boldsymbol{r}_0, \ldots, \boldsymbol{r}_{m-1})$ be the generatrix of a regular m-polytope $\mathcal{P}$, and let $(\boldsymbol{s}_0, \ldots, \boldsymbol{s}_{k-1})$ be that of a regular k-polytope $\mathcal{Q}$. If the mapping Φ, defined by

2C4
$$\boldsymbol{r}_j \Phi := \begin{cases} \boldsymbol{s}_j, & \text{if } j = 0, \ldots, k-1, \\ e, & \text{if } j = k, \ldots, m-1, \end{cases}$$

induces a homomorphism $\Phi\colon \boldsymbol{G}(\mathcal{P}) \to \boldsymbol{G}(\mathcal{Q})$, then we call $\mathcal{Q}$ a *quotient* of $\mathcal{P}$; we also say that $\mathcal{P}$ is a *cover* of $\mathcal{Q}$, and write $\mathcal{P} \searrow \mathcal{Q}$. Observe that we specifically allow $k < m$ here.

If the homomorphism Φ is an isomorphism, then we naturally refer to $\mathcal{P}$ and $\mathcal{Q}$ as *isomorphic*, and write $\mathcal{P} \cong \mathcal{Q}$.

Quotient Criteria

Our main applications of Theorem 2C2 concern quotient relationships. In one direction, we have the *first quotient criterion*, which is

2C5 Theorem *If* $\boldsymbol{G} = \langle \boldsymbol{r}_0, \ldots, \boldsymbol{r}_{m-1} \rangle$ *is an sggi,* $\boldsymbol{H} = \langle \boldsymbol{s}_0, \ldots, \boldsymbol{s}_{m-1} \rangle$ *is a string C-group, and the mapping* $\boldsymbol{r}_j \mapsto \boldsymbol{s}_j$ *for* $j = 0, \ldots, m-1$ *induces a homomorphism* $\Phi\colon \boldsymbol{G} \to \boldsymbol{H}$ *which is one-to-one on* $\boldsymbol{G}_0$*, then* $\boldsymbol{G}$ *is a C-group.*

Proof. We appeal to Theorem 2C2(b). Let $j = 0, \ldots, m-2$, and suppose that $\boldsymbol{g} \in \langle \boldsymbol{r}_0, \ldots, \boldsymbol{r}_j \rangle \cap \boldsymbol{G}_0$. Then

$$\boldsymbol{g}\Phi \in \langle \boldsymbol{s}_0, \ldots, \boldsymbol{s}_j \rangle \cap \langle \boldsymbol{s}_1, \ldots, \boldsymbol{s}_{m-1} \rangle = \langle \boldsymbol{s}_1, \ldots, \boldsymbol{s}_j \rangle,$$

because $\boldsymbol{H}$ is a C-group. Hence $\boldsymbol{g}\Phi$ has a pre-image under Φ in $\langle \boldsymbol{r}_1, \ldots, \boldsymbol{r}_j \rangle$. But Φ is one-to-one on $\boldsymbol{G}_0$, so that $\boldsymbol{g}$ itself is the only pre-image. Thus $\boldsymbol{g} \in \langle \boldsymbol{r}_1, \ldots, \boldsymbol{r}_j \rangle$, as required. □

In the other direction, we should like to know when a quotient of a C-group is also a C-group. There is an extensive discussion of this question in [99, Section 2E]. In theory, the conditions are obvious: if Φ is a homomorphism on a C-group $\boldsymbol{G} = \langle \boldsymbol{r}_0, \ldots, \boldsymbol{r}_{m-1} \rangle$ with kernel $\boldsymbol{N} := \ker \Phi$, then $\boldsymbol{G}\Phi$ is a C-group if and only if

$$\boldsymbol{G}_{\mathsf{J}}\Phi \cap \boldsymbol{G}_{\mathsf{K}}\Phi = \boldsymbol{G}_{\mathsf{J}\cup\mathsf{K}}\Phi \iff \boldsymbol{G}_{\mathsf{J}}\boldsymbol{N} \cap \boldsymbol{G}_{\mathsf{K}}\boldsymbol{N} = \boldsymbol{G}_{\mathsf{J}\cup\mathsf{K}}\boldsymbol{N}$$

for all $\mathsf{J}, \mathsf{K} \subseteq \mathsf{M} := \{0, \ldots, m-1\}$. Of course, in general this condition is not very practical, and usually just comes down to directly verifying the intersection property for the quotient. Indeed, few conditions of this kind are easy to check in practice. Therefore, we will just concentrate on conditions that we shall subsequently find useful.

One of these is the *second quotient criterion.*

2C6 Theorem *With the standard conventions, let* $\boldsymbol{G} = \langle \boldsymbol{r}_0, \ldots, \boldsymbol{r}_{m-1} \rangle$ *be a string C-group, and let* Φ *be a homomorphism on* $\boldsymbol{G}$*. If* $\boldsymbol{G}_0\Phi$ *and* $\boldsymbol{G}_{m-1}\Phi$ *are both C-groups and* $\ker \Phi \leqslant \boldsymbol{G}_{m-1}$*, then* $\boldsymbol{G}\Phi$ *is a C-group.*

Proof. Write $\boldsymbol{N} := \ker \Phi$, $\boldsymbol{H} := \boldsymbol{G}\Phi$, and so on. In view of Theorem 2C2, it suffices to show that $\boldsymbol{H}_0 \cap \boldsymbol{H}_{m-1} = \boldsymbol{H}_{0,m-1}$. One inclusion being obvious, let us suppose that $\boldsymbol{h} \in \boldsymbol{H}_0 \cap \boldsymbol{H}_{m-1}$. Then there is a $\boldsymbol{g} \in \boldsymbol{G}_0$ such that $\boldsymbol{h} = \boldsymbol{g}\Phi$. Since $\boldsymbol{h}\Phi^{-1} = \boldsymbol{g}\boldsymbol{N}$, it follows that $\boldsymbol{g}\boldsymbol{N} \cap \boldsymbol{G}_{m-1} \neq \emptyset$. Since $\boldsymbol{N} \leqslant \boldsymbol{G}_{m-1}$, we deduce that $\boldsymbol{g}\boldsymbol{N} \subseteq \boldsymbol{G}_{m-1}$. Thus $\boldsymbol{g} \in \boldsymbol{G}_{m-1}$ also, and hence $\boldsymbol{g} \in \boldsymbol{G}_{0,m-1}$. We deduce immediately that $\boldsymbol{h} = \boldsymbol{g}\Phi \in \boldsymbol{G}_{0,m-1}\Phi = \boldsymbol{H}_{0,m-1}$, as required. □

One ought, perhaps, to expect the result of Theorem 2F2, which says that (with appropriate conditions) if we identify antipodal elements of a centrally symmetric regular polytope, then we obtain another regular polytope. However, its proof is not as obvious as one might hope. In preparation for the proof, and for general theoretical interest, we need to discuss sparse subgroups. Because we want to know what happens to homomorphic images of C-groups, we phrase things here in terms of normal subgroups, thought of where appropriate as kernels of homomorphisms.

Keeping to the standard conventions, as defined in [99, Section 2E] a normal subgroup $\boldsymbol{N} \lhd \boldsymbol{G}$ is called *sparse* if

2C7 $$\boldsymbol{N} \cap \langle \boldsymbol{r}_0, \ldots, \boldsymbol{r}_{m-2} \rangle \langle \boldsymbol{r}_1, \ldots, \boldsymbol{r}_{m-1} \rangle = \boldsymbol{N} \cap \boldsymbol{G}_{m-1}\boldsymbol{G}_0 = \{\boldsymbol{e}\}$$

(see the notes at the end of the section). The corresponding combinatorial condition is

2C8 Theorem *If $\mathcal{P}$ is a regular m-polytope and $\boldsymbol{N} \lhd \boldsymbol{G} = \boldsymbol{G}(\mathcal{P})$, then $\boldsymbol{N}$ is sparse if and only if any $\boldsymbol{N}$-orbit of vertices of $\mathcal{P}$ meets a given facet of $\mathcal{P}$ at most once.*

Proof. We shall actually show something rather more general. Observe that, for each $0 \leqslant j < k \leqslant n - 1$, we have

$$\langle \boldsymbol{r}_0, \ldots, \boldsymbol{r}_{k-1} \rangle \langle \boldsymbol{r}_{j+1}, \ldots, \boldsymbol{r}_{m-1} \rangle \subseteq \langle \boldsymbol{r}_0, \ldots, \boldsymbol{r}_{m-2} \rangle \langle \boldsymbol{r}_1, \ldots, \boldsymbol{r}_{m-1} \rangle,$$

and hence, bearing Proposition 2B10 in mind, the sparseness condition (2C7) implies that $\boldsymbol{N} \cap \boldsymbol{G}_k\boldsymbol{G}_j = \{\boldsymbol{e}\}$. Indeed, from that proposition we see that, for $\boldsymbol{n} \in \boldsymbol{N}$, if $\boldsymbol{G}_j\boldsymbol{n} \leqslant \boldsymbol{G}_k$, then

$$\boldsymbol{n}^{-1} \in \boldsymbol{N} \cap \boldsymbol{G}_k\boldsymbol{G}_j = \{\boldsymbol{e}\};$$

that is, $\boldsymbol{n} = \boldsymbol{e}$. In other words, no other j-element than $\boldsymbol{G}_j$ in the same $\boldsymbol{N}$-orbit can be incident with $\boldsymbol{G}_k$. But now regularity of $\mathcal{P}$ (that is, transitivity on its flags) and normality of $\boldsymbol{N}$ imply in the same way that at most one j-element in a given $\boldsymbol{N}$-orbit can be incident with a given k-element. With $j = 0$ and $k = m - 1$, this was what the theorem claimed.

It is clear that the argument is reversible; again, by regularity it is enough to consider what happens at the base flag. This establishes the theorem. □

2C9 Remark The reader should not need reminding that the rôles of vertices and cells can be interchanged in Theorem 2C8. Indeed, it is trivial to see that this follows from the condition of the theorem.

The importance of sparse subgroups lies in the following (in effect, [99, Lemma 2E22]), which we may think of as a *third quotient criterion.*

2C10 Theorem *If $\boldsymbol{G}$ is a string C-group and $\boldsymbol{N} \lhd \boldsymbol{G}$ is a sparse normal subgroup, then the quotient $\boldsymbol{G}/\boldsymbol{N}$ is a C-group.*

Proof. Let $\Phi\colon \boldsymbol{G} \to \boldsymbol{G}/\boldsymbol{N}$ be the induced quotient homomorphism, and write $\boldsymbol{H}_{\mathsf{J}} := \boldsymbol{G}_{\mathsf{J}}\Phi$ for each $\mathsf{J} \subseteq \mathsf{M}$. The definition of sparseness implies that $\boldsymbol{N} \cap \boldsymbol{G}_j = \{\boldsymbol{e}\}$ for $j = 0, m-1$, so that Φ is injective on both $\boldsymbol{G}_0$ and $\boldsymbol{G}_{m-1}$. Thus, given any $\boldsymbol{h} \in \boldsymbol{H}_0 \cap \boldsymbol{H}_{m-1}$, there are unique $\boldsymbol{a} \in \boldsymbol{G}_0$ and $\boldsymbol{b} \in \boldsymbol{G}_{m-1}$ such that $\boldsymbol{h} = \boldsymbol{a}\Phi = \boldsymbol{b}\Phi$. But then $\boldsymbol{a}^{-1}\boldsymbol{b} \in \boldsymbol{N} \cap \boldsymbol{G}_0\boldsymbol{G}_{m-1} = \{\boldsymbol{e}\}$, which forces $\boldsymbol{a} = \boldsymbol{b} \in \boldsymbol{G}_0 \cap \boldsymbol{G}_{m-1} = \boldsymbol{G}_{0,m-1}$, and thence $\boldsymbol{h} \in \boldsymbol{G}_{0,m-1}\Phi = \boldsymbol{H}_{0,m-1}$. This is what we wished to show. □

Notes to Section 2C

1. We mostly confine our attention to regular polytopes, rather than pre-polytopes, just because their combinatorial structures (as posets) naturally generalize those of the classical regular polytopes of [27] and convex polytopes of [56, 132]. Grünbaum [57] used the term *polystroma* for regular pre-polytope, but in practice most of his polystromata were actually polytopes.
2. What we have called here the *first* quotient criterion Theorem 2C5 was in [99, Section 2E] just called the quotient criterion, and in earlier papers the quotient lemma.
3. The second quotient criterion of Theorem 2C6 is not previously published.
4. Somewhat surprisingly, while the crucial combinatorial condition for sparseness in Theorem 2C8 is implicit in (2C7), it is not specifically mentioned in [99, Section 2E]. In fact, the relationship stated there is that no proper inface of $\mathcal{P}$ is met by the $\boldsymbol{N}$-orbit of any element more than once; of course, it clearly suffices to take such an inface to be a facet or vertex-figure.

2D Presentations

In Section 1E, we met euclidean reflexion groups, which play so important a part in the whole book. We showed that such a reflexion group is isomorphic to a Coxeter group, namely, an abstract group $\boldsymbol{G} = \langle \boldsymbol{r}_0, \ldots, \boldsymbol{r}_{m-1} \rangle$, whose generators satisfy relations solely of the form $(\boldsymbol{r}_i \boldsymbol{r}_j)^{p_{ij}} = \boldsymbol{e}$, where $p_{jj} = 1$ for each $j = 0, \ldots, m-1$ and $p_{ij} \geqslant 2$ whenever $0 \leqslant i < j \leqslant n-1$. It is a fact that Coxeter groups automatically satisfy the intersection property (with respect to their distinguished generators). In particular, if $\boldsymbol{G}$ is additionally such that $p_{ij} = 2$ whenever $j \geqslant i+2$, then $\boldsymbol{G}$ is a string C-group, and hence is the automorphism group $[p_1, \ldots, p_{m-1}]$ of the *universal* regular m-polytope $\{p_1, \ldots, p_{m-1}\}$, where we write $p_j := p_{j-1,j}$ for $j = 1, \ldots, m-1$.

As we said in Section 2B, the sequence $p_1, \ldots, p_{m-1}$ defines the Schläfli type $\{p_1, \ldots, p_{m-1}\}$ of $\mathcal{P}$; we use the same notation as for the universal polytope, with the understanding that (in general) additional relations will be imposed on the corresponding Coxeter group $[p_1, \ldots, p_{m-1}]$. It therefore follows that the general regular polytope of Schläfli type $\{p_1, \ldots, p_{m-1}\}$ is a quotient of the corresponding universal polytope. We introduce here some notation, which enables us to avoid clumsy circumlocutions of the form "the polytope whose group is the Coxeter group $[p_1, \ldots, p_{m-1}]$, factored out by the relations ...". In fact, the notation will merely be the combinatorial expression of these additional relations; it will also give us some extra flexibility, in that we can then apply the notation to describe a quotient of a general – not necessarily universal – regular polytope.

Flag-Sequences and Relators

In Section 2B we described the automorphism group $\boldsymbol{G} = \boldsymbol{G}(\mathcal{P})$ of a regular m-polytope $\mathcal{P}$, obtaining a direct correspondence between group elements (as products from its generatrix $(\boldsymbol{r}_0, \ldots, \boldsymbol{r}_{m-1})$) and chains of successively adjacent flags:

$$\boldsymbol{\Phi}^{j(1)\cdots j(r)} \longleftrightarrow \boldsymbol{r}_{j(r)} \cdots \boldsymbol{r}_{j(1)}.$$

We have referred to $\mathsf{J} := j(1) \cdots j(r)$ as a flag-sequence; see (2A7). An important observation is that, since $\mathcal{P}$ is regular, we may even forget on which flag the sequence is based, because if we start from a different flag, then we just obtain a conjugate group element. In particular, a relation $\boldsymbol{r}_{j(r)} \cdots \boldsymbol{r}_{j(1)} = \boldsymbol{e}$ gives a relation $\boldsymbol{\Phi}^{j(1)\cdots j(r)} = \boldsymbol{\Phi}$, and conversely. (Note, by the way, that since the $\boldsymbol{r}_j$ are involutions, the group relation is equivalent to $\boldsymbol{r}_{j(1)} \cdots \boldsymbol{r}_{j(r)} = \boldsymbol{e}$ as well.) From a combinatorial point of view, then, the information about the relation is encapsulated in the closed flag-sequence or *flag-cycle* $\mathsf{J} = j(1) \cdots j(r)$. Further, in identifying $\boldsymbol{r}_{j(r)} \cdots \boldsymbol{r}_{j(1)}$ with J, we shall refer to each as a *relator*. This dispenses with any reliance on the notation chosen for the generatrix of the group $\boldsymbol{G}$.

We therefore introduce the following notation. If the regular m-polytope $\mathcal{P}$ is the quotient of the universal polytope $\{p_1, \ldots, p_{m-1}\}$, obtained by imposing the relators $\mathsf{J}_1, \ldots, \mathsf{J}_k$, which are sequences in $0, \ldots, m-1$, then we denote the result in the form of a quotient, namely,

2D1 $$\mathcal{P} := \{p_1, \ldots, p_{m-1}\} / \langle\langle\, \mathsf{J}_1, \ldots, \mathsf{J}_k \,\rangle\rangle.$$

The understanding here, of course, is that none of $\mathsf{J}_1, \ldots, \mathsf{J}_k$ corresponds to the basic relations for the Schläfli type itself, although this restriction is only for convenience (see the notes at the end of the section).

Observe that, in fact, we need not start from a universal polytope. If one regular polytope $\mathcal{P}$ is obtained from another, not necessarily universal, regular m-polytope $\mathcal{Q}$ by the imposition of extra relations on its group, then we may similarly write

2D2 $$\mathcal{P} := \mathcal{Q} / \langle\langle\, \mathsf{J}_1, \ldots, \mathsf{J}_k \,\rangle\rangle,$$

where $\mathsf{J}_1, \ldots, \mathsf{J}_k$ correspond to the *additional* relations imposed on the group $\boldsymbol{G}(\mathcal{Q})$ to get its quotient $\boldsymbol{G}(\mathcal{P})$, or, in other words, to the additional relators which occur in $\mathcal{P}$.

It is appropriate to make some further comments.

2D3 Proposition *If $(\boldsymbol{r}_0, \ldots, \boldsymbol{r}_{m-1})$ is the generatrix of a regular polytope $\mathcal{P}$, then minimal relations determining its group as a quotient of a Coxeter group $[p_1, \ldots, p_{m-1}]$ involve all the $\boldsymbol{r}_i$ in consecutive blocks.*

Proof. This is clear, when we recall from Remark 2B5 that a subgroup $\boldsymbol{G}_j = \langle \boldsymbol{r}_i \mid i \neq j \rangle$ of $\boldsymbol{G} = \boldsymbol{G}(\mathcal{P})$ is an internal direct product

$$\boldsymbol{G}_j = \langle \boldsymbol{r}_0, \ldots, \boldsymbol{r}_{j-1} \rangle \times \langle \boldsymbol{r}_{j+1}, \ldots, \boldsymbol{r}_{m-1} \rangle$$

if $j \neq 0$ or $m-1$. □

What Proposition 2D3 implies is that (for instance) it would be inappropriate to impose a relator $(01542)^5$, since this would be implied by the simpler relators $(012)^5$ and $(45)^5$.

Circuit Criterion

So far, we have discussed the presentation of the group of a regular polytope in very general terms. We now come to a result of much practical importance, which we shall frequently appeal to. If $(\boldsymbol{r}_0, \ldots, \boldsymbol{r}_{m-1})$ is the generatrix of a regular m-polytope $\mathcal{P}$, then a general element $\boldsymbol{g} \in \boldsymbol{G} = \boldsymbol{G}(\mathcal{P})$ can be written in the form

$$\boldsymbol{g} = \boldsymbol{a}_0 \boldsymbol{r}_0 \boldsymbol{a}_1 \boldsymbol{r}_0 \cdots \boldsymbol{r}_0 \boldsymbol{a}_k,$$

where, for $i = 0, \ldots, k$, we have $\boldsymbol{a}_0, \ldots, \boldsymbol{a}_k \in \boldsymbol{G}_0 := \langle \boldsymbol{r}_1, \ldots, \boldsymbol{r}_{m-1} \rangle$, the group of the vertex-figure $\mathcal{V}$ of $\mathcal{P}$ at its base vertex $\boldsymbol{V} = \boldsymbol{G}_0$. (For the moment, it is more convenient to work with group elements, rather than with flag-sequences.) With $\boldsymbol{g}$, we can associate a path in $\mathcal{P}$ with k edges leading from $\boldsymbol{V}$ to $\boldsymbol{V}\boldsymbol{g}$. If $k = 0$, the path consists of $\boldsymbol{V}$ $(= \boldsymbol{V}\boldsymbol{a}_0)$ alone. For $k > 0$, let $(\boldsymbol{E}_1', \ldots, \boldsymbol{E}_{k-1}')$ be an edge-path associated with $\boldsymbol{g}' := \boldsymbol{a}_0 \boldsymbol{r}_0 \boldsymbol{a}_1 \boldsymbol{r}_0 \cdots \boldsymbol{a}_{k-1}$. With $\boldsymbol{g}$ is then associated the path $(\boldsymbol{E}_1, \ldots, \boldsymbol{E}_k)$, given by

$$\begin{aligned} \boldsymbol{E}_1 &:= \boldsymbol{E}\boldsymbol{a}_k \ (= \boldsymbol{E}\boldsymbol{r}_0\boldsymbol{a}_k), \\ \boldsymbol{E}_i &:= \boldsymbol{E}_{i-1}' \boldsymbol{r}_0 \boldsymbol{a}_k \quad \text{for } i = 2, \ldots, k, \end{aligned}$$

where $\boldsymbol{E} := \boldsymbol{G}_1$ is the base edge of $\mathcal{P}$. Of course, this path will not generally be unique, since it depends on the particular expression for $\boldsymbol{g}$.

Conversely, an edge-path $(\boldsymbol{E}_1, \ldots, \boldsymbol{E}_k)$ from $\boldsymbol{V}$ corresponds to an element $\boldsymbol{g} \in \boldsymbol{G}$ of this form, in which $\boldsymbol{r}_0$ occurs k times. If $k > 0$, then there is an $\boldsymbol{a}_k \in \boldsymbol{G}_0$ such that $\boldsymbol{E}_1 = \boldsymbol{E}\boldsymbol{a}_k$. The shorter path $(\boldsymbol{E}_1', \ldots, \boldsymbol{E}_{k-1}')$, given by

$$\boldsymbol{E}_i' := \boldsymbol{E}_{i+1} \boldsymbol{a}_k^{-1} \boldsymbol{r}_0$$

for $i = 1, \ldots, k-1$, also starts at $\boldsymbol{V}$, and we can repeat to obtain $\boldsymbol{g}$ as above, with a free choice of $\boldsymbol{a}_0$.

In the context of group presentations, we deduce what we call the *circuit criterion* (see the notes at the end of the section):

2D4 Theorem *If $\mathcal{P}$ is a regular polytope, then the group $\boldsymbol{G} = \boldsymbol{G}(\mathcal{P})$ of $\mathcal{P}$ is determined by the group of its vertex-figure, and the relations on the generatrix of $\boldsymbol{G}$ induced by the edge-circuits of $\mathcal{P}$ that contain the initial vertex.*

Proof. A relation on $\boldsymbol{G}$ can be written in the form

$$\boldsymbol{a}_0 \boldsymbol{r}_0 \boldsymbol{a}_1 \boldsymbol{r}_0 \cdots \boldsymbol{a}_{k-1} \boldsymbol{r}_0 = \boldsymbol{e},$$

with $\boldsymbol{a}_i \in \boldsymbol{G}_0$ for $i = 0, \ldots, k-1$, which corresponds to an edge-circuit starting and ending at $\boldsymbol{V}$. Conversely, such an edge-circuit is equivalent under $\boldsymbol{G}_0$ to one beginning with $\boldsymbol{E}$, and this gives rise to a relation as above (now the element $\boldsymbol{a}_0$ will be determined by the circuit). This is the result. □

Note, by the way, that among the edge-circuits that determine the group are those of the 2-faces $\{p_1\}$ of $\mathcal{P}$; however, as we have indicated, p_1 is automatically included in the Schläfli type, and so needs no further mention.

We should make a comment on the geometric interpretation of Theorem 2D4. As we saw above, a relation of the form

$$\boldsymbol{a}_0\boldsymbol{r}_0\boldsymbol{a}_1\boldsymbol{r}_0\cdots\boldsymbol{a}_{k-1}\boldsymbol{r}_0 = \boldsymbol{e},$$

with $\boldsymbol{a}_i \in \boldsymbol{G}_0$ for $i = 0, \ldots, k-1$, corresponds to an edge-circuit. However, we do not seem to proceed along the circuit to obtain the relation. In fact, though, we do, if we recall the original association between group elements and flag-chains; this relation corresponds to a sequence

$$s := 0a_{k-1}0\cdots 0a_1 0a_0,$$

with each of $a_{k-1}, \ldots, a_0$ sequences in $1, \ldots, m-1$. Each time we encounter a '0' in the flag-chain, we change vertices in a common edge; thus we do directly obtain a corresponding edge-circuit.

Petrie Polygons and Deep Holes

It is too early to give many examples of how these notions work (but see also the next Section 2E). Instead, we introduce some notation for certain special relations. Theorem 2D4 shows how an automorphism group $\boldsymbol{G}$ is determined by edge-circuits in the corresponding polytope $\mathcal{P}$, once we know its vertex-figure; by transitivity, we may always assume that an edge-circuit under consideration starts at the initial vertex, and then proceeds through the initial edge. A typical such circuit is a regular polygon (2-polytope); this, of course, will be specified by its generatrix $(\boldsymbol{s}, \boldsymbol{t})$ in $\boldsymbol{G}$. Under our assumption, $\boldsymbol{s}\boldsymbol{r}_0$ fixes the initial vertex and the initial edge, and hence $\boldsymbol{s}\boldsymbol{r}_0 \in \langle \boldsymbol{r}_2, \ldots, \boldsymbol{r}_{m-1}\rangle$. Since the converse is clear, we thus have

2D5 Proposition *Up to conjugacy, $(\boldsymbol{s}, \boldsymbol{t})$ is the generatrix of a regular polygon in the edge-graph of a regular m-polytope with generatrix $(\boldsymbol{r}_0, \ldots, \boldsymbol{r}_{m-1})$ exactly when $\boldsymbol{s} \in \boldsymbol{r}_0\boldsymbol{G}_{01} = \boldsymbol{G}_{01}\boldsymbol{r}_0$ and $\boldsymbol{t} \in \boldsymbol{G}_0$ are involutions.*

We allow for the possibility that such a polygon is degenerate.

We first introduce two of these polygons for general rank m, and then say more about the particular case of polyhedra (that is, 3-polytopes). Taking $\boldsymbol{G}$ as usual, we have the *Petrie polygon* with generatrix

2D6 $$(\boldsymbol{s}, \boldsymbol{t}) := (\boldsymbol{r}_0\boldsymbol{r}_2\cdots, \boldsymbol{r}_1\boldsymbol{r}_3\cdots),$$

and *deep hole*, with

2D7 $$(\boldsymbol{s}, \boldsymbol{t}) := (\boldsymbol{r}_0, \boldsymbol{r}_1\boldsymbol{r}_2\cdots\boldsymbol{r}_{m-2}\boldsymbol{r}_{m-1}\boldsymbol{r}_{m-2}\cdots\boldsymbol{r}_2\boldsymbol{r}_1).$$

We write $\{p_1, \ldots, p_{m-1} : r\}$ for the polytope obtained by imposing a Petrie polygon $\{r\}$ (that is, an r-gon) on the universal polytope $\{p_1, \ldots, p_{m-1}\}$, and

$\{p_1, \ldots, p_{m-1} \mid s\}$ for that with a deep hole $\{s\}$. We can combine the notations as $\{p_1, \ldots, p_{m-1} : r \mid s\} = \{p_1, \ldots, p_{m-1} \mid s : r\}$, so imposing both a given Petrie polygon and a given deep hole.

2D8 Remark In terms of the quotient notation, which for obvious reasons we shall not employ in these contexts, we have

$$\{p_1, \ldots, p_{m-1} : r\} = \{p_1, \ldots, p_{m-1}\}/\langle\!\langle (012 \cdots (m-2)(m-1))^r \rangle\!\rangle,$$
$$\{p_1, \ldots, p_{m-1} \mid s\} = \{p_1, \ldots, p_{m-1}\}/\langle\!\langle (012 \cdots (m-2)(m-1)(m-2) \cdots 21)^s \rangle\!\rangle.$$

For the former, see Lemma 2D12 below.

2D9 Example The abstract regular $(m+1)$-apeirotope corresponding to the tiling of $\mathbb{E}^m$ by cubes is $\{4, 3^{m-2}, 4\}$, whose vertex-set we may take to be the integer lattice $\mathbb{Z}^m$. For $r \geqslant 2$, the finite polytope $\{4, 3^{m-2}, 4 \mid r\}$, called a *cubic toroid*, is obtained by identifying points of the sublattice generated by $(r, 0^{m-1})$ and its images under all permutations and changes of sign. Slightly modifying the notation of [99, Section 6D], this is also denoted $\{4, 3^{m-2}, 4\}_{(r,0^{m-1})}$. In a similar way, the cubic toroid $\{4, 3^{m-2}, 4 : rm\} = \{4, 3^{m-2}, 4\}_{(r^m)}$ is obtained from the sublattice generated by all $((\pm r)^m)$.

2D10 Remark For $m \geqslant 3$, there is a third cubic toroid $\{4, 3^{m-2}, 4\}_{(r,r,0^{m-2})}$. It might be thought that there could be more such toroids but, as shown in [97] (see also [99, Section 6D]), for $3 \leqslant k \leqslant m-1$, we have

$$\{4, 3^{m-2}, 4\}_{(r^k,0^{m-k})} = \begin{cases} \{4, 3^{m-2}, 4\}_{(r,0^{m-1})}, & \text{if } k \text{ is odd,} \\ \{4, 3^{m-2}, 4\}_{(r,r,0^{m-2})}, & \text{if } k \text{ is even.} \end{cases}$$

We say a little more about Petrie polygons and deep holes, in particular we give a combinatorial interpretation of the former. We begin with an easy remark, noting that, up to conjugacy, the defining relations are symmetric between dual polytopes.

2D11 Proposition *Dual regular polytopes have the same Petrie polygons and the same deep holes.*

For Petrie polygons, we have an even stronger property.

2D12 Lemma *The different products of the generators of an sggi, in any order, are conjugate, and so have the same period.*

Proof. Let these generators be $\boldsymbol{r}_0, \ldots, \boldsymbol{r}_{m-1}$, in the natural order. We show that any product $\boldsymbol{r}_{j(0)} \cdots \boldsymbol{r}_{j(m-1)}$, with $(j(0), \ldots, j(m-1))$ a permutation of $(0, \ldots, m-1)$, is conjugate to $\boldsymbol{r}_0 \cdots \boldsymbol{r}_{m-1}$. Suppose that, at some stage, we have obtained a conjugate product, with initial string $\boldsymbol{r}_0 \cdots \boldsymbol{r}_{k-1}$ for some k; the inductive argument begins with $k = 0$. Then we have products $\boldsymbol{a}$ and $\boldsymbol{b}$ of $\boldsymbol{r}_{k+1}, \ldots, \boldsymbol{r}_{m-1}$ in some order so that, with $\sim$ denoting conjugacy, our product is now

$$\boldsymbol{r}_0 \cdots \boldsymbol{r}_{k-1}\boldsymbol{a}\boldsymbol{r}_k\boldsymbol{b} = \boldsymbol{a}\boldsymbol{r}_0 \cdots \boldsymbol{r}_{k-1}\boldsymbol{r}_k\boldsymbol{b} \sim \boldsymbol{r}_0 \cdots \boldsymbol{r}_{k-1}\boldsymbol{r}_k\boldsymbol{b}\boldsymbol{a},$$

since $\boldsymbol{a}$ is a product of $\boldsymbol{r}_j$ with $j > k$, each of which commutes with $\boldsymbol{r}_0, \ldots, \boldsymbol{r}_{k-1}$. This is the inductive step, and the lemma follows at once. □

The Petrie polygon of a polygon is clearly that polygon itself. We now set up a recursive definition as an alternative to our original one. For $m \geqslant 3$, a *Petrie polygon* of a regular m-polytope $\mathcal{P}$ is an edge-path $\ldots, \boldsymbol{V}_{-1}, \boldsymbol{E}_{-1}, \boldsymbol{V}_0, \boldsymbol{E}_0, \boldsymbol{V}_1, \boldsymbol{E}_1, \ldots$, with the $\boldsymbol{V}_j$ vertices of $\mathcal{P}$ and $\boldsymbol{E}_j = \{\boldsymbol{V}_j, \boldsymbol{V}_{j+1}\}$ edges of $\mathcal{P}$, such that, for each j, there is a facet $\mathcal{F}_j$ of $\mathcal{P}$ for which $\boldsymbol{V}_j, \boldsymbol{E}_j, \ldots, \boldsymbol{V}_{m+j-2}, \boldsymbol{E}_{m+j-2}, \boldsymbol{V}_{m+j-1}$ belong to a Petrie polygon of $\mathcal{F}_j$, but $\boldsymbol{V}_{m+j}$ is not in $\mathcal{F}_j$. Then we have

2D13 Theorem *The two definitions of a Petrie polygon are equivalent.*

Proof. We choose a particular Petrie polygon given by the second definition. This is such that, for each $k = 0, \ldots, m-1$, that part of the edge-path from $\boldsymbol{V}_0$ to $\boldsymbol{V}_k$ is part of a Petrie polygon of the initial k-face $\mathcal{G}_k$ (say) of $\mathcal{P}$. We make the inductive assumption that, for each such k, the product $\boldsymbol{p}_k := \boldsymbol{r}_{k-1} \cdots \boldsymbol{r}_0$ satisfies $\boldsymbol{V}_j \boldsymbol{p}_k = \boldsymbol{V}_{j+1}$ for $j = 0, \ldots, k-1$. With $k = m-1$, let $\boldsymbol{W}$ be the preceding vertex to $\boldsymbol{V}_0$ in the Petrie polygon of the facet $\mathcal{G}_{m-1}$ of $\mathcal{P}$, and set $\boldsymbol{V}_{-1} := \boldsymbol{W}\boldsymbol{r}_{m-1}$. Then $\boldsymbol{V}_j \boldsymbol{r}_{m-1} = \boldsymbol{V}_j$ for each $j = 0, \ldots, m-2$, because these are vertices of $\mathcal{G}_{m-2}$. It immediately follows that

$$\boldsymbol{V}_j \boldsymbol{p}_m = \boldsymbol{V}_j \boldsymbol{r}_{m-1} \boldsymbol{p}_{m-1} = \begin{cases} \boldsymbol{W}\boldsymbol{p}_{m-1} = \boldsymbol{V}_0, & \text{for } j = -1, \\ \boldsymbol{V}_j \boldsymbol{p}_{m-1} = \boldsymbol{V}_{j+1}, & \text{for } j = 0, \ldots, m-1, \end{cases}$$

and this completes the inductive step. □

There is not so nice a picture of deep holes as we have just seen for Petrie polygons. However, there is one special case. As usual, s^k denotes a string $s, \ldots, s$ of length k.

2D14 Theorem *If the regular m-polytope $\mathcal{P}$ with $m \geqslant 3$ has Schläfli type $\{3^{k-1}, r, 3^{m-k-1}\}$ for some k, then the deep hole of $\mathcal{P}$ is $\{r\}$.*

Proof. The symmetry of deep holes between a polytope and its dual lets us suppose that $k < m-1$. With the usual notation $\boldsymbol{r}_0, \ldots, \boldsymbol{r}_{m-1}$ for the generators of $\boldsymbol{G}(\mathcal{P})$, from $(\boldsymbol{r}_{m-2}\boldsymbol{r}_{m-1})^3 = \boldsymbol{e}$ we have

$$\begin{aligned} &\boldsymbol{r}_0\boldsymbol{r}_1 \cdots \boldsymbol{r}_{m-3}\boldsymbol{r}_{m-2}\boldsymbol{r}_{m-1}\boldsymbol{r}_{m-2}\boldsymbol{r}_{m-3} \cdots \boldsymbol{r}_1 \\ &\quad = \boldsymbol{r}_0\boldsymbol{r}_1 \cdots \boldsymbol{r}_{m-3}\boldsymbol{r}_{m-1}\boldsymbol{r}_{m-2}\boldsymbol{r}_{m-1}\boldsymbol{r}_{m-3} \cdots \boldsymbol{r}_1 \\ &\quad = \boldsymbol{r}_{m-1}\boldsymbol{r}_0\boldsymbol{r}_1 \cdots \boldsymbol{r}_{m-3}\boldsymbol{r}_{m-2}\boldsymbol{r}_{m-3} \cdots \boldsymbol{r}_1\boldsymbol{r}_{m-1} \\ &\quad \sim \boldsymbol{r}_0\boldsymbol{r}_1 \cdots \boldsymbol{r}_{m-3}\boldsymbol{r}_{m-2}\boldsymbol{r}_{m-3} \cdots \boldsymbol{r}_1. \end{aligned}$$

The claim of the theorem now follows by induction on m, since the deep hole of a polygon is that polygon itself. □

Imposed Relators

It may happen that, while certain *explicit* relators specify the group of a regular polytope $\mathcal{P}$, nevertheless additional relators on its facet $\mathcal{P}^f$ or vertex-figure $\mathcal{P}^v$ are consequences of them. Some of these will be *implicit*, and so already implied by the given relators on $\mathcal{P}^f$ or $\mathcal{P}^v$, as appropriate. However, we may also have *imposed* relators, that cannot be deduced in such a way. At present, we lack the machinery to explain why (this will occur in Section 5C), and so we confine ourselves here to giving

2D15 Example There are the following imposed relators:

(a) $\{4,4,3\}/\langle\langle (123)^3 \rangle\rangle = \{4,4,3\}/\langle\langle (0121)^4, (123)^3 \rangle\rangle$,

(b) $\{4,5,3\}/\langle\langle (123)^5 \rangle\rangle = \{4,5,3\}/\langle\langle (0121)^4, (123)^5 \rangle\rangle$.

In each case, a relation on the vertex-figure imposes one on the facet; observe that these facets are ostensibly infinite.

Even and Odd Relators

We end the section with some useful language. If J is a flag-sequence (in particular, a relator), then we call J *even* or *odd* according as it has an even or odd number of terms. Observing that an odd relator merges the maximal simplices of the order complex into a single class, we have

2D16 Proposition *A regular polytope* $\mathcal{P} := \{p_1, \ldots, p_{m-1}\}/\langle\langle \mathsf{J}_1, \ldots, \mathsf{J}_k \rangle\rangle$ *as in (2D1) is orientable if and only if each relator* J_i *is even.*

Naturally, an even relator corresponds to an even product of generators, and so has an even number of indices $0, 1, \ldots, m-1$.

Moreover, if $k = 0, \ldots, m-1$, then we say that J is k*-even* or k*-odd* as J contains k an even or odd number of times. As an example, we have

2D17 Proposition *The edge-graph* $\mathcal{P}_{01}$ *of a regular polytope* $\mathcal{P}$ *is bipartite if and only if every relator on* $\mathcal{P}$ *is* 0*-even. In particular, the entry* p_1 *of the Schläfli type* $\{p_1, \ldots, p_{m-1}\}$ *of* $\mathcal{P}$ *must be even.*

Notes to Section 2D

1. There is no theoretical reason why we should not start from the universal regular m-polytope $\{\infty, \ldots, \infty\}$, and impose the flag-cycles $\mathsf{J}_j = ((j-1)j)^{p_j}$ for $j = 1, \ldots, m-1$. But since our aim is to simplify the notation, it would be somewhat silly to do this.
2. The circuit criterion of Theorem 2D4 was devised by McMullen, in order to find presentations of the groups of two of the 3-dimensional regular apeirohedra found by Grünbaum in [58]; see [98] or Section 10B.
3. The argument of Lemma 2D12 easily extends to general Coxeter groups whose diagrams are trees. In our context, therefore, the only exception is $\mathbf{P}_{d+1}$, whose diagram is a circuit; indeed, it is the case here that the different products of the generating reflexions are not all conjugate, although by Theorem 1D13 all such products have infinite order.
4. The proof of Theorem 2D13 is taken from [27, Section 12.4].

2E Regular Maps

An abstract 3-polytope, or polyhedron, is often called a *map* (*on a surface*). Because regular maps provide us with so many useful examples at an initial stage, we briefly discuss them here.

The intuitive picture of a regular map or polyhedron is of p-gons glued together along their edges, fitting q around each vertex. In this context, p and q are integers or ∞. We usually suppose that $p, q > 2$, but degenerate cases do crop up occasionally. Such a polyhedron is denoted $\{p, q\}$ but, as we shall see, this Schläfli type will generally be modified by additions that yield extra information.

Platonic Polyhedra

The geometric Platonic polyhedra should be familiar; in any event, we shall consider them in detail in Chapter 7. Regarded as regular maps, they are the *tetrahedron* $\{3,3\}$, *octahedron* $\{3,4\}$, *cube* $\{4,3\}$, *icosahedron* $\{3,5\}$ and *dodecahedron* $\{5,3\}$. Since all but the tetrahedron are centrally symmetric (as abstract polyhedra – see the next Section 2F), new regular maps can be formed by identifying antipodal vertices, edges and faces. So, for example, the dodecahedron yields the *hemi-dodecahedron* $\{5, 3 : 5\}$.

Tori

As a special case of the cubic tilings, consider the regular tesselation $\{4,4\}$ of the plane $\mathbb{E}^2$ by squares; we can take its vertex-set to be $\mathbb{Z}^2$. Then we form the *torus* $\{4,4\}_{(r,s)}$ by identifying all vertices of the form $k(r,s) + m(s,-r)$ with $k, m \in \mathbb{Z}$; see Figure 2E1 for the case $(r,s) = (2,1)$. This polyhedron will only be regular if $rs(r-s) = 0$ (that is, $r = 0$, $s = 0$ or $r = s$). Then $\{4,4\}_{(r,0)} = \{4,4 \mid r\}$ will have r^2 vertices and faces and $2r^2$ edges, while $\{4,4\}_{(r,r)} = \{4,4 : 2r\}$ will have twice as many of each.

2E1

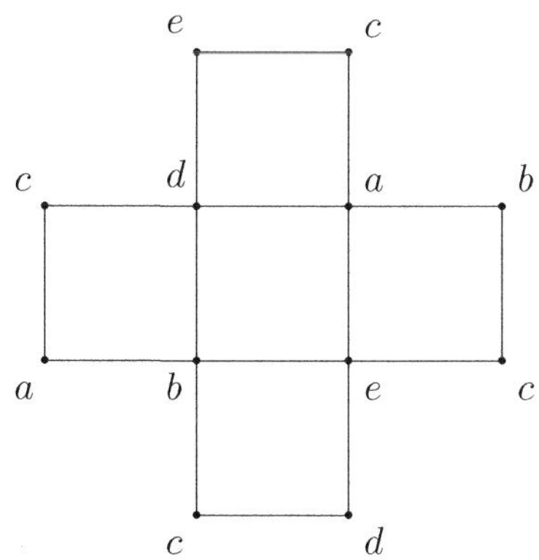

The torus $\{4,4\}_{(2,1)}$

Holes and Zigzags

For polyhedra, we can generalize Petrie polygons and holes (we drop the qualification 'deep' here). The *k-hole* of $\mathcal{P}(\boldsymbol{G})$, with $\boldsymbol{G} = \langle \boldsymbol{r}_0, \boldsymbol{r}_1, \boldsymbol{r}_2 \rangle$, has generatrix

2E2 $$(\boldsymbol{s}, \boldsymbol{t}) := \big(\boldsymbol{r}_0, (\boldsymbol{r}_1\boldsymbol{r}_2)^{k-1}\boldsymbol{r}_1\big),$$

and the *k-zigzag* has generatrix

2E3 $$(\boldsymbol{s}, \boldsymbol{t}) := \big(\boldsymbol{r}_0\boldsymbol{r}_2, (\boldsymbol{r}_1\boldsymbol{r}_2)^{k-1}\boldsymbol{r}_1\big).$$

Thus the hole is actually the 2-hole, while the Petrie polygon is the 1-zigzag.

2E4 Remark If $\mathcal{P}$ is of Schläfli type $\{p, q\}$, then k- and $(q-k)$-holes and zigzags coincide, and so we need only take $k \leqslant \frac{1}{2}q$.

If the regular polyhedron $\mathcal{P}$ is specified by certain of its zigzags and holes, then we write

2E5 $$\mathcal{P} = \{p, q : r_1, r_2, \ldots \mid s_2, s_3, \ldots\}.$$

The convention followed is that each list $r_1, r_2, \ldots$ or $s_2, s_3, \ldots$ terminates with the last specified period; earlier periods unneeded are replaced by $\cdot$; if no zigzags or holes are specified, then the corresponding list and delimiter (: or $|$) are omitted.

We extend this notation further: an entry for a zigzag or hole preceded by an asterisk $*$ indicates the corresponding entry for the *dual* polyhedron. Since Petrie polygons and holes coincide for dual polyhedra, this does not apply to entries r_1 or s_2 in (2E5). Thus, for instance,

$$\{6, 3 : \cdot, *r\} = \{3, 6 : \cdot, r\}^\delta = \{6, 3\}/\langle\langle (01012)^r \rangle\rangle$$

in the notation of Section 2D; we shall meet the case $r = 2$ of this polyhedron in Section 5A.

As we said at the end of Section 2A, while the order complex $\mathcal{C}(\mathcal{P})$ does not usually have a nice topology, in the case of a regular polyhedron we obtain a surface. As an illustration of how our intuition can be helped by this surface, we have

2E6 Theorem *For each $q \geqslant 3$ and $r \geqslant 2$,*

$$\{3, q \mid \cdot, r\} = \{3, q : 2r\}.$$

2E7

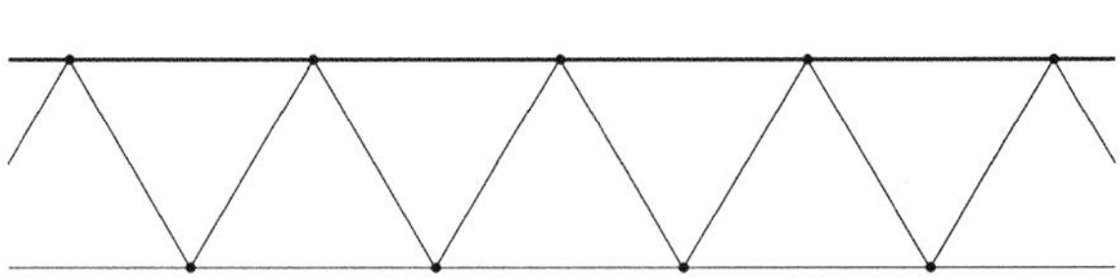

Proof. The situation is depicted in Figure 2E7. With $(\boldsymbol{r}_0, \boldsymbol{r}_1, \boldsymbol{r}_2)$ the generatrix of a regular polyhedron of Schläfli type $\{3, q\}$, we have

$$\boldsymbol{r}_0\boldsymbol{r}_1\boldsymbol{r}_2\boldsymbol{r}_1\boldsymbol{r}_2\boldsymbol{r}_1 \sim \boldsymbol{r}_1\boldsymbol{r}_0\boldsymbol{r}_1\boldsymbol{r}_2\boldsymbol{r}_1\boldsymbol{r}_2 = \boldsymbol{r}_0\boldsymbol{r}_1\boldsymbol{r}_0\boldsymbol{r}_2\boldsymbol{r}_1\boldsymbol{r}_2 = (\boldsymbol{r}_0\boldsymbol{r}_1\boldsymbol{r}_2)^2,$$

where $\sim$ denotes conjugacy. The defining relation for the former polyhedron is

$$(\boldsymbol{r}_0\boldsymbol{r}_1\boldsymbol{r}_2\boldsymbol{r}_1\boldsymbol{r}_2\boldsymbol{r}_1)^r = \boldsymbol{e} \iff (\boldsymbol{r}_0\boldsymbol{r}_1\boldsymbol{r}_2)^{2r} = \boldsymbol{e},$$

which is the defining relation for the latter. □

We shall see applications of Theorem 2E6 in Sections 5A and 7B.

Notes to Section 2E

1. The notational conventions were discussed in detail in the last section, and so we shall not say anything about them here.
2. In spite of their name, the Platonic polyhedra were neither discovered by Plato, nor even first classified by him. The credit for proving that that there are just the five listed apparently goes to Theaetetus (who was killed in battle in 369BCE).

2F Special Polytopes

The last section of this chapter treats various applications of Theorems 2C6 and 2C10 to certain interesting classes of polytopes. It also enables us to introduce some concepts that will be of use later on.

Central Symmetry

We call a regular m-polytope $\mathcal{P}$ *centrally symmetric* if its automorphism group $\boldsymbol{G} = \boldsymbol{G}(\mathcal{P})$ contains a central involution $\boldsymbol{z}$ which does not fix the initial vertex.

2F1 Lemma *The central involution $\boldsymbol{z}$ of a centrally symmetric regular polytope $\mathcal{P}$ does not fix any vertex of $\mathcal{P}$.*

Proof. With the usual conventions for $\boldsymbol{G}$, the condition for $\boldsymbol{z}$ is that $\boldsymbol{z} \notin \boldsymbol{G}_0$, the stabilizer of the initial vertex. But then, since $\boldsymbol{z}$ is central, we also have $\boldsymbol{z} = \boldsymbol{s}^{-1}\boldsymbol{z}\boldsymbol{s} \notin \boldsymbol{s}^{-1}\boldsymbol{G}_0\boldsymbol{s}$ for each $\boldsymbol{s} \in \boldsymbol{G}$, so that $\boldsymbol{z}$ cannot fix the vertex $\boldsymbol{G}_0\boldsymbol{s}$ either. This is the claim of the lemma. □

If $\mathcal{P}$ is a centrally symmetric regular polytope, with central involution $\boldsymbol{z}$, then we write $\mathcal{P}/2$ for the pre-polytope obtained by identifying pairs of elements $\boldsymbol{H}$ and $\widetilde{\boldsymbol{H}} := \boldsymbol{H}\boldsymbol{z}$ of $\mathcal{P}$. The group of $\mathcal{P}/2$ is therefore $\boldsymbol{G}/\boldsymbol{Z}$, where $\boldsymbol{Z} :- \langle \boldsymbol{z} \rangle = \{\boldsymbol{e}, \boldsymbol{z}\}$ is the subgroup of $\boldsymbol{G}$ generated by $\boldsymbol{z}$. Although we need to make little use of the fact, we actually have

2F2 Theorem *If $\mathcal{P}$ is an atomic centrally symmetric regular polytope, then $\mathcal{P}/2$ is a polytope.*

Proof. Recall that $\mathcal{P}$ being atomic means that each element of $\mathcal{P}$ is determined by the set of vertices incident with it. As in the definition, let $\boldsymbol{z} \in \boldsymbol{G} = \boldsymbol{G}(\mathcal{P})$ be the central involution fixing no vertex of $\mathcal{P}$, and let $\boldsymbol{Z} = \langle \boldsymbol{z} \rangle = \{\boldsymbol{e}, \boldsymbol{z}\}$. Then $\boldsymbol{z} \notin \boldsymbol{G}_0 = \langle \boldsymbol{r}_1, \ldots, \boldsymbol{r}_{r-1} \rangle$. We first suppose that $\boldsymbol{z} \notin \boldsymbol{G}_{m-1}$ as well.

For a sparse subgroup $\boldsymbol{Z}$, the result follows from Theorem 2C10. Otherwise, the (initial) facet $\mathcal{F} = \boldsymbol{G}_{m-1}/\boldsymbol{G}_{-1}$ of $\mathcal{P}$ contains some vertex $\boldsymbol{V}$ as well as its antipodal vertex $\widetilde{\boldsymbol{V}} := \boldsymbol{V}\boldsymbol{z}$. If $\widetilde{\mathcal{F}} := \mathcal{F}\boldsymbol{z}$ is the antipodal facet in $\mathcal{P}$ to $\mathcal{F}$, then – applying $\boldsymbol{z}$ – since $\boldsymbol{V}, \widetilde{\boldsymbol{V}} \in \mathcal{F}$ we have $\widetilde{\boldsymbol{V}}, \boldsymbol{V} \in \widetilde{\mathcal{F}}$ also. Applying the automorphisms of $\mathcal{F}$ enables us to assume that $\boldsymbol{V}$ is an arbitrary vertex of $\mathcal{F}$, and thus the vertices of $\mathcal{F}$ fall into antipodal pairs $\boldsymbol{V}, \widetilde{\boldsymbol{V}}$, both in $\mathcal{F}$ and $\widetilde{\mathcal{F}}$. It follows that, if $\mathcal{G} \subset \mathcal{F}$ is any ridge, then – appealing to atomicity – the only other facet to which it can belong is $\widetilde{\mathcal{F}}$, since all its vertices are already in $\widetilde{\mathcal{F}}$. It must now be clear that all proper elements $\boldsymbol{H}$ of $\mathcal{F}$ are also elements of $\widetilde{\mathcal{F}}$, and occur in antipodal pairs $\boldsymbol{H}, \widetilde{\boldsymbol{H}}$. We conclude that $\mathcal{P}$ has just two facets $\mathcal{F}$ and $\widetilde{\mathcal{F}}$, and that $\boldsymbol{G} = \boldsymbol{G}_{m-1} \times \langle \boldsymbol{r}_{m-1} \rangle$, a direct product. Then $\boldsymbol{z}' := \boldsymbol{z}\boldsymbol{r}_{m-1} \in \boldsymbol{G}_{m-1}$ is a central involution, making $\mathcal{F}$ centrally symmetric. Furthermore, with $\boldsymbol{Z}' := \langle \boldsymbol{z}' \rangle$, we see that $\boldsymbol{G}/\boldsymbol{Z} \cong \boldsymbol{G}_{m-1}/\boldsymbol{Z}'$; the obvious induction on rank thus shows that $\boldsymbol{G}/\boldsymbol{Z}$ is a C-group (though of rank $m-1$ rather than m). In this case, therefore, $\mathcal{P}/2$ is actually an $(m-1)$-polytope.

In the other case, we have $\boldsymbol{z} \in \boldsymbol{G}_{m-1}$, so that $\boldsymbol{Z} \leqslant \boldsymbol{G}_{m-1}$; in other words, as an inface $\mathcal{F} = \boldsymbol{F}/\boldsymbol{G}_{-1}$, the (initial) facet $\mathcal{F}$ of $\mathcal{P}$ is centrally symmetric. Now we can appeal to induction on m, and thus claim that $\mathcal{F}/2$ is a polytope; that is, $\boldsymbol{G}_{m-1}/\boldsymbol{Z}$ is a C-group. Since $\boldsymbol{Z} \cap \boldsymbol{G}_0 = \{\boldsymbol{e}\}$, under the quotient map $\Phi \colon \boldsymbol{G} \to \boldsymbol{G}/\boldsymbol{Z}$ we have exactly the situation of Theorem 2C6 (with $\boldsymbol{N} = \boldsymbol{Z}$ and $\boldsymbol{H}_0 \cong \boldsymbol{G}_0$). Therefore, $\boldsymbol{G}/\boldsymbol{Z}$ is a C-group, and consequently $\mathcal{P}/2$ is a polytope.

Thus we have shown that, in all cases, $\mathcal{P}/2$ is polytopal. □

2F3 Remark The first part of this proof shows that a centrally symmetric polytope may have more than one central involution fixing no vertices, though only in unusual circumstances. Note that, in the last case, the facet $\mathcal{F}$ need not contain all the vertices of $\mathcal{P}$, even though $\boldsymbol{Z} \leqslant \boldsymbol{G}_{m-1}$. This part also suggests that the dual of a centrally symmetric polytope need not be centrally symmetric; specific examples are given in Section 8C.

2F4 Remark In Section 5B, we shall see that, if $\mathcal{P}$ is a non-atomic centrally symmetric regular polytope, then $\mathcal{P}/2$ need not be polytopal.

Abstract Blends

Let $\mathcal{P}$ and $\mathcal{Q}$ be abstract regular polytopes, with generatrices $(\boldsymbol{r}_0, \ldots, \boldsymbol{r}_{m-1})$ and $(\boldsymbol{s}_0, \ldots, \boldsymbol{s}_{k-1})$ respectively. We define the (*abstract*) *blend* $\mathcal{P} \# \mathcal{Q}$ by means of its generatrix $(\boldsymbol{t}_0, \boldsymbol{t}_1, \ldots)$ as follows (see the notes at the end of the section). In the product group $\boldsymbol{G}(\mathcal{P}) \times \boldsymbol{G}(\mathcal{Q})$, set

$$\boldsymbol{t}_j := (\boldsymbol{r}_j, \boldsymbol{s}_j)$$

for $j < \max\{m, k\}$, with the convention $\boldsymbol{r}_j = \boldsymbol{e}$ if $j \geqslant m$ and analogously for $\boldsymbol{s}_j$.

It is generally the case that $\mathcal{P} \# \mathcal{Q}$ is *not* polytopal, but there are some particular exceptions. We only prove the one that we most wish to appeal to, and so we shall list others without proof. These all concern the case when $\mathcal{Q}$ is a face of $\mathcal{P}$; recall that $\{2\}$ is the (abstract) 1-polytope or digon.

We first have

2F5 Theorem *Let $\mathcal{P}$ be an abstract regular m-polytope, and let $\mathcal{Q}$ be the k-face of $\mathcal{P}$. Then the blend $\mathcal{P} \# \mathcal{Q}$ is polytopal at least in the cases $k = 1$ and $k = m - 1$.*

Proof. Under the surjective mapping from $\boldsymbol{G}(\mathcal{P} \# \{2\})$ given by $\boldsymbol{t}_j \mapsto \boldsymbol{r}_j$ for $j = 0, \ldots, m-1$, Theorem 2C5 applies. Exactly the same argument applies in the other case, using the dual form of Theorem 2C5. □

We should say a little more about the case $\mathcal{P} \# \{2\}$. Recall that a graph is *bipartite* if its vertex-set V can be partitioned into two disjoint subsets V_1, V_2 such that edges go between V_1 and V_2 only.

2F6 Proposition *If the edge-graph of $\mathcal{P}$ is bipartite, so that each edge-circuit of $\mathcal{P}$ has even length, then $\mathcal{P} \# \{2\} \cong \mathcal{P}$. Otherwise, $\mathcal{P} \# \{2\}$ has automorphism group $\boldsymbol{G}(\mathcal{P}) \times \boldsymbol{C}_2$ and twice as many vertices (in the finite case).*

For the second, we quote a part of [99, Theorem 7A13].

2F7 Proposition *If $\mathcal{P}$ is a regular m-polytope of Schläfli type $\{p_1, \ldots, p_{m-1}\}$ with $p_1, \ldots, p_k$ all odd, and $\mathcal{Q}$ is the k-face of $\mathcal{P}$, then the blend of $\mathcal{P}$ and $\mathcal{Q}$ is polytopal, with automorphism group $\boldsymbol{G} \cong \boldsymbol{G}(\mathcal{P}) \times \boldsymbol{G}(\mathcal{Q})$.*

We adopt a variant notation $\mathcal{P} \otimes \mathcal{Q}$ for the blend in this case; the reason comes from Proposition 4A15.

Amalgamation

In a regular m-polytope $\mathcal{P}$, the $(1, m-1)$-inface is simultaneously the vertex-figure of the facet of $\mathcal{P}$ and the facet of its vertex-figure: $\mathcal{P}^{\boldsymbol{fv}} = \mathcal{P}^{\boldsymbol{vf}}$ (the notation is that introduced in Section 2A). Thus, if $\mathcal{P}$ and $\mathcal{Q}$ are regular m-polytopes, a necessary condition for there to exist a regular $(m+1)$-polytope $\mathcal{R}$ with $\mathcal{R}^{\boldsymbol{f}} \cong \mathcal{P}$ (that is, facets isomorphic to $\mathcal{P}$) and $\mathcal{R}^{\boldsymbol{v}} \cong \mathcal{Q}$ is that $\mathcal{P}^{\boldsymbol{v}} \cong \mathcal{Q}^{\boldsymbol{f}}$; in other words, the vertex-figure of $\mathcal{P}$ must be isomorphic to the facet of $\mathcal{Q}$. The class of such polytopes $\mathcal{R}$ is denoted $\langle \mathcal{P}, \mathcal{Q} \rangle$. The *amalgamation problem* then asks, first, whether $\langle \mathcal{P}, \mathcal{Q} \rangle \neq \emptyset$. If this is so, then the following Theorem 2F8 will show that there is a *universal* member $\{\mathcal{P}, \mathcal{Q}\} \in \langle \mathcal{P}, \mathcal{Q} \rangle$ which covers every other. We then ask, second, whether $\{\mathcal{P}, \mathcal{Q}\}$ is finite if $\mathcal{P}$ and $\mathcal{Q}$ are both finite.

As a consequence of Theorem 2C2, we have

2F8 Theorem *If $\mathcal{P}$ and $\mathcal{Q}$ are regular m-polytopes such that $\langle \mathcal{P}, \mathcal{Q} \rangle \neq \emptyset$, then there is a universal member $\{\mathcal{P}, \mathcal{Q}\} \in \langle \mathcal{P}, \mathcal{Q} \rangle$.*

Proof. Let $(\boldsymbol{s}_0, \ldots, \boldsymbol{s}_{m-1})$ be the generatrix of $\mathcal{P}$ and $(\boldsymbol{t}_0, \ldots, \boldsymbol{t}_{m-1})$ that of $\mathcal{Q}$. Then the generatrix $(\boldsymbol{r}_0, \ldots, \boldsymbol{r}_m)$ of $\{\mathcal{P}, \mathcal{Q}\}$ is defined as follows:

$$\begin{aligned} \langle \boldsymbol{r}_0, \ldots, \boldsymbol{r}_{m-1} \rangle &\cong \boldsymbol{G}(\mathcal{P}) \quad \text{under } \boldsymbol{r}_j \mapsto \boldsymbol{s}_j \text{ for } j = 0, \ldots, m-1, \\ \langle \boldsymbol{r}_1, \ldots, \boldsymbol{r}_m \rangle &\cong \boldsymbol{G}(\mathcal{Q}) \quad \text{under } \boldsymbol{r}_j \mapsto \boldsymbol{t}_{j-1} \text{ for } j = 1, \ldots, m, \\ \boldsymbol{r}_0 &\rightleftharpoons \boldsymbol{r}_m. \end{aligned}$$

In other words, we just impose the minimal relations on a sggi $\langle \boldsymbol{r}_0, \ldots, \boldsymbol{r}_m \rangle$ that are needed. Theorem 2C2 shows that $\boldsymbol{G}$ is a string C-group, because the quotient map onto any member of $\langle \mathcal{P}, \mathcal{Q} \rangle$ is one-to-one on the groups of facet and vertex-figure. Moreover, it is obvious that the corresponding regular $(m+1)$-polytope $\mathcal{R} = \mathcal{P}(\boldsymbol{G})$ covers every member of $\langle \mathcal{P}, \mathcal{Q} \rangle$, so that $\mathcal{R} = \{\mathcal{P}, \mathcal{Q}\}$. □

As it happens, the only known general conditions under which amalgamation is possible lead to polytopes that are usually far from being faithfully realized geometrically; we shall say a little more about this in what immediately follows and in Section 4G. Hence, as a general rule we shall not take much interest in amalgamation here.

On the other hand, when we have a regular polytope $\mathcal{P}$, a natural question to ask is whether it is univeral with facet of type $\mathcal{P}^f$ and vertex-figure of type $\mathcal{P}^v$. Universality in this sense will play a prominent part in what follows, just as it was an important motivation for much of [99].

Flatness

We have a concept which plays a particular rôle when we consider realizations of polytopes in Chapter 4. We call a regular polytope $\mathcal{P}$ (*combinatorially*) *flat* if each of its vertices is incident with each of its cells. It is obvious that the dual of a flat regular polytope is flat. Flatness is also a hereditary property.

2F9 Theorem *If any proper inface of a regular polytope $\mathcal{P}$ is flat, then $\mathcal{P}$ itself is flat.*

Proof. From the symmetry of the definition of flatness between vertices and cells, it is clearly enough to show that, if the facet of $\mathcal{P}$ is flat, then $\mathcal{P}$ is flat. So, let $\boldsymbol{V}$ be any vertex of $\mathcal{P}$ and $\boldsymbol{F}$ any cell. Then, using the connectedness properties of $\mathcal{P}$ of Proposition 2A8, we may choose successive cells $\boldsymbol{F}_j$ and $(m-1)$-elements $\boldsymbol{R}_j$ such that

$$\boldsymbol{V} < \boldsymbol{F}_0 > \boldsymbol{R}_1 < \boldsymbol{F}_1 > \cdots > \boldsymbol{R}_k < \boldsymbol{F}_k = \boldsymbol{F}$$

for some k. Since the facet $\mathcal{F} := \boldsymbol{F}_0/\boldsymbol{G}_{-1}$ is flat, it follows that $\boldsymbol{V} \leqslant \boldsymbol{R}_1$, so that $\boldsymbol{V} < \boldsymbol{F}_1$. We now proceed by induction on k, concluding that $\boldsymbol{V} < \boldsymbol{F}$, as was required. □

2F10 Remark Observe that, as a consequence of Proposition 2B10, a regular m-polytope with group $\boldsymbol{G}$ is flat if and only if $\boldsymbol{G} = \boldsymbol{G}_0\boldsymbol{G}_{m-1}$.

A useful implication of flatness to amalgamation is [99, Theorem 4E5], but the proof was not given there, and so we reproduce it from [96] (see the notes at the end of the section).

2F11 Theorem *If $\mathcal{Q}$ and $\mathcal{R}$ are regular m-polytopes such that $\langle \mathcal{Q}, \mathcal{R} \rangle \neq \emptyset$, and if $\mathcal{Q}$ or $\mathcal{R}$ is flat, then the universal polytope $\{\mathcal{Q}, \mathcal{R}\}$ is flat, and is the only member of $\langle \mathcal{Q}, \mathcal{R} \rangle$.*

Proof. In fact, we need to add little to what we have already said. If $\mathcal{Q}$ is flat, then $\{\mathcal{Q}, \mathcal{R}\}$ is flat by Theorem 2F9. However, any identifications in a flat regular polytope must lead to degeneracy of its facets or vertex-figures, from which the claim follows. □

We shall not give any specific examples of flat regular polytopes here; instead, we shall note flatness when it occurs.

Collapsibility

In the same circle of ideas, we have another concept which is important for realizations. With $(\boldsymbol{r}_0, \ldots, \boldsymbol{r}_{m-1})$ as usual the generatrix of the string C-group $\boldsymbol{G}$, we begin by defining

2F12 $$\boldsymbol{N}_k^+ := \bigcap \{\boldsymbol{N} \lhd \boldsymbol{G} \mid \boldsymbol{r}_k, \ldots, \boldsymbol{r}_{m-1} \in \boldsymbol{N}\},$$

the *normal closure* of $\{\boldsymbol{r}_k, \ldots, \boldsymbol{r}_{m-1}\}$. Then we can express $\boldsymbol{G}$ in the following way.

2F13 Proposition *If $\boldsymbol{G} = \langle \boldsymbol{r}_0, \ldots, \boldsymbol{r}_{m-1} \rangle$ is a string C-group and $\boldsymbol{N}_k^+$ is defined by (2F12), then*

$$\boldsymbol{G} = \langle \boldsymbol{r}_0, \ldots, \boldsymbol{r}_{k-1} \rangle \ltimes N_k^+.$$

Proof. The notation $\boldsymbol{A} \ltimes \boldsymbol{B}$ means that $\boldsymbol{A}$ acts on the group $\boldsymbol{B}$ as a group of automorphisms. Indeed, we can write a general element of $\boldsymbol{G}$ as a product of elements of $\langle \boldsymbol{r}_0, \ldots, \boldsymbol{r}_{k-1} \rangle$ and $\boldsymbol{r}_k, \ldots, \boldsymbol{r}_{m-1}$. We deduce that

$$\boldsymbol{N}_k^+ = \langle \boldsymbol{a}^{-1} \boldsymbol{r}_j \boldsymbol{a} \mid j = k, \ldots, m-1 \text{ and } \boldsymbol{a} \in \langle \boldsymbol{r}_0, \ldots, \boldsymbol{r}_{k-1} \rangle \rangle,$$

from which the claim of the proposition follows. □

Let $\mathcal{P}$ be the regular m-polytope with group $\boldsymbol{G}$ and let $\mathcal{F} = \mathcal{F}_k$ be the (initial) k-face of $\mathcal{P}$. We say that $\mathcal{P}$ is *k-collapsible* if $\mathcal{F}$ is a quotient of $\mathcal{P}$, in the sense of (2C4) (see the notes at the end of the section).

2F14 Theorem *The regular m-polytope $\mathcal{P}$ with group $\boldsymbol{G} = \langle \boldsymbol{r}_0, \ldots, \boldsymbol{r}_{m-1} \rangle$ is k-collapsible if and only if*

$$\langle \boldsymbol{r}_0, \ldots, \boldsymbol{r}_{k-1} \rangle \cap \boldsymbol{N}_k^+ = \{\boldsymbol{e}\},$$

using the notation of (2F12).

Proof. If Φ is any homomorphism on $\boldsymbol{G}$, then $\boldsymbol{r}_k, \ldots, \boldsymbol{r}_{m-1} \in \ker \Phi$ exactly when $\boldsymbol{N}_k^+ \leqslant \ker \Phi$. On the other hand, Φ is one-to-one on $\langle \boldsymbol{r}_0, \ldots, \boldsymbol{r}_{k-1} \rangle$ precisely when $\langle \boldsymbol{r}_0, \ldots, \boldsymbol{r}_{k-1} \rangle \cap \ker \Phi = \{\boldsymbol{e}\}$. The claim of the theorem is an immediate consequence. □

There is a nice picture of what k-collapsibility means in terms of the order complex $\mathcal{C} = \mathcal{C}(\mathcal{P})$. Let $T \in \mathcal{C}$ be the $(m-1)$-simplex associated with the base flag Φ of $\mathcal{P}$, and define

2F15 $$\mathcal{D}_k := \{T\boldsymbol{a} \mid \boldsymbol{a} \in \langle \boldsymbol{r}_0, \ldots, \boldsymbol{r}_{k-1} \rangle\}.$$

Then we have

2F16 Proposition *In the notation above, if $\mathcal{P}$ is a k-collapsible regular m-polytope with order complex $\mathcal{C}$, then $|\mathcal{D}_k|$ is a fundamental region for the action of $\boldsymbol{N}_k^+$ on $|\mathcal{C}|$.*

Proof. This is a direct consequence of Theorem 2F14, if we express the claim of the proposition in terms of flags. We leave the details to the interested reader. □

There are two cases related to k-collapse of particular subsequent interest. As before, we write $\mathcal{E} = \mathcal{E}(\mathcal{P})$ for the edge-graph of the regular polytope $\mathcal{P}$.

2F17 Example The edge-graph $\mathcal{E}$ of a regular polytope $\mathcal{P}$ is bipartite if and only if $\mathcal{P}$ is 1-collapsible; the pre-images of the two vertices of the 1-polytope $\{2\}$ form the partition of the vertices of $\mathcal{P}$.

2F18 Example An *orientation* of a simple graph assigns a direction to each of its edges (thus turning it into a digraph). For the edge-graph $\mathcal{E}$ of a regular polytope $\mathcal{P}$, an orientation is *face-cyclic* if each 2-face is cyclically orientated. The 2-collapsibility of $\mathcal{P}$ will ensure this, but in fact a weaker condition will suffice. Suppose that (in the notation above) $[\langle \boldsymbol{r}_0\boldsymbol{r}_1 \rangle : \langle \boldsymbol{r}_0, \boldsymbol{r}_1 \rangle \cap \boldsymbol{N}_2^+] \geqslant 3$. Then the image $\boldsymbol{G}/\boldsymbol{N}_2^+$ is the group of a proper polygon $\{q\}$, with $3 \leqslant q \mid p_1$, where $\mathcal{P}$ has Schläfli type $\{p_1, \ldots, p_{m-1}\}$. Then a cyclic orientation of $\{q\}$ lifts back into a face-cyclic orientation of $\mathcal{E}$.

The edge-graph of the 3-cube $\{4, 3\}$ has no face-cyclic orientation, which shows that we cannot expect anything better than this. Whether there is a converse result is not clear.

This is another example that occurs frequently. The (standard) *m-staurotope* (*m-cross-polytope* – see the notes at the end of the section) is

2F19 $$\mathsf{X}^m := \operatorname{conv}\{\pm e_1, \ldots, \pm e_m\} \subset \mathbb{E}^m.$$

As an abstract regular polytope, we shall see that it is $\{3^{m-2}, 4\}$. Then we have

2F20 Proposition *For each $m \geqslant 2$, the m-staurotope is $(m-1)$-collapsible.*

Indeed, the natural identification of opposite vertices $\pm e_j$ induces the collapse.

Notes to Section 2F

1. Theorem 2F2 is new. The need for atomicity (or something like it) was pointed out by Daniel Pellicer and Isabel Hubard [106]. Whether the full strength of atomicity is necessary is an open problem.
2. We can say a little more about the general case of central symmetry; we assume atomicity. If the regular m-polytope $\mathcal{P}$ is centrally symmetric, then there is clearly a minimal k such that the k-elements of $\mathcal{P}$ are centrally symmetric (under the central involution $\boldsymbol{z}$, of course). It easily follows that the j-elements of $\mathcal{P}$ are also centrally symmetric for each $j \geqslant k$; that is, the central symmetry of the k-elements induces central symmetry of elements of higher rank. When we come to look at realized regular polytopes, we shall see that examples do occur with $k < m$, or even $k < m-1$.
3. Carrying on from this, we may have (for example) a central symmetry $\boldsymbol{z}_k \in \boldsymbol{G}_k$ induced by $\boldsymbol{z}$ which is not central in $\boldsymbol{G}$ itself.
4. In [99, Section 7A], what we have called the abstract blend is there called the *mix*, and is denoted $\mathcal{P} \diamond \mathcal{Q}$. We avoid this notation, because we wish to reserve it for (5D11). That case may seem needlessly special, but it occurs quite often in our characterizations.
5. Flatness is an important property in the general theory of regular polytopes, in that it facilitates certain types of constructions. However, geometrically the results of these constructions tend to be degenerate, as we shall see in Section 4A. We refer the reader to [99, Sections 4E, 4F] for further details about flatness.
6. Theorem 2F11 is originally due to Egon Schulte; it is [96, Proposition 1].
7. The process of Example 2F18 is called by Coxeter [27] *coherent indexing*.
8. In [99, Chapter 4], k-collapsibility is called the ‘flat amalgamation property with respect to k-elements’, which is why we have introduced it next to flatness. To indicate why, if $\mathcal{Q}$ and $\mathcal{R}$ are regular m-polytopes satisfying the necessary condition for amalgamation, and $\mathcal{Q}^\delta$ and $\mathcal{R}$ are both $(m-1)$-collapsible, then indeed $\langle \mathcal{Q}, \mathcal{R} \rangle \neq \emptyset$, and Theorem 2F11 implies that $\{\mathcal{Q}, \mathcal{R}\}$ is universal.
9. We have long felt the need for a less clumsy term than ‘cross-polytope’. The new term is derived from the Greek ‘$\sigma\tau\alpha\upsilon\rho o\varsigma$’ (meaning ‘cross’ – we have borrowed the prefix from geology).
10. A weaker condition than regularity is chirality. A polytope $\mathcal{P}$ is *chiral* if there are two orbits of flags under its automorphism group, with adjacent flags in different orbits. In this case, the automorphism group (roughly) corresponds to the ‘rotation’ subgroup $\boldsymbol{G}^+ := \langle \boldsymbol{r}_0\boldsymbol{r}_1, \boldsymbol{r}_1\boldsymbol{r}_2, \ldots, \boldsymbol{r}_{m-2}\boldsymbol{r}_{m-1} \rangle$ of the string C-group $\boldsymbol{G}$. There is an extensive theory of chirality, but we shall confine ourselves to referring to the survey by Schulte and Weiss [115].

3

Realizations of Symmetric Sets

This book is concerned with the geometric theory of regular polytopes and apeirotopes. Our next task is therefore to relate this to the abstract theory; the bridging concept is that of realizations. Since the theory for finite polytopes is so much richer, we shall confine our treatment to this case until Section 4F.

We adopt in this chapter a much more general viewpoint than that of [99, Chapter 5] or the related papers [75, 80, 85, 88, 93]. However, the basic theory as treated here largely follows [85, 88], but changing the emphasis to symmetric sets rather than polytopes. After an introductory Section 3A, the core result of Section 3B is that the family of congruence classes of realizations of a fixed pair consisting of a finite group $\boldsymbol{G}$ acting transitively on a point-set $\mathcal{V}$ has the structure of a convex cone, also denoted $\mathcal{V}$. A key idea introduced in Section 3C is that of the inner product and cosine vectors. It is then shown that the cone $\mathcal{V}$ is closed; similarly, the realization domain $\mathcal{N}$ of normalized realizations is a compact convex set. The theory up to this point is illustrated in Section 3D by some examples. We then show in Section 3E that, corresponding to the (tensor) product of representations, there is a product of realizations. Another fundamental notion is that of orthogonality relations for cosine vectors, which we treat in Section 3F. In Section 3G we look more closely at the Wythoff space of realizations derived from the same irreducible representation of $\boldsymbol{G}$; here, we incorporate an important correction from [67] to the earlier theory, particularly as treated in [99, Section 5B] and [85, 88]. The revised theory enables us in Section 3J to define cosine matrices for the general realization domain. Next, Section 3K treats the relationship between actions of $\boldsymbol{G}$ on different subgroups and corresponding cuts. In Section 3L, we consider realizations over subfields of the real numbers. Finally, in Section 3M, we give a brief account of how representations are involved in Section 3G.

3A Transitive Actions

Our starting point is a very general idea. Suppose that we are given a finite set $\mathcal{V}$, which a group $\boldsymbol{G}$ permutes transitively; we shall refer to $\mathcal{V}$ as a *symmetric set*. An alternative viewpoint is to identify $\mathcal{V}$ with the family $\boldsymbol{G}/\boldsymbol{H}$ of right cosets $\boldsymbol{H}\boldsymbol{g}$ of the *stabilizer* $\boldsymbol{H} := \{\boldsymbol{h} \in \boldsymbol{G} \mid \boldsymbol{v}\boldsymbol{h} = \boldsymbol{v}\}$ of a fixed element $\boldsymbol{v} \in \mathcal{V}$ (see the notes at the end of the section). In this sense, we start with a finite group $\boldsymbol{G}$ and a distinguished subgroup $\boldsymbol{H} < \boldsymbol{G}$, thus dispensing with the set $\mathcal{V}$ itself. This is just about what we do when we come to consider realizations of vertex-sets of polytopes, where $\boldsymbol{G}$ will be a string C-group and $\boldsymbol{H} := \boldsymbol{G}_0$ the distinguished subgroup corresponding to its vertices. Moreover, this viewpoint is essential when we come to compare realizations with the same representation of $\boldsymbol{G}$ but different subgroups $\boldsymbol{H}$, for example, when dealing with dual polytopes. However, a concrete picture of an initial set is often useful, as we shall shortly see. Nevertheless, it is sometimes convenient to identify elements of $\mathcal{V}$ with corresponding coset representatives $\boldsymbol{g}$ rather than with the cosets themselves, so that the given fixed element of $\mathcal{V}$ corresponds to the identity $\boldsymbol{e} \in \boldsymbol{G}$.

3A1 Remark Observe that a set $\mathcal{V}$ may be symmetric under the action of (for example) a proper subgroup $\boldsymbol{G}' < \boldsymbol{G}$. For this reason, we should think of a symmetric set as consisting of a pair $(\mathcal{V}, \boldsymbol{G})$, rather than just of $\mathcal{V}$ alone; note as well that we would obtain a different picture in terms of cosets. For various reasons it is convenient to think of $\mathcal{V}$ as an *ordered* set.

Diagonals and Layers

A *diagonal* of $\mathcal{V}$ is a pair $\{\boldsymbol{x}, \boldsymbol{y}\} \subseteq \mathcal{V}$; the diagonals equivalent to a given one under the action of the group $\boldsymbol{G}$ form a *diagonal class*. We label the diagonal classes $\mathcal{D}_s$ for $s = 0, \ldots, r$ (say), with $\mathcal{D}_0$ always consisting of the *trivial* diagonals $\{\boldsymbol{x}, \boldsymbol{x}\}$ (thus r is the number of non-trivial diagonal classes of $\mathcal{V}$). A diagonal $\{\boldsymbol{x}, \boldsymbol{y}\}$ is *symmetric* if the *ordered* pairs $(\boldsymbol{x}, \boldsymbol{y})$ and $(\boldsymbol{y}, \boldsymbol{x})$ are equivalent under $\boldsymbol{G}$; otherwise, the diagonal is *asymmetric*. We use the same terms for the corresponding diagonal classes. Then we have

3A2 Theorem *A diagonal $\{\boldsymbol{H}, \boldsymbol{H}\boldsymbol{g}\}$ containing the initial point $\boldsymbol{v}$ of $\mathcal{V}$ is symmetric just when $\boldsymbol{g}^{-1} \in \boldsymbol{H}\boldsymbol{g}\boldsymbol{H}$.*

Proof. The diagonal opposite to the directed diagonal $(\boldsymbol{H}, \boldsymbol{H}\boldsymbol{g})$ is $(\boldsymbol{H}\boldsymbol{g}, \boldsymbol{H}) \sim (\boldsymbol{H}\boldsymbol{g}, \boldsymbol{H})\boldsymbol{g}^{-1} = (\boldsymbol{H}, \boldsymbol{H}\boldsymbol{g}^{-1})$. If this is equivalent to $(\boldsymbol{H}, \boldsymbol{H}\boldsymbol{g})$, then

$$\boldsymbol{H}\boldsymbol{g}\boldsymbol{h} = \boldsymbol{H}\boldsymbol{g}^{-1} \iff \boldsymbol{g}^{-1} \in \boldsymbol{H}\boldsymbol{g}\boldsymbol{H}$$

for some $\boldsymbol{h} \in \boldsymbol{H}$, as claimed. □

For $s = 0, \ldots, r$, the *layer* $\mathcal{L}_s$ of $\mathcal{V}$ (relative to the initial point $\boldsymbol{v}$) is defined by $\mathcal{L}_s := \big\{\boldsymbol{x} \in \mathcal{V} \mid \{\boldsymbol{v}, \boldsymbol{x}\} \in \mathcal{D}_s\big\}$. The *layer vector* of $\mathcal{V}$ is

3A3 $\quad \Lambda = \Lambda(\mathcal{V}) := (\ell_0, \ldots, \ell_r), \qquad \ell_s := \operatorname{card} \mathcal{L}_s \quad \text{for } s = 0, \ldots, r;$

thus $\ell_0 = 1$ always corresponds to layer $\mathcal{L}_0$ consisting of the initial point $\boldsymbol{e}$ alone. Hence, $\operatorname{card} \mathcal{V} = n = \ell_0 + \cdots + \ell_r =: |\Lambda|$ is the number of points of $\mathcal{V}$. The points in each layer are permuted by $\boldsymbol{H}$. (For polytopes, it is a convention – not universally followed – that $\mathcal{L}_1$ corresponds to the vertex-figure of $\mathcal{V}$.)

3A4 Remark There is nothing mysterious about Λ – it is just an accounting device, to collect together points of $\mathcal{V}$ that behave alike under $\boldsymbol{H}$. This is made clear in Section 3M below.

3A5 Remark We call a layer *symmetric* or *asymmetric* as the corresponding diagonal class is symmetric or asymmetric. We often find it useful to attach an asterisk * to ℓ_s if $\mathcal{D}_s$ is an asymmetric diagonal class.

A useful observation is

3A6 Proposition *The stabilizer $\boldsymbol{H}$ equi-partitions an asymmetric layer of $\mathcal{V}$.*

Proof. Theorem 3A2 implies that we can identify a *half-layer* with a double coset $\boldsymbol{HgH}$. Taking inverses shows that $\boldsymbol{HgH}$ and $\boldsymbol{Hg}^{-1}\boldsymbol{H}$ have the same size, which leads to our claim. □

3A7 Remark As an alternative way of seeing Proposition 3A6, represent an asymmetric diagonal class $\mathcal{D}$ by a directed diagonal $(\boldsymbol{H}, \boldsymbol{Hg})$, say. The images of this diagonal under $\boldsymbol{G}$ then turn $\mathcal{D}$ into a digraph on $\mathcal{V}$, and the transitivity of $\boldsymbol{G}$ on $\mathcal{V}$ ensures that the in- and out-degrees of $\mathcal{D}$ at each point of $\mathcal{V}$ are equal.

Cuts

Symmetric subsets of a symmetric set are also important. If $\boldsymbol{A}, \boldsymbol{B} < \boldsymbol{G}$ and $\boldsymbol{x} \in \mathcal{U} := \boldsymbol{G}/\boldsymbol{A}$, then $\mathcal{C} := \boldsymbol{xB} \subseteq \mathcal{U}$ is clearly itself a symmetric set. We call $(\mathcal{C}, \boldsymbol{B})$ a *cut* of $(\mathcal{U}, \boldsymbol{G})$, or a $\boldsymbol{B}$-*cut* if we wish to specify $\boldsymbol{B}$ (see the notes at the end of the section). We generally suppose $\boldsymbol{B}$ to be chosen maximal such that $\mathcal{C}\boldsymbol{B} = \mathcal{C}$, but occasionally it is useful not to.

A powerful tool for the future is provided by comparison of cuts arising from different subgroups; here, we just consider the combinatorial aspects. As we have said, it is a useful convention to denote the family of right cosets of a subgroup $\boldsymbol{H}$ of a group $\boldsymbol{G}$ by $\boldsymbol{G}/\boldsymbol{H}$; in other words, we can write $\boldsymbol{G}/\boldsymbol{H}$ instead of $\mathcal{V}$.

3A8 Proposition *If $\boldsymbol{A}, \boldsymbol{B} < \boldsymbol{G}$, then there is a one-to-one correspondence between the $\boldsymbol{B}$-cuts of $\boldsymbol{G}/\boldsymbol{A}$ and the $\boldsymbol{A}$-cuts of $\boldsymbol{G}/\boldsymbol{B}$. Moreover, their sizes are in proportion $|\boldsymbol{B}| : |\boldsymbol{A}|$.*

Proof. A $\boldsymbol{B}$-cut of $\boldsymbol{G}/\boldsymbol{A}$ can be identified with a double coset $\boldsymbol{AgB}$; regarded as a union of cosets of $\boldsymbol{A}$, its size is $|\boldsymbol{AgB}|/|\boldsymbol{A}|$. Inverting each element gives

$$\boldsymbol{AgB} \longleftrightarrow \boldsymbol{Bg}^{-1}\boldsymbol{A},$$

and the proposition now easily follows. □

Again, there is a more concrete way of comparing cuts. For $\boldsymbol{A}, \boldsymbol{B} < \boldsymbol{G}$, write $\mathcal{U} := \boldsymbol{G}/\boldsymbol{A}$ and $\mathcal{V} := \boldsymbol{G}/\boldsymbol{B}$. The action of $\boldsymbol{G}$ on $\mathcal{E} := \mathcal{U} \times \mathcal{V}$ partitions it into equivalence classes $\mathcal{E}_1, \ldots, \mathcal{E}_t$, say. With $a := \boldsymbol{A}$ and $b := \boldsymbol{B}$ the initial cosets (or points), for $j = 1, \ldots, t$ we then define $\mathcal{U}_j := \{x \in \mathcal{U} \mid \{x, b\} \in \mathcal{E}_j\}$, and similarly for $\mathcal{V}_j$; we call these the *initial* cuts. Thus $\mathcal{U}_1, \ldots, \mathcal{U}_t$ is a partition of $\mathcal{U}$. Moreover, transitivity under the action of $\boldsymbol{G}$ implies that

3A9 $$\operatorname{card} \mathcal{V} \operatorname{card} \mathcal{U}_j = \operatorname{card} \mathcal{E}_j = \operatorname{card} \mathcal{U} \operatorname{card} \mathcal{V}_j$$

for each k. Since $\operatorname{card} \mathcal{U} = |\boldsymbol{G}|/|\boldsymbol{A}|$ and so on, we have an alternative proof of Proposition 3A8.

3A10 Remark We call $(\mathcal{U}_j, \mathcal{V}_j)$ a *cut pair*; Proposition 3A8 gives a one-to-one correspondence $\mathcal{U}_j \leftrightarrow \mathcal{V}_j$ for $j = 1, \ldots, t$. For any particular j, it is clear that we can replace $\boldsymbol{A}$ or $\boldsymbol{B}$ by suitable conjugates in $\boldsymbol{G}$, to assume that our initial points $a \in \mathcal{U}_j$ and $b \in \mathcal{V}_j$.

An important connexion with what we have just discussed is given by

3A11 Proposition *A symmetric diagonal in a cut $\mathcal{C}$ is symmetric in $\mathcal{U}$ itself.*

Proof. We suppose that the initial vertex $\boldsymbol{u}$ of $\mathcal{U}$ lies in $\mathcal{C}$, and let $\boldsymbol{D} \leqslant \boldsymbol{H}$ be its stabilizer in $\boldsymbol{B}$. If $\{\boldsymbol{D}, \boldsymbol{D}\boldsymbol{g}\}$ is in the given symmetric diagonal class in $\mathcal{C}$, then

$$\boldsymbol{g}^{-1} \in \boldsymbol{D}\boldsymbol{g}\boldsymbol{D} \subseteq \boldsymbol{H}\boldsymbol{g}\boldsymbol{H},$$

showing that the corresponding diagonal class in $\mathcal{V}$ is symmetric. □

3A12 Remark It is clear that symmetric layers are cuts, under the action of the stabilizer $\boldsymbol{H}$. Note that an asymmetric layer is not a cut; however, each of its half-layers is.

Quotients

Quotients play an important rôle in realization theory, and so we make a brief comment about them here. There are two basic ways in which a *quotient* arises (on the abstract level). First, through a homomorphism Φ on $\boldsymbol{G}$, which leads to $\boldsymbol{G}/\boldsymbol{N}$ acting on the cosets of $\boldsymbol{H}\boldsymbol{N}/\boldsymbol{N} = \boldsymbol{H}/(\boldsymbol{H} \cap \boldsymbol{N})$, with $\boldsymbol{N} := \ker \Phi$ the kernel of Φ. Second, $\boldsymbol{G}$ can act on the cosets of $\boldsymbol{H}'$, with $\boldsymbol{H} < \boldsymbol{H}' \leqslant \boldsymbol{G}$ a larger subgroup. Each implies a corresponding identification of points of $\mathcal{V}$. Naturally, a general quotient will combine effects of both kinds of identification; indeed, the first is a special kind of the second, with $\boldsymbol{H}' = \boldsymbol{H}\boldsymbol{N}$. However, at the level of realizations, it is often important to think of them differently; we refer to Remark 3B8 below.

Because quotients may induce identifications, thinking of $\mathcal{V}$ as an ordered set is particularly appropriate here.

3A13 Remark An important instance where quotients play a crucial rôle concerns central symmetry (see the notes at the end of the section). A symmetric set $\mathcal{V}$ is called *centrally symmetric* if there is a central involution $\boldsymbol{z} \in \boldsymbol{G}$ which fixes no point of $\mathcal{V}$. The points of $\mathcal{V}$ thus fall into *antipodal* pairs $\{\boldsymbol{x}, \boldsymbol{x}\boldsymbol{z}\}$; we also call such a pair $\{\boldsymbol{x}, \boldsymbol{x}\boldsymbol{z}\}$ a *diameter*.

Notes to Section 3A

1. Simon [122], for example, uses $\boldsymbol{G}/\boldsymbol{H}$ to denote the family of *left* cosets of $\boldsymbol{H}$ in $\boldsymbol{G}$; we shall not need a separate notation for this family. As we have previously noted, our preference is to follow Coxeter [27] in letting symmetries act on the right.
2. As we have defined the term 'cut', it corresponds quite closely to the usage of [99, Section 6D]. We shall see that cuts of various kinds provide useful tools for investigating realizations.
3. Many regular polytopes are centrally symmetric, so that central symmetry plays an important part in the theory.

3B Realization Cone

The context remains that of the preceding Section 3A; we thus have a finite set $\mathcal{V} = \boldsymbol{G}/\boldsymbol{H}$ acted on transitively by a group $\boldsymbol{G}$, with $\boldsymbol{H}$ the stabilizer of an initial element of $\mathcal{V}$ (which can be identified with the identity $\boldsymbol{e}$ of $\boldsymbol{G}$).

Realizations

A *realization* of a symmetric set $(\mathcal{V}, \boldsymbol{G})$ is a mapping $\varPsi\colon (\mathcal{V}, \boldsymbol{G}) \to (\mathbb{E}, \mathbf{O})$, where $\mathbb{E}$ is some euclidean space and $\mathbf{O} = \mathbf{O}(\mathbb{E})$ is the orthogonal group in $\mathbb{E}$, with the following properties:

- $\varPsi$ induces a homomorphism on $\boldsymbol{G}$,
- $(\boldsymbol{xg})\varPsi = (\boldsymbol{x}\varPsi)(\boldsymbol{g}\varPsi)$ for each $\boldsymbol{x} \in \mathcal{V}$ and $\boldsymbol{g} \in \boldsymbol{G}$.

We write $V := \mathcal{V}\varPsi$ and $\mathbf{G} := \boldsymbol{G}\varPsi$. We shall usually assume that $\mathbb{E} = \operatorname{lin} V$, in which case we call $\mathbb{E}$ the *ambient space* of $\varPsi$; then $d := \dim V = \dim(\operatorname{lin} V) = \dim \mathbb{E}$ is called the *dimension* of the realization. To avoid totally degenerate cases, a further assumption will be that $V \neq \{o\}$, so that $d \geqslant 1$; we refer to the definition of the Wythoff space below. The second property says that $\varPsi$ is symmetric, in that it is compatible with the action of $\boldsymbol{G}$ on $\mathcal{V}$; of course, it is also compatible with our identification of elements of $\mathcal{V}$ with cosets of $\boldsymbol{H}$.

We can look at this in another way. Let $\varPsi\colon \boldsymbol{G} \to \mathbf{O}$ be a real representation, and write $\mathbf{G} := \boldsymbol{G}\varPsi$. Observe that $v := \boldsymbol{v}\varPsi$ is fixed by $\mathbf{H} := \boldsymbol{H}\varPsi$, the image of the stabilizer. Indeed, we obtain a realization by picking as initial point any $v \in \operatorname{axis} \mathbf{H}$, and defining $V := v\mathbf{G}$; then V is a realization of $(\mathcal{V}, \boldsymbol{G})$. In the context of (usually regular) polytopes, this is called Wythoff's construction, and so we shall often use this term loosely in our more general situation. Similarly, we often refer to $\operatorname{axis} \mathbf{H}$ as the *Wythoff space* of the realization $\varPsi$, and employ notation such as $W(\varPsi)$ for it (see the notes at the end of the section).

3B1 Remark The definition of realization says more than $V \sim \mathbf{G}/\mathbf{H}$, since there is added geometric content. Our global assumption that $V \neq \{o\}$ means that realization theory ignores those *trivial* representations $\varPsi$ of $\boldsymbol{G}$ for which $W(\varPsi) = \{o\}$.

3B2 Remark We shall later vary the notion of ambient space. We occasionally need to consider non-symmetric realizations; see, for example, Section 11C.

3B3 Remark We lose no generality in working with the orthogonal group $\mathbf{O}(\mathbb{E})$, rather than with the whole group $\mathbf{M}(\mathbb{E})$ of isometries of $\mathbb{E}$. Recall from Section 1A that a finite group of isometries fixes some point, and conjugating by a suitable translation we can take this point to be the origin o. With this convention, then, the set V is a subset of some sphere centred at the origin o; moreover, the homomorphism Ψ yields an orthogonal representation of $\boldsymbol{G}$. A natural question is whether a given orthogonal representation of $\boldsymbol{G}$ yields realizations of $\mathcal{V}$; this will be very much at the centre of what we do subsequently.

We say that a realization Ψ is *normalized* if $\|x\| = 1$ when $x \in V$. The *unital* realization Ψ_0 maps each $\boldsymbol{x} \in \mathcal{V}$ to $1 \in \mathbb{R}$ and each $\boldsymbol{g} \in \boldsymbol{G}$ to the identity; thus Ψ_0 is normalized.

Just as with representations of groups, we identify two realizations Ψ_1, Ψ_2 if there is an isometry Θ such that $\Psi_2 = \Psi_1\Theta$; in other words, we think of a realization as an equivalence class under congruence, rather than as a single realization. In that sense, employing the convention of [99, Chapter 5], we shall also use $\mathcal{V}$ to denote the family of congruence classes of its realizations. Moreover, if we bear in mind that we think of $\mathcal{V}$ as an ordered set, we see that we lose nothing in identifying the realization Ψ with the image $V = \mathcal{V}\Psi$ of $\mathcal{V}$; identification of points of $\mathcal{V}$ under Ψ will induce a quotient map on $(\mathcal{V}, \boldsymbol{G})$. With this convention, we write $\{1\}$ for the unital realization or *unity*; more generally, particularly in the context of regular polytopes, we shall call $\{1\}$ the *henogon*.

3B4 Remark Observe that, in contrast to the convention of [99, Section 5B], the cone $\mathcal{V}$ now incorporates the henogon $\{1\}$.

We shall write $\mathcal{N}$ for the *realization domain* of $\mathcal{V}$, consisting of the normalized realizations in $\mathcal{V}$.

We call a realization Ψ of $(\mathcal{V}, \boldsymbol{G})$ *faithful* if Ψ is one-to-one on $\mathcal{V}$, and therefore an isomorphism on $\boldsymbol{G}$.

3B5 Remark In the context of realizations of regular polytopes, this concept will be known as vertex-faithful; we shall introduce a more subtle notion in the appropriate place.

As in [99] and earlier papers, we *blend* two realizations Ψ_j in $\mathbb{E}_j$ for $j = 1, 2$ and *scale* a realization Ψ by $\lambda \in \mathbb{R}$ by

3B6 $$\boldsymbol{x}(\Psi_1 \# \Psi_2) := (\boldsymbol{x}\Psi_1, \boldsymbol{x}\Psi_2) \in \mathbb{E}_1 \oplus \mathbb{E}_2,$$

3B7 $$\boldsymbol{x}(\lambda\Psi) := \lambda(\boldsymbol{x}\Psi),$$

for each $\boldsymbol{x} \in \mathcal{V}$. If Ψ_j is identified with the image set V_j for $j = 1, 2$, then we also write the blend as $V_1 \# V_2$, and call each of V_1, V_2 a *component* of the blend; similarly, λV corresponds to $\lambda\Psi$.

It is particularly important to observe that it is the *squares* of scaling factors that are crucial since, if $\mathcal{V}\Psi_j$ is a subset of a sphere of radius ρ_j for $j = 1, 2$, then $\mathcal{V}(\lambda_1\Psi_1 \# \lambda_2\Psi_2)$ lies on a sphere of radius ρ given by

$$\rho^2 = \lambda_1^2\rho_1^2 + \lambda_2^2\rho_2^2.$$

Hence, when working with normalized realizations we need to have $\lambda_1^2 + \lambda_2^2 = 1$.

3B8 Remark Blending and scaling already show that two realizations may induce the same homomorphism on $\boldsymbol{G}$ while differing as mappings on $\mathcal{V}$.

Under blending and squared scaling, $\mathcal{V}$ has the structure of a closed convex cone, called the *realization cone*; this will become clearer when we introduce inner product vectors in the next section. In a similar way, we shall see that the realization domain $\mathcal{N}$ has the structure of a compact convex set.

We call a realization V *centred* if the origin o is the centroid of V. Observe that we do not insist that realizations be centred, even if they are normalized. Thus a non-centred realization will have (a multiple of) $\{1\}$ as a component; again, we refer to the following discussion.

Purity

Because (up to congruence)

3B9 $$\lambda^2 + \mu^2 = \nu^2 \implies (\lambda\Psi) \# (\mu\Psi) = \nu\Psi,$$

there are trivial expressions of realizations as blends. A realization which cannot be expressed as a blend in a non-trivial way is called *pure*. It is clear that pure realizations correspond to irreducible representations of the automorphism group $\boldsymbol{G}$. More generally, for a normalized realization $\Psi \in \mathcal{N}$ we shall have a (possibly non-unique) expression

3B10 $$\Psi = \lambda_0\Psi_0 \# \cdots \# \lambda_r\Psi_r,$$

with $\Psi_0, \ldots, \Psi_r \in \mathcal{N}$ pure, and $\lambda_0^2 + \cdots + \lambda_r^2 = 1$.

We end the section with a partial converse to Theorem 1C6 (see the notes at the end of the section).

3B11 Theorem *If V is a realization of $(\mathcal{V}, \boldsymbol{G})$ whose symmetry group $\mathbf{G}$ is an irreducible representation of $\boldsymbol{G}$, then – up to scaling – V is the image of some orthonormal basis of $\mathbb{E}^n$ under orthogonal projection.*

Proof. If $\mathbb{E}$ is the ambient space and $V = (v_1, \ldots, v_n)$, consider the ellipsoid

$$S := \Big\{x \in \mathbb{E} \;\Big|\; \sum_{j=1}^{n} \langle x, v_j\rangle^2 = 1\Big\}.$$

For each $x \in S$ and $\Phi \in \mathbf{G}$ (which permutes the points of V), we have

$$1 = \sum_{j=1}^{n} \langle x, v_j\rangle^2 = \sum_{j=1}^{n} \langle x\Phi, v_j\Phi\rangle^2 = \sum_{k=1}^{n} \langle x\Phi, v_k\rangle^2,$$

so that $x\Phi \in S$ also. Theorem 1D23 then implies that S is a sphere, of radius ρ, say. Hence, with the usual identification of V and the matrix with rows $v_1, \ldots, v_n$, if $A := \rho V$ has rows $a_1, \ldots, a_n$ and $\|y\| = 1$, then

$$y(A^\mathsf{T} A)y^\mathsf{T} = \sum_{j=1}^{n} \langle y, a_j\rangle^2 = \sum_{j=1}^{n} \langle \rho y, v_j\rangle^2 = 1.$$

Consequently, $A^{\mathsf{T}}A = I_d$, the identity matrix. Thus $(a_1, \dots, a_n)$ is a eutactic star by Theorem 1C6; the claim of the theorem follows at once. □

Notes to Section 3B

1. The term *Wythoff space* is taken from the analogous concept in the theory of hyperplane reflexion groups. In that context, the idea will be treated in detail in Section 4A.
2. For Theorem 3B11, see also [7, Theorem 1]; the proof here is our own, but we have not consulted the original.
3. The degree d of an irreducible unitary representation of a group $\boldsymbol{G}$ divides the order $|\boldsymbol{G}|$ of $\boldsymbol{G}$; see, for example, [122, Theorem III.4.1]. For orthogonal representations, the best that can be claimed is that $d \mid 2|\boldsymbol{G}|$, as the examples of the cyclic groups $\boldsymbol{C}_q$ of odd order q show; see Section 3M.
4. We saw in Proposition 3A6 that the stabilizer splits an asymmetric layer in half. However, it is important to note that, in a realization, these two halves need not be congruent. We do not give an example here; instead, refer to the discussion of the 120-cell in Section 7K.

3C Cosine Vectors

The geometry of a point-set V in a euclidean space is specified by the mutual distances between pairs of points of V. For a realization of a symmetric set, this information can be consolidated in a convenient way.

Inner Product Vectors

Let $V \in \mathcal{V}$ be a realization as before, and write $V_s := \mathcal{L}_s\Psi$ for $s = 0, \dots, r$. If $v \in V_0$ is the initial point and $x \in V_s$, then $\sigma_s := \langle v, x\rangle$ is well-defined; we call

3C1
$$\Sigma = \Sigma(P) := (\sigma_0, \dots, \sigma_r)$$

the *inner product vector* of V. Thus, in particular, $\sigma_0 = \|v\|^2$ is the *squared* radius of the sphere centred at o on which the point-set V lies.

The vector Σ determines the realization up to congruence, since it prescribes the distance between every two points of V (by transitivity, any point can be chosen to be the initial one). Indeed, bearing in mind that Σ is shorthand for the whole $n \times n$ matrix $S = (\sigma_{jk})$ of inner products $\sigma_{jk} := \langle v_j, v_k\rangle$ of pairs of points $v_j, v_k \in V$ (compare Remark 3A4), we see that this claim comes down to

3C2 Proposition *An $n \times n$ matrix S is the matrix of inner products of a set $V = (v_1, \dots, v_n)$ in some euclidean space if and only if S is symmetric and positive semi-definite.*

Proof. Because it is so fundamental, we sketch a proof of this well-known result. First, such a matrix S of inner products is necessarily symmetric. It is also positive semi-definite since, for all $x = (\xi_1, \dots, \xi_n) \in \mathbb{R}^n$, we have

$$0 \leqslant \|\xi_1 v_1 + \cdots + \xi_n v_n\|^2 = xSx^{\mathsf{T}}.$$

Conversely, if $\operatorname{rank} S = d$, then we may reorder the rows (and corresponding columns), and suppose that the leading $d \times d$ submatrix S_d (say) of S is positive definite. The argument is simpler if we first assume that we have already found $v_1, \ldots, v_d$. Let $j > d$. If $s_j := (\sigma_{1j}, \ldots, \sigma_{dj})$, then it may be verified that $x = (\xi_1, \ldots, \xi_d) := s_j S_d^{-1}$ is such that $v_j = \xi_1 v_1 + \cdots + \xi_d v_d$ has $\langle v_j, v_k \rangle = \sigma_{jk}$ for each $k = 1, \ldots, d$. Thus we have found each v_j for $j = d+1, \ldots, n$. Observe that everything is consistent: because $\operatorname{rank} S = d$, linear relations that hold on the first d coordinates of rows of S must hold on the remainder.

We now successively find $v_1, \ldots, v_d \in \mathbb{E}^d$, so that $v_j \in \operatorname{lin}\{e_1, \ldots, e_j\}$ for $j = 1, \ldots, d$; here, $\{e_1, \ldots, e_d\}$ is the standard orthonormal basis of $\mathbb{E}^d$. If, say, we have constructed $v_1, \ldots, v_{k-1}$ for some $k \leqslant d$ (beginning at $k = 1$), then replacing d by $k - 1$ and j by k in the previous calculation now gives the component of v_k parallel to $\operatorname{lin}\{v_1, \ldots, v_{k-1}\}$. Moreover, the complementary component orthogonal to $\operatorname{lin}\{v_1, \ldots, v_{k-1}\}$ has squared norm

$$\sigma_{kk} - s_k S_{k-1}^{-1} s_k^{\mathsf{T}},$$

and putting this component in direction e_k then yields v_k. Note, by the way, that

$$(\sigma_{kk} - s_k S_{k-1}^{-1} s_k^{\mathsf{T}}) \det S_{k-1} = \det S_k > 0,$$

so that the norm of the orthogonal component is indeed positive. Observe that

$$\det S_k = \big(\operatorname{Det}(v_1, \ldots, v_k)\big)^2$$

in terms of the absolute determinant, as defined in (1C13). □

We thus state formally

3C3 Corollary *The inner product vector Σ of a realization V of $\mathcal{V}$ determines V up to congruence.*

In [99, Section 5A], realizations were specified by their diagonal vectors (see the notes at the end of the section). Properties possessed by them have exact analogues for inner product vectors, namely,

3C4 Lemma *Inner product vectors of blends and scalar multiples satisfy*

$$\begin{aligned} \Sigma(P \# Q) &= \Sigma(P) + \Sigma(Q), \\ \Sigma(\lambda P) &= \lambda^2 \Sigma(P). \end{aligned}$$

For the first relation, compare (1C9). We emphasize that the scaling $\Sigma \mapsto \mu\Sigma$ (for $\mu \geqslant 0$) is induced by $V \mapsto \sqrt{\mu} V$ on the original point-sets. Thus we have

3C5 Theorem *The inner product vectors of realizations of $\mathcal{V}$ form a closed convex cone of full dimension in $\mathbb{R}^{r+1}$, naturally to be identified with $\mathcal{V}$ as the realization cone.*

Proof. It remains to check the claims about closure and $\dim \mathcal{V}$; we treat the latter first.

It is clear that $\dim \mathcal{V} \leqslant r+1$, because inner product vectors have $r+1$ coordinates. The *simplex realization* T has its n points at the standard basis vectors $e_1, \ldots, e_n$ of $\mathbb{E}^n$; its inner product vector is thus $\Sigma(T) = (1, 0^r)$ (recall the convention that α^k is shorthand for a string $\alpha, \ldots, \alpha$ of length k). The corresponding inner product matrix S of Proposition 3C2 is just the $n \times n$ identity matrix I_n. This is positive definite, and so any symmetric matrix in a small enough neighbourhood of I_n will also be positive definite. In particular, we may similarly vary the inner product vector Σ freely in a small enough neighbourhood of $(1, 0^r)$, and thus see that $\dim \mathcal{V} \geqslant r+1$. Hence we have equality $\dim \mathcal{V} = r+1$.

The abstract group $\boldsymbol{G}$ acts on T as a permutation group $\mathbf{G}$. A convergent sequence $(\Sigma_j \mid j = 1, 2, \ldots)$ of inner product vectors clearly gives a convergent sequence $(S_j \mid j = 1, 2, \ldots)$ of inner product matrices in Proposition 3C2. Since each S_j is positive semi-definite, so is the limit $\lim_{j\to\infty} S_j$; we immediately deduce that $\lim_{j\to\infty} \Sigma_j$ is also an inner product vector of $\mathcal{V}$. Thus the cone $\mathcal{V}$ is closed. □

The Cosine Vector

When the realization $V \in \mathcal{V}$ has inner product vector $\Sigma = (\sigma_0, \ldots, \sigma_r)$, we define

3C6
$$\Gamma = \Gamma(V) = (\gamma_0, \ldots, \gamma_r) := \sigma_0^{-1}\Sigma,$$

which we call the *cosine vector* of V (recall that $\sigma_0 > 0$ by assumption – see the notes at the end of the section). Thus Γ is the inner product vector of that scalar multiple of V which is normalized; in particular, $\Gamma(V) = \Sigma(V)$ if V is already normalized. We shall always write $\Gamma_0 := \Gamma(\{1\}) = (1^{r+1})$ for the cosine vector of the henogon. While most of our results are phrased in terms of cosine vectors, the more general inner product vectors will have their uses.

If $V \in \mathcal{N}$ is a pure realization other than $\{1\}$, then the centroid of V is the origin o. More generally, recall that V is centred if the centroid of V is the origin o. We clearly have the *layer equation*

3C7 Theorem *If V is a centred realization of $\mathcal{V}$ with cosine vector $\Gamma = (\gamma_0, \ldots, \gamma_r)$, then*

$$\langle \Lambda, \Gamma \rangle = \ell_0\gamma_0 + \cdots + \ell_r\gamma_r = 0.$$

More generally, if Γ is the cosine vector of any normalized realization $V \in \mathcal{N}$ of $\mathcal{V}$, then $\xi_0 := \langle \Lambda, \Gamma \rangle / |\Lambda|$ is the coefficient of $\Gamma(\{1\})$ in the expression of Γ as a convex combination of cosine vectors of pure realizations of $\mathcal{V}$.

We shall also refer to a cosine vector that satisfies the layer equation as *centred*.

Of course, for $V = \{1\}$, where $\Gamma(\{1\})$ has $r+1$ coordinates 1, the sum in Theorem 3C7 will be the number n of points of $\mathcal{V}$. The second claim of the theorem is a straightforward calculation. A weaker form of the claim, which is nevertheless often useful, is the *layer inequality*:

3C8 Corollary *If Σ is the inner product vector of a realization $V \in \mathcal{V}$, then*

$$\xi_0 := \langle \Sigma, \Lambda \rangle / |\Lambda| \geqslant 0$$

is the squared distance of the centroid of V from the origin o.

For the last claim, just recall the relationship between scaling of a realization V and the scaling of its inner product vector Σ, and the fact that blends of realizations correspond to sums of inner product vectors.

We further employ some useful conventions. The simplex realization $T \in \mathcal{N}$ of $\mathcal{V}$ (which we introduced in the proof of Theorem 3C5) has as its point-set an orthonormal basis $\{e_1, \ldots, e_n\}$ of $\mathbb{E}^n$. If $\mathcal{V}$ is centrally symmetric, with (say) $n = 2m$ points, then its *staurotope realization* $X \in \mathcal{N}$ has point-set $\{\pm e_1, \ldots, \pm e_m\}$; its *small simplex realization* $S \in \mathcal{N}$ is the simplex realization of the central quotient $\mathcal{V}/2$ with point-set $\{e_1, \ldots, e_m\}$. As a temporary notation, we write $\mathcal{X}, \mathcal{S} \subseteq \mathcal{N}$ for the subfamilies of components of X, S, respectively.

3C9 Remark We can always take the layer vector of a centrally symmetric point-set $\mathcal{V}$ to be palindromic, except that there may be more than one central layer; the 120-cell $\{5, 3, 3\}$ is a case in point, as [27, Table V(v)] illustrates (see also Section 7F). Realizations of $\mathcal{V}/2$, that is, components of S, will similarly have palindromic cosine vectors, while those of components of X will be anti-palindromic (that is, change sign on reversal of order).

Realization Domain

Just as we identify the realization cone $\mathcal{V}$ with the family of inner product vectors, so we can identify the realization domain $\mathcal{N}$ with the family of cosine vectors. Observe that the analogue of (3B10) for a general cosine vector in terms of those of pure realizations is $\Gamma = \lambda_0 \Gamma_0 + \cdots + \lambda_r \Gamma_r$, with $\lambda_s \geqslant 0$ for $s = 0, \ldots, r$ and $\lambda_0 + \cdots + \lambda_r = 1$, that is, a convex combination. Then we have

3C10 Theorem *The realization domain $\mathcal{N}$ is a compact convex set; it is the convex hull of the cosine vectors of the normalized pure realizations of $\mathcal{V}$, and is a pyramid with apex the henogon $\{1\}$.*

Proof. Since

$$\mathcal{N} = \{\Sigma = (\sigma_0, \ldots, \sigma_r) \in \mathcal{V} \mid \sigma_0 = 1\},$$

and $|\gamma_s| \leqslant 1$ for every coordinate γ_s of a cosine vector $\Gamma = (\gamma_0, \ldots, \gamma_r)$, it follows that $\mathcal{N}$ is bounded. Moreover, it is closed, either by the same argument that showed that $\mathcal{V}$ is closed, or by using the fact that it is the intersection of $\mathcal{V}$ with a hyperplane. Hence $\mathcal{N}$ is compact, and we already know that it is convex.

The next claim is just Carathéodory's Theorem 1B12 applied to $\mathcal{N}$, when we observe that the cosine vectors of pure realizations are exactly the extreme points of $\mathcal{N}$.

Last, if Γ is a general cosine vector, then there are unique λ with $0 \leqslant \lambda \leqslant 1$ and centred cosine vector Γ_1 such that $\Gamma = (1 - \lambda)\Gamma_0 + \lambda \Gamma_1$, where $\Gamma_0 = \Gamma(\{1\})$ as usual. □

We now begin to tie together the different realizations of $\mathcal{V}$. There is an important consequence of Corollary 1C8, which we refer to as the *component equation* (see also [85, Theorem 5.3], where it is called the *dimension* equation – see the notes at the end of the section).

3C11 Theorem *If the simplex realization T of $\mathcal{V}$ with n points is expressed as a blend $T = V_0 \# \cdots \# V_s$ of mutually orthogonal components V_j, then the cosine vectors $\Gamma_j = \Gamma(V_j)$ satisfy*

$$\sum_{i=1}^{s} d_i \Gamma_i = n\Gamma(T) = n(1, 0^r),$$

where $d_j := \dim V_j$ is the linear dimension for each j.

Proof. If $\Sigma_j = (\sigma_{j0}, \ldots, \sigma_{jr}) = \Sigma(V_j)$ for $j = 1, \ldots, s$, then the inner product vectors satisfy

$$\begin{aligned}(1, 0^r) = \Gamma(T) = \Sigma(T) &= \Sigma_1 + \cdots + \Sigma_s \\ &= \sigma_{10}\Gamma_1 + \cdots + \sigma_{s0}\Gamma_s.\end{aligned}$$

However, Corollary 1C8 tells us that the squared circumradius of V_j is $d_j/n = \sigma_{j0}$, and the claim of the theorem is an immediate consequence. □

We shall also refer to a cosine vector Γ_j of Theorem 3C11 as a *component*, here of the simplex realization, but more generally later of any inner product vector.

3C12 Remark An analogue of Theorem 3C11 holds for centrally symmetric realizations of a centrally symmetric set with $n = 2m$ points, as components of its staurotope realization X, except that now the corresponding expression is

$$\sum_{V_i \in \mathcal{X}} d_i \Gamma(V_i) = m\Gamma(X) = m(1, 0^{r-1}, -1),$$

where (conventionally) the last entry of the cosine vector represents the other point of the diameter from the initial point of $\mathcal{V}$ under the central involution $\boldsymbol{z}$. Similarly, there is an analogue for components of the small simplex realization S; here, the expression is

$$\sum_{V_i \in \mathcal{S}} d_i \Gamma(V_i) = m\Gamma(X) = m(1, 0^{r-1}, 1).$$

The pure realizations of $\mathcal{V}$ are thus components of either X or S.

Induced Cosine Vectors

As a symmetric set, a cut $(\mathcal{U}, \boldsymbol{H})$ of $(\mathcal{V}, \boldsymbol{G})$ has its own realization domain and cosine vectors. However, as a subset of $\mathcal{V}$, different diagonal classes of $\mathcal{U}$ may fall together. In such a case, different entries in a general cosine vector Γ of $\mathcal{U}$ will coincide; we then obtain an *induced* cosine vector $\Gamma_{\mathcal{U}}$ of $\mathcal{U}$.

3C13 Remark The symmetric layers $\mathcal{L}_j$ of $\mathcal{V}$ are clearly special sorts of cut. As we have noted, an asymmetric layer falls into two subsets, each of which is a cut; the layer itself is *not* a cut.

In any realization Ψ of $\mathcal{V}$, it it clear that the induced cosine vector $\Gamma_{\mathcal{U}}(U)$ of $U := \mathcal{U}\Psi$ must satisfy the layer inequality with respect to the layer vector $\Lambda(\mathcal{U})$. In fact, we have a stronger result, which is a refined version of Corollary 3C8 for U.

3C14 Theorem *If the cut $\mathcal{U}$ of $\mathcal{V}$ lies in layer s of $\mathcal{V}$ from the initial point, then for any realization $V \in \mathcal{N}$,*

$$\langle \Lambda(\mathcal{U}), U \rangle / m \geqslant \gamma_s(V)^2,$$

where $m :=$ card$\,\mathcal{U}$. Moreover, if $\mathcal{U}$ comprises the whole of layer $\mathcal{L}_s$ and the Wythoff space W has dimension $\dim W = 1$, then equality holds.

Proof. If $\Gamma(V) = (1, \gamma_1, \ldots, \gamma_r)$ is the cosine vector of $V \in \mathcal{V}$, then γ_s is the height (signed distance) of layer s from o, relative to the diameter of V through its initial point v, from which the claimed inequality follows.

For the last part, since the centroid of U is fixed by the stabilizer $\mathbf{H}$ of v, it must lie on the Wythoff space W. Hence, if $\dim W = 1$, then this centroid lies on the diameter of V through v, and so the inequality becomes an equation. $\square$

3C15 Remark More generally, in the latter part of Theorem 3C14, $\mathcal{U}$ could be one half of the equi-partition of an asymmetric layer.

For the moment, we shall not explore these ideas any further; instead, apart from Proposition 3J2, we shall postpone further discussion until Sections 3K and 4D.

Notes to Section 3C

1. We need to revert to diagonal vectors in Section 4F.
2. The inner product of two unit vectors is the cosine of the angle between them, which accounts for the term 'cosine vector'.
3. The term 'dimension equation' for Theorem 3C11 fails to convey what it really says, which is why we renamed it.

3D Examples

At this point, it is appropriate to illustrate the foregoing theory with a few simple examples. For these, only the layer and component equations of Theorems 3C7 and 3C11 will be needed. We make a forward reference to Chapter 7 for the regular polytopes of these examples, although we would be surprised if they were not already familiar to the reader.

3D1 Example The vertex-set $\mathcal{V}$ of the abstract regular d-simplex $\mathcal{T}^d$ has one pure realization V_1 apart from the henogon $\{1\}$; its cosine vector is

$$\Gamma_1 = (1, -\tfrac{1}{d}).$$

To see this most easily, note that the layer vector is clearly $\Lambda(\mathcal{V}) = (1, d)$, whence the expression for Γ_1 follows from the fact that the usual geometric regular d-simplex $\{3^{d-1}\}$ is centred.

3D2 Example We shall encounter several cases of the following example, and so it seems worthwhile to do certain calculations once and for all; this generalizes the previous Example 3D1. We take $\boldsymbol{G} = \boldsymbol{S}_n$, the symmetric group, acting on the symmetric hyperplane $\mathbb{L}^{n-1}$ of (1E15). Let k, m, n be such that $k + m = n$ with $1 \leqslant k \leqslant m$. The (unnormalized) centroids of the $(k-1)$-faces of a centred $(n-1)$-simplex can be taken to be all permutations of $(m^k, (-k)^m)$; this particular vector is the initial point, and is stabilized by $\boldsymbol{H} = \boldsymbol{S}_k \times \boldsymbol{S}_m$ (in the obvious way). For $r = 0, \ldots, k$, the vectors in layer $\mathcal{L}_r$ consist of all permutations in the first k and last m places of $(m^{k-r}, (-k)^r, m^r, (-k)^{m-r})$; hence the entries ℓ_r of the layer vector $\Lambda = (\ell_0, \ldots, \ell_k)$ satisfy

$$\ell_r = \binom{k}{r}\binom{m}{r}.$$

The squared norm of these vectors is

$$k \cdot m^2 + m \cdot (-k)^2 = km(m + k) = kmn.$$

The inner product of the initial vector with a vector in $\mathcal{L}_r$ is

$$(k - r)m^2 + 2rm(-k) + (m - r)(-k)^2;$$

the difference between this and the squared norm is thus

$$rm^2 + 2rmk + rk^2 = r(m + k)^2 = rn^2.$$

From this, it follows at once that the cosine vector $\Gamma = (\gamma_0, \ldots, \gamma_k)$ is such that

$$\gamma_r = 1 - \frac{rn^2}{kmn} = 1 - \frac{rn}{km}.$$

3D3 Remark Observe that various formulæ involving binomial coefficients are incidental consequences of this analysis. For instance,

$$\sum_{r \geqslant 0} \binom{k}{r}\binom{m}{r} = \binom{k+m}{k}$$

just counts the total number of points. Note also the symmetry between k and m, so that the same results hold for the centroids of the $(m-1)$-faces of the $(n-1)$-simplex.

3D4 Example The vertex-set $\mathcal{V}$ of the abstract regular d-staurotope $\mathcal{X}^d$ has two pure realizations V_1, V_2 apart from $\{1\}$; their cosine vectors are

$$\begin{aligned} \Gamma_1 &= (1, -\tfrac{1}{d-1}, 1), \\ \Gamma_2 &= (1, 0, -1). \end{aligned}$$

Here, the layer vector is $\Lambda(\mathcal{V}) = (1, 2(d-1), 1)$; V_2 is the vertex-set of the usual geometric d-staurotope $\mathsf{X}_2 := \{3^{d-2}, 4\}$, while V_1 realizes the $(d-1)$-collapse X_1 of $\mathcal{X}^d$ onto its facet $\mathcal{T}^{d-1}$ (see Proposition 2F20).

3D5 Example Our last example (for the time being) is more elaborate. The *difference body* of a convex body K is $\Delta K := K - K = K + (-K)$. For the d-simplex T^d, the vertex-set V_3 of ΔT^d consists of all $d(d+1)$ permutations of $(1, 0^{d-1}, -1)$ in the symmetric hyperplane $\mathbb{L}^d$, as defined in (1E15). (We shall see this again in Example 4A6.) The symmetry group $\mathbf{G}$ of V_3 consists of all permutations of the $d+1$ coordinates, augmented by the outer automorphism (or *twist*)

$$T\colon x \mapsto -(\xi_d, \ldots, \xi_0);$$

thus the group admits $-I$, which changes all the signs. The stabilizer $\mathbf{H}$ of the initial point $(1, 0^{d-1}, -1)$ consists of the permutations of $\xi_1, \ldots, \xi_{d-1}$, together with T. As a symmetric set $\mathcal{V}$, we thus see that its layer vector is

$$\Lambda = (1, 2(d-1), (d-1)(d-2), 2(d-1), 1).$$

Since $\mathcal{V}$ is centrally symmetric, we can identify opposite points of $\mathcal{V}$ in pairs to form $\mathcal{V}/2$. In general, it is not necessarily easy to see what results from such an identification, but in the present case it is: we just drop the minus signs from the vertices of V_3, and from the twist T. We end up with the $\frac{1}{2}d(d+1)$ permutations of $(1, 0^{d-1}, 1)$ (the mid-points of the edges of a regular d-simplex), and corresponding cosine vector (as a realization of the original point-set)

$$\Gamma = (1, \tfrac{1}{2}, 0, \tfrac{1}{2}, 1).$$

This realization is not centred; there is a component Γ_0 with coefficient

$$\frac{1}{d(d+1)}\langle \Lambda, \Gamma \rangle = \frac{1}{d(d+1)}\bigl(1 + 2(d-1)\cdot\tfrac{1}{2} + 2(d-1)\cdot\tfrac{1}{2} + 1\bigr) = \frac{2}{d+1}.$$

This leads to the centred realization V_1 with cosine vector

$$\Gamma_1 = \bigl(1, \tfrac{d-3}{2(d-1)}, -\tfrac{2}{d-1}, \tfrac{d-3}{2(d-1)}, 1\bigr),$$

so that $\Gamma = \frac{2}{d+1}\Gamma_0 + \frac{d-1}{d+1}\Gamma_1$.

Since $\mathcal{V}/2$ has only two non-trivial diagonals, we deduce that it has a further pure realization V_2, with cosine vector Γ_2 given by

$$\begin{aligned} \Gamma_2 &= \tfrac{2}{(d-2)(d+1)}\bigl(\tfrac{d(d+1)}{2}(1, 0, 0, 0, 1) - \Gamma_0 - d\Gamma_1\bigr) \\ &= \bigl(1, -\tfrac{1}{d-1}, \tfrac{2}{(d-2)(d-1)}, -\tfrac{1}{d-1}, 1\bigr). \end{aligned}$$

We now have just one more pure realization V_4 to find; its cosine vector is similarly given by

$$\begin{aligned}\Gamma_4 &= \tfrac{2}{d(d-1)}\big(\tfrac{d(d+1)}{2}(1,0,0,0,-1) - d\Gamma_3\big)\\ &= \big(1, -\tfrac{1}{d-1}, 0, \tfrac{1}{d-1}, -1\big).\end{aligned}$$

Summarizing, we can describe these pure realizations of $\mathcal{V}$ in the form of a *cosine matrix*, which just expresses the cosine vectors as rows of a matrix (see the notes at the end of the section). We usually state the corresponding layer vector, and also the *dimension vector*

3D6 $$D = D(\mathcal{V}) := (d_0, \dots, d_s),$$

which lists the corresponding dimension d_j of the realization V_j for $j = 0, \dots, s$. Conventionally, $d_0 = 1$ marks the trivial realization. In this case, the cosine matrix of the vertex-set of the abstract difference body of a d-simplex is

3D7 $$\begin{bmatrix} 1 & 1 & 1 & 1 & 1 \\ 1 & \frac{d-3}{2(d-1)} & -\frac{2}{d-1} & \frac{d-3}{2(d-1)} & 1 \\ 1 & -\frac{1}{d-1} & \frac{2}{(d-2)(d-1)} & -\frac{1}{d-1} & 1 \\ 1 & \frac{1}{2} & 0 & -\frac{1}{2} & -1 \\ 1 & -\frac{1}{d-1} & 0 & \frac{1}{d-1} & -1 \end{bmatrix},$$

with layer and dimension vectors

$$\begin{aligned}\Lambda &= (1, 2(d-1), (d-1)(d-2), 2(d-1), 1),\\ D &= (1, d, \tfrac{1}{2}(d-2)(d+1), d, \tfrac{1}{2}d(d-1)).\end{aligned}$$

3D8 Remark This trick of taking absolute values of coordinates is readily applicable when the symmetric group acts on them. We shall see it employed in the future.

Let us look at the particular case $d = 3$ in more detail (see the notes at the end of the section). Here, ΔT^3 is the quasi-regular cuboctahedron, whose vertex-set V_3 we can take (in a different coordinate system) to be all permutations of $(\pm 1, \pm 1, 0)$; it is the left picture in Figure 3D9. The generatrix (R_0, R_1, R_2) of its symmetry group is then that of the cube $\{4, 3\}$, namely,

$$\begin{aligned} R_0 &:= \bar{1}: & x &\mapsto (-\xi_1, \xi_2, \xi_3),\\ R_1 &:= (1\ 2): & x &\mapsto (\xi_2, \xi_1, \xi_3),\\ R_2 &:= (2\ 3): & x &\mapsto (\xi_1, \xi_3, \xi_2).\end{aligned}$$

The Wythoff space is $R_0 \cap R_2$, with the R_j as mirrors. Note, by the way, the abbreviated notation for these symmetries.

3D9

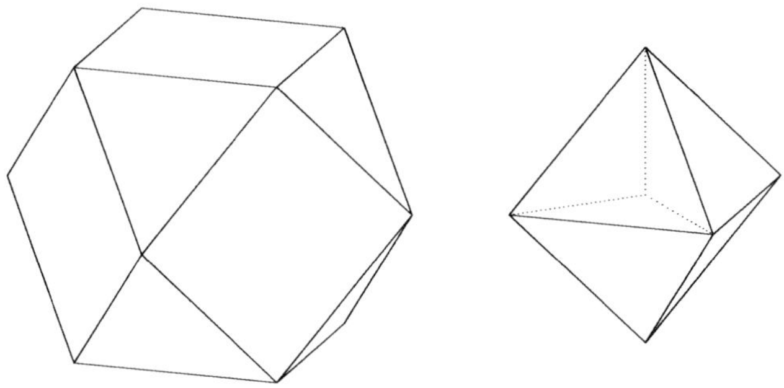

Cuboctahedron and hemi-cuboctahedron

The realization V_2 consists of the vertices of the octahedron, namely, all permutations of $(\pm 1, 0, 0)$; this is the right picture in Figure 3D9. Its generatrix (S_0, S_1, S_2) is given by

$$\begin{aligned} S_0 &:= \overline{2}\,\overline{3} = -R_0, \\ S_1 &:= (1\ 2) = R_1, \\ S_2 &:= (2\ 3) = R_2. \end{aligned}$$

The original tetragons of the cuboctahedron become the diametral squares of the octahedron, while its trigons form alternate triangles of the octahedron. This quasi-regular polyhedron is [32, no. 36], denoted there $\frac{3}{2}\ 3 \mid 2$. For V_1, we have the further identification to the vertices of a trigon.

The final realization V_4 is of much interest. As a set of points, it is the same as the original V_3. However, the generatrix (T_0, T_1, T_2) is now

$$\begin{aligned} T_0 &:= -\overline{1} = -R_0, \\ T_1 &:= R_1, \\ T_2 &:= R_2. \end{aligned}$$

The trigons now have alternate vertices of diametral hexagons of the original cuboctahedron, while the tetragons are now skew polygons, a typical one of which has successive vertices

$$(1, 0, 1),\ (-1, 1, 0),\ (1, 0, -1),\ (-1, -1, 0).$$

Observe here that cyclic permutation of the coordinates gives three such skew polygons abutting the triangle with vertices

$$(-1, 1, 0),\ (1, 0, -1),\ (0, -1, 1).$$

We have thus obtained a 'combinatorial' cuboctahedron quite different from the original.

3D10

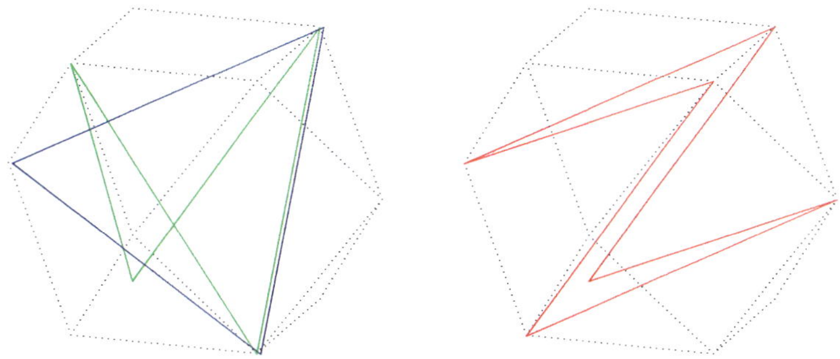

The alternative cuboctahedron

The first picture in Figure 3D10 shows an adjacent trigon (in blue) and tetragon (in green); these determine two hexagons adjacent to both (replacing diametral hexagons in the original), one of which is shown (in red) in the second picture.

Notes to Section 3D

1. The cosine matrix will play a very important rôle in realization theory, as will be seen in Section 3F. We shall extend its definition, and that of the dimension vector, in Section 3J.
2. The curious reader will have checked that the index of **H** in **G** in Example 3D5 is $d(d+1)$, since T belongs neither to the symmetric group $\boldsymbol{S}_{d+1}$ nor to its subgroup $\boldsymbol{S}_{d-1}$. However, on removing the sign, the twist does now belong to $\boldsymbol{S}_{d+1}$, but still not, of course, to the subgroup.
3. The analysis of Example 3D5 extends to the case $d = 2$; the fact that the middle entries of both layer vector Λ and dimension vector D vanish cancels the 'infinity' in the middle entry of Γ_2. Thus the middle row and column of the cosine matrix of (3D7) are absent. We shall look at realizations of regular polygons in general in Section 4G. The cases $d = 4, 5$ will be of interest later, since the vertex-sets are also those of regular polytopes of nearly full rank; for these, see Sections 11B and 13E.
4. The case $d = 5$ of Example 3D5 occurs in Section 15B, where the realizations of the dual to the polytope of Section 13E are also described.

3E Products of Realizations

We introduced tensor products in Section 1A; with that background in mind, we are now ready to define the product of realizations. It may help the reader to bear in mind that, as a consequence of the definition of the tensor product, there are quite close parallels – combinatorially speaking – between the definitions of blend and product; we shall say more about this a little later.

Let Ψ_j be a realization of $(\mathcal{V}, \boldsymbol{G})$ for $j = 1, 2$. The *(tensor) product* of Ψ_1 and Ψ_2 is denoted $\Psi_1 \otimes \Psi_2$, and is defined by

3E1
$$\boldsymbol{x}(\Psi_1 \otimes \Psi_2) := \boldsymbol{x}\Psi_1 \otimes \boldsymbol{x}\Psi_2$$

for each $\boldsymbol{x} \in \mathcal{V}$.

3E2 Remark Since $\mathbb{X} \otimes \mathbb{Y} \cong \mathbb{Y} \otimes \mathbb{X}$ under the twist $x \otimes y \mapsto y \otimes x$, in view of the fact that we identify congruent realizations we see that $\Psi_1 \otimes \Psi_2 = \Psi_2 \otimes \Psi_1$. Similarly, the associativity of tensor products (up to isomorphism) shows that the product of realizations is associative. If we identify a realization Ψ with the image $V = \mathcal{V}\Psi$ in the same spirit as before, then we write the product as $V_1 \otimes V_2$.

3E3 Remark It hardly needs emphasizing that, for symmetry reasons, we can only define the product of finite point-sets lying on spheres centred at the origin; in particular (looking to the future), we cannot take a product of infinite sets.

For scalar multiples and blends, the next result is obvious.

3E4 Lemma *If V, V_1, V_2 are realizations of $\mathcal{V}$ and $\lambda \in \mathbb{R}$, then*

$$\begin{aligned}(\lambda V_1) \otimes V_2 &= \lambda(V_1 \otimes V_2),\\ V \otimes (V_1 \# V_2) &= (V \otimes V_1) \# (V \otimes V_2).\end{aligned}$$

Now we recall Theorem 3C5, which lets us identify a realization V with its inner product vector Σ. Following [85], we define (for our present purposes) the *product* of two real vectors $a = (\alpha_0, \ldots, \alpha_r)$ and $b = (\beta_0, \ldots, \beta_r)$ by

3E5
$$ab := (\alpha_0\beta_0, \ldots, \alpha_r\beta_r);$$

that is, we just multiply coordinates term by term. Bearing in mind that the inner product in $\mathbb{E}_1 \otimes \mathbb{E}_2$ is given by (1C11), namely, if $x_j, y_j \in \mathbb{E}_j$ for $j = 1, 2$, then

$$\langle x_1 \otimes x_2, y_1 \otimes y_2 \rangle = \langle x_1, y_1 \rangle \langle x_2, y_2 \rangle,$$

there follows at once

3E6 Proposition *The inner product and cosine vectors of a product $V_1 \otimes V_2$ are given by*

$$\begin{aligned}\Sigma(V_1 \otimes V_2) &= \Sigma(V_1)\Sigma(V_2),\\ \Gamma(V_1 \otimes V_2) &= \Gamma(V_1)\Gamma(V_2).\end{aligned}$$

We saw in Section 1A that the direct sum $\mathbb{K}_1 \oplus \mathbb{K}_2$ of algebras $\mathbb{K}_j$ over $\mathbb{R}$ for $j = 1, 2$ is such that

$$\begin{aligned}(a_1, a_2) + (b_1, b_2) &= (a_1 + b_1, a_2 + b_2),\\ \lambda(a_1, a_2) &= (\lambda a_1, \lambda a_2),\\ (a_1, a_2)(b_1, b_2) &= (a_1 b_1, a_2 b_2).\end{aligned}$$

It will be immediately clear that we must take the unity or henogon $\{1\}$ into account. Recall that $\{1\}$ consists of the single point $1 \in \mathbb{R}$, and so has linear dimension 1. It is obvious that $\{1\}$ is (as its name indicates) the multiplicative unity with respect to $\otimes$. Putting together the parts of the previous discussion, we arrive at

3E7 Proposition *The algebra of inner product vectors of realizations of $\mathcal{V}$ is that of the direct sum $\mathbb{R} \oplus \cdots \oplus \mathbb{R}$ of $r + 1$ copies of the real numbers $\mathbb{R}$.*

3E8 Remark It is important to bear in mind that the product of cosine vectors is again a cosine vector, and so must satisfy the layer inequality of Corollary 3C8. While non-normalized realizations do occur subsequently, our calculations are mainly carried out on cosine vectors.

The tensor product of two representations of a group splits into irreducible representations with multiplicity. However, (3B9) shows that we may be able to combine pure realizations, so that there is no analogue of 'with multiplicity' for products of realizations. Nevertheless, since an arbitrary realization can be expressed as a blend of pure realizations (not necessarily uniquely), and in view of Lemma 3E4 we may confine our attention to products of pure realizations, there is an immediate implication for the product. Thus we have a *structure theorem*:

3E9 Theorem *For each pair V_1, V_2 of pure realizations of $\mathcal{V}$, for some k there are pure realizations $U_1, \ldots, U_k$ of $\mathcal{V}$ such that*

$$V_1 \otimes V_2 = U_1 \# \cdots \# U_k.$$

3E10 Remark It is important to note that (with some scaling) $\{1\}$ may be a component U_j here. In practical terms, the expression of a product $V_1 \otimes V_2$ in terms of pure realizations $U_1, \ldots, U_k \in \mathcal{N}$ will be by cosine vectors, as a convex combination of the form

$$\Gamma(V_1 \otimes V_2) = \Gamma(V_1)\Gamma(V_2) = \mu_1\Gamma(U_1) + \cdots + \mu_k\Gamma(U_k),$$

for some $\mu_1, \ldots, \mu_k \geqslant 0$ with $\mu_1 + \cdots + \mu_k = 1$.

Theorem 3E9 implies that, in analogy to the character table of a group, we may draw up a table of decompositions of products of pure realizations of a symmetric set $\mathcal{V}$. However, we still have some way to go before we are in a position to do this.

In the centrally symmetric case, the following observation is very useful.

3E11 Theorem *If the symmetric set $\mathcal{V}$ is centrally symmetric, and V_1, V_2 are two pure realizations of $\mathcal{V}$, then $V_1 \otimes V_2$ is a realization of $\mathcal{V}/2$ if V_1 and V_2 are either both or neither centrally symmetric; otherwise, $V_1 \otimes V_2$ is centrally symmetric.*

Proof. The theorem follows from the simple observations that $(-x) \otimes (-y) = x \otimes y$, which identifies opposite points if V_1 and V_2 are centrally symmetric, and that $(\pm x) \otimes y = \pm(x \otimes y)$ gives pairs of opposite points if (say) V_1 is centrally symmetric but V_2 is not. □

It is clear that two realizations $U, V \in \mathcal{V}$ satisfy

3E12 $$\dim(U \otimes V) \leqslant \dim U \cdot \dim V.$$

However, when $U = V$ we have something stronger. Part (b) of the next result is useful in several contexts; we shall give a refined version of part (a) in Lemma 3F3.

3E13 Proposition *If $V \in \mathcal{N}$ is d-dimensional with $d > 1$, then*

(a) *$V \otimes V$ always has a component $\{1\}$ in its expression in terms of pure realizations,*

(b) *the non-trivial component of $V \otimes V$ has dimension at most $\frac{1}{2}(d-1)(d+2)$.*

Proof. For the first claim, note that $\varGamma(V \otimes V) = \varGamma(V)^2$ is a non-negative vector with leading entry 1 for every V; thus $\langle \varGamma(V \otimes V), \varLambda \rangle > 0$, so that $V \otimes V$ indeed has a non-zero component $\{1\}$. For the second claim, note that, if $x \in \mathbb{E}^d$, then $\frac{1}{2}d(d-1)$ pairs of coordinates of $x \otimes x$ coincide. Subtracting the component $\{1\}$, it thus follows that the complementary component of $V \otimes V$ to $\{1\}$ has dimension at most $d^2 - \frac{1}{2}d(d-1) - 1 = \frac{1}{2}(d-1)(d+2)$. □

It is a consequence of the foregoing discussion that, knowing that cosine vectors determine the geometry of realizations, we often just manipulate them without concerning ourselves overmuch with what the corresponding realizations actually look like; this is particularly the case when products are involved. Indeed, we sometimes have little idea of the geometry; the description of the realization domain of the 600-cell $\{3, 3, 5\}$ in Section 7E is a case in point, and that of the 120-cell $\{5, 3, 3\}$ in Section 7K even more so.

3F $\varLambda$-Orthogonality

Retaining the previous conventions, we now introduce a new positive definite *$\varLambda$-inner-product* $\langle \cdot, \cdot \rangle_\varLambda$ on $\mathbb{R}^{r+1}$, defined by

3F1 $$\langle x, y \rangle_\varLambda := \langle \varLambda, xy \rangle / n = \langle \varLambda, xy \rangle / |\varLambda|,$$

where $x = (\xi_0, \ldots, \xi_r)$, $y = (\eta_0, \ldots, \eta_r)$, and $xy = (\xi_0\eta_0, \ldots, \xi_r\eta_r)$ is the product of vectors defined by (3E5). Thus $\langle \cdot, \cdot \rangle_\varLambda$ is just the ordinary inner product with respect to the diagonal matrix $n^{-1} \operatorname{diag}(\ell_0, \ldots, \ell_r)$. There is a corresponding *$\varLambda$-norm*, given by $\|x\|_\varLambda^2 := \langle x, x \rangle_\varLambda$. In addition, we naturally say that two vectors x, y are *$\varLambda$-orthogonal* if $\langle x, y \rangle_\varLambda = 0$.

3F2 Remark A cosine vector Γ induces the linear functional $\langle \cdot, \Gamma \rangle_\Lambda$ on $\mathcal{N}$. Following Remark 1B16, we can talk about Γ separating other cosine vectors; this will be a useful concept later on when we try to determine realization domains.

Our main result will follow directly from the following.

3F3 Lemma *If $V \in \mathcal{V}$ is a realization of $\mathcal{V}$ with cosine vector Γ, then $\|\Gamma\|_\Lambda^2 \geqslant 1/\dim V$.*

Proof. Suppose that $\dim V = d$, so that V sits in $\mathbb{E} := \mathbb{E}^d$ (in this context, recall that $\dim\{1\} = 1$). If $E := \{e_1, \ldots, e_d\}$ is any orthonormal basis of $\mathbb{E}$, then the elements $e_j \otimes e_k$ form an orthonormal basis of $\mathbb{E} \otimes \mathbb{E}$, and the coordinates of $x \otimes y$ with respect to this basis are the products $\xi_j \eta_k$ of the coordinates of x and y with respect to E. In particular, for any $x \in \mathbb{E}$ the sum of the diagonal coordinates of $x \otimes x$ is $\xi_1^2 + \cdots + \xi_d^2 = \|x\|^2$; this is 1 for any point $x \in \operatorname{vert} V$ of a (normalized) realization. This tells us that, if $u := e_1 \otimes e_1 + \cdots + e_d \otimes e_d$, then the points $x \otimes x$ lie in a hyperplane $\{y \in \mathbb{E} \otimes \mathbb{E} \mid \langle y, u \rangle = 1\}$ fixed under the induced action of the realization (it is easy to check this directly); observe that $\|u\| = \sqrt{d}$.

We conclude from this that the points of $V \otimes V$ lie in a hyperplane at distance at least $1/\sqrt{d}$ from o; observe that these points need not be centred on the line $\operatorname{lin}\{u\}$ spanned by u. Considering the component of $\{1\}$ in $V \otimes V$ (see Corollary 3C8), in terms of cosine vectors this says that

$$\|\Gamma\|_\Lambda^2 = \frac{1}{n}\langle \Lambda, \Gamma^2 \rangle \geqslant \frac{1}{d},$$

which is the claim of the lemma. □

Lemma 3F3 bounds the norm in terms of the dimension. Rephrasing the lemma to bound the dimension in terms of the norm can prove useful in analysing a realization domain; we call the resulting expression the *dimension inequality*.

3F4 Corollary *If $V \in \mathcal{V}$ is a realization of $\mathcal{V}$ with cosine vector Γ, then $\dim V \geqslant \|\Gamma\|_\Lambda^{-2}$.*

We now apply Lemma 3F3 to Theorem 3C11 to obtain the *Λ-orthogonality theorem.*

3F5 Theorem *If $\mathcal{V}$ is a symmetric point-set with layer vector Λ whose simplex realization T is expressed as a blend $T = V_1 \# \cdots \# V_s$ of mutually orthogonal components V_j with dimensions $d_j := \dim V_j$ and cosine vectors $\Gamma_j := \Gamma(V_j)$ for $j = 1, \ldots, s$, then*

$$\langle \Gamma_i, \Gamma_j \rangle_\Lambda = \frac{\delta_{ij}}{d_j}$$

for all $i, j = 1, \ldots, s$.

Here, δ_{ij} is the familiar Kronecker delta function.

Proof. Bear in mind here the definition of cosine vector in (3C6). Take the Λ-inner product of the component equation of Theorem 3C11 with Γ_j. If $\mathcal{V}$ has n points, then we get

$$1 = \langle n(1, 0^r), \Gamma_j \rangle_\Lambda = \Big\langle \sum_{i=1}^{s} d_i \Gamma_i, \Gamma_j \Big\rangle_\Lambda \geqslant d_j \cdot \frac{1}{d_j} = 1,$$

from Lemma 3F3 and the fact that $\langle \Gamma_i, \Gamma_j \rangle_\Lambda \geqslant 0$ for all $i \neq j$ (see Remark 3E8). Thus we must have equality throughout, namely, $\|\Gamma_j\|_\Lambda^2 = 1/d_j$ and $\langle \Gamma_i, \Gamma_j \rangle_\Lambda = 0$ for all $i \neq j$. This is just what the theorem asserts. □

Of course, the layer equation of Theorem 3C7 is the particular case $\langle \Gamma_0, \Gamma \rangle_\Lambda = 0$ for the cosine vector Γ of a centred realization; recall that $\Gamma_0 = \Gamma(\{1\}) = (1^{r+1})$, as before.

Observe that we have said nothing yet about purity, although obviously we have in mind the case when the V_j in Theorem 3F5 are pure. As a particular example, we have

3F6 Corollary *If Γ is the cosine vector of a d-dimensional pure realization $V \in \mathcal{N}$ of $\mathcal{V}$, then $\|\Gamma\|_\Lambda^2 = \frac{1}{d}$; moreover, this is the coefficient of Γ_0 in the expression of Γ^2 as a convex combination of pure cosine vectors.*

Recall that, except for the trivial realization $\{1\}$, all pure realizations are centred. Our next observation is thus

3F7 Corollary *If $V_1, V_2 \in \mathcal{V}$ are orthogonal realizations of $\mathcal{V}$ as components of the simplex realization T, then the product $V_1 \otimes V_2$ is centred.*

The condition $\|\Gamma(P)\|_\Lambda^2 = 1/\dim V$ gives a *purity test* for a realization V. However, many blended realizations also pass this test; the precise situation is as follows.

3F8 Proposition *With the assumptions of Theorem 3F5, let $V_1, \ldots, V_k \in \mathcal{N}$ be distinct realizations of $\mathcal{V}$ which are orthogonal as components of the simplex realization T, and for $j = 1, \ldots, k$ let V_j have dimension d_j and cosine vector Γ_j. Set $d := d_1 + \cdots + d_k$. If $\Gamma = \xi_1\Gamma_1 + \cdots + \xi_k\Gamma_k$, for varying $\xi_1, \ldots, \xi_k \geqslant 0$ with $\xi_1 + \cdots + \xi_k = 1$, then $\|\Gamma\|_\Lambda^2$ achieves its minimum $1/d$ just when $\xi_j = d_j/d$ for each j.*

Proof. We have

$$\|\Gamma\|_\Lambda^2 = \sum_{j=1}^{k} \xi_j^2 \|\Gamma_j\|_\Lambda^2 = \sum_{j=1}^{k} \frac{\xi_j^2}{d_j},$$

where we have used Theorem 3F5 to eliminate the cross terms. An easy exercise in calculus yields the unique minimum $1/d$ at $\xi_j = d_j/d$ for $j = 1, \ldots, k$, as claimed. Note that the point-set V with this minimal cosine vector Γ has dimension d. □

We end the section with converses to Theorem 3C11.

3F9 Theorem *If $V_j \in \mathcal{V}$ has dimension d_j and cosine vector Γ_j for $j = 1, \ldots, k$, and*

$$\sum_{i=1}^{k} d_i \Gamma_i = n(1, 0^r),$$

then suitably scaled multiples of $V_1, \ldots, V_k$ are mutually orthogonal components of T. Consequently, $\Gamma_1, \ldots, \Gamma_k$ are mutually Λ-orthogonal.

Proof. Indeed, if we assume that each $V_j \in \mathcal{N}$, then we have $\sqrt{d_1}V_1 \# \cdots \# \sqrt{d_k}V_k = \sqrt{n}T$, with $d_1 + \cdots + d_k = n$. Thus the components V_j must be mutually orthogonal, and an appeal to Theorem 3F5 completes the proof. □

We further have what is called the *complementarity criterion.*

3F10 Theorem *If $V_j \in \mathcal{V}$ has dimension d_j, with the corresponding cosine vectors Γ_j for $j = 1, \ldots, k-1$ mutually Λ-orthogonal, and if and Γ_k d_k are such that*

$$\sum_{i=1}^{k} d_i \Gamma_i = n(1, 0^r), \quad \sum_{i=1}^{k} d_i = n,$$

then Γ_k is the cosine vector of another realization $V_k \in \mathcal{V}$.

Proof. All the theorem says is that (with suitable scaling) V_k is the complement in the simplex realization T of $V_1 \# \cdots \# V_{k-1}$. □

3F11 Remark In practical applications of Theorem 3F10, we usually have extra information; for instance, we may already know that Γ_k is Λ-orthogonal to Γ_j for $j = 1, \ldots, k-1$.

3F12 Remark There are analogous results for components of the staurotope realization X when $\mathcal{V}$ is centrally symmetric, and for those of the small simplex realization S of the central quotient $\mathcal{V}/2$; compare Remark 3C12. Note that cosine vectors of a component of X and a component of S are automatically Λ-orthogonal.

Notes to Section 3F

1. Theorem 3F5 can be proved directly by a suitable choice of coordinates for each V_j. Taking the initial point v_1 (say) to be $v_1 = (1, 0, \ldots, 0)$, the leading entry of each v_j is the corresponding entry of the cosine vector of V_j. The scaling of V_j by ρ_j and orthogonality of the resulting square matrix then results in the theorem.
2. Though there are close parallels with representation theory, it is not clear how Theorem 3F5 can be deduced from (for example) [122, (III.1.2)] when **G** is not real in the sense of Section 3M (see also [122, II.6]).

3G The Wythoff Space

A crucial rôle in the discussion concerns the realizations of $\mathcal{V}$ associated with a fixed irreducible orthogonal representation $\mathbf{G}$ of G. Suppose that the ambient space $\mathbb{E}$ on which $\mathbf{G}$ acts has dimension $d = d_{\mathbf{G}}$ (in other words, d is the degree of the representation). Further, let the Wythoff space W of $\mathbf{G}$ have dimension $w = w_{\mathbf{G}}$. As we have seen, we obtain a realization of $\mathcal{V}$ by applying Wythoff's construction to a point $v \in W$. The realization will be pure unless $v = o$, when it will be trivial. We denote by $\mathcal{V}_{\mathbf{G}}$ the subfamily of $\mathcal{V}$ consisting of all possible blends of these realizations with symmetry group $\mathbf{G}$.

We then have

3G1 Theorem *The cone $\mathcal{V}_{\mathbf{G}}$ is a closed subcone of $\mathcal{V}$. Moreover, $\mathcal{V}$ is the direct sum of the subcones $\mathcal{V}_{\mathbf{G}}$, where $\mathbf{G}$ ranges over the distinct irreducible orthogonal representations of the automorphism group G.*

We already know from Theorem 3F5 that distinct representations of the automorphism group G give rise to realizations whose cosine vectors are Λ-orthogonal, so that our next task is to concentrate on the subcone $\mathcal{V}_{\mathbf{G}}$. As we said earlier, behind Theorem 3C11 lies Theorem 1C7. In the present context, it implies that every pure component in $\mathcal{V}_{\mathbf{G}}$ of the simplex realization T will have squared circumradius d/n. It is thus convenient for the rest of the section tacitly to scale by $\sqrt{n/d}$; we then take pure components of T in $\mathcal{V}_{\mathbf{G}}$ to be normalized, and so to lie in the subdomain $\mathcal{N}_{\mathbf{G}} \subseteq \mathcal{N}$ corresponding to $\mathcal{V}_{\mathbf{G}}$.

We begin with an almost obvious remark.

3G2 Lemma *In any expression of the simplex realization T as a blend of pure realizations, the number $m = m_{\mathbf{G}}$ of these whose symmetry group is the irreducible representation $\mathbf{G}$ of G is independent of the particular expression.*

Proof. Indeed, if (up to scaling) the component of T in $\mathcal{V}_{\mathbf{G}}$ is $V_1 \# \cdots \# V_m$, then this component depends only on $\mathbf{G}$, and so

$$\dim(V_1 \# \cdots \# V_m) = md$$

must be independent of the particular expression of the component. Hence m is independent of it as well. □

We shall refer to $m_{\mathbf{G}}$ as the *multiplicity* of the subcone $\mathcal{V}_{\mathbf{G}}$; see the notes at the end of the section.

The Centralizer

The representation $\mathbf{G}$ of G (in the d-dimensional space $\mathbb{E}$, say) may have a non-trivial centralizer in the corresponding orthogonal group $\mathbf{O} = \mathbf{O}(\mathbb{E})$, that is, other than $\pm I$. It is helpful to generalize this notion a little. So, we define $\mathbf{Z} = \mathbf{Z}(\mathbf{G})$ to consist of all the linear mappings $\Theta \colon \mathbb{E} \to \mathbb{E}$ such that $\Theta\Phi = \Phi\Theta$ for each $\Phi \in \mathbf{G}$. It is clear that the usual addition and multiplication of linear mappings turn $\mathbf{Z}$ into a ring; indeed, $\mathbf{Z}$ is an $\mathbb{R}$-algebra. Then we have the following (compare [122, II.6]).

3G3 Proposition *If* $\mathbf{G}$ *is a real irreducible representation of* $\boldsymbol{G}$*, then there is a division ring* $\mathbb{D} = \mathbb{D}_{\mathbf{G}} = \mathbb{R}, \mathbb{C}$ *or* $\mathbb{H}$ *and a ring isomorphism* $Z\colon \mathbb{D} \to \mathbf{Z}(\mathbf{G})$*. Hence,* $\mathbb{E}$ *has the structure of a* $\mathbb{D}$*-space under scalar multiplication defined by*

$$\boldsymbol{\lambda} x := xZ(\boldsymbol{\lambda})$$

for $\boldsymbol{\lambda} \in \mathbb{D}$ *and* $x \in \mathbb{E}$.

We express things this way in the spirit of [122]; note that our convention of scalar multiplication on the left and application of mappings on the right fits in with the proposition. We have already referred to $\mathbb{D}$ as the *centralizer* of $\mathbf{G}$; we shall call $c = c_{\mathbf{G}} := \operatorname{rank}_{\mathbb{R}} \mathbb{D}$ its *centrality*, so that $c = 1, 2$ or 4.

3G4 Remark It follows from Proposition 3G3 that $w = w_{\mathbf{G}} := \dim W = cm$. However, except for illustrative examples, we shall find that in all cases where we actually describe realization domains we shall have $m = w$.

A core concept is the notion of linear combinations of point-sets with the same symmetry group. More generally, if $U = (u_1, \ldots, u_n)$ and $V = (v_1, \ldots, v_n)$ are two ordered sets in $\mathbb{E}$, we define the *sum* to be $U+V := (u_1+v_1, \ldots, u_n+v_n)$ and *scalar multiple* by $\lambda \in \mathbb{R}$ to be $\lambda V := (\lambda v_1, \ldots, \lambda v_n)$. Let $W = W_{\mathbf{G}} \leqslant \mathbb{E}$ denote the Wythoff space of $\mathbf{G}$. If $U, V \in \mathcal{V}_{\mathbf{G}}$ have initial points $u, v \in W$ (not generally unit vectors) and $\mu, \nu \in \mathbb{R}$, then $\mu u + \nu v \in W$ is the initial point of the *linear combination* $\mu U + \nu V$ with the same group, since we readily see that, if $x \leftrightarrow y$ are corresponding points of U and V, then $\mu x + \nu y$ is that of $\mu U + \nu V$.

Taking linear combinations (with $\mathbf{G}$ fixed) will obviously commute with blends and products, and so we have

3G5 Theorem *If* V_1, V_2 *are realizations of* $\mathcal{V}$ *with the same symmetry group, and hence with the same Wythoff space, and* V *is a further realization of* $\mathcal{V}$*, then*

$$\begin{aligned} V \# (\mu_1 V_1 + \mu_2 V_2) &= \mu_1 (V \# V_1) + \mu_2 (V \# V_2), \\ V \otimes (\mu_1 V_1 + \mu_2 V_2) &= \mu_1 (V \otimes V_1) + \mu_2 (V \otimes V_2), \end{aligned}$$

for all $\mu_1, \mu_2 \in \mathbb{R}$.

The key to the main result of this section employs an orthogonal mapping similar to the reflexion of (1D9); as we know from Theorem 1D10, such reflexions generate an appropriate orthogonal group.

3G6 Lemma *If* U, V *are realizations of* $\mathcal{V}$ *in* $\mathbb{E}$ *with the same symmetry group* $\mathbf{G}$*, and* $\mu, \nu \in \mathbb{R}$ *satisfy* $\mu^2 + \nu^2 = 1$*, then*

$$(\mu U + \nu V) \# (\nu U - \mu V) = U \# V.$$

Proof. First note that $\mathbf{G}$ acts on $\mathbb{E} \oplus \mathbb{E}$ by $(x, y)\Phi := (x\Phi, y\Phi)$ for $x, y \in \mathbb{E}$ and $\Phi \in \mathbf{G}$. The mapping Θ on $\mathbb{E} \oplus \mathbb{E}$ given by

$$(y, z)\Theta := (\mu y + \nu z, \nu y - \mu z)$$

is orthogonal (and involutory). If the two realizations are given for $\boldsymbol{x} \in \mathcal{V}$ by $\boldsymbol{x} \mapsto y, z$ respectively, so that their blend is $\boldsymbol{x} \mapsto (y, z)$, then applying Θ gives a congruent realization, which we therefore identify with the original one. In other words,

$$(\mu U + \nu V) \# (\nu U - \mu V) = (U \# V)\Theta = U \# V,$$

as the lemma claims. □

3G7 Remark As remarked in [99, Section 5B], the irreducibility of $\mathbf{G}$ (or purity of U and V) is unimportant here.

We next explain how we know whether we have a centralizer $\mathbb{D} \neq \mathbb{R}$. Let us suppose that two unit vectors $u, v \in W$ with $u \neq \pm v$ nevertheless give the same realization $V = V(u) = V(v)$ (of course, up to congruence). Lemma 3G6 then says that

$$V(\mu u + \nu v) \# V(\nu u - \mu v) = V(u) \# V(v) = \sqrt{2}V$$

whenever $\mu^2 + \nu^2 = 1$, again up to congruence. But if $w \in \text{lin}\{u, v\}$ is any unit vector, then there are such μ, ν and a suitable λ with $w = \lambda(\mu u + \nu v)$, so that

$$V(\lambda(\mu u + \nu v)) \# V(\lambda(\nu u - \mu v)) = \lambda\sqrt{2}V.$$

In other words, $V(w)$ is a component of the pure realization $\lambda\sqrt{2}V$, from which follows $V(w) = V$ (since both are normalized).

What we have begun to show is

3G8 Proposition *If* $\mathbf{G}$ *is an irreducible representation on* $\mathbb{E}$*, and there are unit vectors* $u, v \in W_{\mathbf{G}}$ *with* $u \neq \pm v$ *such that the corresponding realizations* $V(u)$ *and* $V(v)$ *are congruent, then* $\mathbb{E}$ *admits an action by the complex numbers of absolute value* 1 *which centralizes* $\mathbf{G}$.

Proof. If $\{e_1, e_2\}$ is an orthonormal basis of $W' := \text{lin}\{u, v\}$, then the given conditions imply that every unit vector $v(\vartheta) := \cos\vartheta e_1 + \sin\vartheta e_2 \in W'$ yields a realization $V(\vartheta) \cong V(0)$. Since $\mathbf{G}$ is irreducible, $V(\vartheta) = \{v(\vartheta)\Phi \mid \Phi \in \mathbf{G}\}$ spans $\mathbb{E}$ for each ϑ. On the other hand, since the $V(\vartheta)$ are congruent, the mapping $0 \mapsto \vartheta$ induces a mapping $Z(\vartheta) \in \mathbf{O}(\mathbb{E})$ given by $V(0)Z(\vartheta) = V(\vartheta)$. The orthogonal mapping $Z(\vartheta)$ clearly commutes with each $\Phi \in \mathbf{G}$, and thus exhibits the required centralizing action on $\mathbb{E}$. □

We give a couple of simple examples to illustrate centralizers.

3G9 Example An arbitrary rotation in $\mathbf{SO}_2$ obviously commutes with any (finite) cyclic subgroup of $\mathbf{SO}_2$.

3G10 Example Consider the group $\mathbf{V} = \{\pm 1, \pm\mathbf{i}, \pm\mathbf{j}, \pm\mathbf{k}\}$ of quaternions, acting on the right of $\mathbb{E}^4$ regarded as the space of all quaternions. Then left multiplication $\mathbf{x} \mapsto \mathbf{gx}$ by any quaternion $\mathbf{g} \in \mathbf{Q}$ clearly commutes with the action of $\mathbf{V}$. The abstract situation here is of the right action of $\mathbf{V}$ on itself, so that this is a realization of dimension 4 with $w = 4$ but $m = 1$.

The remaining non-trivial pure realizations are given by the action of $\mathbf{V}$ on its subgroups $\langle \mathbf{g} \rangle$ with $\mathbf{g} = \mathbf{i}, \mathbf{j}, \mathbf{k}$.

If $\mathbb{D}$ is as in Proposition 3G3, then we call two vectors $u, v \in \mathbb{E}$ *strongly orthogonal* if the subspaces $\mathbb{D}u$ and $\mathbb{D}v$ are orthogonal. Similarly, we say that $u_1, \ldots, u_k$ *strongly span* a subspace $L \leqslant \mathbb{E}$ if $L = \mathbb{D}u_1 + \cdots + \mathbb{D}u_k$. By a strong orthonormal basis of L, we thus mean a strongly orthonormal set that strongly spans L.

In what follows, we shall tacitly consider the case $\mathbb{D} = \mathbb{H}$; for the other cases, we just take the obvious simplifications.

3G11 Lemma *If $V_1 \# \cdots \# V_m$ is the component of T in $\mathcal{V}_{\mathbf{G}}$, where $V_j = V(e_j)$ with $e_j \in W$ for $j = 1, \ldots, m$, then $\{e_1, \ldots, e_m\}$ is a strongly orthonormal set.*

Proof. For any $1 \leqslant i < j \leqslant m$ and any $\mu, \nu \in \mathbb{R}$ with $\mu^2 + \nu^2 = 1$, Lemma 3G6 implies that

$$(\mu V_i + \nu V_j) \# (\nu V_i - \mu V_j) = V_i \# V_j,$$

so that we can replace the given components V_i, V_j by $\mu V_i + \nu V_j, \nu V_i - \mu V_j$. But Theorem 1C7 implies that these new pure components of T again have circumradius 1, so that $\mu e_i + \nu e_j$ and $\nu e_i - \mu e_j$ are also unit vectors. However, unless $\langle e_i, e_j \rangle = 0$ this fails to hold whenever $\mu\nu \neq 0$. It follows that $\{e_1, \ldots, e_m\}$ is an orthonormal set.

Moreover, if $\boldsymbol{\lambda}_j \in \mathbb{D}$ with $|\boldsymbol{\lambda}_j| = 1$ for $j = 1, \ldots, m$, then $\boldsymbol{\lambda}_j V_j = V(\boldsymbol{\lambda}_j e_j) \cong V_j$ for each j, so that

$$V(\boldsymbol{\lambda}_1 e_1) \# \cdots \# V(\boldsymbol{\lambda}_m e_m) = V(e_1) \# \cdots \# V(e_m),$$

and hence (by the first part) $\{\boldsymbol{\lambda}_1 e_1, \ldots, \boldsymbol{\lambda}_m e_m\}$ is also an orthonormal set. In other words, $\{e_1, \ldots, e_m\}$ is strongly orthonormal, as claimed. □

3G12 Lemma *If $V_1 \# \cdots \# V_m$ is the component of T in $\mathcal{V}_{\mathbf{G}}$, where $V_j = V(e_j)$ for $j = 1, \ldots, m$, then $\{e_1, \ldots, e_m\}$ strongly spans the Wythoff space W.*

Proof. We first recall by Theorem 3B11 that every pure realization of $\mathcal{V}$ is (as usual up to scaling) a component of T. In particular, if $V \in \mathcal{N}_{\mathbf{G}}$ is pure, then V is a component of $V_1 \# \cdots \# V_m$. If we write $\mathbb{E}_j$ for the copy of $\mathbb{E}$ that contains V_j, then we have an orthogonal projection $\Pi \colon \mathbb{E}_1 \oplus \cdots \oplus \mathbb{E}_m \to \mathbb{E}$ such that $(V_1 \# \cdots \# V_m)\Pi = V$. Since $\mathbb{E}_j = \mathbb{E}$, it follows that Π induces a linear mapping $\Pi_j \colon \mathbb{E} \to \mathbb{E}$ for each j. Moreover, it is clear that Π_j commutes with the action of $\mathbf{G}$, so that there is a $\boldsymbol{\lambda}_j \in \mathbb{D}$ such that $x\Pi_j = \boldsymbol{\lambda}_j x$ for each j. It follows that

$$\begin{aligned} V &= (V_1 \# \cdots \# V_m)\Pi = V_1\Pi_1 + \cdots + V_m\Pi_m \\ &= \boldsymbol{\lambda}_1 V(e_1) + \cdots + \boldsymbol{\lambda}_m V(e_m) = V(\boldsymbol{\lambda}_1 e_1 + \cdots + \boldsymbol{\lambda}_m e_m); \end{aligned}$$

thus the initial point of V lies in $\mathbb{D}e_1 \oplus \cdots \oplus \mathbb{D}e_m$, as we had to show. □

Lemma 3G6 is the core tool we use to investigate what happens. We saw in Theorem 1D10 that the reflexions $R_{st}(\mu, \nu)$ generate the orthogonal group $\mathbf{O}_d$. Here, we apply the theorem in a rather different way. We can regard an ordered set $(u_1, \ldots, u_k)$ of vectors in $\mathbb{E}^d$ as the rows of a $k \times d$ matrix U, which we can multiply by orthogonal $k \times k$ matrices on the left. Such orthogonal matrices are generated by the operations of Lemma 3G6, and so correspond to operations on realizations $V(u_j)$ that do not change the blend $V(u_1) \# \cdots \# V(u_k)$. In addition, since $\mathbb{D}$-equivalent unit vectors u, v give rise to congruent realizations $V(u), V(v)$, we can multiply rows of the matrix U by elements of $\mathbb{D}$ of absolute value 1, again preserving the blend.

We need one more subsidiary result in order to apply Lemma 3G6.

3G13 Lemma *If u and v are strongly orthogonal unit vectors, then so are $\mu u + \nu v$ and $\nu u - \mu v$ whenever $\mu, \nu \in \mathbb{R}$ are such $\mu^2 + \nu^2 = 1$.*

Proof. First observe that, if $\boldsymbol{\lambda} \in \mathbb{D}$, then $\langle u, \boldsymbol{\lambda} u\rangle = \Re\boldsymbol{\lambda}$; moreover, we are told that $\langle u, \boldsymbol{\lambda} v\rangle = \langle v, \boldsymbol{\lambda} u\rangle = 0$. It follows that

$$\begin{aligned}\langle \mu u + \nu v, \boldsymbol{\lambda}(\nu u - \mu v)\rangle &= \mu\nu\langle u, \boldsymbol{\lambda} u\rangle - \mu^2\langle u, \boldsymbol{\lambda} v\rangle + \nu^2\langle v, \boldsymbol{\lambda} v\rangle - \mu\nu\langle v, \boldsymbol{\lambda} v\rangle \\ &= \mu\nu(\Re\boldsymbol{\lambda} - \Re\boldsymbol{\lambda}) = 0,\end{aligned}$$

since the other two terms vanish by the original strong orthogonality. □

Our main result is

3G14 Theorem *If $U_j = V(u_j) \in \mathcal{N}_{\mathbf{G}}$ are pure realizations for $j = 1, \ldots, k$, then – up to scaling – $U_1 \# \cdots \# U_k$ is the component of the simplex realization T in $\mathcal{V}_{\mathbf{G}}$ if and only if $\{u_1, \ldots, u_k\}$ is a strong orthonormal basis of W.*

Proof. We begin by remarking that Lemmas 3G11 and 3G12 together have shown that the condition is necessary, with $k = m$ as in Lemma 3G2.

For the converse, let $\{u_1, \ldots, u_m\}$ be a strong orthonormal basis of W. Thus we have $\boldsymbol{\alpha}_{jk} \in \mathbb{D}$ for $j, k = 1, \ldots, m$ such that $u_j = \boldsymbol{\alpha}_{j1} e_1 + \cdots + \boldsymbol{\alpha}_{jm} e_m$. We now use operations of the kind in Lemma 3G6 and those in the centralizer $\mathbf{Z}$ (that is, multiplication by elements of $\mathbb{D}$) to reduce $\{u_1, \ldots, u_m\}$ to the original strongly orthonormal set $\{e_1, \ldots, e_m\}$.

First, for each j such that $\boldsymbol{\alpha}_{j1} \neq \mathbf{0}$, define $\boldsymbol{\beta}_{j1} := \widetilde{\boldsymbol{\alpha}}_{j1}/|\boldsymbol{\alpha}_{j1}| \in \mathbb{D}$. Thus $|\boldsymbol{\beta}_{j1}| = 1$, so that $V(\boldsymbol{\beta}_{j1} u_j) = V(u_j)$ up to congruence. Moreover, the new coefficient $\gamma_{j1} := \boldsymbol{\alpha}_{j1}\boldsymbol{\beta}_{j1} = \boldsymbol{\alpha}_{j1}\widetilde{\boldsymbol{\alpha}}_{j1}/|\boldsymbol{\alpha}_{j1}| = |\boldsymbol{\alpha}_{j1}|$ of e_1 is real and positive. Write U for the resulting $m \times d$ matrix whose rows are the new u_j.

There is an orthogonal $m \times m$ matrix M (say), such that the leading column of MU is a positive multiple of e_1^{T}. Since the rows of MU remain strongly orthonormal, its first row is e_1 and the multiple is actually 1. We can now appeal to induction on rows $2, \ldots, m$: a further sequence of such operations transforms MU into the matrix with rows $e_1, \ldots, e_m$. Reversing them takes $\{e_1, \ldots, e_m\}$ back into $\{u_1, \ldots, u_m\}$. Since each step preserves the blend $V_1 \# \cdots \# V_m$, we conclude that

$$V(u_1) \# \cdots \# V(u_m) = V(e_1) \# \cdots \# V(e_m)$$

is also the component of T in $\mathcal{V}_{\mathbf{G}}$, as we had to prove. □

In the same spirit, we describe a general realization in $\mathcal{V}_{\mathbf{G}}$; we can think of this as a kind of diagonalization result.

3G15 Theorem *A general element $V \in \mathcal{V}_{\mathbf{G}}$ can be expressed in the form $V = V(a_1) \# \cdots V(a_k)$ for some $k \leqslant m$ and strongly orthogonal $a_1, \ldots, a_k \in W$.*

Proof. By definition of $\mathcal{V}_{\mathbf{G}}$, we know that $V = V(b_1) \# \cdots \# V(b_m)$ for some $b_1, \ldots, b_m \in W$. We apply operations to $b_1, \ldots, b_m$ of the same kind as those in the proof of Theorem 3G14, to produce new vectors $a_1, \ldots$. Among the results is one with $\|a_1\|$ maximal (it need not be unique). We claim that a_1 is strongly orthogonal to the remaining a_j (if there be any). Indeed, if not, then we can change the sign of a_j (if necessary) so that $\langle a_1, a_j \rangle > 0$. Consider $\mu a_1 + \nu a_j$ with $\mu, \nu > 0$ and $\mu^2 + \nu^2 = 1$. Then

$$\begin{aligned}\|\mu a_1 + \nu a_j\|^2 - \|a_1\|^2 &= (\mu^2 - 1)\|a_1\|^2 + 2\mu\nu\langle a_1, a_j \rangle + \nu^2 \|a_j\|^2 \\ &= 2\mu\nu\langle a_1, a_j \rangle - \nu^2(\|a_1\|^2 - \|a_j\|^2) > 0,\end{aligned}$$

whenever ν is small enough, contradicting our assumption. The same conclusion must hold if we replace a_1 by any $\mathbb{D}$-equivalent vector, so leading to our claim.

We now proceed by induction on $a_2, \ldots$, noting that we may end up with fewer vectors than we began with. □

Notes to Section 3G

1. In [85, 88, 93], we wrote $w^* := m$, so that $w^* = w/\dim_{\mathbb{R}} \mathbb{D}$. Indeed, m is a useful symbol in other circumstances, and so we often revert to the notation w^*.
2. We are indebted to Ladisch for pointing out in [67] that our earlier analysis of the case where the centralizer was not just $\{\pm I\}$ was wrong. However, what we do in this section and the next generalizes the techniques of [88], rather than following those of [67].

3H Λ-Orthogonal Bases

Since we have a positive definite inner product $\langle \cdot, \cdot \rangle_\Lambda$, we can apply the Gram–Schmidt orthogonalization process (for example) to turn any basis of $\mathbb{R}^{r+1}$ into a Λ-orthogonal one. However, it is more natural to look for such bases with geometric significance; we shall find such bases here.

We initially work in a general context; in what immediately follows, distinct vectors $u, v, \ldots$ are always assumed to be mutually orthogonal unit vectors in W. Moreover, we write $u \perp_{\mathrm{s}} v$ if u and v are strongly orthogonal. What we already know from Theorems 3F5 and 3G14 is that $\|\Gamma(u)\|_\Lambda^2 = \frac{1}{d}$ and $\langle \Gamma(u), \Gamma(v) \rangle_\Lambda = 0$ for all such u, v with $u \perp_{\mathrm{s}} v$, because u and v can form part of a strong orthonormal basis of W. At each subsequent stage, we shall assume without question what we have already shown.

For any orthogonal u, v, we have an expansion of $\Gamma(\xi u + \eta v)$ as a quadratic in ξ, η with $\xi^2 + \eta^2 = 1$, which we can write as

$$\Gamma(\xi u + \eta v) = \xi^2 \Gamma(u) + 2\xi\eta\Gamma(u, v) + \eta^2 \Gamma(v).$$

We call $\Gamma(u,v)$ a *mixed cosine vector*, even though it is not a cosine vector at all if $u \neq v$. We allow ξ and η to vary freely, subject to $\xi^2+\eta^2=1$. Observe that, if $\Phi_s \in \mathbf{G}$ represents layer $\mathcal{V}_s$, then entry s in $\Gamma(u,v)$ is

3H1 $$\gamma_s(u,v) = \tfrac{1}{2}(\langle u, v\Phi_s\rangle + \langle v, u\Phi_s\rangle);$$

that is, the entries are symmetrized. If $\mathcal{L}_s$ is asymmetric, then it is possible that $\langle u, v\Phi_s\rangle \neq \langle v, u\Phi_s\rangle$.

We now look at various cases in turn. First, if $v \in \mathbb{D}u$, then $\Gamma(\xi u+\eta v) = \Gamma(u) = \Gamma(v)$ for all ξ,η. Hence,

$$\Gamma(\xi u+\eta v) = \xi^2\Gamma(u) + 2\xi\eta\Gamma(u,v) + \eta^2\Gamma(v) = (\xi^2+\eta^2)\Gamma(u) + 2\xi\eta\Gamma(u,v),$$

giving $\Gamma(u,v)=O$.

Next, we expand

$$\langle \Gamma(\xi u+\eta v), \Gamma(\xi u+\eta v)\rangle_\Lambda = \|\Gamma(\xi u+\eta v)\|_\Lambda^2 = \tfrac{1}{d}$$

as a quartic in ξ and η. Substituting for known terms gives

$$\begin{aligned}\tfrac{1}{d} = (\xi^4+\eta^4)\cdot\tfrac{1}{d} &+ 4\xi^2\eta^2\|\Gamma(u,v)\|_\Lambda^2 \\ &+ 4\xi\eta(\xi^2\langle\Gamma(u),\Gamma(u,v)\rangle_\Lambda + \eta^2\langle\Gamma(v),\Gamma(u,v)\rangle_\Lambda).\end{aligned}$$

Changing the sign of ξ and adding the two expressions gives

$$\begin{aligned}8\xi^2\eta^2\|\Gamma(u,v)\|_\Lambda^2 = (2-2\xi^4-2\eta^4)\cdot\tfrac{1}{d} &= 4\xi^2\eta^2\cdot\tfrac{1}{d} \\ \implies \|\Gamma(u,v)\|_\Lambda^2 &= \tfrac{1}{2d}.\end{aligned}$$

Subtracting one from the other and dividing by $8\xi\eta$ then gives

$$\begin{aligned}&\xi^2\langle\Gamma(u),\Gamma(u,v)\rangle_\Lambda + \eta^2\langle\Gamma(v),\Gamma(u,v)\rangle_\Lambda = 0 \\ &\qquad\implies \langle\Gamma(u),\Gamma(u,v)\rangle_\Lambda = \langle\Gamma(v),\Gamma(u,v)\rangle_\Lambda = 0,\end{aligned}$$

since ξ,η are arbitrary.

Now let $u,v \perp_{\mathrm{s}} w$. We first have

$$\begin{aligned}0 &= \langle\Gamma(\xi u+\eta v),\Gamma(w)\rangle_\Lambda = \langle\xi^2\Gamma(u)+2\xi\eta\Gamma(u,v)+\eta^2\Gamma(v),\Gamma(w)\rangle_\Lambda \\ &= 2\xi\eta\langle\Gamma(u,v),\Gamma(w)\rangle_\Lambda,\end{aligned}$$

from which follows

$$\langle\Gamma(u,v),\Gamma(w)\rangle_\Lambda = 0.$$

Since it is clear that $\Gamma(\cdot,\cdot)$ is linear in each variable and $\xi u+\eta v \perp_{\mathrm{s}} w$, we also have

$$\begin{aligned}\tfrac{1}{2d} &= \|\Gamma(\xi u+\eta v, w)\|_\Lambda^2 = \|\xi\Gamma(u,w)+\eta\Gamma(v,w)\|_\Lambda^2 \\ &= (\xi^2+\eta^2)\cdot\tfrac{1}{2d} + 2\xi\eta\langle\Gamma(u,w),\Gamma(v,w)\rangle_\Lambda,\end{aligned}$$

from which we see that

$$\langle \Gamma(u,w), \Gamma(v,w)\rangle_\Lambda = 0.$$

Note that we are only assuming that $\langle u, v\rangle = 0$ here.

Finally, if $u, v \perp_s w, x$, then

$$\begin{aligned} 0 &= \langle \Gamma(\xi u + \eta v), \Gamma(w,x)\rangle_\Lambda = \langle \xi^2\Gamma(u) + 2\xi\eta\Gamma(u,v) + \eta^2\Gamma(v), \Gamma(w,x)\rangle_\Lambda \\ &= 2\xi\eta\langle \Gamma(u,v), \Gamma(w,x)\rangle_\Lambda, \end{aligned}$$

from which follows

$$\langle \Gamma(u,v), \Gamma(w,x)\rangle_\Lambda = 0.$$

We let $\mathbf{B}$ be an orthonormal basis of $\mathbb{D}$ over $\mathbb{R}$. Thus, for instance, we can take

3H2
$$\mathbf{B} = \begin{cases} \{\mathbf{1}\}, & \text{if } \mathbb{D} = \mathbb{R}, \\ \{\mathbf{1}, \mathbf{i}\}, & \text{if } \mathbb{D} = \mathbb{C}, \\ \{\mathbf{1}, \mathbf{i}, \mathbf{j}, \mathbf{k}\}, & \text{if } \mathbb{D} = \mathbb{H}. \end{cases}$$

We write $\mathbf{1}$ and $\mathbf{i}$ (when $\mathbb{D} = \mathbb{C}$) to emphasize that these ought to be thought of as linear mappings.

This previous discussion leads to the core result.

3H3 Theorem *If* $\mathbf{G}$ *is an irreducible representation of* $(\mathcal{V}, \boldsymbol{G})$ *of degree* d *such that that the component of the simplex realization* T *in* $\mathcal{V}_\mathbf{G}$ *has dimension* md*, and if* $E = (e_1, \ldots, e_m)$ *is a strongly orthonormal basis of the Wythoff space* W*, then the following cosine vectors form a* Λ*-orthogonal basis of* $\operatorname{lin} \mathcal{N}_\mathbf{G}$*:*

(a) $\Gamma(e_j, e_j)$ *for* $j = 1, \ldots, m$*;*

(b) $\Gamma(e_j, \boldsymbol{\lambda} e_k)$ *for* $1 \leqslant j < k \leqslant m$ *and* $\boldsymbol{\lambda} \in \mathbf{B}$*.*

Hence, $\mathcal{V}_\mathbf{G}$ *has dimension* $m + \frac{1}{2}m(m-1)c$*, with* $c = c_\mathbf{G}$ *as before.*

Proof. A general unit vector in W is of the form $x = \boldsymbol{\lambda}_1 e_1 + \cdots + \boldsymbol{\lambda}_m e_m$, with $\boldsymbol{\lambda}_1, \ldots, \boldsymbol{\lambda}_m \in \mathbb{D}$ such that $|\boldsymbol{\lambda}_1|^2 + \cdots + |\boldsymbol{\lambda}_m|^2 = 1$. If we write each $\boldsymbol{\lambda}_j$ as a real linear combination in the basis $\mathbf{B}$, then we can expand $\Gamma(x)$ as a quadratic in terms $\Gamma(\boldsymbol{\mu} e_j, \boldsymbol{\nu} e_k)$, with $\boldsymbol{\mu}, \boldsymbol{\nu} \in \mathbf{B}$ and $1 \leqslant j \leqslant k \leqslant m$. However, we can simplify this expression, since

$$\Gamma(\boldsymbol{\mu} e_j, \boldsymbol{\nu} e_j) = \begin{cases} \Gamma(e_j, e_j), & \text{if } j = k \text{ and } \boldsymbol{\mu} = \boldsymbol{\nu}, \\ O, & \text{if } j = k \text{ and } \boldsymbol{\mu} \neq \boldsymbol{\nu}, \\ \Gamma(e_j, \widetilde{\boldsymbol{\mu}} \boldsymbol{\nu} e_k), & \text{if } j < k. \end{cases}$$

Since a general inner product vector in $\mathcal{V}_\mathbf{G}$ is a (non-negative) linear combination of such cosine vectors $\Gamma(x)$, a simple count yields what we claimed. □

From Theorem 3H3 and the previous discussion, we deduce

3H4 Theorem *If the symmetric set* $(\mathcal{V}, \boldsymbol{G})$ *with* $\operatorname{card}\mathcal{V} = n$ *has* r *non-trivial diagonal classes, and the Wythoff space of its simplex realization has dimension* $\overline{w}$*, then*

$$\begin{aligned} n &= \sum_{\mathbf{G}} m_{\mathbf{G}} d_{\mathbf{G}}, \\ r + 1 &= \sum_{\mathbf{G}} m_{\mathbf{G}} + \tfrac{1}{2} m_{\mathbf{G}}(m_{\mathbf{G}} - 1) c_{\mathbf{G}}, \\ \overline{w} &= \sum_{\mathbf{G}} m_{\mathbf{G}} w_{\mathbf{G}}, \end{aligned}$$

where the sum is over the irreducible representations $\mathbf{G}$ *of* $\boldsymbol{G}$*, and the notation is that introduced earlier.*

3H5 Remark Theorem 3H4 can be rewritten in terms of cosets or double cosets of $\boldsymbol{H}$, as in [99, Theorem 5B17]; compare also Theorem 3A2. This also takes us full circle, and explains the connexion between asymmetric diagonal classes and Wythoff spaces W with $w > 1$. Thus, we have

$$\begin{aligned} n &= \operatorname{card}\{\boldsymbol{Hg} \mid \boldsymbol{g} \in \boldsymbol{G}\}, \\ r + 1 &= \operatorname{card}\{\boldsymbol{HgH} \cup \boldsymbol{Hg}^{-1}\boldsymbol{H} \mid \boldsymbol{g} \in \boldsymbol{G}\}, \\ \overline{w} &= \operatorname{card}\{\boldsymbol{HgH} \mid \boldsymbol{g} \in \boldsymbol{G}\}. \end{aligned}$$

The first equation just counts the number of points of $\mathcal{V}$, while Theorem 3A2 shows that the second counts the diagonal classes, including the trivial one. For the third, observe that $\overline{W}$ is spanned by the centroids (or sums) of the distinct orbits of points of T under the group of permutations induced by $\boldsymbol{G}$; such an orbit corresponds to a double coset $\boldsymbol{HgH}$.

It follows from this discussion that each asymmetric diagonal class will result in an extra contribution to the third sum over the second. Comparison with Theorem 3H4 then gives one or more irreducible representations $\mathbf{G}$ of $\boldsymbol{G}$ for which

$$m^2 c = mw > m + \tfrac{1}{2} m(m-1)c \implies (m+1)c > 2,$$

and hence, in particular, $m > 1$ or $c > 1$.

The Real Case

What we shall refer to as the *real* case, when $\mathbf{G}$ is irreducible as a complex representation and so $\mathbb{D} = \mathbb{R}$, turns out to be most important for our purposes; we refer to Section 3M for further details. The distinction is only relevant when $m > 1$; this occurs for realizations of the 120-cell alone among the classical regular polytopes, and even then all are real in this sense (see the notes at the end of the section). In this case, we have $m = w := \dim W$. Since $\mathbf{B} = \{\mathbf{1}\}$, we can simplify the previous notation, and write $\Gamma_{jk} := \Gamma(e_j, e_k)$ for $j, k = 1, \ldots, w$, with $\Gamma_{jk} = \Gamma_{kj}$ throughout.

Before we go further, we make two comments. In calling vectors in W *coplanar*, we mean that they lie in a *linear* plane.

3H6 Proposition *If $u, v \in W$ are unit vectors, then*

$$d\langle \Gamma(u), \Gamma(v)\rangle_\Lambda = \langle u, v\rangle^2.$$

If the coplanar unit vectors $v_1, v_2, v_3 \in W$ are such that $d\langle \Gamma(v_j), \Gamma(v_k)\rangle_\Lambda = \alpha_{jk}$ for each $j \neq k$, then

$$4\alpha_{12}\alpha_{23}\alpha_{31} = (1 - \alpha_{12} - \alpha_{23} - \alpha_{31})^2.$$

Proof. The first claim is clear from the foregoing discussion, if we choose an orthonormal basis $(e_1, e_2, \ldots)$ of W such that $u = e_1$ and $v \in \operatorname{lin}\{e_1, e_2\}$.

For the second, the criterion is just a determinantal one. We have

$$\det(\langle v_j, v_k\rangle) = 1 - \langle v_1, v_2\rangle^2 - \langle v_2, v_3\rangle^2 - \langle v_3, v_2\rangle^2 + 2\langle v_1, v_2\rangle\langle v_2, v_3\rangle\langle v_3, v_2\rangle;$$

equating this to 0 leads immediately to the given condition. □

3H7 Remark The criterion for coplanarity is less useful than one might hope, since its converse is false: just take $\langle v_j, v_k\rangle = \frac{1}{2}$ for each $j \neq k$. Hence it can only be used to disprove coplanarity.

From what we have said, the inner product vector of a general point-set $V \in \mathcal{V}_{\mathbf{G}}$ will be of the form

$$\Sigma(A) := \sum_{j,k=1}^{w} \alpha_{jk}\Gamma_{jk},$$

where $A = (\alpha_{jk})$ is a $w \times w$ symmetric matrix. We call A the *coefficient matrix* of the realization V (see the notes at the end of the section). We now have

3H8 Theorem *If the irreducible representation $\mathbf{G}$ of G is real, with Wythoff space W of dimension w, then the symmetric $w \times w$ matrix A is the coefficient matrix of some realization $V \in \mathcal{V}_{\mathbf{G}}$ if and only if A is positive semi-definite.*

Proof. A coefficient matrix A is the sum of matrices $x^\mathsf{T}x$ corresponding to pure realizations, where $x = \xi_1 e_1 + \cdots \xi_w e_w \in W$ is identified with the row vector $(\xi_1, \ldots, \xi_w)$. These are positive semi-definite; hence A is also.

Conversely, we can always diagonalize a real symmetric matrix A, say $A = U^\mathsf{T}LU$, where L is a diagonal matrix and U is orthogonal. If A is positive semi-definite, then L has non-negative (diagonal) entries $\lambda_1, \ldots, \lambda_w$. It follows that, if U has rows $u_1, \ldots, u_w$, then

$$\sum_{j,k=1}^{w} \alpha_{jk}\Gamma_{jk} = \sum_{j=1}^{w} \lambda_j \Gamma(u_j),$$

which is the inner product vector of $V(\sqrt{\lambda_1}u_1) \# \cdots \# V(\sqrt{\lambda_w}u_w)$. □

3H9 Remark Note that $\Sigma(A)$ is a cosine vector just when the trace $\operatorname{tr} A = 1$.

Recall from Section 1C the definition of the inner product $\langle A, B\rangle = \operatorname{tr} AB^{\mathsf{T}}$ of two square matrices. In terms of inner product vectors, we have

3H10 Proposition *In the real case, if two realizations in $\mathcal{V}_{\mathbf{G}}$ have coefficient matrices A and B with respect to the orthogonal basis $\{\Gamma_{jk}\}$, then*

$$\langle \Sigma(A), \Sigma(B)\rangle_{\Lambda} = \frac{\langle A, B\rangle}{d}.$$

Proof. All we need to observe is that, if $j \neq k$, then the contribution from the (j, k) terms is

$$(2\alpha_{jk})(2\beta_{kj})\|\Gamma_{jk}\|_{\Lambda}^2 = 4\alpha_{jk}\beta_{kj} \cdot \frac{1}{2d} = 2\alpha_{jk}\beta_{kj} \cdot \frac{1}{d};$$

the contribution from the diagonal terms is obviously correct. The claim of the proposition follows at once. □

There is a special case of Proposition 3F8. If $\operatorname{tr} A = 1$, then (by diagonalizing A) we see that

$$\frac{1}{w} \leqslant \|A\|^2 \leqslant 1;$$

more generally, $\|A\|^2 \geqslant 1/s$ if $\operatorname{rank} A = s$.

We may also note the following generalization of Λ-orthogonality.

3H11 Proposition *If $u, v \in W$ are unit vectors, then*

$$\langle \Gamma(u), \Gamma(v)\rangle_{\Lambda} = \frac{1}{d}\langle u, v\rangle^2.$$

Proof. Choose an orthonormal basis $(e_1, e_2, \ldots)$ of W such that $u = e_1$ and $v = \langle u, v\rangle e_1 + \lambda e_2$ (with $\lambda^2 = 1 - \langle u, v\rangle^2$, of course). Then

$$\Gamma(v) = \langle u, v\rangle^2 \Gamma_{11} + 2\langle u, v\rangle \lambda \Gamma_{12} + \lambda^2 \Gamma_{22},$$

and hence, since $\Gamma_{11} = \Gamma(u)$, we have $\langle \Gamma(u), \Gamma(v)\rangle_{\Lambda} = \langle u, v\rangle^2 \|\Gamma(u)\|_{\Lambda}^2$, which is what was claimed. □

3H12 Remark When $w = 2$, there is an alternative parametrization of the normalized pure realizations, namely,

$$\Gamma = \Gamma_m + \cos 2\vartheta \Gamma_c + \sin 2\vartheta \Gamma_s,$$

where $\Gamma_m = \frac{1}{2}(\Gamma_{11} + \Gamma_{22})$, $\Gamma_c = \frac{1}{2}(\Gamma_{11} - \Gamma_{22})$ and $\Gamma_s = \Gamma_{12}$; here, ϑ is an angle parameter in W. This is often useful, particularly since cases with $w > 2$ appear to be less common (but see Section 7K).

3H13 Remark Because we insist on the symmetry $\Gamma_{jk} = \Gamma_{kj}$, following (3H1) we must define its general entry γ_{jks} to be

$$\gamma_{jks} := \tfrac{1}{2}\big(\langle e_j, e_k\Phi_s\rangle + \langle e_k, e_j\Phi_s\rangle\big),$$

where $\Phi_s \in \mathbf{G}$ takes initial vertices in w into layer $\mathcal{L}_s$. However, if $\mathcal{D}_s$ is a symmetric diagonal class, then

$$\langle e_k, e_j\Phi_s\rangle = \langle e_j, e_k\Phi_s^{-1}\rangle = \langle e_j, e_kA\Phi_sB\rangle = \langle e_jB^{-1}, e_kA\Phi_s\rangle = \langle e_j, e_k\Phi_s\rangle$$

for some $A, B \in \mathbf{H}$, so that $\gamma_{jks} = \langle e_j, e_k\Phi_s\rangle$. We shall note at the end of the section that, if $w \geqslant 2$, then there is necessarily some s such that $\langle e_j, e_k\Phi_s\rangle \neq \langle e_k, e_j\Phi_s\rangle$; from what we have just said, the diagonal class $\mathcal{D}_s$ must be asymmetric.

3H14

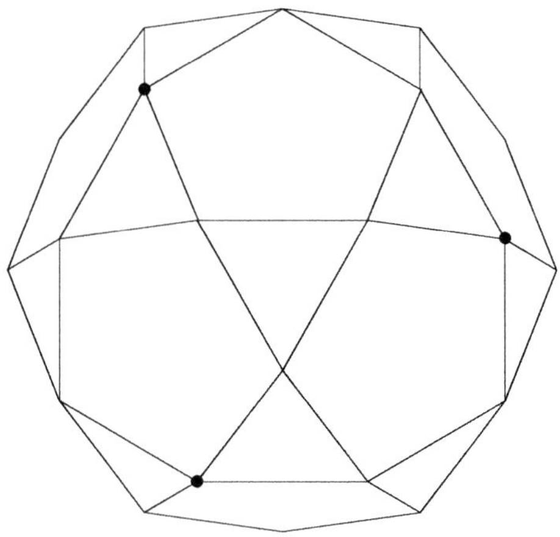

The icosidodecahedron

3H15 Example Let $\mathcal{H}$ be the quasi-regular hemi-icosidodecahedron, as an abstract polytope. This has 15 vertices, and layer vector $(1, 4, 4, 4, 2^*)$, with its automorphisms forming the alternating group $\boldsymbol{A}_5$. We can identify the vertices of $\mathcal{H}$ with the 15 diameters of the original icosidodecahedron. With one diameter as the initial vertex, layer $\mathcal{L}_4$ is formed by the two orthogonal diameters, and the three are permuted by cyclic automorphisms only; see Figure 3H14, where these diameters are picked out. Since the vertices of $\mathcal{H}$ are the mid-points of the edges of the hemi-icosahedron $\{3, 5 : 5\}$ or hemi-dodecahedron $\{5, 3 : 5\}$, and we shall find the realization domains of both in Section 7D, it is easy to determine that of $\mathcal{H}$ itself. Note that $\boldsymbol{A}_5$ has no (non-trivial) representations of degree 2, and that the Wythoff space in degree 3 is trivial.

So, the vertices of the 4-dimensional realization of $\{5, 3 : 5\}$ are the mid-points of the edges of the 4-simplex, and an edge joining two of these vertices goes between opposite edges of a facet of the simplex. Hence 3 such edges have the same mid-point, resulting in the vertices of $\mathcal{H}$ being identified in threes at the vertices of a 4-simplex. Thus we obtain the corresponding cosine vector $\Gamma_1 = (1, -\frac{1}{4}, -\frac{1}{4}, -\frac{1}{4}, 1)$. Another way of thinking about this realization is as that of the group acting on the five triples of three mutually orthogonal diameters. In more abstract terms, this quotient is, in effect, that of $\boldsymbol{A}_5$ acting on its subgroup $\boldsymbol{C}_3$.

Then Λ-orthogonality shows at once that the remaining pure realizations must be 5-dimensional, with cosine vectors of the form $(1, \alpha, \beta, \gamma, -\frac{1}{2})$, where $\alpha + \beta + \gamma = 0$. The dimension equation of Theorem 3F5 implies that

$$3 = \tfrac{15}{5} = 1 + 4\alpha^2 + 4\beta^2 + 4\gamma^2 + 2(-\tfrac{1}{2})^2 \implies \alpha^2 + \beta^2 + \gamma^2 = \tfrac{3}{8}.$$

Thus we have realizations with a 2-dimensional Wythoff space. This can be verified directly from explicit expressions for the generators of the symmetry group as permutations of the coordinates of $\mathbb{E}^6$, with the polytope sitting in the symmetric hyperplane $\mathbb{L}^5$ of (1E15), namely,

$$\begin{aligned} R_0 &:= (0\ 1)(4\ 5)), \\ R_1 &:= (1\ 2)(3\ 4), \\ R_2 &:= (2\ 3)(4\ 5)), \end{aligned}$$

which generate $\boldsymbol{A}_5 = [3, 5]/2$ in the natural way. Then

$$W = R_0 \cap R_2 = \{x \mid \xi_0 = \xi_1,\ \xi_2 = \xi_3,\ \xi_4 = \xi_5\}$$

has dimension 2.

Asymmetric Diagonals

Just until the end of the section, we redefine Γ_{jk} in the non-symmetric way, as $\Gamma_{jk} = (\gamma_{jk0}, \ldots, \gamma_{jkr})$ with $\gamma_{jks} := \langle e_j \Phi_s, e_k \rangle$; otherwise the notation is the same as before. What we called Γ_{jk} earlier is now the *even* part $\Gamma^e_{jk} := \frac{1}{2}(\Gamma_{jk} + \Gamma_{kj})$; we similarly define the *odd* part $\Gamma^o_{jk} := \frac{1}{2}(\Gamma_{jk} - \Gamma_{kj}) = -\Gamma^o_{kj}$. Note that Γ^o_{jk} can have a non-zero entry γ^o_{jks} only if $\mathcal{L}_s$ is an asymmetric layer.

While we shall not prove it here, we actually have (at least in the case when the representation $\mathbf{G}$ is real)

3H16 $$\langle \Gamma_{jk}, \Gamma_{kj} \rangle = 0;$$

this is (in a rather concealed form) a special case of [122, (III.1.2)]. When we substitute $\Gamma_{jk} = \Gamma^e_{jk} + \Gamma^o_{jk}$ and $\Gamma_{kj} = \Gamma^e_{jk} - \Gamma^o_{jk}$, we deduce from Theorem 3F5

3H17 Proposition *With the preceding notation, if $j \neq k$, then*

$$\|\Gamma^o_{jk}\|^2_\Lambda = \|\Gamma^e_{jk}\|^2_\Lambda = \frac{1}{2d}.$$

As an immediate consequence, we have

3H18 Corollary *For all $j \neq k$, there is some asymmetric layer $\mathcal{L}_s$ such that Γ_{jk} and Γ_{kj} differ in entry s.*

3H19 Example We return to the previous example. Since only one layer of the hemi-icosidodecahedron is asymmetric, it follows that, irrespective of the choice of orthonormal basis of W, we shall have $\Gamma^o_{12} = \big(0, 0, 0, 0, \pm\frac{\sqrt{3}}{2}\big)$.

Notes to Section 3H

1. We are indebted to Ladisch [67] for information about the realizations of $\{5,3,3\}$; we shall say more about this in Section 7K.
2. It may be noted that the coefficient matrix A is just a symmetrized version of the canonical matrix of [99, Section 5B], which was introduced in [75]. In retrospect, it is clear that the canonical matrix was far from the best way to describe a realization in $\mathcal{N}_{\mathbf{G}}$; indeed, in several respects it was rather misleading.
3. In case $\mathbf{G}$ is real, we can deduce Theorem 3H3 from [122, (III.1.2)]; the non-symmetric contributions from the half-layers of an asymmetric layer actually cancel. However, in the non-real case, it is unclear how representation theory might lead to the result.

3J Cosine Matrices

For convenience, and because it is the only case that we employ, we continue to assume that all the irreducible representations we encounter are real. Putting together the Γ_{jk} for the different representations $\mathbf{G}$ of $\boldsymbol{G}$ yields a Λ-orthogonal basis of the whole space spanned by the cone $\mathcal{V}$ (as usual identified with the cone of inner product vectors). For a given $\mathbf{G}$, we list the Γ_{jk} by the lexicographic ordering of their indices, namely,

$$\Gamma_{11}, \Gamma_{12}, \Gamma_{13}, \dots, \Gamma_{1m}, \Gamma_{22}, \Gamma_{23}, \dots, \Gamma_{2m}, \dots, \Gamma_{mm}.$$

(The situation for the non-real cases is similar, except that the effect of the centralizer $\mathbb{D}$ needs to be taken into account.) If we write these particular cosine vectors Γ as a block of rows of an $(r+1) \times (r+1)$ matrix in this order, then they are immediately identifiable as a block, since each Γ_{jj} has leading entry 1 (being a cosine vector), while the Γ_{jk} with $j < k$ have leading entry 0. The block is then clearly flanked by the rows Γ_{11} and Γ_{mm}. The resulting matrix with rows so defined is called a *cosine matrix* of $\mathcal{V}$; we must always bear in mind the dependence on a choice of ordered (strongly) orthonormal basis of the essential Wythoff space whenever $m > 1$.

As is well known, if $\{u_1, \dots, u_m\}$ is an orthogonal basis of a vector space $\mathbb{V}$ with respect to an inner product $(\cdot, \cdot)$, then

$$x = \xi_1 u_1 + \dots + \xi_m u_m \iff \xi_j = \frac{(x, u_j)}{(u_j, u_j)}$$

for each j. In the present case of the Λ-inner product, Theorem 3H3 gives the corresponding squared norms. We see that the multiplying factors $1/(u_j, u_j)$ are actual dimensions for those basis elements which are genuine cosine vectors. For a mixed cosine vector, that is, one of the form Γ_{jk} with $j < k$, we have a *notional* dimension $2d$ as the multiplier. We thus obtain a list $D = (d_0, \dots, d_r)$ of dimensions, whether genuine or notional, which we call the *dimension vector*; by convention, $d_0 = 1$ always corresponds to the henogon realization $\{1\}$. We then have

3J1 Theorem *Let $\{\Gamma_0, \ldots, \Gamma_r\}$ be the Λ-orthogonal basis of* $\operatorname{lin} \mathcal{V}$ *given by Theorem 3H3 and the preceding discussion, and let $D = (d_0, \ldots, d_r)$ be the corresponding dimension vector. If $\Sigma = \xi_0 \Gamma_0 + \cdots + \xi_r \Gamma_r$ is the inner product vector of some realization of $\mathcal{V}$, then*

$$\xi_s = d_s \langle \Sigma, \Gamma_s \rangle_\Lambda$$

for $s = 0, \ldots, r$.

At this point, it is worth mentioning a property implied by induced cosine vectors.

3J2 Proposition *Every pure realization of a cut $\mathcal{U}$ of a symmetric set $\mathcal{V}$ is a component of a cut in some pure realization of $\mathcal{V}$.*

Proof. The induced cosine vector of $\mathcal{U}$ in terms of the general one of $\mathcal{V}$ will be $\Gamma^{\mathcal{U}} = (\gamma_{j(0)}, \ldots, \gamma_{j(s)})$ for some subset $\{j(0), \ldots, j(s)\}$ of indices (with $j(0) = 0$, and possibly with repetitions). The component equation Theorem 3C11 implies that the sum of these induced cosine vectors in the pure realizations V_i of $\mathcal{V}$ (*excluding* the mixed ones), scaled by the corresponding dimensions $d_i = \dim V_i$, will be $n(1, 0^s)$. This is just n/k times the component equation for $\mathcal{U}$, where $k := \operatorname{card} \mathcal{U}$. Since each pure realization of $\mathcal{U}$ is a component of its simplex realization, the claim of the proposition follows. □

3J3 Remark Since components of the simplex realization in a given $\mathcal{N}_\mathrm{G}$ can be decomposed in different ways when $m > 1$, we cannot assert that the minimal corresponding realization of $\mathcal{U}$ in Proposition 3J2 is pure; in fact, in general it will not be.

3J4 Remark Since the cosine matrix is non-singular, it is clear that the affine hull of the centred cosine vectors (that is, those that satisfy the layer equation Theorem 3C7) is spanned by the vectors $e_0 - \ell_s^{-1} e_s$ for $s = 1, \ldots, r$. While this observation may not be helpful generally, it can be applied when $\mathcal{V}$ admits a proper quotient $\mathcal{U}$ (say) because, even if the corresponding cosine matrix of $\mathcal{U}$ is not known, nevertheless the Λ-orthogonal complement of the lifts of these vectors into $\mathcal{V}$ will contain the pure cosine vectors in $\mathcal{V} \setminus \mathcal{U}$.

D-Orthogonality

The component equation of Theorem 3C11 relates dimensions and cosine vectors of pure components of the simplex realization T. Moreover, we can write the statement of Theorem 3F5 in the form

$$\frac{1}{n} \sum_{s=0}^{r} \ell_s \gamma_{js} \gamma_{ks} = \frac{\delta_{jk}}{d_k},$$

where $\Gamma_j = (\gamma_{j0}, \ldots, \gamma_{jr})$ for $j = 0, \ldots, r$. We can now generalize the component equation as follows. If $D = (d_0, \ldots, d_r)$ is the dimension vector (including notional dimensions if appropriate), then the *D-inner product* of $x, y \in \mathbb{R}^{r+1}$ is defined by

3J5 $$\langle x, y\rangle_D := \frac{1}{n}\langle D, xy\rangle,$$

in exact analogy to the Λ-inner product; here, xy is the termwise product of x and y as before. We can thus also use the analogous terms *D-norm* $\|\cdot\|_D$ and *D-orthogonal*. Then we have

3J6 Theorem *The columns of the cosine matrix are D-orthogonal. Column s has squared D-norm $1/\ell_s$, where $\Lambda = (\ell_0, \dots, \ell_r)$ is the layer vector.*

Proof. In fact, this is straightforward. If we set

$$\mu_{js} := \sqrt{\frac{d_j \ell_s}{n}}\gamma_{js},$$

then Theorem 3H3 and the definition of the cosine matrix say that the matrix $M = (\mu_{js})$ is orthogonal in the usual sense. The symmetry between rows and columns at once yields the claim of the theorem. □

3J7 Remark Note that the component equation Theorem 3C11 itself, which picks out the genuine cosine vectors, just demonstrates D-orthogonality with respect to the leading column 0, since the leading entries of the mixed cosine vectors make no contribution. What geometric significance (if any) Theorem 3J6 has eludes us.

3J8 Remark It is worth pointing out that the layer vector Λ is the unique vector $(\ell_0, \dots, \ell_r) \in \mathbb{R}^{r+1}$ with $\ell_0 = 1$ which is orthogonal (in the usual sense) to all but row Γ_0 of the cosine matrix. There is no analogous characterization of the dimension vector D if any $w > 1$.

Structure Constants

For each $i, j, k = 0, \dots, r$, we define the *structure constant* α_{ijk} by

3J9 $$\Gamma_i \Gamma_j =: \sum_{k=0}^{r} \alpha_{ijk}\Gamma_k;$$

here, Γ_i is row i of the cosine matrix, whether a genuine or mixed cosine vector. If we write $\Gamma_i = (\gamma_{i0}, \dots, \gamma_{ir})$ for $i = 0, \dots, r$, then using Theorem 3H3 we see that

3J10 $$\alpha_{ijk} = d_k\langle \Gamma_i\Gamma_j, \Gamma_k\rangle_\Lambda = \frac{d_k}{n}\sum_{s=0}^{r} \ell_s\gamma_{is}\gamma_{js}\gamma_{ks}.$$

Of course, this expression is symmetric in i and j, and nearly so in k as well. More exactly,

3J11 Proposition *The structure constants α_{ijk} of (3J9) are such that each $\beta_{ijk} := d_k^{-1}\alpha_{ijk}$ is symmetric in i, j, k.*

This observation considerably shortens the calculations for the multiplication table of the pure normalized realizations of $\mathcal{V}$ (if one should have to do them), since one need only work out those β_{ijk} for $r \geqslant i \geqslant j \geqslant k \geqslant 1$ (the values for $k = 0$ are already given in effect by Theorem 3H3).

3K Cuts and Duality

As we pointed out before Proposition 3A8, an often useful tool to investigate realizations is the relationships between cuts induced by different subgroups of the automorphism group G, in particular their realizations. So, let $\mathbf{G}$ be a representation of G, let $\mathbf{A}, \mathbf{B} < \mathbf{G}$, and write U and V for corresponding realizations with initial points the (as always) unit vectors $u \in \operatorname{axis}\mathbf{A}$ and $v \in \operatorname{axis}\mathbf{B}$. If $E_1, \ldots, E_t$ is the partition of $E := U \times V$ induced by the action of $\mathbf{G}$ and $(x, y) \in E_j$ is any pair, then

3K1 $$\kappa_j := \langle x, y\rangle$$

is well-defined for each $j = 1, \ldots, t$. Of course,

$$\langle u, y\rangle = \kappa_j = \langle x, v\rangle$$

for any $x \in U_j$ and $y \in V_j$, the initial cuts. Though we make little use of it, we can call $K := (\kappa_1, \ldots, \kappa_t)$ the *cut vector*.

We can turn the latter equation into an obvious comment.

3K2 Proposition *For each $j = 1, \ldots, t$, the height of the $\mathbf{B}$-cut U_j of U relative to v is the same as that of the $\mathbf{A}$-cut V_j of V relative to u.*

As we said in Remark 3A10, given any particular $j = 1, \ldots, t$, we can replace $\mathbf{A}$ or $\mathbf{B}$ by suitable conjugates in $\mathbf{G}$ to ensure that $u \in U_j$ and $v \in V_j$. The centroid a_j of U_j is the image of u under orthogonal projection on $\operatorname{axis}\mathbf{G}$, so that

$$\kappa_j = \langle u, v\rangle = \langle a_j, v\rangle \leqslant \|a_j\|$$

because v is a unit vector, with equality if and only if a_j is a non-negative multiple of v. Moreover, if $\eta_j := \eta(U_j)$ is the squared distance from o to the subspace $\operatorname{aff} U_j$ spanned by U_j, then $\eta_j = \|a_j\|^2$. Finally, U_j is itself a symmetric set under the action of $\mathbf{B}$, so that η_j can be calculated from U_j, by feeding its layer vector and induced cosine vector into Corollary 3C8.

Now suppose that the cut U_j is not centred, so that $a_j \neq o$. Then $\widetilde{v} := \|a_j\|^{-1}a_j \in \operatorname{axis}\mathbf{B}$, and hence is a suitable initial point for a new realization $\widetilde{V} := \widetilde{v}\mathbf{G}$. Thinking of $\mathbf{A}$ and $\mathbf{B}$ as (for the moment) fixed up to conjugacy in $\mathbf{G}$, we call $\widetilde{V}$ a *geometric dual* of U. Of course, in general $\widetilde{V}$ depends on j as well as on U. We then obtain corresponding $\mathbf{A}$-cuts of $\widetilde{V}$, and in particular $\widetilde{V}_j := \widetilde{v}\mathbf{A}$, which we also refer to as a geometric dual.

3K3 Remark In a way, we are misusing the term *dual*, since the relationship between U and $\widetilde{V}$ is not symmetric. However, it is convenient, because it points towards the future.

With $\widetilde{v}$ instead of v in the foregoing discussion, what we have achieved here is

$$\widetilde{\kappa}_j := \langle u, \widetilde{v} \rangle = \|a_j\|.$$

Applying the same analysis with $\mathbf{A}$ and $\mathbf{B}$ interchanged, we then deduce the *dual cut criterion*, namely,

3K4 Theorem *If* $\mathbf{A}, \mathbf{B} < \mathbf{G}$ *give rise to realizations* U *and* V*, respectively, such that* V *is the geometric dual of* U *relative to the* $\mathbf{B}$*-cut* U_j *of* U*, then* $\eta(U_j) \leqslant \eta(V_j)$*, with equality if* $\dim(\operatorname{axis} \mathbf{A}) = 1$.

Of course, $\dim(\operatorname{axis} \mathbf{A}) = 1$ forces $\widetilde{u} = \pm u$ in the obvious notation.

There is a further consequence of these ideas; this is the *centred cut criterion.*

3K5 Corollary *If* $\operatorname{axis} \mathbf{B} = \{o\}$*, then the* $\mathbf{B}$*-cuts of* $\mathbf{G}/\mathbf{A}$ *are centred.*

3K6 Remark Similarly, the maximum values of $\eta(U_j)$ and $\eta(V_j)$ are the same; this value is achieved by choosing $u \in U_j$ and $v \in V_j$ so as to maximize $\langle u, v \rangle$. Moreover, if we have a realization for which (say) $\eta(U_j) < \eta(V_j)$ for some j, then we know that $\operatorname{axis} \mathbf{A}$ is at least 2-dimensional.

3L Realizations over Subfields

Recall that we call a geometric symmetric point-set V *rational* if its points have rational coordinates with respect to some (linear or affine) basis. It is important to recognize that rationality is not necessarily with respect to some euclidean coordinate system. For example, the vertex-set of a regular simplex is always rational, but may not have rational *cartesian* coordinates in its ambient space, as in the case of the trigon $\{3\}$. As might be expected, there is a connexion with rational cosine vectors; these turn up frequently, and so are of general interest.

In fact, it makes more sense to treat everything in a general context. If $\mathbb{F}$ is a subfield of $\mathbb{R}$, then we can talk about $\mathbb{F}$-vectors, $\mathbb{F}$-sets (that is, $\mathbb{F}$-point-sets) and $\mathbb{F}$-linear spaces with the obvious meaning. Apart from $\mathbb{Q}$, the field $\mathbb{A}$ of real algebraic numbers is obviously of interest; observe that an $\mathbb{A}$-set actually has coordinates in some finite extension of $\mathbb{Q}$ with respect to a suitable basis.

Our main result here is

3L1 Theorem *Let* $V \in \mathcal{V}$ *have cosine vector* Γ*, and let* $\mathbb{F}$ *be some subfield of* $\mathbb{R}$*. If* Γ *is an* $\mathbb{F}$*-vector, then* V *is an* $\mathbb{F}$*-set. Conversely, if* V *is an* $\mathbb{F}$*-set and* $w = 1$ *for the symmetry group* $\mathbf{G}$ *of* V*, then* Γ *is an* $\mathbb{F}$*-vector.*

Proof. We begin by observing that whether or not V is an $\mathbb{F}$-set is independent of normalization; thus we shall assume that $V \in \mathcal{N}$. Suppose that Γ is an $\mathbb{F}$-vector. That is, if v is the initial point of V and $x \in V$ is any other point, then $\langle x, v \rangle \in \mathbb{F}$. Indeed, since the point v is arbitrary, if $\{v_1, \ldots, v_d\} \subseteq V$ is any basis of the ambient space $\mathbb{E}$, then $\langle x, v_j \rangle \in \mathbb{F}$ for each $j = 1, \dots, d$. Let $\{v_1^*, \dots, v_d^*\}$ be the dual basis of $\mathbb{E}$, so that $\langle v_j, v_k^* \rangle = \delta_{jk}$, the Kronecker delta, for $j, k = 1, \dots, d$. If we write $\mathbb{V} := \mathbb{F}\{v_1, \dots, v_d\}$ for the $\mathbb{F}$-linear space spanned by $\{v_1, \dots, v_d\}$, this says that

$$x = \sum_{j=1}^{d} \langle x, v_j \rangle v_j^* \in \mathbb{V}^* = \mathbb{F}\{v_1^*, \dots, v_d^*\},$$

the $\mathbb{F}$-linear space dual to $\mathbb{V}$. This already shows that V is an $\mathbb{F}$-set, but note that, with $x = v_1, \dots, v_d$, this also implies that $\mathbb{V}$ and $\mathbb{V}^*$ coincide as $\mathbb{F}$-linear spaces; hence, x has coordinates in $\mathbb{F}$ with respect to $\{v_1, \dots, v_d\}$. In any event, we have established the first part of the theorem.

For the converse, suppose that V is an $\mathbb{F}$-set; we clearly lose no generality in taking o to be the centroid of V. Let v be the initial point of V, and extend to a basis $\{v_1, \dots, v_d\} \subseteq V$ of the ambient space $\mathbb{E}$, with $v_1 = v$. Since each point $x \in V$ has coordinates in $\mathbb{F}$ with respect to $\{v_1, \dots, v_d\}$, this says that $\langle x, v_k^* \rangle \in \mathbb{F}$ for each $k = 1, \dots, d$, with $\{v_1^*, \dots, v_d^*\}$ the dual basis of $\mathbb{E}$. Define the $\mathbb{F}$-linear spaces $\mathbb{V}$ and $\mathbb{V}^*$ as before, noting that $\mathbb{V}$ is independent of the choice of a basis of $\mathbb{E}$ in V. If $v^* \in \mathbb{V}^*$ is any vector such that $\langle v, v^* \rangle > 0$, define

$$\overline{v} := \sum_{\Phi \in \mathbf{H}} v^* \Phi \in \mathbb{V}^*,$$

with $\mathbf{H}$ as usual the stabilizer of v. Since $\overline{v}$ is invariant under $\mathbf{H}$, we have $\overline{v} \in W$, the Wythoff space of the group $\mathbf{G}$; because $w = 1$, it follows that $\overline{v} = \lambda v$ for some $\lambda > 0$. However, since $\|v\| = 1$, we have $\lambda = \langle v, \overline{v} \rangle \in \mathbb{F}$. Hence $v \in \mathbb{V}^*$, and the previous argument reverses to show that $\langle x, v \rangle \in \mathbb{F}$ for each $x \in V$, and thus Γ is an $\mathbb{F}$-vector. This completes the proof. □

3L2 Example The two 3-dimensional realizations of the abstract icosahedron $\{3, 5\}$ are the convex icosahedron $\{3, 5\}$ and great icosahedron $\{3, \frac{5}{2}\}$, with cosine vectors $(1, \pm\frac{1}{\sqrt{5}}, \mp\frac{1}{\sqrt{5}}, -1)$; see [85] or Section 7D below. Any non-trivial (normalized) blend of the two is rational, since the points comprise a centrally symmetric set of 12 points which span $\mathbb{E}^6$. However, for rational $0 < \lambda < \frac{1}{2}$ a corresponding cosine vector $(1 - \lambda)\big(1, \frac{1}{\sqrt{5}}, -\frac{1}{\sqrt{5}}, -1\big) + \lambda\big(1, -\frac{1}{\sqrt{5}}, \frac{1}{\sqrt{5}}, -1\big)$ is never rational.

3L3 Remark Example 3L2 shows that some extra condition is needed for the converse in Theorem 3L1. For a pure realization with $w > 1$, the new point $\overline{v}$ (suitably scaled) could give rise to another realization of $\mathcal{V}$ with the same group $\mathbf{G}$ which is not congruent to V. Indeed, even if $m = 1$, the new realization, though necessarily congruent to V, need not coincide with it.

3L4 Remark If $m > 1$, then corresponding cosine vectors vary continuously, and thus some of their entries will take transcendental values. One might guess, though, that when $m = 1$ the cosine vector Γ will be an $\mathbb{A}$-vector. If this is so, then Γ will be rational when the realization is the only one of that dimension.

3M Realizations and Representations

We fill in here a little of the background to what we said in Section 3G. For general references to representation theory in this context, we refer to [34] or, in particular, [122, Section II.6].

We begin by noting that we associate with each unitary representation $\mathbf{H}$ of $\boldsymbol{G}$ a complex conjugate representation $\overline{\mathbf{H}}$. If we think of $\mathbf{H}$ as composed of matrices U, then $\overline{\mathbf{H}}$ is composed of their complex conjugates $\overline{U}$. There are then three possibilities for a real irreducible representation $\mathbf{G}$ of degree d of the automorphism group $\boldsymbol{G}$.

- First, $\mathbf{G}$ is irreducible as a unitary representation; then $\mathbf{G}$ is called *real*. Its degree remains d.
- Otherwise, $\mathbf{G}$ is complex reducible, so that $\mathbf{G} = \mathbf{H} \oplus \overline{\mathbf{H}}$ for some irreducible unitary representation $\mathbf{H}$ of degree $\frac{1}{2}d$; this essentially regards $\mathbf{H}$ as acting on the corresponding euclidean space of twice the dimension. If $\overline{\mathbf{H}}$ is not unitarily equivalent to $\mathbf{H}$, then $\mathbf{H}$ is called *complex*; it is clearly centralized by scalar multiplication by complex numbers of absolute value 1, and this lifts to a corresponding centralizer of $\mathbf{G}$.
- Last, $\mathbf{G}$ is complex reducible, but now $\mathbf{H}$ is self-conjugate (still of degree $\frac{1}{2}d$), so that $\overline{\mathbf{H}}$ is unitarily equivalent to $\mathbf{H}$; one then writes $\mathbf{G} = 2\mathbf{H}$. Note, though, that $\overline{\mathbf{H}}$ does not now yield a different representation from $\mathbf{H}$. In this case, it can be shown that $\mathbf{H}$ can be regarded as a *quaternionic* representation (of degree $\frac{1}{4}d$). If $\mathbf{H}$ acts on the right on its quaternionic vector space, then left (scalar) multiplication by unit quaternions centralizes the action (see Section 1K for details about quaternions).

Referring back to Section 3G, we thus have $m = w/c$, with $c = 1, 2$ or 4. While we cannot expect to recover the orthogonality relations for characters from Λ-orthogonality in general, if only because these involve complex numbers (but there are connexions when the representations are unitarily irreducible), nevertheless we can use a special case of Theorem 3H4 to prove a basic result in representation theory (see, for example, [122, (III.1.1)]).

3M1 Theorem *If $\boldsymbol{G}$ is a finite group, then*

$$\sum_{\mathbf{H}} d(\mathbf{H})^2 = |\boldsymbol{G}|,$$

where the sum is over the unitarily irreducible representations $\mathbf{H}$ of $\boldsymbol{G}$, and $d(\mathbf{H})$ is the degree of $\mathbf{H}$.

Proof. Consider the regular representation of $\boldsymbol{G}$, that is, the permutation action of $\boldsymbol{G}$ on itself; thus the stabilizer $\boldsymbol{H}$ is the trivial subgroup $\{\boldsymbol{e}\}$. For each irreducible real representation $\mathbf{G}$ of $\boldsymbol{G}$, of degree d say, the Wythoff space W is the whole space (in particular, every representation of $\boldsymbol{G}$ will count). The essential Wythoff space m will thus have dimension $m = d/c$, with $c = 1, 2, 4$ as before. We look at the contribution md from $\mathbf{G}$ to $n = |\boldsymbol{G}|$ in the first sum in Theorem 3H4, referring to the listing above:

- $\mathbf{G} = \mathbf{H}$ is real, so that $m = d = d(\mathbf{H})$, giving contribution $md = d^2$ as required;
- $\mathbf{G} = \mathbf{H} \oplus \overline{\mathbf{H}}$ with $\mathbf{H}$ complex, so that $m = \frac{1}{2}d = d(\mathbf{H})$, yielding $md = (\frac{1}{2}d)d = (\frac{1}{2}d)^2 + (\frac{1}{2}d)^2$, again the right contribution;
- $\mathbf{G} = 2\mathbf{H}$, where $\mathbf{H}$ is quaternionic of degree $d(\mathbf{H}) = \frac{1}{2}d$ and $m = \frac{1}{4}d$, giving contribution $md = (\frac{1}{4}d)d = (\frac{1}{2}d)^2$, once more what was wanted.

Thus, in each case, $\mathbf{G}$ makes the correct contribution to the sum. □

It is known that, if $\mathbf{H}$ is an irreducible representation of $\boldsymbol{G}$, then its degree $d(\mathbf{H})$ is a divisor of $|\boldsymbol{G}|$ (see [122, Theorem III.4.1]). When $\mathbf{H}$ is complex, then the corresponding real representation $\mathbf{G}$ has degree $d_{\mathbf{G}} = 2d(\mathbf{H})$; hence, all that we can claim in general is

3M2 Proposition *If* $(V, \mathbf{G})$ *is a pure realization of* $(\mathcal{V}, \boldsymbol{G})$*, then* $\dim V$ *is a divisor of* $2|\mathbf{G}|$.

The almost trivial example of the cyclic group $\boldsymbol{C}_3$ shows that the estimate of Proposition 3M2 cannot be improved.

Finally, we repeat the remark that, while it appears that Theorem 3F5 can be derived (with a little work) from relations like [122, (III.1.2)], geometric properties such as that of Theorem 3C14 – of which we shall make quite extensive use – appear to have nothing to do with representations.

4

Realizations of Polytopes

Now that we have laid the foundations of realization theory in Chapter 3, we move on to the special case of polytopes. In Section 4A, we describe the basic idea of obtaining a polytope from a representation of an automorphism group, by means of Wythoff's construction. We also consider the geometric analogues of some of the basic operations on polytopes. Various connexions between rank and dimension of faithful realizations are then considered in Section 4B, in particular the concept of full rank. Realizations that are degenerate in some respect also play a part; we look at these in Section 4C. In Section 4D we discuss induced cosine vectors with particular reference to polytopes which, in a number of ways, provide additional tools for determining realization domains. A brief account of the alternating product of polytopes is then given in Section 4E. In Section 4F we sketch the theory of realizations of regular apeirotopes; however, some of the complications of the general theory are of little relevance here, and so we shall confine mention of them to the notes. For the most part, we shall postpone description of realization spaces of particular polytopes until we have introduced the polytopes themselves. However, in Section 4G we give several basic examples including polygons; these are needed to formulate the notion of rigidity in Chapter 6.

We note that more extended examples can be found in Part IV. For instance, Chapter 14 illustrates many of the various aspects of the theory by describing the realization domains of those Gosset-Elte polytopes that underpin regular polytopes of nearly full rank, while Chapter 15 treats realizations of certain locally toroidal regular polytopes.

4A Wythoff's Construction

We first remind the reader that, while we leave the consideration of apeirotopes until Section 4F, some of what we say in the initial stages applies to them as well. Initially, therefore, we are looking at realizations of a pair $(\mathcal{V}, \boldsymbol{G})$, where $\boldsymbol{G} = \boldsymbol{G}(\mathcal{P}) = \langle \boldsymbol{r}_0, \ldots, \boldsymbol{r}_{m-1} \rangle$ is the automorphism group of an abstract regular polytope $\mathcal{P}$ and $\mathcal{V} := \mathcal{P}_0$ is its vertex-set. The distinguished subgroup

$\boldsymbol{G}^v = \boldsymbol{G}_0 = \langle \boldsymbol{r}_1, \ldots, \boldsymbol{r}_{m-1} \rangle$ plays the rôle of $\boldsymbol{H}$ in the earlier discussion and, in keeping with it, $\mathcal{V}$ is identified with the set of right cosets of $\boldsymbol{G}_0$ in $\boldsymbol{G}$.

This section is concerned with the way one can construct a geometric regular polytope P from a suitable representation Ψ of the automorphism group $\boldsymbol{G}$. To begin with, the image $R_j := \boldsymbol{r}_j\Psi$ of the distinguished generator $\boldsymbol{r}_j$ of $\boldsymbol{G}$ is an involutory isometry, that is, a reflexion, or the identity. As in Section 1C, we identify R_j with its mirror $\{x \in \mathbb{E} \mid xR_j = x\}$, which is an affine subspace of $\mathbb{E}$. The whole realization can be recovered from these *canonical* mirrors. Since the initial vertex of $\mathcal{P}$ is identified with $\boldsymbol{G}_0$ or, rather, with its coset representative $\boldsymbol{e}$ (the identity), and hence is fixed by each of $\boldsymbol{r}_1, \ldots, \boldsymbol{r}_{m-1}$, we see that its image $v := \boldsymbol{e}\Psi$ lies in each mirror $R_1, \ldots, R_{m-1}$, and hence in their intersection

4A1 $$W = W_{\mathbf{G}} := R_1 \cap \cdots \cap R_{m-1},$$

which is thus the Wythoff space of $\mathbf{G} = \boldsymbol{G}\Psi$ (or of Ψ) in the terminology introduced in Section 3B; moreover, the whole vertex-set of the realization is $V = v\mathbf{G}$. Thus we identify v with the subgroup of $\mathbf{G}$ which fixes it; indeed, if we write $\mathbf{G}_j := \boldsymbol{G}_j\Psi$, then the image of the vertex-set of the initial face $\mathcal{P}_j$ is just $V_j := v\mathbf{G}_j$. (Note that, in effect, $V_j = v\langle R_0, \ldots, R_{j-1} \rangle$, because R_i and R_k commute if $i < j < k$, and R_k fixes v.)

The resulting family P of subsets $V_j\mathbf{G}$ of V, partially ordered by inclusion induced from that of $\mathcal{P}$, yields almost as much of the geometric structure of the latter as we can expect. We call P a *geometric regular polytope*, and the process which leads to it *Wythoff's construction* (see the notes at the end of the section).

As we did with point-sets, we let the same symbol P stand for the realization. We thus use the notation $\mathsf{P}\#\mathsf{Q}$ or $\mathsf{P}\otimes\mathsf{Q}$ for the blend or product of two polytopes $\mathsf{P}, \mathsf{Q} \in \mathcal{P}$; similarly, when P and Q have the same (essential) Wythoff space, we write $\lambda\mathsf{P} + \mu\mathsf{Q}$ for a linear combination. For the blend $\mathsf{P} \# \mathsf{Q}$, the various infaces – particularly faces – inherit their realizations from those of P and Q. The situation for the product $\mathsf{P} \otimes \mathsf{Q}$ is a little more subtle, and so we postpone discussion of this topic until later.

There are two different conventions for the vertex-figure Q of P at its initial vertex v. The strict or *narrow* vertex-figure has initial vertex $w := \frac{1}{2}(v + vR_0)$; observe that w is fixed by the image $\mathbf{G}_1$ of the distinguished subgroup $\boldsymbol{G}_1$ which, we recall, is identified with the initial edge of the abstract polytope $\mathcal{P}$. In contrast, the *broad* vertex-figure has initial vertex vR_0 instead. For most purposes this makes little difference, since the combinatorics are not altered; it is often useful to have the added convenience that the vertices of the vertex-figure occur among those of the polytope itself.

We can carry this through to realizations of more general infaces of $\mathcal{P}$. The iteration of the broad vertex-figure construction is discussed in Section 4D; we shall describe how to generalize the narrow definition very shortly.

4A2 Remark If $(R_0, \ldots, R_{m-1})$ is the generatrix of a geometric string group, with the R_j involutions, it is convenient to write $K_j := R_j \cap \cdots \cap R_{m-1}$, so that $K_1 = W$; see also Section 5D. Thus – leaving aside the intersection property

— to obtain a non-trivial realization, one must be able to find an initial vertex $v \in K_1 \setminus R_0$. We shall pursue this line of investigation further in Section 4C.

Pre-polytopes

A representation of an sggi (string group generated by involutions – see Section 2C) will be a *string group generated by reflexions* or *sggr* for short. The intersection property (2B6) translates directly into

$$\langle R_i \mid i \in \mathsf{J} \rangle \cap \langle R_i \mid i \in \mathsf{K} \rangle = \langle R_i \mid i \in \mathsf{J} \cap \mathsf{K} \rangle.$$

However, a general representation $\mathbf{G}$ of a string C-group $\boldsymbol{G}$ will not necessarily satisfy the intersection property, and so will only be an sggr; this is why we needed to talk about regular pre-polytopes. But Wythoff's construction does not demand that $\mathbf{G}$ satisfy (2B6), and so we can apply it in exactly the same way, and everything continues to go through.

Indeed, it is a crucial feature of realization theory that not all realizations are polytopal; Example 11C17 will show that this can happen even when Ψ is an isomorphism on $\boldsymbol{G}$. Such examples explain why we phrased the definition of realization in the way that we did, with equal emphasis on the point-set and the group acting on it.

Hyperplane Reflexion Groups

The definition of Coxeter diagram, particularly as originally formulated in [18], allows fractional marks p_{jk} on branches, to indicate a dihedral angle π/p_{jk} between the hyperplane mirrors of the corresponding reflexions; we have seen this already in (1E16). There is an important case in which an sggr – for geometric reasons – automatically satisfies the intersection property.

4A3 Theorem *A discrete irreducible string group* $\mathbf{G}$ *generated by hyperplane reflexions is a C-group.*

Proof. Let $\mathbf{G}$ have generatrix $(R_0, \ldots, R_{m-1})$ for some m. Because $\mathbf{G}$ is an irreducible string group, unit normals u_j to the reflexion hyperplanes R_j can be chosen so that $\langle u_{j-1}, u_j \rangle < 0$ for each $j = 1, \ldots, m-1$. The argument of Theorem 1E4 then shows that $\{u_0, \ldots, u_{m-1}\}$ is either linearly independent or is minimally positively dependent, so that the R_j bound either a simple cone or a simplex (possibly with a subspace as a direct sum component).

The obvious inductive argument shows that the intersection property only needs to be verified for $\mathbf{G}_0 \cap \mathbf{G}_{m-1}$. But in this case it is clear; any element in the intersection must fix pointwise the axes of both $\mathbf{G}_0$ and $\mathbf{G}_{m-1}$, and so must fix their affine (or linear) hull. Since this is the axis of $\mathbf{G}_{0,m-1}$, our claim is established. □

4A4 Corollary *Wythoff's construction applied to a discrete irreducible string group generated by hyperplane reflexions yields a regular polytope or apeirotope.*

4A5 Remark The proof of Theorem 4A3 did not depend on discreteness. Of course, the non-discrete cases are generally of lesser interest to us.

Generalizations

Wythoff's construction [131] generalizes in various ways. In principle, we can take any representation $\mathbf{G}$ of a C-group, and pick any point v as initial vertex; then the images of v under the different distinguished subgroups of $\mathbf{G}$ will give the faces of a realization of some polytope, which will not usually be regular. In this generality, the construction is really due to Robinson [111] (see the notes at the end of the section). We do not want such generality here, but it is appropriate to look at some special cases.

First, let $\mathbf{G}$ be a representation of the automorphism group of the abstract regular m-polytope $\mathcal{P}$, with corresponding generatrix $(R_0, \ldots, R_{m-1})$. If, for some k, we pick

$$v_{m-k} \in R_0 \cap \cdots \cap R_{m-k-1} \cap R_{m-k+1} \cap \cdots \cap R_{m-1},$$

then v_{m-k} is the initial vertex for a realization of the k-coface of $\mathcal{P}$ under the subgroup $\langle R_{m-k}, \ldots, R_{m-1} \rangle$. For instance, when $k = m - 1$ we obtain a realization of the vertex-figure of $\mathcal{P}$; thus our initial vertex in this case is $v_1 = vR_0$.

Wythoff's construction is much employed when $\mathbf{G}$ is a discrete hyperplane reflexion group (and sometimes in non-discrete cases as well). One then indicates the choice of v to lie on certain of the canonical mirrors by ringing the nodes of the Coxeter diagram corresponding to the *complementary* set of mirrors. Thus, in particular, a regular polytope is obtained by ringing just an end node of a string diagram.

The corresponding faces are then given by subdiagrams, whose components each contain at least one ringed node. In particular, if just one node of the diagram is ringed, then we obtain a diagram for the vertex-figure by deleting the ringed node, and ringing all the adjacent nodes in the diagram. In general, the derived diagram will then be disconnected, in which case the vertex-figure is a direct product. (The situation is more complicated if two or more nodes are ringed.)

4A6 Example Suppose that we ring the two end nodes of the graph for the group $\mathbf{A}_d$ of the d-simplex T^d, thus:

As a convex set, the resulting polytope is the difference body ΔT^d which we already met in Example 3D5; we shall see why shortly. As a realization of an abstract polytope, we denote it by $\Delta\mathsf{T}^d$. Then the diagrams for its facets are given by deleting a single node, and so will be $(d-1)$-simplices or products of k-simplices and $(d-k-1)$-simplices for $1 \leqslant k \leqslant \frac{1}{2}(d-1)$. Looked at in the hyperplane $\mathbb{L}^d$ of (1E15), the generators of $\mathbf{A}_d$ are the permutations $R_j = (j\ j{+}1)$ of the coordinates for $j = 0, \ldots, d-1$. The initial vertex in $R_1 \cap \cdots \cap R_{d-2}$ can thus be taken to be $v := (1, 0^{d-1}, -1)$, and the remaining vertices are all permutations of $(1, 0^{d-1}, -1)$.

Now we can write

$$v = \tfrac{1}{d+1}\big((d, (-1)^d) + (1^d, -d)\big),$$

with similar expressions for the general vertex, which shows indeed that (up to a scaling factor) we have the vertex-set of $\Delta\mathsf{T}^d$.

4A7 Example Now suppose that we ring a single node of the diagram $\mathbf{P}_{d+1}$, say node 0. The reflexion R_0 can be taken to be

$$R_0 \colon x \mapsto (\xi_d + 1, \xi_1, \ldots, \xi_{d-1}, \xi_0 - 1),$$

with $R_j := (j{-}1\ j)$ for $j = 1, \ldots, d$. We then obtain a honeycomb H with vertex-set $V = \mathbb{L}^d \cap \mathbb{Z}^{d+1}$; that is, V is the set of integer vectors in $\mathbb{E}^{d+1}$ whose coordinates sum to 0. As the rules explained above indicate, the (broad) vertex-figure of H is just $\Delta\mathsf{T}^d$, and the cells (or facets) of H are obtained from the diagram of $\mathbf{A}_d$ by ringing one node in all possible ways.

Different Polytopes

With both blends and products, we can easily generalize the concepts to realizations P and Q of *different* regular polytopes $\mathcal{P}$ and $\mathcal{Q}$; we just think of P and Q as realizations of some common regular cover $\mathcal{R}$ of $\mathcal{P}$ and $\mathcal{Q}$ (see the notes at the end of the section). More exactly, for blends we do the following. If $\mathbf{G}(\mathsf{P})$ is as before, and $\mathbf{G}(\mathsf{Q})$ similarly has generatrix $(S_0, \ldots, S_{k-1})$ for some k, then we formally set $R_j = I$ if $j \geqslant m$, and likewise for Q, and define

4A8 $$T_j := R_j \oplus S_j$$

for $j < \max\{m, k\}$. Then $\mathbf{G} := \langle T_0, T_1, \ldots \rangle$ is a pre-C-group; hence, in general, we only obtain a pre-polytope $\mathsf{P} \# \mathsf{Q}$.

The construction for products is almost exactly analogous. In this case, we actually have

4A9 $$T_j := (R_j \otimes S_j) \oplus (R_j^{\perp} \otimes S_j^{\perp}),$$

with the same conventions as before, but the natural definition of the product $\mathsf{P} \otimes \mathsf{Q}$ has initial vertex $u \otimes v$, where u, v are the initial vertices of P, Q.

Let us comment further about a combinatorial aspect of the construction; this follows from the definition of tensor product.

4A10 Proposition *If* P, Q *are geometric regular polytopes, then the product* $\mathsf{P} \otimes \mathsf{Q}$ *is a quotient of the blend* $\mathsf{P} \# \mathsf{Q}$ *as abstract polytopes. Indeed,* $\mathsf{P} \otimes \mathsf{Q} \cong \mathsf{P} \# \mathsf{Q}$, *except when both are centrally symmetric, with opposite vertices of each corresponding in the blend.*

Of course, the point is that, if (x, y) and $(-x, -y)$ are both corresponding vertices of the blend $\mathsf{P} \# \mathsf{Q}$, then $x \otimes y = (-x) \otimes (-y)$ is the single corresponding vertex of the product $\mathsf{P} \otimes \mathsf{Q}$.

4A11 Remark Blends of different polytopes may be non-polytopal; the close combinatorial connexion between blend and product given by Proposition 4A10 shows that products of realizations of different polytopes may be non-polytopal as well. An example which fails the intersection property on the abstract level comes from $\mathsf{P} \in \{3,3,3\}$ and $\mathsf{Q} \in \{\{5,3:5\},\{3,5:5\}\}$ (for details, see [99, Section 7A]); neither is centrally symmetric, and so the blend and product will be isomorphic.

4A12 Example As illustrations of Proposition 4A10, we have

(a) $\{3,3\} \otimes \{3,5\} \cong \{3,3\} \# \{3,5\}$,

(b) $\{3,4\} \otimes \{3,5\}$ is a quotient of $\{3,4\} \# \{3,5\}$ of index 2.

For (b), note that the index strings for the central symmetries of $\{3,4\}$ and $\{3,5\}$ are $(012)^3$ and $(012)^5$, respectively, which have a common expression as $(012)^{15}$. By the way, it is worth noting that these particular blends and products are all polytopal.

4A13 Remark The case $\mathsf{Q} = \{2\}$ of the digon (geometric 1-polytope), will be of particular importance.

We end the section with three results which relate realizations of polytopes and their faces. The first follows from the notion of quotient.

4A14 Theorem *If $\mathcal{P}$ is a k-collapsible regular polytope and $\mathcal{F}$ is the k-face of $\mathcal{P}$, then every realization of $\mathcal{F}$ induces one of $\mathcal{P}$.*

The second is

4A15 Proposition *If the assumptions of Proposition 2F7 hold, then the pure realizations of $\mathcal{P} \otimes \mathcal{Q}$ are exactly those of the form $\mathsf{P} \otimes \mathsf{Q}$, with $\mathsf{P} \in \mathcal{P}$ and $\mathsf{Q} \in \mathcal{Q}$ pure realizations.*

As we said in Section 2C, it is this result which motivates the abstract notation $\mathcal{P} \otimes \mathcal{Q}$.

Finally, with the notation of Proposition 4A15, we have

4A16 Theorem *If the regular polytope $\mathcal{P}$ is not centrally symmetric, then the realization domain of the abstract blend $\mathcal{P} \# \{2\}$ is $\mathcal{P} \otimes \{2\}$.*

Notes to Section 4A

1. Wythoff [131] only applied the construction named for him to the polytopes related to the 600-cell $\{3,3,5\}$. In the form presented here, it is due to Robinson [111].
2. Since pre-polytopes arise naturally in the theory of realizations, it may be natural to ask why we wish to confine our attention to polytopes. The simple answer (as we have said before) is that there are far too many regular pre-polytopes around which to construct a reasonable theory.

3. The honeycomb H of Example 4A7 is obviously not regular for $d \geqslant 3$ (for $d = 2$ we obtain the tessellation $\{3, 6\}$). However, it plays an important rôle when we come to look at regular apeirotopes of nearly full rank.
4. Monson and Schulte showed in [102] that the (minimal) common regular cover $\mathcal{R}$ of two finite regular polytopes $\mathcal{P}$ and $\mathcal{Q}$ is itself finite.
5. We often appeal to Theorem 4A14, but only meet Proposition 4A15 in Section 15G.

4B Faithful Realizations

There are important restrictions on faithful realizations, which we elucidate in this section. To treat them in a uniform way, it is convenient to think of a *finite* regular polytope P as realized in a sphere $\mathbb{S}$ of dimension $\dim \mathsf{P} - 1$. Thus, in this section (unless we say otherwise) the ambient space $\mathbb{U}$ will be a sphere $\mathbb{S}$, a euclidean space $\mathbb{E}$ or even a hyperbolic space $\mathbb{H}$, as appropriate. Dimensions, and so on, will then refer to the *intrinsic* concepts in $\mathbb{U}$.

We call a realization $\mathsf{P} \in \mathcal{P}$ *faithful* if $\mathbf{G}(\mathsf{P}) \cong \boldsymbol{G}(\mathcal{P})$ and the poset of faces F of P is isomorphic to $\mathcal{P}$.

4B1 Remark Our definition is phrased in this way to treat flat polytopes whose facets are not flat. Thus, while the facets may have the same vertex-sets, they are distinguished by their faces of smaller rank. The previous concept of faithfulness (of realizations of symmetric sets) will now be called *vertex-faithful*; similarly, an *edge-faithful* realization distinguishes edges.

Rank and Dimension

We begin by establishing various connexions between rank and dimension.

4B2 Theorem *Let* P *be a faithful realization of an abstract regular polytope* $\mathcal{P}$*, whose ambient space* $\mathbb{U}$ *is a spherical, euclidean or hyperbolic space. Then* $\dim \mathsf{P} \geqslant \operatorname{rank} \mathcal{P} - 1$.

Proof. The result trivially holds (with equality) when $\operatorname{rank} \mathcal{P} = 1$, because the only 1-polytope is the digon $\{2\}$, which is realized in $\mathbb{S}^0$. When $\operatorname{rank} \mathcal{P} \geqslant 2$, if $\mathcal{Q} := \mathcal{P}^v$ is the vertex-figure of $\mathcal{P}$, then we make the obvious inductive assumption that a faithful realization Q of $\mathcal{Q}$ satisfies $\dim \mathsf{Q} \geqslant \operatorname{rank} \mathcal{Q} - 1 = \operatorname{rank} \mathcal{P} - 2$. In particular, if Q is the (geometric) vertex-figure of P, then $\operatorname{vert} \mathsf{Q}$ lies on a sphere $\mathbb{S}$ in $\mathbb{U}$ with centre the initial vertex v (in the hyperbolic case, $\mathbb{S}$ may be a horosphere). Remembering that we must think of $\operatorname{vert} \mathsf{Q}$ as spanning an appropriate subspace of $\mathbb{S}$ (with the corresponding dimension), we thus see at once that

4B3
$$\begin{aligned} \dim \mathsf{P} &= \dim \mathbb{U} \geqslant \dim \mathbb{S} + 1 \geqslant \dim \mathsf{Q} + 1 \\ &\geqslant (\operatorname{rank} \mathcal{P} - 2) + 1 = \operatorname{rank} \mathcal{P} - 1, \end{aligned}$$

as required. □

If equality holds in Theorem 4B2, then we say that P is of *full rank*. The emphasis is placed this way round, because our later aim will be to classify such polytopes by dimension. In the same way, if $\dim \mathsf{P} = \operatorname{rank} \mathcal{P}$, then we call the realization one of *nearly full rank*. As we said in the foreword to the book, one of our main thrusts is the classification of the regular polytopes and apeirotopes of full or nearly full rank in euclidean spaces.

Because the condition of full rank forces equality in (4B3), an important consequence of the proof of Theorem 4B2 is the following.

4B4 Corollary *Each coface of a regular polytope of full rank is itself of full rank.*

Recall the definitions of vertex- and edge-faithful in Remark 4B1. We make the following comment.

4B5 Theorem *Let $\mathcal{P}$ be an abstract regular polytope with automorphism group $\boldsymbol{G}$. If $\mathbf{G}$ is a faithful representation in $\mathbb{U}$ of $\boldsymbol{G}$, and $v \in \mathbb{U}$ is an initial vertex yielding a vertex-faithful realization $V = v\mathbf{G}$ of $\mathcal{P}$ of full rank in $\mathbb{U}$, then the realization is faithful.*

Proof. It is enough to show that the realization is edge-faithful; we can then proceed by induction on the rank of $\mathcal{P}$. Suppose, to the contrary, that it is not. Then v is joined by multiple edges to the adjacent vertex $w := vR_0$. But this immediately implies that the corresponding representation $\mathbf{G}_0$ of the group $\boldsymbol{G}_0$ of the vertex-figure of $\mathcal{P}$ in its ambient space $\mathbb{S}$ is not faithful, so that there are non-identity elements of $\boldsymbol{G}_0$ whose images under the induced representation are the identity in $\mathbf{G}_0$. Since V is of full rank, it follows that v and $w\mathbf{G}_0$ together span $\mathbb{U}$, and hence there are non-identity elements of $\boldsymbol{G}_0$, and thus also of $\boldsymbol{G}$, whose images act as the identity on $\mathbb{U}$. This contradicts the assumption that the representation $\mathbf{G}$ of $\boldsymbol{G}$ was faithful, and thus establishes the theorem. □

4B6 Remark The recursive argument employed here can be made to show that the groups of realizations of full rank are actually irreducible (affinely, in the case of euclidean realizations). We do not bother to prove this formally, because it emerges in any case from the classifications.

4B7 Remark The extent to which the condition of full rank in the proof of Theorem 4B5 is needed is not clear.

Mirror Vectors

If $(R_0, \ldots, R_{m-1})$ is the generatrix of a realization P of an abstract regular m-polytope, then we call $M(\mathsf{P}) := (\dim R_0, \dim R_1, \ldots, \dim R_{m-1})$ its *mirror vector* (see the notes at the end of the section). It is often the case that a first step in the classification in a fixed dimension d of the faithfully realized regular polytopes or apeirotopes of a fixed rank m is to determine which mirror vectors can occur. We can go further than Theorem 4B2, and place restrictions on the mirror vector of the realization. We treat the finite and infinite cases together by working in the ambient space $\mathbb{U}$, and so initially the mirror vector is interpreted accordingly. We first have a general result.

4B8 Theorem *If* P *is a faithful realization of an abstract regular* m*-polytope, then*

$$M(\mathsf{P}) \geqq (0,1,2,\ldots,m{-}2,m{-}2).$$

Here, $\geqq$ denotes term-by-term inequality.

Proof. This uses the same inductive idea as in the proof of Theorem 4B2; again beginning with the digon $\{2\}$, whose symmetry group has a single generating reflexion R_0, obviously of dimension $\dim R_0 \geqslant 0$. If Q realizes the vertex-figure of P in the sphere $\mathbb{S}$ with centre the initial vertex v of P, then appealing to the natural inductive assumption shows that $M(\mathsf{Q}) \geqq (0,1,\ldots,m{-}3,m{-}3)$. If $(S_0,\ldots,S_{m-2})$ is the generatrix of Q, then R_j is spanned by v and S_{j-1} for $j \geqslant 1$, and hence $\dim R_j \geqslant (j-1)+1 = j$. That $\dim R_0 \geqslant 0$ is trivial, which completes the proof. □

In the case of polytopes of full rank, we can say even more; Theorem 4B9 is the key to the resolution of the case of full rank.

4B9 Theorem *If* $(R_0,\ldots,R_{m-1})$ *is the generatrix of a faithful realization* P *of full rank, then* $\dim R_j = j$ *or* $m-2$ *for* $j = 0,\ldots,m-3$*, and* $\dim R_{m-2} = \dim R_{m-1} = m-2$.

Proof. We first observe that equality in Theorem 4B2 forces the vertex-figure Q of P to be of full rank also; this will then hold for all cofaces (iterated vertex-figures) as well. The core case to be established is that for $j = 0$; the remaining cases will then follow easily. The proof is by induction on m, beginning with the digon $\{2\}$, whose symmetry group is generated by the point-reflexion in the centre of a 0-sphere.

As before, working in the ambient space $\mathbb{U}$ enables us to treat polytopes and apeirotopes at the same time. Suppose, therefore, that P is an m-polytope with $m \geqslant 2$, and that the result holds for polytopes of rank $m-1$. Let the initial vertex of P be v. Then the subset of $\mathbb{U}$ fixed pointwise by the group $\mathbf{G}_0$ of Q consists of the point v (and the antipodal point of $\mathbb{U}$, if $\mathbb{U}$ is a sphere), but contains no higher-dimensional subspace through v; the vertices of Q lie on a sphere (possibly a horosphere) $\mathbb{S}$ in $\mathbb{U}$ centred at v. Let w be the initial vertex of Q, so that w is the mid-point of the edge E through v and vR_0, and E spans the line L joining v and w. Then the group $\mathbf{G}_{01}$ (say) of the edge-figure at E has fixed-point set exactly L (by the same token, since this edge-figure is the vertex-figure of Q), and acts irreducibly on the hyperplane H perpendicular to L through w. Hence, because R_0 commutes with $\mathbf{G}_{01}$, it follows that, if R_0 fixes any point of H distinct from w, then it must fix the whole of H. We conclude at once that either $R_0 = \{w\}$, or $R_0 = H$. This is the required result.

For the remaining assertions, if Q has generatrix $(S_0,\ldots,S_{m-2})$, with the S_j mirrors in $\mathbb{S}$, then R_j is the subspace spanned by S_{j-1} and v for $j = 1,\ldots,m{-}1$, and the obvious inductive assumption yields the result. □

4B10 Remark Observe that the intesection property (2B6) was not appealed to in the proofs of Theorems 4B8 and 4B9; in other words, they apply to regular pre-polytopes as well (see the notes at the end of the section).

4B11 Remark For realizations of finite polytopes we revert to the earlier conventions, and so add 1 to each entry of their mirror vectors. Observe that, if P_j is a realization of $\mathcal{P}$ for $j = 1, 2$, then

$$M(\mathsf{P}_1 \# \mathsf{P}_2) = M(\mathsf{P}_1) + M(\mathsf{P}_2).$$

4B12 Remark In the case of a faithful realization of full rank of a finite regular m-polytope with centre o, if the mirror R_0 is a line, then $-R_0 = (-I)R_0$, the product of R_0 with the central inversion $-I$, is a hyperplane reflexion. If we replace R_0 by $-R_0$, then at worst we have replaced the symmetry group $\mathbf{G}$ by $\mathbf{G} \times \mathbf{Z}_2$, with $\mathbf{Z}_2 = \{\pm I\}$; in any event, we always have another finite group.

For the corresponding geometric polytopes, this process interchanges P and $\mathsf{P} \otimes \{2\}$.

Notes to Section 4B

1. We use the term *mirror vector* (due originally to Pellicer and Schulte [107]) instead of the rather less evocative *dimension vector* of [99] and papers earlier than [86].
2. Remark 4B10 will be useful when we discuss blended vertex-figures in Section 9A.

4C Degenerate Realizations

We next discuss various conditions under which a realization falls short of being faithful. We first observe that a realization may (effectively) have lower rank than the original polytope; this fact is unavoidable. We begin by presenting one important condition.

4C1 Theorem *If the generatrix $(R_0, \ldots, R_{m-1})$ of the realization P of the regular polytope $\mathcal{P}$ is such that $R_k \rightleftharpoons R_{k-1}$ for some $k \geqslant 1$, then P is fixed by each R_j for $j \geqslant k$. In this case, $\mathcal{P}$ is k-collapsible, and the realization is that of the k-collapse of $\mathcal{P}$.*

Proof. Since we now have $R_j \rightleftharpoons R_i$ for each $i \leqslant k-1$ and $j \geqslant k$, we see that R_j fixes each face of P of rank at most k. Moreover, no faces of higher rank than k can be realized, from which we infer that $\mathcal{P}$ must be k-collapsible. □

4C2 Remark The condition of Theorem 4C1 certainly holds if $R_k = I$; in this case, we say that the realization P is *k-trivial.*

In our later discussion of blends, we need a somewhat weaker condition than faithfulness. Let P be a realization of an abstract regular m-polytope $\mathcal{P}$ whose ambient space $\mathbb{U}$ is a sphere or euclidean space according as P is bounded or unbounded. (We shall see in Section 4F that regular apeirotopes can have bounded realizations.) If $\mathbf{G}(\mathsf{P})$ is the group of P, with canonical generators $(R_0, \ldots, R_{n-1})$, then we call P *untrivial* if $R_j \neq I$ for each $j = 0, \ldots, n-1$. We have shown in the preceding Theorem 4C1 that, since we always confine our attention to the ambient space, if $R_k = I$, then $R_j = I$ for each $j \geqslant k$. Now exactly the same inductive argument based on vertex-figures as in Theorem 4B8 shows that $\dim R_j \geqslant j$, at least so long as P is not j-trivial, except that we can have $\dim R_{n-1} = n-2$. Applying this to k-faces (bearing in mind that a realization may collapse onto a k-face, thus replacing m by k in the previous sentence), we deduce the *triviality criterion*:

4C3 Theorem *A realization* P *of an abstract regular polytope* $\mathcal{P}$ *with ambient space* $\mathbb{U}$ *is* k*-trivial for each* $k > \dim \mathbb{U}$.

4C4 Remark In the case of a finite (or bounded) realization P in a euclidean space $\mathbb{E}$, the degeneracy criterion of Theorem 4C3 says that P is k-trivial for each $k \geqslant \dim \mathbb{E}$.

Theorem 4C3 has the following consequence.

4C5 Theorem *Let* P *be an untrivial realization of an abstract regular polytope or apeirotope. If* P *is of full rank, then* P *must be pure. If* P *is of nearly full rank but is not pure, then one component of* P *is* 1*-dimensional (and the other is pure).*

Proof. We take P to be bounded; the proof in the other case is analogous. Let $\dim \mathsf{P} = d$. If P is blended with a component of dimension $r > 0$, then this component must be k-trivial for $k \geqslant r$. Similarly, the other component must be k-trivial for $k \geqslant d - r$. Hence P itself is k-trivial for $k \geqslant \max\{r, d-r\}$, and we immediately deduce the claims of the theorem. □

4C6 Remark It is easy to see that the arguments of Theorems 4B8 and 4B9 really only depend on untriviality rather than faithfulness. Indeed, they do not actually depend on $\mathbf{G}$ being a C-group rather than merely an sggr; we shall be able to make use of this fact in Section 9A.

We shall pay little heed to the hyperbolic case; apart from this being largely beyond the intended scope of the book, its treatment is complicated even in the discrete case. For the finite case, we follow Remark 4B11, and regard a sphere which carries the vertices of a realization P of a finite regular polytope as sitting with centre the origin o in the euclidean space of one larger dimension. The mirrors R_j of its group $\mathbf{G}$ are then thought of as linear subspaces, also of one larger dimension than before.

4D Induced Cosine Vectors

We introduced the concept of induced cosine vectors in a very general way at the end of Section 3C. Since these are the most important cases, we anticipate the more general context of Section 5B by recalling the notation $\mathcal{P}^f$ for the facet of a regular polytope $\mathcal{P}$, and introducing $\boldsymbol{G}^f$ for its group and Γ^f for its induced cosine vector; similarly, $\mathcal{P}^v$, $\boldsymbol{G}^v$ and Γ^v denote the same notions for the vertex-figure of $\mathcal{P}$.

In the context of realizations, we must first make an important observation.

4D1 Lemma *An inface $\mathcal{K}$ of a regular polytope $\mathcal{P}$ induces a cut of $\mathcal{V} := \operatorname{vert} \mathcal{P}$ with full automorphism group $\boldsymbol{G}(\mathcal{K})$.*

Proof. The argument is recursive, since we need only establish two extreme cases. First, for the facet $\mathcal{F} := \mathcal{P}^f$ this is clear: $\operatorname{vert} \mathcal{F} \subseteq \operatorname{vert} \mathcal{P}$, and $\boldsymbol{G}(\mathcal{F}) \leqslant \boldsymbol{G}(\mathcal{P})$. Second, we take the vertex-figure $\mathcal{Q} := \mathcal{P}^v$ at the initial vertex $\boldsymbol{e}$ in the broad sense, so that

$$\operatorname{vert} \mathcal{Q} := \{\boldsymbol{x} \in \operatorname{vert} \mathcal{P} \mid \{\boldsymbol{e}, \boldsymbol{x}\} \text{ an edge of } \mathcal{P}\}.$$

Then $\boldsymbol{G}(\mathcal{Q})$ acts naturally on this copy of $\mathcal{Q}$. For a general inface, we can take successive facets and vertex-figures. This establishes the claim. □

4D2 Remark The cut constructed in this way will not have the full group of automorphisms that preserve an inface $\mathcal{K}$; it will lack those that induce the identity on $\mathcal{K}$. However, we do at least have $\boldsymbol{G}(\mathcal{K})$ acting on $\mathcal{K}$ in a natural way. In fact, other cuts than those of infaces can also prove useful, for instance, those of layers different from the vertex-figure.

Theorem 3C14 and the discussion of Section 3K provide criteria for the existence (or otherwise) of realizations with given cosine vectors. The tests are usually performed on facets or vertex-figures, but clearly more general infaces can be employed. If $\mathcal{K}$ is an inface of $\mathcal{P}$ and K is the induced realization of $\mathcal{K}$ in a realization P of $\mathcal{P}$, then we write

4D3 $$\langle \Gamma_0(\mathcal{K}), \Gamma(\mathsf{K}) \rangle_{\Lambda(\mathcal{K})} =: \begin{cases} \eta_f(\mathsf{P}), & \text{when } \mathcal{K} = \mathcal{P}^f \text{ is the facet,} \\ \eta_v(\mathsf{P}), & \text{when } \mathcal{K} = \mathcal{P}^v \text{ is the vertex-figure;} \end{cases}$$

here, $\Gamma_0(\mathcal{K})$ is just the henogon component of $\mathcal{K}$.

A special case of Theorem 3C14 provides the *vertex-figure criterion*, with the usual convention that layer $\mathcal{L}_1$ of $\mathcal{P}$ consists of the vertex-figure.

4D4 Theorem *If $\mathsf{P} \in \mathcal{P}$ is any realization, then*

$$\eta_v(\mathsf{P}) \geqslant \gamma_1(\mathsf{P})^2,$$

with equality if $w(\mathsf{P}) = 1$.

The converse of this result is often useful.

4D5 Corollary *If $\mathsf{P} \in \mathcal{P}$ is a realization such that $\eta_v(\mathsf{P}) > \gamma_1(\mathsf{P})^2$, then $w(\mathsf{P}) > 1$.*

If $\mathcal{P}$ is a regular polytope with automorphism group $\boldsymbol{G}$, then a $\boldsymbol{G}^{\boldsymbol{f}}$-cut of $\mathcal{P}$ is called a *stratum*. We have a generalization of a classical result which we shall call the *dual facet criterion* (see the notes at the end of the section); it is the special case of Theorem 3K4 with $\boldsymbol{A} = \boldsymbol{G}^{\boldsymbol{v}}$ and $\boldsymbol{B} = \boldsymbol{G}^{\boldsymbol{f}}$.

4D6 Theorem *If* $\mathsf{P} \in \mathcal{N}$ *is a realization of* $\mathcal{P}$ *such that* $\eta_{\boldsymbol{f}}(\mathsf{P}) > 0$, *then there is a realization* $\mathsf{P}^\delta \in \mathcal{N}(\mathcal{P}^\delta)$ *of a geometric dual in the same space. Moreover, if* $w(\mathsf{P}) = 1$, *then such a geometric dual* Q *can be chosen so that*

$$\eta_{\boldsymbol{f}}(\mathsf{Q}) = \eta_{\boldsymbol{f}}(\mathsf{P}).$$

4D7 Remark There are obvious generalizations of Theorem 4D6. For instance, suppose that $w(\mathsf{P}) = 1 = w(\mathsf{Q})$, with $\mathsf{Q} := \mathsf{P}^\delta$. Then Proposition 3K2 tells us that we can rearrange P and Q so that their strata fall into the same set of parallel hyperplanes.

A simple example of this is the dual pair of icosahedron and dodecahedron. An even more striking case consists of the 600-cell and 120-cell; compare the coordinates of [27, Table V(iv)].

4D8 Remark In most cases, the first part of the theorem is more useful, since the mere fact that $\eta_{\boldsymbol{f}}(\mathsf{P}) > 0$ tells us little about the corresponding realization P^δ of the dual. For the second part of the theorem, observe that there are examples when $w^*(\mathsf{P}) > 1$ for which the second dual $\mathsf{P}^{\delta\delta}$ does not coincide with P (see the notes at the end of the section). Furthermore, if $\eta_{\boldsymbol{f}}(\mathsf{Q}) > 0$, and Q yields a geometric dual P for which $\eta_{\boldsymbol{f}}(\mathsf{Q}) < \eta_{\boldsymbol{f}}(\mathsf{P})$, then we immediately deduce that $w^*(\mathsf{Q}) > 1$.

Notwithstanding this remark, a special case is worth noting.

4D9 Corollary *If* $\mathcal{P}$ *is a simplicial* m*-polytope that is not* $(m-1)$*-collapsible, then to every pure realization* P *of* $\mathcal{P}$ *corresponds a pure realization* P^δ *of* $\mathcal{P}^\delta$ *of the same dimension. Moreover, if* $\Gamma = (1, \gamma_1, \ldots)$ *is the cosine vector of* P, *then* $\gamma_1 > -1/(m-1)$.

Proof. Indeed, the facet F of P cannot be centred, so that $\eta_{\boldsymbol{f}}(\mathsf{P}) > 0$. Since the induced cosine vector of the facet $\mathcal{F}$ of $\mathcal{P}$ is $(1, \gamma_1)$, the second claim follows from Corollary 3C8. □

A converse of Theorem 4D6 is also very useful; this is a restatement of Corollary 3K5, and is called the *centred facet criterion.*

4D10 Corollary *If* P *is a pure realization of* $\mathcal{P}$, *such that there is no pure realization of the dual* $\mathcal{P}^\delta$ *of dimension* $\dim \mathsf{P}$, *then the facets of* P *are centred.*

While we are looking at duals, we can make a general comment. Recall from Section 2F that a polytope is flat if each of its vertices belongs to every facet. It is obvious that no realization of a flat polytope can have a geometric dual. By contrast, we have

4D11 Proposition *If the regular polytope $\mathcal{P}$ is not flat, then there is some pure realization of $\mathcal{P}$ that has a geometric dual.*

Proof. It is clear from its definition in (4D3) that $\eta_{\boldsymbol{f}}$ is a linear function of the cosine vector. If T' is the complementary component of the simplex realization T to the henogon $\{1\}$, then T' is centred, while the corresponding realization F' of the facet $\mathcal{F}$ of $\mathcal{P}$ is not; that is, $\eta_{\boldsymbol{f}}(\mathsf{T}') > 0$. It follows at once that $\eta_{\boldsymbol{f}}(\mathsf{P}) > 0$ for at least one pure realization $\mathsf{P} \in \mathcal{P}$, and the claim of the proposition is then a consequence of Theorem 4D6. □

4D12 Remark Induced cosine vectors can be used in various other ways; for an application of a quite different kind, we make a forward reference to Sections 7C and 7F.

At this point, we introduce some useful notation. If Γ is the cosine vector of a polytope P, then we write $\Gamma(\mathsf{P}) := \Gamma$ and $\mathsf{P}(\Gamma) := \mathsf{P}$; furthermore, we define $d(\Gamma) := \dim \mathsf{P}$. Thus we can express Corollary 3F4 in the form

4D13 $$d(\Gamma) \geqslant 1/\|\Gamma\|_{\Lambda}^2,$$

without mentioning the corresponding polytope P.

As we saw in Section 3H, in the real case when $w > 1$ the cosine vector of a corresponding pure realization $\mathsf{P} = \mathsf{P}(x)$ is a quadratic expression of the form

$$\Gamma(\mathsf{P}) = \sum_{j,k=1}^{w} \xi_j \xi_k \Gamma_{jk},$$

where the Γ_{jk} are fixed and $x = (\xi_1, \ldots, \xi_w)$ is a unit vector. The same is therefore true of $\eta_{\boldsymbol{f}}(\mathsf{P})$, which we can thus write in the form $\eta_f(\mathsf{P}) = xHx^{\mathsf{T}}$ for some fixed symmetric matrix $H = (\eta_{jk})$. Since necessarily $\eta_{\boldsymbol{f}}(\mathsf{P}) \geqslant 0$, we deduce

4D14 Proposition *The quadratic form xHx^{T}, such that*

$$\eta_{\boldsymbol{f}}(\mathsf{P}(x)) = xHx^{\mathsf{T}},$$

is positive semi-definite.

It seems appropriate at this point to comment on the case of centred facets. We have

4D15 Proposition *If the ridges of a geometric regular polytope* P *are centred in its facets, then the facets are centred in* P.

Proof. If F, G are adjacent facets, meeting on the ridge R, then their centres satisfy

$$c(\mathsf{F}) = c(\mathsf{R}) = c(\mathsf{G}).$$

It follows at once that all facets of P have the same centre, which must be that of P itself. □

Notes to Section 4D

1. Theorem 4D6 generalizes an observation made (according to Hypsicles in Book XIV of Euclid's *Elements*) by Apollonius of Perga, which says, in effect, that regular icosahedra and dodecahedra with the same circumsphere have the same insphere. In the notation used by Coxeter throughout [27], $\eta_f(\mathsf{P}) = \cos^2\chi(\mathsf{P})$.
2. There are 10-dimensional realizations P of the regular polyhedron $\{5, 5 : 5\}$ with $w = 2$ for which $\mathsf{P}^{\delta\delta} \neq \mathsf{P}$. Indeed, as noted in [88], successively taking second duals leads to an infinite sequence of pairwise incongruent realizations.

4E Alternating Products

In Section 1A we introduced a different product of vector spaces, namely, the alternating product, and in Section 1C we described the corresponding euclidean structure. As we suggested there, it does play a minor part in realization theory, but under such restrictive circumstances that it would have been inappropriate to have mentioned it earlier.

We can only apply the alternating product to a single space, $\mathbb{E}$ say, although we may apply it to several terms. However, in order to apply it more than twice, it is clear that we must obtain something non-trivial in $\mathbb{E} \wedge \mathbb{E}$ first. If $\boldsymbol{G}$ is the automorphism group of the abstract regular polytope $\mathcal{P}$, then the alternating property of the product implies that the representations $\mathbf{G}$ and $\mathbf{H}$ of $\boldsymbol{G}$ in $\mathbf{O} = \mathbf{O}(\mathbb{E})$ must be very closely related. This can happen in one of two ways.

First, $\mathbf{G} = \mathbf{H}$, with Wythoff space W^* of dimension $w^* \geqslant 2$. Then initial vertices u of P and v of Q (say) must be chosen so that $u \neq \pm v$ (assuming that we take normalized realizations for simplicity); the resulting realization $\mathsf{P} \wedge \mathsf{Q}$ depends only on the subspace $\operatorname{lin}\{u, v\}$, and so the natural choice is to take $\langle u, v\rangle = 0$ (this also ensures that $\|u \wedge v\| = 1$). In practice, we have found no cases where such products make a useful contribution.

Naturally, for general $w^* > 1$ we can take an alternating product of up to w^* polytopes, whose initial vertices are linearly independent. It is not too hard to see that, if P_j has initial vertex $e_j \in W^*$ for $j = 1, \ldots, s$, with $\{e_1, \ldots, e_s\}$ an orthonormal set, then the cosine vector of $\mathsf{P}_1 \wedge \cdots \wedge \mathsf{P}_s$ is just

$$\varGamma(\mathsf{P}_1 \wedge \cdots \wedge \mathsf{P}_s) = \det(\varGamma_{jk}).$$

Here, $\varGamma_{jk}$ is the non-symmetric mixed cosine vector (as defined at the end of Section 3G) when $j \neq k$, so that – as we saw – $\varGamma_{jk} \neq \varGamma_{kj}$; moreover, the multiplication of cosine vectors in expanding the determinant is just the usual one. For instance, if $s = 2$, then we have

$$\varGamma(\mathsf{P}_1 \wedge \mathsf{P}_2) = \varGamma_{11}\varGamma_{22} - \varGamma_{12}\varGamma_{21};$$

observe that, since $\langle \varGamma_{11}, \varGamma_{22}\rangle_\Lambda = \langle \varGamma_{12}, \varGamma_{21}\rangle_\Lambda = 0$, the alternating product is centred. (By the way, notice that the symmetrized $\varGamma^e_{12}$ cannot stand here instead of $\varGamma_{12}$ and $\varGamma_{21}$, since that would make the layer inequality negative.)

Second, $\mathbf{G} \neq \mathbf{H}$, when the generatrices $(R_0, \ldots, R_{m-1})$ and $(S_0, \ldots, S_{m-1})$ must be such that $R_j = \pm S_j$ for $j = 0, \ldots, m-1$, again because of the alternating nature of the product. Moreover, since the initial vertices u and v (as before) must satisfy $u \neq \pm v$, having just $R_0 = -S_0$ alone will not generally suffice. When we come to such instances as there are, it turns out that $R_1 = -S_1$ is the productive case; see the notes at the end of the section. Moreover, we have never genuinely needed to use the alternating product to find missing realizations; only subsequently does it become apparent that certain realizations can be obtained in this way.

For now, we confine ourselves to a single case. In Example 3D5, the only difference between $\mathcal{V}$ and $\mathcal{V}/2$ is that the group of the former contains a central involution while that of the latter does not. In particular, this shows up in the otherwise close relationship between the two d-dimensional realizations V_1 and V_3 (in the notation of the example). Using the coordinate vectors given in the example and the definition of the inner product in (1C12), we can easily check that $V_1 \wedge V_3 = V_4$.

Notes to Section 4E

1. It happens that the pure faithful 6-dimensional realization of $\{4, 6 \mid 3\}$ is actually $\{4, \frac{6}{1,3} \mid 3\} \wedge \{4, \frac{6}{2,3} \mid 3\}$; see Remark 11C8 for some details.
2. In Section 14C it can be seen retrospectively that the alternating product of two 6-dimensional pure realizations is a pure realization of dimension 15. Appealing to the alternative construction thus provides independent evidence that the description of the realization domain is correct.

4F Apeirotopes

Much of the material expounded in the early part of the last chapter and the first sections of this carry over to apeirotopes with no more than obvious changes; indeed, the treatment in Section 4B explicitly covered apeirotopes as well. We now have a countable group $\boldsymbol{G}$ – the automorphism group of a regular apeirotope $\mathcal{P}$, and so of the form $\boldsymbol{G} = \langle \boldsymbol{r}_0, \ldots, \boldsymbol{r}_{m-1} \rangle$ – acting transitively on the set $\mathcal{V} = \operatorname{vert} \mathcal{P}$, which is therefore also countable.

The definition of a realization is much the same, except that we must replace the orthogonal group $\mathbf{O}(\mathbb{E})$ by the isometry group $\mathbf{M}(\mathbb{E})$. The operations of scaling and blending of realizations can be defined as before, and hence the realization space, also denoted $\mathcal{P}$ as in our previous convention, again has the structure of a convex cone.

However, inner product and cosine vectors are now clearly inappropriate, and so we must revert to diagonal vectors as in the earlier account of the theory (see [99, Chapter 5]). The vertices of $\mathcal{P}$ still fall into layers $\mathcal{V}_0, \mathcal{V}_1, \ldots$ (now infinitely many), with $\mathcal{V}_0 = \{v\}$ consisting of the initial vertex $v \in R_1 \cap \cdots \cap R_{m-1}$ alone. If Ψ is a realization of $\mathcal{P}$, $v_s \in V_s := \mathcal{V}_s \Psi$ and $\delta_s := \|v_s - v\|^2$ for $s \geqslant 1$, then

4F1 $$\Delta = \Delta(\Psi) := (\delta_1, \delta_2, \ldots)$$

is called the *diagonal vector* of Ψ (or of the corresponding apeirotope $\mathsf{P} = \mathcal{P}\Psi$). We may alternatively define $\delta_s = \|x - y\|^2$, with $\{x, y\} \subseteq V$ representing the sth diagonal class $\mathcal{D}_s$ of $\mathcal{P}$.

Blending and scalar multiplication behave for diagonal vectors in exactly the same way as for inner product vectors:

4F2
$$\Delta(\mathsf{P} \# Q) = \Delta(\mathsf{P}) + \Delta(Q),$$
$$\Delta(\lambda P) = \lambda^2 \Delta(\mathsf{P}).$$

4F3 Remark It is obvious that we have no product structure for realizations of an apeirotope; thus the henogon $\{1\}$ plays no rôle here. Correspondingly, the diagonal vector has no entry $\delta_0 = 0$ for the trivial diagonal.

The Wythoff space W of a realization is now to be thought of as an affine subspace of the ambient space $\mathbb{E}$; if W is not a singleton set, then we can take affine combinations of initial vertices of different realizations of $\mathcal{P}$, and so define affine combinations of them. However, except in the somewhat restricted context which we shall consider in Section 5E, we do not seem to find much use for this concept.

Discreteness

There is a restriction on realizations of infinite sets which we subsequently wish to impose, that of discreteness. For the moment, this just means that $V = \mathcal{V}\Psi$ is discrete. Observe though that, in general, a discrete infinite realization that is a blend can have non-discrete components; we shall see an example at the end of the section. When the ambient space $\mathbb{E} = \operatorname{conv} V$, this will imply that the symmetry group $\mathbf{G}(V)$ is crystallographic. It is usual in these circumstances to take the initial vertex v to be the origin o of coordinates, so that the stabilizer subgroup $\mathbf{H} = \mathbf{G}_0$ is an orthogonal group. If $\mathcal{Q}$ is the vertex-figure of $\mathcal{P}$, on which the discreteness condition will impose finiteness, it will be the realizations of $\mathcal{Q}$ rather than of $\mathcal{P}$ itself which will largely govern any investigations.

The example of $\mathbb{Z}$ acted on by itself (as a group of 'translations'), realized as points on a circle, with successive points subtending a fixed irrational angle at the centre, shows that infinite sets can have faithful bounded realizations, though clearly these cannot be discrete.

Our treatment of realizations of apeirotopes will be rather more cursory than that of finite polytopes; for further details, we refer the reader to [80]. For the most part, this is because the more bizarre features of their realization cones will not be exhibited by any of the examples presented in this book. In particular, irrational angles (that is, irrational multiples of π) do not play a rôle when the corank is small.

The Apeirogon

Some of the problems which we wish to avoid meeting subsequently can be found in a simple example. As we have seen, the *apeirogon* is the infinite (regular) polytope $\{\infty\}$ of rank 2; we discuss some features of its realization cone.

The group $\mathbf{G} = \mathbf{G}(\mathsf{P})$ of a realization P of $\{\infty\}$ is generated by two reflexions R_0 and R_1. Then $T := R_1R_0$ acts as a (combinatorial) translation, moving one vertex of P to the next, and the diagonal vector $\Delta(\mathsf{P})$ of P is clearly determined by T and the initial vertex of P.

The discussion of isometries in Section 1C shows that we can express T, in a unique way, as a commuting product of a translation (by a vector t, say, which may be o) and an orthogonal transformation Θ acting on a hyperplane orthogonal to t (if $t \neq o$). Further, we can decompose Θ, again uniquely, into a product of rotations, by distinct angles $\vartheta_1, \ldots, \vartheta_k$ say, with $0 < \vartheta_j < \pi$ for $j = 1, \ldots, k$, and a reflexion, corresponding to an eigenvalue -1; some or all of these may be absent. Taking the ambient space $\mathbb{E}$ to be the affine hull of P, we see that eigenvalues $\exp(\pm i\vartheta_j)$ for $j = 1, \ldots, k$ or -1 of Θ are not repeated. Thus P is a blend of pure polygons and apeirogons, with at least one of the latter if we confine our attention to the case when P has infinitely many vertices. We have a polygon corresponding to each rational angle ϑ_j. To a non-zero translation vector t, and each irrational angle ϑ_j, corresponds an apeirogon, in the latter case bounded and non-discrete.

4F4 Example If R_0 and R_1 are two non-intersecting lines in $\mathbb{E}^3$ with angle φ between them (or, rather, between suitable translates which meet), then a point on R_1, but not on the common perpendicular of the lines, is the initial vertex of a regular helical apeirogon P with symmetry group $\mathbf{G}(\mathsf{P}) = \langle R_0, R_1 \rangle$ (thus R_0 and R_1 are reflexions identified with their mirrors), whose twist-angle is 2φ. When $\varphi = \pi/r$ for some integer $r > 2$ we have an *r-helix*; the case $r = 2$ gives a *zigzag* or *skew apeirogon*.

4F5 Remark A particular case is that some angle ϑ_j is irrational; with a translational component the apeirogon is discrete, but it does not actually have any translations in its symmetry group.

In fact, this discussion shows that the algebraic dimension of the realization cone of $\{\infty\}$ is uncountably infinite, and so it cannot be closed. To be more precise, let $\vartheta_1, \vartheta_2, \ldots$ be distinct angles with $0 < \vartheta_j < \pi$, let P_j be the apeirogon with vertices on the unit circle in $\mathbb{E}^2$ corresponding to the rotation though ϑ_j (thus if ϑ_j is rational, the apeirogon covers the appropriate polygon). Writing $\mathsf{Q}_n := P_1 \# \frac{1}{2}P_2 \# \cdots \# \frac{1}{n}P_n$, we see that the limit $\Delta := \lim_{n\to\infty} \Delta(Q_n)$ of the diagonal vectors certainly exists, since each of its coordinates is a sum of non-negative terms which is bounded above by

$$\sum_{j=1}^{\infty} \left(\frac{2}{j}\right)^2 = \frac{2\pi^2}{3}.$$

But this vector Δ cannot be that of a realization. The reason is simple: the Cayley–Menger criterion for embedding a set of points in some euclidean space (see [99, Section 5B]) shows that the limit apeirogon, were it to exist, would have an infinite affinely independent set of vertices. Notice also that, if we blend this example with the ordinary linear apeirogon, then we have a sequence of discrete realizations whose diagonal vectors converge pointwise, but with limit which is not a diagonal vector of any realization. In other words, in contrast to the finite case, the realization cone is not closed.

Discrete Realizations

We cannot completely avoid considering non-discrete realizations, because (as we observe below) these naturally occur as components under blending of discrete realizations. However, for regular apeirotopes, the focus of the book will be on discrete realizations.

A bounded discrete set of points is finite, and hence for bounded discrete realizations the previous analysis can be followed with minor modifications. We can thus suppose that our discrete realization P of a regular apeirotope $\mathcal{P}$ is infinite; there are then two cases, according as the ambient space $\mathbb{E}$ (a euclidean space now, as we recall) is or is not the convex hull of the vertex-set V of P. For our first result, discreteness is actually irrelevant.

4F6 Theorem *Let P be an unbounded realization of a regular apeirotope $\mathcal{P}$ with vertex-set V and ambient space $\mathbb{E}$. If $\operatorname{conv} V \neq \mathbb{E}$, then P is a non-trivial blend, one of whose components is bounded.*

Proof. The convex hull $K := \operatorname{conv} V$ is full-dimensional, since $\mathbb{E}$ is the ambient space of P. All points of V are equivalent under $\mathbf{G}$, and hence $V \subseteq \operatorname{bd} K$, the boundary of K; otherwise, if, say, $v \in V$ is such that $B(v, \rho) \subseteq K$, where $B(x, \sigma)$ denotes the ball in $\mathbb{E}$ with centre x and radius $\sigma > 0$, then the transitivity of $\mathbf{G}$ on V shows that $B(v', \rho) \subseteq K$ for every $v' \in V$, which would contradict $K = \operatorname{conv} V$.

The recession cone $\operatorname{rec} K$ of K, which (as defined in Section 1B) consists of the vectors $a \in \mathbb{E}$ such that $x + a \in K$ for every $x \in \operatorname{int} K$, is the orthogonal sum of a linear subspace L, say, and a pointed convex cone C, say; we show that $\operatorname{rec} K = L$. Certainly, L must be fixed by $\mathbf{G}$ (as a set of directions), since $\mathbf{G}$ takes K into itself. Hence, if we project orthogonally along L, then we shall obtain another realization of P, with corresponding recession cone C.

Were $C \neq \{o\}$, we could replace P by its projection as above, and then take $\operatorname{rec} K = C$. Now C, again as a set of directions, is invariant under $\mathbf{G}$. If we let g be the centre of gravity of the intersection $C \cap B$ of C with the unit ball of $\mathbb{E}$, then $g \in \operatorname{int} C$ because C is pointed, and so is contained in some half-ball of B. Since C is invariant under $\mathbf{G}$, the same is true of g, and thus the hyperplane H through o orthogonal to g is also invariant under $\mathbf{G}$. Since C is pointed, it follows that $-g \notin C$, so that H must support C (in $\{o\}$ alone). Then K has a support hyperplane H' parallel to H, which must meet V since $K = \operatorname{conv} V$. Using the transitivity of $\mathbf{G}$ on V, we thus have $V = (V \cap H')\mathbf{G} \subset H'\mathbf{G} = H'$. But since $C = \operatorname{rec} K \neq \{o\}$ by our assumption, K must also have vertices outside H'. This is the required contradiction, so that $C = \{o\}$ as claimed. $\square$

Theorem 4F6 shows that we can translate P so that the points of V are equidistant from some linear subspace L of $\mathbb{E}$. The orthogonal projection of V on L yields another realization of $\mathcal{P}$.

4F7 Lemma *The projection of V on L is a discrete set.*

Proof. The points of V are at a fixed distance $\rho > 0$ from L. Since the sphere is compact, any cluster point of the projection lifts to a cluster point of V itself. But V is discrete; hence no such cluster point can exist. □

Observe that the projection along L is certainly bounded, but need not be discrete, as the examples of realizations of the apeirogon which we discussed at the beginning of the section demonstrate.

Before we discuss the other case, let us recall Theorems 4B2 and 4B8: if P is a discrete faithful realization of a regular m-apeirotope, with group $\mathbf{G}(\mathsf{P}) = \langle R_0, \ldots, R_{m-1} \rangle$, then

- $\dim \mathsf{P} \geqslant m - 1$,
- $\dim R_j \geqslant j$ for $j = 0, \ldots, m-2$, and $\dim R_{m-1} \geqslant m - 2$,
- if $\dim \mathsf{P} = m - 1$, then $\dim R_{m-1} = \dim R_m = m - 2$.

Note, by the way, that the last equality forces P (or, rather, its vertex-set) to be infinite, since otherwise we have a contradiction to Theorem 4B2. In fact, we can go further.

4F8 Theorem *Let P be a realization of a regular polytope $\mathcal{P}$, whose vertex-figure Q is such that $\dim \mathsf{Q} = \dim \mathsf{P}$. If P has vertex-set V and ambient space $\mathbb{E}$, then $\operatorname{conv} V = \mathbb{E}$. In particular, V is infinite and $\mathcal{P}$ is an apeirotope.*

Proof. Notice that we are not assuming discreteness or faithfulness here. The proof is straightforward. As we saw in the proof of Theorem 4F6, if $\operatorname{conv} V \neq \mathbb{E}$, then $V \subseteq \operatorname{bd} \operatorname{conv} V$. However, since $\dim \mathsf{Q} = \dim \mathbb{E}$ ($= \dim \mathsf{P}$ by definition), for the initial vertex v we have $v \in \operatorname{int} \operatorname{conv} \mathsf{Q} \subseteq \operatorname{int} \operatorname{conv} V$. Thus each vertex lies in the convex hull of its vertex-figure, which contradicts $V \subseteq \operatorname{bd} \operatorname{conv} V$. □

Resuming our discussion, we may now suppose that $\operatorname{conv} V = \mathbb{E}$. Once more, of course, we are assuming that our realization P is discrete and faithful. The symmetry group $\mathbf{G}$ of P is transitive on V, and so, by the famous theorem of Bieberbach (see Theorem 1D21), $\mathbf{G}$ contains a subgroup T of translations of full rank $\dim \mathbb{E}$. This subgroup T is discrete, and so can be identified with a lattice in $\mathbb{E}$. Note that $\mathbf{G}$ will then act as a group of automorphisms (that is, symmetries) of T.

Non-Discrete Realizations

As we remarked earlier, we cannot entirely avoid non-discrete realizations. For instance, in the plane $\mathbb{E}^2$, the tessellation $\{8, \frac{8}{3}\}$ is a non-discrete realization of the hyperbolic honeycomb $\{8, 8\}$, as is its allomorph (and dual) $\{\frac{8}{3}, 8\}$. If we blend these two realizations together, with an appropriate choice of scaling of each, the resulting vertex-set of the blended realization $\{8, \frac{8}{3}\} \# \{\frac{8}{3}, 8\}$ in $\mathbb{E}^4$ is $\mathbb{Z}^4$. Indeed, every non-discrete regular tessellation $\{p, q\}$ in $\mathbb{E}^2$ (with p and q rational, and at least one fractional) may be lifted to a discrete realization in an analogous way, although more than two components (isomorphic, but not similar) are generally required. Thus, as another example, we need to blend all three of the non-discrete planar tessellations $\{7, \frac{14}{5}\}$, $\{\frac{7}{2}, \frac{14}{3}\}$ and $\{\frac{7}{3}, 14\}$ in order to obtain a discrete realization (actually, if we wish to think of it in this way, of the hyperbolic tiling $\{7, 14\}$).

Notes to Section 4F

1. A curiosity about discrete realizations is the following. Call a realization P of a regular apeirotope $\mathcal{P}$ *translation-free* if its symmetry group $\mathbf{G} = \mathbf{G}(\mathsf{P})$ contains no translations. Of course, a bounded realization must be translation-free, but we also observed above that we have unbounded translation-free realizations of the regular apeirogon. Somewhat surprisingly, if a regular apeirotope $\mathcal{P}$ has a discrete realization, then it has a discrete translation-free realization; see [80] or [99, Section 5C].
2. Another curiosity is that regular hyperbolic honeycombs can have faithful bounded realizations, naturally non-discrete. We specifically constructed an example in [80], but it is likely that most cases of that kind provide appropriate isomorphisms. For instance, the proof of Proposition 17B3 shows that $\{3, 3, 3, \frac{5}{2}\} \cong \{3, 3, 3, 5\}$. Moreover, since we have sections like $\{5, 5\} \prec \{3, 3, 3, 5\}$, we obtain isomorphisms such as $\{\frac{5}{2}, \frac{5}{2}\} \cong \{5, 5\}$, $\{\frac{5}{2}, 3, \frac{5}{2}\} \cong \{5, 3, 5\}$ and $\{3, \frac{5}{2}, 3\} \cong \{3, 5, 3\}$.

4G Examples of Realizations

We shall meet many examples of realizations in the rest of the book, and so what we mainly need from this section are some basic realization domains to which we can subsequently refer. However, it would be unfair to leave illustration of the various techniques of the theory until much later (indeed, we confine consideration of many cases to Part IV, in order not to break up the main narrative), and so we also show how they work on some examples that do not turn up again.

Polygons

We begin with polygons (see also [30, Chapter 1], [45] and [58, Section 2]); in geometric terms, these are our basic building blocks, and so we need to know exactly what polygons look like. This also gives us the opportunity to introduce notation that describes – to a certain extent – the geometry of realizations of polytopes of higher rank.

For each integer $p \geqslant 3$, the (abstract) regular p-gon $\{p\}$ has $\lfloor \frac{1}{2}p \rfloor$ diagonal classes, and for each $k = 1, \ldots, \lfloor \frac{1}{2}p \rfloor$, there is a planar regular polygon $\{\frac{p}{k}\}$, which gives a degenerate realization if the greatest common divisor $(p, k) > 1$, reducing to a digon $\{2\}$ when $k = \frac{1}{2}p$. It is easy to see directly that a general regular p-gon is a blend of these pure realizations. For each irreducible representation $\mathbf{G}$, we have $w_{\mathbf{G}} = 1$ in the notation used in Chapter 3.

To be more specific, we can take the unit normal u_j to the generating mirror R_j to be

$$u_j := \begin{cases} (-\sin \frac{k\pi}{p}, \cos \frac{k\pi}{p}), & \text{if } j = 0, \\ (0, -1), & \text{if } j = 1. \end{cases}$$

With initial vertex $v_0 = (1, 0)$, we see that the vertices of $\{\frac{p}{k}\}$ are $v_j = \big(\cos \frac{2jk\pi}{p}, \sin \frac{2jk\pi}{p}\big)$ for $j = 0, \ldots, p-1$.

With the usual abbreviation for number strings, the layer vector of $\{p\}$ is

4G1 $$\Lambda(\{p\}) = \begin{cases} (1, 2^{(p-2)/2}, 1), & \text{if } p \text{ is even}, \\ (1, 2^{(p-1)/2}), & \text{if } p \text{ is odd}, \end{cases}$$

while the cosine vector of $\{\frac{p}{k}\}$ is

4G2 $$\Gamma_k = \Gamma(\{\tfrac{p}{k}\}) = \big(1, \cos(\tfrac{2k\pi}{p}), \cos(\tfrac{4k\pi}{p}), \ldots\big),$$

for $k = 0, \ldots, \lfloor \frac{1}{2}p \rfloor$. Bear in mind that the cosine vectors are those of p-gons, which may be degenerate.

So far as products are concerned, it is straightforward to see that, if Θ, Ψ are rotations of $\mathbb{E}^2$ through ϑ, φ, respectively, then $\Theta \otimes \Psi$ is (in general) a double rotation of $\mathbb{E}^4 = \mathbb{E}^2 \otimes \mathbb{E}^2$ through angles $\vartheta \pm \varphi$. Bear in mind that, since we identify by congruence, we regard rotations through $\pm\vartheta$ as the same. Indeed,

$$\cos\vartheta \cos\varphi = \tfrac{1}{2}(\cos(\vartheta - \varphi) + \cos(\vartheta + \varphi)).$$

The cases where $\varphi = \vartheta$ or $\varphi = \pi - \vartheta$ are therefore special. We thus have

4G3 Example For rational $p \geqslant q \geqslant 2$,

$$\{p\} \otimes \{q\} = \begin{cases} \left\{\dfrac{pq}{p-q}\right\} \# \left\{\dfrac{pq}{p+q}\right\}, & \text{if } p > q, \\ \left\{\dfrac{p}{2}\right\} \# \{1\}, & \text{if } p = q. \end{cases}$$

Note the special case where $1/p + 1/q = 1/2$, which gives $\{2\}$ as one of the components. Strictly speaking, if all the polygons are normalized, then the pure components of the product should carry multiplicative scalars $1/\sqrt{2}$.

The ideas here extend in an natural way to ‘nice’ realizations of the abstract apeirogon $\{\infty\}$, that is, those whose bounded components are actual polygons. This lets us introduce notation which we shall find invaluable in everything that follows.

A general realization of the finite polygon $\{p\}$ is of the form

$$\{p_1\} \# \cdots \# \{p_k\},$$

where each p_j is a fraction $p_j = r/s_j$ for some $1 \leqslant s_1 < \cdots < s_k \leqslant \frac{1}{2}p$, not necessarily in its lowest terms. When s_j is not a divisor of p, we have a regular star-polygon; observe that $s_k = \frac{1}{2}p$ gives the digon $\{2\}$. We can similarly denote the linear apeirogon by $\{\infty\} = \{\frac{1}{0}\}$; the general geometric regular polygon (or apeirogon) can thus be denoted by

4G4 $$\{p\} = \Bigl\{\frac{r}{s_0, \ldots, s_k}\Bigr\} = \Bigl\{\frac{r}{s_0}\Bigr\} \# \cdots \# \Bigl\{\frac{r}{s_k}\Bigr\},$$

where now $0 \leqslant s_0 < s_1 < \cdots < s_k \leqslant \frac{1}{2}p$. This notation takes no account of the relative scaling of the components, but we shall see that it is exactly what we need when we come to discuss rigidity in Chapter 6, and will subsequently lead to simplified notation for certain regular polytopes of higher rank. The entry

$$p := \frac{r}{s_0, \ldots, s_k}$$

is called a *generalized fraction*.

For $2 \leqslant q \leqslant \infty$, we define q'' by

4G5 $$\frac{1}{q} + \frac{1}{q''} = \frac{1}{2},$$

with the obvious conventions $2'' = \infty$ and $\infty'' = 2$. For a general regular polygon (or apeirogon) $\{p\} = \{q_1\} \# \cdots \# \{q_k\}$, with the $\{q_j\}$ planar polygons (including $\{2\}$ and $\{\infty\}$), we then define

4G6 $$\{p''\} := \{q_k''\} \# \cdots \# \{q_1''\}.$$

We refer to $\{p''\}$ as the *supplement* of $\{p\}$; the order of the components of the blend is reversed in accord with (4G4).

Simplex and Staurotope

We already described the realization domains of the simplex and staurotope in Section 3C, and so we merely make brief remarks about products.

For the (geometric) d-simplex T^d with cosine vector $\Gamma_1 = (1, -\frac{1}{d})$, there is only one non-trivial product, namely,

$$\Gamma_1{}^2 = \tfrac{1}{d}\Gamma_0 + \tfrac{d-1}{d}\Gamma_1;$$

the coefficient of Γ_0 comes from Corollary 3F6.

There is a $(d-1)$-collapse of the abstract d-staurotope $\mathcal{X}^d$ onto its facet $\mathcal{T}^{d-1}$. With $\Gamma_1 = \Gamma(\mathsf{T}^{d-1}) = (1, -\frac{1}{d-1}, 1)$ (as a cosine vector of $\mathcal{X}^d$) and $\Gamma_2 = \Gamma(\mathsf{X}^d) = (1, 0, -1)$, we have

$$\begin{aligned} \Gamma_1\Gamma_2 &= \Gamma_2, \\ \Gamma_2{}^2 &= \tfrac{1}{d}\Gamma_1 + \tfrac{d-1}{d}\Gamma_1. \end{aligned}$$

It is worthwhile introducing some more general examples, all of non-regular polytopes. The aim here is to illustrate a number of the ideas which have been discussed in this chapter and the last, without encroaching on the formal classification of regular polytopes of full or nearly full rank which occupy us from Chapter 7 onward.

4G7 Example Let $\mathcal{P}$ be the product $\{3^{d-1}\} \times \{3^{d-1}\}$ of two d-simplices, with group $([3^{d-1}] \times [3^{d-1}]) \rtimes \boldsymbol{Z}_2$ with $\boldsymbol{Z}_2 = \langle \boldsymbol{z} \rangle$; the involutory automorphism $\boldsymbol{z}$ interchanges the two copies of the simplex. It is not hard to see that the layer vector of $\mathcal{P}$ is $\Lambda = (1, 2d, d^2)$. One non-trivial pure realization is the natural product $\mathsf{T}^d \times \mathsf{T}^d$ of two d-simplices T^d in $\mathbb{E}^{2d} = \mathbb{E}^d \oplus \mathbb{E}^d$, with vertices $\frac{1}{\sqrt{2}}(x_j, y_k)$, where the two simplices have (normalized) vertex-sets $\{x_0, \ldots, x_d\}$ and $\{y_0, \ldots, y_d\}$. We need one more pure realization of dimension $(d+1)^2 - 1 - 2d = d^2$, and it should not be particularly surprising that it is the tensor product $\mathsf{T}^d \otimes \mathsf{T}^d$, whose vertices are all $x_j \otimes y_k \in \mathbb{E}^d \otimes \mathbb{E}^d$. The cosine matrix thus takes on the nice symmetric form

$$\begin{bmatrix} 1 & 1 & 1 \\ 1 & \frac{d-1}{2d} & -\frac{1}{d} \\ 1 & -\frac{1}{d} & \frac{1}{d^2} \end{bmatrix}$$

Some Unitary Polytopes

The last two examples concern highly symmetric polytopes whose basic symmetry groups are unitary (possibly with twists – for these, see Section 5C).

4G8 Example Consider the diagrams

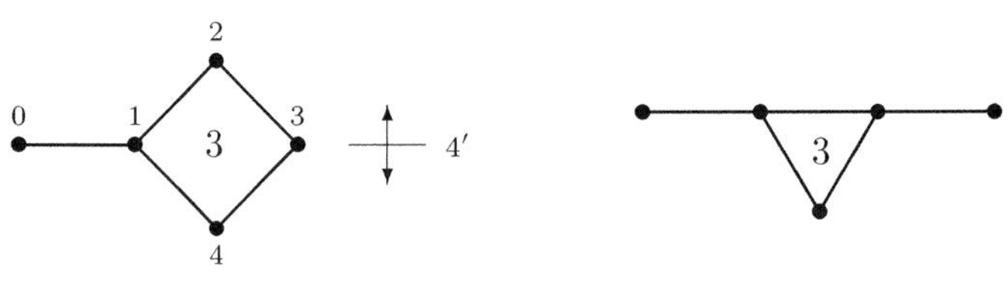

These are of the same finite subgroup $\mathbf{G}_5$ of $\mathbf{U}_5$, with different generators; the relationship between the generatrices is as in Figures 1E23 and 1E25 related by (1E24) (there is a general discussion of such unitary groups in Section 1E). What we will recall is that a central mark q in a circuit of k nodes and unmarked branches, corresponding to (abstract) generators $\boldsymbol{s}_1, \ldots, \boldsymbol{s}_k$ in cyclic order, induces an extra relation

$$(\boldsymbol{s}_1 \boldsymbol{s}_2 \cdots \boldsymbol{s}_k \boldsymbol{s}_{k-1} \boldsymbol{s}_{k-2} \cdots \boldsymbol{s}_2)^q = \boldsymbol{e}.$$

Taking the first diagram, the geometric generators are (complex involutory) reflexions R_j in the hyperplanes with unit normals u_j for $j = 0, \ldots, 4$, where

$$\begin{aligned} u_0 &:= \tfrac{1}{\sqrt{6}}(-\sqrt{2}, 1, 1, 1, 1), \\ u_1 &:= \tfrac{1}{\sqrt{2}}(0, \omega, -\overline{\omega}, 0, 0), \\ u_2 &:= \tfrac{1}{\sqrt{2}}(0, 0, 1, -1, 0), \\ u_3 &:= \tfrac{1}{\sqrt{2}}(0, 0, 0, 1, -1), \\ u_4 &:= \tfrac{1}{\sqrt{2}}(0, -1, 0, 0, 1); \end{aligned}$$

we also have the twist

$$R_4' \colon z \mapsto -(\overline{\zeta}_1, \overline{\zeta}_3, \overline{\zeta}_2, \overline{\zeta}_5, \overline{\zeta}_4),$$

which is such that $R_4' R_3 R_4' = R_4$. Here, $\omega = \frac{1}{2}(-1 + i\sqrt{3})$ is a cube root of 1, so that $\overline{\omega} = \omega^2 = \frac{1}{2}(-1 - i\sqrt{3})$.

We now set $\mathbf{H} := \langle R_1, \ldots, R_4 \rangle$, so that we can take the initial vertex to be $v := (3, 0, 0, 0, 0)$. We then find that the 80 vertices of the realization $\mathsf{P}_3 \in \mathcal{P}_5$ are all

$$\pm(3, 0, 0, 0, 0), \ \pm(1, \sqrt{2}\omega^{\kappa}, \sqrt{2}\omega^{\lambda}, \sqrt{2}\omega^{\mu} \sqrt{2}\omega^{\nu}), \ \pm(\omega - \overline{\omega})(1, \sqrt{2}\omega^{\lambda}, 0, 0, 0),$$

where in the second block $\kappa, \lambda, \mu, \nu = 0, 1, 2$ with $\kappa + \lambda + \mu + \nu \equiv 0 \pmod 3$, and in the third $\lambda = 0, 1, 2$ with all permutations of $\zeta_2, \ldots, \zeta_5$. The resulting polytope is semi-regular; its facets are 4-staurotopes (with full symmetry if the twist is included) and 4-simplices.

From this, we conclude that the layer vector of the abstract polytope $\mathcal{P}_5$ of rank 5 is $\Lambda = (1, 27, 24^*, 27, 1)$. Note that the diagonal class $\mathcal{D}_2$ is asymmetric; it is clear that the 10-dimensional realization just described has $w = 2$ but $w^* = 1$, and so we can multiply our initial vertex v by any (non-zero) complex number. On the other hand, if we replace R_4 by the twist R_4' (which we shall do here), then $\mathcal{D}_2$ becomes symmetric and (strictly speaking) our initial vertex should be $\pm iv$ up to real scalar multiple.

While the coordinates given for this realization do not demonstrate the full symmetry (see the notes at the end of the section), they do enable us easily to write down the induced cosine vector of the vertex-figure. This is $\Gamma^{\boldsymbol{v}} = (1, \gamma_1, \gamma_2, \gamma_3)$ in terms of the general cosine vector $\Gamma = (1, \gamma_1, \ldots, \gamma_4)$ of $\mathcal{P}_5$ itself; the corresponding layer vector is $\Lambda^{\boldsymbol{v}} = (1, 12, 8, 6)$.

We now readily find the remaining pure realizations of $\mathcal{P}_5$. Since $\mathcal{P}_5$ is centrally symmetric, we first look at $\mathcal{P}_5/2$, with cosine vector of the form $(1, \alpha, \beta, \alpha, 1)$ such that

$$1 + 27\alpha + 24\beta + 27\alpha + 1 = 0 \implies \beta = -\tfrac{1}{12}(1 + 27\alpha).$$

Applying Theorem 3C14, we conclude that

$$27\alpha^2 = 1 + 12\alpha + 8\beta + 6\alpha = 1 + 18\alpha - \tfrac{2}{3}(1 + 27\alpha) = \tfrac{1}{3},$$

giving $\alpha = \pm\frac{1}{9}$. We thus have

$$\begin{aligned} \Gamma_1 &= (1, \tfrac{1}{9}, -\tfrac{1}{3}, \tfrac{1}{9}, 1), \\ \Gamma_2 &= (1, -\tfrac{1}{9}, \tfrac{1}{6}, -\tfrac{1}{9}, 1), \end{aligned}$$

with corresponding dimensions given by

$$\begin{aligned} \tfrac{1}{d_1} &= \tfrac{1}{80}\bigl(1 + 27(\tfrac{1}{9})^2 + 24(-\tfrac{1}{3})^2 + 27(\tfrac{1}{9})^2 + 1\bigr) = \tfrac{16}{3\cdot 80} = \tfrac{1}{15}, \\ \tfrac{1}{d_2} &= \tfrac{1}{80}\bigl(1 + 27(-\tfrac{1}{9})^2 + 24(\tfrac{1}{6})^2 + 27(-\tfrac{1}{9})^2 + 1\bigr) = \tfrac{10}{3\cdot 80} = \tfrac{1}{24}, \end{aligned}$$

from Theorem 3F5.

We already have one component P_3 of the staurotope realization X, whose cosine vector is $\Gamma_3 = (1, \frac{1}{3}, 0, -\frac{1}{3}, -1)$ with dimension $d_3 = 10$, and so there is just one more component to find. The cosine vector is of the form $\Gamma_4 = (1, \alpha, 0, -\alpha, -1)$; there are at least three ways of finding it. We could apply Theorem 3C14 again, but it is easier to use either the component equation of Theorem 3C11 or Λ-orthogonality of Theorem 3F5. For the former, $d_4 = 40 - 10 = 30$, and hence

$$\Gamma_4 = \tfrac{1}{30}\bigl(40(1, 0, 0, 0, -1) - 10(1, \tfrac{1}{3}, 0, -\tfrac{1}{3}, -1)\bigr) = (1, -\tfrac{1}{9}, 0, \tfrac{1}{9}, -1).$$

For the latter,

$$\begin{aligned} 0 = \langle \Gamma_3, \Gamma_4 \rangle_\Lambda &= \tfrac{1}{80}\bigl(1 + 27(\tfrac{1}{3})\alpha + 27(-\tfrac{1}{3})(-\alpha) + (-1)(-1)\bigr) \\ &= \tfrac{1}{80}(2 + 18\alpha), \end{aligned}$$

from which follows $\alpha = -\frac{1}{9}$ just as before. One must then use Theorem 3F5 to verify that $d_4 = 30$.

In conclusion, we have shown that $\mathcal{P}_5$ has cosine matrix

$$\begin{bmatrix} 1 & 1 & 1 & 1 & 1 \\ 1 & \frac{1}{9} & -\frac{1}{3} & \frac{1}{9} & 1 \\ 1 & -\frac{1}{9} & \frac{1}{6} & -\frac{1}{9} & 1 \\ 1 & \frac{1}{3} & 0 & -\frac{1}{3} & -1 \\ 1 & -\frac{1}{9} & 0 & \frac{1}{9} & -1 \end{bmatrix},$$

with layer and dimension vectors

$$\Lambda = (1, 27, 24, 27, 1), \qquad D = (1, 15, 24, 10, 30).$$

4G9 Remark The alternative diagram for the group $\mathbf{G}_5$ admits a vertical twist. This would lead to dual regular polytopes of rank 4 in $\mathbb{E}^{10}$, which are even more remote from the focus of the book.

4G10 Remark We have performed the calculations here in some detail as an illustration. In future, we shall often leave such details to the interested reader.

4G11 Remark It is often the case that the non-trivial component of $\mathsf{P} \otimes \mathsf{P}$ is pure, where P is the basic geometric polytope from which the abstract polytope is generalized. However, for P_3 this is not the case; in fact

$$\Gamma_3^2 = \tfrac{1}{10}\Gamma_0 + \tfrac{1}{2}\Gamma_1 + \tfrac{3}{5}\Gamma_2,$$

as one can calculate using Theorem 3J1 or (in this case) merely verifying directly.

4G12 Example We finally consider the diagrams

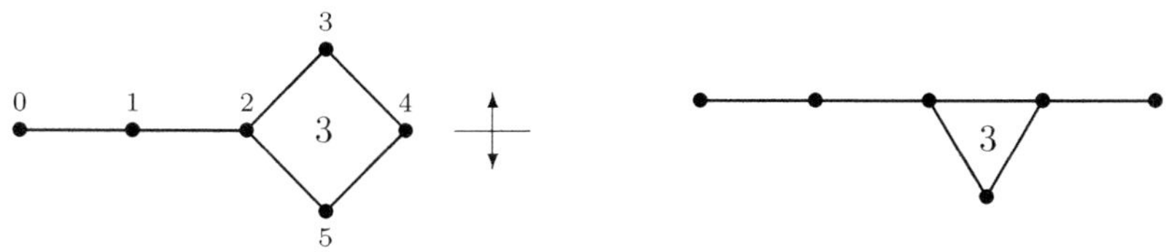

4G13

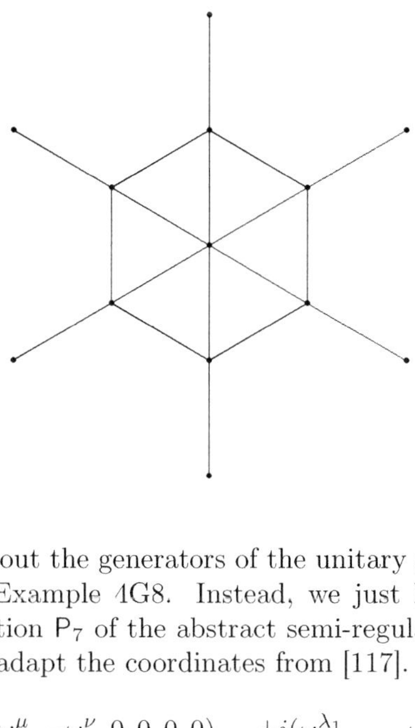

We shall not write out the generators of the unitary group $\mathbf{G}_6 \leqslant \mathbf{U}_6$, which admits a twist as in Example 4G8. Instead, we just list the vertices of the 12-dimensional realization P_7 of the abstract semi-regular polytope $\mathcal{P}_6$ derived from the diagram; we adapt the coordinates from [117]. These are all

$$\sqrt{3}(\omega^{\mu}, -\omega^{\nu}, 0, 0, 0, 0), \quad \pm i(\omega^{\lambda_1}, \ldots, \omega^{\lambda_6}),$$

with free permutations of coordinates in the first block, and $\mu, \nu, \lambda_1, \ldots, \lambda_6 = 0, 1, 2$ and $\lambda_1 + \cdots + \lambda_6 \equiv 0 \pmod 3$. We thus have $30 \cdot 9 + 2 \cdot 243 = 756$ vertices in all. The layer vector is

$$\Lambda = (1, 80, 2^*, 160, 270, 160, 2^*, 80, 1),$$

with asymmetric diagonal classes $\mathcal{D}_2$ and $\mathcal{D}_6$. The latter classes arise from the cyclic subgroup $\mathbf{Z}_6$ of symmetries generated by scalar multiplication by

$-\overline{\omega} = \exp(i\pi/3)$. If we adjoin the diagram twist, as we implicitly shall in what follows, then these diagonal classes become symmetric.

Figure 4G13 shows the projection of P_7 on the (complex) line through one (and hence six) of its vertices. The outer points are single, each point of the inner ring comes from 80 vertices of P_7, and the centre is the projection of the remaining 270.

The vertex-figure of $\mathcal{P}_6$ is just the semi-regular polytope $\mathcal{P}_5$ of Example 4G8. The realization P_7 (with index chosen to fit into the final classification) tells us that the induced cosine vector of $\mathcal{P}_5$ is $\Gamma^{\boldsymbol{v}} = (1, \gamma_1, \gamma_3, \gamma_4, \gamma_7)$ in terms of the general cosine vector $\Gamma = (1, \gamma_1, \ldots, \gamma_8)$ of $\mathcal{P}_6$.

The action on $\mathcal{P}_6$ of the abstract subgroup $\boldsymbol{Z}_6 = \langle \boldsymbol{z} \rangle$ corresponding to $\mathbf{Z}_6$ enables us to identify the vertices of $\mathcal{P}_6$ in twos, threes or sixes. This suggests the line of approach: we first tackle $\mathcal{P}_6/6$, then what remain of $\mathcal{P}_6/3$ and $\mathcal{P}_6/2$, and finally the rest of $\mathcal{P}_6$ itself. Moreover, the picture shows exactly how the identifications manifest themselves.

Thus $\mathcal{P}_6/6$ has (intrinsic) layer vector $(1, 80, 45)$, and its general cosine vector $(1, \alpha, \beta)$ embeds in that of $\mathcal{P}_6$ by

$$(1, \alpha, \beta) \hookrightarrow (1, \alpha, 1, \alpha, \beta, \alpha, 1, \alpha, 1).$$

So far as the vertex-figure $\mathcal{P}_5$ is concerned, we can think of the identification as collapsing its layer vector to $(1, 52, 27)$ (with the same cosine vectors). Applying Theorem 3C14 yields the two equations

$$\begin{aligned} 80\alpha^2 &= 1 + 52\alpha + 27\beta, \\ 0 &= 1 + 80\alpha + 45\beta. \end{aligned}$$

Eliminating β leads to

$$0 = 200\alpha^2 - 10\alpha - 1 = (10\alpha - 1)(20\alpha + 1),$$

and thus to the two cosine vectors

$$\begin{aligned} \Gamma_1 &= (1, \tfrac{1}{10}, -\tfrac{1}{5}), \\ \Gamma_2 &= (1, -\tfrac{1}{20}, \tfrac{1}{15}). \end{aligned}$$

The corresponding dimensions (calculated using Theorem 3F5 – we omit the details) are $d_1 = 35$ and $d_2 = 90$; with $d_0 = 1$ for the henogon $\{1\}$, we have $d_0 + d_1 + d_2 = 126$, as expected.

We now move on to $\mathcal{P}_6/3$. This is centrally symmetric, with $\boldsymbol{z}^3$ as its central involution, and so its intrinsic layer vector is $(1, 80, 90, 80, 1)$, with corresponding cosine vector of the form

$$(1, \alpha, 0, -\alpha, -1) \hookrightarrow (1, \alpha, -1, -\alpha, 0, \alpha, 1, -\alpha, -1)$$

as embedded in a general cosine vector. When we apply Theorem 3C14, we obtain

$$80\alpha^2 = 1 + 27\alpha - 24\alpha - \alpha = 1 + 2\alpha,$$

so that

$$0 = 80\alpha^2 - 2\alpha - 1 = (8\alpha - 1)(10\alpha + 1).$$

We thus have the cosine vectors

$$\varGamma_3 = (1, \tfrac{1}{8}, -1, -\tfrac{1}{8}, 0, \tfrac{1}{8}, 1, -\tfrac{1}{8}, -1),$$
$$\varGamma_4 = (1, -\tfrac{1}{10}, -1, \tfrac{1}{10}, 0, -\tfrac{1}{10}, 1, \tfrac{1}{10}, -1);$$

the dimensions given by Theorem 3F5 are $d_3 = 56$ and $d_4 = 70$, again summing to 126.

We next consider $\mathcal{P}_6/2$, whose instrinsic cosine vectors will be of the form $(1, \alpha, \beta, \gamma, \delta)$, with layer vector $(1, 80, 2, 160, 135)$. However, we also have $\varLambda$-orthogonality with respect to

$$(1, \tfrac{1}{10}, -\tfrac{1}{5}) \hookrightarrow (1, \tfrac{1}{10}, 1, \tfrac{1}{10}, -\tfrac{1}{5}),$$
$$(1, -\tfrac{1}{20}, \tfrac{1}{15} \hookrightarrow (1, -\tfrac{1}{20}, 1, -\tfrac{1}{20}, \tfrac{1}{15}).$$

As a result, we have the three equations

$$1 + 80\alpha + 2\beta + 160\gamma + 135\delta = 0,$$
$$1 + 8\alpha + 2\beta + 16\gamma - 27\delta = 0,$$
$$1 - 4\alpha + 2\beta - 8\gamma + 9\delta = 0,$$

the first being the layer equation; these imply that $\beta = -\frac{1}{2}$ (which we could have guessed), $\gamma = -\frac{1}{2}\alpha$ and $\delta = 0$.

We now apply Theorem 3C14; after substituting for β, γ and δ, we find that

$$80\alpha^2 = 1 + 27\alpha - 12\alpha + \alpha = 1 + 16\alpha,$$

so that

$$0 = 80\alpha^2 - 16\alpha - 1 = (4\alpha - 1)(20\alpha + 1).$$

This results in the cosine vectors

$$\varGamma_5 = (1, \tfrac{1}{4}, -\tfrac{1}{2}, -\tfrac{1}{8}, 0, -\tfrac{1}{8}, -\tfrac{1}{2}, \tfrac{1}{4}, 1),$$
$$\varGamma_6 = (1, -\tfrac{1}{20}, -\tfrac{1}{2}, \tfrac{1}{40}, 0, \tfrac{1}{40}, -\tfrac{1}{2}, -\tfrac{1}{20}, 1);$$

once again, we use Theorem 3F5 to find the dimensions $d_5 = 42$ and $d_6 = 210$.

We can finish off $\mathcal{P}_6$ in several ways, just as we could with $\mathcal{P}_5$. However, for variety, let us introduce a fourth method. We first note that we already have P_7 with cosine vector

$$\varGamma_7 = (1, \tfrac{1}{2}, \tfrac{1}{2}, \tfrac{1}{4}, 0, -\tfrac{1}{4}, -\tfrac{1}{2}, -\tfrac{1}{2}, -1)$$

and dimension $d_7 = 12$. We then define

$$\varGamma_8 = \varGamma_2\varGamma_7 = \varGamma_4\varGamma_5 = (1, -\tfrac{1}{40}, \tfrac{1}{2}, -\tfrac{1}{80}, 0, \tfrac{1}{80}, -\tfrac{1}{2}, -\tfrac{1}{40}, -1).$$

It is routine to check that Γ_8 is Λ-orthogonal to each Γ_j with $j = 0, \ldots, 7$; actually, we need only check this for $j = 3, 4, 7$, because the rest correspond to components of the small simplex realization S. Since a count of diagonals shows that we lack a single component of the simplex realization T, it is clear that P_8 must be it. Indeed, we find that $d_8 = 240$ from Theorem 3F5, and then that the component equation of Theorem 3C11 holds.

We can summarize this discussion as follows: the cosine matrix of $\mathcal{P}_6$ is

$$\begin{bmatrix} 1 & 1 & 1 & 1 & 1 & 1 & 1 & 1 & 1 \\ 1 & \frac{1}{10} & 1 & \frac{1}{10} & -\frac{1}{5} & \frac{1}{10} & 1 & \frac{1}{10} & 1 \\ 1 & -\frac{1}{20} & 1 & -\frac{1}{20} & \frac{1}{15} & -\frac{1}{20} & 1 & -\frac{1}{20} & 1 \\ 1 & \frac{1}{8} & -1 & -\frac{1}{8} & 0 & \frac{1}{8} & 1 & -\frac{1}{8} & -1 \\ 1 & -\frac{1}{10} & -1 & \frac{1}{10} & 0 & -\frac{1}{10} & 1 & \frac{1}{10} & -1 \\ 1 & \frac{1}{4} & -\frac{1}{2} & -\frac{1}{8} & 0 & -\frac{1}{8} & -\frac{1}{2} & \frac{1}{4} & 1 \\ 1 & -\frac{1}{20} & -\frac{1}{2} & \frac{1}{40} & 0 & \frac{1}{40} & -\frac{1}{2} & -\frac{1}{20} & 1 \\ 1 & \frac{1}{2} & \frac{1}{2} & \frac{1}{4} & 0 & -\frac{1}{4} & -\frac{1}{2} & -\frac{1}{2} & -1 \\ 1 & -\frac{1}{40} & \frac{1}{2} & -\frac{1}{80} & 0 & \frac{1}{80} & -\frac{1}{2} & -\frac{1}{40} & -1 \end{bmatrix},$$

with layer and dimension vectors

$$\Lambda = (1, 80, 2, 160, 270, 160, 2, 80, 1), \quad D = (1, 35, 90, 56, 70, 42, 210, 12, 240).$$

4G14 Remark We can also see that $\Gamma_6 = \Gamma_4 \Gamma_7$, but it is less easy to work this fact into the analysis. Recall that this means that $\mathsf{P}_6 = \mathsf{P}_4 \otimes \mathsf{P}_7$. Relationships of this kind among pure realizations turn up not infrequently.

Notes to Section 4G

1. Quaternions (see Section 1K) provide an alternative picture for the tensor product of two planar rotations.
2. In notes dating from 1986, McMullen was already using the notation of (4G4) for generalized polygons. Norman Johnson [65, Section 5.2] independently came up with a notation for Petrie polygons of regular polytopes like (4G4), but with an additional entry to give some indication of the geometry. Our notation encompasses apeirogonal components as well.
3. More symmetric coordinates for vertices of the 10-dimensional realization P_3 of $\mathcal{P}_5$ come from its rôle as vertex-figure of the 12-dimensional realization P_7 of $\mathcal{P}_6$; compare [117] again. However, these are less easy to work with than the ones that we gave above.

5

Operations and Constructions

In this chapter, we describe various operations and constructions on groups, both abstract and geometric, which lead – at least putatively – from old regular polytopes to new ones.

Two basic kinds of operations – mixing and twisting – are of fundamental importance. However, there is a certain amount of overlap between them, and some string C-groups can be derived from other groups by both methods. There is a wide range of mixing operations; rather than having a very long section, we have split the discussion into the two Sections 5A and 5B. They work on an abstract as well as geometric level, and so we shall treat them abstractly. One important case for the future is Petrie contraction, but other operations such as faceting, halving and so on will occur frequently. On the other hand, some twisting operations considered in Section 5C will not work geometrically, or will only work in restricted circumstances; since we are more concerned with the geometric aspect, some care will need to be taken.

The next types of operation are purely geometric; while some may have abstract analogues, it is often hard to describe them in algebraic terms. In Section 5D, we look at ways in which one or more generating reflexions of a symmetry group can be replaced by their products with commuting (geometric) involutions; centriversion and eversion are important special cases.

In Section 5E, we consider extensions in general, particularly introducing the abelian extension and its relationship with eversion and Petrie contraction. Finally, in Section 5F, we consider the general notion of constructing regular polytopes in a recursive way suggested by our initial definition of a polytope. This idea concentrates on restrictions on vertex-figures as basic building blocks for regular polytopes of one higher rank when the corank is small.

The treatment in this chapter will be largely theoretical, and so while we shall mention some applications as we go along, for the most part we postpone discussion of practical examples until we have introduced suitable polytopes to which the operations can be applied. The notes after each section will often point to places where the techniques are used.

5A Operations on Polyhedra

Many mixing operations were originally applied to polyhedra (3-polytopes), and so in this section we confine our attention to this special case. With certain restrictions (which we shall discuss in Section 5B) they actually apply to 3-cofaces of general regular polytopes. Indeed, one of them – faceting – should really be thought of as applying to polygons. We then use the same notation for the corresponding operation applied to the coface of appropriate rank of a general regular polytope. The crucial feature of many of these operations is that they lead to new regular polytopes whose vertices form a subset of those of old ones, or even the same set if the operations are invertible.

Faceting

The first operation applies to finite regular polygons, and thence to 2-cofaces of polytopes of higher rank. This replaces $\boldsymbol{s}_0$ by some other reflexion (conjugate of $\boldsymbol{s}_0$ or $\boldsymbol{s}_1$) in $\langle \boldsymbol{s}_0, \boldsymbol{s}_1 \rangle$. More specifically, the kth *faceting* operation φ_k is given by

5A1 $$\varphi_k \colon (\boldsymbol{s}_0, \boldsymbol{s}_1) \mapsto (\boldsymbol{s}_0(\boldsymbol{s}_1\boldsymbol{s}_0)^{k-1}, \boldsymbol{s}_1) =: (\boldsymbol{r}_0, \boldsymbol{r}_1).$$

We shall suppose that $2 \leqslant k < \frac{1}{2}q$, since φ_{q-k} has the same effect as φ_k up to isomorphism (actually, conjugation of the whole group by $\boldsymbol{s}_1$), and the case $k = \frac{1}{2}q$ yields the segment $\{2\}$. We shall also find it convenient to allow the case $k = 1$ as well, where $\varphi_1 = \boldsymbol{\iota}$ is just the identity mixing operation. When the highest common factor $(k, q) = 1$, then $\boldsymbol{r}_0\boldsymbol{r}_1$ has the same period q as $\boldsymbol{s}_0\boldsymbol{s}_1$; indeed, the groups are the same, and φ_k is inverted by φ_j, where $jk \equiv \pm 1 \pmod{q}$. In fact, we have

5A2 Lemma *The faceting operations satisfy $\varphi_j\varphi_k = \varphi_{jk}$, where the suffix is to be read as the number between 0 and $\frac{1}{2}q$ that is congruent to $\pm jk$ modulo q.*

Proof. We apply φ_j and φ_k in succession to the group. Noting that only $\boldsymbol{s}_0$ changes, and writing $\boldsymbol{r}_0 = \boldsymbol{s}_0(\boldsymbol{s}_1\boldsymbol{s}_0)^{j-1} = (\boldsymbol{s}_0\boldsymbol{s}_1)^j\boldsymbol{s}_1$, we obtain

$$(\boldsymbol{r}_0\boldsymbol{r}_1)^k\boldsymbol{r}_1 = ((\boldsymbol{s}_0\boldsymbol{s}_1)^j\boldsymbol{s}_1^2)^k\boldsymbol{s}_1 = (\boldsymbol{s}_0\boldsymbol{s}_1)^{jk}\boldsymbol{s}_1,$$

as required. □

This lemma covers all possible j and k. Generally speaking, we shall be rather less interested in the case $(j, q) > 1$, although it will occasionally be useful. For instance, we shall employ φ_2 with q even (actually, $q = 6$) in Section 10B; we also discuss the effect of φ_2 in Proposition 5A4, which deals with the theory we shall need.

We now consider the effect of faceting on polyhedra; indices of distinguished generators are thus raised by 1. As usual, we shall take our regular polyhedron $\mathcal{Q}$ to have $\boldsymbol{G}(\mathcal{Q}) = \boldsymbol{H} = \langle \boldsymbol{s}_0, \boldsymbol{s}_1, \boldsymbol{s}_2 \rangle$, and we suppose that $\mathcal{Q}$ is of Schläfli type $\{p, q\}$. Thus each operation $\boldsymbol{\mu}$ will lead to a new group $\boldsymbol{G}$, and a new polyhedron $\mathcal{P} := \mathcal{Q}^{\boldsymbol{\mu}}$ with $\boldsymbol{G}(\mathcal{P}) = \boldsymbol{G}$.

As we did in Section 2E, it is often helpful to think of polyhedra as (finite or infinite) regular maps. The underlying surface for such a map $\mathcal{Q}$ is, of course, the order complex $\mathcal{C}(\mathcal{Q})$ of $\mathcal{Q}$, or, more exactly, its underlying (topological) polyhedron $|\mathcal{C}(\mathcal{Q})|$. Recall that a triangle $T(\boldsymbol{\Psi})$ of $\mathcal{C}(\mathcal{Q})$ is associated with each flag $\boldsymbol{\Psi}$ of $\mathcal{Q}$, with each vertex of $T(\boldsymbol{\Psi})$ associated with a face of $\boldsymbol{\Psi}$, and two triangles share an edge precisely when the corresponding flags are adjacent. Here, many of the operations have direct geometric interpretations.

The effect of a mixing operation on a polyhedron $\mathcal{Q}$ can often be pictured geometrically by applying the abstract analogue of Wythoff's construction in the order complex $\mathcal{C}(\mathcal{Q})$. However, observe that the new faces (that is, 2-faces) which are obtained, regarded as circuits of vertices and edges, will not usually bound discs in $|\mathcal{C}(\mathcal{Q})|$.

Combinatorially, φ_k has the following effect when $(k, q) = 1$. The new polyhedron $\mathcal{P} := \mathcal{Q}^{\varphi_k}$ has the same vertices and edges as $\mathcal{Q}$. However, a typical face of $\mathcal{P}$ is a k-hole of $\mathcal{Q}$, as defined by (2E2). Recall that this is formed by the edge-path which leaves a vertex by the kth edge from which it entered, in the same sense (that is, keeping always to the left, say, in some local orientation of $\mathcal{C}(\mathcal{Q})$). The faces of $\mathcal{P}$ then comprise all the k-holes of $\mathcal{Q}$. Hence, if such a k-hole is an r-gon, so that r is the period of

$$\boldsymbol{r}_0\boldsymbol{r}_1 = \boldsymbol{s}_0 \cdot (\boldsymbol{s}_1\boldsymbol{s}_2)^{k-1}\boldsymbol{s}_1,$$

then $\mathcal{Q}^{\varphi_k}$ is of type $\{r, q\}$. If $\mathcal{Q}$ is infinite, then it is possible that $r = \infty$, even if p is finite.

Of course, we must not forget to verify the intersection property, but in this case it is much easier to do this 'geometrically', thinking of $\mathcal{P} = \mathcal{Q}^{\varphi_k}$ as embedded in a surface. Naturally, this will not generally be the same as the original surface $|\mathcal{C}(\mathcal{Q})|$, although in practice we shall be able to work with $|\mathcal{C}(\mathcal{Q})|$ instead of the new surface $|\mathcal{C}(\mathcal{P})|$. Noting that $\boldsymbol{G}(\mathcal{Q})$ is transitive on the k-holes of $\mathcal{Q}$, the actual condition is just the usual interpretation of the intersection property.

5A3 Lemma *Let $\mathcal{Q}$ be a regular polyhedron of type $\{p, q\}$, and let $2 \leqslant k < \frac{1}{2}q$ with $(k, q) = 1$. Then the kth faceting operation φ_k applied to $\mathcal{Q}$ yields a regular polyhedron $\mathcal{P} = \mathcal{Q}^{\varphi_k}$ if and only if each vertex of $\mathcal{Q}$ is visited by a given k-hole at most once.*

When $(k, q) > 1$, the situation is similar, except that now we may obtain a polyhedron $\mathcal{P}$ of type $\{r, s\}$ inscribed in $\mathcal{Q}$, where $s := q/(k, q)$; the copies of $\mathcal{P}$ under the action of $\boldsymbol{H}$ will then form a compound of several polyhedra. Even if $k = \frac{1}{2}q$, when only a polygon $\{r\}$ can be obtained (or, more strictly perhaps, a dihedron $\{r, 2\}$), the number r may be of interest. However, $\mathcal{Q}^{\varphi_k}$ will often fail to be polytopal, when the vertex-figure at a vertex is no longer connected; that is, several sets of s r-gons will share the same (original) vertex of $\mathcal{Q}$. The following result illustrates a case of particular interest.

5A4 Proposition *Let $\mathcal{Q}$ be a regular polyhedron of type $\{3, 2s\}$ for some $s \geqslant 3$. If $\mathcal{P} := \mathcal{Q}^{\varphi_2}$ is a regular polyhedron, then it is of type $\{2s, s\}$. Moreover, $\boldsymbol{G}(\mathcal{P})$ is then a subgroup of $\boldsymbol{G}(\mathcal{Q})$ of index 3.*

Proof. We appeal to the geometry of the underlying surface $|\mathcal{C}(\mathcal{Q})|$ of $\mathcal{Q}$; we shall see that $|\mathcal{C}(\mathcal{P})| = |\mathcal{C}(\mathcal{Q})|$. Let $\mathcal{Q}$ have automorphism group $\boldsymbol{H}$ with generatrix $(\boldsymbol{s}_0, \boldsymbol{s}_1, \boldsymbol{s}_2)$, and let $\boldsymbol{\Psi}$ be the base flag of $\mathcal{Q}$. If $T := T(\boldsymbol{\Psi})$ is the fundamental region of $\boldsymbol{H}$ in $|\mathcal{C}(\mathcal{Q})|$ (that is, the base triangle in $\mathcal{C}(\mathcal{Q})$), then the corresponding fundamental region of $\boldsymbol{G} = \boldsymbol{G}(\mathcal{P})$ is $T \cup T\boldsymbol{s}_1 \cup T\boldsymbol{s}_0\boldsymbol{s}_1$. In $|\mathcal{C}(\mathcal{Q})|$, we can think of the $2s$-gonal faces of $\mathcal{P}$ as unions of $2s$ trigons of $\mathcal{Q}$ containing a common vertex. From either point of view, we see that the index $[\boldsymbol{H} : \boldsymbol{G}] = 3$. □

5A5 Remark The alternative possibility is that $\boldsymbol{G} = \boldsymbol{H}$, in which case we would only obtain a pre-polyhedron $\mathcal{P}$. For, since $\mathcal{P}$ would have $|\boldsymbol{G}|/2s = 2 \cdot |\boldsymbol{H}|/4s$ vertices, each occurring among those of $\mathcal{Q}$, we see that $\mathcal{P}$ would actually have the same vertices as $\mathcal{Q}$, each doubly-covered by the $2s$-gons through it. Thus $\mathcal{P}$ must degenerate.

5A6 Example As applied to the abstract icosahedron, $\{3,5\}^{\varphi} = \{5,5 \mid 3\}$; recall from Theorem 2D14 that $\{3,5\} = \{3,5 \mid 5\}$.

5A7 Example It is not often the case that the intersection property fails to carry over under faceting, but here is an instance where it does. Let

$$\mathcal{Q} := \{5,3,5,5\}/\langle\langle\, (01213)^4, (2343)^3 \,\rangle\rangle,$$
$$\mathcal{P} := \{5,3,3,5\}/\langle\langle\, (0123214)^4 \,\rangle\rangle;$$

the notation is that introduced in Section 2D. Then $\mathcal{Q}$ is polytopal, while $\mathcal{P}$ is not (see the notes at the end of the section); however, $\mathcal{P} = \mathcal{Q}^{\varphi_2}$.

Petrie operations

Initially applied to polyhedra, the *Petrie operation* $\boldsymbol{\pi}$ is defined by

5A8 $$\boldsymbol{\pi} \colon (\boldsymbol{s}_0, \boldsymbol{s}_1, \boldsymbol{s}_2) \mapsto (\boldsymbol{s}_0\boldsymbol{s}_2, \boldsymbol{s}_1, \boldsymbol{s}_2) =: (\boldsymbol{r}_0, \boldsymbol{r}_1, \boldsymbol{r}_2).$$

The resulting polyhedron $\mathcal{Q}^{\pi}$ is often called the *Petrie dual* or, more briefly, the *Petrial* of $\mathcal{Q}$. It has the same vertices and edges as $\mathcal{Q}$; however, each face is a Petrie polygon of $\mathcal{Q}$ as defined by (2D6), whose defining property is that two successive edges, but not three, are edges of a face of $\mathcal{Q}$. Thus a face of $\mathcal{Q}^{\pi}$ is a zigzag, leaving a face of $\mathcal{Q}$ after traversing two of its edges.

The polyhedra obtained from a given one by iterating the Petrie operation and duality form a family of up to six; that is, we have

5A9 Proposition *The Petrie operation and duality on polyhedra satisfy* $(\boldsymbol{\pi\delta})^3 = \boldsymbol{\iota}$, *the identity operation.*

Proof. Indeed, considering the groups, we have

$$\begin{aligned}(\boldsymbol{s}_0, \boldsymbol{s}_1, \boldsymbol{s}_2) &\overset{\pi}{\longmapsto} (\boldsymbol{s}_0\boldsymbol{s}_2, \boldsymbol{s}_1, \boldsymbol{s}_2) \overset{\delta}{\longmapsto} (\boldsymbol{s}_2, \boldsymbol{s}_1, \boldsymbol{s}_0\boldsymbol{s}_2) \overset{\pi}{\longmapsto} (\boldsymbol{s}_0, \boldsymbol{s}_1, \boldsymbol{s}_0\boldsymbol{s}_2)\\ &\overset{\delta}{\longmapsto} (\boldsymbol{s}_0\boldsymbol{s}_2, \boldsymbol{s}_1, \boldsymbol{s}_0) \overset{\pi}{\longmapsto} (\boldsymbol{s}_2, \boldsymbol{s}_1, \boldsymbol{s}_0) \overset{\delta}{\longmapsto} (\boldsymbol{s}_0, \boldsymbol{s}_1, \boldsymbol{s}_2),\end{aligned}$$

as claimed. □

It is clear that the Petrie operation $\boldsymbol{\pi}$ is involutory, so that $\boldsymbol{\pi}^2 = \boldsymbol{\iota}$; that is, $(\mathcal{Q}^{\pi})^{\pi} = \mathcal{Q}$. If $\mathcal{Q}^{\pi}$ is isomorphic to $\mathcal{Q}$, then we call $\mathcal{Q}$ *self-Petrie*. Observe, however, that a self-Petrie polyhedron and its Petrial do not coincide, since while they share the same vertices and edges, their faces are different.

5A10 Remark The Petrial of a realized self-Petrie regular polyhedron need not be congruent to the original. Again, the 10-dimensional realizations $P_{10}(\vartheta)$ of $\{5, 5 : 5\}$ of [88, Remark 14.7] are instances: Petriality is induced by $\vartheta \mapsto \alpha - \vartheta$ (with $\alpha = \arctan\sqrt{11}$ as before), so that $\mathsf{P}_{10}(\vartheta)$ and its Petrial are congruent only if $\vartheta = \frac{1}{2}\alpha$ or $\frac{1}{2}(\alpha + \pi)$.

Another self-Petrie polyhedron is the hemi-dodecahedron $\{5, 3 : 5\}$, obtained from the dodecahedron $\{5, 3\}$ by identifying its faces of each dimension under the central involutory symmetry (thus $\{5, 3 : 5\} = \{5, 3\}/2$). Recall from Section 2D that a regular polyhedron of type $\{p, q\}$ is denoted $\{p, q : r\}$ if the length r of its Petrie polygons determines its combinatorial type; observe that $\{p, q : r\}^{\pi} = \{r, q : p\}$. We shall meet further examples below. The group of $\{p, q : r\}$ is denoted by $[p, q : r]$ or (following [25] and [33, Section 8.6]) $G^{p,q,r}$, where the notation is now symmetric in p, q, r.

For polyhedra, there is a close connexion between self-duality and self-Petriality. To explain this in combinatorial terms, let $\mathcal{P}$ be a regular polyhedron with group $\boldsymbol{G}$ and generatrix $(\boldsymbol{r}_0, \boldsymbol{r}_1, \boldsymbol{r}_2)$. For self-duality, we have $\mathcal{P}^{\delta} \cong \mathcal{P}$, so that the mapping $\boldsymbol{r}_j \mapsto \boldsymbol{r}_{2-j}$ for $j = 0, 1, 2$ induces an automorphism of $\boldsymbol{G}$. Similarly, since the Petrial $\mathcal{P}^{\pi}$ of $\mathcal{P}$ is given by the operation

$$(\boldsymbol{r}_0, \boldsymbol{r}_1, \boldsymbol{r}_2) \mapsto (\boldsymbol{r}_0\boldsymbol{r}_2, \boldsymbol{r}_1, \boldsymbol{r}_2) =: (\boldsymbol{s}_0, \boldsymbol{s}_1, \boldsymbol{s}_2),$$

we see that $\mathcal{P}$ is self-Petrial provided that the mapping $\boldsymbol{r}_0 \mapsto \boldsymbol{r}_0\boldsymbol{r}_2$ and $\boldsymbol{r}_j \mapsto \boldsymbol{r}_j$ for $j = 1, 2$ induces an automorphism of $\boldsymbol{G}$.

We find it convenient to write $\boldsymbol{\pi}^* := \boldsymbol{\delta\pi\delta}$ $(= \boldsymbol{\pi\delta\pi}$ by Proposition 5A9), which is the dual operation

5A11 $$\boldsymbol{\pi}^* : (\boldsymbol{s}_0, \boldsymbol{s}_1, \boldsymbol{s}_2) \mapsto (\boldsymbol{s}_0, \boldsymbol{s}_1, \boldsymbol{s}_0\boldsymbol{s}_2) =: (\boldsymbol{r}_0, \boldsymbol{r}_1, \boldsymbol{r}_2).$$

We can usefully think of $\boldsymbol{\pi}^*$ as the conjugate of $\boldsymbol{\pi}$ by duality $\boldsymbol{\delta}$. Then we have

5A12 Proposition *The abstract regular polyhedron $\mathcal{P}$ is self-dual if and only if $\mathcal{P}^{\pi^*}$ is self-Petrie.*

Proof. Just notice that the Petrie operation on $\mathcal{P}^{\pi^*}$ is

$$(\boldsymbol{r}_0, \boldsymbol{r}_1, \boldsymbol{r}_0\boldsymbol{r}_2) \mapsto (\boldsymbol{r}_2, \boldsymbol{r}_1, \boldsymbol{r}_0\boldsymbol{r}_2);$$

the claim is now obvious. □

5A13 Remark We could use $\boldsymbol{\pi\delta}$ instead of $\boldsymbol{\pi}^*$ here. The latter is marginally preferable, because it is involutory.

Of course, Proposition 5A12 immediately translates into geometric terms; from a suitable realized regular polyhedron which is combinatorially but not geometrically self-dual, we can obtain another which is combinatorially but not geometrically self-Petrie. However, when we try to apply the idea to, for example, $\{5, \frac{5}{2}\} \cong \{5, 5 \mid 3\}$, we find that we have no corresponding realization of $\{5, 5 \mid 3\}^{\pi^*}$ in $\mathbb{E}^3$, and the remaining realization in $\mathbb{E}^5$ is actually that of $\{5, 5 : 3\}$, in which the Petrie polygons $\{6\}$ of $\{5, 5 \mid 3\}$ collapse onto trigons.

5A14 Remark A consequence of this example is that, in considering rigidity in the next Chapter 6, we may not assume that combinatorial self-Petriality, even with faces and Petrie polygons of the same kind, will necessarily imply that faces and Petrie polygons are themselves congruent. Thus other conditions will have to be brought into play if we need to use any additional symmetry.

The notation $\{p, q : r_1; r_2, \ldots \mid s_2, s_3, \ldots\}$ for a regular polyhedron specified by certain of its zigzags and holes was introduced in (2E5). The following is a straightforward observation.

5A15 Proposition *Applying the Petrie operation* π *to the general regular polyhedron of (2E5) switches k-zigzags and k-holes, and so yields*

$$\{p, q : r_1, r_2, r_3, \ldots \mid s_2, s_3, \ldots\}^{\pi} = \{r_1, q : p, s_2, s_3, \ldots \mid r_2, r_3, \ldots\}.$$

5A16

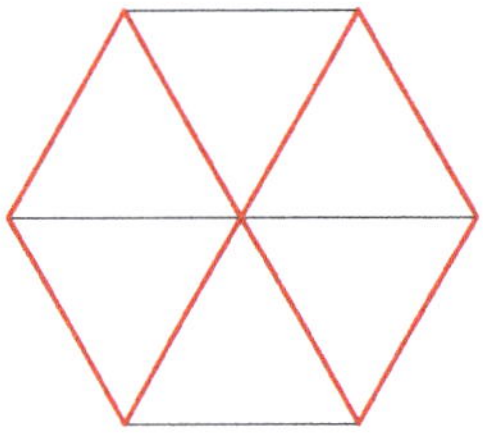

The regular polyhedron $\{3, 6 : \cdot, 2\}$.

There are rare cases in which the Petrial of a polyhedron is not polytopal (that is, is not itself a polyhedron). One instance is the flat toroidal polyhedron $\{3, 6 : \cdot, 2\}$; again refer to Section 2D for the notation. This has six triangles with a common vertex, which form a hexagon with opposite edges identified (see Figure 5A16), and so has 3 vertices, 9 edges and 6 faces. It is easy to check its polytopality, which is not vitiated by (for example) its edges falling into three triples with the same vertices. But its Petrial is not a polyhedron; each of the vertices of a Petrie polygon of $\{3, 6 : \cdot, 2\}$ (illustrated by the red lines in Figure 5A16) belongs to four of its edges, rather than two.

What is happening here indicates when polytopality is retained.

5A17 Lemma *The Petrial* $\mathcal{Q}^{\pi}$ *of a regular polyhedron* $\mathcal{Q}$ *is polytopal if and only if a Petrie polygon of* $\mathcal{Q}$ *visits any given vertex at most once.*

Proof. Of course, this is the familiar interpretation of the intersection property, as we previously saw in the case of faceting. □

5A18

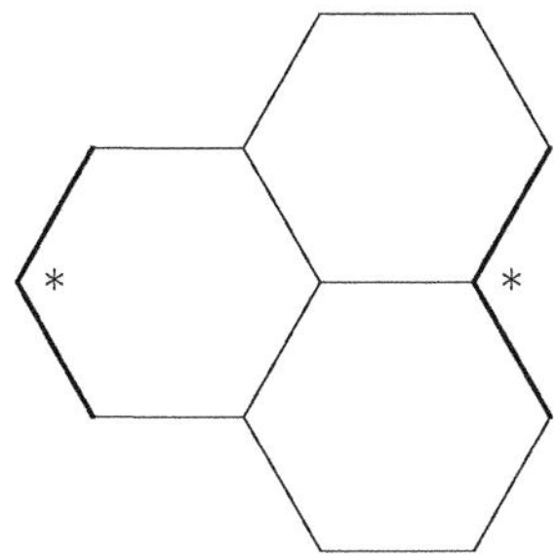

The regular polyhedron $\{6, 3 : \cdot, *2\}$.

Curiously, since the dual polyhedron $\{6, 3 : \cdot, *2\} = \{6, 3\}/\langle\langle (01012)^2 \rangle\rangle$ is self-Petrie, its Petrial is polytopal; this may be verified alternatively by (for example) looking at Figure 5A18, where the asterisks indicate a typical identification to create the torus.

In general, though, the Petrial of a regular polyhedron will also be a regular polyhedron (that is, it will also be polytopal).

The Petrie operation and faceting are related as follows.

5A19 Proposition *The operations π and φ_k commute.*

Proof. This is easily verified algebraically. However, it is even more instructive to look at the geometry. Whether we apply π or φ_k first, the result will be (assuming that it exists) a polyhedron $\mathcal{P}$ whose vertices and edges are those of $\mathcal{Q}$, but whose typical face is a k-zigzag (as defined in (2E3)), given by an edge-path which (as for φ_k) leaves a vertex at the kth edge from the one by which it entered, but in the oppositely oriented sense at alternate vertices (see, for example, [129]). Thus the Petrie polygons themselves are 1-zigzags, which accounts for our nomenclature. □

In view of Proposition 5A19, we can define the *generalized Petrie operation* π_k by

5A20 $$\pi_k := \pi\varphi_k = \varphi_k\pi.$$

Thus we think of $\varphi_1 := \iota$ as before and $\pi_1 := \pi$.

Halving

The *halving* operation η applies initially to a regular polyhedron $\mathcal{Q}$ of type $\{4, q\}$ for some $q \geqslant 3$, and turns it into a self-dual polyhedron $\mathcal{P} := \mathcal{Q}^\eta$ of type $\{q, q\}$. We define η on its generatrix (s_0, s_1, s_2) by

5A21 $$\eta\colon (s_0, s_1, s_2) \mapsto (s_0 s_1 s_0, s_2, s_1) =: (r_0, r_1, r_2).$$

The intersection property is easily checked for $\boldsymbol{G} := \langle r_0, r_1, r_2 \rangle$; it will become clear from the discussion below. When we think of $\boldsymbol{H} = \boldsymbol{G}(\mathcal{Q})$ acting on the surface $|\mathcal{C}(\mathcal{Q})|$, as we remarked in at the end of Section 2B the triangle $T = T(\boldsymbol{\Phi})$ associated with the base flag $\boldsymbol{\Phi}$ of $\mathcal{Q}$ is a fundamental region for $\boldsymbol{H}$, and $\boldsymbol{s}_0$, $\boldsymbol{s}_1$ and $\boldsymbol{s}_2$ act as reflexions in the sides of T. Then $T' := T \cup T\boldsymbol{s}_0$ is the fundamental region for $\boldsymbol{G}$, and $\boldsymbol{G}$ is similarly generated by the reflexions in the sides of T'.

If we now apply Wythoff's construction (or, rather, its abstract analogue, in the underlying surface $|\mathcal{C}(\mathcal{Q})|$), then we see that there are two possibilities.

First, if the edge-graph of $\mathcal{Q}$ is bipartite, so that the edge-circuits of $\mathcal{Q}$ have even length, then $\mathcal{P}$ is a map on the same surface $|\mathcal{C}(\mathcal{Q})|$ with half as many vertices as $\mathcal{Q}$, namely, those in the same partition of the vertex-set $\mathcal{Q}_0$ of $\mathcal{Q}$ as the initial vertex in $\boldsymbol{\Phi}$. Further, $\boldsymbol{G}$ has index 2 in $\boldsymbol{H}$. As we asserted above, $\mathcal{P}$ is self-dual, since $\boldsymbol{s}_0 \in \boldsymbol{H}$ acts as an automorphism of $\boldsymbol{G}$, interchanging $\boldsymbol{r}_0$ and $\boldsymbol{r}_2$ and leaving $\boldsymbol{r}_1$ fixed. The vertices of the dual $\mathcal{P}^\delta$ are then those in the other partition of $\mathcal{Q}_0$. We may observe that $\boldsymbol{H}$ can be recovered from $\boldsymbol{G}$ by a twisting operation, of the kind that we shall consider in Section 5C.

Otherwise the edge-graph of $\mathcal{Q}$ is not bipartite and, unless $q = 4$ also, $\mathcal{P}$ is a map on a different surface from $|\mathcal{C}(\mathcal{Q})|$. Indeed, $|\mathcal{C}(\mathcal{P})|$ will be a double cover of $|\mathcal{C}(\mathcal{Q})|$ in every case; we must be careful to note that $\mathcal{Q}$ itself does not cover $\mathcal{P}$ in general. We now have $\boldsymbol{G} = \boldsymbol{H}$, and $\mathcal{P}$ has the same vertex-set $\mathcal{Q}_0$ as $\mathcal{Q}$. Finally, $\mathcal{P}$ is still self-dual, although now the conjugating element $\boldsymbol{s}_0$ is in $\boldsymbol{G}$.

In either case, we have $\boldsymbol{r}_0\boldsymbol{r}_1\boldsymbol{r}_2\boldsymbol{r}_1 = \boldsymbol{s}_0\boldsymbol{s}_1\boldsymbol{s}_0\boldsymbol{s}_2\boldsymbol{s}_1\boldsymbol{s}_2 = (\boldsymbol{s}_0\boldsymbol{s}_1\boldsymbol{s}_2)^2$. This shows that, if the original polyhedron $\mathcal{Q}$ has Petrie polygons of length h, then the new polyhedron $\mathcal{P}$ will have 2-holes of length h or $h/2$ according as h is odd or even. Note that the latter will be the case when the graph of $\mathcal{Q}$ is bipartite (but possibly in other cases also); then $[\boldsymbol{H} : \boldsymbol{G}] = 2$.

In this spirit, in certain cases the combinatorial type of a polyhedron $\mathcal{P} = \mathcal{Q}^\eta$ is easily determined from that of $\mathcal{Q}$.

5A22 Theorem *If $q \geqslant 4$ and $s \geqslant 2$, then*

(a) $\{4, q \mid 2s\}^\eta = \{q, q : 2s\}$,

(b) $\{4, q : 2s\}^\eta = \{q, q \mid s\}$.

Proof. The graphs of the two polyhedra $\{4, q \mid 2s\}$ and $\{4, q : 2s\}$ are bipartite, since their defining circuits are 4-gonal faces and $2s$-gonal holes or zigzags. The operation $\mathcal{Q} \mapsto \mathcal{Q}^\eta =: \mathcal{P}$ is given by (5A21). In case (a), we therefore have

$$\begin{aligned} \boldsymbol{e} &= (\boldsymbol{s}_0\boldsymbol{s}_1\boldsymbol{s}_2\boldsymbol{s}_1)^{2s} = (\boldsymbol{s}_0\boldsymbol{s}_1\boldsymbol{s}_2\boldsymbol{s}_1\boldsymbol{s}_0\boldsymbol{s}_1\boldsymbol{s}_2\boldsymbol{s}_1)^s = (\boldsymbol{s}_0\boldsymbol{s}_1\boldsymbol{s}_2 \cdot \boldsymbol{s}_0\boldsymbol{s}_1\boldsymbol{s}_0\boldsymbol{s}_1\boldsymbol{s}_0 \cdot \boldsymbol{s}_2\boldsymbol{s}_1)^s \\ &\sim (\boldsymbol{s}_1\boldsymbol{s}_2 \cdot \boldsymbol{s}_0\boldsymbol{s}_1\boldsymbol{s}_0 \cdot \boldsymbol{s}_1\boldsymbol{s}_2 \cdot \boldsymbol{s}_0\boldsymbol{s}_1\boldsymbol{s}_0)^s = (\boldsymbol{s}_1\boldsymbol{s}_2 \cdot \boldsymbol{s}_0\boldsymbol{s}_1\boldsymbol{s}_0)^{2s} = (\boldsymbol{r}_2\boldsymbol{r}_1\boldsymbol{r}_0)^{2s}, \end{aligned}$$

where $\sim$ as usual denotes conjugacy. Similarly, in case (b), we have

$$\boldsymbol{e} = (\boldsymbol{s}_0\boldsymbol{s}_1\boldsymbol{s}_2)^{2s} = (\boldsymbol{s}_0\boldsymbol{s}_1\boldsymbol{s}_2\boldsymbol{s}_0\boldsymbol{s}_1\boldsymbol{s}_2)^s = (\boldsymbol{s}_0\boldsymbol{s}_1\boldsymbol{s}_0 \cdot \boldsymbol{s}_2\boldsymbol{s}_1\boldsymbol{s}_2)^s = (\boldsymbol{r}_0\boldsymbol{r}_1\boldsymbol{r}_2\boldsymbol{r}_1)^s.$$

In each case, the defining relation of the original polyhedron $\mathcal{Q}$ is equivalent to the corresponding defining relation for the new polyhedron $\mathcal{P}$. □

We now consider the effect of applying $\boldsymbol{\eta}$ to the 3-coface of a suitable regular polytope of higher rank; the following exhibits some extra features of $\boldsymbol{\eta}$.

5A23 Theorem *Let $\mathcal{P}$ be a regular m-polytope of Schläfli type $\{p_1, \ldots, p_{m-1}\}$ with $m \geqslant 4$ and $p_{m-2} = 4$. If $\mathcal{P}^\eta$ is polytopal, then*

- *the $(m-3)$-faces of $\mathcal{P}^\eta$ and $\mathcal{P}$ are the same,*
- *faces of $\mathcal{P}^\eta$ are holes of facets of $\mathcal{P}$,*
- *holes of the facets of $\mathcal{P}^\eta$ are deep holes of $\mathcal{P}$,*
- *if, in addition, $p = 3$, then facets of $\mathcal{P}^\eta$ are vertex-figures of $\mathcal{P}$.*

Proof. All the claims except for the last follow easily. For the last assertion, if $\mathcal{P}$ has generatrix $(\boldsymbol{r}_0, \ldots, \boldsymbol{r}_3)$, then conjugating $(\boldsymbol{r}_0, \boldsymbol{r}_1, \boldsymbol{r}_2)$ by $\boldsymbol{r}_1\boldsymbol{r}_0$ (which moves the initial vertex of $\mathcal{P}$ to that of the vertex-figure of $\mathcal{P}$), and using the fact that $\boldsymbol{r}_0\boldsymbol{r}_1\boldsymbol{r}_0\boldsymbol{r}_1\boldsymbol{r}_0 = \boldsymbol{r}_1$, with generatrix $(\boldsymbol{s}_0, \ldots, \boldsymbol{s}_3)$ of $\mathbf{G}(\mathcal{P}^\eta)$ we obtain

$$\boldsymbol{r}_0\boldsymbol{r}_1(\boldsymbol{s}_0, \boldsymbol{s}_1, \boldsymbol{s}_2)\boldsymbol{r}_1\boldsymbol{r}_0 = \boldsymbol{r}_0\boldsymbol{r}_1(\boldsymbol{r}_0, \boldsymbol{r}_1\boldsymbol{r}_2\boldsymbol{r}_1, \boldsymbol{r}_3)\boldsymbol{r}_1\boldsymbol{r}_0 = (\boldsymbol{r}_1, \boldsymbol{r}_2, \boldsymbol{r}_3),$$

since $\boldsymbol{r}_1$ commutes with $\boldsymbol{r}_3$ and $\boldsymbol{r}_0$ commutes with $\boldsymbol{r}_2$ and $\boldsymbol{r}_3$. □

5A24 Example We shall meet many examples of halving as we go on. For now, we give a few simple ones using Theorem 5A22, all on the abstract level.

- The inscription of the tetrahedron in the cube corresponds to $\{4,3\}^\eta = \{3,3\}$; this fits in, because $\{4,3\} = \{4,3:6\}$ and $\{3,3\} = \{3,3 \mid 3\}$.
- In Sections 11B and 11C, we consider the example $\{4,5:6\}^\eta = \{5,5 \mid 3\}$. In view of Example 5A6, namely, $\{5,5 \mid 3\} = \{3,5\}^{\varphi_2}$, this shows that the automorphism group of $\{4,5:6\}$ (and the other members of its family) has order 240.
- For the tori, $\{4,4 \mid 2r\}^\eta = \{4,4:2r\}$ and $\{4,4:2r\}^\eta = \{4,4 \mid r\}$ for each $r \geqslant 3$. On the other hand, if r is odd, then $\{4,4 \mid r\}^\eta = \{4,4 \mid r\}$.

Skewing

We next define the *skewing* operation $\boldsymbol{\sigma}$ on a regular polyhedron $\mathcal{Q}$ of Schläfli type $\{4, q\}$ (for some q), again on the generatrix $(\boldsymbol{s}_0, \boldsymbol{s}_1, \boldsymbol{s}_2)$, by

5A25 $$\boldsymbol{\sigma} := \boldsymbol{\pi}^*\boldsymbol{\eta}\boldsymbol{\pi}^* \colon (\boldsymbol{s}_0, \boldsymbol{s}_1, \boldsymbol{s}_2) \mapsto (\boldsymbol{s}_1, \boldsymbol{s}_0\boldsymbol{s}_2, (\boldsymbol{s}_0\boldsymbol{s}_1)^2) =: (\boldsymbol{r}_0, \boldsymbol{r}_1, \boldsymbol{r}_2),$$

modulo conjugation by $\boldsymbol{s}_0$, with $\boldsymbol{\pi}^*$ and $\boldsymbol{\eta}$ as defined previously (see the notes at the end of the section). Hence, just as we could think of $\boldsymbol{\pi}^*$ as the conjugate of $\boldsymbol{\pi}$ by $\boldsymbol{\delta}$, so we can think of $\boldsymbol{\sigma}$ as the conjugate of $\boldsymbol{\eta}$ by $\boldsymbol{\pi}^*$.

The definition indicates that $\boldsymbol{\sigma}$ halves the order of the group just when $\boldsymbol{\eta}$ does, modulo the double of application of $\boldsymbol{\pi}^*$, that is, just when the graph of $\mathcal{Q}^{\pi^*}$ is bipartite. If this is the case, then $\boldsymbol{s}_0 \notin \boldsymbol{G} := \boldsymbol{G}(\mathcal{Q}^\sigma) = \langle \boldsymbol{r}_0, \boldsymbol{r}_1, \boldsymbol{r}_2 \rangle$, but acts on it as an automorphism. In any event, $\mathcal{P} := \mathcal{Q}^\sigma$ is self-Petrie; the isomorphism between $\mathcal{P}$ and $\mathcal{P}^\pi$ is given by conjugation of $\boldsymbol{G}(\mathcal{P})$ by $\boldsymbol{s}_0$, since

$$\boldsymbol{s}_0\boldsymbol{s}_1\boldsymbol{s}_0 = \boldsymbol{s}_1 \cdot (\boldsymbol{s}_0\boldsymbol{s}_1)^2.$$

The type $\{s, t\}$ of $\mathcal{P} = \mathcal{Q}^\sigma$ is determined by the periods of

$$\boldsymbol{r}_0\boldsymbol{r}_1 = \boldsymbol{s}_1 \cdot \boldsymbol{s}_0\boldsymbol{s}_2 \sim \boldsymbol{s}_0\boldsymbol{s}_1\boldsymbol{s}_2, \quad \boldsymbol{r}_1\boldsymbol{r}_2 = \boldsymbol{s}_0\boldsymbol{s}_2 \cdot (\boldsymbol{s}_0\boldsymbol{s}_1)^2 \sim \boldsymbol{s}_0\boldsymbol{s}_1\boldsymbol{s}_2\boldsymbol{s}_1.$$

Hence s is the length of the Petrie polygons of $\mathcal{Q}$, while t is the length of its 2-holes.

5A26 Remark Since $\mathcal{C}(\mathcal{Q}^{\sigma})$ is not in general closely related to $\mathcal{C}(\mathcal{Q})$, we do not usually attempt to apply $\boldsymbol{\sigma}$ to m-polytopes with $m \geqslant 4$. Thus skewing mainly applies only in rank 3.

We shall postpone discussion of specific examples of skewing until later.

Quartering

We finally come to a new operation $\boldsymbol{\psi}$, which we shall call *quartering*; see the notes at the end of the section. Initially, this is applicable to a polyhedron $\mathcal{Q}$ of type $\{p, q : 4\}$, and on its generatrix $(\boldsymbol{s}_0, \boldsymbol{s}_1, \boldsymbol{s}_2)$ is

5A27 $$\boldsymbol{\psi}\colon (\boldsymbol{s}_0, \boldsymbol{s}_1, \boldsymbol{s}_2) \mapsto (\boldsymbol{s}_0\boldsymbol{s}_1\boldsymbol{s}_0, \boldsymbol{s}_1, \boldsymbol{s}_2\boldsymbol{s}_1\boldsymbol{s}_2) =: (\boldsymbol{r}_0, \boldsymbol{r}_1, \boldsymbol{r}_2).$$

Observe that $\boldsymbol{r}_0\boldsymbol{r}_2 = \boldsymbol{s}_0\boldsymbol{s}_1\boldsymbol{s}_0 \cdot \boldsymbol{s}_2\boldsymbol{s}_1\boldsymbol{s}_2 = (\boldsymbol{s}_0\boldsymbol{s}_1\boldsymbol{s}_2)^2$ does have period 2.

We have the following geometric picture of what $\boldsymbol{\psi}$ does. Suppose first that $p \geqslant 6$ is even, so that the edge-graph of the universal polyhedron $\{p, q : 4\}$ is bipartite. In each polygonal face $\{p\}$ is inscribed a polygon $\{\frac{1}{2}p\}$. Because the Petrie polygon has length 4, we see that adjacent edges of alternate such polygons $\{\frac{1}{2}p\}$ around a given vertex actually coincide, so that these polygons fit together to form a polyhedron. If q is odd, then q such polygons fit round the vertex, going around twice. If q is also even, then there is a further split into two polyhedra. Just as with halving $\boldsymbol{\eta}$, the same construction works if p and q are odd, but now there is no splitting.

5A28 Example As a simple example, we have

$$\{6, 5 : 4\}^{\psi} = \{3, 5\}.$$

A more complicated one is the application to $\{6, 7 : 4\}$. This polyhedron has automorphism group $\mathrm{PGL}(2, 13) \times C_2$, with 312 vertices. We can then make the identification

$$\{6, 7 : 4\}^{\psi} = \{3, 7 : 12\};$$

the latter polyhedron has automorphism group $\mathrm{PGL}(2, 13)$, and has 156 vertices.

Notes to Section 5A

1. Proposition 5A4 and Remark 5A5 correct the assertion in [99, Lemma 7B10] that $\mathcal{P}$ can be polytopal if $\boldsymbol{G} = \boldsymbol{H}$.
2. The data for Example 5A7 were provided by Marston Conder using the computer algebra program MAGMA [5]; see further Section 17F. In fact, if the generatrix of $\mathcal{Q}$ is $(\boldsymbol{s}_0, \ldots, \boldsymbol{s}_4)$, then $\langle \boldsymbol{s}_0, \ldots, \boldsymbol{s}_3\rangle \cap \langle \boldsymbol{s}_1, \ldots, \boldsymbol{s}_4\rangle$ has order 96, rather than 24 as it would have if it had the intersection property.
3. As we have defined it, $\boldsymbol{\sigma}$ is not the same as the operation so denoted in [99, (7B19)]. Instead, largely for geometric reasons, we prefer instead to define $\boldsymbol{\sigma}$ on polyhedra of type $\{4, q\}$, thus preceeding the former $\boldsymbol{\sigma}$ by duality $\boldsymbol{\delta}$. It is worth noting that $\boldsymbol{\sigma}$ is applied to several polyhedra of type $\{4, q\}$ in Section 11F that do not have geometric duals.
4. Quartering is so called because – potentially at least – two halving operations are involved. The name itself was suggested by a biochemist friend, Iain Mowbray [104], after several clumsier terms had been rejected.

5B General Mixing

A general *mixing* operation $\boldsymbol{\mu}$ on a group $\boldsymbol{H}$ with generatrix $(\boldsymbol{s}_0, \ldots, \boldsymbol{s}_{k-1})$, with the $\boldsymbol{s}_j \in \boldsymbol{H}$ involutions, is denoted by

5B1 $$\boldsymbol{\mu}\colon (\boldsymbol{s}_0, \ldots, \boldsymbol{s}_{k-1}) \mapsto (\boldsymbol{r}_0, \ldots, \boldsymbol{r}_{m-1})$$

with each $\boldsymbol{r}_j \in \boldsymbol{H}$ an involution, yielding the generatrix $(\boldsymbol{r}_0, \ldots, \boldsymbol{r}_{m-1})$ of an sggi (string pre-C-group) $\boldsymbol{G}$. The general notation for such an operation is $\boldsymbol{G} := \boldsymbol{H}^{\boldsymbol{\mu}}$; moreover, if $\boldsymbol{H} = \boldsymbol{G}(\mathcal{Q})$ for some regular k-polytope $\mathcal{Q}$, then we set $\mathcal{P} := \mathcal{Q}^{\boldsymbol{\mu}}$ for the regular pre-polytope with group $\boldsymbol{G}$. It is therefore obviously important to find conditions under which $\boldsymbol{G}$ is a C-group; in fact, it is more often the case that we can describe circumstances under which $\boldsymbol{G}$ fails to have the intersection property (2B6), and so is not a C-group.

Duality

Our first mixing operation is not commonly thought of as such. This is *duality*, denoted throughout by $\boldsymbol{\delta}$, and given by

5B2 $$\boldsymbol{\delta}\colon (\boldsymbol{s}_0, \ldots, \boldsymbol{s}_{m-1}) \mapsto (\boldsymbol{s}_{m-1}, \ldots, \boldsymbol{s}_0) =: (\boldsymbol{r}_0, \ldots, \boldsymbol{r}_{m-1}).$$

The *dual* of $\mathcal{Q}$ is thus denoted $\mathcal{P} := \mathcal{Q}^{\boldsymbol{\delta}}$.

5B3 Remark We shall observe in Section 5D that the corresponding operation

$$\boldsymbol{\delta}\colon (S_0, \ldots, S_{m-1}) \mapsto (S_{m-1}, \ldots, S_0)$$

on a geometric group may fail to lead to a geometric dual. We shall see some examples as soon as Section 8C, where quasi-duality comes into play.

5B4 Remark Even with a self-dual polytope, a geometric dual may exist, but need not be congruent (or similar) to the original; for instance, we have the great dodecahedron $\{5, \frac{5}{2}\}$ and its geometrically dual small stellated dodecahedron $\{\frac{5}{2}, 5\}$ to which it is isomorphic. Indeed, in [88, Remark 14.7] it was noted that the 10-dimensional realizations $\mathsf{P}_{10}(\vartheta)$ (with $0 \leqslant \vartheta < \pi$ and $\mathsf{P}_{10}(\vartheta) = \mathsf{P}_{10}(\vartheta + \pi)$) of the self-dual, self-Petrie regular polyhedron $\{5, 5 : 5\}$ are such that geometric duality is induced by $\vartheta \mapsto \vartheta + \alpha$, where $\alpha = \arctan \sqrt{11}$; hence, successively taking the geometric dual leads to an infinite sequence of mutually incongruent polyhedra.

5B5 Remark In the same spirit, the *facet* $\mathcal{Q}^{\boldsymbol{f}}$ and *vertex-figure* $\mathcal{Q}^{\boldsymbol{v}}$ of $\mathcal{Q}$ can be thought of as given by mixing operations $\boldsymbol{f}$ and $\boldsymbol{v}$, namely,

$$\boldsymbol{f}\colon (\boldsymbol{s}_0, \ldots, \boldsymbol{s}_{m-1}) \mapsto (\boldsymbol{s}_0, \ldots, \boldsymbol{s}_{m-2}),$$
$$\boldsymbol{v}\colon (\boldsymbol{s}_0, \ldots, \boldsymbol{s}_{m-1}) \mapsto (\boldsymbol{s}_1, \ldots, \boldsymbol{s}_{m-1}),$$

with $\boldsymbol{G}(\mathcal{P}) = \langle \boldsymbol{s}_0, \ldots, \boldsymbol{s}_{m-1} \rangle$ as before. We extend this notation to the *ridge* $\boldsymbol{r} := \boldsymbol{ff}$ and *edge-figure* $\boldsymbol{e} := \boldsymbol{vv}$. More general notions of this kind will not be needed, but observe that $\boldsymbol{fv} = \boldsymbol{vf}$.

Vertex-Figure Replacement

An operation that will prove very useful – abstractly and geometrically – is that of *vertex-figure replacement*; see particularly Sections 7G, 16F and 17F. If $(\boldsymbol{r}_0, \ldots, \boldsymbol{r}_{m-1})$ is the generatrix of an m-polytope $\mathcal{P}$ and its automorphism group $\boldsymbol{G} = \boldsymbol{G}(\mathcal{P})$, then the operation is of the form $(\boldsymbol{r}_0, \ldots, \boldsymbol{r}_{m-1}) \mapsto (\boldsymbol{s}_0, \ldots, \boldsymbol{s}_{m-1})$, where $(\boldsymbol{s}_1, \ldots, \boldsymbol{s}_{m-1})$ is the generatrix of a string C-group, and

5B6
$$\begin{aligned} \boldsymbol{s}_0 &= \boldsymbol{r}_0, \\ \langle \boldsymbol{s}_1, \ldots, \boldsymbol{s}_{m-1} \rangle &= \langle \boldsymbol{r}_1, \ldots, \boldsymbol{r}_{m-1} \rangle, \\ \langle \boldsymbol{s}_2, \ldots, \boldsymbol{s}_{m-1} \rangle &= \langle \boldsymbol{r}_2, \ldots, \boldsymbol{r}_{m-1} \rangle. \end{aligned}$$

If $(\boldsymbol{s}_0, \ldots, \boldsymbol{s}_{m-1})$ also satisfies the intersection property, then we obtain a new regular m-polytope $\mathcal{Q}$, say, with the same automorphism group $\boldsymbol{G}(\mathcal{Q}) = \boldsymbol{G}$ and the same initial vertex and edge as $\mathcal{P}$. However, vertex-figure replacement can lead from a polytope to a non-polytope, as Example 5A7 shows (faceting is a particular case of vertex-figure replacement).

A special case of vertex-figure replacement often proves useful, as we shall see particularly in Chapters 16 and 17. This replaces $\mathcal{P}^{\boldsymbol{v}}$ by an allomorph (which is generally unique in the contexts where it is applied); we denote such an operation by $\boldsymbol{\lambda}\colon \mathcal{P} \mapsto \mathcal{P}^{\boldsymbol{\lambda}}$.

Petrie Polygons and Deep Holes

We introduced Petrie polygons and deep holes in Section 2D. On an m-polytope $\mathcal{Q}$, they correspond quite naturally to mixing operations $\boldsymbol{p}$ given by

5B7
$$\boldsymbol{p}\colon (\boldsymbol{s}_0, \ldots, \boldsymbol{s}_{m-1}) \mapsto (\boldsymbol{s}_0\boldsymbol{s}_2\boldsymbol{s}_4 \cdots, \boldsymbol{s}_1\boldsymbol{s}_3\boldsymbol{s}_5 \cdots)$$

for the Petrie polygon (with the natural convention $\boldsymbol{s}_j := \boldsymbol{e}$ for $j \geqslant m$), and $\boldsymbol{h}$ defined by

5B8
$$\boldsymbol{h}\colon (\boldsymbol{s}_0, \ldots, \boldsymbol{s}_{m-1}) \mapsto (\boldsymbol{s}_0, \boldsymbol{s}_1\boldsymbol{s}_2 \cdots \boldsymbol{s}_{m-1}\boldsymbol{s}_{m-2}\boldsymbol{s}_{m-3} \cdots \boldsymbol{s}_1)$$

for the deep hole. We thus write $\mathcal{Q}^{\boldsymbol{p}}$ and $\mathcal{Q}^{\boldsymbol{h}}$ for these polygons. In each case, the new generators are chosen so that the initial vertex and edge of the polygon are the same as those of $\mathcal{Q}$.

For our purposes, the most frequent and important case of faceting of an m-polytope of Schläfli type $\{q_1, \ldots, q_{m-1}\}$ is that when $q_{m-1} = 5$; we often write $\boldsymbol{\varphi} := \boldsymbol{\varphi}_2$ in such a case. We cannot make any general claims about Petrie polygons; however, we do have

5B9 Proposition *If $\mathcal{Q}$ is a regular m-polytope with $m \geqslant 3$ and $\boldsymbol{\varphi} = \boldsymbol{\varphi}_2$, then $\mathcal{Q}^{\boldsymbol{\varphi f h}} = \mathcal{Q}^{\boldsymbol{h}}$. Moreover, if the Schläfli type of $\mathcal{Q}$ is $\{q_1, \ldots, q_{m-1}\}$ with $q_{m-1} = 5$, then $\mathcal{Q}^{\boldsymbol{\varphi h}} = \mathcal{Q}^{\boldsymbol{f h}}$.*

Proof. Under $\boldsymbol{\varphi}$, we have $(\ldots, \boldsymbol{s}_{m-2}, \boldsymbol{s}_{m-1}) \mapsto (\ldots, \boldsymbol{s}_{m-2}\boldsymbol{s}_{m-1}\boldsymbol{s}_{m-2}, \boldsymbol{s}_{m-1})$. It is clear from this that the generatrices of $\mathcal{Q}^{\boldsymbol{\varphi f h}}$ and $\mathcal{Q}^{\boldsymbol{h}}$ are the same. The second claim then follows, since $\boldsymbol{\varphi}$ is invertible here. □

Petrie Operations

We next extend the Petrie operations π_k to higher ranks. If $\mathcal{Q}$ is a regular m-polytope with $m \geqslant 4$, then $\mathcal{Q}^{\pi_k}$ is obtained by applying π_k to the 3-coface of $\mathcal{Q}$. As we said earlier, sometimes there are restrictions on applications of certain operations; these particularly concern the π_k. We give here a very general such restriction, and then draw attention to some particular cases; the terminology used here was introduced at the end of Section 2D.

5B10 Theorem *For $m \geqslant 4$, the Petrie operations π_k on a regular m-polytope $\mathcal{Q}$ with generatrix $(s_0, \ldots, s_{m-1})$ are invalid if there is an $(m-3)$-odd relator on the ridge $\mathcal{Q}^r$ of $\mathcal{Q}$.*

Proof. It suffices to consider π; the further application of φ_k will be irrelevant. Let the given relator $\boldsymbol{a}$ in $(\boldsymbol{s}_0, \ldots, \boldsymbol{s}_{m-3}) \subset \boldsymbol{H} := \boldsymbol{G}(\mathcal{Q})$ contain $2k+1$ (say) occurrences of $\boldsymbol{s}_{m-3}$, and let $\boldsymbol{b} \in \boldsymbol{G} := \boldsymbol{H}^{\pi}$ be obtained from $\boldsymbol{a}$ when $\boldsymbol{s}_{m-3}$ is replaced by $\boldsymbol{s}_{m-3}\boldsymbol{s}_{m-1}$. Since $\boldsymbol{s}_{m-1}$ commutes with each of $\boldsymbol{s}_0, \ldots, \boldsymbol{s}_{m-3}$, we see that $\boldsymbol{s}_{m-1} = \boldsymbol{a}\boldsymbol{s}_{m-1} = \boldsymbol{a}\boldsymbol{s}_{m-1}^{2k+1} = \boldsymbol{b} \in \boldsymbol{G}_{m-2,m-1}$, violating the intersection property (2B6). This is the claim of the theorem. □

5B11 Remark The converse of the theorem says that, if $\mathcal{Q}^{\pi}$ is polytopal, then the edge-circuits of the dual $\mathcal{G}^{\delta}$ of the original ridge $\mathcal{G} := \mathcal{Q}^r$ of $\mathcal{Q}$ are even; hence the facets of $\mathcal{G}$ are 2-colourable.

5B12 Corollary *For $m \geqslant 4$, the Petrie operations π_k on a regular m-polytope $\mathcal{Q}$ are invalid whenever, for some $j = 0, \ldots, m-2$, the deep hole or Petrie polygon of its $(j, m-2)$-inface is odd.*

5B13 Remark In the context of Corollary 5B12, bear in mind that both the deep hole and the Petrie polygon of a polygon are that polygon itself.

For more details about how the Petrie operation works, we have

5B14 Theorem *If $\mathcal{Q}$ is a regular polytope of rank at least 4 such that $\mathcal{Q}^{\pi}$ is polytopal, then π has the following effects:*

(a) *the new vertex-figure is $\mathcal{Q}^{\pi v} = \mathcal{Q}^{v\pi}$;*

(b) *for the ridge, $\mathcal{Q}^{\pi r} = \mathcal{Q}^r$;*

(c) *for the Petrie polygons, $\mathcal{Q}^{\pi p} = \mathcal{Q}^{fp}$, and hence $\mathcal{Q}^{\pi fp} = \mathcal{Q}^p$;*

(d) *for the deep holes, $\mathcal{Q}^{\pi h} = \mathcal{Q}^h$.*

Proof. The first claim is obvious. For the second, since $\mathcal{G}$ is fixed by $\boldsymbol{s}_{m-1}$ (in the previous notation), it is unaffected by π (note that this will hold on the geometric level as well). The exchange of the Petrie polygon relators is clear, while the new deep hole relator is the old one conjugated by $\boldsymbol{s}_{m-1}$, which fixes the initial vertex. □

Schläfli Operations

We can generalize Petrie operations in a different way. Let $\mathcal{Q}$ be a regular m-polytope of type $\{3^{m-2}, q\}$ for some q, with generatrix $(\boldsymbol{s}_0, \ldots, \boldsymbol{s}_{m-1})$. Then the *Schläfli operation* $\boldsymbol{\chi}_k \colon (\boldsymbol{s}_0, \ldots, \boldsymbol{s}_{m-1}) \mapsto (\boldsymbol{r}_0, \ldots, \boldsymbol{r}_{m-1})$ is defined by

5B15
$$\boldsymbol{r}_j := \begin{cases} \boldsymbol{s}_{m-k+j}, & \text{for } j = 0, \ldots, k-1, \\ \boldsymbol{s}_0\boldsymbol{s}_1 \cdots \boldsymbol{s}_{m-2}\boldsymbol{s}_{m-1}\boldsymbol{s}_{m-2} \cdots \boldsymbol{s}_0, & \text{for } j = k, \\ \boldsymbol{s}_{j-k-1} & \text{for } j = k+1, \ldots, m-1. \end{cases}$$

The indices are chosen so that it is $\boldsymbol{s}_k$ that is the new generator; the others are all original, but relabelled. With different indices, this is the group-theoretic equivalent of Theorem 1H12(a) (see the notes at the end of the section).

5B16 Remark The pre-polytope $\mathcal{Q}^{\boldsymbol{\chi}_k}$ is of type $\{3^{k-2}, q, r, q, 3^{d-k-2}\}$, with

$$r = \begin{cases} q, & \text{if } q \text{ is odd}, \\ \frac{1}{2}q, & \text{if } q \text{ is even}. \end{cases}$$

We shall not address polytopality in the general situation, since we have little occasion to use these operations.

Debrunner Operations

In higher ranks, there are further operations, of which halving $\boldsymbol{\eta}$ is the first. Suppose that $\mathcal{P}$ is an abstract regular m-polytope of type $\{3^{m-3}, 4, q\}$ for some q, with group $\boldsymbol{G}$ and generatrix $(\boldsymbol{s}_0, \ldots, \boldsymbol{s}_{m-1})$. The simplex dissection result Theorem 1H12(b) of Debrunner [40] can be interpreted as a mixing operation $\boldsymbol{\mu}_k \colon (\boldsymbol{s}_0, \ldots, \boldsymbol{s}_{m-1}) \mapsto (\boldsymbol{r}_0, \ldots, \boldsymbol{r}_{m-1})$, where

5B17
$$\boldsymbol{r}_j := \begin{cases} \boldsymbol{s}_j, & \text{for } j = 0, \ldots, k-1, \\ \boldsymbol{s}_k\boldsymbol{s}_{k+1} \cdots \boldsymbol{s}_{n-2}\boldsymbol{s}_{m-3} \cdots \boldsymbol{s}_k, & \text{for } j = k, \\ \boldsymbol{s}_{m+k-j} & \text{for } j = k+1, \ldots, m-1. \end{cases}$$

We shall call $\boldsymbol{\mu}_k$ a *Debrunner operation*. The range is $0 \leqslant k \leqslant m-2$; formally, we can take $\boldsymbol{\mu}_{-1} = \boldsymbol{\delta}$ to be duality. Observe that $\boldsymbol{\mu}_{n-2} = \boldsymbol{\iota}$ is the identity, while $\boldsymbol{\mu}_{n-3} = \boldsymbol{\eta}$. We used the same operation (in a dual form, with different indices) in [79]; see also [99, Section 14A]. The new group $\boldsymbol{G}_k := \langle \boldsymbol{s}_0, \ldots, \boldsymbol{s}_{n-1} \rangle$ is a subgroup of $\boldsymbol{G}_{n-2} := \boldsymbol{G}$ of index a divisor of $\binom{m-1}{k+1}$, and the operation $\boldsymbol{\mu}_k$ leads (potentially, at least) to a new regular polytope of type $\{3^{k-2}, 4, q, q, 4, 3^{m-k-3}\}$.

5B18 Remark In the case of the Coxeter group $[3^{m-3}, 4, q]$, we obtain the Coxeter group $[3^{k-2}, 4, q, q, 4, 3^{m-k-3}]$ on applying $\boldsymbol{\mu}_k$. It is routine, if tedious, to check that $[3^{m-3}, 4, q]^{\boldsymbol{\mu}_k}$ does satisfy the relations of $[3^{k-2}, 4, q, q, 4, 3^{m-k-3}]$. The simplex dissection result of [40] shows that the index is indeed $\binom{m-1}{k+1}$.

Petrie Contraction

We now come to a further construction, which generalizes one introduced in [94] (see also [83]); unlike most of those considered above (apart, of course, from $\boldsymbol{f}$ and $\boldsymbol{v}$), this operation reduces the rank by 1.

The mixing operation ϖ involved here is called *Petrie contraction*, and is given by

5B19 $$\varpi\colon (\boldsymbol{s}_0, \ldots, \boldsymbol{s}_m) \mapsto (\boldsymbol{s}_1, \boldsymbol{s}_0\boldsymbol{s}_2, \boldsymbol{s}_3, \ldots, \boldsymbol{s}_m) =: (\boldsymbol{r}_0, \ldots, \boldsymbol{r}_{m-1});$$

when ϖ is a valid operation, an $(m+1)$-polytope $\mathcal{Q}$ yields an m-polytope $\mathcal{P} := \mathcal{Q}^{\varpi}$.

5B20 Remark Observe that, in the notation of (5B19), $\mathcal{P}^{\delta}$ is the facet of $\mathcal{Q}^{\delta\pi}$, whether or not the latter is actually polytopal (which in most applications – compare the next Proposition 5B21 – it would not be). This accounts for our nomenclature.

5B21 Proposition *If $\mathcal{Q}$ is an $(m+1)$-polytope of Schläfli type $\{q_1, \ldots, q_m\}$ with $q := q_3$ odd, then $\boldsymbol{G}(\mathcal{Q}^{\varpi}) = \boldsymbol{G}(\mathcal{Q})$, and Petrie contraction is reversible.*

Proof. With the notation of (5B19), since

$$(\boldsymbol{r}_1\boldsymbol{r}_2)^q = (\boldsymbol{s}_0\boldsymbol{s}_2\boldsymbol{s}_3)^q = \boldsymbol{s}_0^q(\boldsymbol{s}_2\boldsymbol{s}_3)^q = \boldsymbol{s}_0,$$

we can recover $\boldsymbol{s}_0$, and hence the whole group $\boldsymbol{G}(\mathcal{Q})$. □

When we come to applying ϖ to a realized regular polytope Q, the case which is obviously of most interest to us, we can give a partial description of Q^{ϖ} in terms of Q:

- the vertices of Q^{ϖ} are the mid-points of the edges of Q,
- the edges of the 2-faces of Q^{ϖ} join mid-points of successive edges of the Petrie polygons of the 3-faces of Q,
- the vertex-figure of Q^{ϖ} is the blend $(\mathsf{Q}/\mathsf{G}) \# \{2\}$, with G the 2-face of Q,
- for $k \geqslant 3$, the k-face of Q^{ϖ} is G^{ϖ}, where G is the $(k+1)$-face of Q,
- if G is a 4-face of Q, then the hole of G^{ϖ} is that of the vertex-figure of G.

5B22 Remark Note also that the Petrie polygon of Q^{ϖ} is (broadly) of the same kind as that of Q.

Branch Contraction

Another form of contraction is occasionally useful. If $\boldsymbol{s}, \boldsymbol{t} \in \boldsymbol{G}$ are involutions with $\boldsymbol{st}$ of odd period q, then we define

$$(\boldsymbol{st})^{q/2} := \underbrace{\boldsymbol{sts}\cdots\boldsymbol{s}}_{q} = \underbrace{\boldsymbol{tst}\cdots\boldsymbol{t}}_{q};$$

if q is even, then $(\boldsymbol{st})^{q/2}$ has its usual meaning. Observe that $(\boldsymbol{st})^{q/2}$ is again an involution in either case. If $(\boldsymbol{s}_0, \ldots, \boldsymbol{s}_{m-1})$ is the generatrix of a regular (pre-)polytope $\mathcal{Q}$ of Schläfli type $\{q_1, \ldots, q_{m-1}\}$ and $1 \leqslant k \leqslant m-1$, then the mixing operation $\boldsymbol{\beta}_k$ is defined by

5B23
$$\begin{aligned}\boldsymbol{\beta}_k \colon (\boldsymbol{s}_0, \ldots, \boldsymbol{s}_{m-1}) &\mapsto (\boldsymbol{s}_0, \ldots, \boldsymbol{s}_{k-2}, (\boldsymbol{s}_{k-1}\boldsymbol{s}_k)^{q_k/2}, \boldsymbol{s}_{k+1}, \ldots, \boldsymbol{s}_{m-1}) \\ &=: (\boldsymbol{r}_0, \ldots, \boldsymbol{r}_{m-2})\end{aligned}$$

is called k*-branch contraction* (see the notes at the end of the section). We write $\mathcal{P} = \mathcal{Q}^{\boldsymbol{\beta}_k}$ for the effect of $\boldsymbol{\beta}_k$ on $\mathcal{Q}$.

For the new Schläfli type $\{p_1, \ldots, p_{m-2}\}$ of $\mathcal{P}$, obviously

$$p_j = \begin{cases} q_j, & \text{if } j = 1, \ldots, k-2, \\ q_{j+1}, & \text{if } j = k+1, \ldots, m-2. \end{cases}$$

The entries p_{k-1} and p_k will depend on the infaces of types $\{q_{k-1}, q_k\}$ and $\{q_k, q_{k+1}\}$ (they are the lengths of holes or zigzags for p_{k-1} and their duals for p_k). However, in certain cases that are of most interest, we can say a little more:

- if $q_k = 3$, then $p_{k-1} = q_{k-1}$, $p_k = q_{k+1}$;
- if $q_k = 5$, then p_{k-1} is the hole of the inface $\{q_{k-1}, q_k\}$ and p_k is the hole of the inface $\{q_k, q_{k+1}\}$.

5B24 Example To illustrate what can happen in a simple case with $k = 2$ and $p_2 = 4$, we have $\{3, 4, 3\}^{\boldsymbol{\beta}_2} = \{4, 4 \mid 4\}$.

Edge Replacement

One more operation of this kind has limited applicability, although it does provide useful bridges between pairs of polytopes in Chapter 16 in particular. If $\mathcal{P}$ is a regular m-polytope with generatrix $(\boldsymbol{r}_0, \ldots, \boldsymbol{r}_{m-1})$ whose facet $\mathcal{F} := \mathcal{P}^f$ is centrally symmetric with central involution $\boldsymbol{z} = \boldsymbol{z}(\mathcal{F})$, then we define the *edge replacement* operation $\boldsymbol{\varepsilon}$ by

$$\boldsymbol{\varepsilon} \colon (\boldsymbol{r}_0, \ldots, \boldsymbol{r}_{m-1}) \mapsto (\boldsymbol{z}, \boldsymbol{r}_{m-1}, \ldots, \boldsymbol{r}_1) =: (\boldsymbol{s}_0, \ldots, \boldsymbol{s}_{m-1});$$

see the notes at the end of the section. The reversal of $\boldsymbol{r}_1, \ldots, \boldsymbol{r}_{m-1}$ makes sense here, because the new edge can be thought of as pointing in the direction of the initial vertex of the dual $\mathcal{P}^{v\delta}$ of the vertex-figure of $\mathcal{P}$.

Generalized Quartering

Quartering admits generalizations to an m-polytope $\mathcal{P}$. If $1 \leqslant k \leqslant m-2$, then we define $\boldsymbol{\psi}_k$ on its generatrix by

5B25
$$\begin{aligned}&\boldsymbol{\psi}_k \colon (\boldsymbol{s}_0, \ldots, \boldsymbol{s}_{m-1}) \\ &\quad\mapsto (\boldsymbol{s}_0, \ldots, \boldsymbol{s}_{k-2}, \boldsymbol{s}_{k-1}\boldsymbol{s}_k\boldsymbol{s}_{k-1}, \boldsymbol{s}_k, \boldsymbol{s}_{k+1}\boldsymbol{s}_k\boldsymbol{s}_{k+1}, \boldsymbol{s}_{k+2}, \ldots, \boldsymbol{s}_{m-1});\end{aligned}$$

that is, we apply ψ to the appropriate 3-inface. Observe that the string property is preserved; of course, in each instance the intersection property will need to be checked. Further, $\mathcal{P}^{\psi_k}$ has the same vertex-set as $\mathcal{P}$.

Orientability

Certain mixing operations preserve orientability of a regular polytope, while others do not. As in (5B1), let $\boldsymbol{\mu}\colon (\boldsymbol{s}_0, \ldots, \boldsymbol{s}_{k-1}) \mapsto (\boldsymbol{r}_0, \ldots, \boldsymbol{r}_{m-1})$ be a general mixing operation, so that each $\boldsymbol{r}_j$ is a product of original involutions $\boldsymbol{s}_i$; we call $\boldsymbol{r}_j$ *even* or *odd* according as the number of such $\boldsymbol{s}_i$ in the product is even or odd. Further, if each $\boldsymbol{r}_j$ is even, then we say that $\boldsymbol{\mu}$ is even. Then we have

5B26 Proposition *An even mixing operation $\boldsymbol{\mu}$ preserves orientability.*

By this, we mean that, if $\mathcal{P}$ has generatrix $(\boldsymbol{s}_0, \ldots, \boldsymbol{s}_{k-1})$, then $\mathcal{P}$ and $\mathcal{P}^{\boldsymbol{\mu}}$ are orientable or not together.

5B27 Example Many mixing operations are even, including duality, faceting, halving, quartering and the Schläfli and Debrunner operations. However, the Petrie operation is not.

Notes to Section 5B

1. We have called the operations χ_k after Schläfli, because he implicitly used them for $m = 4$ in his calculation of certain integrals related to the group $[3, 3, 5]$.
2. The main applications of branch contraction will be in Chapters 16 and 17, but it also has relevance in Section 7G.
3. It might be more informative to talk about *edge-diameter replacement*, but we prefer the shorter term.
4. The generalized quartering operations ψ_k play a rôle in relating various polytopes of Chapters 16 and 17. Note that ψ_k induces faceting φ_2 on $(k+1)$-faces, and the dual operation $\delta\varphi_2\delta$ on $(m-k)$-cofaces.

5C Twisting

In talking about twisting, we must distinguish between the abstract and the geometric. So, while we may be able to extend an abstract group (which need not itself be a C-group) by an (outer) automorphism to obtain the group of an abstract regular polytope, it may not be possible correspondingly to extend a representation of that group or, if it is, such an extension may lead to degeneracy of some kind. This section will therefore deal largely with the general principles of twisting, both abstract and geometric, leaving most particular applications until later.

The notion of extending a group $\boldsymbol{H}$ by a subgroup $\boldsymbol{A}$ of its group $\mathrm{Aut}(\boldsymbol{H})$ of automorphisms to a group $\boldsymbol{G} := \boldsymbol{H} \rtimes \boldsymbol{A} = \boldsymbol{A} \ltimes \boldsymbol{H}$ should be familiar to the reader; indeed, we have already met several instances of the idea. In cases considered here, $\boldsymbol{H}$ will usually be a C-group, perhaps with generators other than standard ones, and $\boldsymbol{A}$ will similarly be generated by involutions.

To set the scene, we begin with a simple example, which ties in with what we did in the previous section.

5C1 Example Let $\mathcal{Q}$ be a self-dual regular polyhedron of type $\{q, q\}$, with generatrix $(\boldsymbol{s}_0, \boldsymbol{s}_1, \boldsymbol{s}_2)$. Then there is a *polarity* $\boldsymbol{\omega}$ which fixes the base flag of $\mathcal{Q}$, so that $\boldsymbol{\omega}^2 = \iota$ and $\boldsymbol{s}_j^{\boldsymbol{\omega}} = \boldsymbol{s}_{2-j}$ for each j. If $\boldsymbol{G} := \langle \boldsymbol{s}_0, \boldsymbol{s}_1, \boldsymbol{s}_2 \rangle \rtimes \langle \boldsymbol{\omega} \rangle = \boldsymbol{G}(\mathcal{Q}) \rtimes \boldsymbol{C}_2$, then $\boldsymbol{G}$, with generatrix $(\boldsymbol{r}_0, \boldsymbol{r}_1, \boldsymbol{r}_2) := (\boldsymbol{\omega}, \boldsymbol{s}_2, \boldsymbol{s}_1)$, is the group of a polyhedron $\mathcal{P}$ of type $\{4, q\}$, with $[\boldsymbol{G} : \boldsymbol{G}(\mathcal{Q})] = 2$, where we now write $\boldsymbol{\omega g \omega} := \boldsymbol{g}^{\boldsymbol{\omega}}$. Note that $(\boldsymbol{\omega s}_2 \boldsymbol{s}_1 \boldsymbol{s}_2)^2 = (\boldsymbol{s}_0 \boldsymbol{s}_1 \boldsymbol{s}_2)^2$. (It is straightforward to verify the intersection property in this case.)

We notice that the edge-graph of $\mathcal{P}$ is bipartite, and that $\mathcal{P}^\eta = \mathcal{Q}$. But it must also be observed that, if $\mathcal{P}$ is a regular polyhedron of type $\{4, q\}$ whose edge-graph is *not* bipartite, then $\mathcal{P}$ is not recovered from $\mathcal{P}^\eta$ by twisting in this way; instead, we obtain a double-cover of $\mathcal{P}$. In this case, there can be a parting of the ways between the abstract and geometric, since an abstract twist may not be realizable geometrically.

In most cases, our basic examples will be obtained by twisting diagrams of hyperplane reflexion groups. These groups will, of course, be among those classified in Section 1E, with diagrams not necessarily in standard form. In this situation, we distinguish four kinds of twisting operation or *twist* for short. First, a twist can be *outer* or *inner*, according as the induced automorphism is outer or inner. (Strictly speaking, twisting by an inner automorphism is a case of mixing, but we find it convenient to include the notion here.) Second, while a twist must permute the mirrors of the diagram reflexions, it need not permute the corresponding (unit) normal vectors to those mirrors – it may change some signs, while preserving others. If the normals can be chosen to be permuted by the twist, then we call it *proper* (see the notes at the end of the section); otherwise the twist is *improper*.

We give some simple examples to illustrate these ideas; we shall revisit these examples later, so for now we merely sketch various points of interest. The transpositions $\boldsymbol{s}_j := (j\ j{+}1)$ for $j = 0, \ldots, 3$ generate the symmetric group $\boldsymbol{S}_5$ on $\{0, \ldots, 4\}$. This acts by permutations of the basic coordinates vectors $e_0, \ldots, e_4$ of $\mathbb{E}^5$, and these preserve the symmetric hyperplane $\mathbb{L}^4$ of (1E15). We can use the same notation $(j\ j{+}1)$ for the representative S_j of $\boldsymbol{s}_j$ in the orthogonal group $\mathbf{O}_4$, regarded as acting on $\mathbb{L}^4$, in which case (as we saw previously) it is the reflexion in the hyperplane with unit normal vector $u_j := \frac{1}{\sqrt{2}}(e_j - e_{j+1})$.

Now we can twist this group in two ways. First, the inner automorphism $T_+ := (0\ 4)(1\ 3)$ acts as $u_j T_+ = -u_{4-j}$. This is not actually improper, as we see by changing the initial signs of u_2 and u_3. Then the operation

$$(S_0, \ldots, S_3, T_+) \mapsto (S_0, T_+, S_2) =: (R_0, R_1, R_2)$$

leads to the group of the regular polyhedron $\{4, \frac{6}{2,3} \mid 3\} \cong \{4, 6 : 5 \mid 3\}$. Here, we make a forward appeal to the rigidity criterion of Theorem 6D5 to see that we need not mention the type $\{\frac{5}{1,2}\}$ of the Petrie polygon to determine its geometry.

Alternatively, we can apply the outer automorphism $T_- := -(0\ 4)(1\ 3)$ following the previous twist by the central inversion. The new twist T_- respects the original normal vectors u_j. With the analogous operation to that used before, we now obtain $\{4, \frac{6}{1,3} \mid 3\} \cong \{4, 6 \mid 3\}$, the universal polyhedron of its kind.

We now extend the basic group $\boldsymbol{S}_5$ to $\boldsymbol{S}_6$, by adjoining the transposition $\boldsymbol{s}_5 := (2\ 5)$, and let it act on the corresponding hyperplane $\mathbb{L}^5 \leqslant \mathbb{E}^6$. We naturally represent $\boldsymbol{s}_5$ by the reflexion S_5 in the hyperplane with unit normal $u_5 := \frac{1}{\sqrt{2}}(e_2 - e_5)$. We can define $T_\pm$ as before, noting that $T_\pm$ mentions neither e_2 nor e_5; however, now the inner T_+ is proper (change the signs of u_2 and u_3 as before), but the outer T_- is improper, as we can see by looking at its effect on u_1, u_2, u_5. At this stage, we will not discuss any related regular polytopes; we shall meet this twist again in Chapter 13.

While we postpone most examples of twisting until later, the way in which twisting relates the various hyperplane reflexion groups is of interest, since it results in the need to calculate rather fewer group orders in the treatment of Section 1H. We begin with a general construction. Actually, it is a special case of one even more general, which we shall not need; what we do is mainly taken from [99, Section 8D]. We refer the interested reader to [99, Chapter 8] for the full range of possibilities.

5C2 Example Let $\mathcal{Q}$ be a finite regular $(m-1)$-polytope with k vertices and group $\boldsymbol{H}$. Then $\boldsymbol{H}$ acts as a group of permutations on $\boldsymbol{C}_2^k = \boldsymbol{C}_2 \times \cdots \times \boldsymbol{C}_2$, giving the group $\boldsymbol{G} := \boldsymbol{C}_2^k \rtimes \boldsymbol{H}$. We identify the involutory generators $\boldsymbol{z}_1, \ldots, \boldsymbol{z}_k$ of the cyclic components with the k vertices of $\mathcal{Q}$, where $\boldsymbol{z}_1$ represents the initial vertex. If $\boldsymbol{H} = \langle \boldsymbol{s}_0, \ldots, \boldsymbol{s}_{m-2} \rangle$, then we write

$$(\boldsymbol{r}_0, \ldots, \boldsymbol{r}_{m-1}) := (\boldsymbol{z}_1, \boldsymbol{s}_0, \ldots, \boldsymbol{s}_{m-2}),$$

so that $\boldsymbol{G} = \langle \boldsymbol{r}_0, \ldots, \boldsymbol{r}_{m-1} \rangle$ is clearly an sggi. We denote the corresponding (pre-)polytope by $\mathbf{2}^{\mathcal{Q}}$ (see the notes at the end of the section).

In fact, it is clear that $\boldsymbol{G}$ is a C-group, and that $\mathbf{2}^{\mathcal{Q}}$ is a polytope. We can see this by induction on the rank m: if $\mathcal{G}$ is the facet of $\mathcal{Q}$, then the facet of $\mathbf{2}^{\mathcal{Q}}$ is clearly $\mathbf{2}^{\mathcal{G}}$.

As a special case of this, we have $\mathcal{Q} = \{3^{m-2}\}$, the $(m-1)$-simplex, giving

$$\mathbf{2}^{\{3^{m-2}\}} = \{4, 3^{m-2}\},$$

the (abstract) m-cube. Indeed, the natural geometric realization represents the involutions by reflexions in the m coordinate hyperplanes of $\mathbb{E}^m$, with the symmetry group $\mathbf{A}_{m-1} = [3^{m-2}]$ acting in the obvious way as permutations of the coordinates. With a change of index, this therefore represents the symmetry group of the geometric d-cube as

5C3 $$\mathbf{C}_d \cong \mathbf{Z}_2^d \rtimes \mathbf{A}_{d-1},$$

with $\mathbf{C}_d$ as in Table 1E12. We write $\mathbf{Z}_2$ for a general geometric reflexion; this will often (but not always) be a reflexion in a hyperplane.

Twisting certain hyperplane reflexion groups produces others; this provides an alternative approach to some of the results of Section 1E. The diagrams of Tables 1E12 and 1E14 indicate obvious cases where twisting is applicable, but we shall later see that these are far from the only ones. So, we clearly have

$$\mathbf{C}_d = \mathbf{B}_d \rtimes \mathbf{Z}_2$$

for each $d \geqslant 3$; here, we have $\mathbf{B}_3 = \mathbf{A}_3$ in the obvious way. However, we have additional twists if $d = 4$, namely,

$$\mathbf{F}_4 = \mathbf{B}_4 \rtimes \mathbf{A}_2;$$

moreover, this gives subgroup relationships

$$\mathbf{B}_4 < \mathbf{C}_4 < \mathbf{F}_4,$$

with indices 2 and 3, respectively.

In a similar way, for the infinite groups we have

$$\mathbf{S}_{d+1} = \mathbf{Q}_{d+1} \rtimes \mathbf{Z}_2, \quad \mathbf{R}_{d+1} = \mathbf{S}_{d+1} \rtimes \mathbf{Z}_2.$$

For $d = 3$, we replace $\mathbf{S}_4$ by $\mathbf{P}_4$. Just as in the finite case, the dimension $d = 4$ is special:

$$\mathbf{U}_5 = \mathbf{Q}_5 \rtimes \mathbf{A}_2 = \mathbf{Q}_5 \rtimes \mathbf{A}_3,$$

since we clearly have two more ways of twisting $\mathbf{Q}_5$ than in the general case. This leads to subgroup relationships

$$\mathbf{R}_5 < \mathbf{U}_5, \quad \mathbf{U}_5 < \mathbf{U}_5,$$

with indices 6 and 4, respectively.

5C4 Remark It is clear that each infinite group of Table 1E14 acting on $\mathbb{E}^d$ is a subgroup of itself of index 2^d. That $\mathbf{U}_5$ should be a subgroup of itself of index 4 is thus a quirk; however, see also Remark 5B18 and the following discussion.

We can generalize these specific constructions. There is a nice family of examples of twisting coming from the Coxeter diagram $\mathcal{D}_{m,0}(q)$, which consists of a central node 0 joined to nodes $1, \ldots, m$ by branches marked q, and with no other branches. Thus the symmetric group $\mathbf{A}_{m-1}$ (thought of as before as a geometric group) acts on nodes $1, \ldots, m$ by diagram automorphisms. However, we can split these nodes into subsets, and allow subgroups of $\mathbf{A}_{m-1}$ to act just as groups of permutations of these nodes. In our case, we wish to obtain string groups, and so we take a split $m = r + s$, say, with $r \geqslant s \geqslant 0$; we then relabel the diagram $\mathcal{D}_{r,s}(q)$. Figure 5C5 illustrates the case $q = 3$, $m = 5$ with $r = 3$, from which we obtain $[3, 4, 3, 3, 4]$ (with generators corresponding in the natural way to the labels on the nodes and twists).

5C5 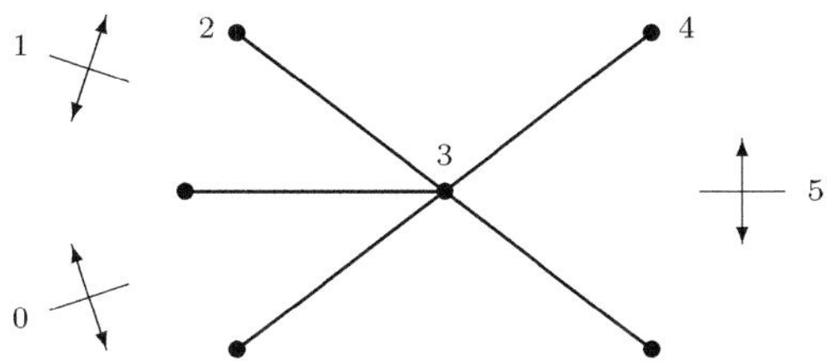

Twisting $\mathcal{D}_{3,2}(3)$

This provides an alternative proof of Remark 5B18; in case (for example) $s = 1$, this is to be interpreted as omitting the last entry 4 as well.

We end the section by returning to Example 2D15. Call a regular polytope *neighbourly* if all its diagonals are edges (see the notes at the end of the section). We refer to [99, Corollary 8E6] for

5C6 Theorem *If the regular polytope $\mathcal{Q}$ is neighbourly with facet $\mathcal{G}$, then $\mathbf{2}^{\mathcal{Q}} = \{\mathbf{2}^{\mathcal{G}}, \mathcal{Q}\}$ is universal.*

Theorem 5C6 explains the imposed relators of Example 2D15. For the first, in effect applying π in reverse, we have

$$\begin{aligned}\{4,4,3\}/\langle\langle(123)^3\rangle\rangle = \{4,3,3\}^{\pi} = (\mathbf{2}^{\{3,3\}})^{\pi} = (\mathbf{2}^{\{4,3\}/\langle\langle(123)^3\rangle\rangle})^{\pi}\\ = \{\{\mathbf{2}^{\{4\}}\},\{4,3\}\}/\langle\langle(123)^3\rangle\rangle = \{\{4,4\mid 4\},\{4,3:3\}\}.\end{aligned}$$

The proof of the other is just the same, with the observation that

$$\mathbf{2}^{\{5\}} = \{4,5\mid 4\}.$$

Notes to Section 5C

1. We could allow a proper twist to change the signs of all the normal vectors, but in practice we need not let this happen.
2. In the general context of [99, Chapter 8], we should perhaps write $\{2\}^{\mathcal{Q}}$ rather than $\mathbf{2}^{\mathcal{Q}}$. However, we follow the conventions of [99, Section 8D] here.
3. As we have said, the twisting applied to $\mathcal{D}_{m.0}(q)$ is equivalent to Theorem 1H12(b). Indeed, the simplex dissections exactly correspond to the twists induced by the actions of the subgroups of $\mathbf{A}_{m-1}$; Figure 5C5 gives a good illustration of this.
4. In the context of *convex* polytopes, what we have called here 'neighbourly' is really '2-neighbourly'. In fact, Theorem 5C6 holds more generally, if $\mathcal{Q}$ is *weakly neighbourly*, meaning that every two of its vertices belong to some common facet.

5D Modifying Mirrors

Although the idea of modifying the mirrors of the generating reflexions of the group $\mathbf{H}$ of a geometric regular polytope Q is one of the last of the general techniques discussed here to be formally recognized, nevertheless it proves to be one of the more versatile. The modification is usually by multiplying some of the reflexions in the generatrix by other involutions. It is important to bear in mind our convention of identifying a reflexion R with its mirror; the context will usually make clear how R is to be regarded.

5D1 Remark It is worth emphasizing that the operations described here are geometric, and so we should not necessarily expect that they have abstract analogues; indeed, in many cases they will not.

Axiversion

Let $(S_0, \ldots, S_{m-1})$ be the generatrix of a fixed geometric string C-group $\mathbf{H}$. For each $r = 0, \ldots, m-1$, define $K_r := S_r \cap \cdots \cap S_{m-1}$, so that K_r is the axis of the subgroup $\langle S_r, \ldots, S_{m-1} \rangle$. We define *axiversion* $\boldsymbol{\kappa}_{jr}$ by

5D2
$$\begin{aligned}\boldsymbol{\kappa}_{jr} \colon (S_0, \ldots, S_{m-1}, K_r) \mapsto (S_0, \ldots, S_{j-1}, S_j K_r, S_{j+1}, \ldots, S_{m-1}) \\ =: (R_0, \ldots, R_{m-1}).\end{aligned}$$

There are various restrictions on the validity of $\boldsymbol{\kappa}_{jr}$, like those on $\boldsymbol{\pi}$ that we met in Section 5A; indeed, we can see that $\boldsymbol{\pi} = \boldsymbol{\kappa}_{m-3,m-1}$. However, we first have

5D3 Proposition *The operation $\boldsymbol{\kappa}_{jr}$ can only yield an sggi if $j = r-2$ or r.*

Proof. Since $K_r \subseteq S_j$ for $j \geqslant r$, we see that $K_r \rightleftharpoons S_j$ for these j. Moreover, Corollary 1D16 implies that $K_r \rightleftharpoons S_j$ for $j \leqslant r-2$. However, it is clear that K_r does not commute with S_{r-1}. Hence we must have $j = (r-1) \mp 1 = r-2$ or r, as claimed. $\square$

There is a further restriction, whose proof we shall omit because it follows the lines of that of Theorem 5B10 (in an abstract context); it is

5D4 Proposition *If $j = r-2$ or r, then the operation $\boldsymbol{\kappa}_{jr}$ is invalid if $\langle S_0, \ldots, S_j \rangle$ has a relator which contains S_j an odd number of times.*

We write $\mathsf{P}^{\boldsymbol{\kappa}_{jr}}$ for the result of applying axiversion $\boldsymbol{\kappa}_{jr}$ to P. In view of Proposition 5D3, the case $j = 0$, $r = 2$ will be of more importance than the others; we therefore adopt a special notation $\boldsymbol{\tau} := \boldsymbol{\kappa}_{02}$.

If $\mathbf{H}$ is the symmetry group of an apeirotope Q, then $K_0 = \emptyset$, and so no corrresponding operation K_{00} can arise. In this context, an important case is $j = r = 1$, for which we abbreviate the notation to $\boldsymbol{\kappa} := \boldsymbol{\kappa}_{11}$. We then have

5D5 Proposition *If Q is a regular apeirotope with facet F and vertex-figure G, then the facet of $\mathsf{Q}^{\boldsymbol{\kappa}}$ is $\mathsf{F}^{\boldsymbol{\kappa}}$, and its vertex-figure is $\mathsf{G}^{\boldsymbol{\zeta}}$.*

The operation $\boldsymbol{\zeta}$ is defined immediately below. As we shall see, when we shortly come to extend the use of $\boldsymbol{\kappa}$ in an important way, the claim about the vertex-figure will need to be interpreted carefully; see the notes at the end of the section. We therefore postpone further discussion of it until then.

Centriversion

The other operations of this kind only work in the finite case. As usual, we think of a finite regular polytope Q as being centred at the origin o. We then set $Z := -I = \{o\}$, the latter indentifying Z with its mirror. In this event, *centriversion* $\boldsymbol{\zeta}_k$ is defined by

5D6
$$\begin{aligned}\boldsymbol{\zeta}_k\colon (S_0,\dots,S_{m-1},Z) &\mapsto (S_0,\dots,S_{k-1},S_kZ,S_{k+1},\dots,S_{m-1})\\ &=: (R_0,\dots,R_{m-1}).\end{aligned}$$

As applied to the corresponding regular m-polytope Q, we denote the resulting new polytope by $\mathsf{P} := \mathsf{Q}^{\boldsymbol{\zeta}_k}$.

5D7 Remark Observe that, if any relator in $\langle S_0,\dots,S_{m-1}\rangle$ contains S_k an odd number of times, then $Z \in \langle R_0,\dots,R_{m-1}\rangle$. In particular, this will happen if either of the entries q_k or q_{k+1} in the Schläfli type $\{q_0,\dots,q_{m-1}\}$ of Q is odd.

5D8 Remark Recall that, for a faithful realization P, there is the lower bound $\dim R_k \geqslant k+1$ of Theorem 4B8, which excludes the deployment of $\boldsymbol{\zeta}_k$ in many cases.

5D9 Remark Centriversion has a natural analogue for centrally symmetric abstract regular polytopes; we denote it by the same symbol $\boldsymbol{\zeta}_k$.

The most frequent case is $k = 0$; we therefore abbreviate the notation here, write $\boldsymbol{\zeta} := \boldsymbol{\zeta}_0$, and call $\mathsf{P}^{\boldsymbol{\zeta}}$ the *centrivert* of P. There is a nice alternative way of looking at $\boldsymbol{\zeta}$; it follows simply from the fact that the symmetry group of the digon $\{2\}$ is generated by the reflexion $\{o\}$ in $\mathbb{R}$.

5D10 Theorem *If* Q *is a finite regular polytope, then* $\mathsf{Q}^{\boldsymbol{\zeta}} = \mathsf{Q} \otimes \{2\}$.

Because $\{2\} \otimes \{2\} = \{1\}$, we see from Theorem 5D10 that the application of $\boldsymbol{\zeta}$ to blends with $\{2\}$ is a little anomalous; what happens is made clear in the proof of Theorem 5D13. Since we are not employing the notation $\diamond$ for the abstract mixing operation (in contrast to the usage of [99]), it is free for use alternatively as

5D11
$$\mathsf{Q} \diamond \{2\} := \mathsf{Q}^{\boldsymbol{\zeta}} \# \{2\}.$$

Since the term is also free, we refer to $\mathsf{Q} \diamond \{2\}$ as the *mix* of Q and a digon. Then we have

5D12 Proposition *If the regular polytope* Q *is not already a blend with the digon* $\{2\}$, *then* $\mathsf{Q} \diamond \{2\} \cong \mathsf{Q} \# \{2\}$.

Proof. The isomorphism is induced by

$$(x, \varepsilon) \leftrightarrow (\varepsilon x, \varepsilon)$$

for $x \in \operatorname{vert} \mathsf{Q}$, where $\varepsilon = \pm 1$. $\square$

For the next part, it is convenient to change notation, and let P be a regular m-polytope with vertex-figure $\mathsf{Q} := \mathsf{P}^{\boldsymbol{v}}$. The effect of $\boldsymbol{\zeta}$ on Q is straightforward: the (broad) vertex-figure of P^{ζ} is $-\mathsf{Q} = \mathsf{Q}Z$ (we could replace the initial vertex v by vZ to preserve Q, but it is more convenient not to). The effect of $\boldsymbol{\zeta}$ on a proper (initial) face $\mathsf{F} < \mathsf{P}$ depends on two things: whether the centre $c = c(\mathsf{F})$ of F coincides with o or not, and whether F is a blend with one component $\{2\}$ or not.

5D13 Theorem *Let* F *be an initial (proper) face with centre* c *of a finite regular polytope* P *with centre* o*. Then, under* $\boldsymbol{\zeta}$*, the corresponding initial face* $\widehat{\mathsf{F}}$ *of* P^{ζ} *with centre* $\widehat{c}$ *is as follows:*

(a) *if* F *is not a blend with component* $\{2\}$ *and* $c = o$*, then* $\widehat{\mathsf{F}} = \mathsf{F}^{\zeta}$ *with* $\widehat{c} = o$*;*

(b) *if* F *is not a blend with component* $\{2\}$ *and* $c \neq o$*, then* $\widehat{\mathsf{F}} = \mathsf{F}^{\zeta} \# \{2\}$ *with* $\widehat{c} = o$*;*

(c) *if* $\mathsf{F} = \mathsf{G} \# \{2\}$ *is a blend with component* $\{2\}$ *and* $c = o$*, then* $\widehat{\mathsf{F}} = \mathsf{G}^{\zeta}$ *with* $\widehat{c} \neq o$*;*

(d) *if* $\mathsf{F} = \mathsf{G}\#\{2\}$ *is a blend with component* $\{2\}$ *and* $c \neq o$*, then* $\widehat{\mathsf{F}} = \mathsf{G}^{\zeta}\#\{2\}$ *with* $\widehat{c} \neq o$ *and* $\langle c, \widehat{c} \rangle = 0$*.*

Proof. Except for the last claim that $\langle c, \widehat{c} \rangle = 0$, for which see Figure 5D14, this is all fairly obvious. In the picture, G' is a translate of the polytope G involved in the blend, and edges of F go from a vertex of G to the corresponding adjacent vertex of G'. As indicated, $\mathsf{G}'' = -\mathsf{G}' = \mathsf{G}'Z$, and edges of $\widehat{\mathsf{F}}$ similarly go from a vertex of G to the corresponding adjacent vertex of G''. Observe, by the way, that this illustrates the pairing $\mathsf{G} \diamond \{2\} = \mathsf{G}^{\zeta} \# \{2\}$ of (5D11). $\square$

5D14

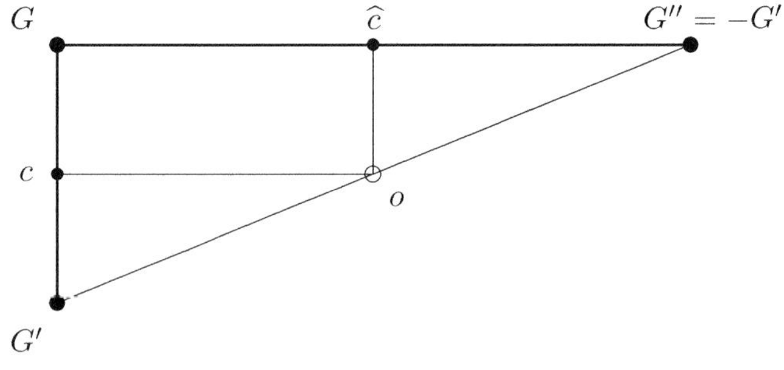

As we have seen, blends with the digon behave somewhat anomalously under $\boldsymbol{\zeta}$. Otherwise, we have the following.

5D15 Theorem *If* P *and* Q *are finite regular polytopes, neither of which is a blend with one component a digon, then* $(\mathsf{P} \# \mathsf{Q})^{\varsigma} = \mathsf{P}^{\varsigma} \# \mathsf{Q}^{\varsigma}$. *In particular, if* $\{p\}$ *is a finite regular polygon, then* $\{p\}^{\varsigma} = \{p''\}$.

Proof. Only the latter claim needs to be checked, and for this it is clear that, if $\{q\}$ is a (finite) planar regular polygon, then $\{q\}^{\varsigma} = \{q''\}$. Recall the definition of the supplement $\{p''\}$ of $\{p\}$ in (4G6). □

Eversion

We next observe that a particular case of axiversion as introduced above actually yields a construction if suitably generalized. To set the scene, we first suppose that P is a regular apeirotope with initial vertex o at the centre of a blended vertex-figure; we can therefore write this vertex-figure in the form $\mathsf{G} \diamond \{2\}$ $(= \mathsf{G}^{\varsigma} \# \{2\})$. The discussion of Theorem 5D13(c) shows that the vertex-figure of P^{κ} is just G, with now o not at its centre. We then see that P^{κ}, whether polytopal or not, will be finite (see the notes at the end of the section).

To reverse this procedure, let Q be a finite regular polytope whose group $\mathbf{H} = \mathbf{G}(\mathsf{Q}) = \langle S_0, \ldots, S_{d-1} \rangle$ is crystallographic (recall that this means that $\mathbf{H}$ is a subgroup of the point-group of an irreducible discrete infinite group), and let w be the initial vertex of Q. The point-reflexion in w is $\{w\}$ (regarded as a mirror, as previously mentioned); the construction is then given by the new generatrix

$$(S_0, \ldots, S_{d-1}) \mapsto (S_0, S_1\{w\}, S_2, \ldots, S_{d-1}) =: (R_0, \ldots, R_{d-1}).$$

Were Q an apeirotope with $S_1 \cap \cdots \cap S_{d-1}$ a point (as would often be the case), we would have designated the corresponding operation by $\boldsymbol{\kappa}$ (and recall that we confined its usage to apeirotopes); here, $\boldsymbol{\kappa}$ would just be the operation $\boldsymbol{\zeta}$ applied to the group $\langle S_1, \ldots, S_{d-1} \rangle$ of the vertex-figure Q^v of Q. In keeping with what we did in [84], we use the same symbol $\boldsymbol{\kappa}$ for this construction; however, just in this context we give it the alternative name *eversion*, and talk about Q^{κ} as the *evert* of Q. We therefore have

5D16 Theorem *If* Q *is a finite crystallographic regular polytope of full rank, then its evert* Q^{κ} *is a putative discrete regular apeirotope of nearly full rank. Conversely, if* P *is a regular apeirotope of nearly full rank whose vertex-figure is blended, then* P^{κ} *is a putative regular polytope of full rank.*

5D17 Remark We say 'putative' here, because in many cases eversion or its inverse fails to yield an apeirotope or polytope (whose group satisfies the intersection property – see the notes at the end of the section again). When $d \geqslant 4$, observe that – whether or not Q is infinite – facets are related under $\boldsymbol{\kappa}$ by $\mathsf{Q}^{\kappa f} = \mathsf{Q}^{f \kappa}$.

There is a parallel to Proposition 5D5. Since $\boldsymbol{\kappa}$ is involutory, this indicates how that proposition can go wrong when the vertex-figure of an apeirotope is blended. In the case of apeirotopes of nearly full rank, we shall go into things in detail in Section 9A.

5D18 Proposition *If* Q *is a finite regular polytope with facet* F *and vertex-figure* G*, then the facet of* $\mathsf{Q}^{\boldsymbol{\kappa}}$ *is* $\mathsf{F}^{\boldsymbol{\kappa}}$*, and its vertex-figure is* $\mathsf{G} \diamond \{2\}$.

Because we apply $\boldsymbol{\kappa}$ fairly widely, we need to record its effect on faces. In fact, we have a general result.

5D19 Theorem *If* P *is a regular polytope or apeirotope, and* $\{p\}$ *is a regular polygon in the edge-graph of* P *with generatrix* (J, K) *(and* J, K *as flag-sequences), then the corresponding polygon in the edge-graph of* $\mathsf{P}^{\boldsymbol{\kappa}}$ *is* $\{p\}$ *or* $\{p''\}$*, according as* K *is* 1*-even or* 1*-odd. In particular, this holds for zigzags and holes of the* 3*-faces of* P.

Proof. Assuming (as usual) that the initial vertex of P is o, the generatrix of the new polygon is $(S, \pm T)$, with S, T the geometric reflexions whose flag-sequences are J, K. It is enough to verify what happens for a planar polygon $\{p\}$, in which case it is fairly obvious. □

5D20 Remark Note that Theorem 5D19 continues to hold for the degenerate polygons, namely,

$$\{2\}^{\boldsymbol{\kappa}} = \{\infty\}, \quad \{\infty\}^{\boldsymbol{\kappa}} = \{2\}.$$

We end the present discussion with the following obvious result.

5D21 Theorem *If* P, Q *are regular polytopes for which* $\boldsymbol{\kappa}$ *is valid, then*

$$(\mathsf{P} \# \mathsf{Q})^{\boldsymbol{\kappa}} = \mathsf{P}^{\boldsymbol{\kappa}} \# \mathsf{Q}^{\boldsymbol{\kappa}}.$$

Quasi-Duality

The idea of modifying mirrors plays a rôle in a rather different context. As we know from Section 2B, the generatrix of the dual $\mathcal{P}^{\delta}$ of an abstract regular polytope $\mathcal{P}$ is obtained by reversing the order of the generators in the generatrix of $\mathcal{P}$. When this carries over to a realization P of $\mathcal{P}$, so that Wythoff's construction applies to the reversed generatrix, we have the geometric dual P^{δ} of P. However, we shall meet many cases where the geometric dual degenerates (of course, the generatrix will still engender a C-group). In some of these cases, though, a combinatorial dual of P can be obtained by modifying the reversed generatrix of P by changing the signs of some of its generators or, equivalently, replacing mirrors by their orthogonal complements. This yields what we shall call a *quasi-dual* of P. Thus, quasi-duality is weaker than geometric duality $\boldsymbol{\delta}$, but stronger than mere combinatorial duality. We shall not give examples here, but instead make reference to the foregoing, particularly to Section 11B.

Notes to Section 5D

1. We shall see in Section 8B that applying $\boldsymbol{\kappa}$ to a regular apeirotope whose vertex-figure is a blend with a segment will not generally result in a polytope unless, of course, it was itself constructed by means of $\boldsymbol{\kappa}$.

5E Extensions

There is another general construction that we need to introduce. Let X be a point-set in a euclidean space $\mathbb{E}$. As in Section 3E, we call X *rational* if the points of X can be chosen to have rational coordinates with respect to some (linear or affine) coordinate system in $\mathbb{E}$. The following remark is obvious.

5E1 Lemma *Let $\mathbb{E}$ be a euclidean space, and let X be a finite point-set in $\mathbb{E}$. Then the group $\mathbf{R}(X)$ generated by the point-reflexions (inversions) in the points of X is discrete if and only if X is rational.*

Proof. The product of the point-reflexions in $t, u \in X$ (in this order) is the translation by $2(u - t)$. Then $\mathbf{R}(X)$ is discrete just when these translations generate a discrete lattice, which, in turn, is equivalent to the condition of the lemma. □

If Q is a regular polytope with ambient space $\mathbb{E}$, then we similarly call Q *rational* if its vertex-set is rational. We have the following very general construction.

5E2 Theorem *Let Q be a rational regular m-polytope in the euclidean space $\mathbb{E}$, with generatrix $(S_0, \ldots, S_{m-1})$, vertex-set W and initial vertex $w \in W$, and suppose that $v \in (S_0 \cap \cdots \cap S_{m-1}) \setminus \operatorname{aff} W$. Let $R_0 = \{w\}$ be the point-reflexion in the point w, and define $R_j := S_{j-1}$ for $j = 1, \ldots, m$. Then $(R_0, \ldots, R_m)$ is the generatrix of a discrete regular $(m+1)$-apeirotope P, with 2-faces apeirogons $\{\frac{2}{0,1}\}$ and vertex-figure Q at the initial vertex v.*

Proof. We first observe that the 2-faces of P are indeed apeirogons $\{\frac{2}{0,1}\}$, because $v \notin \operatorname{aff} \mathsf{Q}$. Indeed this initial choice of v ensures that P must be a blend, with its vertices lying in two translates of $\operatorname{aff} W$.

Note next that $xR_0 = 2w - x$, and that, if $G \in \mathbf{H} = \langle S_0, \ldots, S_{m-1} \rangle$, then

$$xGR_0 = 2w - xG = (2w - x)G + 2(w - wG).$$

We deduce from this that a general element of $\mathbf{G} = \langle R_0, \ldots, R_m \rangle$ is uniquely expressible in the form $x \mapsto xG + t$ or $x \mapsto (2w - x)G + t$, with $G \in \mathbf{H}$ and $t \in \mathbf{T}$, the translation subgroup $\mathbf{T}$ of $\mathbf{R}(W)$, namely,

$$\mathbf{T} = \langle 2(x - y) \mid x, y \in W \rangle.$$

Hence, $\mathbf{G}$ is the semidirect product $\mathbf{G} = \mathbf{R}(W) \ltimes \mathbf{G}_0$, and $V := v\mathbf{G}$ is the vertex-set of P.

Now consider a subgroup $\langle R_0, \ldots, R_k \rangle$ of $\mathbf{G}$ for some k. The uniqueness ensures that an element of this which also belongs to $\mathbf{H}$ must be of the form $x \mapsto xG$ with $G \in \mathbf{H}$, and therefore actually lies in $\langle R_1, \ldots, R_k \rangle$. By Theorem 2C2, this shows that $\mathbf{G}$ is a C-group, so that a genuine polytope results. □

We call P a *free abelian apeirotope* on Q, or *with vertex-figure* Q and *base vertex* v. We denote the family of such apeirotopes by $\operatorname{apeir} \mathsf{Q}$, although we often loosely write $\mathsf{P} = \operatorname{apeir} \mathsf{Q}$. The construction is not quite the dual of the free extension of [99, Theorem 4D4], because its geometric nature imposes commutativity conditions on it; additionally, $\operatorname{apeir} \mathsf{Q}$ depends on the particular realization Q, rather than on the original abstract regular polytope $\mathcal{Q}$, say. In fact, there is nothing in the construction which depends on Q being rational; however, if it is not, then the resulting apeirotope will not be discrete, which is contrary to the ground rules established in Section 4F.

The same apeir construction also applies in the special case when $v \in \operatorname{aff} W$, for which we use the affix $\boldsymbol{\alpha}$. Thus we write $\mathsf{P} = \mathsf{Q}^{\alpha}$, and call Q^{α} the *free abelian extension* of Q. A general apeirotope in $\operatorname{apeir} \mathsf{Q}$ is blended, of the form $\mathsf{Q}^{\alpha} \# \{2\}$; thus the following consequence of the construction is clear.

5E3 Corollary *Let* Q *be a finite rational regular polytope. If every edge-circuit of* Q^{α} *is even, then* Q^{α} *is polytopal.*

5E4 Remark An odd edge-circuit can be assumed to start at the initial vertex; we lose no generality in taking Q to be centred, so that $v = o$. If $w = w_0, w_1, \ldots, w_{2k}$ are the successive mid-points of these edges, then

$$w_0 - w_1 + w_2 - \cdots + w_{2k} = o \implies 2w \in \mathbf{T}.$$

In practice, we can use either criterion (or, rather, its negative) to show that all edge-circuits are actually even. For example, if (say) the jth coordinate of each vertex of Q is odd, then Q^{α} can have no odd edge-circuits.

5E5 Remark Under the following circumstances, Q^{α} fails to be polytopal:

- Q^{α} has odd edge-circuits;
- Q is not centrally symmetric, or $\dim \mathsf{Q}$ is odd and Q is handed (has a purely rotational symmetry group – see, for instance, Section 13B).

The crucial point here is that such an edge-circuit brings the centre v of Q back to its starting point, but reverses the orientation of Q.

5E6 Remark The set of centres of point-reflexions in $\mathbf{G}$, which is just $W + \mathbf{T}$, is closed under all combinations $(t, u) \mapsto (k+1)t - ku$ with $k \in \mathbb{Z}$. However, this set need not be closed under *all* affine combinations with integer coefficients; the following example illustrates this.

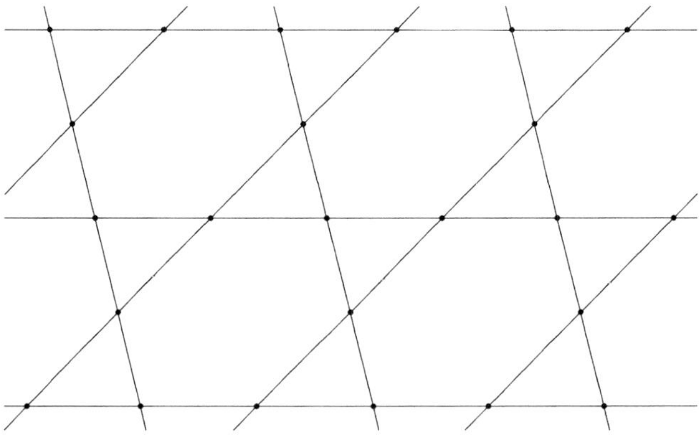

The missing centres of the hexagons are affine combinations of the form $a-b+c$, where a, b, c are any three successive vertices of such a hexagon.

Before we go on to discuss individual cases, we have two further general results about this construction; the first is related to those of Section 5D in an important way.

5E7 Theorem *If* Q *is a finite rational regular polytope such that* Q^{α} *and* Q^{κ} *are polytopal, then* $\mathsf{Q}^{\alpha\varpi} = \mathsf{Q}^{\kappa}$.

Proof. We use the notation in the definition of Q^{α}. Bearing in mind that w is the initial vertex of Q (and so of Q^{κ} also), the generatrix of $\mathsf{Q}^{\alpha\varpi}$ is

$$(\{w\}, S_0, S_1, S_2, \ldots,) \overset{\varpi}{\longmapsto} (S_0, \{w\}S_1, S_2, \ldots) = (R_0, R_1, R_2, R_3, \ldots),$$

which replaces S_1 by its image under the point-reflexion in $w \in S_1$. This just defines Q^{κ}, as claimed. □

5E8 Remark We shall see in Section 8B that, if Q is of full rank, then Q^{α} and Q^{κ} are always polytopal.

In general terms, we also have the following.

5E9 Proposition *Let* Q *be a finite regular* m*-polytope with* $m \geqslant 5$. *Then* $\mathsf{Q}^{\zeta\varpi} = \mathsf{Q}^{\varpi\kappa}$, *if the polytope exists.*

Proof. Indeed, since $\boldsymbol{\zeta}$ changes the sign of S_0, which in turn changes the sign of R_1, this combined operation is $\boldsymbol{\varpi\kappa}$ by definition. □

Notes to Section 5E

1. In fact, we know of no examples where the abelian extension falls foul of the conditions of Remark 5E5 or, indeed, fails for any other reason to yield a genuine apeirotope when applied to a rational regular polytope.

5F Vertex-Figures

The recursive definition of a polytope that we gave in Section 2A was symmetric between facets and vertex-figures. However, the various constructions – abstract and geometric – that we have described in the foregoing sections of this chapter concentrate on vertex-figures, making little or no mention of facets. This is also a feature of realization theory, as we saw particularly in Section 4D.

Indeed, even for the classical regular polytopes of [27], which we shall discuss in Chapter 7, there is no way that we can just list them without reference to the solution of the corresponding classification problem in lower rank or dimension. For the non-classical polytopes, and more so for the polytopes of larger corank, we must appeal to recursion of some kind or other. Since we have much less control over what the facets of new polytopes may look like, we find that we have to concentrate on the vertex-figures; see the notes at the end of the section.

Subgroups

It is unnecessary to point out that the symmetry group $\mathbf{H} = \mathbf{G}(\mathsf{Q})$ of the vertex-figure Q of a regular polytope (or apeirotope) P is a subgroup of the group $\mathbf{G} = \mathbf{G}(\mathsf{P})$ of P, whose index $[\mathbf{G} : \mathbf{H}]$ is the number of vertices of P. However, for a polytope of full or nearly full rank in $\mathbb{E}^d$ with $d \geqslant 5$ which is not a blend, $\mathbf{G}$ will be a subgroup of a hyperplane reflexion group, possibly with outer automorphisms; this will even hold for polytopes whose groups consist of direct isometries only, as we shall see in Section 13B. This severely restricts what $\mathbf{H}$ can be. Moreover, because $\mathbf{H}$ will (in effect) be a subgroup of $\mathbf{G}$ in only one way, it will immediately determine the vertex-set of the polytope P; in turn, since $\operatorname{vert} \mathsf{Q} \subseteq \operatorname{vert} \mathsf{P}$ (in the obvious way), this places further restrictions on Q.

As we shall see in Section 13A, this is particularly important for the finite polytopes in the 'gateway' dimension $d = 5$ (see the notes at the end of the section), because here $\mathbf{G}$ must be a subgroup of $[3, 3, 3, 3] \rtimes \mathbf{Z}_2 = [3, 3, 3, 3] \times \mathbf{Z}_2$ or $[3, 3, 3, 4]$; by this stage, we will already have classified the regular polytopes and apeirotopes of full or nearly full rank in lower dimensions.

So, let us make the general procedure explicit. The assumption is that we are testing a given (finite) geometric regular polytope Q with symmetry group $\mathbf{H}$ as a potential vertex-figure of a regular polytope (or apeirotope) P. We thus do the following:

- identify suitable groups $\mathbf{G}$ for which $\mathbf{H} < \mathbf{G}$;
- from the Wythoff space (of $\mathbf{H}$ in $\mathbf{G}$), identify the vertex-set V of P;
- check whether the vertex-set of Q is an appropriate subset of V;
- check whether $\mathbf{G}$ has a suitable reflexion interchanging the initial vertices of P and Q.

Of course, the procedure applies most readily when the Wythoff space is 1-dimensional; this will happen when the corank is small, but not usually more generally. The last condition particularly applies to apeirotopes P, when there may be more than one choice of this reflexion.

Nearest Vertices

Another test for vertex-figures or their groups is what we call the *nearest vertices criterion*; this provides an easy way to eliminate many possibilities in (for instance) Section 13B. If X is a finite set of points in an ambient space $\mathbb{U}$ (a sphere or euclidean space) and $v \notin X$ is a further point, then we denote by $\alpha(v, X)$ the smallest angle subtended at v in $\mathbb{U}$ by some pair of points of X. The following claim should be obvious.

5F1 Proposition *If* P *is a geometric regular polytope with initial vertex* v*, and* X *is the set of vertices of* P *nearest to* v *apart from* v *itself, then* $\alpha(v, X) > \pi/3$*. Similarly, if* P *is a regular apeirotope, then* $\alpha(v, X) \geqslant \pi/3$*.*

Proof. If the given condition failed to hold, then two points of X would be closer to each other than they are to v, a contradiction to the definition of X and the regularity of P. (Note that the angle of a spherical equilateral triangle is greater than $\pi/3$.) □

5F2 Remark Observe that the example $\{5,3\}$ as the vertex-figure of $\{\frac{5}{2},5,3\}$ shows that Proposition 5F1 need not hold when X is, instead, the vertex-set of the vertex-figure of P at v. See Section 7G for $\{\frac{5}{2},5,3\}$ and the other 4-dimensional regular star-polytopes.

Notes to Section 5F

1. The term 'gateway' comes from genealogy. A 'gateway' ancestor is one through whom many lines of descent are traced from notable ancestors; in English history, Edward III is often cited as an instance. As another example, all the kings and queens of England after Edgar II was deposed in 1066 (none actually English by male-line descent) can trace their ancestry to the usurper William the Bastard. (All except William himself are descended from both King Alfred the Great and the Frankish Emperor Charles the Great through his wife Matilda of Flanders.)
2. As is pointed out in [27, 7.x], Gosset [53] was one of many who rediscovered higher-dimensional regular polytopes; however, few of them also came across the Schläfli symbol. In fact, Gosset used an interesting variant of it, writing $|p_1|p_2|\cdots|p_{d-1}|$ instead of $\{p_1,p_2,\ldots,p_{d-1}\}$. Moreover, he regarded $|p_1$ as an operator that can be applied to $|p_2|\cdots|p_{d-1}|$, which chimes in exactly with the approach described in this section.

6

Rigidity

In this chapter, we discuss the relatively new topic of rigidity. The term 'rigidity' is used in several different areas of mathematics, most notably (for connexions with the present interests) concerning frameworks and convexity. (There is a useful survey of rigidity in its various aspects in [14]; more recently, several articles in [15] concern rigidity.) The latter is closest in concept to what we do here, so let us begin by recalling the classical rigidity theorem, proved (almost) by Cauchy [9, 10], and later extended by other authors to higher dimensions. This states that, if $d \geqslant 3$ and corresponding facets of two isomorphic convex d-polytopes P and Q are congruent, then P and Q are themselves congruent. In the context of regular polytopes, what we ask of rigidity is very much in the same spirit: to what extent does the specification of certain geometric data about a regular polytope P determine P? Under normalization, we could talk about congruence classes here as well; however, particularly in the context of regular apeirotopes, it is perhaps more natural merely to look at shapes.

In detail, the chapter goes as follows. Section 6A gives a little historical background to the subject. In Section 6B the central notion of fine Schläfli symbols is described, and then in Section 6C the formal concepts of shape (or similarity class) and rigidity are defined, and some (to be hoped) familiar examples are introduced. Finally, general criteria for rigidity are given in Section 6D.

6A Basic concept

Before we go into the concept of rigidity in formal detail, it is appropriate to introduce it in a way that would have been understood by the contributors to Euclid's *Elements*. We can generalize how Theaetetus classified the (misnamed) Platonic solids, and replace the convex polygons by arbitrary planar regular star-polygons. If we wish to construct a regular polyhedron (in ordinary euclidean space $\mathbb{E}^3$) with faces $\{p\}$ and vertex-figures $\{q\}$, then we must be able to fit p-gons in a q-gonal fashion around a vertex with something to spare. This leads at once to a very familiar relationship, that is, in terms of normalized angles (with sum 1 rather than 2π), we must have

$$q\left(\frac{1}{2} - \frac{1}{p}\right) < 1 \iff \frac{1}{p} + \frac{1}{q} > \frac{1}{2}.$$

At this point, though, if we carry over our historical comparisons, the Greeks are stalled; they lack the combined geometric and algebraic tools to determine what pairs $\{p, q\}$ yield finite (or, equivalently, discrete) polyhedra. Curiously, Plato (who was no mathematician) had in [109] what might be called half an insight into the solution. In his description of the three regular polyhedra with trigonal faces (tetrahedron, octahedron and icosahedron), he divided their faces into six triangles by their angle bisectors or altitudes, thus producing what is essentially the kaleidoscope of reflexion planes of the polyhedra. It was only half an insight, because he did not do the analogous construction for the cube or dodecahedron, otherwise he might possibly have spotted the underlying method of expressing the symmetries of the polyhedra.

Nevertheless, the basic idea of the argument sketched above can be made to illustrate the general principle. Irrespective of finiteness, such a Schläfli symbol $\{p, q\}$ uniquely determines a 3-dimensional polyhedron, at least in the bounded case, in the following way. Fix a vertex v on, say, the unit sphere S in $\mathbb{E}^3$. In sections of S by planes perpendicular to the direction v, two great circles on S through v subtending a normalized angle $1/q$ induce planar angles from varying from $1/q$ to 0. In particular, since the internal angle of a p-gon is $\frac{1}{2} - \frac{1}{p}$, and $\frac{1}{2} - \frac{1}{p} < \frac{1}{q}$, at one stage p-gons with vertices on S will fit q-gonally around v. This provides us with the vertices adjacent to v. The construction then iterates, and we end up with a (not necessarily finite) regular polyhedron $\{p, q\}$. As we have said, quite different methods need to be deployed to determine finiteness.

6B Fine Schläfli symbols

Since we shall be concerned in this chapter with similarity classes of realizations of polytopes (or apeirotopes), the henogon $\{1\}$ will play no part in the present discussion. In other words, we are actually looking at realization cones in the earlier sense of [99, Section 5A].

A *fine Schläfli symbol* determines a realization subcone $\mathcal{P}$ by specifying to a certain extent the geometry of some regular polygons (that is, 2-polytopes) that occur among the vertices of $\mathcal{P}$. In other words, $\mathcal{P}$ results from the imposition of certain geometric conditions on an abstract regular polytope $\mathcal{Q}$ which are not necessarily indicated by its abstract type (as given, for instance, by its Schläfli type and relators).

In this sense, the realization subcone of a regular polygon $\mathcal{P}$ is denoted by

$$\{p\} := \left\{\frac{s}{t_1, \ldots, t_k}\right\},$$

with p now a generalized fraction, so that s is a positive integer and $t_1, \ldots, t_k$ are non-negative integers such that $0 \leqslant t_1 < \cdots < t_k \leqslant \frac{1}{2}s$; moreover, their greatest common divisor is $(s, t_1, \ldots, t_k) = 1$. As in (4G4), this means that

6B1 $$\{p\} = \{p_1\} \# \cdots \# \{p_k\},$$

with $p_j := s/t_j$ in lowest terms; such a $\{p_j\}$ is a *component* of $\{p\}$. In this context, the linear apeirogon $\{\infty\}$ is sometimes written $\{\frac{1}{0}\}$, particularly when there is a possibility of ambiguity; as previously, $\{2\}$ is the digon (1-polytope). Then, for $p_j \neq \infty, 2$, each $\{p_j\}$ is a planar regular polygon, which will be a star-polygon if p_j is not an integer. Observe that the family $\mathcal{P}$ of polygons with this fine Schläfli symbol forms a k-dimensional subcone of the whole realization cone (which is $\{\infty\}$ if $t_1 = 0$, and otherwise is $\{s\}$), because there are k degrees of freedom for the sizes of the k components.

We shall write $|p| := s$ when $t_1 > 0$, so that $\{p\}$ is then a finite $|p|$-gon, and $|p|$ is the period of the product of its generating reflexions.

We should say a few more words here about certain special regular polygons or apeirogons. First, we have *skew* or *zigzag* polygons, of the form $\{\frac{2k}{t,k}\}$; the most frequently occurring cases are $k = 2$ or 3 and $t = 1$. Note that the polygon $\{\frac{6}{1,3}\}$ is centrally symmetric. Second, we have the *r-helix* $\mathcal{H}_r = \{\frac{r}{0,1}\}$ with r an integer; its fundamental property is that, if the vertices of some $\mathsf{H} \in \mathcal{H}_r$ are $\ldots, a_{-1}, a_0, a_1, a_2, \ldots$, then $t = t(\mathsf{H}) := a_{r+j} - a_j$ (for any j) is a vector which yields a translational symmetry of H, though not necessarily of a regular apeirotope whose specification involves helices in $\mathcal{H}_r$. (More general helices will also occur, and we shall draw attention to them in the appropriate place.)

6B2 Remark It is important to bear in mind that, while the fine Schläfli symbol will tell us, for example, that a zigzag is a blend of a planar polygon and a digon, it does not specify the relative sizes of the components.

We specify a realization subcone more exactly by introducing generalized fractions into the abstract notation of Section 2D. Suppose that J and K are index strings corresponding to involutory elements (reflexions) S and T in the realization P. Then $\langle S, T\rangle$ is the symmetry group of a polygon $\{q\}$ with initial vertex a vertex of P lying in T. It is important to bear in mind that the order (S, T) is crucial here; while the periods of the products ST and TS are the same, reversing the order may yield a different regular polygon $\{r\}$. An example to illustrate this is the Petrie–Coxeter apeirohedron $\{4, \frac{6}{1,3} \mid 4\}$, whose (geometric) dual $\{6, \frac{4}{1,2} \mid 4\}$ is obtained by reversing the order of the canonical generating reflexions (R_0, R_1, R_2); thus (R_1, R_2) gives a skew hexagon $\{\frac{6}{1,3}\}$, while (R_2, R_1) gives a planar hexagon $\{6\}$. If J, K are flag sequences corresponding to S, T, then the resulting fine Schläfli symbol is of the form

6B3 $$\{p_1, \ldots, p_{n-1}\}/\langle\langle \ldots, (\mathsf{J} \cdot \mathsf{K})^q, \ldots\rangle\rangle,$$

where $p_1, \ldots, p_{n-1}, q$ are now generalized fractions, and $\cdot$ is used as an indicatory separator. Abstractly, of course, we have a corresponding relator $(\mathsf{JK})^{|q|}$ when appropriate.

6B4 Remark We continue to use notation such as $\{p_1, \ldots, p_{n-1} : q\}$ or $\{p_1, \ldots, p_{n-1} \mid q\}$ to specify Petrie polygons or deep holes. Further, it is still often the case the notation of (6B3) is clumsier than employing concatenations, and so far as we are concerned it is of more theoretical rather than practical importance.

6C Shapes

At the very least, a fine Schläfli symbol will look like $\{p_1, \ldots, p_{n-1}\}$, where the p_j are generalized fractions; in general, further geometric data are given. This particularly applies to discrete realizations with infinite 2-faces; even so, in a fine Schläfli symbol the geometric type of these apeirogons will be specified.

We should emphasize from the outset that a given regular polytope may have radically different fine Schläfli symbols, which nevertheless completely determine its similarity class or *shape*, as we shall more usually call it. We address this topic here, although we postpone giving significant examples until Section 10C.

6C1 Remark We can restate Cauchy's theorem in a way that is even closer to the spirit of this chapter: *for each $d \geqslant 3$, if corresponding facets of two isomorphic d-polytopes have the same shape, then the polytopes themselves have the same shape.*

A fine Schläfli symbol therefore determines a subcone $\mathcal{P}$ of the realization cone of some abstract regular polytope, assuming (of course) that the implied group relations yield a string C-group $\boldsymbol{G}$. If this cone is a ray (in other words, 1-dimensional), so that $\mathcal{P}$ consists of a single similarity class, then we say that $\mathcal{P}$ is *rigid*. Thus the polytopes in $\mathcal{P}$ must be pure (and faithful) realizations, but the converse need not hold; rigidity is a geometric rather than an algebraic property. For convenience, we shall usually talk about subcones $\mathcal{P}$ rather than their individual polytopes.

More generally, we can ask what subcone $\mathcal{P}$ is specified by a fine Schläfli symbol and, in particular (in the case of apeirotopes) whether $\mathcal{P}$ is finite-dimensional. In this context, it is worth reminding ourselves that the realization cones of (abstract) regular apeirotopes will generally have uncountably infinite algebraic dimension, as was shown in [80].

With the understanding that they are fine Schläfli symbols, we shall write $\mathcal{P} \preccurlyeq \mathcal{Q}$ to mean that the cone specified by $\mathcal{P}$ is a subcone of that specified by $\mathcal{Q}$, and $\mathcal{P} \approx \mathcal{Q}$ to mean that the two fine Schläfli symbols specify the same cone.

Before we go on to establish some general criteria, we give a couple of simple examples. We shall see in Section 7D that the realization cone of the regular icosahedron is 4-dimensional; apart from the henogon $\{1\}$, which we ignore, a general realization will have three pure components: the usual 3-dimensional icosahedron $\{3, 5\}$ and great icosahedron $\{3, \frac{5}{2}\}$, and the 5-dimensional hemi-icosahedron $\{3, \frac{5}{1,2} : \frac{5}{1,2}\}$. The first two are isomorphic (in fact, allomorphic), while the third is (abstractly) $\boldsymbol{\{3, 5 : 5\}}$. The usual Schläfli symbols – their fine Schläfli symbols in our terms – distinguish the first two. The abstract Schläfli symbol of the third is enough to determine the geometry of a realization, because each pair of its six vertices forms an edge, thus determining a regular 5-simplex.

The case of the regular dodecahedron is rather different. As we shall see in Section 7D, the realization cone is 6-dimensional; there are five pure components apart from $\{1\}$. Two of these are the familiar 3-dimensional dodecahedron $\{5,3\}$ and great stellated dodecahedron $\{\frac{5}{2},3\}$. There is also a 4-dimensional faithful realization, whose fine Schläfli symbol is $\{\frac{5}{1,2},3 : \frac{10}{1,3}\}$; because we specify the Petrie polygon $\{\frac{10}{1,3}\}$ (preventing it from having components $\{\frac{5}{1,2}\}$, $\{\frac{10}{1,5}\}$ or $\{\frac{10}{3,5}\}$ arising from the other pure realizations), this again completely determines the geometry.

There are also two realizations – 4- and 5-dimensional, respectively – of the hemi-dodecahedron $\{5,3 : 5\}$. However, these both have fine Schläfli symbol $\{\frac{5}{1,2},3 : \frac{5}{1,2}\}$, with no apparent way of separating them by regular edge-circuits and the like. In fact, they can only be distinguished by an even more detailed description of their faces. The angle at a vertex of the general regular pentagon varies between $\pi/5 = \arccos(\tau/2)$ for the pentagram and $3\pi/5 = \arccos(\tau^{-1}/2)$ for the (convex) pentagon; for the face of the general $\{\frac{5}{1,2},3 : \frac{5}{1,2}\}$ the angle varies from $\arccos(3/4)$ for the 4-dimensional pure realization to $\pi/2$ for the 5-dimensional one (in between, the realization is 9-dimensional and blended). The crucial point is that $\{\frac{5}{1,2},3 : \frac{5}{1,2}\}$ does not determine a ray in the realization cone of the dodecahedron.

6C2 Remark The fine Schläfli type $\mathcal{P} = \{\frac{5}{1,2},3 : \frac{5}{1,2}\}$ is self-Petrie, as its notation indicates. While $\mathcal{P}$ has certain regular hexagonal edge-circuits, these do not have their full symmetries in the group $\boldsymbol{G}$ of $\mathcal{P}$. However, we obtain all their symmetries, if we adjoin an outer automorphism of $\boldsymbol{G}$ which interchanges the faces and Petrie polygons of $\mathcal{P}$. In the 4-dimensional realization these hexagons are $\{\frac{6}{2,3}\}$, while in the 5-dimensional one they are $\{\frac{6}{1,3}\}$, so that specifying either hexagon does now determine the shape.

It should be noted as well that more than geometry is involved here. For instance, the edges and diagonals of the vertex figure of the hemi-icosahedron $\{3,\frac{5}{1,2} : \frac{5}{1,2}\}$ are equal. However, this is also true of the realization of the icosahedron whose vertices are those of the regular 6-cross-polytope; this will have fine Schläfli symbol $\{3,\frac{5}{1,2}\}$, illustrating the fact that $\{3,\frac{5}{1,2} : \frac{5}{1,2}\} \preccurlyeq \{3,\frac{5}{1,2}\}$ (the latter, of course, not being rigid).

6D Rigidity Criteria

We now describe a number of general conditions which will ensure rigidity of a regular polytope. Bear in mind here that rigidity depends on a given fine Schläfli symbol, so that one that ensures rigidity may be rather different from one that describes its abstract type.

By definition, a rigid regular polytope $\mathcal{P}$ determines a single similarity class. The same is then true of all sections of $\mathcal{P}$, even though some of these may not be rigid themselves. Thus many of our arguments really depend less on rigidity itself than on working within fixed similarity classes. In this context, it is natural to raise

6D1 Question Suppose that $\mathcal{P}$ is a geometric regular n-polytope such that, for some $2 \leqslant k \leqslant n-1$, the k-face and $(n-k+1)$-coface of $\mathcal{P}$ are constrained to lie in fixed similarity classes. Does $\mathcal{P}$ then consist of a single similarity class?

This is the best that we could expect, since we clearly need some overlap of face and coface.

What we shall do in this section is establish rigidity in a variety of different general cases. The first covers the classical regular polytopes.

6D2 Theorem *If* $\operatorname{rank} \mathcal{P} \geqslant 2$ *and the fine Schläfli type* $\mathcal{P}$ *has planar* 2*-faces and rigid vertex-figure* $\mathcal{Q}$*, then* $\mathcal{P}$ *is rigid.*

Proof. This is clear, because the ratio $1 : 2\cos(\pi/p)$ between the edge-length and next diagonal of the planar regular polygon $\{p\}$ fixes the relative distance of the initial vertex from the adjacent vertices. Since $\mathcal{Q}$ is itself rigid, this fixes the whole geometry of $\mathcal{P}$. □

6D3 Remark It is worth pointing out that even more is determined in this case. Considering the actual realizations $\mathsf{P} \in \mathcal{P}$ and $\mathsf{Q} \in \mathcal{Q}$, if the distance forces the initial vertex v to be at the centre of Q, then P is necessarily infinite. If the distance is greater, then P must be finite, and its centre and circumradius can immediately be found.

6D4 Corollary *The classical regular polytopes and honeycombs are rigid.*

We recall that the symmetry groups of the regular polytopes of [27] are generated by hyperplane reflexions (see also Chapter 7); all the entries in their Schläfli symbols are simple fractions (and thus correspond to planar polygons).

Observe, in fact, that all 18 of the finite regular polyhedra in $\mathbb{E}^3$ are rigid, if we specify them appropriately. For instance, as we have seen, the great dodecahedron is $\{\frac{5}{2}, 3\}$. Its Petrial is $\{\frac{10}{3,5}, 3 : \frac{5}{2}\}$, which distinguishes it from the Petrial $\{\frac{10}{1,5}, 3 : 5\}$ of the Platonic dodecahedron $\{5, 3\}$. (Indeed, we could denote these Petrials by $\{\cdot, 3 : \frac{5}{2}\}$ and $\{\cdot, 3 : 5\}$, respectively, since the designations of the faces is actually redundant.) Thus we have another family of regular polytopes which are automatically rigid.

6D5 Theorem *Suppose that* $\operatorname{rank} \mathcal{P} \geqslant 3$ *and that the fine Schläfli type of* $\mathcal{P}$ *is such that the type of the vertex-figure is a blend* $\mathcal{Q} \# \{2\}$ *with* $\mathcal{Q}$ *rigid, and the* 2*-face and the hole of the* 3*-face of* $\mathcal{P}$ *are planar. Then* $\mathcal{P}$ *is rigid.*

Proof. Again, the geometry of the 2-face and hole fix the shape of the zigzag 2-faces of the vertex figure; in turn (as in the proof of the previous Theorem 6D2), this fixes the geometry of $\mathcal{P}$. □

Theorem 6D5 enables us to deal with another wide range of examples in a categorical fashion, and so no individual treatment of such cases will be needed. Observe that we have already mentioned the theorem in Section 5C.

II

Polytopes of Full Rank

7

Classical Regular Polytopes

Classical regular polytopes were the subject of Coxeter's immensely influential monograph *Regular Polytopes* [27], which first appeared in 1948. This part of the book concerns polytopes of full rank; because all polytopes of full rank are intimately related to the classical ones, we begin in this chapter by classifying the latter, and describing the relationships among them.

The symmetry groups of the classical polytopes are the hyperplane reflexion groups that we classified in Section 1E. We saw already in Section 4B that the cofaces of a polytope of full rank are themselves of full rank. In Section 7A, we give a new condition to be a classical polytope, which restores the duality between facets and vertex-figures. In Section 7B we describe the three sequences of the corresponding polytopes – simplex, staurotope (as we have renamed the cross-polytope) and cube – which occur in all dimensions. Section 7C is devoted to studying the regular 24-cell; here, we begin to see the useful rôle played by quaternions. In Section 7D we describe the icosahedron and dodecahedron. In Section 7E we correspondingly construct the regular 600-cell and in the next Section 7F we treat its dual 120-cell; the description of the realization domain of the latter is postponed until Section 7K. In Section 7G we find all the regular star-polytopes and their (abstract) automorphism groups. In Section 7H we classify the discrete regular apeirotopes or honeycombs. Then, in Section 7J we describe the regular compounds of polytopes, and make some brief comments on compounds of honeycombs.

To avoid constant repetition, we take all polytopes in the chapter to be regular, unless specified otherwise. Moreover, we assume throughout that all polytopes are faithfully realized and finite or discrete as appropriate, again except when specifically stated otherwise.

7A Faces of Full Rank

Followwing what we said in the preamble to the chapter about [27], we are motivated to call a geometric polytope P of rank $m \geqslant 3$ *classical* if its generatrix consists of hyperplane reflexions. We also recall that we showed in Corollary 4B4 that the cofaces of a polytope of full rank – in particular its vertex-figures – are themselves of full rank. The main Theorem 7A1 of the section gives an equivalent definition that uses the dual condition as well; this is somewhat in the spirit of amalgamation as discussed in Section 2F, and corresponds in a way to the definition used by Coxeter in [27].

We begin by recalling what the classical polytopes of low rank m are:

- for $m = 0$, the henogon or point-set $\{1\}$;
- for $m = 1$, the digon or (line-)segment $\{2\}$;
- for $m = 2$, the finite polygons $\{p\}$ for $2 < p \in \mathbb{Q}$, and the (linear) apeirogon $\{\infty\}$.

Note that these are just the polytopes of full rank $m \leqslant 2$.

The new characterization is given by

7A1 Theorem *A polytope* P *of rank* $m \geqslant 3$ *is classical if and only if its facets and vertex-figures are of full rank.*

We shall refer to the condition of the theorem, namely, that P has facets and vertex-figures of full rank, as the *FV-criterion.* On the surface, the FV-criterion appears to be comparatively weak; indeed, neither of the equivalent conditions actually assumes that P itself is of full rank. The proof will be accomplished by a sequence of lemmas.

First, from the definition and the obvious inductive assumption we clearly have

7A2 Lemma *A classical polytope satisfies the FV-criterion.*

The next step is

7A3 Lemma *A polytope of rank* $m \geqslant 3$ *satisfying the FV-criterion is itself of full rank.*

Proof. Let P be such a polytope, with initial vertex v, facet $\mathsf{P}^{\boldsymbol{f}} =: \mathsf{F} \ni v$ and vertex-figure $\mathsf{Q} := \mathsf{P}^{\boldsymbol{v}}$ at v; we take the latter in the broad sense, so that

$$\operatorname{vert} \mathsf{Q} = \{w \in \operatorname{vert} \mathsf{P} \mid \{v, w\} \text{ is an edge of } \mathsf{P}\}.$$

We first observe that the argument of Theorem 4B2 shows that, if $\mathsf{G} := \mathsf{F}^{\boldsymbol{v}}$ is the vertex-figure of F at v, then $\operatorname{aff} \mathsf{F} = \operatorname{aff}(\{v\} \cup \mathsf{G})$.

There are two possibilities. If $v = c(\mathsf{Q})$ (the centre of Q), then $\operatorname{aff} \mathsf{F} \subseteq \operatorname{aff}(\{v\} \cup \mathsf{Q}) = \operatorname{aff} \mathsf{Q}$. Of course, this holds for each facet of P through v, and then for each vertex w adjacent to v, and so on; the connectedness of P then implies that P must be infinite, and that $\mathsf{P} \subseteq \operatorname{aff} \mathsf{Q}$. Thus P is an apeirotope of full rank.

Alternatively, if $v \neq c(\mathsf{Q})$ and S is the sphere spanned by v and $\operatorname{vert} \mathsf{Q}$, then $\operatorname{vert} \mathsf{F}$ similarly lies in the sphere spanned by v and $\operatorname{vert} \mathsf{G}$, which is contained in S; once again, this holds for all facets of P through v. It is clear that all symmetries of Q (as a subgroup of $\mathbf{G}(\mathsf{P})$) preserve S. Similarly, the symmetries of P also preserve S – indeed, they preserve the line though $c(\mathsf{F})$ perpendicular to aff F, which passes through the centre of S. But these symmetries generate $\mathbf{G}(\mathsf{P})$, and consequently vert $\mathsf{P} \subset S$. It follows that P is a (finite) polytope of full rank, as was claimed. □

We expand on Lemma 7A3 in

7A4 Lemma *Each inface of a polytope of rank $m \geqslant 3$ satisfying the FV-criterion is itself of full rank, and hence satisfies the FV-criterion.*

Proof. Let F be a facet of such a polytope P of rank $m \geqslant 3$, and let G be a ridge (face of corank 2) of P contained in F. Then G must also be of full rank, since otherwise span G = span F, and then the reflexion R_{m-1} that fixes G and interchanges F with the other facet that contains G will be the identity. However, this contradicts our blanket assumption that P is faithfully realized.

We now appeal to induction on the corank, to see that all faces of P are of full rank. Finally, since Corollary 4B4 tells us that cofaces of polytopes of full rank themselves have full rank, it follows that all infaces of P have full rank, as claimed. □

To avoid constant repetition, from now on we take the generatrix of a regular d-dimensional polytope or apeirotope P of full rank to be $(R_0, \ldots, R_{m-1})$, with $m = d$ or $d+1$ as appropriate; that is, we fix the dimension rather than the rank. What Lemma 7A4 has shown is that, if F is a j-face of P satisfying the FV-criterion, then dim $\mathsf{F} = j$, except when P is infinite and $\mathsf{F} = \mathsf{P}$.

We next have a geometrical lemma concerning the *centre* $c(\mathsf{P})$ of a polytope P, that is, the centroid of its vertices.

7A5 Lemma *If P is a finite polytope of rank at least 3 satisfying the FV-criterion and F is a facet of P, then $c(\mathsf{P}) \notin$ aff F.*

Proof. We employ an induction argument on $d :=$ dim P, noting that the result already holds for polytopes of lower dimension. If $d > 1$, let $\mathsf{G} < \mathsf{F}$ be a ridge of P; in such circumstances, we may always assume that faces under consideration belong to the base flag of P. By the inductive assumption, $c(\mathsf{F}) \notin$ aff G, since F itself satisfies the FV-criterion. Were we to have $c(\mathsf{P}) \in$ aff F, then $c(\mathsf{F}) = c(\mathsf{P})$, by the uniqueness of the centre. But then the generating reflexion R_{d-1} of $\mathbf{G}(\mathsf{P})$ would have to fix G and $c(\mathsf{F})$, and so would fix F itself, since aff$(\mathsf{G} \cup \{c(\mathsf{F})\}) =$ aff F. Since P is faithful, this is impossible. The contradiction thus yields the result. □

The final step is

7A6 Lemma *If P is a polytope of rank at least 3 satisfying the FV-criterion, then its symmetry group $\mathbf{G}(\mathsf{P})$ is generated by hyperplane reflexions.*

Proof. First suppose that P is finite. Once again, we may use induction on $d :=$ dim P, the case $d = 1$ being trivial. Let F and G be as in the proof of Lemma 7A5. Within aff F, by the inductive assumption, (the mirrors of) $R_0, \ldots, R_{d-2}$ are hyperplanes; that is, $\dim(R_j \cap \text{aff}\, \mathsf{F}) = d-2$ for each $j = 0, \ldots, d-2$. Since

$c(\mathsf{P}) \notin \operatorname{aff} \mathsf{F}$ by Lemma 7A5, we see that $R_j = \operatorname{aff}((R_j \cap \operatorname{aff} \mathsf{F}) \cup \{c(\mathsf{P})\})$ is a hyperplane for $j = 0, \ldots, n-2$. Finally, since R_{d-1} fixes G, with $\dim \mathsf{G} = d-2$ and $c(\mathsf{P}) \notin \operatorname{aff} \mathsf{G} \subset \operatorname{aff} \mathsf{F}$, or by an appeal to Theorem 4B8, we conclude that $\dim R_{d-1} \geqslant d-1$. But $R_{d-1} \neq \mathbb{E}^d$, so that the opposite inequality is trivial, which proves the result in this case.

When P is infinite, say of rank $d+1$ in $\mathbb{E}^d$, exactly the same argument yields the reflexion hyperplanes $R_0, \ldots, R_{d-1}$ as before, except that now they all pass through the centre $c(\mathsf{F})$ of a facet of P. Since the remaining generating symmetry mirror R_d must contain the initial ridge G of P, and since $\dim \mathsf{G} = d-1$, this shows that $R_d = \operatorname{aff} \mathsf{G}$ is also a hyperplane; again, we could quote Theorem 4B8. Thus the result holds here as well. □

7A7 Remark It should be obvious that the facet of a non-classical regular polytope of full rank must itself be of nearly full rank.

Since the classical polytopes turn up in many later contexts, in later sections we shall attempt a description of their geometry. For the moment, though, we can make an appeal to geometry and Wythoff's construction, to see that a suitable hyperplane reflexion group must give rise to a polytope, which actually has a geometric dual.

So, until the end of the section, $\mathbf{G}$ is a string group generated by hyperplane reflexions in $\mathbb{E}^d$, which is either finite (and then thought of as an orthogonal group) or infinite and discrete. Moreover, we lose no generality in supposing that $\mathbf{G}$ is indecomposable and acts *effectively* on $\mathbb{E}^d$, and so has no invariant (affine) subspaces. All this means that the mirrors of the generating reflexions bound a simple cone (in the finite case) or a simplex (otherwise).

7A8 Remark Referring to Table 1E14 for the discrete infinite case, we see that, for $d \geqslant 3$, no marks greater than 4 occur on any Coxeter diagram, and only 6 occurs for $d = 2$. Further, only for $d = 3$ or 4 do we find a mark 5 in the finite cases in Table 1E12.

We thus have a very general result. In view of the specific descriptions of the hyperplane reflexion groups in Tables 1E12 and 1E14, it could be used to find the vertices of the corresponding polytopes as well as other details, but at this stage we shall leave it on a theoretical level. Thus, except for the rather important first part, the proof will be a mere sketch.

7A9 Theorem *Each indecomposable finite or discrete infinite string group generated by hyperplane reflexions in $\mathbb{E}^d$ is the symmetry group of a faithfully realized polytope or apeirotope. Moreover, each such polytope has a geometric dual.*

Proof. The astute reader will have noticed that we have not yet mentioned the intersection property in this chapter. However, now that we know that we are dealing with a group generated by hyperplane reflexions, we can appeal to Theorem 4A3. With the intersection property, everything sails through with no problems. Nothing stands in the way of a direct application of Wythoff's construction, with initial vertex $v \in (R_1 \cap R_2 \cap \cdots) \setminus R_0$ as in (4A1). Induction shows that all proper faces are as to be expected. Moreover, the initial vertex $w := \frac{1}{2}(v + vR_0)$ yields the vertex-figure (and its faces).

Finally, reversing the order of the generating reflexions in this argument produces the dual. □

Schläfli Symbol

It is often the case that an abstract regular polytope can only be described combinatorially by considerable modification of the symbol denoting its Schläfli type. For classical polytopes, on the other hand, we have a much more concise notation. We treat the finite case here, but the situation for apeirotopes needs little modification.

Let P be a classical regular d-polytope with generatrix $(R_0, \ldots, R_{d-1})$ and symmetry group $\mathbf{G} = \mathbf{G}(\mathsf{P})$, where the R_j are hyperplane reflexions associated with a base flag of P. Taking the ambient space of P to be $\mathbb{E}^d$ and $c(\mathsf{P}) = o$, for $j = 0, \ldots, d-1$, we may write

$$R_j = \{x \in \mathbb{E}^d \mid \langle x, u_j \rangle = 0\},$$

with each u_j a unit vector. Further, we may change signs of the u_j, if necessary, to ensure that $\langle u_{j-1}, u_j \rangle < 0$ for $j = 1, \ldots, d-1$ (we have $\langle u_j, u_k \rangle = 0$ if $j \leqslant k-2$). Since $\langle R_{j-1}, R_j \rangle$ is a finite group, for each $j = 1, \ldots, d-1$ there is a rational number $p_j > 2$ (written as a fraction in its lowest terms), such that $\langle u_{j-1}, u_j \rangle = -\cos(\pi/p_j)$. Then P is denoted by its *Schläfli symbol*

7A10 $$\mathsf{P} := \{p_1, \ldots, p_{d-1}\}.$$

This new notation naturally accords with our informal earlier usage, such as $\{\frac{5}{2}\}$ for the regular pentagram, and $\{\frac{5}{2}, 5\}$ for the small stellated dodecahedron. We shall show in subsequent sections that the classical polytopes whose Schläfli symbols contain only integer entries are precisely the convex polytopes, which are isomorphic to the finite universal polytopes $\{p_1, \ldots, p_{d-1}\}$. Those whose symbols contain at least one non-integer entry are therefore *starry*, or are *star-polytopes*, so called because their faces or cofaces are not those of the convex hull of the vertices.

The proof of Lemma 7A6 shows

7A11 Proposition *The dual of classical polytope* $\mathsf{P} = \{p_1, \ldots, p_{d-1}\}$ *is the classical polytope* $\mathsf{P}^\delta = \{p_{d-1}, \ldots, p_1\}$.

Similarly, we have

7A12 Proposition *If* $d \geqslant 2$, *then the regular* d*-polytope* $\mathsf{P} = \{p_1, \ldots, p_{d-1}\}$ *has facet* $\mathsf{P}^f = \{p_1, \ldots, p_{d-2}\}$ *and vertex-figure* $\mathsf{P}^v = \{p_2, \ldots, p_{d-1}\}$.

Proof. This is clear either directly, or by applying Wythoff's construction, in the first case to the subgroup $\langle R_0, \ldots, R_{d-2} \rangle$ with the same initial vertex v of P, and in the second to $\langle R_1, \ldots, R_{d-1} \rangle$ with initial vertex the other vertex w of the initial (base) edge E of P. □

7A13 Remark The description of classical regular polytopes, particularly with Proposition 7A12 in mind, shows that the Schläfli symbol defined here from the viewpoint of reflexion groups coincides with the fine Schläfli symbol introduced in Section 6B. We have already observed in Theorem 6D2 that the classical polytopes, as determined by their Schläfli symbols, are rigid.

Schläfli Determinant

When we come to discuss regular star-polytopes in Section 7G, we need to appeal to certain features of the geometry of a classical polytope. Let $\mathsf{P} = \{p_1, \ldots, p_{d-1}\}$ be such a d-polytope. The *Schläfli determinant* $\Sigma(\mathsf{P}) = \Sigma(p_1, \ldots, p_{d-1})$ of P is defined to be

7A14 $$\Sigma(\mathsf{P}) := \begin{vmatrix} 1 & -\cos\frac{\pi}{p_1} & 0 & \cdots & 0 & 0 \\ -\cos\frac{\pi}{p_1} & 1 & -\cos\frac{\pi}{p_2} & \cdots & 0 & 0 \\ 0 & -\cos\frac{\pi}{p_2} & 1 & \cdots & 0 & 0 \\ \vdots & \vdots & \vdots & \ddots & \vdots & \vdots \\ 0 & 0 & 0 & \cdots & 1 & -\cos\frac{\pi}{p_{d-1}} \\ 0 & 0 & 0 & \cdots & -\cos\frac{\pi}{p_{d-1}} & 1 \end{vmatrix}$$

(see the notes at the end of the section). Thus the appropriate definition for the digon (1-polytope) $\{2\}$ is

7A15 $$\Sigma(\{2\}) := 1.$$

If we write U for the $d \times d$ matrix whose rows are the unit vectors $u_0, \ldots, u_{d-1}$, then we see that

7A16 $$\Sigma(\mathsf{P}) = \det(UU^{\mathsf{T}}) > 0,$$

since clearly $\{u_0, \ldots, u_{d-1}\}$ spans $\mathbb{E}^d$.

In view of the use to which we put the Schläfli determinant in Section 7G, it is of interest to see that we can make a recursive definition that does not mention determinants. So, suppose instead that we *define*

7A17 $$\Sigma(\mathsf{P}) := \Sigma(\mathsf{P}^{\boldsymbol{v}}) - \cos^2 \frac{\pi}{p_1} \Sigma(\mathsf{P}^{\boldsymbol{vv}})$$

for $d \geqslant 2$, with $\Sigma(\{1\}) = \Sigma(\{2\}) := 1$, the latter as in (7A15). Recall that the commuting operators $\boldsymbol{v}$ and $\boldsymbol{f}$ give the vertex-figure and facet, respectively, and that $\boldsymbol{\delta v} = \boldsymbol{f \delta}$. In the proof of Theorem 7G1 we need the following

7A18 Proposition *If* P *is a classical polytope, then* $\Sigma(\mathsf{P}^{\boldsymbol{\delta}}) = \Sigma(\mathsf{P})$.

Proof. There is clearly nothing to prove if $d \leqslant 2$, so we suppose that $d \geqslant 3$, and appeal to whatever inductive assumptions we need. Then we have

$$\begin{aligned}\Sigma(\mathsf{P}) &= \Sigma(\mathsf{P}^{\boldsymbol{v}}) - \cos^2 \frac{\pi}{p_1} \Sigma(\mathsf{P}^{\boldsymbol{vv}}) \\ &= \Sigma(\mathsf{P}^{\boldsymbol{vf}}) - \cos^2 \frac{\pi}{p_{d-1}} \Sigma(\mathsf{P}^{\boldsymbol{vff}}) - \cos^2 \frac{\pi}{p_1} \Sigma(\mathsf{P}^{\boldsymbol{vvf}}) \\ &\qquad + \cos^2 \frac{\pi}{p_{d-1}} \cos^2 \frac{\pi}{p_1} \Sigma(\mathsf{P}^{\boldsymbol{vvff}}) \\ &= \Sigma(\mathsf{P}^{\boldsymbol{\delta}}),\end{aligned}$$

by symmetry of the expression. Note that the last term is absent if $d = 3$. □

The Schläfli determinant also has a geometric interpretation. Following [27, Section 7.7], we let $\varphi(\mathsf{P})$ be such that an edge E of P subtends the angle $2\varphi(\mathsf{P})$ at the centre $c(\mathsf{P})$ of P (we usually suppose that $c(\mathsf{P}) = o$). Then we have

7A19 Theorem *If $d \geqslant 2$ and P is a classical regular d-polytope, then*

$$\Sigma(\mathsf{P}) = \Sigma(\mathsf{P}^{\boldsymbol{v}}) \sin^2 \varphi(\mathsf{P}).$$

Proof. We use the idea of absolute determinant introduced in (1C13), noting that $\Sigma(\mathsf{P}^{\boldsymbol{v}}) = \operatorname{Det}(u_1, \ldots, u_{d-1})^2$, and so on. Let v be the initial vertex of P; we may suppose that v is a unit vector. Since $\langle v, u_j \rangle = 0$ for $j \geqslant 1$, and bearing in mind that Det gives volume of the appropriate dimension, it follows that

$$\operatorname{Det}(u_0, \ldots, u_{d-1}) = \mu \operatorname{Det}(u_1, \ldots, u_{d-1}),$$

where μ is the length of the projection of u_0 in direction v. But clearly μ is the same as the length of the projection of v in direction u_0, namely, $\sin \varphi(\mathsf{P})$. This establishes the result. □

There is another angle associated with a polytope P which will later play a useful part. Again following [27, Section 7.7], let $\chi(\mathsf{P})$ be the angle subtended at the centre of P by a radius of a facet F of P. In other words, if the circumradii of P and F are ρ and σ, respectively, then

7A20
$$\sin \chi(\mathsf{P}) = \frac{\sigma}{\rho}.$$

From Theorem 7A19, we deduce

7A21 Theorem *If $d \geqslant 3$ and P is a classical regular d-polytope, then*

$$\sin^2 \chi(\mathsf{P}) = \frac{\Sigma(\mathsf{P})\Sigma(\mathsf{P}^{\boldsymbol{vf}})}{\Sigma(\mathsf{P}^{\boldsymbol{v}})\Sigma(\mathsf{P}^{\boldsymbol{f}})}.$$

Proof. We calculate the edge-length 2λ (say) of P in two different ways. First, from P itself, Theorem 7A19 says that

$$\lambda^2 = \rho^2 \sin^2 \varphi(\mathsf{P}) = \rho^2 \frac{\Sigma(\mathsf{P})}{\Sigma(\mathsf{P}^{\boldsymbol{v}})}.$$

On the other hand, from its facet F the analogous formula gives

$$\lambda^2 = \sigma^2 \frac{\Sigma(\mathsf{P}^{\boldsymbol{f}})}{\Sigma(\mathsf{P}^{\boldsymbol{vf}})}.$$

The result then follows from eliminating λ and using the definition of $\chi(\mathsf{P})$ in (7A20). □

We have a consequence of Theorem 7A21, which follows from the symmetry of its expression; see the notes at the end of the section.

7A22 Corollary *If* P *is a classical regular d-polytope with* $d \geqslant 3$, *then*

$$\chi(\mathsf{P}^{\delta}) = \chi(\mathsf{P}),$$

with P^{δ} *the geometric dual of* P.

7A23 Remark Since $\cos^2 \chi(\mathsf{P}) = \eta_{\boldsymbol{f}}(\mathsf{P})$, Corollary 7A22 is just a special case of Theorem 4D6 expressed in a different way; see the notes at the end of the section.

Petrie Polygons

We end the section with two general observations; the first is about Petrie polygons.

7A24 Theorem *The Petrie polygon of a classical polytope or honeycomb is full-dimensional.*

Proof. This is just a straightforward consequence of Theorem 1D13, making use of Lemma 2D12, because the product of the generating reflexions has no fixed point (in the sphere or whole space, as appropriate). Note that, in the infinite case, the inductive argument shows that a Petrie polygon of a cell (or facet) is already full-dimensional. □

7A25 Remark In later sections, we shall find the types of the Petrie polygons of all the classical polytopes by synthetic methods.

Central symmetry

Finally, we have a useful observation.

7A26 Theorem *If* P *is a finite classical polytope of rank at least* 3 *whose vertex-figure* Q *is centrally symmetric, then*

(a) *the edge-graph of* P *contains diametral planar regular polygons,*

(b) P *itself is centrally symmetric.*

Proof. The central symmetry of Q, taken to be the vertex-figure at the initial vertex v of P, induces a reflexion S, say, with mirror the line through v and the centre c of P. Then $\mathbf{H} := \langle R_0, S\rangle$ is the symmetry group of a regular polygon D with initial vertex v, whose vertices and edges lie in $\mathcal{E}(\mathsf{P})$. Since the two vertices of D adjacent to v are opposite vertices of Q, it follows that D lies in a plane through c. Further, since R_0 and S cannot be conjugate (this is where $\operatorname{rank}\mathsf{P} \geqslant 3$ comes in), we see that D is itself centrally symmetric. Thus to each vertex v of P corresponds a diametrically opposite vertex v', say.

At this point, a little care is needed. If D is a $2r$-gon with r odd, then the central symmetry of D automatically induces the (unique) central symmetry of P. On the contrary, if D is a $4s$-gon for some s, then the central symmetry of D does not induce one of P itself. However, special pleading (looking at individual cases) shows that this happens only for the staurotopes (see the next section), which we know to be centrally symmetric. □

7A27 Remark In practice, we only ever appeal to Theorem 7A26(a).

Notes to Section 7A

1. Schläfli introduced the determinant named after him in [113], together with the recursive formula. The notation $\Sigma(\mathsf{P})$ is confined to this chapter.
2. In [27, 7·74], Coxeter introduces the Schläfli determinant by the recursive formula (7A17), and then observes that an easy induction argument equates it with the actual determinant.
3. Corollary 7A22 was observed by Apollonius of Perga in classical times for the convex regular polyhedra; see the historical remarks in [27, Section 2.9].
4. Further to Remark 7A25, we point out that Coxeter [27] finds the abstract types of the Petrie polygons, but only in the finite cases, by solving certain trigonometric equations.
5. Theorem 7A26(b) is a kind of analogue of the Aleksandrov–Shephard theorem for convex polytopes: *a convex d-polytope with $d \geqslant 3$ which has centrally symmetric facets is itself centrally symmetric.* See [1] for the 3-dimensional case, [118] for the general case, and [72] for a simpler proof.

7B Polytopes in All Dimensions

In this section, we shall introduce the three sequences of finite polytopes which occur in all dimensions, namely, the simplex, staurotope and cube. It may be assumed that the reader is very familiar with these; one main purpose is to describe the realization space of the cube, but we shall also show how to find the geometric types of all the Petrie polygons by direct methods.

Simplex

The group of the abstract *d-simplex* $\mathcal{T}^d = \{3^{d-1}\}$ is $\boldsymbol{G}(\mathcal{T}^d) = \boldsymbol{S}_{d+1}$, the symmetric group on $\mathsf{D} := \{0, \ldots, d\}$ (the reason for this choice will soon become apparent), with $\boldsymbol{r}_j := (j\ j{+}1)$ for $j = 0, \ldots, d-1$. Recall that, in a vector or the like, α^r means a sequence $\alpha, \ldots, \alpha$ of length r. Thus the initial vertex is 0; in fact, the initial j-face has vertex-set $\{0, \ldots, j\}$. It should be clear that the facet and vertex-figure of $\mathcal{T}^d$ are both $(d-1)$-simplices $\mathcal{T}^{d-1}$; we can take the initial vertex of the latter to be 1.

There is an obvious realization, namely, that as the standard simplex whose vertices form the standard basis $\{e_0, \ldots, e_d\}$ of $\mathbb{E}^{d+1}$. However, the centroid of these vertices is not the origin o; a choice for vertices for which it is o consists of the permutations of $(d, (-1)^d)$ in the hyperplane of (1E15), namely,

$$\mathbb{L}^d := \{(\xi_0, \ldots, \xi_d) \in \mathbb{E}^{d+1} \mid \xi_0 + \cdots + \xi_d = 0\};$$

this gives the geometric regular simplex $\mathsf{T}^d = \{3^{d-1}\}$. In a natural way, the symmetric group acts on $\mathbb{L}^d$ as an orthogonal group $\mathbf{A}_d$ in the notation of Table 1E12; just as for the abstract group, we usually denote its elements by permutations (here of the coordinate vectors).

Either directly or by calculating the product of the group generators, we see that the Petrie polygon of the d-simplex has length $d+1$. The direct method uses induction on d: a Petrie polygon of the initial facet has successsive vertices $e_0, \ldots, e_{d-1}$ and, since the Petrie polygon of $\mathcal{T}^d$ itself must now leave that facet, it can only go to e_d (and then back again to e_0, of course). From the group, we have the product

$$\boldsymbol{r}_{d-1}\boldsymbol{r}_{d-2}\cdots\boldsymbol{r}_0 = (d{-}1\ d)(d{-}2\ d{-}1)\cdots(0\ 1) = (0\ 1\ \ldots\ d),$$

of period $d+1$.

However, neither calculation gives the fine Schläfli symbol for the Petrie polygon. If we take it to be

$$\left\{\frac{d+1}{t_1, \ldots, t_m}\right\},$$

with $1 \leqslant t_1 < \cdots < t_m \leqslant \frac{1}{2}(d+1)$, then $m = \lfloor \frac{1}{2}(d+1) \rfloor$ because of its full-dimensionality, which shows the polygon to be $\{\frac{d+1}{1,2,\ldots,m}\}$. Observe that, as we might expect, there is a component $\{2\}$ when d is odd.

We have already described the realization space of $\mathcal{T}^d$ in Example 3D1. It is, as to be expected, very simple; we briefly recall the details. The layer vector is $\Lambda = \Lambda(\mathcal{T}^d) = (1, d)$, as is the dimension vector $D = (1, d)$. This is more than enough information to show that the cosine matrix of $\mathcal{T}^d$ is

7B1 $$\begin{bmatrix} 1 & 1 \\ 1 & -\frac{1}{d} \end{bmatrix}.$$

As we saw in Section 4G, there is only one non-trivial product.

Staurotope

As we know, the *d-staurotope* (see Section 2F and its notes) $\mathcal{X}^d = \{3^{d-2}, 4\}$ has a single pure faithful realization $\mathsf{X}^d = \{3^{d-2}, 4\}$ in $\mathbb{E}^d$. We can take the generators of its symmetry group to be

7B2 $$R_j = \begin{cases} (j{+}1\ j{+}2), & \text{for } j = 0, \ldots, d-2, \\ \overline{d}, & \text{for } j = d-1, \end{cases}$$

where the permutations are of the cartesian coordinates of $\mathbb{E}^d$ (or of the basis vectors $e_1, \ldots, e_d$) and

$$\overline{d}\colon x \mapsto (\xi_1, \ldots, \xi_{d-1}, -\xi_d),$$

where $x = (\xi_1, \ldots, \xi_d)$. We shall often find it convenient to employ such shorthand for changing signs of coordinates. We easily find that X^d has vertices $\pm e_1, \ldots, \pm e_d$. Moreover, the facet of $\mathcal{X}^d$ is a $(d-1)$-simplex $\mathcal{T}^{d-1}$, while the vertex-figure (geometrically with vertices $\pm e_2, \ldots, \pm e_d$, at the initial vertex e_1) is a $(d-1)$-staurotope $\mathcal{X}^{d-1}$.

Recall that we saw in Proposition 2F20 that the d-staurotope $\mathcal{X}^d$ is $(d-1)$-collapsible.

As with the simplex, we can use geometry or algebra to find the length of the Petrie polygon. By either method, the Petrie polygon can be taken to have successive vertices $e_1, \ldots, e_d, -e_1, \ldots, -e_d$, and so is of length $2d$. Note that $e_1, \ldots, e_d$ give d successive vertices of a Petrie polygon of the initial facet, while $e_2, \ldots, e_d, -e_1$ give those of a Petrie polygon of the adjacent facet.

Again, though, for the fine Schläfli symbol, we need a little more work. Assume the type to be

$$\left\{\frac{2d}{t_1, \ldots, t_m}\right\},$$

with $1 \leqslant t_1 < \cdots < t_m \leqslant d$; then, as before, $m = \lfloor \frac{1}{2}(d+1) \rfloor$ from the full-dimensionality. Further, d steps along the Petrie polygon will give its central symmetry $x \mapsto -x$, and hence that of X^d itself. We conclude that each t_i must be odd, from which follows that the type is $\{\frac{2d}{1,3,\ldots,r}\}$, with $r = d-1$ or d as d is even or odd.

We have already described the realization space of $\mathcal{X}^d$ in Example 3D4. Here, the layer vector is $\Lambda = (1, 2(d-1), 1)$ and the dimension vector is $D = (1, d-1, d)$. The staurotope realization is $\mathsf{X} = \mathsf{X}^d$ itself, with cosine vector $(1, 0, -1)$. There is one other pure realization, namely, the (small) $(d-1)$-simplex S (realizing $\mathcal{X}^d/2$) with cosine vector $(1, -\frac{1}{d-1}, 1)$ (as a realization of $\mathcal{X}^d$, of course). This ties in with the fact that Theorem 2F14 implies that $\mathcal{X}^d$ is $(d-1)$-collapsible: adapting the notation of (2F12), the normal closure $\mathbf{N}_{d-1}^{+}$ of R_{d-1} consists of all changes of sign of the coordinates of $\mathbb{E}^d$. Thus the cosine matrix of $\mathcal{X}^d$ is

7B3 $$\begin{bmatrix} 1 & 1 & 1 \\ 1 & -\frac{1}{d-1} & 1 \\ 1 & 0 & -1 \end{bmatrix}$$

The first two rows are therefore the cosine vectors of the realizations of $\mathcal{X}^d/2$.

Cube

We now move on to the *d-cube* $\mathcal{C}^d = \{4, 3^{d-2}\}$, which is, of course, the dual of the staurotope $\mathcal{X}^d$. Before we look at its geometry, it is instructive to see how we can construct it abstractly. Indeed, in the terminology of Section 5C, we have

$$\mathcal{C}^d = \mathbf{2}^{\{3^{d-2}\}};$$

that is, we let the group S_d of the (abstract) $(d-1)$-simplex act by permutations in the natural way on the product $\mathrm{C}_2^d = \mathrm{C}_2 \times \cdots \times \mathrm{C}_2$ of d copies of the cyclic group C_2 of order 2.

Geometrically, we can take the jth copy C_2 acting on $\mathbb{E}^d$ to be $\overline{\jmath}$, which we recall changes the sign of the jth coordinate. In other words, we can take the generators of the symmetry group of the geometric $\{4, 3^{d-2}\}$ to be

7B4 $$R_j = \begin{cases} \overline{1}, & \text{for } j = 0, \\ (j\ j{+}1), & \text{for } j = 1, \ldots, d-1. \end{cases}$$

We see that $\{4, 3^{d-2}\}$ has vertices $(\pm 1, \ldots, \pm 1)$, with all 2^d changes of sign; for the initial j-face, the last $d - j$ signs are all positive.

The Petrie polygon of $\{4, 3^{d-2}\}$ has the same fine Schläfli symbol as that of $\{3^{d-2}, 4\}$. As with the staurotope $\mathcal{X}^d$, we can find its length geometrically or algebraically; we then apply full-dimensionality as before.

There are various relationships among the three polytopes that we have considered in this section. However, we shall postpone discussion of these until Section 7C, when we can treat them all together.

7B5

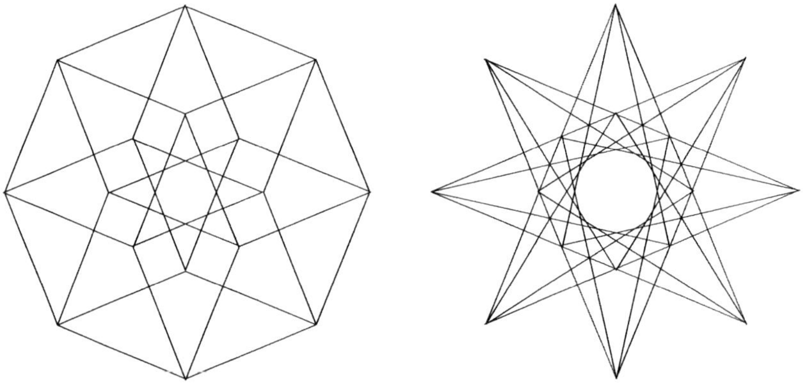

The two 4-dimensional realizations of $\{4, 3, 3\}$

In Figure 7B5, we depict symmetric projections of the two 4-dimensional realizations of the 4-cube $\{4, 3, 3\}$; these appeared on the dust-jacket of [99].

Note that they have the same vertices, and that the edges of the second are the long diagonals of the cubical facets of the first.

We shall see that, while we cannot find closed-form expressions for the pure cosine vectors of realizations of $\mathcal{C}^d$, nevertheless we can explicitly construct their multiplication table. Our investigation also demonstrates how powerful generating functions can be for encapsulating complicated expressions involving binomial coefficients.

We described the pure realizations of $\mathcal{C}^d$ in [75] (see also [99, Section 5B]). Including the trivial one, $\mathcal{C}^d$ has $d+1$ pure realizations, of dimensions $\binom{d}{r}$ for $r = 0, \ldots, d$, with corresponding cosine vectors $\Gamma_{dr} = (\gamma_{dr0}, \ldots, \gamma_{drn})$. Here, $\Gamma_{d0} = (1^{d+1})$ gives the trivial realization as the henogon $\{1\}$, Γ_{d1} is that of the standard d-cube $\{4, 3^{d-2}\}$, and Γ_{dd} – given by $\gamma_{ddk} = (-1)^k$ – is that of the 1-collapse of $\mathcal{C}^d$ onto the digon $\{2\}$.

7B6 Remark The 1-collapsibility of $\mathcal{C}^d$ is, in a sense, the dual property to the $(d-1)$-collapsibility of $\mathcal{X}^d$ shown in Proposition 2F20.

The layer vector and dimension vector both take the same form:

7B7 $$\Lambda = D = \left(1, d, \binom{d}{2}, \binom{d}{3}, \ldots, \binom{d}{d-1}, 1\right).$$

The matrices of cosine vectors of $\mathcal{C}^d$ have an interesting form; we illustrate this by giving the cases $d \leqslant 5$ (the first two are for $\{1\}$ and $\{2\}$).

7B8

$$\begin{bmatrix} 1 \end{bmatrix} \quad \begin{bmatrix} 1 & 1 \\ 1 & -1 \end{bmatrix} \quad \begin{bmatrix} 1 & 1 & 1 \\ 1 & 0 & -1 \\ 1 & -1 & 1 \end{bmatrix} \quad \begin{bmatrix} 1 & 1 & 1 & 1 \\ 1 & \frac{1}{3} & -\frac{1}{3} & -1 \\ 1 & -\frac{1}{3} & -\frac{1}{3} & 1 \\ 1 & -1 & 1 & -1 \end{bmatrix}$$

$$\begin{bmatrix} 1 & 1 & 1 & 1 & 1 \\ 1 & \frac{1}{2} & 0 & -\frac{1}{2} & -1 \\ 1 & 0 & -\frac{1}{3} & 0 & 1 \\ 1 & -\frac{1}{2} & 0 & -\frac{1}{2} & 1 \\ 1 & -1 & 1 & -1 & 1 \end{bmatrix} \quad \begin{bmatrix} 1 & 1 & 1 & 1 & 1 & 1 \\ 1 & \frac{3}{5} & \frac{1}{5} & -\frac{1}{5} & -\frac{3}{5} & -1 \\ 1 & \frac{1}{5} & -\frac{1}{5} & -\frac{1}{5} & \frac{1}{5} & 1 \\ 1 & -\frac{1}{5} & -\frac{1}{5} & \frac{1}{5} & \frac{1}{5} & -1 \\ 1 & -\frac{3}{5} & \frac{1}{5} & \frac{1}{5} & -\frac{3}{5} & 1 \\ 1 & -1 & 1 & -1 & 1 & -1 \end{bmatrix}$$

Various features of the matrices are easily explained. Multiplication by the digon $\{2\}$ (interchanging two pure realizations of the same dimension) implies that $\gamma_{d,d-r,k} = (-1)^k \gamma_{drk}$ for each r and k – numbering of rows and columns begins with 0. Similarly, the symmetry or antisymmetry of the kth and $(d-k)$th diagonals (acccording as the realization is one of $\mathcal{C}^d/2$ or not – that is, whether or not opposite vertices of $\mathcal{C}^d$ are identified) explains why $\gamma_{d,r,d-k} = (-1)^r \gamma_{drk}$

for each r and k. However, there is also the symmetry by reflexion in the main diagonal of the matrix, which follows directly from

7B9 Theorem *The kth entry γ_{drk} in the cosine vector Γ_{dr} of the rth realization of the d-cube of dimension $\binom{d}{r}$ is given by*

$$\gamma_{drk} = \binom{d}{r}^{-1} \sum_{j \geqslant 0} (-1)^j \binom{k}{j} \binom{d-k}{r-j}.$$

In particular, $\gamma_{drk} = \gamma_{dkr}$ for all d, r and k.

Proof. The ambient space of the rth pure realization C_r^d (say) is spanned by the symmetric r-tensors

$$f_{j(1)\dots j(r)} := e_{j(1)} \otimes \dots \otimes e_{j(r)}$$

on *distinct* basis elements $e_1, \dots, e_d$ of $\mathbb{E}^d$, so that $1 \leqslant j(1) < \dots < j(r) \leqslant d$. Thus, before normalization, the initial vertex is their sum, and the typical kth diagonal goes from this to the vertex obtained by changing the signs of $e_1, \dots, e_k$. Thus there is a contribution ± 1 to the (r, k)-entry of the (unnormalized) inner product vector from a (possible) change of sign of $f_{j(1)\dots j(r)}$ according as $\operatorname{card}\{i \mid j(i) \leqslant k\}$ is even or odd. Counting these signs for each $j \leqslant k$, and dividing by the squared circumradius $\binom{d}{r}$ to normalize, leads to the formula stated.

The last claim follows from the easily verified

$$\binom{d}{r}^{-1} \binom{k}{j} \binom{d-k}{r-j} = \binom{d}{k}^{-1} \binom{r}{j} \binom{d-r}{k-j},$$

which completes the proof. □

7B10 Remark The symmetry of the matrix of cosine vectors provides a nice illustration of Theorem 3C11 in case of cubes; it also follows from (7B7). As a consequence, if $d = 2r$, then alternate entries of the middle row c_{dr} are 0. This tells us that alternate vertices of this pure realization of $\mathcal{C}^d$ fall into two mutually orthogonal subspaces. For example, the vertices of the 6-dimensional realization of $\mathcal{C}^4$ (actually of $\mathcal{C}^4/2$) are those of two tetrahedra in orthogonal 3-spaces.

A straightforward argument shows the following; we use sans serif letters to denote indeterminates.

7B11 Lemma *The inner sum in the formula of Theorem 7B9 is the coefficient of t^r in $(1-\mathsf{t})^k(1+\mathsf{t})^{d-k}$. In other words,*

$$\sum_{r \geqslant 0} \binom{d}{r} \gamma_{drk} \mathsf{t}^r = (1-\mathsf{t})^k (1+\mathsf{t})^{d-k}.$$

Proof. Indeed, taking out a term t^j and summing first over r, we have

$$\begin{aligned}\sum_{r\geqslant 0}\sum_{j\geqslant 0}(-1)^j\binom{k}{j}\binom{d-k}{r-j}\mathsf{t}^r &= \sum_{j\geqslant 0}(-1)^j\binom{k}{j}\mathsf{t}^j(1+\mathsf{t})^{d-k}\\ &= (1-\mathsf{t})^k(1+\mathsf{t})^{d-k},\end{aligned}$$

as claimed. □

7B12 Remark Observe that, either directly or using the relation

$$\begin{aligned}\binom{d}{r}^{-1}(1-\mathsf{t})^k(1+\mathsf{t})^{d-k} &= \frac{r}{d}\mathsf{t}\cdot\binom{d-1}{r-1}^{-1}(1-\mathsf{t})^k(1+\mathsf{t})^{d-k-1}\\ &\quad+\frac{d-r}{d}\cdot\binom{d-1}{r}^{-1}(1-\mathsf{t})^k(1+\mathsf{t})^{d-k-1},\end{aligned}$$

we obtain the recurrence

$$\gamma_{drk} = \frac{r}{d}\gamma_{d-1,r-1,k} + \frac{d-r}{d}\gamma_{d-1,rk}$$

for $0 \leqslant k < d$. We originally found the cosine matrices for the d-cube up to $d = 4$ directly, and then used this recurrence to calculate that for $d = 5$.

While the expression of Theorem 7B9 does not help very much in calculating the entries γ_{drk}, it is the starting point for the remaining analysis. Thus our next step shows how to express a general cosine vector of a realization of $\mathcal{C}^d$ in terms of the c_{dt}.

Let us write C_d for the $(d+1) \times (d+1)$ matrix with rows $\Gamma_{d0}, \ldots, \Gamma_{dn}$, and B_d for its inverse, so that, if $x = (\xi_0, \ldots, \xi_d)$ is an appropriate cosine vector, then $x = \eta_0\Gamma_{d0} + \cdots + \eta_d\Gamma_{dn}$, where $y = (\eta_0, \ldots, \eta_d)$ is given by $y = xB_d$.

7B13 Theorem *The inverse matrix $B_d = (\beta_{dij})$ of C_d is given by*

$$\beta_{dij} = \frac{1}{2^d}\binom{d}{i}\binom{d}{j}\gamma_{dij}.$$

Proof. This is a straightforward calculation: Theorem 7B9 and Lemma 7B11 imply that

$$\begin{aligned}\sum_{i\geqslant 0}\mathsf{t}^i\sum_{j\geqslant 0}\beta_{dij}\gamma_{djk} &= \sum_{j\geqslant 0}\frac{1}{2^d}(1-\mathsf{t})^j(1+\mathsf{t})^{d-j}\cdot\sum_{h\geqslant 0}(-1)^h\binom{k}{h}\binom{d-k}{j-h}\\ &= \frac{1}{2^d}\sum_{h\geqslant 0}(-1)^h\binom{k}{h}(1-\mathsf{t})^h(1+\mathsf{t})^{k-h}\\ &\quad\times\sum_{j\geqslant 0}\binom{d-k}{j-h}(1-\mathsf{t})^{j-h}(1+\mathsf{t})^{d-k-j+h}\\ &= \frac{1}{2^d}\sum_{h\geqslant 0}(-1)^h\binom{k}{h}(1-\mathsf{t})^h(1+\mathsf{t})^{k-h}\cdot 2^{d-k}\\ &= \frac{1}{2^d}(2\mathsf{t})^k\cdot 2^{d-k} = \mathsf{t}^k.\end{aligned}$$

That is, the (i,k)-entry in $B_d C_d$ is the coefficient of t^i in the polynomial t^k, namely, the Kronecker delta δ_{ik}; therefore, indeed $B_d = (C_d)^{-1}$. □

7B14 Remark We could have employed Theorem 3J1 here, since we know that $w = 1$ for all the pure realizations of $\mathcal{C}^d$. In other words, if $\Gamma = (\gamma_0, \ldots, \gamma_d)$ is a general cosine vector in the realization domain $\mathcal{N}$ of $\mathcal{C}^d$, then the coefficient ξ_r in the expression $\Gamma = \xi_0 \Gamma_{d0} + \cdots + \xi_d \Gamma_{dd}$ is given by

$$\xi_r = 2^{-d}\binom{d}{r}\sum_{s=0}^{d}\binom{d}{s}\gamma_{drs}\gamma_s.$$

The symmetry $\gamma_{drs} = \gamma_{dsr}$ then yields Theorem 7B13. However, we have kept the direct calculation, as an illustration of the fact that we do not always have to use the most sophisticated techniques.

According to Theorem 4D6, for each $r \geqslant 2$ the facet of the realization C_r^d of $\mathcal{C}^d$ should be centred. It is therefore a useful illustration of the theorem to check this. As before, we use generating functions. From the expression of Lemma 7B11, we have

$$\begin{aligned}
\sum_{r\geqslant 0}\binom{d}{r}\Bigl(\sum_{k\geqslant 0}\binom{d-1}{k}\gamma_{drk}\Bigr)\mathsf{t}^r &= \sum_{k\geqslant 0}\binom{d-1}{k}\Bigl(\sum_{r\geqslant 0}\binom{d}{r}\gamma_{drk}\mathsf{t}^r\Bigr)\\
&= \sum_{k\geqslant 0}\binom{d-1}{k}(1-\mathsf{t})^k(1+\mathsf{t})^{d-k}\\
&= (1+\mathsf{t})\sum_{k\geqslant 0}\binom{d-1}{k}(1-\mathsf{t})^k(1+\mathsf{t})^{d-k-1}\\
&= 2^{d-1}(1+\mathsf{t}).
\end{aligned}$$

This tells us that

$$\eta_{\boldsymbol{f}}(\mathsf{C}_r^d) = \begin{cases} 1, & \text{if } r = 0,\\ \frac{1}{d}, & \text{if } r = 1,\\ 0, & \text{if } r \geqslant 2; \end{cases}$$

which is rather more than we asked for.

We complete the discussion (in the spirit of Remark 7B14) by finding the multiplication table for the pure realizations of $\mathcal{C}^d$. We do this purely as an exercise, since – as we have said – we are more interested in products as an aid to determining realization domains rather than for their own sake. We have

7B15 Theorem *In the multiplication table for the d-cube $\mathcal{C}^d$, the coefficient α_{drst} in the expression $\Gamma_{dr}\Gamma_{ds} - \sum_t \alpha_{drst}\Gamma_{dt}$ is given by*

$$\alpha_{drst} = \begin{cases} \dbinom{d}{r}^{-1}\dbinom{s}{u}\dbinom{d-s}{r-u}, & \textit{if } r+s-t = 2u \textit{ is even},\\[2ex] 0, & \textit{otherwise.} \end{cases}$$

Proof. The coefficients α_{drst} are given by

$$\begin{aligned}\alpha_{drst} &= \sum_{k\geqslant 0} \gamma_{drk}\gamma_{dsk}\beta_{dtk} \\ &= \frac{1}{2^d}\sum_{k\geqslant 0}\binom{d}{t}\binom{d}{k}\gamma_{drk}\gamma_{dsk}\gamma_{dtk}.\end{aligned}$$

Of course, this is symmetric in r and s as it must be, but observe that $\binom{d}{r}\binom{d}{s}\alpha_{drst}$ is actually symmetric in r, s, t.

This points towards what we should do. With indeterminates s, t and calculations like those previously using Lemma 7B11, we have

$$\begin{aligned}&\sum_{r,t\geqslant 0}\binom{d}{r}\alpha_{drst}\mathsf{s}^r\mathsf{t}^t \\ &\quad = \frac{1}{2^d}\sum_{k\geqslant 0}\sum_{i\geqslant 0}(-1)^i\binom{s}{i}\binom{d-s}{k-i}(1-\mathsf{s})^k(1+\mathsf{s})^{d-k}(1-\mathsf{t})^k(1+\mathsf{t})^{d-k} \\ &\quad = \frac{1}{2^d}\sum_{i\geqslant 0}(-1)^i\binom{s}{i}(1-\mathsf{s})^i(1+\mathsf{s})^{s-i}(1-\mathsf{t})^i(1+\mathsf{t})^{s-i}\big(2(1+\mathsf{st})\big)^{d-s} \\ &\quad = \frac{1}{2^d}\big(2(\mathsf{s}+\mathsf{t})\big)^s\big(2(1+\mathsf{st})\big)^{d-s} \\ &\quad = (\mathsf{s}+\mathsf{t})^s(1+\mathsf{st})^{d-s}.\end{aligned}$$

Here, we have written alternate exponents in the first expression as $k = i+(k-i)$ and $d-k = (s-i)+(d-s-k+i)$, and summed first over k. We also use

$$\begin{aligned}(1+\mathsf{s})(1+\mathsf{t}) + (1-\mathsf{s})(1-\mathsf{t}) &= 2(1+\mathsf{st}), \\ (1+\mathsf{s})(1+\mathsf{t}) - (1-\mathsf{s})(1-\mathsf{t}) &= 2(\mathsf{s}+\mathsf{t}).\end{aligned}$$

There is only one term in $\mathsf{s}^r\mathsf{t}^t$ in the ultimate polynomial; if we have a contribution $\mathsf{s}^i\mathsf{t}^{s-i}$ from the first factor and $(\mathsf{st})^j$ from the second, then $i+j=r$ and $(s-i)+j=t$ imply that $r+s-t=2i$ and $r-s+t=2j$ are even. We immediately deduce the claimed result. □

7B16 Remark Observe particularly that the usual convention for binomial coefficients holds, namely, $\binom{m}{k} = 0$ unless $0 \leqslant k \leqslant m$.

7B17 Remark Expanding the binomial coefficients shows that this expression for α_{drst} is symmetric in r and s, though superficially it may not seem to be.

We end the section with an illustration of Theorem 7B15.

7B18 Example Taking $d = 5$, $r = 3$ and $s = 2$, the coefficients we are interested in are

$$\binom{5}{3}^{-1}\binom{2}{u}\binom{3}{3-u};$$

the possible values of u are $0, 1, 2$, giving $t = 5, 3, 1$, respectively. Reading off the cosine vectors $\mathit{\Gamma}_{5t}$ from (7B8), we then have

$$\begin{aligned}\tfrac{3}{10}\mathit{\Gamma}_{51} + \tfrac{3}{5}\mathit{\Gamma}_{53} + \tfrac{1}{10}\mathit{\Gamma}_{55} &= \tfrac{3}{10}(1, \tfrac{3}{5}, \tfrac{1}{5}, -\tfrac{1}{5}, -\tfrac{3}{5}, -1) \\ &\quad + \tfrac{3}{5}(1, -\tfrac{1}{5}, -\tfrac{1}{5}, \tfrac{1}{5}, \tfrac{1}{5}, -1) + \tfrac{1}{10}(1, -1, 1, -1, 1, -1) \\ &= (1, -\tfrac{1}{25}, \tfrac{1}{25}, -\tfrac{1}{25}, \tfrac{1}{25}, -1) \\ &= (1, -\tfrac{1}{5}, -\tfrac{1}{5}, \tfrac{1}{5}, \tfrac{1}{5}, -1)(1, \tfrac{1}{5}, -\tfrac{1}{5}, -\tfrac{1}{5}, \tfrac{1}{5}, 1) \\ &= \mathit{\Gamma}_{53}\mathit{\Gamma}_{52},\end{aligned}$$

as Theorem 7B15 claims.

7C The 24-Cell

There are several different ways of approaching the 24-cell $\{3, 4, 3\}$. Since one of them uses quaternions, it is appropriate to begin with the quaternion groups involved. As we said in Section 1K, we would show how to obtain the expressions for these groups that we presented there. And to do this, we first need to look more closely at the related regular polyhedra in $\mathbb{E}^3$, and various connexions among them.

We begin with the 3-cube $\{4, 3\}$. The edge-graph here is bipartite; as we have observed, the d-cube $\{4, 3^{d-2}\}$ is 1-collapsible. Thus, when we apply the halving operation $\boldsymbol{\eta}$, we halve the number of vertices. Indeed, we have

$$\{4, 3\}^{\eta} = \{3, 3\}$$

as abstract polytopes, and so the same relationship holds for the geometric ones. Taking $\{4, 3\}$ with the standard vertices all $(\pm 1, \pm 1, \pm 1)$ as in the previous section, we see that we can take the vertices of $\{3, 3\}$ to be all $(\pm 1, \pm 1, \pm 1)$ with an even number of minus signs.

Further, since $\{3, 3\}$ is self-dual, its group $\langle \boldsymbol{s}_0, \boldsymbol{s}_1, \boldsymbol{s}_2 \rangle$ admits a twist $\boldsymbol{t}$ such that $\boldsymbol{t}\boldsymbol{s}_j\boldsymbol{t} = \boldsymbol{s}_{2-j}$ for each j. Then the operation

$$(\boldsymbol{s}_0, \boldsymbol{s}_1, \boldsymbol{s}_2, \boldsymbol{t}) \mapsto (\boldsymbol{s}_1, \boldsymbol{s}_0, \boldsymbol{t}) =: (\boldsymbol{r}_0, \boldsymbol{r}_1, \boldsymbol{r}_2)$$

yields the group of the octahedron $\{3, 4\}$. When we perform the corresponding geometric operation, we find that the vertices of $\{3, 4\}$ are the mid-points of the edges of $\{3, 3\}$.

In terms of Wythoff's general construction, we have

7C1 $\{3, 4\}$ = 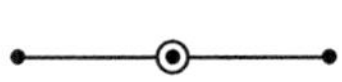 = 

Observe that this exhibits twisting as well.

We now see how to lift the rotation groups $\mathbf{A}_3^+ = [3, 3]^+$ and $\mathbf{C}_3^+ = [3, 4]^+$ into the corresponding subgroups of quaternions. The former has half-turns about each coordinate axis and 3-fold rotations about the axes in directions

$(\pm 1, \pm 1, \pm 1)$. Together with ± 1, we obtain exactly $\mathbf{T}$. The latter group has, in addition, 4-fold rotations about the coordinate axes and half-turns about axes whose directions are permutations of $(\pm 1, \pm 1, 0)$. These lift to quaternions $\nu_0 + \nu_1 \mathbf{i} + \nu_2 \mathbf{j} + \nu_3 \mathbf{k}$, where $(\nu_0, \nu_1, \nu_2, \nu_3)$ are permutations of

$$\tfrac{1}{\sqrt{2}}(\pm 1, \pm 1, 0, 0),$$

forming the set which we denoted $\mathbf{U}$; the 2-fold rotations are those that lift with first coordinate 0. Thus $\mathbf{O} = \mathbf{T} \cup \mathbf{U}$, with $\mathbf{T}$ a subgroup of index 2 in $\mathbf{O}$ and $\mathbf{U}$ its other coset in $\mathbf{O}$.

We are now ready to describe how to construct $\mathsf{F} := \{3, 4, 3\}$ (the notation – here and below – is chosen with Section 7J in mind). There are three ways available. Two are geometric and, indeed, dual. The third is more algebraic, and yields extra information.

First, consider the convex polytope obtained from the 4-staurotope $\mathsf{X} := \{3, 3, 4\}$ by complete truncation at its vertices. By this, we mean that the vertices of the truncate are the mid-points of the edges of the staurotope. The (convex) polytope F thus obtained has two kinds of facet. First, we have the complete truncates of the tetrahedral facets of $\{3, 3, 4\}$, which we have just seen to be octahedra. Second, replacing each original vertex of $\{3, 3, 4\}$ is its (narrow) octahedral vertex-figure $\{3, 4\}$. In other words, our new polytope F has $16 + 8 = 24$ regular octahedral facets $\{3, 4\}$. On the other hand, at each vertex of F we have 4 octahedra arising from the facets of $\{3, 3, 4\}$ and two (opposite ones) arising from the two original vertices of $\{3, 3, 4\}$ of the edge of F which contains the new vertex. That is, the vertex-figure has 6 tetragonal faces, and so must be a 3-cube $\{4, 3\}$. We conclude that F itself is regular, of type $\{3, 4, 3\}$. Observe that, if we start from the staurotope with vertices $\pm\sqrt{2} e_j$ for $j = 1, \ldots, 4$, then – identified with quaternions – the vertex-set of F is $\mathbf{U}$.

In terms of diagrams, we have the analogue of (7C1):

7C2 $\{3, 4, 3\}$ =

For the dual construction, we begin with the 4-cube C in the standard form of the previous Section 7B, with vertices $(\pm 1, \pm 1, \pm 1, \pm 1)$ and edge-length 2. The point $2e_1 = (2, 0, 0, 0)$ is at distance 2 from each of the vertices $(1, \pm 1, \pm 1, \pm 1)$, and the mid-point of the digon joining $2e_1$ to $2e_2$ is the centre of the tetragonal face $\{4\}$ of the cube $\{4, 3, 3\}$ with vertices $(1, 1, \pm 1, \pm 1)$. We see at once that these six points are the vertices of a regular octahedron with edge-length 2. Each of the 24 tetragonal faces of the cube gives rise to such an octahedron, so that we have obtained a convex 4-polytope with 24 regular octahedral facets $\{3, 4\}$. Observe that we have placed a pyramid with apex $\pm 2e_j$ and base a 3-cube over each of the 8 corresponding cubical facets of $\{4, 3, 3\}$. It is clear that the vertex-figure at each of the new vertices is a cube $\{4, 3\}$; while it is not quite so transparent, the vertex-figures at the old vertices are again bounded by six tetragons, and so must be cubes $\{4, 3\}$. Thus we have another copy of

$\mathsf{F} = \{3,4,3\}$; halving the vertices and regarding them as quaternions gives the set $\mathbf{T}$.

7C3 Remark This last comment, showing that the vertex-set as constructed was $2\mathbf{T}$, actually obviates any need for special pleading about the symmetry: the group structure of $\mathbf{T}$ implies that what is true geometrically about one vertex must be true of them all. In a similar way, since $\mathbf{U}$ is the other coset of $\mathbf{T}$ in $\mathbf{O}$, exactly the same considerations apply to the previous construction.

We are now poised to describe the symmetry group of the 24-cell. First, we know that the group $\mathbf{F}_4$ of Table 1E12 is the universal group $[3,4,3]$, whose order we found in Section 1H to be 1152. This accords with our calculation $24 \cdot 48 = 1152$ from the group $[4,3]$ of the vertex-figure and the number 24 of vertices. Taking $\mathbf{T}$ to be the vertex-set of $\{3,4,3\}$, we can now collect together various pieces of information:

- the hyperplane reflexion $\mathbf{x} \mapsto -\widetilde{\mathbf{x}}$ is a symmetry;
- multiplication on left or right by an element of $\mathbf{T}$ is a symmetry;
- multiplication on left or right by an element of $\mathbf{U}$ interchanges $\mathbf{T}$ and $\mathbf{U}$.

Counting the symmetries of different kinds, we conclude that we have proved

7C4 Theorem *The common symmetry group of the* 24*-cells with vertex-sets* $\mathbf{T}$ *or* $\mathbf{U}$ *is* $(\mathbf{O}/\mathbf{T};\mathbf{O}/\mathbf{T})^*$, *consisting of all mappings of the form* $\mathbf{x} \mapsto \widetilde{\mathbf{g}}\mathbf{x}\mathbf{h}$ *and* $\mathbf{x} \mapsto \widetilde{\mathbf{g}}\widetilde{\mathbf{x}}\mathbf{h}$, *with* $\mathbf{g}, \mathbf{h} \in \mathbf{O}$ *such that* $\widetilde{\mathbf{g}}\mathbf{h} \in \mathbf{T}$.

We leave to the reader the minor task of ensuring that we do indeed obtain 1152 symmetries in this way; recall the identification $(-1)\mathbf{x}(-1) = \mathbf{x}$ (and further see the notes at the end of the section).

Thus far, we have not mentioned the fact that $\mathbf{T}$ has a subgroup $\mathbf{V}$ of index 3, which as a set of coordinate vectors comprises $\{\pm e_1, \pm e_2, \pm e_3, \pm e_4\}$. The various symmetries lead us at once to the following compounds of copies of one polytope with its vertices a subset of the vertices of another:

- $\{3,4,3\}$ contains three copies of $\{3,3,4\}$;
- $\{3,4,3\}$ contains three copies of $\{4,3,3\}$, with each vertex of the former belonging to two copies of the latter;
- $\{4,3,3\}$ contains two copies of $\{3,3,4\}$.

In the last case, it is clear that the inscribed staurotope cannot have its full symmetry group as a subgroup of that of the cube. However, this leads on neatly to our algebraic approach, using the general Wythoff construction.

7C5 $\{3,4,3\} \quad =$ 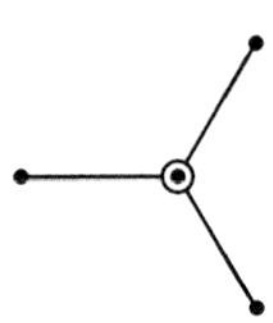

The diagram of $\mathbf{B}_d$ has an obvious twist, which shows that it is a subgroup of index 2 in $\mathbf{C}_d$. This even holds, of course, for $d = 3$, when $\mathbf{B}_3 = \mathbf{A}_3$ (which was what we saw in (7C1)). The case $d = 4$, though, is special, in that the diagram may be acted on by the dihedral or symmetric group $\mathbf{A}_2 \cong \boldsymbol{S}_3$. This gives the alternative diagram of (7C5), and thus yet another way of proving

7C6 Proposition *The group* $[3, 3, 4]$ *is a subgroup of* $[3, 4, 3]$ *of index* 3.

7C7 Remark The proposition also illustrates Remark 5B18.

7C8

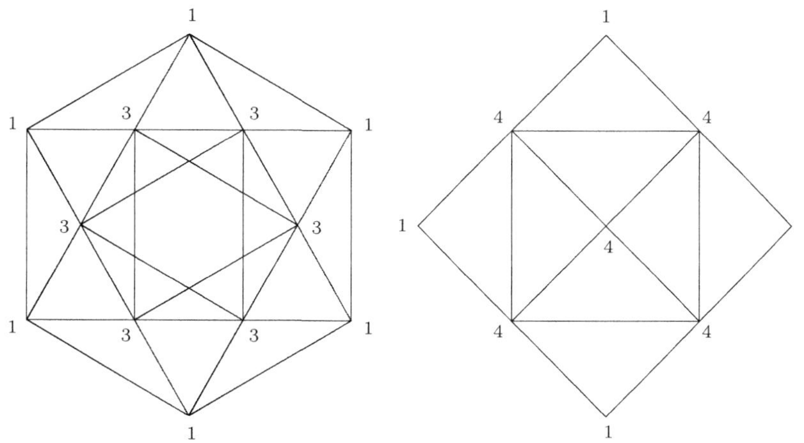

Two projections of the 24-cell $\{3, 4, 3\}$

From the second geometric construction, we easily see how the vertices of $\{3, 4, 3\}$ fall into layers from an initial one, say $(2, 0, 0, 0)$. At height 1 we have a cube (the broad vertex-figure) with edge-length 2. The central layer is an octahedron of edge-length $2\sqrt{2}$. We then have another cube, finishing with the antipodal vertex $(-2, 0, 0, 0)$. These layers are the horizontal ones in both pictures of Figure 7C8.

The slanted lines in the second picture in Figure 7C8 show successively an octahedral facet $\{3, 4\}$, a central cuboctahedron, and then the opposite facet. The labels give the numbers of vertices falling together in the projections: in the first, we have trigons of edge-length $2\sqrt{2}$ (edge-figures, that is, vertex-figures of the cubical vertex-figures), and in the second we have tetragons (namely, central squares of octahedra) of edge-length 2 on the periphery and edge-length $2\sqrt{2}$ at the centre.

7C9 Remark The second picture also shows rather clearly the alternative construction of $\{3, 4, 3\}$, erecting pyramids on the facets of the 4-cube $\{4, 3, 3\}$.

Petrie Polygon

We next look at the Petrie polygon of $\{3, 4, 3\}$. Theorem 7A26 tells us that the edge-graph of $\{3, 4, 3\}$ contains diametral planar polygons, which is a consequence of its centrally symmetric vertex-figure; as we said earlier, this observation goes back to van Oss (see also [27, §14.6]). In fact, it follows directly from the construction that these planar diametral polygons are hexagonal; one such in our second construction has (successive) vertices

$$2e_1,\ (1,1,1,1),\ (-1,1,1,1),\ -2e_1,\ -(1,1,1,1),\ -(-1,1,1,1).$$

This is illustrated by the outer hexagon in the first picture of Figure 7C8.

7C10

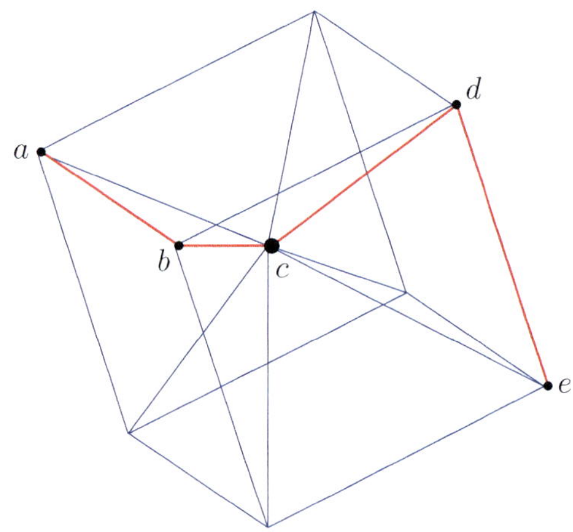

Now it is easy to see directly that four steps along a Petrie polygon a, b, c, d, e results in going two steps a, c, e along a diametral hexagon; see Figure 7C10, which for clarity shows only half of each octahedral facet of $\{3, 4, 3\}$ containing the vertex c. (Note that, as they should be, a, b, c, d, and b, c, d, e are four successive vertices of Petrie polygons of the two octahedral facets which share the 2-face with vertices b, c, d.) This implies at once that the Petrie polygon is dodecagonal, of the form

$$\left\{\frac{12}{t_1, t_2}\right\},$$

with $1 \leqslant t_1 < t_2 < 6$ (with strict inequality $t_2 < 6$ because the Petrie polygon is 4-dimensional). However, since two steps along it yields a planar hexagon $\{6\}$, we see that $t_1, t_2 \equiv \pm 1 \pmod 6$, and we conclude at once that $t_1 = 1,\ t_2 = 5$. That is,

7C11 Proposition *The Petrie polygon of the 24-cell* $\{3, 4, 3\}$ *is of type* $\{\frac{12}{1,5}\}$.

Projections

We have just seen that the Petrie polygon of $\{3, 4, 3\}$ is of type $\{\frac{12}{1,5}\}$. In fact, the vertices of $\{3, 4, 3\}$ are those of two complementary Petrie polygons (in several ways). Writing the Petrie polygon explicitly as a blend, namely, $\{\frac{12}{1,5}\} = \{12\} \# \{\frac{12}{5}\}$, suggests a natural orthogonal projection of the 24-cell into the plane of one of these two component polygons $\{12\}$ or $\{\frac{12}{5}\}$. Indeed, it is reasonable to expect (and is actually the case) that, if we project one Petrie polygon into a dodecagon $\{12\}$, then the complementary Petrie polygon is projected into a dodecagram $\{\frac{12}{5}\}$.

We can employ induced cosine vectors, as described in Section 4D, to work out the sizes of these planar polygons. Alternate vertices of a Petrie polygon are those of a diametral planar hexagon, and so the first picture of Figure 7C8 shows that the cosine vector of the Petrie polygon induced from that of $\{3,4,3\}$ will be $(1,\gamma_1,\gamma_1,\gamma_2,\gamma_3,\gamma_3,\gamma_4)$. Thus, since $\Gamma(\{12\}) = (1,\frac{\sqrt{3}}{2},\frac{1}{2},0,-\frac{1}{2},-\frac{\sqrt{3}}{2},-1)$, the radius ρ_1 of the projection of the Petrie polygon of a general realization on its component $\{12\}$ is given by

$$\begin{aligned}\rho_1^2 &= \tfrac{2}{12}\big(1+2\tfrac{\sqrt{3}}{2}\gamma_1+2\tfrac{1}{2}\gamma_1+\cdots+2(-\tfrac{\sqrt{3}}{2})\gamma_3+(-1)\gamma_4\big)\\ &= \tfrac{1}{6}\big(1+(\sqrt{3}+1)\gamma_1-(\sqrt{3}+1)\gamma_3-\gamma_4\big),\end{aligned}$$

in terms of the cosine vector $\Gamma = (1,\gamma_1,\gamma_2,\gamma_3,\gamma_4)$.

So, from $\{3,4,3\}$ itself, we find that

$$\rho_1^2 = \tfrac{1}{6}\big(1+(\sqrt{3}+1)\tfrac{1}{2}-(\sqrt{3}+1)(-\tfrac{1}{2})-(-1)\big) = \tfrac{1}{6}(3+\sqrt{3}).$$

The analogous calculation for the projection on the component $\{\frac{12}{5}\}$ gives $\rho_2^2 = \frac{1}{6}(3-\sqrt{3})$; note that $\rho_1^2+\rho_2^2 = 1$ as a check on our working.

Of course, knowing these radii alone does not lead directly to the projection; however, Figure 7C12 almost draws itself.

7C12

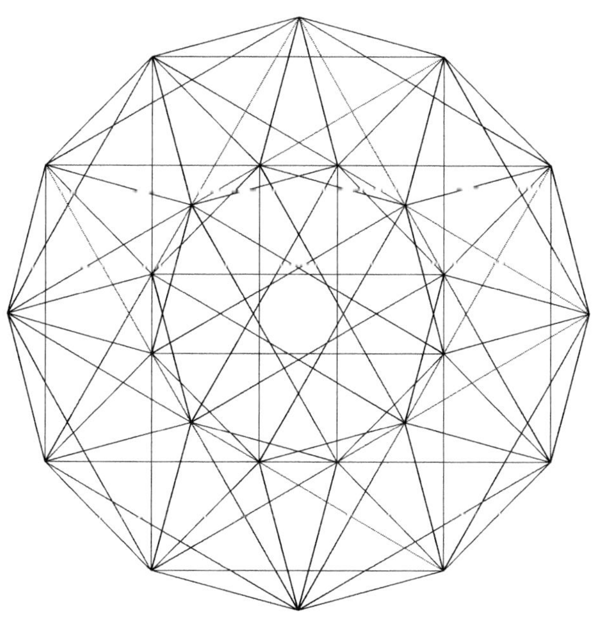

The dodecagonal projection of the 24-cell

The two different kinds of octahedral facet of $\{3,4,3\}$ can be picked out in the projection, although the whole structure is rather less easy to visualize than in the two projections of Figure 7C8.

To draw Figure 7C12, we can appeal to the geometry of $\{3,4,3\}$. Three successive edge of the Petrie polygon are edges of an octahedral facet $\{3,4\}$. Moreover, we note that

$$\frac{5\pi}{6} = \frac{\pi}{2} + \frac{\pi}{3};$$

that is, the internal angle of the dodecagon $\{12\}$ is the sum of those of the tetragon $\{4\}$ and trigon $\{3\}$. Thus, we can start the drawing with

7C13

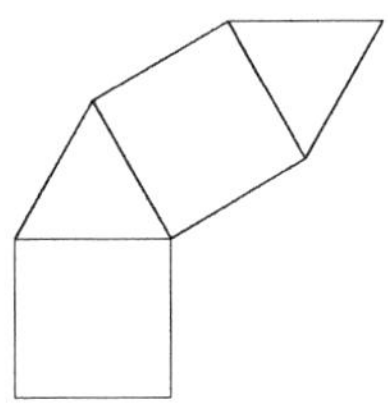

We already recognize the outline of an octahedron in Figure 7C13.

Realization space

We end the section by finding the realization space of the (abstract) 24-cell $\mathcal{F}^4 = \{3,4,3\}$. Surprisingly, perhaps, it is not too difficult to construct the cosine matrix, and thus the multiplication table for the pure realizations. As we shall see, $\mathsf{F}_0 = \{1\}$, F_1 corresponds to the 2-collapse of $\{3,4,3\}$ onto $\{3\}$, F_2 is the 9-dimensional realization of $\{3,4,3\}/2$ which we shall identify below, $\mathsf{F}_3 = \{3,4,3\}$ itself (the standard convex 4-polytope, so that F_3 is what we initially labelled F), and F_4 is an 8-dimensional faithful realization.

7C14 Remark If $(R_0, \ldots, R_3)$ is the generatrix of $\{3,4,3\}$, then, adapting the notation of Theorem 2F14 concerning the 2-collapse of $\{3,4,3\}$, the normal closure $\mathbf{N}_2^+$ of the subgroup $\langle R_2, R_3 \rangle$ of $\mathbf{G}(\{3,4,3\}) = \mathbf{F}_4$ is $\mathbf{B}_4 = [3^{1,1,1}]$. In fact, as is clear from the symmetry between the 24-cell and its dual, there are two copies of $\mathbf{B}_4$ in $\mathbf{F}_4$.

We originally described the realization space of the 24-cell in [99, Section 5B], and the cosine matrix in [85]. However, here we proceed differently, to illustrate an approach that could prove useful in other contexts. We pretend that, while we know a lot about the combinatorial structure of $\{3,4,3\}$, we know nothing about its geometry. To be specific, we suppose that we have the data that a computer algebra program such as GAP [50] can easily provide. So, imagine that we are told that all diagonals of $\{3,4,3\}$ are symmetric (and hence $w(\mathsf{P}) = 1$ for each pure realization $\mathsf{P} \in \{3,4,3\}$), that it is centrally symmetric with layer vector $\Lambda = (1,8,6,8,1)$ and that, in terms of the cosine vector $\Gamma = (1, \gamma_1, \gamma_2, \gamma_3, \gamma_4)$ of a general realization, the induced cosine vector of its vertex-figure $\{3,4,3\}^v =$

$\{4,3\}$ is $\Gamma^v = (1, \gamma_1, \gamma_2, \gamma_3)$. The layers of $\{3,4,3\}$ are depicted in Figure 7C8; the first of the two projections also shows how we get the induced cosine vector of the vertex-figure.

For the pure components of the small simplex realization S (that is, of the quotient $\{3,4,3\}/2$), we have $\gamma_3 = \gamma_1$ and $\gamma_4 = 1$, so that, with $\gamma_1 = \alpha$ and $\gamma_2 = \beta$, from the layer equation of Theorem 3C7 we see that $1 + 8\alpha + 3\beta = 0$. Then Theorem 3C14 reads

$$8\alpha^2 = 1 + 3\alpha + 3\beta + \alpha = -4\alpha \implies \alpha = -\tfrac{1}{2} \text{ or } 0.$$

For the pure components of the staurotope realization X, we have $\gamma_2 = 0$, $\gamma_3 = -\gamma_1$ and $\gamma_4 = -1$, so that, with $\gamma_1 = \gamma$, Theorem 3C14 reads

$$8\gamma^2 = 1 + 3\gamma + 3 \cdot 0 + 1(-\gamma) = 1 + 2\gamma \implies \gamma = \tfrac{1}{2} \text{ or } -\tfrac{1}{4}.$$

We conclude from this discussion the following.

7C15 Theorem *The cosine matrix of* $\{3,4,3\}$ *is*

$$\begin{bmatrix} 1 & 1 & 1 & 1 & 1 \\ 1 & -\frac{1}{2} & 1 & -\frac{1}{2} & 1 \\ 1 & 0 & -\frac{1}{3} & 0 & 1 \\ 1 & \frac{1}{2} & 0 & -\frac{1}{2} & -1 \\ 1 & -\frac{1}{4} & 0 & \frac{1}{4} & -1 \end{bmatrix},$$

with layer and dimension vectors

$$\Lambda = (1,8,6,8,1), \quad D = (1,2,9,4,8).$$

Proof. We still have to find the dimensions of these pure realizations; observe that we have not assumed anything about them in the foregoing. Indeed, we now use Theorem 3F5 to calculate that the dimension vector D is as claimed. □

Various well-known things about the pure realizations $\mathsf{F}_0 = \{1\}, \mathsf{F}_1, \ldots, \mathsf{F}_4$ now follow. First, as we have said, F_1 realizes the 2-collapse of $\{3,4,3\}$ onto its face $\{3\}$; this can be thought of as the action of the group $[3,4,3]$ on its subgroup $[3,3,4]$ of index 3, which identifies the 8 vertices of each inscribed 4-staurotope. Second, F_2 is the (unique) pure faithful realization of the quotient $\{3,4,3\}/2$; from the cosine vector, it follows that the vertices fall into three tetrahedra in mutually orthogonal subspaces of $\mathbb{E}^9$, with each tetrahedron corresponding to identification of antipodal vertices of the staurotopes just alluded to; see also [79] or [99, Section 14A], where F_2 is the first in a certain family of locally projective polytopes. Of course, $\mathsf{F}_3 = \{3,4,3\}$, the classical convex realization.

Since $\mathsf{F}_4 = \mathsf{F}_1 \otimes \mathsf{F}_3$, its geometry is determined as well; the expresssion of [99, 5B27] already suggests this, although it does not fully reflect the symmetry. If we use complex coordinates, and work in $\mathbb{C}^4$ rather than $\mathbb{E}^8$, then we can write down the vertices in a way that immediately illustrates the product structure, as well as the subgroup relationship $[3,3,4] < [3,4,3]$. So, we have

7C16
$$\begin{array}{lll}(\pm 1, \pm 1, 0, 0), & \omega(\pm 1, 0, \pm 1, 0), & \overline{\omega}(\pm 1, 0, 0, \pm 1), \\ (0, 0, \pm 1, \pm 1), & \omega(0, \pm 1, 0, \pm 1), & \overline{\omega}(0, \pm 1, \pm 1, 0).\end{array}$$

Here, $\omega = \frac{1}{2}(-1 + i\sqrt{3})$ is a cube root of 1, and $\overline{\omega} = \omega^2$ is its complex conjugate. As we can see from (7C16), the component F_1 shows up in the mapping

7C17
$$z \mapsto \omega(\zeta_1, \zeta_4, \zeta_2, \zeta_3),$$

which permutes the three copies of $\{3, 3, 4\}$ inscribed in $\{3, 4, 3\}$.

It is easy to verify the Λ-orthogonality relations of Theorem 3F5. Further, F_2 is (up to scaling) the non-trivial component of $\mathsf{F}_3 \otimes \mathsf{F}_3$, since

$$\Gamma_3{}^2 = \tfrac{1}{4}\Gamma_0 + \tfrac{3}{4}\Gamma_2$$

(the coefficient of Γ_0 also comes from Theorem 3F5).

7C18 Remark Since six steps along a Petrie polygon of $\{3, 4, 3\}$ gives its central symmetry, we see that we can write $\{3, 4, 3\}/2 = \{3, 4, 3 : 6\}$.

Notes to Section 7C

1. Further to Theorem 7C4, note that we obtain $\frac{1}{2}(24{\cdot}24) = 288$ rotational symmetries $\mathbf{x} \mapsto \widetilde{\mathbf{g}}\mathbf{x}\mathbf{h}$ with $\mathbf{g}, \mathbf{h} \in \mathbf{T}$. Counting the three other possibilities (with $\mathbf{x}$ replaced by $\widetilde{\mathbf{x}}$ or $\mathbf{T}$ replaced by $\mathbf{U}$) – each with the same number of associated symmetries – gives group order $4 \cdot 288 = 1152$, as required.
2. In the first edition of [27], Figure 8.2A (that is, our Figure 7C12) was drawn incorrectly. In [27, Figure 8.2B], Coxeter has depicted a 4-cube and a 4-staurotope inscribed in the 24-cell.

7D Pentagonal Polyhedra

In this section, we show how to construct the two regular (convex) polyhedra with pentagonal symmetries. Of course, we could also work directly from the reflexion group $\mathbf{G}_3 = [3, 5]$; what we did in Section 1E will readily produce a suitable generatrix. However, the geometry of the polyhedra feeds into the binary icosahedral group $\mathbf{I}$ of (1K13). In addition, we find the Petrie polygons of the polyhedra, and describe their realization domains.

Construction

We have already remarked in Proposition 2F20 that the octahedron $\mathcal{X}^3$ is 2-collapsible, so that its edge-graph admits a face-cyclic orientation. Indeed, in the proposition, we collapsed the pair $\pm e_j$ of vertices of the geometric X^3 onto e_j for $j = 1, 2, 3$. If $\lambda, \mu > 0$ and we define $v_j := \lambda e_j + \mu e_{j+1}$, with indices taken modulo 3 (so that $e_0 = e_3$), and lift back to X^3, then we obtain 12 vertices of the form $\pm\lambda e_j \pm \mu e_{j+1}$ on the edges of the scaled copy $(\lambda+\mu)\mathsf{X}^3$ of the octahedron.

Assuming that $\lambda > \mu$, the convex hull I of these 12 points has two edge-lengths ρ and σ, given by

$$\begin{aligned}\rho^2 &= \lambda^2 + (\lambda-\mu)^2 + \mu^2 = 2(\lambda^2 - \lambda\mu + \mu^2),\\ \sigma^2 &= 4\mu^2.\end{aligned}$$

Here, ρ is the edge-length of the trigons inscribed in the faces of $(\lambda+\mu)\mathsf{X}^3$, while σ is the distance between (for example) the two points $\lambda e_1 \pm \mu e_2$. When $\lambda = \tau := \frac{1}{2}(1+\sqrt{5})$ (the golden number) and $\mu = 1$, we see that

$$\rho^2 = \tau^2 + \tau^{-2} + 1 = 4 = \sigma^2,$$

since $\tau - 1 = \tau^{-1}$, and so all the edge-lengths are equal. Each original vertex of X^3 has been replaced by two trigons, and we then see that each vertex of I belongs to 5 trigons: $1+2$ from the vertices of X^3 of the corresponding edge, and 2 from the trigons inscribed in its faces; this is depicted in Figure 7D1. The vertex-set of I thus consists of all *even* (that is, cyclic) permutations of $(\pm\tau, \pm1, 0)$, with all changes of sign.

7D1

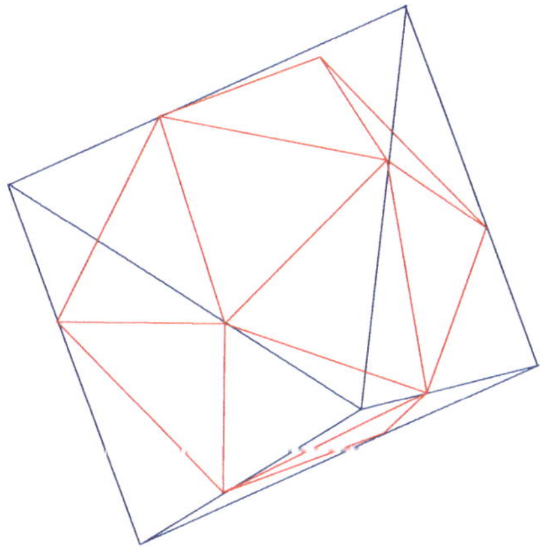

Icosahedron constructed from octahedron

To see that I is actually regular – the regular *icosahedron* – we should check that we have all the required symmetries. In fact, it is enough to verify that the vertices of I adjacent to a given one are those of a regular pentagon. Taking $\tau e_1 + e_2 = (\tau, 1, 0)$ as the given vertex, those adjacent to it are (in cyclic order)

$$(\tau,-1,0),\ (1,0,\tau),\ (0,\tau,1),\ (0,\tau,-1),\ (1,0,-\tau);$$

this is a planar cyclic pentagon with equal edge-lengths and equal diagonals, and so is regular. As a check, note that the diagonals of this pentagon have squared length

$$\tau^2 + \tau^4 + 1 = 4\tau^2,$$

since $\tau + 1 = \tau^2$. Thus I has 5-fold rotational symmetry about the axis through $(\tau, 1, 0)$, and the reflexion in the plane with normal e_3 gives the full dihedral symmetry.

For reference, we describe various symmetries of I.

7D2 Theorem *Among the symmetries of the regular icosahedron* $\mathsf{I} = \{3, 5\}$ *are*

(a) 5-*fold rotations about axes* $(\tau, \pm 1, 0)$ *(even permutations),*
(b) 3-*fold rotations about axes* $(1, \pm 1, \pm 1)$ *and* $(\tau^{-1}, \pm\tau, 0)$ *(even permutations),*
(c) 2-*fold rotations about axes* e_j $(j = 1, 2, 3)$ *and* $(\tau, \pm\tau^{-1}, \pm 1)$ *(even permutations),*
(d) *reflexions in planes with unit normals* e_j $(j = 1, 2, 3)$ *and* $\frac{1}{2}(\tau, \pm\tau^{-1}, \pm 1)$ *(even permutations).*

There are also rotatory-reflexions that are not listed.

In listing the 3-fold rotational axes of the icosahedron I, we have almost listed the vertices of the dual *dodecahedron* D. These are all $(\pm 1, \pm 1, \pm 1)$ and all even permutations of $(\pm\tau^{-1}, \pm\tau, 0)$.

7D3 Remark It is worth noting here that the 30 points $\pm 2e_j$ (for $j = 1, 2, 3$) and even permutations of $(\pm\tau, \pm\tau^{-1}, \pm 1)$ (that is, the scaled normals to the reflexion planes) are the vertices of the *icosidodecahedron*; these points are (up to scaling) the common mid-points of the 30 edges of the icosahedron or of the dodecahedron.

Petrie Polygons

We can easily find the Petrie polygons of the icosahedron and dodecahedron using Theorem 2E6, which states that

$$\{3, q \mid \cdot, r\} = \{3, q : 2r\}.$$

We recall that r is the length of the 3-hole, while $2r$ is the length of the Petrie polygon; at this stage, we work on the abstract level. For the icosahedron $\{3, 5\}$, the 3-hole coincides with the 2-hole, and by Theorem 2D14 (which is obvious in this case) the 2-hole of $\{3, 5\}$ is a pentagon $\{5\}$. In other words, $\{3, 5\} = \{3, 5 : 10\}$, so that the Petrie polygon of $\{3, 5\}$ is a decagon $\{10\}$.

Geometrically, of course, Theorem 1D13 tells us that the Petrie polygon is full-dimensional, with a component $\{2\}$. Since two steps along a Petrie polygon of $\{3, 5\}$ gives a pentagon $\{5\}$, we see immediately that

7D4 Proposition *The Petrie polygons of both the regular icosahedron and the dodecahedron are of type* $\{\frac{10}{1,5}\}$.

Realization Spaces

Again, the realization domain of the icosahedron $\{3,5\}$ is known from [85]. However, we found it there by assuming that we knew the two 3-dimensional realizations $\{3,5\}$ and $\{3,\frac{5}{2}\}$. Suppose, to the contrary, that we just assume the basic combinatorics of $\{3,5\}$, in particular, that all its diagonals are symmetric, that its layer vector is $\Lambda = (1,5,5,1)$, and that it is centrally symmetric.

Bearing in mind Remark 3C9, we then see that the central quotient $\{3,5\}/2$ is neighbourly, and so there is a single non-trivial pure component of the small simplex realization S, with cosine vector $\Gamma_1 = (1,-\frac{1}{5},-\frac{1}{5},1)$. This gives the hemi-icosahedron $\mathsf{I}_1 = \{3,\frac{5}{1,2} : \frac{5}{1,2}\} \cong \{3,5:5\}$ (compare Remark 7C18 for the notation), whose dimension is $d_1 = 5$.

Applying the Λ-orthogonality of Theorem 3F5 then shows that any other cosine vector $\Gamma = (1,\gamma_1,\gamma_2,\gamma_3)$ must satisfy

$$\begin{aligned} 1 + 5\gamma_1 + 5\gamma_2 + \gamma_3 &= 0, \\ 1 - \gamma_1 - \gamma_2 + \gamma_3 &= 0; \end{aligned}$$

these lead at once to $\gamma_2 = -\gamma_1$ and $\gamma_3 = -1$, so that $\Gamma = (1,\gamma,-\gamma,-1)$ for some $\gamma = \gamma_1$ (of course, we can appeal to Remark 3C9 again, since we are here looking at components of the staurotope realization X). The induced cosine vector of the vertex-figure $\{3,5\}^v = \{5\}$ is $\Gamma^v = (1,\gamma_1,\gamma_2)$, so that Theorem 3C14 yields

$$5\gamma_1^2 = 1 + 2\gamma_1 + 2\gamma_2 \implies 5\gamma^2 = 1 \implies \gamma = \pm\tfrac{1}{\sqrt{5}}.$$

In other words, we have proved

7D5 Theorem *An inner product matrix of* $\{3,5\}$ *is*

$$\begin{bmatrix} 1 & 1 & 1 & 1 \\ 5 & -1 & -1 & 5 \\ \sqrt{5} & 1 & 1 & \sqrt{5} \\ \sqrt{5} & -1 & 1 & -\sqrt{5} \end{bmatrix},$$

with corresponding layer and dimension vectors

$$\Lambda = (1,5,5,1), \quad D = (1,5,3,3).$$

Note particularly that we have found the cosine matrix of $\{3,5\}$ without knowing anything in advance about its geometry.

7D6 Remark From now on, we shall frequently aid readability and save space by writing inner product matrices instead of cosine matrices. To obtain the latter from the former, just divide by the leading entries.

If we write $\mathsf{I}_0 = \{1\}$, $\mathsf{I}_1 = \{3, \frac{5}{1,2} : \frac{5}{1,2}\}$ (as before), $\mathsf{I}_2 = \{3, 5\}$ (the usual convex icosahedron), and $\mathsf{I}_3 = \{3, \frac{5}{2}\}$ (the great icosahedron), and Γ_j for the cosine vector of I_j, then we have (for example)

$$\Gamma_2{}^2 = \Gamma_3{}^2 = \tfrac{1}{3}\Gamma_0 + \tfrac{2}{3}\Gamma_1,$$
$$\Gamma_2\Gamma_3 = \Gamma_1.$$

For the last, we know that $\mathsf{I}_2 \otimes \mathsf{I}_3$ is centred and realizes $\{3, 5\}/2$; thus it can only be I_1.

In preparation for Theorem 7D8, we list the values of η_f for each of these three pure realizations I_j:

7D7
$$\eta_f(\mathsf{I}_j) = \begin{cases} \dfrac{1}{5}, & \text{if } j = 1, \\ \dfrac{\tau^3}{3\sqrt{5}}, & \text{if } j = 2, \\ \dfrac{\tau^{-3}}{3\sqrt{5}}, & \text{if } j = 3. \end{cases}$$

There remains the dodecahedron $\mathcal{D}^3 := \{5, 3\}$. For this, we shall prove

7D8 Theorem *An inner product matrix of* $\{5, 3\}$ *is*

$$\begin{bmatrix} 1 & 1 & 1 & 1 & 1 & 1 \\ 6 & -4 & 1 & 1 & -4 & 6 \\ 3 & 1 & -1 & -1 & 1 & 3 \\ 3 & \sqrt{5} & 1 & -1 & -\sqrt{5} & -3 \\ 3 & -\sqrt{5} & 1 & -1 & \sqrt{5} & -3 \\ 2 & 0 & -1 & 1 & 0 & -2 \end{bmatrix},$$

with corresponding layer and dimension vectors

$$\Lambda = (1, 3, 6, 6, 3, 1), \quad D = (1, 4, 5, 3, 3, 4).$$

Proof. Our proof will illustrate various of the techniques at our disposal, rather than being the most straightforward. The five non-trivial diagonal classes of $\mathcal{D}^3 = \{5, 3\}$ are symmetric, so that $w(\mathsf{P}) = 1$ for each pure realization P. The layer vector is $\Lambda = (1, 3, 6, 6, 3, 1)$, and the induced cosine vector of the vertex-figure $\{3\}$ is $\Gamma^v = (1, \gamma_2)$ (in terms of the general cosine vector $\Gamma = (1, \gamma_1, \ldots, \gamma_5)$ of $\mathcal{D}^3$).

The hemi-dodecahedron $\{5, 3\}/2 \cong \{5, 3 : 5\}$ has two non-trivial diagonal classes, so that its cosine vectors are of the form $\Gamma = (1, \alpha, \beta, \beta, \alpha, 1)$. Moreover, the induced cosine vector of its facet $\{5\}$ is $(1, \alpha, \beta)$. Combining Theorem 4D6 with the layer equation gives

$$\begin{aligned} 1 + 3\alpha + 6\beta &= 0, \\ 1 + 2\alpha + 2\beta &= 5\eta_{\boldsymbol{f}}(\mathsf{D}_j), \end{aligned}$$

with $\eta_{\boldsymbol{f}}(\mathsf{D}_1) = 0$ and $\eta_{\boldsymbol{f}}(\mathsf{D}_2) = \eta_{\boldsymbol{f}}(\mathsf{I}_1)$ given by (7D7). This gives the cosine vectors

$$\begin{aligned} 6\Gamma_1 &= (6, -4, 1, 1, -4, 6), \\ 3\Gamma_2 &= (3, 1, -1, -1, 1, 3); \end{aligned}$$

we can then use Theorem 3F5 to find the corresponding dimensions $d_1 = 4$ and $d_2 = 5$. (Note that we already knew that $d_2 = 5$, but we have not needed to appeal to this.) We shall describe these realizations D_1 and D_2 below.

For the components of the staurotope realization X, where the cosine vector takes the form $(1, \alpha, \beta, -\beta, -\alpha, -1)$, we can begin by applying Theorem 4D6 again. Here, we need to feed in a little more information than we did with $\{5, 3 : 5\}$, because the layer equation tells us nothing in the centrally symmetric case. However, we have Theorem 3C14 in reserve, and so now we solve (for the dual of I_2 above)

$$\begin{aligned} 1 + 2\alpha + 2\beta &= 5 \cdot \frac{\tau^3}{3\sqrt{5}}, \\ 1 + 2\beta &= 3\alpha^2. \end{aligned}$$

Eliminating β gives

$$9\alpha^2 + 6\alpha = \tau^3\sqrt{5} \implies \alpha = \frac{\sqrt{5}}{3} \text{ or } -\frac{\tau^3}{3}.$$

Since $\tau^3 > 3$, the second value does not yield a cosine vector, and so we obtain

$$3\Gamma_3 = (3, \sqrt{5}, 1, -1, -\sqrt{5}, -3).$$

We can now use the Λ-orthogonality of Theorem 3F5; thus the remaining cosine vectors satisfy

$$\begin{aligned} 1 + \sqrt{5}\alpha + 2\beta &= 0, \\ 1 + 2\alpha + 2\beta &= 5\eta_{\boldsymbol{f}}(\mathsf{D}_j) \end{aligned}$$

for $j = 4, 5$, with $\eta_{\boldsymbol{f}}(\mathsf{D}_4) = \eta_{\boldsymbol{f}}(\mathsf{I}_3)$ and $\eta_{\boldsymbol{f}}(\mathsf{D}_5) = 0$. We then easily obtain

$$\begin{aligned} 3\Gamma_4 &= (3, -\sqrt{5}, 1, -1, \sqrt{5}, -3), \\ 2\Gamma_5 &= (2, 0, -1, 1, 0, -2). \end{aligned}$$

We can easily calculate the dimensions $d_3 = d_4 = 3$ and $d_5 = 4$ using Theorem 3F5, although we already know d_3 and d_4 from the analysis of the icosahedron $\{3, 5\}$, and $d_5 = 10 - d_3 - d_4$. In any event, this now completes the proof. □

7D9 Remark The product structure of the realization domain of $\{5,3\}$ is more complicated than that of $\{3,5\}$. In parallel to that, we do have

$$\Gamma_3{}^2 = \Gamma_4{}^2 = \tfrac{1}{3}\Gamma_0 + \tfrac{2}{3}\Gamma_2.$$

However, now we have

$$\Gamma_3\Gamma_4 = \tfrac{2}{3}\Gamma_1 + \tfrac{1}{3}\Gamma_2,$$

as we can see directly or by using Theorem 3J1, which is quite a different relation from the equivalent one for the icosahedron. (Observe, though, that $\mathsf{D}_3 \otimes \mathsf{D}_4$ now does have full dimension $4+5 = 3 \cdot 3$.) Let us give just one more example, which is very easy to see:

$$\Gamma_2\Gamma_5 = \tfrac{1}{2}(\Gamma_3 + \Gamma_4).$$

Notice, though, that $\dim(\mathsf{D}_3 \# \mathsf{D}_4) = 3+3 = 6$, as opposed to $\dim \mathsf{D}_2 \cdot \dim \mathsf{D}_5 = 5 \cdot 4 = 20$.

We give a brief description of these realizations, except for the 3-dimensional convex dodecahedron and great stellated dodecahedron, with which we assume the reader is familiar (for a picture of the former, see Figure 7J6). We begin by pointing out that there is a mistake in [99, Section 5B (p. 137)]. The simplest coordinates of the vertices of the 4-dimensional faithful realization D_5 can be taken to be all permutations of $(1,-1,0,0,0)$ in the hyperplane $\mathbb{L}^4$ of (1E15); these are not the mid-points of the edges of a 4-simplex T^4 and their opposites, but instead the vertices of the difference body $\Delta\mathsf{T}^4 = \mathsf{T}^4 - \mathsf{T}^4$ (Minkowski sum) of the simplex.

However, we can use a familiar trick by looking at D_5 first; see Remark 3D8. Replacing the coordinates of the vertices of D_5 by their absolute values gives the 10 permutations of $(1,1,0,0,0)$. This vertex-set is not centred, but if we centre it, to obtain (after scaling as well) all permutations of $(3,3,-2,-2,-2)$, we now have the vertices of D_1; these are the mid-points of the edges of a regular 4-simplex.

We last have D_2. Its vertices are half of those of the form $(1,1,1,-1,-1,-1)$ in the hyperplane $\mathbb{L}^5$. All 20 points of this form are the vertices of the polytope 0_{22}, which is a precursor of the Gosset–Elte family that will play a prominent rôle in Chapter 13. In layers from a vertex, we have the product $\mathsf{T}^2 \times \mathsf{T}^2$ of two trigons, then its negative, and last the opposite vertex. If we pick in the second product a trigon which looks like the blend $\mathsf{T}^2 \# \mathsf{T}^2$, then this forms the vertex-figure of the realization D_2; the complementary set of six vertices in the first product then completes the vertex-set.

7E The 600-Cell

In this section and the next, we construct the pentagonal 4-polytopes $\{3,3,5\}$ (the 600-cell) and $\{5,3,3\}$ (the 120-cell), and briefly describe them. However, the intimate relationship between these polytopes and the quaternion group **I** of order 120 (the binary icosahedral group) behoves us to recall the latter first.

Binary Icosahedral Group

We said in Section 1K that **I** consists of the quaternions $\nu_0 + \nu_1\mathbf{i} + \nu_2\mathbf{j} + \nu_3\mathbf{k}$, with the vectors $(\nu_0, \nu_1, \nu_2, \nu_3) \in \mathbb{E}^4$ comprising all even permutations of

$$(\pm 1, 0, 0, 0), \quad \tfrac{1}{2}(\pm 1, \pm 1, \pm 1, \pm 1), \quad \tfrac{1}{2}(\pm\tau, \pm 1, \pm\tau^{-1}, 0),$$

where $\tau = \frac{1}{2}(1 + \sqrt{5})$ is the golden number. To verify that these do provide appropriate coordinates, we refer to the description in the previous Section 7D of the rotational symmetries of the icosahedron $\{3, 5\}$. The identity and half-turns about the coordinate axes of $\mathbb{E}^3$ lift to the first block of quaternions; the remaining half-turns lift to the pure imaginary quaternions (that is, with $\nu_0 = 0$) in the third block. The third-turns about the diameters of the dodecahedron lift to the quaternions in the second block and those in the third with $\nu_0 = \pm\frac{1}{2}$. The rotations of period 5 lift to the remaining quaternions with $\nu_0 = \pm\frac{1}{2}\tau$ or $\pm\frac{1}{2}\tau^{-1}$, all in the third block. We need no more than this to see that **I** is indeed a group (it would be tedious in the extreme to have to check this directly).

Geometric Construction

The starting point for the geometric construction of $\{3, 3, 5\}$ is the 24-cell $\mathsf{F} = \{3, 4, 3\}$ with vertices all permutations of $(\pm 1, \pm 1, 0, 0)$. We know that $\{3, 4, 3\}$ is 2-collapsible (as an abstract polytope); indeed, if we take the trigonal face of $\{3, 4, 3\}$ with vertices $a_j := e_1 + e_{j+2}$ for $j = 0, 1, 2$, then the 2-collapse can be given by

$$\begin{aligned}
(\pm 1, \pm 1, 0, 0), (0, 0, \pm 1, \pm 1) &\mapsto a_0 = (1, 1, 0, 0),\\
(\pm 1, 0, \pm 1, 0), (0, \pm 1, 0, \pm 1) &\mapsto a_1 = (1, 0, 1, 0),\\
(\pm 1, 0, 0, \pm 1), (0, \pm 1, \pm 1, 0) &\mapsto a_2 = (1, 0, 0, 1);
\end{aligned}$$

compare also (7C16). The three sets of vertices are, of course, those of the three 4-staurotopes with the vertices of the 24-cell.

We now utilize the fact that the graph of F is face-cyclic. We take new vertices $b_j := a_j + \tau^{-1}a_{j+1}$ for $j = 0, 1, 2$ (with indices modulo 3), and lift back to $\mathbb{E}^4$, in essentially the same way that we did in Section 7D. (Note that we lift in all possible ways *before* applying the various sign changes.) In other words, in each trigonal face of $\{3, 4, 3\}$ (scaled up by τ) we inscribe a smaller trigon with its vertices on the original edges. We thus obtain 96 points, which are all the even permutations of $(\pm\tau, \pm 1, \pm\tau^{-1}, 0)$.

The resulting (convex) 4-polytope is called the *snub* 24-cell, and is denoted $s\{3, 4, 3\}$. It has 24 regular icosahedral facets, corresponding to the 24 original facets of F (compare the construction of the icosahedron in Section 7D), and a cluster of 5 regular tetrahedral facets corresponding to each original vertex of F, giving $24 \cdot 5 = 120$ tetrahedra in all. For example, corresponding to the vertex $(1, 1, 0, 0)$ of $\{3, 4, 3\}$ is the tetrahedron with vertices $(\tau, 1, \pm\tau^{-1}, 0)$ and $(1, \tau, 0, \pm\tau^{-1})$, with the four vertices $(\tau, \tau^{-1}, 0, \pm 1)$ and $(\tau^{-1}, \tau, \pm 1, 0)$ each joined to three of the first four to form four additional tetrahedra. (Note that only half the cubical symmetry at the vertices of F is preserved.) But now each vertex $\pm 2e_j$ $(j = 1, \ldots, 4)$ and $(\pm 1, \pm 1, \pm 1, \pm 1)$ of the dual 24-cell lies above an icosahedral facet, forming with it a pyramid whose lateral facets are also regular tetrahedra. Thus taking the convex hull of these $96 + 24 = 120$ vertices yields a 4-polytope in which the 24 icosahedra are each replaced by 20 tetrahedra, yielding $120 + 24 \cdot 20 = 600$ tetrahedral facets in all.

7E1

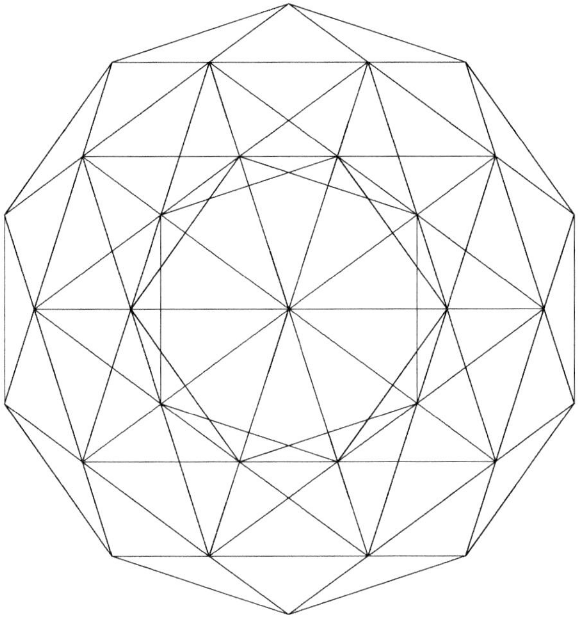

The 600-cell $\{3, 3, 5\}$

The vertex-set G of this new polytope G, regarding 4-vectors as quaternions, is just $2\mathbf{I}$; this already suggests that we do indeed have a regular polytope. Thus

7E2 $$G := \{(2, 0, 0, 0)^s, (1, 1, 1, 1)^s, (\tau, 1, \tau^{-1}, 0)^s\}.$$

The notation $(\cdots)^s$ means that we take all even permutations with arbitrary changes of sign of the coordinates within the brackets $(\cdots)$; this convention will be useful with more complicated point-sets. A depiction of G is given in Figure 7E1. In fact, we have

7E3 Theorem *The symmetry group of the* 600*-cell* G *is* $(\mathbf{I}/\mathbf{I}; \mathbf{I}/\mathbf{I})^*$, *consisting of all mappings* $\mathbf{x} \mapsto \widetilde{\mathbf{a}}\mathbf{x}\mathbf{b}$ *and* $\mathbf{x} \mapsto \widetilde{\mathbf{a}}\widetilde{\mathbf{x}}\mathbf{b}$, *with* $\mathbf{a}, \mathbf{b} \in \mathbf{I}$.

Proof. It is clear that these mappings are indeed symmetries of the point-set $\mathbf{I}$, and counting them yields a total $2 \cdot (120 \cdot 120/2) = 14400 = |[3, 3, 5]|$ from the calculations of Section 1H. □

7E4 Remark The description in Theorem 7E3 of the symmetry group of the 600-cell has a remarkable implication: if P is *any* 4-polytope whose symmetry group is $[3,3,5]$ or its rotational subgroup $[3,3,5]^+$, then the vertices of P fall into disjoint copies of the vertices of $\{3,3,5\}$ (usually in two ways). These arise from the right (or left) action of $\mathbf{I}$ on the vertex-set of P.

The preceding description of the vertex-set of $\{3,3,5\}$ shows how the vertices fall into layers from an initial one, say $(2,0,0,0)$. The vertex-figure, at height τ, is (of course) an icosahedron of edge-length $2\tau^{-1}$. At height 1 we have a dodecahedron, of the same edge-length $2\tau^{-1}$. At height τ^{-1} we have another icosahedron, this time of edge-length 2 (naturally, its edges do not appear in Figure 7E1). The central layer (at height 0) is an icosidodecahedron of edge-length $2\tau^{-1}$ again. The remaining layers then repeat these, in reverse order, to the antipodal vertex $(-2,0,0,0)$.

The original description in Section 1K shows that $\mathbf{T}$ is a subgroup of $\mathbf{I}$ of index 5. If $\mathbf{p} \in \mathbf{I}$ is any quaternion of period 5 and $\mathbf{p}_j := \mathbf{p}^j$ for $j = 0,\dots,4$, then $\mathbf{p}_0,\dots,\mathbf{p}_4$ are right (and left) coset representatives of $\mathbf{T}$ in $\mathbf{I}$. This implies that the vertices of $\{3,3,5\}$ fall into sets of five copies of the vertices of $\{3,4,3\}$. Bear in mind, though, that the coordinates given for $\mathbf{T}$ are fairly arbitrary, and so a conjugate of $\mathbf{T}$ in $\mathbf{I}$ could be chosen as the initial one instead. If we wish to preserve the full symmetry, then we should actually write

$$\mathbf{I} = \bigcup\{\mathbf{p}_{-j}\mathbf{T}\mathbf{p}_k \mid j,k = 0,\dots,4\},$$

thus expressing the vertices of $\{3,3,5\}$ as those of 25 copies of $\{3,4,3\}$, each vertex of the former corresponding to 5 vertices of the latter. We refer to Section 7J for further details.

For future reference, we fix on

7E5 $$\mathbf{p} := -\tfrac{1}{2}(\tau - \mathbf{i} + \tau^{-1}\mathbf{j})$$

once and for all. We know from Section 1K that, if $\mathbf{u} \in \mathbf{B}$ (the other coset from $\mathbf{A}$ in $\mathbf{O}$), then $\tilde{\mathbf{u}}\mathbf{I}\mathbf{u} = \mathbf{I}^\ddagger$; we shall also fix

7E6 $$\mathbf{u} := \tfrac{1}{\sqrt{2}}(1 - \mathbf{k}).$$

The point of these choices is

7E7 Proposition *The elements* $\mathbf{p}$ *of (7E5) and* $\mathbf{u}$ *of (7E6) are compatible, in the sense that, if* $\mathbf{q} := \mathbf{p}^\ddagger$, *then* $\tilde{\mathbf{u}}\mathbf{p}\mathbf{u} = \mathbf{q}^2$ *and* $\tilde{\mathbf{u}}\mathbf{q}\mathbf{u} = \mathbf{p}^2$.

Proof. This interaction between $\mathbf{p}$ and $\mathbf{q}$ is central in Sections 7F and 7J. To prove it, we have

$$\mathbf{q} = \tfrac{1}{2}(\tau^{-1} + \mathbf{i} + \tau\mathbf{j}) \implies \mathbf{q}^2 = \tfrac{1}{2}(-\tau + \tau^{-1}\mathbf{i} + \mathbf{j}) = \tilde{\mathbf{u}}\mathbf{p}\mathbf{u},$$

as is easily seen (note that $\tilde{\mathbf{u}}\mathbf{i} = \mathbf{i}\mathbf{u}$, and so on). The other calculation is similar. □

Petrie Polygons

We can find the Petrie polygon of $\{3, 3, 5\}$ by a similar argument to that applied to $\{3, 4, 3\}$; indeed, we can find the Petrie polygons of all the pentagonal 4-polytopes by direct methods; see Section 7G for the others. In the case of $\{3, 3, 5\}$, we appeal to Theorem 7A26(a) to see that it has diametral planar polygons and, as the coordinates which we found above show, these will be decagons $\{10\}$ (the peripheral decagon in Figure 7E1 is a projection of one such). We then have

7E8 Theorem *The Petrie polygons of the* 600-*cell* $\{3, 3, 5\}$ *and its dual* 120-*cell* $\{5, 3, 3\}$ *are of type* $\{\frac{30}{1,11}\}$.

7E9

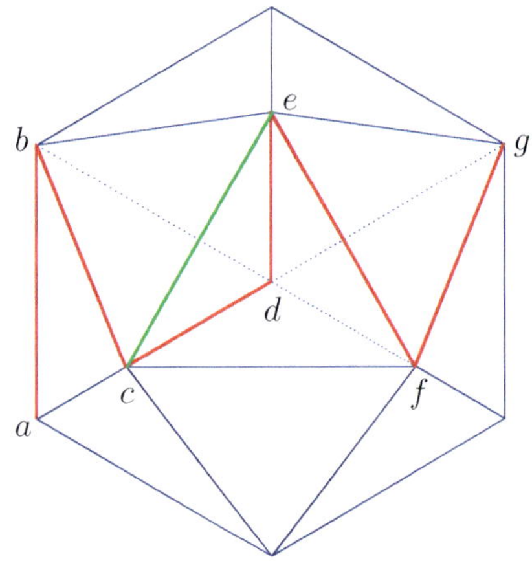

Petrie polygon of $\{3, 3, 5\}$

Proof. Let a, b, c, d, e, f, g be seven successive vertices of a Petrie polygon of $\{3, 3, 5\}$. It is not hard to see that a, b, c, e, f, g are six successive vertices of a Petrie polygon of the broad vertex-figure at the middle vertex d; see Figure 7E9. Then a and g are the opposite ends of a diameter of this vertex-figure; as a result, these six edges with successive vertices a, b, c, d, e, f, g go two steps (that is, a, d, g) along a diametral decagon of $\{3, 3, 5\}$. It follows that the Petrie polygon is of the form

$$\left\{\frac{30}{t_1, t_2}\right\},$$

with $1 \leqslant t_1 < t_2 < 15$. Since three steps along it gives one along the decagon, we conclude that $t_1, t_2 \equiv \pm 1 \pmod{10}$. Thus certainly $t_1, t_2 \in \{1, 9, 11\}$; we must show that $t_j \neq 9$ for either j. (Of course, we should expect that $(t_j, 30) = 1$ for both j, but we cannot assume this.) To show this, we look at the rotation group of $\{3, 3, 5\}$, consisting of the mappings $\mathbf{x} \mapsto \tilde{\mathbf{a}}\mathbf{x}\mathbf{b}$ with $\mathbf{a}, \mathbf{b} \in \mathbf{I}$. First, recall that, if $\mathbf{a} = \cos\vartheta + \sin\vartheta\mathbf{u}$ and $\mathbf{b} = \cos\varphi + \sin\varphi\mathbf{v}$ for some $0 \leqslant \vartheta, \varphi \leqslant \pi$ and pure imaginary unit quaternions $\mathbf{u}, \mathbf{v}$, then $\mathbf{x} \mapsto \tilde{\mathbf{a}}\mathbf{x}\mathbf{b}$ is a double rotation through $\vartheta \mp \varphi$. Second, notice that, if $\mathbf{a}, \mathbf{b} \in \mathbf{I}$, then ϑ and φ can only be multiples of $\frac{\pi}{2}$, $\frac{\pi}{3}$ or $\frac{\pi}{5}$. We thus see that (with these restrictions) $\{\vartheta, \varphi\} = \{\frac{2\pi}{5}, \frac{\pi}{3}\}$ is essentially the only choice which yields one of $\frac{\pi}{15}$ or $\frac{11\pi}{15}$, and then this choice automatically gives the other as well.

Since dual classical polytopes have Petrie polygons of the same type, this proves our claim. □

Realizations of $\{3, 3, 5\}$

Our next task is to describe the realization domain of the abstract 600-cell $\mathcal{G} := \{3, 3, 5\}$. In contrast to what we did previously, we shall not write out (even in outline) the multiplication table for its pure realizations. Our intention here is rather to demonstrate how the theory we have developed, including the multiplicative structure, can be used to determine the pure realizations of $\mathcal{G}$, at least in principle (see the notes at the end of the section).

Observe that the hemi-600-cell or *300-cell* (as we shall call it) $\mathcal{G}/2$ has four non-trivial diagonal classes, while $\mathcal{G}$ itself has eight. Moreover, all diagonal classes are symmetric, so that $w = 1$ for each pure realization of $\mathcal{G}$. It is clear that all our cosine vectors will have entries in the field $\mathbb{Q}[\sqrt{5}] = \mathbb{Q}[\tau]$; as previously, we make use of the involutory automorphism ‡ of $\mathbb{Q}[\sqrt{5}]$ given by changing the sign of $\sqrt{5}$, which also corresponds to $\tau \leftrightarrow -\tau^{-1}$. For the layer vector, we can refer to the preceding description of the vertex-set of $\{3, 3, 5\}$ and Figure 7E1. We shall then show

7E10 Theorem *An inner product matrix of the abstract* 600-*cell* $\{3, 3, 5\}$ *is*

$$\begin{bmatrix} 1 & 1 & 1 & 1 & 1 & 1 & 1 & 1 & 1 \\ 3 & \tau & 0 & -\tau^{-1} & -1 & -\tau^{-1} & 0 & \tau & 1 \\ 3 & -\tau^{-1} & 0 & \tau & -1 & \tau & 0 & -\tau^{-1} & 1 \\ 4 & -1 & 1 & -1 & 0 & -1 & 1 & -1 & 1 \\ 5 & 0 & -1 & 0 & 1 & 0 & -1 & 0 & 1 \\ 2 & \tau & 1 & \tau^{-1} & 0 & -\tau^{-1} & -1 & -\tau & -1 \\ 2 & -\tau^{-1} & 1 & -\tau & 0 & \tau & -1 & \tau^{-1} & -1 \\ 4 & 1 & -1 & -1 & 0 & 1 & 1 & -1 & -1 \\ 6 & -1 & 0 & 1 & 0 & -1 & 0 & 1 & -1 \end{bmatrix},$$

with layer and dimension vectors

$$\Lambda = (1, 12, 20, 12, 30, 12, 20, 12, 1), \quad D = (1, 9, 9, 16, 25, 4, 4, 16, 36).$$

Proof. We assume as given the standard convex 600-cell $\mathsf{G}_5 = \{3, 3, 5\}$ and the stellated 600-cell $\mathsf{G}_6 = \{3, 3, \frac{5}{2}\}$ (for this, we make a forward reference to Section 7G), with dimensions $d_5 = d_6 = 4$ and cosine vectors given by

$$2\Gamma_5 = (2, \tau, 1, \tau^{-1}, 0, -\tau^{-1}, -1, -\tau, -2),$$

$$2\Gamma_6 = (2, -\tau^{-1}, 1, -\tau, 0, \tau, -1, \tau^{-1}, -2).$$

For the components of the small simplex realization S, we follow [88] by showing that Γ_1, Γ_2 and Γ_3, given by

$$\begin{aligned}4{\Gamma_5}^2 &= \Gamma_0 + 3\Gamma_1,\\ 4{\Gamma_6}^2 &= \Gamma_0 + 3\Gamma_2,\\ \Gamma_5\Gamma_6 &= \Gamma_3,\end{aligned}$$

that is,

$$\begin{aligned}3\Gamma_1 &= (3, \tau, 0, -\tau^{-1}, -1, -\tau^{-1}, 0, \tau, 3),\\ 3\Gamma_2 &= (3, -\tau^{-1}, 0, \tau, -1, \tau, 0, -\tau^{-1}, 3),\\ 4\Gamma_3 &= (4, -1, 1, -1, 0, -1, 1, -1, 4),\end{aligned}$$

are mutually Λ-orthogonal. Observe that the component $\frac{1}{4}\Gamma_0$ of ${\Gamma_5}^2$ and ${\Gamma_6}^2$ is given by Theorem 3F5. Using Corollary 3F4 to bound $\dim \mathsf{G}_j$ from below, and (3E12) and Proposition 3E13 to bound it from above, we obtain

$$\begin{aligned}9 = \|\Gamma_j\|_\Lambda^{-2} &\leqslant \dim \mathsf{G}_j \leqslant \tfrac{1}{2}(4-1)(4+2) = 9, \quad \text{for } j = 1, 2,\\ 16 = \|\Gamma_3\|_\Lambda^{-2} &\leqslant \dim \mathsf{G}_3 \leqslant 4 \cdot 4 = 16;\end{aligned}$$

hence we have equality throughout. Since S can have at most four non-trivial components, whose dimensions sum to $60 - 1 = 59$, we see that $\mathsf{G}_1, \mathsf{G}_2, \mathsf{G}_3$ must all be pure, with dimensions $d_1 = d_2 = 9$ and $d_3 = 16$. We then use the component equation of Theorem 3C11 to show that the remaining pure realization G_4 has dimension $d_4 = 60 - 1 - 9 - 9 - 16 = 25$, with cosine vector given by

$$5\Gamma_4 = \tfrac{1}{5}\big(60(1, 0^7, 1) - \Gamma_0 - 9\Gamma_1 - 9\Gamma_2 - 16\Gamma_3\big) = (5, 0, -1, 0, 1, 0, -1, 0, 5).$$

Note that we can check the dimension using Theorem 3F5; moreover, we can see that Γ_4 is Λ-orthogonal to each of $\Gamma_1, \Gamma_2, \Gamma_3$.

We now carry out an exactly analogous procedure to find the remaining two pure components of the staurotope realization X. We already have Γ_5 and Γ_6; we then observe that

$$6\Gamma_8 := 6\Gamma_1\Gamma_6 = 6\Gamma_2\Gamma_5 = (6, -1, 0, 1, 0, -1, 0, 1, -6)$$

is Λ-orthogonal to Γ_5, Γ_6. Just as before, we estimate

$$36 = \|\Gamma_8\|_\Lambda^{-2} \leqslant \dim \mathsf{G}_8 \leqslant 9 \cdot 4 = 36,$$

again giving equality. Since we have at most four components of X, and $4 + 4 + 36 < 60$, it follows that G_8 must be pure with dimension $d_8 = 36$. We finally use Theorem 3C11 again to calculate

$$4\Gamma_7 = (4, 1, -1, -1, 0, 1, 1, -1, -4)$$

from

$$4\Gamma_5 + 4\Gamma_6 + 16\Gamma_7 + 36\Gamma_8 = 60(1, 0^7, -1),$$

incidentally verifying the corresponding dimension $d_7 = 16$ and Λ-orthogonality to Γ_5, Γ_6 and Γ_8 using Theorem 3F5. Alternatively, we have

$$3\Gamma_1\Gamma_5 = \Gamma_5 + 2\Gamma_7,$$

and the same relation with Γ_6 replacing Γ_5. □

7E11 Remark The last relation can also be written in the form $2{\Gamma_s}^3 = \Gamma_s + \Gamma_7$, with $s = 5, 6$. We shall find this useful in Section 7K.

7E12 Remark A point to bear in mind is that, from its construction, among the vertices of $\mathcal{G}$ are those of the 24-cell $\{3, 4, 3\}$. It does not occur with full symmetry (in fact, a certain subgroup of $[3, 4, 3]$ of index 2 is a subgroup of $[3, 3, 5]$ of index 25 – see Proposition 7F10 below), but it is symmetrical enough that every realization of $\mathcal{G}$ induces a corresponding one of $\{3, 4, 3\}$.

7E13 Remark It should be emphasized that our constructions only yield the realizations $\mathsf{G}_1, \mathsf{G}_2, \mathsf{G}_3$ and G_8 in principle, and give no hint at all about the geometry of G_4 and G_7. (We say a little more about G_3 and G_4 in Section 7F.) It is, perhaps, a notable feature of our approach that we do not need to know what these latter polytopes look like.

7E14 Remark There are two curiosities about the pure realizations of $\mathcal{G}$, the first of which ties in with Theorem 3F5 (see the notes at the end of the section):

(a) for each j, there is a $q \in \mathbb{N}$ such that, apart from 0 or ± 1, each coordinate of Γ_j is of the form ε/q, where $\varepsilon \in \mathbb{Z}[\tau]$ is a unit (in practice, $\pm 1, \pm\tau$ or $\pm\tau^{-1}$);

(b) for each j, $\dim \mathsf{G}_j = q^2$, with q as in (a).

Because of the latter, the rows of the matrix of Theorem 7E10 actually form a Λ-orthonormal basis of the realization cone. The fact that each $\dim \mathsf{G}_j$ is a square is related to the way that $[3, 3, 5]$ is expressible by quaternionic multiplication from both sides. Even more significant is the fact that the central quotient $[3, 3, 5]/2$ is isomorphic to $(\mathrm{A}_5 \times \mathrm{A}_5) \rtimes \mathrm{C}_2$.

7E15 Remark As we shall see, the two expressions $\mathsf{G}_8 := \mathsf{G}_1 \otimes \mathsf{G}_6 = \mathsf{G}_2 \otimes \mathsf{G}_5$ have implications for realizations of the 120-cell in Section 7K.

Notes to Section 7E

1. The construction of the 600-cell given here basically goes back to Gosset [53], with a refinement due to Alicia Boole Stott; see [23, p. 338]. Further historical information may be found in [27, Section 8.9].
2. The treatment of realizations of $\{3, 3, 5\}$ in [88] was already superior to that of the earlier [85]; what we have done here improves on it further.
3. The curiosities mentioned in Remark 7E14 have been explained by Ladisch [67].

7F The 120-Cell

We now show how to construct the dual 120-cell $\{5,3,3\}$, at least in principle. The construction also yields several relationships among the various regular 4-polytopes. In particular, we have the rather striking

7F1 Proposition *Among the vertices of the* 120*-cell* $\{5,3,3\}$ *are to be found the vertices of each of the other five convex regular* 4*-polytopes.*

Before we do anything more complicated, an appeal to Remark 7E4 shows that the vertices of $\{5,3,3\}$ already contain copies of those of $\{3,3,5\}$, and hence also those of $\{3,3,4\}$, $\{4,3,3\}$ and $\{3,4,3\}$. Thus, in fact, only the 4-simplex remains to be found; we shall see this shortly in Proposition 7F10 .

7F2

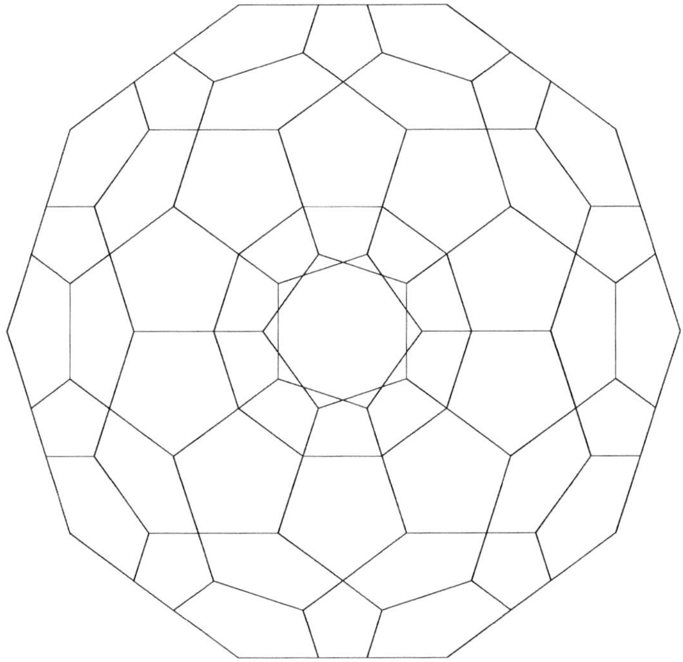

The regular 120-cell $\{5,3,3\}$

We can express the vertex-set of the 120-cell in two rather different ways. To begin with, we look at it as the geometric dual of the 600-cell G of Section 7E. A typical facet of G has vertices

$$(\tau,1,\pm\tau^{-1},0),\ (1,\tau,0,\pm\tau^{-1})$$

in G of (7E2), and so the corresponding vertex of $\{5,3,3\}$ is $(2,2,0,0)$. As a quaternion, this is $\sqrt{8}\mathbf{u}$, with $\mathbf{u}=(1+\mathbf{i})/\sqrt{2}\in\mathbf{U}$. It follows at once that (up to scaling), we can take vert$\{5,3,3\} = \mathbf{IuI} = \mathbf{IUI}$. Here, this copy of $\{5,3,3\}$ has the original symmetry group $(\mathbf{I}/\mathbf{I};\mathbf{I}/\mathbf{I})^*$ of the 600-cell. Reversing the scaling to remove fractions, we call the new set of vertices H:

7F3
$$(2,2,0,0)^s,\quad (\sqrt{5},1,1,1)^s,\quad (\tau^2,\tau^{-1},\tau^{-1},\tau^{-1})^s,\quad (\tau^{-2},\tau,\tau,\tau)^s,$$
$$(\sqrt{5},\tau^{-1},\tau,0)^s,\quad (\tau^2,\tau^{-2},1,0)^s,\quad (2,\tau,\tau^{-1},1)^s;$$

recall the convention of (7E2). In fact, we do not need this detail. All we have to remember is that $v \in H$ just when $v = x + y$ for some orthogonal $x, y \in G$ (that is, $\langle x, y\rangle = 0$). Observe that each $v \in H$ is representable as such a sum in three ways, for example, $(2,2,0,0) = (2,0,0,0) + (0,2,0,0) = (1,1,1,\pm1) + (1,1,-1,\mp1)$; hence their number is indeed $120 \cdot 30/2 \cdot 3 = 600$.

The picture of the 120-cell in Figure 7F2 gives a fair idea of its geometry. The periphery is formed of a ring of ten dodecahedra $\{5,3\}$, corresponding to the diametral decagon of $\mathsf{G} = \{3,3,5\}$ in Figure 7E1. Dodecahedra corresponding to the other rings of vertices of G are clearly visible; at the centre are ten more superimposed dodecahedra, corresponding to the diametral decagon of G that projects to the centre of Figure 7E1.

One problem with this picture of $\{5,3,3\}$ is that it is not easy to identify layers. To get around this, we consider instead

$$\mathbf{IuI} = \mathbf{u}\cdot\widetilde{\mathbf{u}}\mathbf{IuI} = \mathbf{uI}^{\ddagger}\mathbf{I},$$

and multiplying by $\widetilde{\mathbf{u}}$ on the left we arrive at

7F4
$$\mathrm{vert}\{5,3,3\} = \mathbf{I}^{\ddagger}\mathbf{I} =: \mathbf{H}.$$

A disadvantage of this expression is that the symmetries take on an awkward form. If $\mathbf{g} \in \mathbf{I}$, then we write

7F5
$$\mathbf{g}^* := \widetilde{\mathbf{g}}^{\ddagger} = (\mathbf{g}^{\ddagger})^{-1};$$

since the involution ‡ applies equally to $\mathbf{I}^{\ddagger}$, it is clear that * is also involutory. Then we have

7F6 Theorem *If the vertex-set of* $\{5,3,3\}$ *is taken in the form* $\mathbf{H}$ *of (7F4), then its symmetries are those mappings of the form*

$$\mathbf{x} \mapsto \mathbf{a}^*\mathbf{x}\mathbf{b} \text{ or } \mathbf{a}^*\widetilde{\mathbf{u}}\widetilde{\mathbf{x}}\mathbf{u}\mathbf{b},$$

with $\mathbf{a}, \mathbf{b} \in \mathbf{I}$ *and* $\mathbf{u} \in \mathbf{U}$ *any fixed element.*

As a consequence, we have

7F7 Proposition *The coordinate vectors of the vertices of* $\{5,3,3\}$ *consist of all permutations with an even number of changes of sign of*

$$(4,0,0,0),\ (2,2,2,\pm2),\ (2\tau,2,2\tau^{-1},0),$$
$$(-\sigma,\tau^{-1},\tau^{-1},\tau^{-1}),\ (-\sigma^{\ddagger},\tau,\tau,\tau),$$
$$(-1,\sqrt{5},\sqrt{5},\sqrt{5}),\ (3,\sqrt{5},1,1),$$
$$(\tau\sqrt{5},-\tau,\tau^{-2},\tau^{-2}),\ (\tau^{-1}\sqrt{5},-\tau^{-1},\tau^2,\tau^2),$$

where $\sigma := \frac{1}{2}(3\sqrt{5}+1) = \sqrt{5}+\tau$.

7F8 Remark The coordinates of Proposition 7F7 differ from those of Coxeter [27, Table V(v)], which are obtained by an *odd* number of changes of sign from ours. We recognize that the vertices in the first line include those of an inscribed copy of the 600-cell $\{3,3,5\}$ (in fact, there are two copies); in turn, among these are the vertices of a 24-cell $\{3,4,3\}$.

7F9 Remark The change of sign of $\sqrt{5}$ yields a copy of the star-polytope $\{\frac{5}{2},3,3\}$, for which see the next Section 7G (then changing the sign of the last coordinate gives the same vertex-set). We shall need this observation shortly.

At this point, it is appropriate to refer to the claims in Proposition 1F3 that $\mathbf{A}_4, \mathbf{B}_4 < \mathbf{G}_4 = [3,3,5]$. Recalling that $[3,4,3] = (\mathbf{O}/\mathbf{T};\mathbf{O}/\mathbf{T})^*$ and $[3,3,4] = (\mathbf{O}/\mathbf{V};\mathbf{O}/\mathbf{V})^*$, we see that Theorem 7F6 implies the second and third parts of

7F10 Proposition *The group* $[3,3,5]$ *has subgroups*

(a) $\mathbf{A}_4 = [3,3,3]$ *of index* 120*,*
(b) $\mathbf{B}_4 = [3^{1,1,1}]$ *of index* 75*,*
(c) *a subgroup* $[3,4,3]^*$ *of* $[3,4,3]$ *of index* 2 *whose index is* 25*.*

Proof. We can prove the first two parts directly: defining reflexion mirrors in $[3,3,5]$ in its original form by their normals

$$u_0 := \tfrac{1}{2}(1,1,1,1), \quad u_j := -e_j \text{ for } j=1,\ldots,4, \quad u_5 = \tfrac{1}{2}(-\tau,1,\tau^{-1},0),$$

with $R_0,\ldots,R_5$ the corresponding reflexions, we see that

$$(R_5,R_2,R_0,R_4), \quad (R_1,R_0,R_2,R_4)$$

are generatrices of $\mathbf{A}_4$ and $\mathbf{B}_4$, respectively. Moreover, with $(2,0,-2,0)$ as initial vertex, we obtain a 4-simplex and 4-staurotope inscribed in $\{5,3,3\}$. □

The form $\operatorname{vert}\{5,3,3\} = \mathbf{H} = \mathbf{I}^{\ddagger}\mathbf{I}$ shows at once that copies of $\operatorname{vert}\{3,3,4\}$ and $\operatorname{vert}\{3,4,3\}$ are subsets, since $\mathbf{V},\mathbf{T} \subset \mathbf{H}$. Indeed, a useful observation at this point is the following.

7F11 Remark With $\mathbf{p}$ as in (7E5), write $\mathbf{p}_j = \mathbf{p}^j$ as before. Since the powers of $\mathbf{p}$ are the right coset representatives of $\mathbf{T}$ in $\mathbf{I}$, we see that

$$\operatorname{vert}\{5,3,3\} = \mathbf{H} = \bigcup\{\mathbf{p}_j^*\mathbf{T}\mathbf{p}_k \mid j,k=0,\ldots,4\}.$$

Our next task is to identify a copy of $\operatorname{vert}\{3,3,3\}$ in $\mathbf{H}$. With the same notation, we have

7F12 Lemma *If* $\mathbf{g} \in \mathbf{I}$*, then* $\mathbf{g}^*\mathbf{g} = \mathbf{p}_j^*\mathbf{p}_j$ *for some* $j=0,\ldots,4$*.*

Proof. Indeed, if $\mathbf{g} \in \mathbf{T}\mathbf{p}_j$ (with $j=0,\ldots,4$), then we have $\mathbf{g} = \mathbf{a}\mathbf{p}_j$ for some $a \in \mathbf{T}$, and hence

$$\mathbf{g}^*\mathbf{g} = \mathbf{p}_j^*\mathbf{a}^*\mathbf{a}\mathbf{p}_j = \mathbf{p}_j^*\mathbf{p}_j,$$

since $\mathbf{a}^{\ddagger} = \mathbf{a}$, so that $\mathbf{a}^* = \mathbf{a}^{-1}$. This is the result claimed. □

Lemma 7F12 actually tells us that the same set $\mathbf{P} := \{\mathbf{p}_j^*\mathbf{p}_j \mid j = 0, \ldots, 4\}$ will be obtained from *any* family $\{\mathbf{p}_0, \ldots, \mathbf{p}_4\}$ of coset representatives of $\mathbf{T}$ in $\mathbf{I}$. Observe that, with $\mathbf{p} = \mathbf{p}_1 = -\frac{1}{2}(\tau - \mathbf{i} + \tau^{-1}\mathbf{j})$ as in (7E5), the mapping $\mathbf{x} \mapsto \mathbf{p}^*\mathbf{x}\mathbf{p}$ induces a cyclic permutation of $\mathbf{P}$. We should (perhaps) not be surprised to find

7F13 Proposition *The set* $\mathbf{P}$ *forms the vertices of a regular* 4*-simplex.*

Proof. We have, for instance,

$$\mathbf{p}^*\mathbf{p} = -\tfrac{1}{4}(\tau^{-1} - \mathbf{i} - \tau\mathbf{j})(\tau - \mathbf{i} + \tau^{-1}\mathbf{j}) = \tfrac{1}{4}\big(-1 + \sqrt{5}(\mathbf{i} + \mathbf{j} + \mathbf{k})\big);$$

one reason for choosing $\mathbf{p}$ in (7E5) was to make this so. Similarly, the $\mathbf{p}_j^*\mathbf{p}_j$ with $j = 2, 3, 4$ are all of the form $\frac{1}{4}(-1 \pm \sqrt{5}\mathbf{i} \pm \sqrt{5}\mathbf{j} \pm \sqrt{5}\mathbf{k})$, with an even number of minus signs. The claim of the lemma is an easy consequence. □

In fact, we can recover the subgroup relationship $[3, 3, 3] < [3, 3, 5]$ from Proposition 7F13, when we observe that the hyperplane reflexions $\mathbf{x} \mapsto -\mathbf{u}\tilde{\mathbf{x}}\mathbf{u}$ with $\mathbf{u} \in \mathbf{U}$ pure imaginary preserve the initial vertex $1 \in \mathbf{H}$ and generate the group $\mathbf{A}_3 = [3, 3]$ acting on coordinates ξ_2, ξ_3, ξ_4 in the natural way.

Notes to Section 7F

1. Our treatment of the classical polytopes has been inspired (at least in part) by that of Du Val in [43]. We have drawn quite heavily on [43] for the description of the 120-cell $\{5, 3, 3\}$ and its relationships.
2. The projection of $\{5, 3, 3\}$ depicted in the Figure 7F2 is drawn inaccurately in [30, p. 41]. Of course, our picture is still only an approximation, but a much better one. There is a very symmetric projection of $\{5, 3, 3\}$ into a truncate of the rhombic triacontahedron in $\mathbb{E}^3$. A very large model of this hangs in the Fields Institute in Toronto; from a certain angle, it is just possible to see how the planar projection arises from it.

7G Star-Polytopes

In this section, we shall complete the classification of the classical polytopes by describing the regular star-polytopes. Among other things, we shall demonstrate that there are no regular star-polytopes in dimensions larger than four, and no regular star-honeycombs at all. In fact, we shall establish a result first observed by Hess [62], and proved using a case by case enumeration by van Oss [124]. Several proofs are also to be found in [27, Chapter 14]. However, to enumerate the star-polytopes, we follow here the systematic approach of McMullen in [71] (see also [99, Section 7D]), which does not rely on any prior knowledge of the actual polytopes. For what the *density* of a regular star-polytope means we refer the reader to Coxeter [17, 19, 20] and [27, Section 14.8]; see also Section 1H.

Hess's Theorem

The observation made by Hess was

7G1 Theorem *A classical regular star-polytope has the same vertices as some convex polytope.*

Proof. By definition, the generatrix of a classical regular polytope P consists of hyperplane reflexions; Theorem 1E11 identifies each reflexion group with a Coxeter diagram $\mathcal{D}$ in Table 1E12. The groups of the facet P^f and vertex-figure P^v cannot be decomposable; they must therefore correspond to connected string subdiagrams of $\mathcal{D}$ obtained by deleting a single node (here, we can appeal to the obvious inductive assumption, starting with regular polygons). Recalling how Wythoff's construction works, we see that $\mathcal{D}$ itself is a string diagram, with the deleted node an end node. This corresponds to a convex polytope Q, say, and the initial vertex of P coincides with that of Q or its dual Q^δ, as we had to show. □

7G2 Remark This independent proof of Hess's theorem justifies the approach to finding the star-polytopes called by Coxeter in [27, Section 14.3] *systematic faceting* (see the notes at the end of the section).

Vertex-Figure Replacement

We shall construct the classical regular star-polytopes as a consequence of a proof of Theorem 7G1 in the following stronger form (again, see the notes at the end of the section).

7G3 Theorem *If* P *is a starry regular d-polytope, then there is a unique regular convex polytope* $\overline{\mathsf{P}}$ *such that*

(a) $\operatorname{vert}\mathsf{P} = \operatorname{vert}\overline{\mathsf{P}}$,
(b) $\mathbf{G}(\mathsf{P}) = \mathbf{G}(\overline{\mathsf{P}})$,
(c) $\Sigma(\mathsf{P}) > \Sigma(\overline{\mathsf{P}})$.

Here, Σ *is the Schläfli determinant of (7A14).*

Proof. The induction begins with the case $d = 2$. If $\mathsf{P} = \{\frac{s}{t}\}$ is a regular star-polygon, with s and t integers satisfying $2 \leqslant t < \frac{1}{2}s$ and $(s, t) = 1$, then $\overline{\mathsf{P}}$ is the convex regular s-gon $\{s\}$ with the same vertices. Clearly, $\mathbf{G}(\mathsf{P}) = \mathbf{G}(\overline{\mathsf{P}})$; further

$$\Sigma(\mathsf{P}) = \sin^2 \frac{t\pi}{s} > \sin^2 \frac{\pi}{s} = \Sigma(\overline{\mathsf{P}}).$$

Thus the conditions hold here.

Now suppose that $d \geqslant 3$, and that $\mathsf{P} = \{p_1, \ldots, p_{d-1}\}$ is a regular star-polytope – that is, *not* convex. Then at least one of $p_1, \ldots, p_{d-1}$ is not an integer; appealing to Proposition 7A11, we may dualize, if necessary, and suppose that it is one of $p_2, \ldots, p_{d-1}$. (Note, by the way, that Corollary 4A4 ensures that we do always obtain a polytope by such dualization.) By Proposition 7A12, the vertex figure $\mathsf{P}^v = \{p_2, \ldots, p_{d-1}\}$ of P is a classical regular $(n-1)$-polytope, which is starry. Our inductive assumption then enables us to replace $\mathsf{P}_0 := \mathsf{P}^v$ by the unique regular convex polytope $\mathsf{Q}_0 = \{q_2, \ldots, q_{d-1}\}$, say, with the same vertices and the same symmetry group $\mathbf{G}(\mathsf{Q}_0) = \mathbf{G}(\mathsf{P}_0)$ (thus $\mathsf{Q}_0 = \overline{\mathsf{P}}_0$ in the inductive scheme). Note that $q_2, \ldots, q_{d-1} \geqslant 3$ are integers.

Let $(R_0, \ldots, R_{d-1})$ be the generatrix of P, v its initial vertex and $\mathsf{E} = \{v, w\}$ its initial edge. Choose any generatrix $(S_1, \ldots, S_{d-1})$ of $\mathbf{G}(\mathsf{Q}_0)$, with $S_1, \ldots, S_{d-1} \in \mathbf{G}(\mathsf{P}^v)$ hyperplane mirrors in $\mathbb{E}^d$ containing v, associated with a base flag of Q_0 containing w. Define $S_0 := R_0$. Since $v \in S_j$ for each $j = 1, \ldots, d-1$, and $w \in S_j$ for $j = 2, \ldots, d-1$, we see that $\mathsf{E} \subseteq S_j$ for $j = 2, \ldots, d-1$. But $S_0 = R_0$ is the perpendicular bisector of E, so that $S_0 \perp S_j$ for $j = 2, \ldots, d-1$.

Our inductive assumption says that $\mathbf{G}(\mathsf{Q}_0) = \mathbf{G}(\mathsf{P}_0) = \mathbf{G}(\mathsf{P}^v)$, and hence

$$\langle S_0, \ldots, S_{d-1} \rangle = \langle R_0, \ldots, R_{d-1} \rangle.$$

If we apply Wythoff's construction, with the same initial vertex v, we thus obtain a polytope

$$\mathsf{Q} = \{q_1, \ldots, q_{d-1}\}$$

with vertex-figure $\mathsf{Q}^v = \mathsf{Q}_0$, for some q_1 giving the new 2-face; once again, we can appeal to Corollary 4A4 here. By construction, Q will have the same vertices and edges as P; observe that E is also the initial edge of Q. We call this process of obtaining Q from P *vertex-figure replacement*.

We now have the first of two subsidiary results; this clearly follows from Theorem 7A19, since P and Q have the same edges.

7G4 Lemma *If* Q *is obtained from* P *by vertex-figure replacement, then*

$$\frac{\Sigma(\mathsf{Q})}{\Sigma(\mathsf{P})} = \frac{\Sigma(\mathsf{Q}^v)}{\Sigma(\mathsf{P}^v)}.$$

Since $\Sigma(\mathsf{Q}^v)$ is known, Lemma 7G4 gives the new Schläfli determinant $\Sigma(\mathsf{Q})$; moreover, in view of the inductive assumption $\Sigma(\mathsf{Q}^v) < \Sigma(\mathsf{P}^v)$, we conclude that

$$\Sigma(\mathsf{Q}) < \Sigma(\mathsf{P}).$$

We next need to calculate q_1. For this, we have

7G5 Lemma *If* P *yields* $\mathsf{Q} = \{q_1, \ldots, q_{d-1}\}$ *by vertex-figure replacement, then*

$$\frac{\cos(\pi/q_1)}{\cos(\pi/p_1)} = \frac{\sin \varphi(\mathsf{Q}^v)}{\sin \varphi(\mathsf{P}^v)}.$$

Proof. If the vertices of P lie on a sphere of radius $\rho(\mathsf{P})$ about o and P has edge-length $\lambda(\mathsf{P})$, then

$$\lambda(\mathsf{P}) = 2\rho(\mathsf{P}) \sin \varphi(\mathsf{P}).$$

The vertices of the broad vertex-figure P^v are the other vertices of the edges of P through the initial vertex v, so that a typical edge of P^v has length

$$\lambda(\mathsf{P}^v) = 2\lambda(\mathsf{P}) \cos \frac{\pi}{p_1},$$

the edge-length of the vertex-figure of the 2-face $\{p_1\}$ of P. Replacing P by $\mathsf{P}^{\boldsymbol{v}}$ in the first expression leads to

$$\rho(\mathsf{P}^{\boldsymbol{v}}) \sin \varphi(\mathsf{P}^{\boldsymbol{v}}) = \tfrac{1}{2}\lambda(\mathsf{P}^{\boldsymbol{v}}) = \lambda(\mathsf{P}) \cos \frac{\pi}{p_1}.$$

Since $\lambda(\mathsf{P}) = \lambda(\mathsf{Q})$ and $\rho(\mathsf{P}^{\boldsymbol{v}}) = \rho(\mathsf{Q}^{\boldsymbol{v}})$, the claim of the lemma follows at once. □

Of course, using Theorem 7A19 we can now express $\sin \varphi(\mathsf{P}^{\boldsymbol{v}})$ and $\sin \varphi(\mathsf{Q}^{\boldsymbol{v}})$ in terms of Schläfli determinants.

If q_1 is also an integer, then we are done. Otherwise, we dualize and replace the new vertex-figure, repeating the process as often as necessary. Bear in mind Proposition 7A18, which states that $\Sigma(\mathsf{P}^{\boldsymbol{\delta}}) = \Sigma(\mathsf{P})$. Now $\mathbf{G}(\mathsf{P})$ is finite, and so there are only finitely many choices of generatrices in $\mathbf{G}(\mathsf{P})$. Since $\Sigma(\mathsf{Q}) < \Sigma(\mathsf{P})$, the Schläfli determinant strictly decreases at each stage. The only way the process can terminate is thus at a convex polytope $\overline{\mathsf{P}}$, say. We complete the argument as in the proof of Theorem 7G1, and conclude that – possibly after dualizing $\overline{\mathsf{P}}$ – we have $\operatorname{vert} \mathsf{P} = \operatorname{vert} \overline{\mathsf{P}}$. This completes the inductive step of the argument. □

7G6 Remark A simple manipulation of expressions in the preceding proof also leads to

$$\cos \varphi(\mathsf{P}) = \frac{\cos(\pi/p_1)}{\sin \varphi(\mathsf{P}^{\boldsymbol{v}})}.$$

In the case of a regular polyhedron $\mathsf{P} = \{p, q\}$, this gives the formula of [27, 2.44], namely, $\cos \varphi(\mathsf{P}) = \cos(\pi/p) \operatorname{cosec}(\pi/q)$.

Theorem 7G1 shows how a regular star-polytope is related to some regular convex polytope. But the method of proof of Theorem 7G1 is, if anything, even more instructive. For it says that we may find all the regular star-polytopes, by reversing the vertex-figure replacement process in all possible ways, with starting-point a suitable regular convex polytope. Thus – in the classical case – we may replace any vertex-figure by another polytope with the same vertices. This process is somewhat akin to Coxeter's systematic faceting of [27, Section 14.3].

We can take for granted the regular star-polygons, since we have already considered them in the proof of Theorem 7G1. Therefore, the first genuine application of the procedure is to 3-polytopes. In this case, Lemma 7G5 says that

$$\frac{\cos(\pi/q_1)}{\cos(\pi/p_1)} = \frac{\sin(\pi/q_2)}{\sin(\pi/p_2)}.$$

The only regular convex polyhedron whose vertex-figure allows replacement by a star-polygon is $\mathsf{Q} = \{3, 5\}$; here, we may replace $\{5\}$ by $\{\frac{5}{2}\}$. Iterating the dualization and vertex-figure replacement procedure then results in the family of Table 7G7 (see the notes at the end of the section).

7G7

$\{5,3\}$	$\{3,5\}$		
	$\{5,\frac{5}{2}\}$	$\{\frac{5}{2},5\}$	
		$\{3,\frac{5}{2}\}$	$\{\frac{5}{2},3\}$

The regular star-polyhedra

Polyhedra in the same row of the table are dual, while those in the same column are obtained by vertex-figure replacement. Indeed, in each column, the polytopes are related in pairs by $\varphi := \varphi_2$ (following the convention of Section 5A). Considering the symmetry groups of the vertex-figures shows that, of the star-polyhedra, $\{\frac{5}{2},3\}$ has the same vertices as $\{5,3\}$, while the others have the same vertices as $\{3,5\}$. Observe also that interchanging 5 and $\frac{5}{2}$, which gives the table a half-turn, takes each polytope into its allomorph.

7G8

$\{5,3,3\}$	$\{3,3,5\}$			
	$\{3,5,\frac{5}{2}\}$	$\{\frac{5}{2},5,3\}$		
	$\{5,\frac{5}{2},5\}$			
	$\{5,3,\frac{5}{2}\}$		$\{\frac{5}{2},3,5\}$	
			$\{\frac{5}{2},5,\frac{5}{2}\}$	
		$\{5,\frac{5}{2},3\}$	$\{3,\frac{5}{2},5\}$	
			$\{3,3,\frac{5}{2}\}$	$\{\frac{5}{2},3,3\}$

The regular star-polytopes in $\mathbb{E}^4$.

For a regular 4-polytope $\mathsf{P} = \{p_1, p_2, p_3\}$, Remark 7G6 yields

$$\cos\varphi(\mathsf{P}^v) = \cos\frac{\pi}{p_2}\operatorname{cosec}\frac{\pi}{p_3}$$

for its vertex-figure P^v (see also [27, 2.44]). Then Lemma 7G5 implies that

$$\frac{\cos^2(\pi/q_1)}{\cos^2(\pi/p_1)} = \frac{1-\cos^2(\pi/q_2)\operatorname{cosec}^2(\pi/q_3)}{1-\cos^2(\pi/p_2)\operatorname{cosec}^2(\pi/p_3)}.$$

7G9

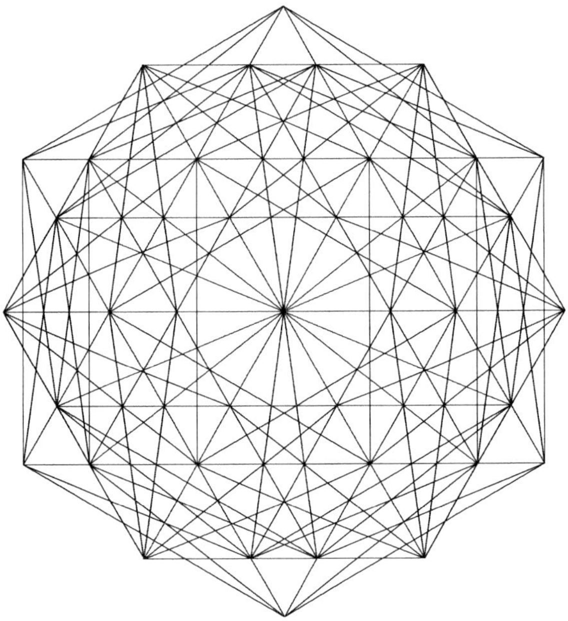

The star-polytopes $\{\frac{5}{2}, 5, 3\}$ and $\{5, \frac{5}{2}, 3\}$

7G10

The remaining star-polytopes in $\mathbb{E}^4$

The only possible starting point for applying the procedure is the 600-cell

$Q = \{3, 3, 5\}$. With the same conventions as before, the resulting family is given in Table 7G8.

Considering the groups of the vertex-figures, we see that, among the star-polytopes, $\{\frac{5}{2}, 3, 3\}$ will have the vertices of $\{5, 3, 3\}$, while the remainder will have the vertices of $\{3, 3, 5\}$. Note that, in the second and fourth columns, the first two polytopes share the same 2-faces (as well as vertices and edges), as do the last two. Moreover, diametrically opposite polytopes in the table are allomorphic, as in Table 7G7.

Since the procedure is not applicable to any regular convex d-polytope with $d \geqslant 5$, we see that the lists in Tables 7G7 and 7G8 complete the enumeration of the classical polytopes.

We shall not draw any pictures of the star-polytope $\{\frac{5}{2}, 3, 3\}$; even the simplest one, with the vertices projected in the same way as in Figure 7F2, is too complicated (see the notes at the end of the section). However, we do give pictures of all the others. Indeed, since our pictures can only depict vertices and edges, from Table 7G8 we see that Figure 7E1 serves as pictures of $\{3, 5, \frac{5}{2}\}$, $\{5, \frac{5}{2}, 5\}$ and $\{5, 3, \frac{5}{2}\}$ in the second column as well. Next, in Figure 7G9, we see the polytopes in the centre column of Table 7G8 (see the notes at the end of the section). Finally, in Figure 7G10, we see the four star-polytopes $\{\frac{5}{2}, 3, 5\}$, $\{\frac{5}{2}, 5, \frac{5}{2}\}$, $\{3, \frac{5}{2}, 5\}$ and $\{3, 3, \frac{5}{2}\}$ in the penultimate column of Table 7G8.

Petrie Polygons

We now look at Petrie polygons, beginning with $\{3, 5, \frac{5}{2}\}$, whose facets are the broad vertex-figures at the vertices of $\{3, 3, 5\}$. We employ the notation introduced in Section 7E, so that two adjacent facets of $\{3, 5, \frac{5}{2}\}$ are the vertex-figures of $\{3, 3, 5\}$ at b and f, which meet on the trigon with vertices c, d, e; see Figure 7G11. Then a, c, d, e, g are successive vertices of a Petrie polygon of $\{3, 5, \frac{5}{2}\}$, so that four steps along it give two steps along a diametral decagon, and thus two steps along it give an edge of the diametral decagon. It follows that the Petrie polygon must be of type $\{\frac{20}{s,t}\}$, with $1 \leqslant s < t < 10$ and $s, t \equiv \pm 1 \pmod{10}$; since Theorem 7A24 tells us that the Petrie polygon is full-dimensional, the only possibility is $\{\frac{20}{1,9}\}$.

The facets of $\{5, \frac{5}{2}, 5\}$ are great dodecahedra $\{5, \frac{5}{2}\}$ with the same vertices and edges as those of the facets $\{3, 5\}$ of $\{3, 5, \frac{5}{2}\}$; thus we see that a, c, e, g are four successive vertices of a Petrie polygon of the facet in the vertex-figure at d. Hence the vertices of a Petrie polygon of $\{5, \frac{5}{2}, 5\}$ are alternate vertices of one of $\{3, 3, 5\}$; its type is therefore $\{\frac{15}{1,4}\}$.

Without doing any calculations at all, we can argue that the Petrie polygon of $\{5, 3, \frac{5}{2}\}$ and $\{\frac{5}{2}, 3, 5\}$ must be of type $\{\frac{12}{1,5}\}$, because – among the possibilities listed in the discussion of the group $[3, 3, 5]$ in terms of quaternions – the double rotation though $\frac{\pi}{6}$ and $\frac{5\pi}{6}$ is the only feasible one which is unaffected by the interchange of 5 and $\frac{5}{2}$. Moreover, we can see that alternate vertices subtend angle $\frac{\pi}{3}$ at the centre.

7G11

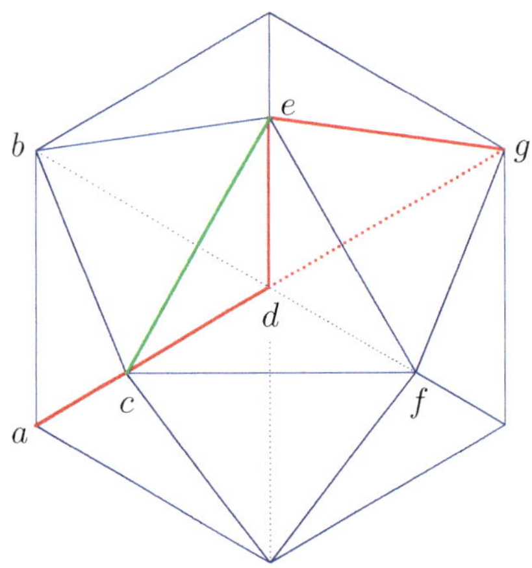

Petrie polygon of $\{3, 5, \frac{5}{2}\}$

7G12 Remark In fact, this is not quite the case, since double rotations $(\frac{\pi}{5}, \frac{3\pi}{5})$ and $(\frac{2\pi}{5}, \frac{4\pi}{5})$ also occur. However, there are no corresponding polygons with vertices in $\{3, 3, 5\}$; moreover, the minimal edge-length of a regular polygon of radius 1 of either type is greater than that of $\{5, 3, \frac{5}{2}\}$.

Leaving out the duals, whose Petrie polygons have to be of the same type (because the double rotations are the same), the remaining cases can be found either by interchanging 5 and $\frac{5}{2}$ and then looking at the effect on the Petrie polygons, or by carrying out the same analysis as before, but starting from $\{3, 3, \frac{5}{2}\}$ which has diametral decagrams $\{\frac{10}{3}\}$ instead of decagons. Thus we easily find that $\{3, 3, \frac{5}{2}\}$ itself has Petrie polygons $\{\frac{30}{7,13}\}$, $\{3, \frac{5}{2}, 5\}$ has Petrie polygons $\{\frac{20}{3,7}\}$, and $\{\frac{5}{2}, 5, \frac{5}{2}\}$ has Petrie polygons $\{\frac{15}{2,7}\}$.

Group Presentations

We next describe the groups of the regular star-polytopes in abstract terms (see also [81]). To some extent, we shall employ the mixing operations which we introduced in Section 5A, and variants of them. However, in many cases it is much quicker to appeal to the circuit criterion Theorem 2D4. We begin with a result that we have already quoted; it occurs first in [24, p.52].

7G13 Theorem *The regular star-polyhedra* $\{5, \frac{5}{2}\}$ *and* $\{\frac{5}{2}, 5\}$ *are isomorphic to the abstract polyhedron* $\{5, 5 \mid 3\}$.

Proof. The vertex-figure replacement procedure shows that $\{5, \frac{5}{2}\}$ has the same edge-graph as the original icosahedron $\{3, 5\}$. Let $\boldsymbol{G}(\{5, \frac{5}{2}\}) = \langle \boldsymbol{r}_0, \boldsymbol{r}_1, \boldsymbol{r}_2 \rangle$, in terms of its distinguished generators as an abstract group. Since $\{5, \frac{5}{2}\} = \{3, 5\}^{\varphi}$ (with $\varphi = \varphi_2$ as is usual in such a context), we see that the required group is obtained from the Coxeter group $[3,5] = \langle \boldsymbol{s}_0, \boldsymbol{s}_1, \boldsymbol{s}_2 \rangle$ by the mixing operation

$$(\boldsymbol{s}_0, \boldsymbol{s}_1, \boldsymbol{s}_2) \mapsto (\boldsymbol{s}_0, \boldsymbol{s}_1\boldsymbol{s}_2\boldsymbol{s}_1, \boldsymbol{s}_2) = (\boldsymbol{r}_0, \boldsymbol{r}_1, \boldsymbol{r}_2).$$

Since we have

$$\boldsymbol{r}_0\boldsymbol{r}_1\boldsymbol{r}_2\boldsymbol{r}_3 = \boldsymbol{s}_0 \cdot \boldsymbol{s}_1\boldsymbol{s}_2\boldsymbol{s}_1 \cdot \boldsymbol{s}_2 \cdot \boldsymbol{s}_1\boldsymbol{s}_2\boldsymbol{s}_1 = \boldsymbol{s}_0\boldsymbol{s}_1\boldsymbol{s}_1\boldsymbol{s}_2 \sim \boldsymbol{s}_0\boldsymbol{s}_1,$$

we see that the hole of $\{5, \frac{5}{2}\}$ is a trigon, and therefore $\{5, \frac{5}{2}\}$ is a quotient of $\{5, 5 \mid 3\}$. Indeed, this trigon is a face of the icosahedron $\{3, 5\}$; since each edge-circuit can be contracted over such trigons, Theorem 2D4 yields the required isomorphism. □

7G14 Remark It was, perhaps, unnecessary to appeal to the circuit criterion here, since the isomorphism is easy to establish by reversing the operation φ_2. However, this points out the best approach subsequently.

We now move on to the 4-polytopes. We shall soon see that the interchange of 5 and $\frac{5}{2}$ induces isomorphism of polytopes here as well. If we anticipate this result, and also confine our attention to one of each dual pair, we conclude that we need only consider the three polytopes $\{3, 5, \frac{5}{2}\}$, $\{5, \frac{5}{2}, 5\}$ and $\{5, 3, \frac{5}{2}\}$. For the last, we must recall a concept introduced in Section 2D. If the automorphism group $\boldsymbol{G}(\mathcal{P})$ of the abstract regular d-polytope $\mathcal{P}$ is obtained from the Coxeter group $[p_1, \ldots, p_{d-1}] = \langle \boldsymbol{r}_0, \ldots, \boldsymbol{r}_{d-1} \rangle$ by imposing the single extra relation

$$(\boldsymbol{r}_0\boldsymbol{r}_1 \cdots \boldsymbol{r}_{d-1}\boldsymbol{r}_{d-2} \cdots \boldsymbol{r}_1)^r = \boldsymbol{e},$$

then we write

$$\mathcal{P} = \{p_1, \ldots, p_{d-1} \mid r\};$$

the extra relation determines the deep hole. Recall also that the group of the latter is generated by the involutions $\boldsymbol{r}_0$ and $\boldsymbol{r}_1 \cdots \boldsymbol{r}_{d-1}\boldsymbol{r}_{d-2} \cdots \boldsymbol{r}_1$; the former interchanges the two vertices of the initial edge in the base flag of $\mathcal{P}$, and the latter is a conjugate of $\boldsymbol{r}_{d-1}$ which fixes the initial vertex. The relation for a deep hole is preserved under duality, so that

$$\mathcal{P}^\delta = \{p_{d-1}, \ldots, p_0 \mid r\}.$$

We remark that Theorem 2D14 showed that, for $k \geqslant 3$, a polytope of type $\{3^{k-2}, q\}$ has deep holes of type $\{q\}$.

The starry regular 4-polytopes are then given by

7G15 Theorem *The following isomorphisms hold:*

(a) $\{3, 5, \frac{5}{2}\} \cong \{\{3, 5\}, \{5, 5 \mid 3\}\}$,
(b) $\{5, \frac{5}{2}, 5\} \cong \{\{5, 5 \mid 3\}, \{5, 5 \mid 3\}\}$,
(c) $\{5, 3, \frac{5}{2}\} \cong \{5, 3, 5 \mid 3\}$.

Proof. Note that (a) and (b) say that the two polytopes are universal of their type, while (c) says that the type of $\{5, 3, \frac{5}{2}\}$ is determined by trigonal deep holes. Of course, $\{5, 3, \frac{5}{2}\}$ cannot be universal of type $\{\mathbf{5}, \mathbf{3}, \mathbf{5}\}$, since the latter is infinite. The theorem also justifies the earlier remark about the interchange of 5 and $\frac{5}{2}$; in particular, it follows from (c) that $\{5, 3, \frac{5}{2}\}$ is isomorphic to its dual $\{\frac{5}{2}, 3, 5\}$.

As with Theorem 7G13, we appeal to the circuit criterion Theorem 2D4. The crucial observation is that, as we saw in the classification in Table 7G8, all three of these star-polytopes have the same edge-graph as $\{3, 3, 5\}$, and their vertex-figures have the same symmetry group. Since each edge-circuit can be contracted over the trigonal faces of $\{3, 3, 5\}$, it is consequently enough to show that a typical such trigon is induced for each polytope.

For (a), since $\{3, 5, \frac{5}{2}\} = \{3, 3, 5\}^{\varphi}$, we see from Theorem 7G13 that the vertex-figure must be $\{5, \frac{5}{2}\} \cong \{\mathbf{5}, \mathbf{5} \mid \mathbf{3}\}$, and the facet must be $\{3, 5\} \cong \{\mathbf{3}, \mathbf{5}\}$. Since the trigonal faces of the facets are faces of $\{3, 3, 5\}$, (a) follows from the circuit criterion.

For (b), the vertex-figure must be $\{\frac{5}{2}, 5\} \cong \{\mathbf{5}, \mathbf{5} \mid \mathbf{3}\}$, and the facet must be $\{5, \frac{5}{2}\} \cong \{\mathbf{5}, \mathbf{5} \mid \mathbf{3}\}$. A trigonal hole of a facet will be a face of $\{3, 3, 5\}$, and then the circuit criterion yields the claim of (b).

For (c), we only have to work a little harder. We have $\{5, 3, \frac{5}{2}\} = \{5, \frac{5}{2}, 5\}^{\varphi}$, yielding a corresponding operation

$$(\boldsymbol{s}_0, \ldots, \boldsymbol{s}_3) \mapsto (\boldsymbol{s}_0, \boldsymbol{s}_1, \boldsymbol{s}_2\boldsymbol{s}_3\boldsymbol{s}_2, \boldsymbol{s}_3) =: (\boldsymbol{r}_0, \ldots, \boldsymbol{r}_3)$$

on their abstract automorphism groups. For the deep hole of $\{5, 3, \frac{5}{2}\}$, we have

$$\begin{aligned}\boldsymbol{r}_0\boldsymbol{r}_1\boldsymbol{r}_2\boldsymbol{r}_3\boldsymbol{r}_2\boldsymbol{r}_1 &= \boldsymbol{s}_0 \cdot \boldsymbol{s}_1 \cdot \boldsymbol{s}_2\boldsymbol{s}_3\boldsymbol{s}_2 \cdot \boldsymbol{s}_3 \cdot \boldsymbol{s}_2\boldsymbol{s}_3\boldsymbol{s}_2 \cdot \boldsymbol{s}_1 \\ &= \boldsymbol{s}_0\boldsymbol{s}_1\boldsymbol{s}_3\boldsymbol{s}_2\boldsymbol{s}_3\boldsymbol{s}_1 \\ &\sim \boldsymbol{s}_0\boldsymbol{s}_1\boldsymbol{s}_2\boldsymbol{s}_1,\end{aligned}$$

giving the trigonal hole of the facet, which is (as already noted) a face of $\{3, 3, 5\}$. Once again, the claim in (c) follows from the circuit criterion. □

7G16 Remark Since $\{\{\mathbf{5}, \mathbf{5} \mid \mathbf{3}\}, \{\mathbf{5}, \mathbf{5} \mid \mathbf{3}\}\}^{\varphi} = \{\mathbf{5}, \mathbf{3}, \mathbf{5} \mid \mathbf{3}\}$, it follows from Theorem 5B14 that the former has deep hole $\{\mathbf{3}\}$, which is perhaps not obvious. Similarly, $\{\{\mathbf{3}, \mathbf{5}\}, \{\mathbf{5}, \mathbf{5} \mid \mathbf{3}\}\}^{\varphi} = \{\mathbf{3}, \mathbf{3}, \mathbf{5}\}$, so that the former has deep hole $\{\mathbf{3}\}$, which can easily be seen directly.

Allomorphs

When we come to vertex-figure replacement in Chapter 17, we shall want to know the effect of replacing one realization of $\{\mathbf{3}, \mathbf{3}, \mathbf{5}\}$ by its allomorph with the same vertices and initial vertex. In other words, we need a mixing operation that passes from one to the other. We can do this geometrically, by reference to the realization $\{3, 3, 5\}$ that we have discussed above.

In fact, there are two different ways to achieve this.

7G17 Proposition *The mixing operations*

$$(\boldsymbol{r}_0, \ldots, \boldsymbol{r}_3) \mapsto (\boldsymbol{s}_0, \boldsymbol{r}_1\boldsymbol{r}_2\boldsymbol{r}_3\boldsymbol{r}_2\boldsymbol{r}_1, \boldsymbol{r}_3\boldsymbol{r}_2\boldsymbol{r}_3, \boldsymbol{r}_2) =: (\boldsymbol{s}_0, \ldots, \boldsymbol{s}_3),$$

where

$$\boldsymbol{s}_0 = \boldsymbol{r}_0(\boldsymbol{r}_1\boldsymbol{r}_2\boldsymbol{r}_3)^5\boldsymbol{r}_0(\boldsymbol{r}_1\boldsymbol{r}_2\boldsymbol{r}_3)^5\boldsymbol{r}_0 \quad \text{or} \quad \boldsymbol{r}_0(\boldsymbol{r}_1\boldsymbol{r}_2\boldsymbol{r}_3)^5\boldsymbol{r}_0(\boldsymbol{r}_0\boldsymbol{r}_1\boldsymbol{r}_2\boldsymbol{r}_3)^{15},$$

are allomorphisms of $\{3, 3, 5\}$.

Proof. The first choice of $\boldsymbol{s}_0$ is a conjugate of $\boldsymbol{r}_0$ that sends the initial vertex into one of layer $\mathcal{L}_3$, while for the second choice bear in mind that multiplication by the central inversion interchanges reflexions in lines and those in their normal hyperplanes. □

For our applications, we shall be more interested in the central quotient $\{3, 3, 5\}/2 = \{3, 3, 5 : 15\}$, for which $(\boldsymbol{r}_0\boldsymbol{r}_1\boldsymbol{r}_2\boldsymbol{r}_3)^{15} = \boldsymbol{e}$.

7G18 Corollary *The mixing operation*

$$(\boldsymbol{r}_0, \ldots, \boldsymbol{r}_3) \mapsto (\boldsymbol{r}_0(\boldsymbol{r}_1\boldsymbol{r}_2\boldsymbol{r}_3)^5\boldsymbol{r}_0, \boldsymbol{r}_1\boldsymbol{r}_2\boldsymbol{r}_3\boldsymbol{r}_2\boldsymbol{r}_1, \boldsymbol{r}_3\boldsymbol{r}_2\boldsymbol{r}_3, \boldsymbol{r}_2) =: (\boldsymbol{s}_0, \ldots, \boldsymbol{s}_3)$$

is an allomorphism of $\{3, 3, 5\}/2 = \{3, 3, 5 : 15\}$.

An operation that replaces $\{5, 3, 3\}$ by its allomorph is straightforward. We apply edge replacement ε of Section 5B, observing that the vertex-figure at the opposite vertex in $\mathcal{L}_{30}$ to the initial one $\boldsymbol{v} \in \mathcal{L}_0$ forms layer $\mathcal{L}_5$. In turn, a typical such vertex is obtained from $\boldsymbol{v}$ by $\boldsymbol{z}(\mathcal{F}) = (\boldsymbol{r}_0\boldsymbol{r}_1\boldsymbol{r}_2)^5$, with $\mathcal{F} = \{5, 3\}$ the initial facet. If we further apply the central inversion $\boldsymbol{\zeta}$, we arrive at

7G19 Proposition *An allomorphism of* $\{5, 3, 3\}$ *is given by the operation* $\varepsilon\zeta$, *namely,*

$$(\boldsymbol{r}_0, \ldots, \boldsymbol{r}_3) \mapsto \big((\boldsymbol{r}_0\boldsymbol{r}_1\boldsymbol{r}_2)^5(\boldsymbol{r}_0\boldsymbol{r}_1\boldsymbol{r}_2\boldsymbol{r}_3)^{15}, \boldsymbol{r}_3, \boldsymbol{r}_2, \boldsymbol{r}_1\big) =: (\boldsymbol{s}_0, \ldots, \boldsymbol{s}_3).$$

The operation is even simpler when we pass to the central quotient.

7G20 Corollary *Edge replacement*

$$\varepsilon\colon (\boldsymbol{r}_0, \ldots, \boldsymbol{r}_3) \mapsto \big((\boldsymbol{r}_0\boldsymbol{r}_1\boldsymbol{r}_2)^5, \boldsymbol{r}_3, \boldsymbol{r}_2, \boldsymbol{r}_1\big) =: (\boldsymbol{s}_0, \ldots, \boldsymbol{s}_3)$$

induces an allomorphism of $\{5, 3, 3\}/2 = \{5, 3, 3 : 15\}$.

Notes to Section 7G

1. It is curious that Coxeter failed to observe the straightforward direct proof of Theorem 7G1, even though he tacitly made the assumption that the symmetry groups that he worked with were generated by hyperplane reflexions. It has to be said, though, that this proof appears for the first time here – it was also missed in [71] and in [99, Section 7D].

2. Vertex-figure replacement was introduced by McMullen as his first original result (though published second as [71]). This result, for which he retains a natural affection, was formulated as Theorem 7G3; in fact, he found it before he formally became a research student.
3. In the original paper [81], which was followed in [99, Section 7D], Theorems 7G13 and 7G15 were proved by explicitly finding the changes of generators leading from the convex polytopes to the starry ones, and then reversing them. The use of the circuit criterion considerably shortens the proofs of the results, and probably makes the results themselves more transparent as well.
4. All the regular polyhedra are depicted in the anaglyphs of [47].
5. Figure 7G9 is projected in a different way from those of Figures 7E1 and 7G10; the motive is to display the layers of the vertex-figures $\{5, 3\}$ or $\{\frac{5}{2}, 3\}$.
6. McMullen's picture of $\{\frac{5}{2}, 3, 3\}$ projected into the plane of a component $\{\frac{30}{7}\}$ of its Petrie polygon $\{\frac{3}{7,13}\}$ is reproduced in [30, Figure 4.7D].

7H Honeycombs

We have already found the classical regular apeirotopes in $\mathbb{R}$ and $\mathbb{E}^2$; recall that they are the *apeirogon* $\{\infty\}$ in $\mathbb{R}$ and the three regular tesselations $\{4, 4\}$, $\{3, 6\}$ and $\{6, 3\}$ in $\mathbb{E}^2$. We now complete the enumeration of the classical regular apeirotopes or *honeycombs* in higher dimensions. The most important observation to make is that the symmetry group of such a honeycomb must be given by one of the string diagrams of Table 1E12, namely, $\mathbf{R}_{d+1}, \mathbf{U}_5, \mathbf{V}_3$ or $\mathbf{W}_2$, of which we have already dealt with the latter two.

All Dimensions

In view of this observation, the only infinite family of suitable groups consists of the $\mathbf{R}_{d+1}$. In $\mathbb{E}^d$, we thus have the self-dual honeycomb $\mathsf{R}^{d+1} := \{4, 3^{d-2}, 4\}$ for $d \geqslant 2$; this is the familiar *cubic tiling*. The symmetry group has generators (regarded as hyperplanes)

7H1
$$\begin{aligned} R_0\colon\ & \xi_1 = \tfrac{1}{2}, \\ R_j\colon\ & \xi_j = \xi_{j+1} \qquad \text{for } j = 1, \ldots, d-1, \\ R_d\colon\ & \xi_d = 0, \end{aligned}$$

in terms of cartesian coordinates $\xi_1, \ldots, \xi_d$. Thus the vertex-set is $\mathbb{Z}^d$, the set of points of $\mathbb{E}^d$ with integer cartesian coordinates.

This shows that the family can be thought of as beginning with $\{\infty\}$. In fact, this is very natural, when we note that

7H2
$$\mathbf{R}_{d+1} \cong (\mathbf{W}_2)^d \rtimes \mathbf{A}_{d-1},$$

with the symmetric group $\mathbf{A}_{d-1} \cong \boldsymbol{S}_d$ acting in the obvious way by permuting coordinates. Compare the corresponding expression (5C3) for the symmetry group of the d-cube.

Geometric Construction

It remains for us to treat the dual honeycombs $\{3,3,4,3\}$ and $\{3,4,3,3\}$ in $\mathbb{E}^4$; we observe that Wythoff's construction of Section 4A already ensures their existence.

We begin by describing a geometric construction for $\{3,4,3,3\}$; as might be expected, it is closely related to one of our constructions for $\{3,4,3\}$ in Section 7C. We saw that the mid-points of the edges of the 4-staurotope $\{3,3,4\}$ are the vertices of $\{3,4,3\}$. Equally, we may take these vertices to be the centres of the tetragonal 2-faces $\{4\}$ of the 4-cube $\{4,3,3\}$; it is this alternative view which generalizes.

Now consider the cubic tiling $\mathsf{R} := \{4,3,3,4\}$. The centres of its tetragonal 2-faces are also the centres of the 2-faces of the dual cubic tiling R^δ, whose vertices are the centres of the cubic cells $\{4,3,3\}$ of R. So, within each cell of R and of R^δ we inscribe a 24-cell $\{3,4,3\}$. We initially colour these 24-cells black or white, according as they arise from cells of R or of R^δ. Then two adjacent black cells meet on common facets with normals $\pm e_j$ for some $j = 1, \ldots, 4$, as do two adjacent white cells. But adjacent black and white cells meet on common facets with normals $\frac{1}{2}(\pm1,\pm1,\pm1,\pm1)$. It is clear that each 24-cell is completely surrounded by others in this way; each vertex belongs to four black and four white cells.

7H3 Remark As a subgroup of $\mathbf{G}(\mathsf{R})$, the new vertex-figure has the symmetry of $\{4\} \times \{4\}$, rather than the whole symmetry of $\{4,3,3\}$. The partition into black and white cells respects this symmetry.

If we take R to have vertex-set $2\mathbb{Z}^4$, consisting of all quadruples of even integers, then R^δ has vertex-set $2\mathbb{Z} + (1,1,1,1)$, which consists of all quadruples of odd integers. We then find that the vertex-set of $\{3,4,3,3\}$ can be taken to consist of all $(\zeta_1, \ldots, \zeta_4) \in \mathbb{Z}^4$ with two out of the four ζ_j even and the other two odd. The vertices of the dual honeycomb $\{3,3,4,3\}$ then consist of the centres of the cells of $\{3,4,3,3\}$, which are the vertices of R and R^δ together, namely, all $(\zeta_1, \ldots, \zeta_4) \in \mathbb{Z}^4$ with $\zeta_1 \equiv \cdots \equiv \zeta_4 \pmod 2$.

Bearing in mind the two different forms of the vertices of $\{3,4,3\}$, it is easy to see that another choice for the vertex-set of $\{3,3,4,3\}$ consists of all $(\zeta_1, \ldots, \zeta_4) \in \mathbb{Z}^4$ with $\zeta_1 + \cdots + \zeta_4 \equiv 0 \pmod 2$. We find this geometrically using an alternative construction for $\{3,4,3,3\}$. Begin with R and R^δ as before, but with the vertex-set of R now $\mathbb{Z}^4$. Since $\{4,3,3,4\}$ is 3-collapsible, we may colour the cubical cells of R^δ alternately black and white, with adjacent cells carrying different colours; suppose that the cell corresponding to the initial vertex o of R is black. We repeat the second construction of $\{3,4,3\}$ from $\{4,3,3\}$ on each black cell. Then each white cell is divided into eight pyramids with bases its facets $\{4,3\}$, and the new cells $\{3,4,3\}$ clearly fit together exactly to form the new honeycomb $\{3,4,3,3\}$. Notice, however, that the vertices of this $\{3,4,3,3\}$ do not take on such a tidy form as before, although it is easy to see that the vertex-set of the corresponding dual $\{3,3,4,3\}$ is as asserted.

7H4 Remark When halved, namely,

$$V := \{(\xi_1, \ldots, \xi_4) \in \tfrac{1}{2}\mathbb{Z}^4 \mid \xi_1 \equiv \cdots \equiv \xi_4 \pmod 1\},$$

the vertex-set of the first copy of $\{3, 3, 4, 3\}$ consists of the integer quaternions (in the usual representation as coordinate vectors). This set and the second,

$$V^* := \{(\eta_1, \ldots, \eta_4) \in \mathbb{Z}^4 \mid \eta_1 + \cdots + \eta_4 \equiv 0 \pmod 2\},$$

are then *reciprocal* lattices, in the sense that

$$V^* = \{y \in \mathbb{E}^4 \mid \langle x, y\rangle \in \mathbb{Z} \text{ for all } x \in V\}.$$

Algebraic Constructions

If we take the unit normals to the fundamental simplex of the group $\mathbf{U}_5 = [3, 3, 4, 3]$ in the form given in the proof of Theorem 1E13, then we can write down the group generators (as usual identifying reflexions with hyperplanes) as

7H5
$$\begin{aligned} R_0 &: \xi_1 + \xi_2 = 1,\\ R_1 &: \xi_2 = \xi_3,\\ R_2 &: \xi_3 = \xi_4,\\ R_3 &: \xi_4 = 0,\\ R_4 &: \xi_1 = \xi_2 + \xi_3 + \xi_4, \end{aligned}$$

in terms of cartesian coordinates $\xi_1, \ldots, \xi_4$). This gives the vertices of $\{3, 3, 4, 3\}$ in the second form, namely, all $(\xi_1, \ldots, \xi_4) \in \mathbb{Z}^4$ with $\xi_1 + \cdots + \xi_4$ even.

The construction also shows that the vertices of the cubic tiling $\{4, 3, 3, 4\}$ occur among those of $\{3, 3, 4, 3\}$, and this indicates that $\mathbf{R}_5 = [4, 3, 3, 4]$ is a subgroup of $\mathbf{U}_5 = [3, 3, 4, 3]$. In fact, that construction strongly suggests how this happens. The diagrams of $\mathbf{Q}_5$ and $\mathbf{R}_5$ admit obvious twists, and suitable applications of Wythoff's construction yield

7H6 $\{3, 4, 3, 3\}$ = 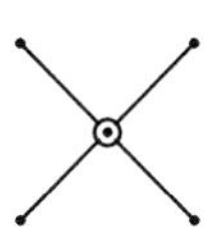 = 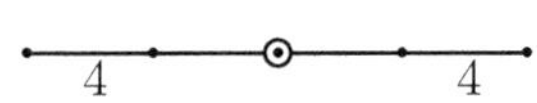

Compare here the corresponding expressions for the 24-cell $\{3, 4, 3\}$ in (7C2).

Further note that, as the alternative forms of the vertex-set indicate, $\mathbf{U}_5$ is a subgroup of itself of index 4 (we expect index 16 through self-symmetry). In terms of Wythoff's construction, this relationship is illustrated by

7H7 $\{3, 4, 3, 3\}$ = 

Collapsibility

Collapsibility (see Section 2F) casts another light on these relationships. The 1-collapsibility of the cubic tiling $\{4, 3^{d-2}, 4\}$ for each $d \geqslant 2$ merely reflects the obvious fact that its edge-graph is bipartite, while its d-collapsibility yields the 2-colouring of the tiles. However, we obtain rather more from $\{3,3,4,3\}$ and its dual.

First, since $\{3,3,4,3\}$ is 3-collapsible, we can 4-colour its vertices. The vertices carrying the same colour as a given one are obtained by moving along diameters of staurotopal cells $\{3,3,4\}$. If we take the vertices in the second form, namely, all $(\xi_1,\xi_2,\xi_3,\xi_4) \in \mathbb{Z}^4$ with $\xi_1+\cdots+\xi_4$ even, then the ones of the same colour as the origin o are all those with $\xi_1 \equiv \cdots \equiv \xi_4 \pmod 2$; in other words, they are the vertices of the first kind. The four colours then give the subgroup relationship just mentioned.

In a similar way, $\{3,4,3,3\}$ is 2-collapsible. Just as with the 24-cell, we conclude that the graph of $\{3,4,3,3\}$ is face-cyclic. Of course, we should not expect the corresponding construction of the 600-cell $\{3,3,5\}$ to generalize; however, see Remark 7H8 below. But we do see that the vertices of $\{3,4,3,3\}$ can be 3-coloured. The three colour classes form three copies of the vertex-set of the dual tiling $\{3,3,4,3\}$. Indeed, this fits in with what we have just said about $\{3,3,4,3\}$: any three out of the four colour classes of its vertices form the vertex-set of an inscribed copy of $\{3,4,3,3\}$, the vertices of whose dual form the fourth class.

7H8 Remark We can generalize the construction of the 600-cell so far as obtaining a highly symmetric but non-regular tiling of $\mathbb{E}^4$ by snub 24-cells and regular 4-staurotopes and 4-simplices. What this tiling does show by careful inspection is that the dihedral angles of the three regular convex 4-polytopes with tetrahedral facets (namely, $\{3,3,3\}$, $\{3,3,4\}$ and $\{3,3,5\}$) sum to 2π.

Petrie Apeirogons

We end the section by calculating the types of the Petrie apeirogons of the regular honeycombs. We begin by recalling two facts about them:

- they are indeed apeirogons (that is, infinite);
- they are full-dimensional.

Hence a d-dimensional classical regular honeycomb has a Petrie apeirogon of type

$$\left\{\frac{s}{0,t_2,\ldots,t_m}\right\},$$

for some $0 < t_2 < \cdots \leqslant \frac{1}{2}s$, with $m := \lfloor \frac{1}{2}d \rfloor$. Morover, if d is even, then necessarily s is also even and $t_m = \frac{1}{2}s$.

There is nothing to say about the cases $d \leqslant 2$, since the linear apeirogon $\{\infty\} = \{\frac{1}{0}\}$ is its own Petrie apeirogon, and the planar tessellations $\{4,4\}$, $\{3,6\}$ and $\{6,3\}$ must have Petrie apeirogons $\{\frac{2}{0,1}\}$. We are thus left with the cubic tilings $\{4,3^{d-2},4\}$ for each $d \geqslant 3$ and the dual 4-dimensional honeycombs $\{3,3,4,3\}$ and $\{3,4,3,3\}$.

There is no problem about the cubic tiling $\{4,3^{d-2},4\}$, which we take in the standard form as above. Each d successive edges of a Petrie apeirogon belong to a Petrie polygon of a cubical cell $\{4, 3^{d-2}\}$; we lose no generality in assuming these to be in directions $e_1, \ldots, e_d$ (in that order). Since the Petrie apeirogon now leaves this cube to move to the adjacent one which shares the previous $d-1$ edges, its next edge must therefore be in direction e_1, then e_2, and so on. Thus d steps along the Petrie apeirogon yields a translation, and so we conclude

7H9 Proposition *For each $d \geqslant 2$, the Petrie apeirogon of the d-dimensional cubic tiling $\{4, 3^{d-2}, 4\}$ is of type $\{\frac{d}{0,1,\ldots,m}\}$, with $m := \lfloor \frac{1}{2}d \rfloor$.*

For the remaining cases, we shall prove

7H10 Proposition *The Petrie apeirogons of the dual 4-dimensional regular honeycombs $\{3, 3, 4, 3\}$ and $\{3, 4, 3, 3\}$ are of type $\{\frac{6}{0,2,3}\}$.*

Proof. As we have already pointed out, the Petrie apeirogon G, say, is of type $\{\frac{2s}{0,t,s}\}$ for some $0 < t < s$. Let $T := R_1R_3 \cdot R_0R_2R_4$ be the 'translational' symmetry of G, so that we take $[3, 3, 4, 3] = \langle R_0, \ldots, R_4 \rangle$. Then the image $\widetilde{T}$ of T in the point (or special) group $[3, 4, 3]$ is the rotational symmetry of a polygon $\widetilde{H} = \{\frac{2s}{t,s}\}$. Since $\widetilde{T}$ leaves the line parallel to the axis of G invariant, it must belong to a subgroup $[3, 4]$ or $[3] \times [2]$ of $[3, 4, 3]$, and it then follows that $\widetilde{H} = \{\frac{4}{1,2}\}$ or $\{\frac{6}{2,3}\}$. Our task is thus to eliminate the former possibility.

Suppose that the Petrie apeirogon G were of type $\{\frac{4}{0,1,2}\}$. In $\{3, 3, 4, 3\}$, four successive edges of G are those of a Petrie polygon of a cell $C = \{3, 3, 4\}$, and so go between diametrically opposite vertices v and w, say, of C. Then $w - v$ is the basic translational symmetry of G, so that the next four edges of G are the translates of the first four by $w - v$. Now the intermediate sets of four successive edges of G should belong to cells of $\{3, 3, 4, 3\}$ as well, which form adjacent ones containing w. However, it takes six such adjacent cells (at least) to get from C to its translate $C + (w - v)$; that is, six steps through adjacent facets are needed to get from one facet of the vertex-figure $\{3, 4, 3\}$ to its opposite. We have thus obtained a contradiction, and so established the proposition. □

7J Regular Compounds

If P, Q are regular polytopes (or apeirotopes – at this stage we do not make any distinction), then we say that Q is *inscribed* in P if $\operatorname{vert} \mathsf{Q} \subseteq \operatorname{vert} \mathsf{P}$, and write $\mathsf{Q} \preccurlyeq \mathsf{P}$. A *vertex-regular compound* is a family $\mathcal{C}$ of polytopes satisfying the following conditions:

- there is a polytope P such that $\mathsf{Q} \preccurlyeq \mathsf{P}$ for each $\mathsf{Q} \in \mathcal{C}$,
- the symmetry group $\mathbf{G}(\mathcal{C})$ of the compound is transitive on $\mathcal{C}$,
- further, $\mathbf{G}(\mathcal{C})$ is transitive on $\operatorname{vert} \mathsf{P}$.

Our definition is sufficiently general that it does not (for instance) demand that $\operatorname{rank} \mathsf{P} = \operatorname{rank} \mathsf{Q}$ for $\mathsf{Q} \in \mathcal{C}$, although we shall not want that level of generality here. If n copies of Q are each inscribed in P, and together cover $\operatorname{vert} \mathsf{P}$ m times, then we employ the notation $m\mathsf{P}[n\mathsf{Q}] := \mathcal{C}$.

When R is classical and $\mathcal{C}$ consists of classical polytopes of the same rank, then it can make sense to talk about the dual concept of a *cell-regular compound*. The notation for such a compound is $[n\mathsf{Q}]k\mathsf{R}$. Naturally, a *regular* compound is one which is both vertex- and cell-regular; the notation will be $m\mathsf{P}[n\mathsf{Q}]k\mathsf{R}$. In such a case, it is usual that P and R be the same or dual polytopes.

Until the end of this section, we shall mostly confine our attention to the case where both P and Q are finite and convex; we shall also concentrate on vertex-regular compounds, often leaving the dual cell-regular compounds to be understood. The many regular compounds of 4-polytopes can only be found by systematic enumeration; quaternions provide the key to this, and so extensive reference will be made to Section 1K.

Inscription Criteria

Before we describe the regular compounds, we need some preliminary results. For the first of these, let P, Q be two (finite) polytopes in $\mathbb{E}^d$. We give here three criteria which P and Q must satisfy if $\mathsf{Q} \preccurlyeq \mathsf{P}$; the notation $\varphi(\mathsf{P})$ and $\chi(\mathsf{P})$ was introduced in Section 7A.

7J1 Theorem *If* P, Q *are finite regular d-polytopes such that* $\mathsf{Q} \preccurlyeq \mathsf{P}$, *then*

(a) $\varphi(\mathsf{P}) \leqslant \varphi(\mathsf{Q})$,

(b) $\chi(\mathsf{P}) \leqslant \chi(\mathsf{Q})$,

(c) $\Gamma(\mathsf{Q})$ *is a subvector of* $\Gamma(\mathsf{P})$.

Proof. Part (a) is clear, since edges of Q are diagonals of P; recall that $\gamma_1(\mathsf{P}) = \cos 2\varphi(\mathsf{P})$ is the first entry of the cosine vector of P. For part (b), consider the inspheres of P and Q. If P and Q have circumradius ρ, then P has inradius $\sigma(\mathsf{P}) = \rho \cos \chi(\mathsf{P})$, and similarly for $\sigma(\mathsf{Q})$. Since $\mathsf{Q} \subseteq \mathsf{P}$, it is clear that $\sigma(\mathsf{Q}) \leqslant \sigma(\mathsf{P})$. It follows that $\chi(\mathsf{Q}) \geqslant \chi(\mathsf{P})$, as claimed. For (c), just observe that each layer of Q must be inscribed in some layer of P. □

There is a special case to which we shall appeal later.

7J2 Corollary *If* P, Q *are two dual regular d-polytopes such that* $\mathsf{Q} \preccurlyeq \mathsf{P}$, *then each facet of* P *contains an inscribed copy of a facet of* Q.

Proof. We now have $\chi(\mathsf{P}) = \chi(\mathsf{Q}) =: \chi$, say; thus P and Q have a common insphere S, say. Let $\mathsf{F} \lhd \mathsf{P}$ be any facet. Then F touches S at its centre p, say, and is separated from S by $\operatorname{bd} \mathsf{Q}$. The only way that this can happen is that some facet $\mathsf{G} \lhd \mathsf{Q}$ touches S at the same point p. Then p is also the centre of G, and since $\operatorname{vert} \mathsf{G} \subset \operatorname{vert} \mathsf{Q} \subseteq \operatorname{vert} \mathsf{P}$, it follows that G must be inscribed in F, as claimed. □

7J3 Remark Analogous arguments apply to regular compounds of hyperbolic honeycombs. If correctly interpreted, Corollary 7J2 still holds for geometrically dual euclidean honeycombs, but we cannot use the same proof.

Common Subgroups

We next look at symmetry groups. Suppose that we are given two regular d-polytopes P, Q such that $\mathsf{Q} \preccurlyeq \mathsf{P}$. We consider here the *common subgroup*

$$\mathbf{K}(\mathsf{P}, \mathsf{Q}) := \mathbf{G}(\mathsf{P}) \cap \mathbf{G}(\mathsf{Q}).$$

We emphasize that we have two *geometric* groups, rather than abstract ones (which may even be isomorphic). We write

$$k(\mathsf{P}, \mathsf{Q}) := [\mathbf{G}(\mathsf{P}) : \mathbf{K}(\mathsf{P}, \mathsf{Q})]$$

for the index of $\mathbf{K}(\mathsf{P}, \mathsf{Q})$ in $\mathbf{G}(\mathsf{P})$, with $k(\mathsf{Q}, \mathsf{P})$ defined analogously. We then have a general result about compounds.

7J4 Theorem *If* P, Q *are two regular d-polytopes such that* $\mathsf{Q} \prec \mathsf{P}$, *then there exists a vertex-regular compound* $c\mathsf{P}[k(\mathsf{P}, \mathsf{Q})\mathsf{Q}]$ *for some* c.

Proof. Indeed, if we write $k := k(\mathsf{P}, \mathsf{Q})$, then each (right) coset representative of $\mathbf{K}(\mathsf{P}, \mathsf{Q})$ in $\mathbf{G}(\mathsf{P})$ will give rise to a distinct copy of Q inscribed in P. It is clear that the resulting compound is vertex-regular, which means that there will be an appropriate number c. □

7J5 Remark The number c in Theorem 7J4 will depend on the relative orders of $\mathbf{G}(\mathsf{P})$ and $\mathbf{G}(\mathsf{Q})$ and the relative numbers of their vertices. We shall not give the general calculation here; it will be sufficiently illustrated by the particular examples.

Compounds of Polygons

It should be obvious that $\{q\}$ can be inscribed in $\{p\}$ only if $p = kq$ for some k. In this case, $k(\mathsf{P}, \mathsf{Q}) = k$ and $k(\mathsf{Q}, \mathsf{P}) = 1$, and we have a vertex- and cell-regular compound

$$\{kq\}[k\{q\}]\{kq\}.$$

Compounds of Polyhedra

The subgroup relationship $\mathbf{A}_3 < \mathbf{C}_3$ (or $[3, 3] < [3, 4]$) leads immediately to the vertex- and cell-regular compound

$$\{4, 3\}[2\{3, 3\}]\{3, 4\};$$

here, we have $k(\mathsf{P}, \mathsf{Q}) = 2$ and $k(\mathsf{Q}, \mathsf{P}) = 1$. Of course, this compound is closely connected with the fact that $\{3, 3\} = \{4, 3\}^{\eta}$ (see the notes at the end of the section).

The construction in Section 7D of the icosahedron and dodecahedron shows that we can inscribe the cube in the dodecahedron, and hence inscribe the tetrahedron also. (More precisely, the vertices of $\{3, 5\}$ lie in the edge of $\{3, 4\}$, so that the edges of $\{4,3\}$ lie in the faces of $\{5,3\}$.) The case of the cube is depicted in Figure 7J6 (see the notes at the end of the section).

7J6

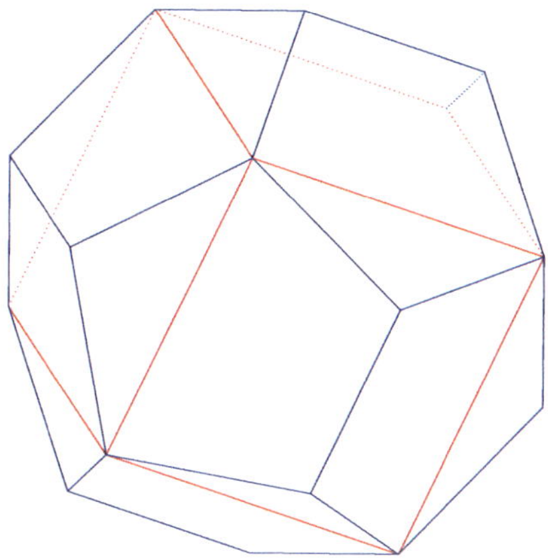

Cube inscribed in dodecahedron

Here, it is useful to look at the common subgroup $\mathbf{K}(\mathsf{P},\mathsf{Q})$ in detail. For the tetrahedron Q in the dodecahedron P, we have common subgroup

$$\mathbf{K}(\mathsf{P},\mathsf{Q}) = \mathbf{A}_3^+ = [3,3]^+,$$

its rotation subgroup. Hence we have $k(\mathsf{P},\mathsf{Q}) = 10$ and $k(\mathsf{Q},\mathsf{P}) = 2$, and in view of Theorem 7J4 we therefore have a vertex- and cell-regular compound

$$2\{5,3\}[10\{3,3\}]2\{3,5\}.$$

However, we can allow just the rotation subgroup $\mathbf{G}_3^+ = [3,5]^+$ to act on an inscribed tetrahedron, and this yields the compound

$$\{5,3\}[5\{3,3\}]\{3,5\}.$$

In fact, there are two such compounds in a given dodecahedron, related by any opposite symmetry, such as the central inversion Z (see the notes at the end of the section).

For the cube $\mathsf{Q} = \{4,3\}$ in the dodecahedron P, we have common subgroup

$$\mathbf{K}(\mathsf{P},\mathsf{Q}) = \mathbf{A}_3^+ \times \mathbf{Z}_2 = [3,3]^+ \times \mathbf{Z}_2,$$

with $\mathbf{Z}_2 = \langle Z \rangle$ the subgroup generated by the central inversion Z. We now have $k(\mathsf{P},\mathsf{Q}) = 5$ and $k(\mathsf{Q},\mathsf{P}) = 2$, which leads to the vertex-regular compound

$$2\{5,3\}[5\{4,3\}].$$

Naturally, we have the dual cell-regular compound $[5\{3,4\}]2\{3,5\}$; we can see how this arises, just from the way that we initially constructed the icosahedron from the octahedron in Section 7D.

The 24-Cell

In effect, in Section 7C we have already described the compounds

7J7 $\{4,3,3\}[\mathit{2}\{3,3,4\}], \quad [\mathit{2}\{4,3,3\}]\{3,3,4\},$

7J8 $\{3,4,3\}[\mathit{3}\{3,3,4\}]\mathit{2}\{3,4,3\}, \quad \mathit{2}\{3,4,3\}[\mathit{3}\{4,3,3\}]\{3,4,3\},$

that involve only the rational polytopes, but we shall briefly recall some salient features.

The vertex-set of the 24-cell $\mathsf{F} := \{3,4,3\}$ can be identified with $\mathbf{T}$, and that of the dual 24-cell F^{δ} with $\mathbf{U}$. The symmetry group $[3,4,3]$ of these dual polytopes is $(\mathbf{O}/\mathbf{T};\mathbf{O}/\mathbf{T})^*$ (consisting of the mappings (1K5) with $\mathbf{a},\mathbf{b} \in \mathbf{O}$ such that $\widetilde{\mathbf{a}}\mathbf{b} \in \mathbf{T}$), and its rotation subgroup is $[3,4,3]^+ = (\mathbf{O}/\mathbf{T};\mathbf{O}/\mathbf{T})$.

The compounds of (7J8) arise in the following way. First, $\mathbf{V}$ is a normal subgroup of $\mathbf{T}$ of index 3. Regarded as vertex-sets, this inscribes three copies of the staurotope $\mathsf{X} = \{3,3,4\}$ in F. Since the vertex-set of the dual cube $\mathsf{C} = \{4,3,3\}$ is $\operatorname{vert}\mathsf{C} = \operatorname{vert}\mathsf{F} \setminus \operatorname{vert}\mathsf{X} = \mathbf{T} \setminus \mathbf{V} =: \mathbf{W}$, say, we immediately obtain the dual pair of fully regular compounds of (7J8).

This construction has also inscribed two copies of X in C (that is, the other two cosets of $\mathbf{V}$ in $\mathbf{T}$ as subsets of $\mathbf{W}$), giving the first (only) vertex-regular compound of (7J7); the second is its facet-regular dual, and again both are fully regular.

The symmetry group of X and C is $[3,3,4] = (\mathbf{O}/\mathbf{V};\mathbf{O}/\mathbf{V})^*$ of order 384, while the common subgroup of a copy of X inscribed in C is isomorphic to $(\mathbf{T}/\mathbf{V};\mathbf{T}/\mathbf{V})^*$ of order 192, and is $\mathbf{B}_4 = [3^{1,1,1}]$.

7J9 Remark A useful observation is that a given staurotope X is inscribed in a unique 24-cell F. The key to this is the relationship of the vertex-sets of these polytopes to the finite groups of quaternions. We lose no generality in taking as vertex-set of X the subset $\mathbf{V} = \{\pm 1, \pm\mathbf{i}, \pm\mathbf{j}, \pm\mathbf{k}\}$ of unit quaternions. The only set in $\mathbf{Q}$ which contains $\mathbf{V}$ and is congruent to $\mathbf{T}$ is $\mathbf{T}$ itself, establishing the claim.

The same result holds for a cube C, whose vertex-set we can identify with $\mathbf{W}$. Since $\mathbf{W}$ consists of the other two cosets of $\mathbf{V}$ in $\mathbf{T}$, it similarly follows that the staurotope X is inscribed in exactly two cubes.

The action of $\mathbf{F}_4$ on its subgroup $\mathbf{B}_4$ induces the 2-collapse of $\{3,4,3\}$ onto its face $\{3\}$; compare Remark 7C14. This collapse manifests itself in another way, as the action of $\mathbf{G}(\{3,4,3\})$ on the dual compounds of (7J8).

The 600-Cell

We now come to the rich family of compounds involving the pentagonal polytopes. We begin with the well-known compounds in the 600-cell $\{3,3,5\}$, which are fairly straightforward to handle.

Since $\mathbf{T} \subset \mathbf{I}$, we can obviously inscribe $\mathsf{F} = \{3,4,3\}$ in G. However, it is worth noting that only half its symmetry group survives into the common subgroup, namely, $[3,4,3]^* := (\mathbf{T}/\mathbf{T};\mathbf{T}/\mathbf{T})^*$. As groups, we have $\mathbf{T} < \mathbf{I}$. Recall that we defined $\mathbf{p} := -\frac{1}{2}(\tau - \mathbf{i} + \tau^{-1}\mathbf{j})$ in (7E5) as a typical element of $\mathbf{I}$ of period 5, so that the powers of $\mathbf{p}$ act as both left and right coset representatives of $\mathbf{T}$ in $\mathbf{I}$. Indeed, the 25 subsets $\mathbf{p}^{-r}\mathbf{T}\mathbf{p}^s = \widetilde{\mathbf{p}}^r\mathbf{T}\mathbf{p}^s \subset \mathbf{I}$ are distinct, and yield a compound

7J10 $$\mathit{5}\{3,3,5\}[\mathit{25}\{3,4,3\}]\{3,3,5\};$$

we shall justify the facet-regularity a little later.

At this point, it is worth introducing a binary dihedral group that will appear quite often. Observe that $\widetilde{\mathbf{kpk}} = \widetilde{\mathbf{p}}$, and since $\mathbf{k}^2 = -1$, it should be clear that $\langle \mathbf{p}, \mathbf{k} \rangle$ is binary dihedral of order 20, and so a copy of $\mathbf{D} = \mathbf{D}_5$.

Now we can clearly consider the (say) right cosets $\mathbf{T}\mathbf{p}^s$ alone, which must give rise to a compound of type $\mathsf{G}[\mathit{5}\mathsf{F}]$. If we dualize, keeping track of the vertices as quaternions, then for the typical copy of F we have

$$\delta\colon \mathbf{T}\mathbf{p}^s \mapsto \mathbf{U}\mathbf{p}^s = \widetilde{\mathbf{u}}\mathbf{T}\mathbf{p}^s = \mathbf{T}\widetilde{\mathbf{u}}\mathbf{p}^s,$$

with $\mathbf{u} \in \mathbf{U}$ as before. The last two expressions tell us two things. First, the compound is also facet-regular, with the facets touching those of $\widetilde{\mathbf{u}}\mathsf{G}^\delta$, a copy of the 120-cell $\{5,3,3\}$; hence it is of kind

7J11 $$\{3,3,5\}[\mathit{5}\{3,4,3\}]\{5,3,3\}.$$

Second, note that we can only apply elements of the subgroup $\mathbf{T}$ on the left; of course, elements of $\mathbf{I}$ applied on the right just permute the cosets of $\mathbf{T}$. Following Section 1K, therefore, the symmetry group of the compound is $(\mathbf{T}/\mathbf{T};\mathbf{I}/\mathbf{I})$ of order $24 \cdot 120/2 = 1440$, while the common subgroup is the rotation subgroup $(\mathbf{T}/\mathbf{T};\mathbf{T}/\mathbf{T})$ of $[3,4,3]^*$ of order $24 \cdot 24/2 = 288$, which accounts for there being $1440/288 = 5$ copies of F in the compound.

Dealing with the left cosets of $\mathbf{T}$ in $\mathbf{I}$ instead results in an enantiomorphic copy of the compound; the two will be swapped by any opposite symmetry of the 600-cell.

The compounds of the staurotope and cube in the 24-cell lead to further compounds

7J12 $$\mathit{5}\{3,3,5\}[\mathit{75}\{3,3,4\}]\mathit{10}\{5,3,3\}, \quad \mathit{10}\{3,3,5\}[\mathit{75}\{4,3,3\}]\mathit{5}\{5,3,3\},$$

7J13 $$\{3,3,5\}[\mathit{15}\{3,3,4\}]\mathit{2}\{5,3,3\}, \quad \mathit{2}\{3,3,5\}[\mathit{15}\{4,3,3\}]\{5,3,3\}.$$

The symmetry group of the dual pair of (7J12) remains the whole group $[3,3,5]$; the common subgroup is $[3^{1,1,1}] = (\mathbf{T}/\mathbf{V};\mathbf{T}/\mathbf{V})^*$ of order $2 \cdot 24 \cdot 8/2 = 192$. The group of the two compounds of (7J13) remains $(\mathbf{T}/\mathbf{T};\mathbf{I}/\mathbf{I})$, but this is now acting on the common subgroup $(\mathbf{T}/\mathbf{V};\mathbf{T}/\mathbf{V})$.

7J14 Remark The edges of $\{4,3,3\}$ and $\{3,4,3\}$ inscribed in $\{3,3,5\}$ are among those of $\{\frac{5}{2},5,3\}$ and $\{5,\frac{5}{2},3\}$; we thus see that Figure 7G9 also serves as a projection of the corresponding two compounds with full symmetry group $\mathbf{G}_4 = [3,3,5]$. Figure 7G9 is a little confusing, in part because $\{\frac{5}{2},5,3\}$ and $\{5,\frac{5}{2},3\}$ have 1200 edges, rather than the 720 of $\{3,3,5\}$. Nevertheless, it should be possible to make out a copy of $\{3,4,3\}$ inscribed in $\{3,3,5\}$ in the first projection of Figure 7C8.

The* 120*-Cell* – 600*-Cells

There are many compounds with the vertices of $\mathsf{H} := \{5,3,3\}$; throughout the rest of the section, we take vert H as in Theorem 7F6 or Proposition 7F7, whichever is more convenient. For the compounds of $\mathsf{G} = \{3,3,5\}$ in H, we first appeal to Corollary 7J2. In each facet of the 120-cell must be inscribed a 3-simplex, and this can be done in exactly ten ways; of course, we have here tetrahedra inscribed in dodecahedra, and so we can refer to the two compounds

$$\{5,3\}[5\{3,3\}]\{3,5\}, \quad 2\{5,3\}[10\{3,3\}]2\{3,5\},$$

which we described earlier. These lead immediately to the two 4-dimensional compounds

$$\{5,3,3\}[5\{3,3,5\}], \quad 2\{5,3,3\}[10\{3,3,5\}].$$

These are only vertex-regular; the first occurs in two enantiomorphic pairs, which together form the second. In keeping with what we said before, we shall not mention the dual cell-regular compounds.

Although it is not of immediate importance, it is appropriate at this point to describe the common subgroup of a 600-cell inscribed in a 120-cell.

7J15 Proposition *The common subgroup of* $\{3,3,5\}$ *inscribed in its dual* $\{5,3,3\}$ *has order* 1440, *and hence index* 10.

Proof. For this, quaternions provide the straightforward way of getting to the answer. The relationship between the two polytopes is symmetric, and so we fix $\{3,3,5\}$ with vert$\{3,3,5\} = \mathbf{I}$, and escribe $\{5,3,3\}$ to it by vert$\{5,3,3\} = \mathbf{I}\mathbf{u}\mathbf{I}\widetilde{\mathbf{u}}$, with $\mathbf{u} = (1-\mathbf{k})/\sqrt{2}$ as in (7E6). (The last multiplication by $\widetilde{\mathbf{u}}$ is to bring the vertex 1 back to its original position.) We first notice that mappings involving $\mathbf{x} \mapsto \widetilde{\mathbf{x}}$ do not preserve vert$\{5,3,3\}$. Of the rotations $\mathbf{x} \mapsto \widetilde{\mathbf{a}}\mathbf{x}\mathbf{b}$ with $\mathbf{a}, \mathbf{b} \in \mathbf{I}$, we can allow $\mathbf{a}$ to take any value. However, it is only those $\mathbf{b} \in \mathbf{T}$ which we can permit on the right, since $\widetilde{\mathbf{u}}\mathbf{b}\mathbf{u} \in \mathbf{I}^{\ddagger} \setminus \mathbf{T}$ if $\mathbf{b} \in \mathbf{I} \setminus \mathbf{T}$. Thus the subgroup of allowable elements has order $120 \cdot 24/2 = 1440$, as claimed. □

We have further compounds of this kind in $\{5,3,3\}$, since we can replace the 600-cell $\{3,3,5\}$ by any one of the seven regular star-polytopes of Table 7G8 with the same vertices; except for those with $\{3,3,\frac{5}{2}\}$, these new compounds will now be cell-regular as well. Indeed, we even have self-dual compounds such as

$$\{5,3,3\}[5\{5,\tfrac{5}{2},5\}]\{3,3,5\}.$$

We do not bother to write out the complete list; instead, we refer the reader to [27, Table VII].

The* 120*-Cell* – *Simplices

We first recall from Proposition 7F10 that we have two reflexion subgroups of $\mathbf{G}_4 = [3,3,5]$, namely, $\mathbf{A}_4 = [3,3,3]$ of index $14400/120 = 120$ and $\mathbf{B}_4 = [3^{1,1,1}]$ of index $14400/192 = 75$. The first immediately induces a compound

7J16 $$\{5,3,3\}[120\{3,3,3\}]\{3,3,5\};$$

however, we shall shortly see that this compound is not the only possible one involving the simplex. For this compound, compare the description of the 120-cell in Section 7F in terms of unit quaternions, in particular the action of $\mathbf{I}$ on the vertex-set $\mathbf{P}$ of the initial 4-simplex. It should be recalled at this point that among these symmetries are all permutations of the coordinates with an even number of changes of sign; in particular, we can freely permute the points $\mathbf{p}_j^*\mathbf{p}_j$ of $\mathbf{P}$ for $j = 1, \ldots, 4$. In addition, referring to Proposition 7K2, we see that the hyperplane reflexion with (unscaled) normal $(\sqrt{5}, -1, -1, -1)$ is a symmetry of the 120-cell H (say) which interchanges $(4, 0, 0, 0)$ and $(-1, \sqrt{5}, \sqrt{5}, \sqrt{5})$ while fixing each of the other three points; this shows directly why the corresponding five vertices of $\{5, 3, 3\}$ are those of a regular 4-simplex A, say.

There are 24 more vertices of H at the same distance from the initial vertex v as these. Consider the symmetries given by

$$S\colon \mathbf{x} \mapsto \mathbf{qxp},$$
$$T\colon \mathbf{x} \mapsto \widetilde{\mathbf{u}}\widetilde{\mathbf{x}}\mathbf{u},$$

with $\mathbf{p}$ given as before by (7E5), $\mathbf{u} := \frac{1}{\sqrt{2}}(1 - \mathbf{k})$ as in (7E6) and $\mathbf{q} = \mathbf{p}^\ddagger$. Recall from Proposition 7E7 that $\widetilde{\mathbf{u}}\mathbf{pu} = \mathbf{q}^2$ and $\widetilde{\mathbf{u}}\mathbf{qu} = \mathbf{p}^2$, so that

$$\widetilde{\mathbf{u}}\widetilde{\mathbf{qp}}\mathbf{u} = \widetilde{\mathbf{u}}\widetilde{\mathbf{p}}\mathbf{u} \cdot \widetilde{\mathbf{u}}\widetilde{\mathbf{p}}\mathbf{u} = \mathbf{q}^{-2}\mathbf{p}^{-2}.$$

We thus have a subgroup $\langle S, T\rangle$ of order 20 acting on the five vertices $v_k := 4\mathbf{q}^k\mathbf{p}^k$ for $k = 0, \ldots, 4$, where (in terms of coordinates) $v_0 = (4, 0, 0, 0)$ and

7J17 $$v_1, v_4 = (-1, \pm(1, 3), -\sqrt{5}), \quad v_2, v_3 = (-1, \pm(3, -1), \sqrt{5}),$$

with the positive sign taken first in each pair. It is now clear that the points $v_0, \ldots, v_4$ are the vertices of another regular 4-simplex B. Moreover, we have seen that the common subgroup is

7J18 $$\mathbf{K}(\mathsf{B}, \mathsf{H}) = \langle S, T\rangle.$$

Now the form of (7J17) shows that there are five other tetrahedra with vertices of this kind, permuted by the symmetries of the vertex-figure of H at v_0. These give 24 points which, together with v_0, form the vertices of six copies of B. Transforming these six under the action of $[3, 3, 5]$ therefore yields a self-dual regular compound

7J19 $$\mathit{6}\{5, 3, 3\}[\mathit{720}\{3, 3, 3\}]\mathit{6}\{3, 3, 5\}.$$

The self-duality results from the polar of B being $-\mathsf{B}$. Observe as well that $720 = 14400/20$, as we should expect.

We must next check whether any further compounds arise as subcompounds of that of (7J19). In fact, there is (up to symmetry) just one more, as we shall now show.

The symmetry group $\mathbf{G}$ of such a compound must have order $600k$ for some k, since it is transitive on $\operatorname{vert} \mathsf{H}$. Of course, it is also a subgroup of $[3, 3, 5]$ of order 14400, so that k must be a divisor of 24. Looked at as a group of both left- and right-acting quaternions, we see that one of the two components has to be the whole group $\mathbf{I}$ of 120 icosians; we lose no generality in taking this to give the right action. The left action will be a subgroup of $\mathbf{I}$ or rather (as we have things) of the isomorphic group $\mathbf{I}^{\ddagger}$. If $\mathbf{G}$ contains an opposite isometry, then the left group must be the whole of $\mathbf{I}^{\ddagger}$, so that $\mathbf{G} = [3, 3, 5]$; this yields the compound of (7J19).

Let us denote the left group by $\mathbf{D}$. We first note that $\mathbf{x}T^2 = -\mathbf{kxk}$, and then observe that, for any $\mathbf{g} \in \mathbf{I}$, we have

$$\begin{aligned}\mathbf{q}\mathsf{B}\mathbf{g} &= \mathbf{q}\mathsf{B}\mathbf{p} \cdot \widetilde{\mathbf{p}}\mathbf{g} = \mathsf{B}S\mathbf{g}' = \mathsf{B}\mathbf{g}',\\ \mathbf{k}\mathsf{B}\mathbf{g} &= \mathbf{k}\mathsf{B}(-\mathbf{k}) \cdot \mathbf{k}\mathbf{g} = \mathsf{B}T^2\mathbf{k}\mathbf{g} = \mathsf{B}\mathbf{g}'',\end{aligned}$$

for some $\mathbf{g}', \mathbf{g}'' \in \mathbf{I}$. In other words, $\mathbf{D} \geqslant \langle \mathbf{q}, \mathbf{k} \rangle \cong \mathbf{D}_5$, which is binary dihedral of order 20. But $\mathbf{D}_5$ is maximal in $\mathbf{I}^{\ddagger}$, and since $\mathbf{D} \neq \mathbf{I}^{\ddagger}$, we must have $\mathbf{D} = \mathbf{D}_5$.

It follows that $\mathbf{G} = (\mathbf{D}/\mathbf{D}; \mathbf{I}/\mathbf{I})$, of order $20 \cdot 120/2 = 1200$, and our compound therefore consists of the 120 copies $\mathsf{B}\mathbf{g}$ of B, with $\mathbf{g} \in \mathbf{I}$, namely,

7J20 $\{5, 3, 3\}[\mathit{120}\{3, 3, 3\}]\{3, 3, 5\}^{(\mathrm{var})}$;

once again, the compound is self-dual, and actually coincides with its dual since $-I \in \mathbf{G}$. Since the common subgroup of the compound is now only $\langle S, T^2 \rangle$ of order 10, this accounts for it containing $1200/10 = 120$ copies of B.

7J21 Remark Observe that this new compound is distinct from the one of (7J16) (that listed in [27, Table VII(i)]). Formally, it is designated by a similar symbol, and so we have distinguished it by $^{(\mathrm{var})}$. Note that its symmetries are all direct, so that it actually occurs in two enantiomorphic forms, each its own dual. In some sense, it is an analogue of the compound of five tetrahedra in a dodecahedron, except that the dual of the latter is its enantiomorph.

7J22 Remark In [27, p. 47], Coxeter stated conditions for regular compounds (of polyhedra there, but applied generally) that do not include our transitivity conditions. They would thus permit taking any of the six copies of our basic 4-simplex B containing $(4, 0, 0, 0)$ and their images under right multiplication by $\mathbf{I}$. However, as we have seen, only in case of one or all six copies do we obtain a compound for which all vertices of the 120-cell and all copies of B look alike.

The 120-Cell – Staurotopes

For the remaining compounds of the three polytopes $\{3, 3, 4\}$, $\{4, 3, 3\}$ and $\{3, 4, 3\}$ inscribed in H the situation is also complicated, because in each of these the same polytope can be inscribed in different ways. We base our treatment on the 4-staurotope X, and derive the other compounds from it.

The central section from the vertex $1 = (1, 0, 0, 0)$ has vertices obtained from $(0, 1, 0, 0)$ and $\frac{1}{2}(0, \tau, 1, \tau^{-1})$ by permutations in the last three coordinates and arbitrary changes of sign. There are then two essentially different ways –

up to symmetries of the vertex-figure $\{3,3\}$ (all permutations of the last three coordinates with an even number of changes of sign) – to complete $(1,0,0,0)$ to an orthonormal basis using some of these vectors: we apply cyclic permutations in the last three coordinates to either $\mathbf{i} = (0,1,0,0)$ or $\frac{1}{2}(0,\tau,-\tau^{-1},1)$.

In the first case, the inscribed 4-staurotope that results exhibits the subgroup relationship $[3^{1,1,1}] < [3,3,5]$ (the common subgroup is the former here, of course). Since the index is $14400/192 = 75$, we quickly arrive at the compounds

$$\{5,3,3\}[75\{3,3,4\}]2\{3,3,5\},$$
$$2\{5,3,3\}[75\{4,3,3\}]\{3,3,5\},$$
$$\{5,3,3\}[25\{3,4,3\}]\{3,3,5\}.$$

The second is the dual of the first, and the third arises from the compound $\{3,4,3\}[3\{3,3,4\}]2\{4,3,3\}$ encountered earlier, which gathers the staurotopes in threes into a 24-cell.

In the second case, once again we find that quaternions provide the quickest way of arriving at the solution. We define

$$\mathbf{a} := \tfrac{1}{2}(\tau\mathbf{i} - \tau^{-1}\mathbf{j} + \mathbf{k}),\ \ \mathbf{b} := \tfrac{1}{2}(\mathbf{i} + \tau\mathbf{j} - \tau^{-1}\mathbf{k}),\ \ \mathbf{c} := \tfrac{1}{2}(-\tau^{-1}\mathbf{i} + \mathbf{j} + \tau\mathbf{k}),$$

noting that $\mathbf{a},\mathbf{b},\mathbf{c} \in \mathbf{I}$. In fact, $(\mathbf{a},\mathbf{b},\mathbf{c}) = \tilde{\mathbf{u}}(\mathbf{i},-\mathbf{k},\mathbf{j})\mathbf{u}$, with $\mathbf{u}$ again as in (7E6), so that they behave like $\mathbf{i},\mathbf{j},\mathbf{k}$, in that $\mathbf{a}^2 = -1$, $\mathbf{ab} = \mathbf{c}$, and so on. The vertices of the second kind of staurotope Y, say, can therefore be identified with $\pm 1, \pm\mathbf{a}, \pm\mathbf{b} \pm \mathbf{c}$. It is now easy to find the common subgroup. Certainly, we can multiply on the right by elements of $\{\pm 1, \pm\mathbf{a}, \pm\mathbf{b} \pm \mathbf{c}\}$. We can also permute cyclically the last three coordinates by $\mathbf{x} \mapsto \tilde{\mathbf{v}}\mathbf{x}\mathbf{v}$ with $\mathbf{v} = \frac{1}{2}(-1 + \mathbf{i} + \mathbf{j} + \mathbf{k})$. Thus the common subgroup has order at least $8 \cdot 3 = 24$. (We can see as well that mappings involving $\mathbf{x} \mapsto \mathbf{u}\tilde{\mathbf{x}}\mathbf{u}$ do not preserve Y, but we do not need to appeal to that fact.)

Note that 1 belongs to exactly eight such staurotopes, obtained by arbitrary permutation of the last three coordinates of $\frac{1}{2}(0,\tau,-\tau^{-1},1)$ with all changes of sign. Moreover, these fall into two sets of four, related by the even permutations. Under the rotation group $[3,3,5]^+$, we obtain the dual compounds

$$4\{5,3,3\}[300\{3,3,4\}]8\{3,3,5\},\quad 8\{5,3,3\}[300\{4,3,3\}]4\{3,3,5\};$$

these each occur in enantiomorphic pairs. Allowing the whole symmetry group $[3,3,5]$ to act gives the dual compounds

$$8\{5,3,3\}[600\{3,3,4\}]16\{3,3,5\},\quad 16\{5,3,3\}[600\{4,3,3\}]8\{3,3,5\};$$

since $14400/600 = 24$, the latter must be the order of the common subgroup.

We can now group staurotopes (or cubes) together in threes to make the vertices of $\{3,4,3\}$, and these lead to the self-dual compounds

$$4\{5,3,3\}[100\{3,4,3\}]4\{3,3,5\},\quad 8\{5,3,3\}[200\{3,4,3\}]8\{3,3,5\},$$

respectively; again, the first occurs in enantiomorphic pairs.

However, just as in the case of simplices, there exist subcompounds. For a change, let us concentrate on $\mathsf{S} = \{3,4,3\}$. Corresponding to Y, we have the copy $\mathsf{T} := \widetilde{\mathbf{p}}\mathsf{S}\mathbf{p}$; the common subgroup $\mathbf{K}(\mathsf{T},\mathsf{H})$ now has order $3 \cdot 24 = 72$. Let $\mathbf{C}$ be the cyclic subgroup of $\mathbf{I}^{\ddagger}$ generated by $-\mathbf{q}$, with $\mathbf{q} = \mathbf{p}^{\ddagger}$ as usual. If we let $\mathbf{I}$ act on T on the right, but only $\mathbf{C}$ on the left, and recalling that the $\mathbf{p}^k$ are the coset representatives of $\mathbf{T}$ in $\mathbf{I}$, we see that the resulting set of vertices can be identified with

$$\bigcup\{\mathbf{q}^j\widetilde{\mathbf{p}}\mathbf{T}\mathbf{p}^k \mid j,k = 0,\ldots,4\}.$$

We thus obtain all the vertices of H, and a new compound

$$\{5,3,3\}[25\{3,4,3\}]^{(\mathrm{var})}$$

with symmetry group $(\mathbf{C}/\mathbf{C};\mathbf{I}/\mathbf{I})$. Inspection of the dual shows that this is only vertex-regular.

7J23 Remark Even though $\mathbf{C}$ is not maximal in $\mathbf{I}^{\ddagger}$, there are no further compounds of this kind. We cannot extend $\mathbf{C}$ by $\mathbf{k}$ to obtain a binary dihedral group $\mathbf{D}_5$, since doing so would lead to a new copy of S as starting point, ultimately giving the whole compound $4\{5,3,3\}[100\{3,4,3\}]4\{3,3,5\}$.

Replacing the 24-cell by the inscribed copies of the staurotope or cube then yields the dual pair

$$\{5,3,3\}[75\{3,3,4\}]2\{3,3,5\}^{(\mathrm{var})}, \quad 2\{5,3,3\}[75\{4,3,3\}]\{3,3,5\}^{(\mathrm{var})},$$

which are vertex- and facet-regular.

Compounds of Honeycombs

We shall not treat compounds of regular honeycombs of euclidean spaces systematically; instead, we confine ourselves to indicating a few possibilities. We shall see by examples that it is very complicated; even in the planar case, it is not altogether straightforward. The general reference to the background here is [99, Sections 1D, 6D and 6E].

First, consider the tiling $\{4,4\}$ of the plane by squares (tetragons). There is no way to inscribe a trigon $\{3\}$ in the square lattice, and so it is only compounds of larger copies of $\{4,4\}$ itself that we need look at. Up to symmetry, such a copy is defined by a non-zero vector $(a,b) \in \mathbb{Z}^2$ with $a,b \geqslant 0$; the sublattice $\Lambda_{(a,b)} \leqslant \mathbb{Z}^2$ spanned by (a,b) and $(b,-a)$ then forms the vertex-set of the initial copy, and then translates of this copy by $\mathbb{Z}^2$ give a^2+b^2 copies in all.

If $ab(a-b) \neq 0$, then $\Lambda_{(a,b)}$ only has the rotational symmetries of $\{4,4\}$. Thus the vector (b,a) gives a different lattice $\Lambda_{(b,a)}$, and hence a different compound. These two compounds can then be combined to give a compound of $2(a^2+b^2)$ congruent copies of the larger tiling.

The situation for inscribing larger copies of the triangular tessellation $\{3,6\}$ in itself is very similar. However, we can also inscribe copies of $\{6,3\}$ in $\{3,6\}$, as well as inscribing the geometric dual $\{3,6\}$ in $\{6,3\}$ (where Corollary 7J2 holds), which thus leads to inscribing larger copies of $\{6,3\}$ in itself. We resile from going into the details here.

There are many different types of regular compounds whose vertices are those of cubic honeycombs $\{4,3^{d-2},4\}$ in $\mathbb{E}^d$ with $d \geqslant 3$, with different kinds of common subgroup. Let us just give two examples with $d = 3$. The orthogonal vectors $(3,4,0),(4,-3,0),(0,0,5)$ have the same length 5, and so span a larger copy of the cubic lattice. Just by translations of $\mathbb{Z}^3$ alone we obtain a compound of 125 copies of the corresponding larger cubic tiling, but we clearly find more compounds under the action of larger subgroups of the whole symmetry group $[4,3,4]$. In the same way, the vectors $(-2,3,6),(6,-2,3),(3,6,-2)$ are mutually orthogonal; they result in a compound of $7^3 = 343$ copies of a larger version of the cubic tiling, and once again we can apply bigger subgroups of $[4,3,4]$ than $\mathbb{Z}^3$ to obtain further compounds. But we observe that the common subgroups in these two cases are quite distinct; if we expand the first copy by 7 and the second by 5, then we have two families of compounds of 35 times larger copies of $\{4,3,4\}$ which are essentially unrelated.

Furthermore, the symmetry group $\mathbf{G}$ of a regular honeycomb P of $\mathbb{E}^d$ is a subgroup of itself of index k^d for every k, because we can scale up a fundamental region of $\mathbf{G}$ by k within the group (just scale up from a vertex corresponding to the point group of $\mathbf{G}$). In certain cases, though, there are other possibilities. For example, $\mathbf{V}_3 = [3,6]$ has an isomorphic subgroup of index 3 as well as 4, and – as we have already seen – $\mathbf{U}_5 = [3,3,4,3]$ has an isomorphic subgroup of index 4. We leave to the interested reader the exploration of these cases.

Notes to Section 7J

1. The definition adopted in this section is more rigorous than that usually found, although in practice it does seem to be what is expected of a regular compound. All the compounds described by Coxeter [27, 14.3] satisfy our definition; it might be guessed that something like it was what he had in mind.
2. Du Val [43, Section 2.11] uses the term *maximum common subgroup*.
3. The compounds of tetrahedra are depicted in anaglyphs in [47].
4. The two new compounds of 4-simplices in the 120-cell come from [90]. It is curious that Coxeter missed these compounds, by failing to notice that the section 12_0 of $4+24$ points in his Table V(v) should be identified as $1+6$ tetrahedra.
5. The new compounds of 4-staurotopes, 4-cubes and 24-cells also come from [90]; here, Coxeter did not observe that some of his compounds had subcompounds.

7K Realizations of $\{5,3,3\}$

We have left the description of the realization domain of the 120-cell $\{5,3,3\}$ until last, because we need part of the previous Section 7J as background. The compounds described there arise from the action of $[3,3,5]$ on various subgroups. Apart from the obvious components of the small simplex realization S, we have the subgroups $[3,3,3]$, $[3,3,3] \times \mathbf{Z}_2$ with $\mathbf{Z}_2 = \{\pm I\}$, $[3^{1,1,1}]$ and the subgroup $[3,4,3]^*$ of Proposition 7F10(c), of index 2 in $[3,4,3]$. The vertices of $\mathcal{H}$ can thus be identified in sets of k for $k = 2, 5, 8, 10, 24$; we write $\mathcal{H}_k$ for the corresponding quotient. Thus $\mathcal{H}_2 = \{5, 3, 3\}/2$ identifies antipodal vertices. Being able to describe $\mathcal{H}_k$ for $k = 5, 8, 10, 24$ is very helpful.

We may also observe that Corollary 4D9 has the implication that $\mathcal{H}$ has pure realizations of dimensions $9, 9, 16, 25, 4, 4, 16, 36$, with the first four realizations of the *60-cell* or hemi-120-cell $\mathcal{H}/2$. Of course, the two 4-dimensional realizations are just $\{5, 3, 3\}$ and $\{\frac{5}{2}, 3, 3\}$. It is also clear that the 9-dimensional realizations are the non-trivial components of $\{5, 3, 3\} \otimes \{5, 3, 3\}$ and $\{\frac{5}{2}, 3, 3\} \otimes \{\frac{5}{2}, 3, 3\}$ (these have $w = 1$), while we obtain a 16-dimensional realization $\{5, 3, 3\} \otimes \{\frac{5}{2}, 3, 3\}$ (with $\eta_f = \frac{1}{64}$ – we shall bear this notion in mind). We can similarly identify some 36-dimensional realizations; compare Remark 7E15.

7K1 Remark Because of the length of this section, we omit many of the calculations that go into the analysis; however, we have given more details than in [92], which this section draws on. We must also mention here the invaluable help we received from Frieder Ladisch; see the notes at the end of the section.

Layer Vector

Proposition 7F7 will yield the layer vector Λ of $\mathcal{H} = \{5, 3, 3\}$, and tell us which diagonal classes are asymmetric. Since $\mathcal{H}$ has many layers, we compromise by writing down only the first half of Λ. For the diagonals classes, we have

7K2 Proposition *The matrices below, scaled by $\frac{1}{4}$, are involutory symmetries of the copy of $\{5, 3, 3\}$ with the vertices of Proposition 7F7:*

$$\begin{bmatrix} -1 & \sqrt{5} & \sqrt{5} & \sqrt{5} \\ \sqrt{5} & 3 & -1 & -1 \\ \sqrt{5} & -1 & 3 & -1 \\ \sqrt{5} & -1 & -1 & 3 \end{bmatrix}, \quad \begin{bmatrix} \sigma & \tau^{\ddagger} & \tau^{\ddagger} & \tau^{\ddagger} \\ \tau^{\ddagger} & \varphi^{\ddagger} & -\tau^2 & -\tau^2 \\ \tau^{\ddagger} & -\tau^2 & \varphi^{\ddagger} & -\tau^2 \\ \tau^{\ddagger} & -\tau^2 & -\tau^2 & \varphi^{\ddagger} \end{bmatrix}, \quad \begin{bmatrix} -\sigma^{\ddagger} & \tau & \tau & \tau \\ \tau & -\varphi & \tau^{-2} & \tau^{-2} \\ \tau & \tau^{-2} & -\varphi & \tau^{-2} \\ \tau & \tau^{-2} & \tau^{-2} & -\varphi \end{bmatrix}.$$

Here, $\varphi := \tau\sqrt{5}$, so that $\varphi^{\ddagger} = \tau^{-1}\sqrt{5}$.

If we permute the rows of a matrix in some way, then for its inverse we must permute the columns in the same way. So, for instance, if we take $(4, 0, 0, 0)$ to be the initial vertex of $\{5, 3, 3\}$, then, switching the first and second rows in the first matrix, we obtain the vertices $(\sqrt{5}, 3, -1, -1)$ and $(\sqrt{5}, -1, \sqrt{5}, \sqrt{5})$ in the same layer, but inequivalent under the symmetries of the vertex-figure (all permutations and an even number of changes of sign in coordinates $2, 3, 4$). In other words, this shows that the $12 + 12$ vertices in that layer ($\mathcal{L}_7$, as it happens) form a single asymmetric diagonal class with $(4, 0, 0, 0)$.

The diagonal classes to $(2, 2, 2, \pm 2)$, $(-2, 2, 2, \pm 2)$ and $(0, 2\tau, 2\tau^{-1}, \pm 2)$ are similarly asymmetric, but this is more easily seen using quaternions. For the first, $\mathbf{x} \mapsto \mathbf{xg}$, with $\mathbf{g} = \frac{1}{2}(1 + \mathbf{i} + \mathbf{j} + \mathbf{k})$, takes $(4, 0, 0, 0)$ into $(2, 2, 2, 2)$, while $\mathbf{g}^{-1}$ has image $(2, -2, -2, -2)$; the second is similar. The mapping induced by $\tau\mathbf{i} + \mathbf{j} + \tau^{-1}\mathbf{k}$ acts in the same way for $\pm 2(0, \tau, 1, \tau^{-1})$.

7K3 Remark In the initial realization $\mathsf{H} := \{5,3,3\}$ itself, there are several cases where two layers have the same value for the first coordinate ξ_1. However, the reader will probably have [27, Table V(v)] to hand, and so for convenience of reference to it we have labelled such layers by the same main index, differentiated by adding 'a' or 'b'. To be specific, we have

$$\begin{aligned}
\mathcal{L}_{8a} &= \{(2,2,2,\pm 2),\ldots\},\\
\mathcal{L}_{12a} &= \{1,\sqrt{5},\sqrt{5},-\sqrt{5}),\ldots\},\\
\mathcal{L}_{15a} &= \{(0,4,0,0),\ldots\},\\
\mathcal{L}_{18a} &= \{-1,\sqrt{5},\sqrt{5},\sqrt{5}),\ldots\},\\
\mathcal{L}_{22a} &= \{(-2,2,2,\pm 2),\ldots\},
\end{aligned}$$

where the remaining vertices in each set are those obtained by applying the symmetries of the vertex-figure, and the layers $\mathcal{L}_{8b}$ and so on consist of the other vertices with the same first coordinate (see Proposition 7F7).

We can summarize the foregoing discussion in

7K4 Proposition *The layer vector of* $\{5,3,3\}$ *is*

$$\Lambda = (1,4,12,24,12;4,24,24^*,8^*,24;24^*,12,24,4,24;24^*,24,6,48^*,\ldots),$$

where $6, 48^*$ *are the middle terms, the rest repeats the first half in reverse order, and* * *denotes an asymmetric diagonal class. Thus* $\{5,3,3\}$ *has* 36 *diagonal classes, including the trivial one, of which* 9 *are asymmetric.*

Semicolons rather than commas in three places separate the entries into blocks for easier identification.

7K5 Remark Regarding $\mathcal{H}_2 = \{5,3,3\}/2$ as a polytope in its own right, its layer vector is now

$$\Lambda = (1,4,12,24,12;4,24,24^*,8^*,24;24^*,12,24,4,24;24^*,24,3,24);$$

note that $\mathcal{L}_{15b}$ is no longer asymmetric.

Centred Strata

The strata of a pure realization of $\mathcal{H}$ in any dimension except those for which there are corresponding realizations of the 600-cell $\{3,3,5\}$ (namely, 9, 16, 25 for $\mathcal{H}_2$ and 4, 16, 36 otherwise) must be centred; compare Theorem 4D6. Indeed, whenever $w > 1$, a subspace of the Wythoff space W of codimension 1 will also yield realizations with centred strata. The following list is not comprehensive, but does provide workable criteria. In practice, we have made only limited appeal to it.

7K6 Proposition *The following are conditions for centred strata:*

$$\begin{aligned}
\mathcal{S}_1 &: 1 + 3\gamma_1 + 6\gamma_2 + 6\gamma_3 + 3\gamma_4 + \gamma_5 = 0, \\
\mathcal{S}_6 &: 1 + 3\gamma_{25} + 6\gamma_{10} + 6\gamma_{20} + 3\gamma_4 + \gamma_{29} = 0, \\
\mathcal{S}_2 &: 1 + 3\gamma_2 + 6\gamma_4 + 6\gamma_{8b} + 3\gamma_{10} + \gamma_{12a} = 0, \\
\mathcal{S}_3 &: 1 + 4\gamma_2 + 4\gamma_4 + 4\gamma_6 + 4\gamma_{8a} + 4\gamma_{10} + 4\gamma_{12b} + 4\gamma_{14} + \gamma_{15a} = 0.
\end{aligned}$$

Proof. The strata $\mathcal{S}_1, \mathcal{S}_2, \mathcal{S}_6$ are dodecahedra, the first being the facet of $\{5, 3, 3\}$ and the last the facet of its allomorph; the stratum $\mathcal{S}_3$ is an icosidodecahedron. There are four copies of $\mathcal{S}_2$ and six copies of $\mathcal{S}_3$ through each vertex.

The derivation of $\mathcal{S}_6$ from $\mathcal{S}_1$ needs no comment. The vertex-figures of $\mathcal{S}_2$ and $\mathcal{S}_3$ lie in layer $\mathcal{L}_2$, which is a truncated tetrahedron. We can choose the copy $\mathcal{S}_2$, whose vertex-figure is a trigonal face of $\mathcal{L}_2$, to be preserved by the permutations of ξ_2, ξ_3, ξ_4; we need six vertices in $\mathcal{L}_{8ab}$ and just one in $\mathcal{L}_{12ab}$, which forces us to take $\mathcal{L}_{8b}$ and $\mathcal{L}_{12a}$ rather than the alternatives. Similarly, one copy of $\mathcal{S}_3$ is preserved by the permutation $(3\ 4)$, and this obliges us to pick $\mathcal{L}_{12b}$ and $\mathcal{L}_{15a}$ rather than their alternatives. □

Quotients of $\{5, 3, 3\}$

As we have said, a crucial feature of the realization domain $\mathcal{N}$ of $\mathcal{H} = \{5, 3, 3\}$ is the existence of the families of quotients $\mathcal{H}_k := \mathcal{H}/k$ for $k = 5, 10, 8, 24$. It is natural to treat each quotient in its own right, that is, with its own subdomain $\mathcal{N}_k$ and layer vector Λ_k, say, with $k = 2, 5, 10, 8, 24$. Except for $k = 2$, it is not straightforward to work out how Λ_k embeds in Λ; this will be our first task.

The Subfamily $\mathcal{H}_8$

We begin with

7K7 Lemma *Layers in* $\mathcal{H}_8$ *are identified as follows:*

0	1	1	4	2	12	4	12	3	24	8*a*	8
15*a*	6	6	24	5	4	7	24	8*b*	24	22*a*	8
30	1	10	12	9	24	12*a*	4	11	24		
		13	24	14	24	12*b*	24	15*b*	48		
		17	24	16	24	18*a*	4	19	24		
		20	12	21	24	18*b*	24	22*b*	24		
		24	24	25	4	23	24	27	24		
		29	4	28	12	26	12				
	8		128		128		128		192		16

The further identifications to give $\mathcal{H}_{24}$ *are* $\{\mathcal{L}_0, \mathcal{L}_{8a}\}$ *and* $\{\mathcal{L}_1, \mathcal{L}_2, \mathcal{L}_4\}$.

7K8 Remark The entry in the second half of a column counts the number of points in that layer. The order is chosen to emphasize the (somewhat surprising) symmetry among $\{\mathcal{L}_1, \mathcal{L}_2, \mathcal{L}_4\}$. Even though layers $12a$ and $12b$ fall together, as do layers $18a$ and $18b$, we have separated them to emphasize the correspondence with the previous two identifications. We shall see why shortly.

For our first step, we have

7K9 Proposition *The quotient $\mathcal{H}_{24}$ of $\{5,3,3\}$ has two non-trivial diagonal classes. Combinatorially, it is the product $\{3,3,3\} \times \{3,3,3\}$ of two 4-simplices with group $([3,3,3] \times [3,3,3]) \rtimes C_2$, with edges of $\mathcal{H}$ corresponding to non-edges of the product. Consequently, the cosine matrix of $\mathcal{H}_{24}$ is*

$$\begin{bmatrix} 1 & 1 & 1 \\ 1 & -\frac{1}{4} & \frac{3}{8} \\ 1 & \frac{1}{16} & -\frac{1}{4} \end{bmatrix},$$

with layer and dimension vectors

$$\Lambda = (1, 16, 8), \quad D = (1, 8, 16).$$

Proof. The key to this is to note that the identification $\mathcal{H} \searrow \mathcal{H}_{24}$ preserves inscribed 4-simplices $\{3,3,3\}$. Think, as well, of the description of the vertex-set of $\{5,3,3\}$ in Remark 7F11, as $\mathbf{T}$ acted on by powers of $\mathbf{p}^*$ on the left and $\mathbf{p}$ on the right.

However, a more formal approach is the following (for reasons that will become clear, we cannot appeal to the vertex-figure criterion of Theorem 3C14 here). Since $\mathcal{H}_{24}$ has $600/24 = 25$ vertices, the dimensions of its two non-trivial pure components sum to 24, which is not a sum of dimensions of realizations of $\{3,3,5\}/2$. Thus Theorem 4D6 implies that the facet of at least one component must be centred, leading us to solve

$$\begin{aligned} 1 + 16\alpha + 8\beta &= 0, \\ 1 + 13\alpha + 6\beta &= 0, \end{aligned}$$

for the cosine vector $(1, \alpha, \beta)$, where we have collected terms in the second equation using the table of Lemma 7K7. This yields $(1, -\frac{1}{4}, \frac{3}{8})$, with dimension 8 from Theorem 3F5, and then Theorem 3C11 gives the other cosine vector $(1, \frac{1}{16}, -\frac{1}{4})$ with dimension 16. □

Observe that, for the 16-dimensional realization, we have $\eta_f = \frac{1}{64}$, whereas we have $\eta_f = \frac{1}{16}$ for the 16-dimensional realization of $\{3,3,5\}/2$. Remark 3K6 thus already tells us that $w^* > 1$ in this case, and so explains why we cannot appeal to Theorem 3C14.

We now move on to $\mathcal{H}_8$; we shall prove the following.

7K10 Proposition *The pure cosine vectors in $\mathcal{H}_8$ that are Λ-orthogonal to $\mathcal{N}_{24}$ are of the form*

$$\Gamma = (1, \alpha, \beta, \gamma, 0, -\tfrac{1}{2}), \qquad \text{with } \alpha + \beta + \gamma = 0, \quad \alpha^2 + \beta^2 + \gamma^2 = \tfrac{3}{32}.$$

The common dimension is 25.

Proof. The layer vector of $\mathcal{H}_8$ is $(1, 16, 16, 16, 24, 2^*)$; the diagonal class $\mathcal{D}_{8a}$ remains asymmetric, because it is still the case that no automorphism fixes one copy of $\{3, 3, 4\}$ inscribed in $\{3, 4, 3\}$ while interchanging the other two. The injection of cosine vectors of $\mathcal{H}_{24}$ into $\mathcal{H}_8$ is given by

$$(1, \alpha, \beta) \hookrightarrow (1, \alpha, \alpha, \alpha, \beta, 1).$$

Using the layer equation Theorem 3C7, and applying Theorem 3F5 to what we know from Proposition 7K9 (bearing in mind Remark 3J4), shows that the remaining cosine vectors of $\mathcal{H}_8$ satisfy

$$\begin{aligned} 1 + 16\gamma_1 + 16\gamma_2 + 16\gamma_3 + 24\gamma_4 + 2\gamma_5 &= 0, \\ 1 - \gamma_1 - \gamma_2 - \gamma_3 + 2\gamma_5 &= 0, \\ 1 - 3\gamma_4 + 2\gamma_5 &= 0. \end{aligned}$$

This shows that the pure cosine vectors are of the form $(1, \alpha, \beta, \gamma, 0, -\frac{1}{2})$, with $\alpha + \beta + \gamma = 0$. The asymmetric diagonal class suggests that we might expect common dimension 25 with 2-dimensional Wythoff space, which would account for the dimension deficit 50 and the three remaining diagonal classes.

In the discussion of the groups, we have seen various operations – vertex-figure replacements – that lead from one polytope to another. We also have Proposition 7G19 which replaces $\mathcal{H} = \{5, 3, 3\}$ by its allomorph, and the same for its central quotient $\mathcal{H}_2$ comes from Corollary 7G20. In particular, the latter holds for the quotient $\mathcal{H}_8$, where it swaps layers $\mathcal{L}_1$ and $\mathcal{L}_2$.

Now a typical automorphism of $\mathcal{H}_8$ of period 3 that permutes the points of $\mathcal{L}_0$ and $\mathcal{L}_5$ cyclically corresponds on H to the (double) rotation of $\mathbb{E}^4$ given by right multiplication by $\mathbf{u} := -\frac{1}{2}(1 + \mathbf{i} + \mathbf{j} + \mathbf{j})$. If we multiply $-1 + \sqrt{5}(\mathbf{i} + \mathbf{j} + \mathbf{k})$ by $\mathbf{u}$ and $\mathbf{u}^2 = \widetilde{\mathbf{u}}$, then we obtain in turn

$$\sigma - \tau^{-1}(\mathbf{i} + \mathbf{j} + \mathbf{k}), \quad \sigma^{\ddagger} - \tau(\mathbf{i} + \mathbf{j} + \mathbf{k}).$$

In other words, the multiplication permutes representatives of layers $\mathcal{L}_1$, $\mathcal{L}_2$ and $\mathcal{L}_3$, and so induces a corresponding automorphism $\boldsymbol{\varrho}$, say, of $\mathcal{H}_8$.

It follows that, if we conjugate the operation $\boldsymbol{\tau}$ of Corollary 7G20 by $\boldsymbol{\varrho}$, then we are enabled to swap $\mathcal{L}_1$ or $\mathcal{L}_2$ with $\mathcal{L}_3$ as well, thus making the corresponding diagonal class $\mathcal{D}_3$ the edge-class of realizations. For instance, $\boldsymbol{\varrho}^{-1}\boldsymbol{\tau}\boldsymbol{\varrho}$ fixes the initial vertex and interchanges $\mathcal{L}_2$ and $\mathcal{L}_3$.

This symmetry makes it clear that the Wythoff space W of any realization of $\mathcal{H}_8$ must contain $\mathcal{L}_5$ also. In the present case, this implies that $\dim W = 2$, and the relation of the theorem results from applying Theorem 3F5 to calculate the dimension 25:

$$\tfrac{1}{75}\big(1 + 16(\alpha^2 + \beta^2 + \gamma^2) + 2(-\tfrac{1}{2})^2\big) = \tfrac{1}{25}.$$

This completes the proof. □

7K11 Remark We can now see from Proposition 7K10 that

$$|\beta| \leqslant \tfrac{1}{4} \implies 0 \leqslant \eta_{\boldsymbol{f}} \leqslant \tfrac{1}{10}.$$

This is still short of the value $\eta_{\boldsymbol{f}} = \frac{1}{4}$ for the 25-dimensional realization of $\{3, 3, 5\}/2$, and so we know from Remark 3K6 that we must have yet more 25-dimensional realizations of $\{5, 3, 3\}$.

7K12 Remark The symmetry among the three diagonal classes $\mathcal{D}_1, \mathcal{D}_2, \mathcal{D}_3$ of $\{5, 3, 3\}/8$ implies that (for a given cosine vector) there may be as many as six distinct allomorphs sharing the same vertices. This will have bearing on vertex-figure replacements in Chapter 17 (further, see the notes at the end of the section).

The Subfamily $\mathcal{H}_5$

What is perhaps surprising is that we can describe $\mathcal{H}_5$ as well. As with $\mathcal{H}_8$, we begin with

7K13 Lemma *Layers of $\mathcal{H}$ are identified in $\mathcal{H}_5$ as follows:*

0	1	1	4	3	24	2	12
18*a*	4	4	12	12*b*	24	7	24
		9	24	16	24	10	12
		17	24	19	24	15*ab*	54
		22*ab*	32	24	24	20	12
		25	4			23	24
						28	12
	5		100		120		150
12*a*	4	5	4	6	24		
30	1	8*ab*	32	11	24		
		13	24	14	24		
		21	24	18*b*	24		
		26	12	27	24		
		29	4				
	5		100		120		

The identifications to give $\mathcal{H}_{10}$ are in the same columns, which are arranged to suggest the central symmetry of $\mathcal{H}_5$.

As before, the entries in the second half of each column count the number of points in that layer. Note that corresponding layers of $\{5,3,3\}$ and $\{\frac{5}{2},3,3\}$ (that is, under the change of sign of $\sqrt{5}$) fall into the same layers of $\mathcal{H}_5$.

The technique used to describe $\mathcal{H}_5$ is quite different from the previous one. Recall that the basic geometric 4-simplex $\{3,3,3\}$ inscribed with full symmetry in $\{5,3,3\}$ has vertex-set $\mathbf{P} = \{\mathbf{p}_k^*\mathbf{p}_k \mid k = 0,\ldots,4\}$. We next see that the 600 points $\mathbf{Pg}$ with $\mathbf{g} \in \mathbf{I}$ are distinct, because $\arccos(-\frac{1}{4})$ is not among the angles subtended by pairs of vertices of $\{3,3,5\}$, and from this we conclude that each copy of $\mathbf{P}$ inscribed in the 120-cell is uniquely determined by the point $\mathbf{g} = 1\mathbf{g}$.

Exactly the same argument applies to left multiplication by elements of $\mathbf{I}^\ddagger$. However, if $\mathbf{h}^\ddagger\mathbf{P}$ and $\mathbf{Pg}$ have any point in common, then they coincide. Since we can write

$$\mathbf{g} = \mathbf{g}^\ddagger \cdot \mathbf{g}^*\mathbf{g} = \mathbf{g}^\ddagger \cdot \mathbf{p}_k^*\mathbf{p}_k$$

for some k, it follows that $\mathbf{g} \in \mathbf{I}$ and $\mathbf{g}^\ddagger \in \mathbf{I}^\ddagger$ determine the same copy of the simplex. As a consequence, we have

7K14 Lemma *Any realization of* $\{3,3,5\}$ *that is invariant under the change of sign of* $\sqrt{5}$ *induces a realization of* $\{5,3,3\}/5$ *of the same dimension.*

Confining our attention to the pure realizations of $\{3,3,5\}$, Lemma 7K14 permits $\Gamma_3, \Gamma_4, \Gamma_7, \Gamma_8$. However, we must replace two pairs by half their sums:

$$\Gamma_{12} := \tfrac{1}{2}(\Gamma_1 + \Gamma_2), \quad \Gamma_{56} := \tfrac{1}{2}(\Gamma_5 + \Gamma_6).$$

We can justify this in another way by an appeal to the component equation of Theorem 3C11, since $\Gamma_{1,2}$ and $\Gamma_{5,6}$ are the complements of the remaining components in the small simplex S and staurotope X, respectively. Effectively, then, the cosine vectors are of the form $(1, \gamma_6, \gamma_{15}, \gamma_4, \gamma_{37}, \gamma_2, \gamma_8)$, with the double indices for entries that are forced to be equal; the re-ordering is deliberate. To see how these are injected into $\mathcal{H}_5$, we just look for the layers of $\{5,3,3\}$ containing $\mathbf{I}$; these are $\mathcal{L}_0, \mathcal{L}_3 = \mathcal{L}_{19}, \mathcal{L}_{8ab}, \mathcal{L}_{11} = \mathcal{L}_{27}, \mathcal{L}_{15ab}, \mathcal{L}_{22ab}, \mathcal{L}_{30}$. Taking the layer vector of $\mathcal{H}_5$ to be as implied by Lemma 7K13, we see that the induced cosine vector is exactly of the form we chose, so confirming our initial claim.

The description of $\mathcal{H}_5$ is thus given by

7K15 Proposition *The cosine matrix of* $\{5,3,3\}/5$ *is*

$$\begin{bmatrix} 1 & 1 & 1 & 1 & 1 & 1 & 1 \\ 1 & 0 & \frac{1}{6} & -\frac{1}{3} & \frac{1}{6} & 0 & 1 \\ 1 & \frac{1}{4} & -\frac{1}{4} & 0 & -\frac{1}{4} & \frac{1}{4} & 1 \\ 1 & -\frac{1}{5} & 0 & \frac{1}{5} & 0 & -\frac{1}{5} & 1 \\ 1 & -\frac{1}{2} & \frac{1}{4} & 0 & -\frac{1}{4} & \frac{1}{2} & -1 \\ 1 & \frac{1}{4} & \frac{1}{4} & 0 & -\frac{1}{4} & -\frac{1}{4} & -1 \\ 1 & 0 & -\frac{1}{6} & 0 & \frac{1}{6} & 0 & -1 \end{bmatrix},$$

with layer and dimension vectors

$$\Lambda = (1, 20, 24, 30, 24, 20, 1), \quad D = (1, 18, 16, 25, 8, 16, 36).$$

7K16 Remark Putting Propositions 7K10 and 7K15 together confirms the observation of Remark 7K11 that the 25-dimensional realizations must have Wythoff space of dimension at least 3.

Small Dimensions

We now begin working through dimension by dimension. It is useful to bear in mind one thing. Implicitly, for each dimension d our calculations yield a lower bound w for the dimension of the Wythoff space, giving minimal contributions wd and $\frac{1}{2}w(w+1)$ to the total dimension and number of diagonal classes. The upper bounds provided by the latter enforce equality throughout.

Apart from $\Gamma_0 = (1^{36})$, we have the 4-dimensional starting points $\Gamma_{19} = \Gamma(\{5,3,3\})$ itself, and $\Gamma_{20} = \Gamma_{19}{}^{\ddagger} = \Gamma(\{\frac{5}{2},3,3\})$.

We have already seen the 8-dimensional cases $\Gamma_1 \in \mathcal{N}_{24} \subset \mathcal{N}_8$ and $\Gamma_{21} \in \mathcal{N}_5$. Though quite unrelated, they play surprisingly parallel rôles in the analysis.

Next, we have the 9-dimensional Γ_2 and Γ_3, given by

$$4\Gamma_{19}{}^2 = \Gamma_0 + 3\Gamma_2,$$
$$4\Gamma_{20}{}^2 = \Gamma_0 + 3\Gamma_3.$$

Finally, there is the 18-dimensional $\Gamma_7 \in \mathcal{N}_{10}$, again already noted.

Dimension 16 – Small Simplex

So far, we know three 16-dimensional realizations of $\mathcal{H}_2$. First, we have $\mathsf{Q}_1 := \{5,3,3\} \otimes \{\frac{5}{2},3,3\}$. Second, we have the 16-dimensional component Q_2 in $\mathcal{H}_{24}$. Third, there is the 16-dimensional component Q_3 in $\mathcal{H}_{10}$. In temporary notation (which we appeal to later), their respective cosine vectors are given by

$$16\Delta_1 = (16, -11, 5, -4, 9; -11, 1, -5, 4, 4; -1, 5, -4, 1, 1; -1, 1, 0, 0),$$

7K17 $$16\Delta_2 = (16, 1, 1, -4, 1; 1, 1, 1, 16, -4; 1, 1, -4, 1, 1; 1, 1, 16, -4),$$

$$4\Delta_3 = (4, 1, 0, -1, 1; 1, -1, 0, 1, 1; 1, 0, -1, 4, -1; 1, -1, 0, 0).$$

The corresponding values of η_f are $\frac{1}{64}, \frac{1}{64}, \frac{1}{16}$.

7K18 Remark The last tells us that $\mathsf{Q}_3 \in \mathcal{H}_{10}$ is the geometric dual of the 16-dimensional realization of $\{3,3,5\}/2$; it is interesting that, in this dual, the vertices are identified in tens. It is also worth noting that, even though Q_2 and Q_3 lie in disjoint quotients of $\mathcal{H}$ (except for $\{1\}$), their cosine vectors Γ_2 and Γ_3 are not Λ-orthogonal; in fact, $16\langle \Delta_j, \Delta_k \rangle_\Lambda = \frac{1}{4}$ whenever $j \neq k$.

This then leads us to suspect that $w = 2$ (which we shall confirm shortly – it also follows from direct construction of the symmetry group as that of $\{5,3,3\} \otimes \{\frac{5}{2},3,3\}$). Expecting that there should be a 16-dimensional realization with $\eta_f = 0$, it is straightforward to see that we have

7K19 Proposition *With the immediately previous notation, the general cosine vector Δ of a pure* 16*-dimensional realization of* $\{5,3,3\}/2$ *is*

$$\Delta = \lambda_1\Delta_1 + \lambda_2\Delta_2 + \lambda_3\Delta_3, \quad \textit{with } \lambda_1+\lambda_2+\lambda_3 = 1,\ \lambda_1^2+\lambda_2^2+\lambda_3^2 = 1.$$

The realization Q_4 with $\eta_f = 0$ has cosine vector $\Delta_4 := \frac{2}{3}\Delta_1 + \frac{2}{3}\Delta_2 - \frac{1}{3}\Delta_3$, namely,

$$12\Delta_4 = (12,-6,3,-3,4;-6,2,-2,9,-1;-1,3,-3,-3,2;-1,2,8,-2).$$

Moreover, if we set $\Gamma_4 := \Delta_3$, $\Gamma_6 := \Delta_4$ and define Γ_5 by

$$\begin{aligned} 8\sqrt{3}\Gamma_5 &:= 8(\Delta_2 - \Delta_1) \\ &= (0,6,-2,0,-4;6,0,3,6,-4;1,-2,0,0,0;1,0,8,-2), \end{aligned}$$

then we have the Λ-orthogonal basis vectors $\Gamma_4, \Gamma_5, \Gamma_6$ in standard form (the indices refer to the final table of Theorem 7K36).

7K20 Remark The entries in Δ_4 are certainly nicer than those in Δ_1, and arguably nicer than the ones in Δ_2. It is curious, therefore, that Q_4 seems not to arise in a natural way; rewriting the definition of Δ_4 as $3\Delta_4 + \Delta_3 = 2\Delta_1 + 2\Delta_2$ does not shed any more light on the matter.

7K21 Remark Observe that Δ_2, Δ_3 occur in

$$\begin{aligned} 8{\Gamma_1}^2 &= \Gamma_0 + 3\Gamma_1 + 4\Delta_2, \\ 8{\Gamma_{21}}^2 &= \Gamma_0 + 4\Delta_3 + 3\Gamma_7. \end{aligned}$$

Moreover, Propositions 7K9 and 7K15 imply that

$$\begin{aligned} 16{\Delta_2}^2 &= \Gamma_0 + 6\Gamma_1 + 9\Delta_2, \\ 16{\Delta_3}^2 &= \Gamma_0 + 4\Delta_3 + 6\Gamma_7 + 5\Psi, \end{aligned}$$

where Ψ is the 25-dimensional component of $\mathcal{H}_{10}$.

We could tackle the 16-dimensional realizations of $\{5,3,3\}/2$ in the same way that we shall treat the 25-dimensional realizations, except that we appeal to the 4-dimensional realization of $\{5,3:5\}$, with a corresponding action on 5×5 matrices. However, what we have done here already provides us with all the information that we need later.

Dimension 16 – Staurotope

We already have the cosine vector Γ_{22} of the 16-dimensional component of $\mathcal{H}_5$ in X, namely, $\Phi_1 := \Gamma_{22}$ as in the table of Theorem 7K37 (the notation Φ_1 is temporary). If $x \in \mathbb{E}^4$, then $x^3 = x \otimes x \otimes x$ has 4 components of type ξ_i^3, 12 of type $\xi_i^2\xi_j$ and 4 of type $\xi_i\xi_j\xi_k$, giving total dimension $4+12+4 = 20$. The crucial observation is then that $d({\Gamma_{19}}^3) \leqslant 20$ and, since

$$4\langle {\Gamma_{19}}^3, \Gamma_{19}\rangle_\Lambda = 4\langle {\Gamma_{19}}^2, {\Gamma_{19}}^2\rangle_\Lambda = \tfrac{1}{4}\langle \Gamma_0 + 3\Gamma_2, \Gamma_0 + 3\Gamma_2\rangle_\Lambda = \tfrac{1}{4}(1+1) = \tfrac{1}{2},$$

we can write

7K22 $$\Gamma_{19}{}^3 = \tfrac{1}{2}\Gamma_{19} + \tfrac{1}{2}\Phi_2,$$

where $d(\Phi_2) \leqslant 16$; compare here Remark 7E11. In exactly the same way, we have

$$\Gamma_{20}{}^3 = \tfrac{1}{2}\Gamma_{20} + \tfrac{1}{2}\Phi_3,$$

with $\Phi_3 := \Phi_2{}^\ddagger$ and $d(\Phi_3) \leqslant 16$.

Suppressing some messy calculations, we find that

$$16\langle \Phi_1, \Phi_2\rangle_\Lambda = \tfrac{3}{8} = 16\langle \Phi_1, \Phi_3\rangle_\Lambda;$$

this shows that Φ_2, Φ_3 have components with the same group as Φ_1, so that $d(\Phi_j) \geqslant 16$ for $j = 2, 3$. Hence, in fact, $d(\Phi_j) = 16$ for $j = 2, 3$; moreover, each Φ_j is also pure. Since $\Gamma_{19}\Gamma_{20} = \Delta_1$, we also see that

$$16\langle \Phi_2, \Phi_3\rangle_\Lambda = 16\langle \Phi_2 + \Gamma_{19}, \Phi_3 + \Gamma_{20}\rangle_\Lambda = 64\langle \Gamma_{19}{}^3, \Gamma_{20}{}^3\rangle_\Lambda = \tfrac{1}{16},$$

by direct calculation of the last term; note that the remaining three terms in the expression vanish, by an appeal to Λ-orthogonality. The Λ-inner product here involves terms $N(\alpha) := \alpha\alpha^\ddagger$, with $\alpha \in \mathbb{Q}[\sqrt{5}]$. If $\alpha = p + q\sqrt{5}$, then the norm $N(\alpha) = p^2 - 5q^2$ (as it is called) is rational.

7K23 Remark The following observation uses very similar calculations, and so we perform them here; they will be needed in dimension 36. We have

$$\begin{aligned} 3\Gamma_2\Gamma_{19} &= 4\Gamma_{19}{}^3 - \Gamma_{19} \\ &= 2\Gamma_{19} + 2\Phi_2 - \Gamma_{19} = 2\Phi_2 + \Gamma_{19}, \end{aligned}$$

with an analogous expression for $\Gamma_3\Gamma_{20}$. It follows that

$$\begin{aligned} 36\langle \Gamma_2\Gamma_{19}, \Gamma_3\Gamma_{20}\rangle_\Lambda &= 4\langle 2\Phi_2 + \Gamma_{19}, 2\Phi_3 + \Gamma_{20}\rangle_\Lambda \\ &= 16\langle \Phi_2, \Phi_3\rangle_\Lambda = \tfrac{1}{16}, \end{aligned}$$

by the previous calculation, with cross terms again vanishing.

The remaining calculations are routine. The mixed cosine vector is the appropriate multiple $\Gamma_{23} - \frac{2}{\sqrt{15}}(\Phi_2 - \Phi_3)$ with $\|\Gamma_{23}\|_\Lambda^2 = \frac{1}{32}$, while the required pure cosine vector that is Λ-orthogonal to Γ_{22} is $\Gamma_{24} = \frac{1}{5}(4\Phi_2 + 4\Phi_3 - 3\Phi_1)$; again, refer to the table for details.

7K24 Remark The 16-dimensional component of $\{3, 3, 5\}$ in X has $\eta_f = \frac{7}{16}$. The corresponding values in $\mathcal{H}$ are $\eta_f = (\frac{3}{16}, \frac{\sqrt{3}}{8}, \frac{1}{4})$ for $(\Gamma_{22}, \Gamma_{23}, \Gamma_{24})$. The realization achieving the maximum $\eta_f = \frac{7}{16}$ (that of the geometric dual of the corresponding realization of $\{3, 3, 5\}$) has cosine vector $\alpha^2\Gamma_{22} + 2\alpha\beta\Gamma_{23} + \beta^2\Gamma_{24}$ when $(\alpha, \beta) = \frac{1}{\sqrt{7}}(\sqrt{3}, 2)$, which leads to cosine vector beginning

$$\tfrac{1}{28}(28, 23, 16, 7, 3; 1, \ldots)$$

(we have given enough to check that the correct value of η_f is attained). It is curious that this realization is not nicer.

Dimension 24

We obtain the realizations in $\mathbb{E}^{24}$ by methods which are curiously parallel, even though the 8-dimensional realizations employed – one in $\mathcal{H}_5$ and the other in $\mathcal{H}_{24}$ – seem completely unrelated. We have

7K25
$$4\Gamma_{20}\Gamma_{22} = \Gamma_1 + 3\Gamma_8, \qquad 4\Gamma_1\Gamma_{20} = \Gamma_{22} + 3\Gamma_{25},$$
$$4\Gamma_{19}\Gamma_{22} = \Gamma_1 + 3\Gamma_9, \qquad 4\Gamma_1\Gamma_{19} = \Gamma_{22} + 3\Gamma_{26}.$$

Observe that we have exact counts of dimensions: $4 \cdot 8 = 32 = 8 + 24$. More specifically, if the dimension is d, then $d + 8 \leqslant 32$ giving $d \leqslant 24$. In the other direction, we have (for instance)

$$\|\Gamma_{19}\Gamma_{21}\|_\Lambda^2 = \langle {\Gamma_{19}}^2, {\Gamma_{21}}^2 \rangle_\Lambda = \langle \tfrac{1}{4}\Gamma_0 + \tfrac{3}{4}\Gamma_2, \tfrac{1}{8}\Gamma_0 + \tfrac{3}{8}\Gamma_7 + \tfrac{1}{2}\Delta_3 \rangle_\Lambda = \tfrac{1}{4} \cdot \tfrac{1}{8} = \tfrac{1}{32},$$

where we have used Remark 7K21 (the Λ-orthogonality of the other terms is due to their different dimensions). But then

$$\tfrac{1}{32} = \|\Gamma_{19}\Gamma_{21}\|_\Lambda^2 = \|\tfrac{1}{4}\Gamma_1 + \tfrac{3}{4}\Gamma_9\|_\Lambda^2 = \tfrac{1}{16} \cdot \tfrac{1}{8} + \tfrac{9}{16} \cdot \|\Gamma_9\|_\Lambda^2$$

by Λ-orthogonality of the construction gives $\|\Gamma_9\|_\Lambda^2 = \frac{1}{24}$, which leads to $d \geqslant 1/\|\Gamma_9\|_\Lambda^2 = 24$ by the dimension inequality Corollary 3F4.

The purity of these new realizations can be checked by their Λ-orthogonality to all the lower-dimensional realizations.

Dimension 25

To settle this case, we use the fact that $[3, 3, 5]/2 \cong (\mathrm{A}_5 \times \mathrm{A}_5) \rtimes \mathrm{C}_2$, and mimic the description of the geometric group $[3, 3, 5]$ in terms of quaternions. The context is that of 6×6 matrices Z with row and column sums 0, acted on by transposition and left and right multiplication by orthogonal matrices. In particular, we observe that the mapping

7K26
$$Z \mapsto AZ^{\mathsf{T}}A$$

is involutory, because $A^{-1} = A^{\mathsf{T}}$. In the present case, A will be a permutation matrix; the permutations are easily obtained from the 5-dimensional realization of $\{5, 3 : 5\}$. In effect, we are in $\mathbb{L}^5 \otimes \mathbb{L}^5$, with $\mathbb{L}^5$ as usual the symmetric hyperplane of (1E15). The generatrix $(R_0, \ldots, R_3)$ of the symmetry group is given by

7K27
$$R_0 \colon \sigma_0 = (1\ 2\ 3\ 4\ 5),$$
$$R_1 \colon \sigma_1 = \iota \text{ (the identity)},$$
$$R_2 \colon \sigma_2 = (0\ 1\ 2)(3\ 5\ 4),$$
$$R_3 \colon \sigma_3 = (1\ 5)(2\ 4).$$

A little work establishes

7K28 Lemma *The Wythoff space* W *of the realizations of* $\mathcal{H}_2$ *with generatrix (7K27) consists of those matrices*

$$\begin{bmatrix} \alpha & \beta & \gamma & \delta & \gamma & \beta \\ \beta & \gamma & \alpha & \beta & \delta & \gamma \\ \gamma & \alpha & \beta & \gamma & \beta & \delta \\ \delta & \beta & \gamma & \alpha & \gamma & \beta \\ \gamma & \delta & \beta & \gamma & \beta & \alpha \\ \beta & \gamma & \delta & \beta & \alpha & \gamma \end{bmatrix},$$

with $\alpha + 2\beta + 2\gamma + \delta = 0$.

In particular, it follows that $w := \dim W = 3$, thus confirming the observation of Remark 7K16.

7K29 Remark The analogous calculation for $\{3, 3, 5\}$ leads to initial vertex

$$\begin{bmatrix} 5 & -1 & -1 & -1 & -1 & -1 \\ -1 & -1 & 5 & -1 & -1 & -1 \\ -1 & 5 & -1 & -1 & -1 & -1 \\ -1 & -1 & -1 & -1 & -1 & 5 \\ -1 & -1 & -1 & -1 & 5 & -1 \\ -1 & -1 & -1 & 5 & -1 & -1 \end{bmatrix}.$$

We adopt the temporary notation A for the matrix with $\alpha = 1$ and $\beta = \gamma = \delta = 0$, with B, C, D defined analogously in the obvious way; thus A, B, C, D are mutually orthogonal, with 6×6 matrices regarded as vectors in $\mathbb{E}^{36} = \mathbb{E}^6 \otimes \mathbb{E}^6$. The condition for W leads to an initial orthogonal (but not orthonormal) basis consisting of

$$A - D, \quad B - C, \quad \tfrac{1}{3}(2A - B - C + 2D) = A + D - \tfrac{1}{3}J,$$

where J is the 6×6 matrix all of whose entries are 1; we scale the last in this way for future convenience. The square norms of these matrices are 12, 24 and 8, respectively.

We start with the third initial vertex. A possible image $A' + D'$ of $A + D$ (we can ignore $-\frac{1}{3}J$ for the moment) is determined by three disjoint pairs of row indices, and three such pairs of column indices, giving at most

$$15 \cdot 15 = 225 < 300.$$

This tells us at once that we must be in a further quotient of $\mathcal{H}_2$, and since 75 is the only plausible divisor of 225 in the frame, we must actually be in $\mathcal{H}_8$. Indeed, the corresponding entry of the cosine vector will be

$$\tfrac{1}{8}\langle A + D - \tfrac{1}{3}J, A' + D' - \tfrac{1}{3}J\rangle = \tfrac{1}{8}\langle A + D, A' + D'\rangle - \tfrac{1}{2},$$

and working out the first few entries applying powers of the product $R_1 \cdot R_0R_2$ (which goes along a Petrie polygon of the initial facet), or using calculations like those following, leads to

$$8\Gamma_{13} = (8,-1,-1,0,2;-1,-1,2,-4,0;-1,-1,0,2,2;-1,-1,8,0),$$

which we recognize from Proposition 7K10. (Our labelling is again looking to the future.)

Though the situation for $B - C$ is not quite so clear, the case is similar. There are 225 choices for B, and each choice of B seems to allow two choices for C (the other, in the initial matrix, being $A + D$), thus giving $2 \cdot 225 = 450$ possibilities. (In fact, it is more likely that B determines C uniquely; this should come out of a deeper look at how the group acts.) Since 300 is not a divisor of 450, again we must be in a quotient, which will be $\mathcal{H}_8$. This is confirmed by direct calculation, leading to the cosine vector

$$8\Gamma_{15} = (8,1,1,0,-2;1,1,-2,-4,0;1,1,0,-2,-2;1,1,8,0),$$

exactly as to be expected from Λ-orthogonality.

We can use the matrix representation to work out the corresponding mixed cosine vector Γ_{14} as well, and find that

$$8\Gamma_{14} = \sqrt{3}(0,-1,1,0,0;1,-1,0,0,0;1,-1,0,0,0;-1,1,0,0,0).$$

Once again, knowing that we are in $\mathcal{H}_8$ here means that we need only calculate the first few terms to obtain all of them.

There remain the cosine vectors derived from $A - D$; note that, with this as initial vertex, we do indeed obtain all 300. We approach this case in three steps. First, for each layer $\mathcal{L}_s$, we choose $\mathbf{a} \in \mathbf{I}^{\ddagger}$ and $\mathbf{b} \in \mathbf{I}$ such that (regarded as a quaternion) $4\tilde{\mathbf{a}}\mathbf{b} \in \mathcal{L}_s$. In fact, as we shall see, we can choose either $\mathbf{a} = 1$ or $\mathbf{a}$ as a fixed pure imaginary quaternion; this simplifies the later calculations. Second, we relate these quaternions $\mathbf{a}$, $\mathbf{b}$ to permutations in A_5 acting on antipodal vertices of the icosahedron $\{3,5\}$, with the identifications

$$\begin{aligned} 0 &:= \pm(\tau,-1,0), & 3 &:= \pm(\tau,1,0), \\ 1 &:= \pm(0,\tau,1), & 4 &:= \pm(1,0,\tau), \\ 2 &:= \pm(-1,0,\tau), & 5 &:= \pm(0,\tau,-1). \end{aligned}$$

In this context, recall from (1K10) that the quaternion $\cos\vartheta + \sin\vartheta\mathbf{u}$, with $\mathbf{u}$ a pure imaginary unit quaternion identified with a unit vector in $\mathbb{E}^3$, induces a rotation of -2ϑ about the axis $\mathbf{u}$. Third, we lift these permutations to actions on our 6×6 matrices or, rather, their basis elements. As a check on the workings, we can use Λ-orthogonality with respect to $\mathcal{H}_8$ (we cannot use $\mathcal{H}_{10}$, because we still have a component there).

First, $\mathbf{a} = 1$; the corresponding $\mathbf{b}$ with induced permutations are as follows, with the asymmetric layers coming second:

$$\begin{array}{rll}
\mathcal{L}_0\colon & 1 & \iota, \\
\mathcal{L}_3\colon & \frac{1}{2}(\tau + \mathbf{i} + \tau^{-1}\mathbf{j}) & (0\ 5\ 2\ 1\ 3), \\
\mathcal{L}_{8b}\colon & \frac{1}{2}(1 + \tau^{-1}\mathbf{i} + \tau\mathbf{j}) & (0\ 4\ 3)(1\ 2\ 5), \\
\mathcal{L}_{11}\colon & \frac{1}{2}(\tau^{-1} + \tau\mathbf{i} + \mathbf{j}) & (0\ 2\ 3\ 5\ 1), \\
\mathcal{L}_{15a}\colon & \mathbf{i} & (1\ 5)(2\ 4), \\
\mathcal{L}_{8a}\colon & \frac{1}{2}(1 + \mathbf{i} + \mathbf{j} + \mathbf{k}) & (0\ 2\ 5)(1\ 3\ 4), \\
\mathcal{L}_{15b}\colon & \frac{1}{2}(\tau\mathbf{i} + \tau^{-1}\mathbf{j} + \mathbf{k}) & (1\ 4)(2\ 3).
\end{array}$$

We have listed $\mathcal{L}_{15b}$ here, even though it becomes symmetric in $\mathcal{H}_2$.

For the remainder, we can take $\mathbf{a} = \frac{1}{2}(\tau\mathbf{i} + \mathbf{j} + \tau^{-1}\mathbf{k}) \in \mathbf{I}^{\ddagger}$ a fixed involution, with corresponding permutation (1 4)(2 3):

$$\begin{array}{rll}
\mathcal{L}_1\colon & \frac{1}{2}(\tau\mathbf{i} + \tau^{-1}\mathbf{j} + \mathbf{k}) & (1\ 4)(2\ 3), \\
\mathcal{L}_2\colon & \frac{1}{2}(\mathbf{i} + \tau\mathbf{j} + \tau^{-1}\mathbf{k}) & (0\ 1)(3\ 4), \\
\mathcal{L}_4\colon & \frac{1}{2}(\tau^{-1}\mathbf{i} + \mathbf{j} + \tau\mathbf{k}) & (1\ 3)(4\ 5), \\
\mathcal{L}_5\colon & \frac{1}{2}(\mathbf{i} + \tau\mathbf{j} - \tau^{-1}\mathbf{k}) & (0\ 4)(3\ 5), \\
\mathcal{L}_6\colon & \frac{1}{2}(\tau\mathbf{i} + \tau^{-1}\mathbf{j} - \mathbf{k}) & (0\ 2)(4\ 5), \\
\mathcal{L}_{10}\colon & \frac{1}{2}(\tau\mathbf{i} - \tau^{-1}\mathbf{j} - \mathbf{k}) & (2\ 5)(3\ 4), \\
\mathcal{L}_{12a}\colon & \frac{1}{2}(\tau^{-1}\mathbf{i} - \mathbf{j} + \tau\mathbf{k}) & (0\ 1)(2\ 5), \\
\mathcal{L}_{12b}\colon & \frac{1}{2}(\tau^{-1}\mathbf{i} + \mathbf{j} - \tau\mathbf{k}) & (0\ 5)(1\ 4), \\
\mathcal{L}_{14}\colon & \frac{1}{2}(\mathbf{i} - \tau\mathbf{j} + \tau^{-1}\mathbf{k}) & (0\ 2)(1\ 3), \\
\mathcal{L}_7\colon & \frac{1}{2}(\tau + \mathbf{i} + \tau^{-1}\mathbf{j}) & (0\ 5\ 2\ 1\ 3), \\
\mathcal{L}_9\colon & \frac{1}{2}(\tau^{-1} + \tau\mathbf{i} - \mathbf{j}) & (0\ 4\ 3\ 1\ 5), \\
\mathcal{L}_{13}\colon & \frac{1}{2}(\tau + \mathbf{j} - \tau^{-1}\mathbf{k}) & (1\ 5\ 4\ 3\ 2).
\end{array}$$

We shall not give the details of calculating the entries of the cosine vectors. However, it is worth illustrating why we may need to find two inner products for an asymmetric diagonal class. For instance, the unsymmetrized mixed vectors corresponding to Γ_{14} are actually (before scaling)

$$(0,-1,1,0,0;1,-1,0,\pm 4,0;1,-1,0,0,0;-1,1,0,0),$$

illustrating the asymmetry of $\mathcal{L}_{8a}$. We thus obtain

$$\begin{array}{rll}
12\Gamma_{10} &= (12,3,3,0,-2;3,-1,-2,0,-4;-1,3,0,6,2;-1,-1,-4,4), & 6, \\
4\sqrt{6}\Gamma_{11} &= (0,3,-1,0,2;3,-1,-2,0,0;1,-1,0,-6,2;1,-1,0,0), & \sqrt{6}, \\
4\sqrt{2}\Gamma_{12} &= (0,3,1,0,0;-3,-1,0,0,0;-1,-1,0,0,0;1,1,0,0), & 3\sqrt{2}.
\end{array}$$

The final column gives $40\eta_{\boldsymbol{f}}$ (notionally in case of the mixed cosine vectors); the corresponding values for $\Gamma_{13}, \Gamma_{14}, \Gamma_{15}$ are $1, \sqrt{3}, 3$, respectively. As usual, the indices are as they appear in the final table of Theorem 7K36.

It follows that, for the general cosine vector, $\eta_{\boldsymbol{f}} = xHx^{\mathsf{T}}$, where

$$H = \frac{1}{40}\begin{bmatrix} 6 & \sqrt{6} & 3\sqrt{2} \\ \sqrt{6} & 1 & \sqrt{3} \\ 3\sqrt{2} & \sqrt{3} & 3 \end{bmatrix}$$

and x is a unit vector. Generally speaking, one might have problems finding the maximum and minimum of $\eta_{\boldsymbol{f}}$. However, here it can be done by inspection. Indeed, the unit vectors

$$\begin{aligned} u_1 &:= \tfrac{1}{\sqrt{10}}(\sqrt{6}, 1, \sqrt{3}), \\ u_2 &:= \tfrac{1}{2\sqrt{5}}(2\sqrt{2}, -\sqrt{3}, -3), \\ u_3 &:= \tfrac{1}{2}(0, \sqrt{3}, -1) \end{aligned}$$

are mutually orthogonal, and yield $\eta_{\boldsymbol{f}} = \frac{1}{4}, 0, 0$, respectively.

7K30 Remark The fact that $\eta_{\boldsymbol{f}} = 0$ for u_2 and u_3 implies that $\eta_{\boldsymbol{f}} = 0$ for the unit vectors in a whole 2-plane in the essential Wythoff space. We shall see the same phenomenon in the 36-dimensional realizations.

Dimensions 30 *and* 40

We treat these cases together, because they come out of the same stable. At this stage, in $\mathcal{H}_2$ (that is, components of the small simplex S) we have three diagonal classes to account for, and a dimension deficit of 100. The remaining cases $\Gamma_{16}, \Gamma_{17}, \Gamma_{18}$ are given by

$$\begin{aligned} 12\Gamma_{19}\Gamma_{26} &= 3\Gamma_1 + 3\Gamma_7 + \Gamma_9 + 5\Gamma_{16}, \\ 12\Gamma_{20}\Gamma_{25} &= 3\Gamma_1 + 3\Gamma_7 + \Gamma_8 + 5\Gamma_{17}, \\ 12\Gamma_{19}\Gamma_{25} &= 4\Delta_6 + 3\Gamma_8 + 5\Gamma_{18}, \\ 12\Gamma_{20}\Gamma_{26} &= 4\Delta_6 + 3\Gamma_9 + 5\Gamma_{18}, \end{aligned}$$

where

$$\Delta_6 := \tfrac{3}{4}\Gamma_4 + \tfrac{\sqrt{3}}{2}\Gamma_5 + \tfrac{1}{4}\Gamma_6$$

is pure of dimension 16. (We give both constructions for Γ_{18}, to complete the pattern.)

Observe that the dimensions do not match up here. For the first two,

$$8 + 18 + 24 + 30 = 80 < 96 = 4 \cdot 24,$$

while for the others we have

$$16 + 24 + 40 = 80 < 96.$$

Thus these do not help to pin down the dimensions. However, we have

$$\|\Gamma_{16}\|_\Lambda^2 = \|\Gamma_{17}\|_\Lambda^2 = \tfrac{1}{30}, \quad \|\Gamma_{18}\|_\Lambda^2 = \tfrac{1}{40},$$

giving $d_{16}, d_{17} \geqslant 30$ and $d_{18} \geqslant 40$, both by appeal to the dimension inequality Corollary 3F4. But now $d_{16} + d_{17} + d_{18} = 100$ imposes equality.

Dimension 48

We treat dimension 48 before dimension 36, because we need the results of the former case to feed into the latter. However, it is worth referring to the initial remarks in that case about what is already known. First note that the present dimension deficit of components of X is 204. We shall see immediately that the Wythoff space in $\mathbb{E}^{48}$ is at least 2-dimensional, while that in $\mathbb{E}^{36}$ has dimension at least 3. Since $2\cdot 48+3\cdot 36 = 204$, these dimensions must be exactly 2 and 3, respectively.

We define $K_1, \ldots, K_4$ by

$$\begin{aligned} 3\Gamma_7\Gamma_{19} &= \Gamma_{26} + 2K_1, \\ 3\Gamma_7\Gamma_{20} &= \Gamma_{25} + 2K_2, \\ 3\Gamma_2\Gamma_{21} &= \Gamma_{26} + 2K_3, \\ 3\Gamma_3\Gamma_{21} &= \Gamma_{25} + 2K_4. \end{aligned}$$

In each case, we have an exact dimension count: $24 + 48 = 72 = 18 \cdot 4 = 9 \cdot 8$. Indeed, each K_j has dimension 48. We have $\|K_j\|_\Lambda^2 = \frac{1}{48}$ for each j, yielding $48 \leqslant d_j \leqslant 48$ as for dimension 24. Alternatively, for instance,

$$\|\Gamma_7\Gamma_{19}\|_\Lambda^2 = \langle \Gamma_7{}^2, \Gamma_{19}{}^2 \rangle_\Lambda = \tfrac{1}{18} \cdot \tfrac{1}{4} = \tfrac{1}{72},$$

since

$$\Gamma_7{}^2 = \tfrac{1}{18}\Gamma_0 + \tfrac{1}{6}\Gamma_7 + \tfrac{2}{9}\Delta_2 + \tfrac{5}{9}H, \quad \Gamma_{19}{}^2 = \tfrac{1}{4}\Gamma_0 + \tfrac{3}{4}\Gamma_2,$$

with Δ_2 pure of dimension 16 as before and H pure of dimension 25, and the other inner products vanish because the dimensions are different. On the other hand,

$$\tfrac{1}{72} = \|\Gamma_7\Gamma_{19}\|_\Lambda^2 = \|\tfrac{1}{3}\Gamma_7 + \tfrac{2}{3}K_1\|_\Lambda^2 = \tfrac{1}{9}\|\Gamma_7\|_\Lambda^2 + \tfrac{4}{9}\|K_1\|_\Lambda^2 - \tfrac{1}{9} \cdot \tfrac{1}{24} - \tfrac{4}{9} \cdot \tfrac{1}{d},$$

yielding $d = 48$, as claimed.

In a similar way, the various inner products of the K_j can be found:

$$\langle \Gamma_7\Gamma_{19}, \Gamma_7\Gamma_{20} \rangle_\Lambda = \langle \Gamma_7{}^2, \Gamma_{19}\Gamma_{20} \rangle_\Lambda = \tfrac{2}{9}\langle \Delta_2, \Delta_1 \rangle_\Lambda = \tfrac{2}{9} \cdot \tfrac{1}{64} = \tfrac{1}{288},$$

from which it follows that $48\langle K_1, K_2 \rangle_\Lambda = \frac{3}{8}$. Appealing to similar calculations, or directly, we find that the matrix of inner products $48\langle K_i, K_j \rangle_\Lambda$ is

$$\begin{bmatrix} 1 & \frac{3}{8} & \frac{1}{16} & \frac{3}{8} \\ \frac{3}{8} & 1 & \frac{3}{8} & \frac{1}{16} \\ \frac{1}{16} & \frac{3}{8} & 1 & \frac{27}{32} \\ \frac{3}{8} & \frac{1}{16} & \frac{27}{32} & 1 \end{bmatrix}.$$

The fact that these inner products are positive shows that the K_j belong to the same family. We know that the initial vertices in the Wythoff space must be coplanar, which chimes in with Proposition 3H6.

Defining K_0 by

$$24K_0 = (24, -6, -2, 1, -2; 6, 2, 0, -6, -1; 0, -2, -1, 6, 2; 0, 2, 0, 0),$$

so that $\|K_0\|_\Lambda^2 = \frac{1}{96}$, we find that

$$K_0 = \tfrac{3}{10}(K_1 + K_2) + \tfrac{1}{5}(K_3 + K_4).$$

The general pure cosine vector Γ in the family must be an affine combination of the K_j which satisfies $\|\Gamma - K_0\|_\Lambda^2 = \frac{1}{96}$ A little work then shows that

$$\Gamma = (1 - \lambda - \mu)K_0 + \lambda K_1 + \mu K_2,$$

for some λ, μ such that $\lambda^2 - \frac{1}{2}\lambda\mu + \mu^2 = 1$.

The best choice for a basis seems to be to take $\lambda_1 = -\lambda_2 = \pm\sqrt{2/5}$ to obtain Γ_{33}, Γ_{35}, while $K_1 + K_2 - 2K_0$, appropriately normalized, yields Γ_{34} (even so the entries are not very nice). We shall not write out these vectors here, but instead refer to the table of Theorem 7K37.

Dimension 36

We begin by listing what we already know. First, Proposition 7K15 gives the 36-dimensional component $\Theta_1 := \Gamma_{27}$ in $\mathcal{H}_5$, namely,

$$6\Theta_1 = (6, 0, 0, -1, 0; 0, 1, 0, 0, 0; 0, 0, 1, -6, -1; 0, 1, 0, 0).$$

(The notation is, of course, temporary.) Further, $\Theta_2 := \Gamma_2\Gamma_{20}$ and $\Theta_2{}^\ddagger = \Gamma_3\Gamma_{19}$ are pure 36-dimensional, in analogy to the corresponding construction for $\{3, 3, 5\}$; however, note here that $\Theta_2 \neq \Theta_2{}^\ddagger$. We find that

$$36\langle\Theta_1, \Theta_2\rangle_\Lambda = 36\langle\Theta_1, \Theta_2{}^\ddagger\rangle_\Lambda = \tfrac{1}{4}, \quad 36\langle\Theta_2, \Theta_2{}^\ddagger\rangle_\Lambda = \tfrac{1}{16}.$$

The first two calculations are direct. For the third, we have

$$36\langle\Theta_2, \Theta_2{}^\ddagger\rangle_\Lambda = 36\langle\Gamma_2\Gamma_{19}, \Gamma_3\Gamma_{20}\rangle_\Lambda = 16\langle\Phi_2, \Phi_2{}^\ddagger\rangle_\Lambda,$$

as we saw in Remark 7K23.

There follow two things. First, all three realizations are indeed pure with the same group. Second, Proposition 3H6 shows that the initial vertices of these three polytopes are not coplanar in the Wythoff space W, which is thus at least 3-dimensional. As we remarked at the beginning of the previous case, we thus see that $\dim W = 3$.

The strongest tool is provided by the products $\Delta\Gamma_{19}$ (and $\Delta\Gamma_{20}$), with Δ a pure 16-dimensional realization of $\mathcal{H}_2$; these must all have the same group. We first have

$$4\langle\Delta\Gamma_{19}, \Gamma_{20}\rangle_\Lambda = 4\langle\Delta, \Gamma_{19}\Gamma_{20}\rangle_\Lambda = 4\langle\Delta, \Delta_1\rangle_\Lambda,$$

which takes the values $\frac{1}{4}, \frac{1}{16}, \frac{1}{16}$ for $\Delta = \Delta_1, \Delta_2, \Delta_3$, respectively.

Next, recall that $\Gamma_1, \Delta_2 \in \mathcal{H}_{24}$, with $4\Gamma_1\Delta_2 = \Gamma_1 + 3\Delta_2$. Hence, with $\Delta = \Delta_2$, from the 24-dimensional cases we have

$$\begin{aligned} 24\langle \Delta_2\Gamma_{19}, \Gamma_{25}\rangle &= 8\langle \Delta_2\Gamma_{19}, 4\Gamma_1\Gamma_{20} - \Gamma_{21}\rangle_\Lambda = 8\langle \Delta_2, 4\Gamma_1\Gamma_{19}\Gamma_{20} - \Gamma_{19}\Gamma_{21}\rangle_\Lambda \\ &= 32\langle \Delta_2, \Gamma_1\Delta_1\rangle_\Lambda - 2\langle \Delta_2, \Gamma_1 + 3\Gamma_9\rangle_\Lambda \\ &= 32\langle \Gamma_1\Delta_2, \Delta_1\rangle_\Lambda = 8\langle \Gamma_1 + 3\Delta_2, \Delta_1\rangle_\Lambda \\ &= 8 \cdot 3 \cdot \tfrac{1}{64} = \tfrac{3}{8}; \end{aligned}$$

remember that pure components of different dimensions are orthogonal.

Taking $\Delta = \Delta_1$ and recalling that $\Delta_1 = \Gamma_{19}\Gamma_{20}$, we then have

$$4\Delta_1\Gamma_{19} = 4{\Gamma_{19}}^2\Gamma_{20} = 3\Gamma_2\Gamma_{20} + \Gamma_{20}.$$

We know that $\Gamma_2\Gamma_{20}$ is pure of dimension 36 from the beginning of the section, and the coefficient of Γ_{20} is that given above. We have thus shown

7K31 Lemma *If* Δ *is a pure* 16*-dimensional component of* $\mathcal{H}_2$*, then* $\Delta\Gamma_{19}$ *can only have components* Γ_{20}, Γ_{25} *and a pure component of dimension* 36.

Proof. We have seen that we have found such components among the products $\Delta\Gamma_{19}$. Since $\Delta\Gamma_{19}$ has dimension at most $16 \cdot 4 = 64 = 4 + 24 + 36$, giving an exact count, it follows that there can be no others. □

We now consider $\Delta_3 \in \mathcal{H}_5$. Since $\Theta_1 \in \mathcal{H}_5$ also, we see that

$$\langle \Delta_3\Gamma_{19}, \Theta_1\rangle_\Lambda = \langle \Gamma_{19}, \Delta_3\Theta_1\rangle_\Lambda = 0,$$

since no component of $\mathcal{H}_5$ has dimension 4. In other words,

7K32 Lemma *The pure* 36*-dimensional components of* $\Delta_3\Gamma_{19}$ *and* $\Delta_3\Gamma_{20}$ *are* Λ*-orthogonal to* Θ_1*.*

We calculate directly that

$$24\langle \Delta_3\Gamma_{19}, \Gamma_{25}\rangle_\Lambda = \tfrac{3}{8},$$

and since we already know that $4\langle \Delta_3\Gamma_{19}, \Gamma_{20}\rangle_\Lambda = \frac{1}{16}$, it follows that the pure 36-dimensional component Θ_3, say, is given by

7K33 $$16\Delta_3\Gamma_{19} = \Gamma_{20} + 6\Gamma_{25} + 9\Theta_3;$$

similarly,

7K34 $$16\Delta_3\Gamma_{20} = \Gamma_{19} + 6\Gamma_{26} + 9{\Theta_3}^{\ddagger}.$$

Now,

$$\begin{aligned} 256\langle \Delta_3\Gamma_{19}, \Delta_3\Gamma_{20}\rangle_\Lambda &= 256\langle \Gamma_{19}\Gamma_{20}, {\Delta_3}^2\rangle_\Lambda \\ &= 16\langle \Delta_1, \Gamma_0 + 4\Delta_3 + 6\Gamma_7 + 5\Pi\rangle_\Lambda \\ &= 4 \cdot \tfrac{1}{4} = 1, \end{aligned}$$

where we have appealed to Remarks 7K18 and 7K21. If we substitute $\Delta_3\Gamma_{19}$ and $\Delta_3\Gamma_{20}$ using (7K33) and (7K34), we obtain

$$1 = \langle \Gamma_{20} + 6\Gamma_{25} + 9\Theta_3, \Gamma_{19} + 6\Gamma_{26} + 9\Theta_3{}^{\ddagger} \rangle_\Lambda = 81\langle \Theta_3, \Theta_3{}^{\ddagger} \rangle_\Lambda,$$

since the remaining terms in the Λ-inner product vanish. In other words, we have shown that

7K35 $$36\langle \Theta_3, \Theta_3{}^{\ddagger} \rangle_\Lambda = \tfrac{36}{81} = \tfrac{4}{9}.$$

We now appeal to what we did earlier. The complement of the remaining terms in the component equation Theorem 3C11 for the staurotope X is $108\Theta_0$, with Θ_0 the centre of the realization domain $\mathcal{N}$, or (then multiplied by 3) the sum of any three mutually Λ-orthogonal pure cosine vectors in $\mathcal{N}$; compare here Remark 3C12. The entry is

$$36\Theta_0 = (36, 3, 3, -6, 2; -3, -1, 0, 0, 0; -1, 3, 6, -6, 0; 1, -1, 0, 0).$$

It follows that the centre of the Λ-circle in $\mathcal{N}$ that contains $\Delta_3\Gamma_{19}, \Delta_3\Gamma_{20}$ and is Λ-orthogonal to Θ_1 is $\Theta_4 := \frac{1}{2}(3\Theta_0 - \Theta_1)$, whose square Λ-norm is $\frac{1}{72}$, as should be expected.

Since $\langle \Theta_3, \Theta_4 \rangle_\Lambda = \langle \Theta_3{}^{\ddagger}, \Theta_4 \rangle_\Lambda = \frac{1}{72}$ also, the Λ-orthogonal sub-basis is formed by $\lambda(\Theta_3 + \Theta_3{}^{\ddagger}) + (1 - 2\lambda)\Theta_4$ and $\mu(\Theta_3 - \Theta_3{}^{\ddagger})$ for suitable λ and μ. For λ, we must solve

$$\begin{aligned} 1 &= 36\|\lambda(\Theta_3 + \Theta_3{}^{\ddagger}) + (1 - 2\lambda)\Theta_4\|_\Lambda^2 \\ &= \lambda^2(1 + \tfrac{8}{9} + 1) + 2\lambda(1 - 2\lambda)(\tfrac{1}{2} + \tfrac{1}{2}) + (1 - 2\lambda)\tfrac{1}{2}, \end{aligned}$$

whence $\lambda = \pm\frac{3}{4}$. Similarly, for μ, we need

$$1 = 72\|\mu(\Theta_3 - \Theta_3{}^{\ddagger})\|_\Lambda^2 = \mu^2(2 - \tfrac{16}{9} + 2) = \tfrac{20}{9},$$

giving $\mu = \pm\frac{3}{2\sqrt{5}}$ (either sign will do). In other words, setting

$$\begin{aligned} 4\Gamma_{30} &:= 10\Theta_4 - 3(\Theta_3 + \Theta_3{}^{\ddagger}), \\ 2\sqrt{5}\Gamma_{31} &:= 3(\Theta_3 - \Theta_3{}^{\ddagger}), \\ 4\Gamma_{32} &:= 3(\Theta_3 + \Theta_3{}^{\ddagger}) - 2\Theta_4, \end{aligned}$$

gives us a Λ-orthogonal basis of the subdomain of $\mathcal{N}$ that is Λ-orthogonal to Γ_{27}.

We next bring Δ_2 into play. Exactly parallel to what we had for Δ_3, we find that

$$16\Delta_2\Gamma_{19} = \Gamma_{20} + 6\Gamma_{26} + 9\Theta_5,$$

where Θ_5 is pure 36-dimensional. Then

$$36\Theta_5 = (36, -3, 5, -6, 2; 3, 1, 0, 18, -12; -3, 5, 6, -6, 4; 3, 1, 0, 0).$$

Since this is rational, we get the same if we substitute Γ_{20} for Γ_{19}. We calculate that

$$36\langle \Gamma_{27}, \Theta_5\rangle_\Lambda = \tfrac{1}{3}, \quad 36\langle \Gamma_{30}, \Theta_5\rangle_\Lambda = \tfrac{2}{3}, \quad \langle \Gamma_{31}, \Theta_5\rangle_\Lambda = \langle \Gamma_{32}, \Theta_5\rangle_\Lambda = 0.$$

The appropriate mixed cosine vector Γ_{28} is thus given by

$$3\Theta_5 = \Gamma_{27} + \sqrt{8}\Gamma_{28} + 2\Gamma_{30}.$$

For the last step, we return to the pure $\Theta_2 := \Gamma_2\Gamma_{20}$ and $\Gamma_3\Gamma_{19}$. Again after suppressed calculations, we have

$$\begin{aligned} 36\langle \Theta_2, \Gamma_{30}\rangle_\Lambda &= \tfrac{1}{8}, \quad 36\langle \Theta_2, \Gamma_{32}\rangle_\Lambda = \tfrac{5}{8}, \\ 72\langle \Theta_2, \Gamma_{28}\rangle_\Lambda &= \tfrac{1}{\sqrt{8}}, \quad 72\langle \Theta_2, \Gamma_{30}\rangle_\Lambda = -\tfrac{\sqrt{5}}{4}. \end{aligned}$$

Since Θ_2 is pure, we know that the remaining mixed cosine vector Γ_{29} must satisfy (up to sign)

$$8\Theta_2 = 2\Gamma_{27} + \sqrt{8}\Gamma_{28} + 2\sqrt{10}\Gamma_{29} + \Gamma_{30} - 2\sqrt{5}\Gamma_{31} + 5\Gamma_{32},$$

and so we have completed the basis. We shall not write out these basis vectors, but instead refer to the table of Theorem 7K37.

The Classification

In Theorems 7K36 and 7K37, we list the realizations of $\mathcal{H}$. We have adopted the following abbreviations, again for reasons of space:

$$\varphi := \tau\sqrt{5}, \quad \rho := 1 + 2\sqrt{5}, \quad \psi := \tfrac{1}{2}(7 + 5\sqrt{5}).$$

Note that $\psi = -\tau^2\rho^\ddagger$; we continue to employ $\tau := \frac{1}{2}(1+\sqrt{5})$ (but write $\tau^\ddagger$ rather than $-\tau^{-1}$) and $\sigma := \frac{1}{2}(1 + 3\sqrt{5}) = \tau + \sqrt{5}$. Further, we write

$$\alpha := 12 + 6\sqrt{2}, \quad \beta := -4 + 4\sqrt{2}, \quad \gamma := 4 + 5\sqrt{2};$$

we denote changing the sign of $\sqrt{2}$ by ‡ as well, but no confusion should arise.

Instead of cosine vectors, as we have been doing all along we actually have inner product vectors in the tables, with scaling of mixed vectors indicated by brackets replacing what should be 0; this has been done so as to fit each table into a single page. (Alternatively, the tables can be viewed as providing a merely Λ-orthogonal basis.) Therefore, to obtain the actual (mixed) cosine vector, divide by the scaling factor in the second column. The first column gives the dimension d of the realization; for the mixed vectors, recall that the notional dimension is $2d$, indicated by †.

7K36 Theorem *The components of* $\{5,3,3\}$ *in the small simplex* S *are as follows:*

	0	1	2	3	4	5	6	7	$8a$	$8b$	9	10	11	$12a$	$12b$	13	14	$15a$	$15b$
d	1	4	12	24	12	4	24	24*	8*	24	24*	12	24	4	24	24*	24	3	24
1	1	1	1	1	1	1	1	1	1	1	1	1	1	1	1	1	1	1	1
8	8	-2	-2	3	-2	-2	-2	-2	8	3	-2	-2	3	-2	-2	-2	-2	8	3
9	12	3φ	ψ	4τ	5	$3\varphi^\ddagger$	$-\sigma^\ddagger$	1	0	0	$-\varphi^\ddagger$	$\psi^\ddagger$	$4\tau^\ddagger$	-3	-3	$-\varphi$	$-\sigma$	-4	-4
9	12	$3\varphi^\ddagger$	$\psi^\ddagger$	$4\tau^\ddagger$	5	3φ	$-\sigma$	1	0	0	$-\varphi$	ψ	4τ	-3	-3	$-\varphi^\ddagger$	$-\sigma^\ddagger$	-4	-4
16	4	1	0	-1	1	1	-1	0	1	1	1	0	-1	4	-1	1	-1	0	0
$32^\dagger$	$(8\sqrt{3})$	6	-2	0	-4	6	0	3	6	-4	1	-2	0	0	0	1	0	8	-2
16	12	-6	3	-3	4	-6	2	-2	9	-1	-1	3	-3	-3	2	-1	2	8	-2
18	6	0	-2	1	0	0	1	-2	0	0	0	-2	1	6	1	0	1	-2	-2
24	24	$6\sqrt{5}$	2	$-\rho$	-10	$-6\sqrt{5}$	$-2\tau^\ddagger$	2	0	5	$2\sqrt{5}$	2	$-\rho^\ddagger$	-6	4	$-2\sqrt{5}$	-2τ	-8	-3
24	24	$-6\sqrt{5}$	2	$-\rho^\ddagger$	-10	$6\sqrt{5}$	-2τ	2	0	5	$-2\sqrt{5}$	2	$-\rho$	-6	4	$2\sqrt{5}$	$-2\tau^\ddagger$	-8	-3
25	12	3	3	0	-2	3	-1	-2	0	-4	-1	3	0	6	2	-1	-1	-4	4
$50^\dagger$	$(4\sqrt{6})$	3	-1	0	2	3	-1	-2	0	0	1	-1	0	-6	2	1	-1	0	0
$50^\dagger$	$(4\sqrt{2})$	3	1	0	0	-3	-1	0	0	0	-1	-1	0	0	0	1	1	0	0
25	8	-1	-1	0	2	-1	-1	2	-4	0	-1	-1	0	2	2	-1	-1	8	0
$50^\dagger$	$(\frac{8}{\sqrt{3}})$	-1	1	0	0	1	-1	0	0	0	1	-1	0	0	0	-1	1	0	0
25	8	1	1	0	-2	1	1	-2	-4	0	1	1	0	-2	-2	1	1	8	0
30	12	$-3\tau^{-2}$	$-\sigma$	2τ	1	$-3\tau^2$	$-\tau$	1	0	-2	τ^2	$-\sigma^\ddagger$	$2\tau^\ddagger$	-3	-1	τ^{-2}	$-\tau^\ddagger$	-4	0
30	12	$-3\tau^2$	$-\sigma^\ddagger$	$2\tau^\ddagger$	1	$-3\tau^{-2}$	$-\tau^\ddagger$	1	0	-2	τ^{-2}	$-\sigma$	2τ	-3	-1	τ^2	$-\tau$	-4	0
40	24	6	-6	-3	2	6	4	2	0	1	-2	-6	-3	-6	-2	-2	4	-8	5

7K37 Theorem *The components of* $\{5, 3, 3\}$ *in the staurotope* X *are as follows:*

	0	1	2	3	4	5	6	7	$8a$	$8b$	9	10	11	$12a$	$12b$	13	14	$15a$	$15b$
d	1	4	12	24	12	4	24	24^*	8^*	24	24^*	12	24	4	24	24^*	24	6	48^*
4	4	σ	φ	2τ	3	$-\sigma^\ddagger$	τ^2	$\sqrt{5}$	2	2	τ	$\varphi^\ddagger$	$-2\tau^\ddagger$	1	1	$-\tau^\ddagger$	τ^{-2}	0	0
4	4	$\sigma^\ddagger$	$\varphi^\ddagger$	$2\tau^\ddagger$	3	$-\sigma$	τ^{-2}	$-\sqrt{5}$	2	2	$\tau^\ddagger$	φ	-2τ	1	1	$-\tau$	τ^2	0	0
8	4	-2	0	1	-2	2	-1	0	2	2	-2	0	-1	-4	1	2	-1	0	0
16	4	1	0	1	1	-1	-1	0	-1	-1	1	0	-1	-4	-1	-1	-1	0	0
$32^\dagger$	$(\frac{8}{\sqrt{3}})$	2	2	0	0	2	0	-1	0	0	-1	-2	0	0	0	-1	0	0	0
16	4	2	1	1	0	-2	0	0	-1	-1	-1	1	-1	1	-2	1	0	0	0
24	12	3τ	$-\varphi^\ddagger$	$\sigma^\ddagger$	-1	$-3\tau^\ddagger$	$-\tau^\ddagger$	$\sqrt{5}$	6	1	τ^2	$-\varphi$	$-\sigma$	3	-2	$-\tau^{-2}$	$-\tau$	0	0
24	12	$3\tau^\ddagger$	$-\varphi$	σ	-1	-3τ	$-\tau$	$-\sqrt{5}$	6	1	τ^{-2}	$-\varphi^\ddagger$	$-\sigma^\ddagger$	3	-2	$-\tau^2$	$-\tau^\ddagger$	0	0
36	6	0	0	-1	0	0	1	0	0	0	0	0	1	-6	-1	0	1	0	0
$72^\dagger$	$(12\sqrt{2})$	-6	0	0	4	6	0	0	6	-4	-1	0	0	0	0	1	0	0	0
$72^\dagger$	$(12\sqrt{2})$	-6	4	0	0	-6	0	-2	0	0	3	-4	0	0	0	3	0	0	0
36	24	9	5	-4	-6	-9	-1	0	6	-4	-1	5	4	6	6	1	-1	0	0
$72^\dagger$	(24)	9	-1	0	0	9	-5	-2	0	0	3	1	0	0	0	3	-5	0	0
36	24	-3	1	-4	10	3	-5	0	-6	4	-1	1	4	6	-2	1	-5	0	0
48	48	$-\alpha$	β	2	-4	$\alpha^\dagger$	$\gamma^\dagger$	$8\sqrt{2}$	-12	-2	$-2\sqrt{2}$	$\beta^\dagger$	-2	-12	4	$-2\sqrt{2}$	γ	0	0
$96^\dagger$	$(8\sqrt{6})$	6	4	0	-4	6	-1	0	0	0	2	4	0	0	-2	-2	-1	0	0
48	48	$-\alpha^\dagger$	$\beta^\dagger$	2	-4	α	γ	$-8\sqrt{2}$	-12	-2	$2\sqrt{2}$	β	-2	-12	4	$2\sqrt{2}$	$\gamma^\dagger$	0	0

Final Remarks

The quotient $\mathcal{H}_8 = \{5, 3, 3\}/8$ plays an important rôle in Chapter 17, where it forms layers in (abstract) regular 5-polytopes belonging to an interesting family related to the simple group $S_4(4)$. We postpone further discussion of properties of the quotient until then, when we have more motivation.

At an earlier stage in our investigations of the realization cone $\mathcal{H}$, we made use of the alternating product, though in the end we were able to make do without it. Nevertheless, one rather striking case does merit mention. Consider the 25-dimensional realizations in $\mathcal{H}_8$; the relevant cosine vectors are given by

$$\begin{aligned} 8\Gamma_{13} &= (8, -1, -1, 0, 2, -1, -1, 2, -4, 0, -1, -1, 0, 2, 2, -1, -1, 8, 0), \\ 8\Gamma_{14}{}^{\pm} &= \sqrt{3}(0, -1, 1, 0, 0, 1, -1, 0, \pm 4, 0, 1, -1, 0, 0, 0, -1, 1, 0, 0), \\ 8\Gamma_{15} &= (8, 1, 1, 0, -2, 1, 1, -2, -4, 0, 1, 1, 0, -2, -2, 1, 1, 8, 0), \end{aligned}$$

where (as earlier) we have replaced the mixed cosine vector Γ_{14} by its asymmetric versions. Writing $\Gamma := \Gamma(\mathsf{P}_{13} \wedge \mathsf{P}_{15})$, a little calculation then gives

$$\begin{aligned} 16\Gamma &= 16(\Gamma_{13}\Gamma_{15} - \Gamma_{14}{}^{+}\Gamma_{14}{}^{-}) \\ &= (16, -1, -1, 0, -1; -1, -1, -1, 16, 0; -1, -1, 0, -1, -1; -1, -1, 16, 0), \end{aligned}$$

from which follows

$$\Gamma = \tfrac{2}{5}\Gamma_1 + \tfrac{3}{5}\Delta_2,$$

namely, a combination of the 8- and 16-dimensional realizations in $\mathcal{H}_{24}$. At first glance, this may seem surprising – a product of two 25-dimensional polytopes is only 24-dimensional. However, the alternating product $\mathsf{P}_{13} \wedge \mathsf{P}_{15}$ must remain in $\mathcal{H}_8$ and should be expected to be Λ-orthogonal to P_{13} and P_{15}, and so can only lie in $\mathcal{H}_{24}$.

Notes to Section 7K

1. A shorter account of the contents of this section is to be found in [92].
2. The table provided to us by Ladisch [68] was invaluable in finding many of these realizations. For instance, we had been unaware of the existence of 24-dimensional realizations, although it turned out that they are relatively easy to construct. The direct construction of the 30- and 40-dimensional realizations of $\mathcal{H}_2$ also relied on our knowing them from [68]. Similarly, we appealed to [68] to complete the descriptions of the 36- and 48-dimensional realizations; in practice, as well, we did not actually calculate the remaining component Θ_0 in the 36-dimensional case using complementarity, but instead took the vector from [68].
3. As we have said, we shall look into the quotient $\mathcal{H}_8$ in more depth in Chapter 17, where knowing further details about its geometry will be relevant.
4. A specific representation of [3, 3, 5] in dimension 36, like those we found for [3, 3, 5]/2 in $\mathbb{E}^{16}$ and $\mathbb{E}^{25}$, would be very helpful. It is just about possible to find one, though the details are so messy as to make it well nigh impossible to work with (the space is the tensor product of $\mathbb{E}^4$ with a certain 9-dimensional quotient of $\mathbb{E}^{16}$).

8

Non-Classical Polytopes

This chapter is devoted to completing the classification of the regular polytopes and apeirotopes of full rank, by treating the non-classical cases. The material is largely taken from [82], though we have considerably reorganized it. Naturally, we take for granted the listing of the classical regular polytopes and apeirotopes in Chapter 7.

To avoid repetition of the same arguments in subsequent sections, we begin in Sections 8A and 8B by dealing with the cases of full rank derived from the four infinite families of regular polytopes and honeycombs which occur in all dimensions, both describing those which do occur, and eliminating others that do not. There are two interesting non-polytopes in dimensions six and eight, the first particularly so since its facet is genuinely polytopal, as we shall see in Section 13F.

The exceptional cases are to be found in dimensions two, three and four only. In Section 8C we look at the remaining regular polyhedra in $\mathbb{E}^3$, after remarking that the apeirohedra in $\mathbb{E}^2$ need little comment. The other examples are in $\mathbb{E}^4$, and are covered in Section 8D.

8A Polytopes in All Dimensions

A core result about finite regular polytopes of full rank, which at present we cannot prove except on a case-by-case basis, is the following.

8A1 Theorem *If* P *is a finite faithfully realized regular polytope of full rank, then its symmetry group is either a Coxeter group (generated by hyperplane reflexions), or the direct product of a Coxeter group with a cyclic group of order two (generated by the reflexion in a point). Moreover,* P *is classical, or is obtained from a classical regular polytope by operations* π *and* ζ *alone.*

However, in view of Theorem 7A1, what we do have is

8A2 Theorem *The facets of a non-classical regular polytope or apeirotope of full rank are themselves of nearly full rank. More generally, the faces are of full or nearly full rank.*

The term *nearly full rank* was defined after Theorem 4B2.

Our approach differs from that in [82], in that we work with the ideas of Section 5F more closely in mind. In other words, in this section we consider each regular $(d-1)$-polytope of full rank that we have already found as a possible candidate to be a vertex-figure of a regular d-polytope of full rank. In the next section, we treat apeirotopes in the same way.

In brief, we first prove that, for each classical regular polytope P in one of the three infinite families – simplex, staurotope and cube – its centrivert $\mathsf{P}^{\varsigma} = \mathsf{P} \otimes \{2\}$ is polytopal. We then show that, for $d \geqslant 5$, no such centrivert can be a vertex-figure of a finite regular d-polytope of full rank. This means that Theorem 4B9 considerably overstates the range of possibilities for the mirror vector; indeed, as we shall see in Sections 8C and 8D, only when $d \leqslant 4$ do all the possible mirror vectors occur. It is a curious fact that the arguments have to be tailored to each individual case; we shall see that the cube provides the most interest here. As we proceed, we shall describe – to a certain extent – the new polytopes and their automorphism groups as abstract polytopes.

We assume in the following that Theorem 8A1 holds in each dimension lower than the one under consideration. We write $(S_0, \ldots, S_{d-1})$ for the generatrix of a putative such polytope P of rank and dimension d, as usual centred at o. Consider the narrow vertex-figure $\mathsf{Q} := \mathsf{P}^{\boldsymbol{v}}$, whose generatrix is $(S_1, \ldots, S_{d-1})$. Our assumption implies that $\mathbf{G}(\mathsf{Q}) = \langle S_1, \ldots, S_{d-1} \rangle$ contains a subgroup of index at most 2 generated by hyperplane reflexions $R_1, \ldots, R_{d-1}$ in $\mathbb{E}$ whose mirrors contain the initial vertex w of Q. The distinguished generator S_0 is either a hyperplane or a line. In the former case set $R_0 := S_0$, and in the latter $R_0 := -S_0$ (an application of centriversion $\boldsymbol{\zeta}$). Then $\mathbf{K} := \langle R_0, \ldots, R_{d-1} \rangle$ is a finite Coxeter group, although not necessarily with the standard generating reflexions. We therefore look among such groups $\mathbf{K}$ for ones to which operations such as Petriality $\boldsymbol{\pi}$ and axiversion $\boldsymbol{\kappa}_j$ can be applied (for some j, although only $j = 0$ and, initially at least, $j = 1$ are allowed); in this context, we also bear in mind Theorem 5B10, which eliminates most potential applications of $\boldsymbol{\pi}$.

Thus, for general $d \geqslant 5$, with the two exceptions just mentioned, we begin with a string group $\mathbf{K} := \langle R_0, \ldots, R_{d-1} \rangle$ generated by hyperplane reflexions, and attempt to apply one or more of $\boldsymbol{\pi}$, $\boldsymbol{\zeta}$ and $\boldsymbol{\kappa}_1$ to it. However, we only have the possibilities $\mathbf{K} = [3^{d-1}]$, $[3^{d-2}, 4]$ and $[4, 3^{d-2}]$, with the latter two the same group in dual form. Theorem 5B10 excludes any application of $\boldsymbol{\pi}$. The only cases, therefore, are the three universal polytopes

8A3 $$\{3^{d-1}\}, \quad \{3^{d-2}, 4\}, \quad \{4, 3^{d-2}\},$$

namely, the d-simplex, d-staurotope and d-cube, respectively, and the results of applying $\boldsymbol{\zeta}$ or $\boldsymbol{\kappa}_1$ to them (if a polytope is obtained). We treat them in turn.

Simplices

8A4 Theorem *For $d \geqslant 3$, the centrivert $\{3^{d-1}\}^{\zeta} \cong \{3^{d-1}\} \# \{2\}$ is polytopal; it has group $[3^{d-1}] \times \mathbf{Z}_2 = \mathbf{A}_d \times \mathbf{Z}_2$, and $2(d+1)$ vertices (namely, those of the simplex and its opposite).*

Proof. The assertion for $\boldsymbol{\zeta}$ is straightforward. We remark that we should verify the intersection property (2B6) in this case and those following but, because we have explicit geometric descriptions of the groups, this is largely unnecessary. Note, by the way, that we have $Z = (S_0S_1)^3$, with $(S_0, \ldots, S_{d-1})$ the generatrix of the centrivert. □

8A5 Theorem *The centrivert $\{3^{d-2}\}^{\zeta}$ of $\{3^{d-2}\}$ is not the vertex-figure of any finite regular d-polytope of full rank.*

Proof. Suppose the claim to be false, so that Q is a finite regular d-polytope of full rank whose vertex-figure is $\{3^{d-2}\}^{\zeta}$. If the symmetry group $\mathbf{H}$ of Q has generatrix $(S_0, \ldots, S_{d-1})$, then $K = (S_1S_2)^3 \in \mathbf{H}$ is the half-turn about the diameter of Q through its initial vertex v, say.

We now define $R_0 := S_0$ or $S_0^{\perp}$ according as S_0 is a hyperplane or line reflexion. With $R_1 := S_1K$ and $R_j := S_j$ for $j \geqslant 2$, we now have a finite string group $\mathbf{G}$ with generatrix $(R_0, \ldots, R_{d-1})$ consisting of hyperplane reflexions; Wythoff's construction then leads to a classical regular d-polytope P with the same initial vertex v as Q. Now $\mathbf{G}$ must also contain the half-turn $L := R_0KR_0$ about the other vertex $w := vR_0$ of the initial edge of P. For P to exist, the product of K and L must have finite period; this can be so only if the angle between K and L (identified now with their mirrors as lines) is a rational multiple of π. Since $d \geqslant 5$ and P has vertex-figure $\{3^{d-2}\}$, it can only be the simplex $\{3^{d-1}\}$ or the cube $\{4, 3^{d-2}\}$. In the first case, the angle between K and L is $\arccos(-1/d)$, which is an irrational multiple of π for each $d \geqslant 3$. In the other case, the angle is $\arccos((d-2)/d)$, again an irrational multiple of π for $d \geqslant 5$. Hence in neither case is the resulting group $\mathbf{G}$ actually finite, and this establishes the theorem. □

For $\boldsymbol{\zeta}$, we can go down to $d = 2$ with the hexagon $\{6\}$, while for $d = 3$ we obtain the Petrial $\{\frac{6}{1,3}, 3 : 4\}$ of the cube $\{4, 3\}$.

We can characterize $\{3^{d-1}\}^{\zeta}$ as an abstract polytope in the following way.

8A6 Theorem *For each $d \geqslant 4$,*

$$\{3^{d-1}\}^{\zeta} \cong \{3^{d-1}\} \# \{2\} \cong \{6, 3^{d-2}\}/\langle\langle (101012)^2 \rangle\rangle.$$

Proof. The first isomorphism is clear. Certainly, $\{3^{d-1}\}^{\zeta}$ is of Schläfli type $\{6, 3^{d-2}\}$. However, this polytope is obviously infinite (since its 3-face is). The operation which yields the group of $\{3^{d-1}\} \# \{2\}$ from that of $\{3^{d-1}\}$ is

$$(\boldsymbol{r}_0, \ldots, \boldsymbol{r}_{d-1}, \boldsymbol{z}) \mapsto (\boldsymbol{r}_0\boldsymbol{z}, \ldots, \boldsymbol{r}_{d-1}) =: (\boldsymbol{s}_0, \ldots, \boldsymbol{s}_{d-1}),$$

where $\boldsymbol{z}$ is an involution with $\boldsymbol{z} \rightleftharpoons \boldsymbol{r}_j$ for each $j = 0, \dots, d-1$. Now we first see that $\boldsymbol{z} = (\boldsymbol{r}_0\boldsymbol{r}_1)\boldsymbol{z}^3 = (\boldsymbol{s}_0\boldsymbol{s}_1)^3$, and next that $\boldsymbol{r}_0 = \boldsymbol{s}_0\boldsymbol{z} = \boldsymbol{s}_1\boldsymbol{s}_0\boldsymbol{s}_1\boldsymbol{s}_0\boldsymbol{s}_1$. This then implies that $\boldsymbol{r}_0\boldsymbol{r}_1 = (\boldsymbol{s}_1\boldsymbol{s}_0)^2$, with correct period. Hence the operation

$$(\boldsymbol{s}_0, \dots, \boldsymbol{s}_{d-1}) \mapsto (\boldsymbol{s}_1\boldsymbol{s}_0\boldsymbol{s}_1\boldsymbol{s}_0\boldsymbol{s}_1, \dots, \boldsymbol{s}_{d-1}) =: (\boldsymbol{r}_0, \dots, \boldsymbol{r}_{d-1})$$

leads back to $[3^{d-1}]$, as long as we ensure that $\boldsymbol{r}_0 \rightleftharpoons \boldsymbol{r}_j$ for each $j \geqslant 2$. Now this is automatic if $j \geqslant 3$, since $\boldsymbol{r}_j = \boldsymbol{s}_j$ for $j \neq 0$. Thus we only need this for $j = 2$, and then the required relation is exactly $(\boldsymbol{s}_1\boldsymbol{s}_0\boldsymbol{s}_1\boldsymbol{s}_0\boldsymbol{s}_1 \cdot \boldsymbol{s}_2)^2 = \boldsymbol{e}$, as the notation of the theorem implies. □

Staurotopes

We have the exact analogues of Theorems 8A4 and 8A5, whose proofs we shall give together; see the notes at the end of the section.

8A7 Theorem *If $d \geqslant 3$, then the centrivert $\{3^{d-2}, 4\}^\varsigma$ of the d-staurotope is polytopal; it has the same group $[3^{d-2}, 4]$ and the same vertices as $\{3^{d-2}, 4\}$.*

8A8 Theorem *If $d \geqslant 4$, then the centrivert $\{3^{d-2}, 4\}^\varsigma$ of the $(d-1)$-staurotope is not the vertex-figure of a finite regular d-polytope of full rank.*

Proof. The argument in this case is rather different from that for the simplex. For Theorem 8A7, we have

$$K_0 = (R_0 \cdots R_{d-1})^d$$

in the original group $\langle R_0, \dots, R_{d-1} \rangle = [3^{d-2}, 4]$. However, once again, we have $K_0 = (S_0S_1)^3$ in the group $\mathbf{G} = \langle S_0, \dots, S_{d-1} \rangle$. It follows that the operation which yields the group is invertible. The new group does give a polytope, whose k-faces (for $k = 2, \dots, d-1$) are blends $\{3^{k-1}\} \diamond \{2\} \cong \{3^{k-1}\} \# \{2\}$ with the vertices of the $(k+1)$-dimensional staurotope $\{3^{k-1}, 4\}$ (that is, they are not copies of $\{3^{k-1}\}^\varsigma$). Since the abstract blend $\{3^{d-2}, 4\} \# \{2\}$ (with $4d$ vertices) is the universal regular polytope with facet $\{3^{d-2}\} \# \{2\}$ and vertex-figure $\{3^{d-3}, 4\}$, we see that $\{3^{d-2}, 4\}^\varsigma$ is not universal of its kind. Notice that $\{3^{d-2}, 4\}^\varsigma$ is actually flat (meaning, we recall, that each vertex belongs to each facet), with 2^{d-1} facets $\{3^{d-2}\} \diamond \{2\}$.

For Theorem 8A8, we first observe that a putative polytope P with $\{3^{d-2}, 4\}^\varsigma$ as vertex-figure can only have the vertices of the staurotope $\{3^{d-2}, 4\}$. Using the notation of signed permutations introduced in Section 1D, this implies that its generatrix $(S_0, \dots, S_{d-1})$ is of the form

$$\begin{aligned} S_0 &= \varepsilon(1\ 2), \\ S_1 &= -\overline{1}(2\ 3), \\ S_j &= (j\ j{+}1) \quad \text{for } j = 3, \dots, d-2, \\ S_{d-1} &= \overline{d}, \end{aligned}$$

where $\varepsilon = \pm 1$. Note that $-\overline{1}$ changes the sign of all coordinates except the first. By assunption, $d \geqslant 4$, so that $(S_1 S_2)^3 = -\overline{1}$; we immediately recover $R_1 = (2\,3)$. Then similarly $(S_1 R_2)^3 = \varepsilon I$ gives R_1 if $\varepsilon = -1$. Now conjugating $-\overline{1}$ by $R_1, R_2, \ldots, R_{d-2}$ in turn gives $-\overline{k}$ for each k, and hence $\overline{j}\,\overline{k}$ for all $j \neq k$. We now readily see that $\langle S_0, \ldots, S_{d-2} \rangle \geqslant \mathbf{B}_d$, the group consisting of all permutations of the coordinates $\xi_1, \ldots, \xi_d$, together with all even numbers of changes of sign (see Table 1E12). Indeed, since $-\overline{1}$ changes an odd number of signs if d is even, in that case we actually obtain the whole group $\mathbf{C}_d$. Thus the intersection property (2B6) is violated, and hence a polytope does not result. □

8A9 Remark There is an abstract description of $\{3^{d-2}, 4\}^{\zeta}$, analogous to that of $\{3^{d-1}\}^{\zeta}$ in Theorem 8A6. Basically the same operation

$$(\boldsymbol{r}_0, \ldots, \boldsymbol{r}_{d-1}) \mapsto (\boldsymbol{r}_1\boldsymbol{r}_0\boldsymbol{r}_1\boldsymbol{r}_0\boldsymbol{r}_1, \ldots, \boldsymbol{r}_{d-1}) =: (\boldsymbol{s}_0, \ldots, \boldsymbol{s}_{d-1})$$

leads from the group of $\{3^{d-2}, 4\}^{\zeta}$ back to that of $\{3^{d-2}, 4\}$; however, in addition to the relation $(\boldsymbol{r}_1\boldsymbol{r}_0\boldsymbol{r}_1\boldsymbol{r}_0\boldsymbol{r}_1\boldsymbol{r}_2)^2 = \boldsymbol{e}$ expressing the fact that we have a string group, we also need to insist that $\boldsymbol{z} = (\boldsymbol{s}_0 \cdots \boldsymbol{s}_{d-1})^d$ or, in other words,

$$(\boldsymbol{r}_0\boldsymbol{r}_1)^3 = (\boldsymbol{r}_1\boldsymbol{r}_0\boldsymbol{r}_1\boldsymbol{r}_0\boldsymbol{r}_2 \cdots \boldsymbol{r}_{d-1})^d,$$

which is not a very nice relation.

Cubes

8A10 Theorem *If $d \geqslant 3$, then $\{4, 3^{d-2}\}^{\zeta}$ is polytopal.*

Proof. The result of applying $\boldsymbol{\zeta}$ to the cube depends on whether d is even or odd. First note that the central inversion is

$$Z = (R_0 \cdots R_{d-1})^d$$

in the original Coxeter group $[4, 3^{d-2}]$ with generatrix $(R_0, \ldots, R_{d-1})$. Thus the generatrix $(S_0, \ldots, S_{d-1})$ of $\{4, 3^{4-2}\}^{\zeta}$ is given by $S_0 = R_0 Z$ and $S_j = R_j$ for $j = 1, \ldots, d-1$. We also have

$$(S_0 \cdots S_{d-1})^d = (R_0 \cdots R_{d-1})^d Z^d = Z^{d+1} = \begin{cases} Z, & \text{if } d \text{ is even}, \\ I, & \text{if } d \text{ is odd}. \end{cases}$$

Thus $\boldsymbol{\zeta}$ is an involutory mixing operation if d is even. In this case, we obtain an isomorphic copy of the cube with the same vertices (that is, an allomorph). Indeed, we can specify the isomorphism in terms of the coordinates $(\pm 1, \ldots, \pm 1)$ of the vertices of $\{4, 3^{d-2}\}$:

$$(\varepsilon_1, \ldots, \varepsilon_d) \mapsto \varepsilon_1 \cdots \varepsilon_d(\varepsilon_1, \ldots, \varepsilon_d),$$

where $\varepsilon_j = \pm 1$ for $j = 1, \ldots, d$.

When d is odd, this operation on the generatrix identifies opposite vertices of the cube, and produces a polytope isomorphic to $\{4, 3^{d-2}\}/2 = \{4, 3^{d-2} : d\}$; we can appeal to Theorem 2F2 to see that this is polytopal. The vertices are alternate vertices of those of the cube, say all $(\varepsilon_1, \ldots, \varepsilon_d)$ with $\varepsilon_j = \pm 1$ for $j = 1, \ldots, d$ and $\varepsilon_1 \cdots \varepsilon_d = 1$, and the symmetry group is the group $\mathbf{B}_d$ just mentioned. Indeed, we can see that the mixing operation

8A11 $$(S_0, \ldots, S_{d-1}) \mapsto (S_0 S_1 S_0, S_1, \ldots, S_{d-1}) =: (T_0, \ldots, T_{d-1})$$

yields the standard generatrix $(T_0, \ldots, T_{d-1})$ of $\mathbf{B}_d$. Moreover,

$$\{4, 3^{d-2}\} = \{4, 3^{d-2} : d\} \# \{2\}$$

as abstract polytopes, because the digon $\{2\}$ is not a realization of $\{4, 3^{d-2} : d\}$ when d is odd. □

8A12 Remark In fact, the operation (8A11) yields $\mathbf{B}_d$ for $\{4, 3^{d-2}\}$ and $\{4, 3^{d-2}\}^\varsigma$, whether d is even or odd.

The geometric polytope $\{4, 3^{d-2}\}^\varsigma$, with d odd, does not just have the same vertices as the half-cube ($h\gamma_d$ in the notation of [27]), but has the same symmetry group; this is what leads to the anomalous cases which we have to eliminate in Theorem 8A16. There is another way of looking at this, which ties in with that theorem. We write the Coxeter diagram of the reflexion group $\mathbf{B}_m$ as in Figure 8A14, taking its generatrix $(R_0, \ldots, R_{m-1})$ in the form

8A13
$$\begin{aligned} R_0 &:= (1\ \overline{2}), \\ R_j &:= (j\ j{+}1) \quad \text{for } j = 1, \ldots, m-1; \end{aligned}$$

the labels on Figure 8A14 correspond to these generators. We use a neutral index m, because this will form a subdiagram of other diagrams for various values of m.

8A14

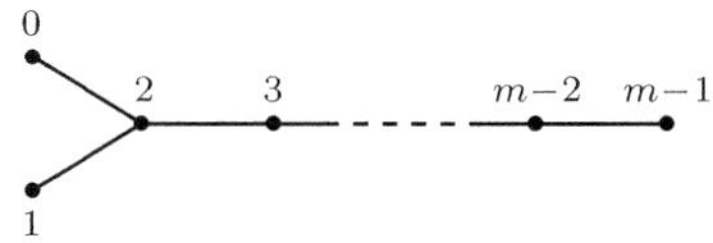

First, we go over the construction of $\{4, 3^{d-2}\}^\varsigma$ again. Indeed, we can twist Figure 8A14 with $m = d$ by $T := \overline{1}$ (change the sign of ξ_1), which gives the operation

$$(R_0, \ldots, R_{d-1}, T) \mapsto (T, R_1, \ldots, R_{d-1});$$

this produces the generatrix of the convex cube $\{4, 3^{d-2}\}$. On the other hand, if we change the sign, with the new twist $T := -\overline{1}$ (which changes the sign of $\xi_2, \ldots, \xi_d$), then the same operation yields the generatrix of $\{4, 3^{d-2}\}^\varsigma$.

However, there is yet another approach. If m is even, then (in terms of (8A13)) $(R_0 \cdots R_{m-1})^{m-1} = -I$, the central inversion. We thus see that, if d is odd, then the operation

8A15 $$(R_0, \ldots, R_{d-1}) \mapsto \bigl((R_0 \cdots R_{d-2})^{d-2}, R_{d-1}, \ldots, R_1\bigr) =: (S_0, \ldots, S_{d-1})$$

leads to a different copy of $\{4, 3^{d-2}\}^\varsigma$. The initial edge is now the long diagonal of the original initial facet, rather than that of the facet through the initial vertex that does not contain the original initial edge. Note the corresponding reversal of the order of the generatrix of the vertex-figure $\{3^{d-2}\}$, which accords with the fact that $R_j \rightleftharpoons (R_0 \cdots R_{d-2})^{d-2}$ for $j = 1, \ldots, d-2$.

With this in mind, we next come to

8A16 Theorem *For $d \geqslant 5$, neither $\{4, 3^{d-3}\}$ nor $\{4, 3^{d-3}\}^\varsigma$ is the vertex-figure of a finite regular d-polytope of full rank.*

Proof. We shall treat $\{4, 3^{d-3}\}^\varsigma$ first, because in the end we return to $\{4, 3^{d-3}\}$. There are two different cases to consider here, according to the parity of d. The easier case is d odd. In this event, $\{4, 3^{d-3}\}^\varsigma$ has the same vertices and symmetry group as $\{4, 3^{d-3}\}$ itself. Thus, if the regular d-polytope Q has vertex-figure $\{4, 3^{d-3}\}^\varsigma$, then we can apply axiversion $\boldsymbol{\kappa}_1$ (which induces a symmetry of Q) to replace $\{4, 3^{d-3}\}^\varsigma$ by $\{4, 3^{d-3}\}$. Similarly, if the generator S_0 of the generatrix $(S_0, \ldots, S_{d-1})$ of Q is a line, then we can apply $\boldsymbol{\zeta}$, and replace it by the orthogonal hyperplane $S_0^\perp$. In other words, by applying $\boldsymbol{\kappa}_1$ and (possibly) $\boldsymbol{\zeta}$, we have arrived at a hyperplane reflexion group, and so a corresponding classical regular d-polytope of the form $\{p, 4, 3^{d-3}\}$ for some p. However, since $d \geqslant 5$ there is no such polytope.

Observe now that the same argument applies to $\{4, 3^{d-3}\}$, for all $d \geqslant 4$.

As we saw in the proof of Theorem 8A10, when d is even, $\{4, 3^{d-3}\}^\varsigma$ has the vertices and symmetry group of the half-d-cube $1_{d-4,1} = \{3, 3^{d-4,1}\}$; indeed, (8A11) showed how to recover the generatrix of $\mathbf{B}_{d-1}$. As with the first case, we can apply $\boldsymbol{\zeta}$ to Q, if necessary, so that $R_0 = S_0$ or $S_0^\perp$ is a hyperplane reflexion. With the new generatrix $(R_0, \ldots, R_{d-1})$, we conclude that we have the symmetry group of a (non-regular) polytope $\mathsf{P} = \{p, 3, 3^{d-4,1}\}$ for some p. (Here, we should make a forward reference to the discussion of the Gosset–Elte polytopes in Section 13C and Chapter 14.) Observe as well that (8A15) enables one to return to Q (or Q^ς).

We now see at once that we can only have $p = 3$ and $d = 6$ or 8, so that $\mathsf{P} = 2_{d-4,1}$. We address the two subcases in turn, and show that neither yields a regular polytope of full rank d.

First, let $d = 6$, and take 2_{21} to have unit edge-length. The edge-length of the vertex-figure $\{4, 3, 3, 3\}^\varsigma$ of our putative 6-polytope Q_6 inscribed in 1_{21} is $\sqrt{2}$, and so we expect Q_6 to have square 2-faces. Indeed, the 3-faces are now going to be tori $\{4, \frac{4}{1,2} \mid 3\}$, whose 2-holes (of length 3) correspond to 2-faces of $1_{2,1}$. Moreover, the facets will then have to be quotients of 5-polytopes of type $\{4, 4, 3, 3\}$, with these toroidal 3-faces. Consider the mixing operation

$$(S_0, \ldots, S_4) \mapsto (S_0, S_1 S_2 S_1, S_1 S_0 S_1, S_2, S_3, S_4) =: (U_0, \ldots, U_5)$$

on the generatrix $(S_0, \ldots, S_4)$ of the facet of Q_6. We recover the original group if we twist by S_1 (thus we have a semidirect product), and so the subgroup is of index 1 or 2. (In fact, the group is the same, because S_1 is induced by an inner automorphism.) Then $\langle U_0, \ldots, U_5 \rangle$ is a group generated by hyperplane reflexions which satisfies the relations for the group $[3^{2,2,1}]$ of 2_{21} itself, and so must actually be that group. Obviously this contradicts the intersection property for C-groups, since the trivial relation $S_5 \notin \langle S_0, \ldots, S_4 \rangle$ is violated. The situation is illustrated by the diagram of Figure 8A17, with a label j on a node (or on the twist) given by the suffix of the corresponding generator S_j.

8A17

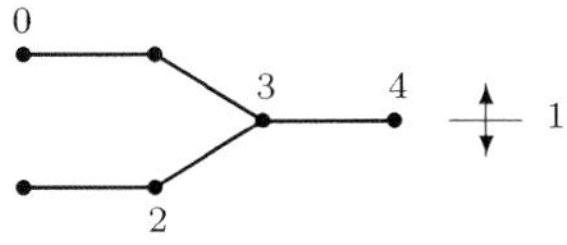

Second, let $d = 8$. If we take $2_{41} = \{3, 3, 3^{4,1}\}$ to have edge-length 1, then the edge-length of the vertex-figure $\{4, 3, 3, 3, 3, 3\}^{\varsigma}$ of our putative polytope Q_8 inscribed in $1_{42} = \{3, 3^{4,1}\}$ is $\sqrt{3}$; thus we expect the 2-faces of Q_8 to be hexagons. Indeed, we can see that the 3-faces of Q_8 would be regular polyhedra $\{6, \frac{4}{1,2} \mid 3\}$, with the holes trigonal faces of 1_{41}. We perform the analogue of the operation used in the previous case on the facet, namely,

$$(S_0, \ldots, S_6) \mapsto (S_0, S_1S_2S_1, S_1S_0S_1, S_2, \ldots, S_6) =: (U_0, \ldots, U_7).$$

Once again, we obtain a hyperplane reflexion group $\langle U_0, \ldots, U_7 \rangle$; the supposed twist is actually by the inner automorphism S_1. The situation is illustrated in the diagram of Figure 8A18, with the same conventions as before. The form of the group is a little concealed, but a change of generators as in (1E24) enables us to recognize it as the whole group $[3^{4,2,1}]$ of 2_{41}; compare Figures 1E23 and 1E25 and the discussion at the end of Section 1E for the mark 2 on the pentagonal circuit. In other words, we have $S_7 \in \langle S_0, \ldots, S_6 \rangle$, contrary to the intersection property. □

8A18

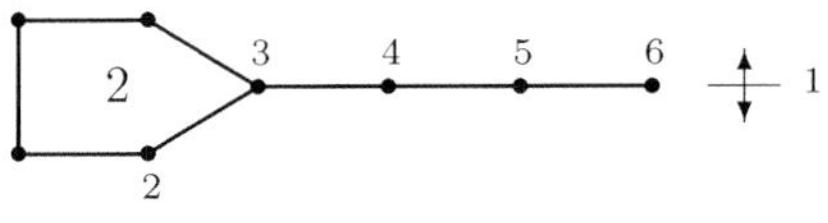

8A19 Remark We shall see in Section 13F that the facet of P_6 is a genuine regular 5-polytope of nearly full rank. However, we shall see in Section 13D that, while the ridge of P_8 is a regular 6-polytope of nearly full rank, its facet degenerates.

8A20 Remark The cases when $d = 4$ can also be excluded, but we shall postpone the discussion of these until Section 8D.

The foregoing discussion produces the list in Table 8A21 of the non-classical regular polytopes and apeirotopes of full rank in dimensions at least 5; in the last generalized fraction, $m = \frac{1}{2}(d-1)$. For the applications of ζ to the d-cube, as we have already noted the fine Schläfli symbol cannot ensure rigidity, and so merely gives an indication of part of the geometry. However, the cosine vector $(1, -\frac{2}{d}, \frac{4}{d}, -\frac{6}{d}, \ldots)$ of $\{4, 3^{d-2}\}^\zeta$ does specify its geometry.

8A21

$$\{3^{d-1}\} \quad \overset{\zeta}{\longleftrightarrow} \quad \{3^{d-1}\} \otimes \{2\}$$

$$\{3^{d-2}, 4\} \quad \overset{\zeta}{\longleftrightarrow} \quad \{3^{d-2}, 4\} \otimes \{2\}$$

$$\delta \updownarrow$$

$$\{4, 3^{d-2}\} \quad \overset{\zeta}{\longleftrightarrow} \quad \begin{cases} \{\frac{4}{1,2}, 3^{d-2} : \frac{2d}{1,3,\ldots,d-1}\}, & \text{if } d \text{ is even,} \\ \{\frac{4}{1,2}, 3^{d-2} : \frac{d}{1,2,\ldots,m}\}, & \text{if } d \text{ is odd.} \end{cases}$$

Finite d-polytopes for $d \geqslant 5$

Notes to Section 8A

1. Theorem 8A8 and its proof correct the statement of [82, Theorem 5.3], which omitted the case $\zeta\kappa_1$, and add the detail about the case d even. Of course, the theorem excludes the possiblity of applying κ_1 or $\zeta\kappa_1$ to the d-staurotope when $d \geqslant 4$.

8B Apeirotopes in All Dimensions

We now move on to the apeirotopes. We first recall that the basic generator R_0 must be a point or a hyperplane. In the latter case, inspection shows readily that (for $d = 3$ or $d \geqslant 5$) the only possible vertex-figures (among those considered in Section 8A) are the staurotope $\{3^{d-2}, 4\}$ and its image $\{3^{d-2}, 4\}^\zeta$ under ζ. Indeed, the same arguments as in the proofs of Theorems 8A4 and 8A10 show that otherwise $(R_0R_1)^2$ is a rotation through an irrational angle. We thus only obtain $\mathsf{R}_{d+1} := \{4, 3^{d-2}, 4\}$ (the cubic tessellation) and its image $\mathsf{R}_{d+1}{}^\kappa := \{4, 3^{d-2}, 4\}^\kappa$ under $\kappa := \kappa_1$ (this is just ζ applied to the vertex-figure – as explained in Section 5D, we use the abbreviated notation because of the frequency of its occurrence).

The apeirotope $\mathsf{R}_{d+1}{}^\kappa$ is unfamiliar, and so deserves a little more attention. A purely geometric approach will tell us much of what we need to know. In fact, we begin (to avoid repetition of essentially the same arguments in Section 8D) by working in a rather more general context than the present one.

We thus suppose that $d \geqslant 3$, and that we have a universal $(d+1)$-apeirotope $\{p_1, \ldots, p_d\}$ in $\mathbb{E}^d$, whose (Coxeter) group is $[p_1, \ldots, p_d] = \langle R_0, \ldots, R_d \rangle$ in the usual way; thus each R_j is a hyperplane reflexion. Write $K = K_1$ for the point-reflexion in the initial vertex, which as usual we take to be o. Then the group of $\{p_1, \ldots, p_d\}^{\kappa}$ is given by

$$(R_0, \ldots, R_d) \mapsto (R_0, R_1 K, R_2, \ldots, R_d) =: (S_0, \ldots, S_d).$$

By definition, the new vertex-figure is $\{p_2, \ldots, p_d\}^{\varsigma}$. If the new apeirohedron is of Schläfli type $\{q_1, \ldots, q_d\}$, then $\{q_3, \ldots, q_d\} = \{p_3, \ldots, p_d\}$, because κ does not affect any of the relevant group generators $R_2, \ldots, R_d$.

Now K clearly preserves the affine (actually, linear) hull L_k of the base k-face, so that $\dim L_k = k$ for each $k \leqslant d$. In particular, for $k = 3$, write $T_j := S_j \cap L_3$ for $j = 0, 1, 2$. Then $T_1 = S_1 \subset L_3$, while each of T_0 and T_2 is a plane; hence the corresponding mirror vector is $\dim(T_0, T_1, T_2) := (\dim T_0, \dim T_1, \dim T_2) = (2, 1, 2)$. Reference to [24, 98] or [99, Section 7E] then suggests that the new 3-face F_3 is a Petrie–Coxeter apeirohedron; we shall describe these apeirohedra in greater detail in Section 10A.

To verify this claim, first observe that the polygonal face $\mathsf{F}_2 = \{q_1\}$ is planar, with $q_1 = p_1''$ given by

$$\frac{1}{q_1} + \frac{1}{p_1} = \frac{1}{2},$$

because the edge $\mathsf{F}_1 S_1$ through o in F_3 adjacent to the base edge F_1 goes along the same line as before, but in the opposite direction. In contrast, the section $\{q_2\}$ is skew; we have

$$\{q_2\} = \{p_2\} \diamond \{2\} = \{p_2''\} \# \{2\},$$

with (as for q_1)

$$\frac{1}{p_2''} + \frac{1}{p_2} = \frac{1}{2}.$$

Thus the 2-faces are planar, and the vertex-figures of 3-faces are skew, which is appropriate for some Petrie–Coxeter apeirohedron $\{q_1, q_2 \mid h\}$, with a suitable h. Indeed, the polygon $\{h\}$ has generatrix

$$(S_0, S_1 S_2 S_1) = (R_0, R_1 K \cdot R_2 \cdot R_1 K) = (R_0, R_1 R_2 R_1),$$

which just gives the (planar) hole of the original 3-face; this is that $\{h\}$ for the Petrie–Coxeter apeirohedron.

After these general remarks, we now return to the specific case $\{4, 3^{d-2}, 4\}$, with $d \geqslant 3$. The 3-faces are thus Petrie–Coxeter apeirohedra $\{4, \frac{6}{1,3} \mid 4\}$, whose vertex-sets form copies of the cubic lattice $\mathbb{Z}^3$. To see what the general face looks like, we give suitable generators for the symmetry group. In terms of the usual cartesian coordinates $x = (\xi_1, \ldots, \xi_d)$, they are

8B1
$$\begin{aligned} S_0 &:= \bar{1}T(e_1), \\ S_1 &:= -(1\ 2), \\ S_j &:= (j\ j{+}1) \quad \text{for } j := 2, \ldots, d-1, \\ S_d &:= \bar{d}, \end{aligned}$$

where – with our usual convention – permutations are those of coordinates, $T(e_1)$ is the translation by e_1 (here following the change of sign of ξ_1), and $-(1\ 2)$ indicates the transposition combined with changing signs of all coordinates. Of course, (8B1) just gives the standard generatrix of the cubic tiling $\{4, 3^{d-2}, 4\}$ after the application of $\boldsymbol{\kappa}$. It is easy to check similarly that, for $3 \leqslant k \leqslant d$, the k-faces are k-dimensional apeirotopes, whose vertices form cubic lattices $\mathbb{Z}^k$. Since

$$xS_{k-1} \cdots S_0 = (\xi_k + 1, -\xi_1, \ldots, -\xi_{k-1}, -\xi_{k+1}, \ldots, -\xi_d),$$

it is tedious, but routine, to verify that

$$x(S_{k-1} \cdots S_0)^k = (-1)^{k-1}(\xi_1, \ldots, \xi_k, -\xi_{k+1}, \ldots, -\xi_d) + v_k,$$

where $v_k := (1, -1, \ldots, (-1)^{k-1}, 0^{d-k})$. In particular, for the facet F_d it follows that $(S_{d-1} \cdots S_0)^d$ is the translation by v_d if d is odd, but the point-reflexion in $\frac{1}{2}v_d$ if d is even. When d is even, we can see that, if $Q := (S_{d-2} \cdots S_0)^{d-1}$, then $S_0 Q^{-1} S_0 Q$ is the translation by $2e_1$. In other words, denoting by $\boldsymbol{\Lambda}(v)$ the subgroup of translations generated by the vector $v \in \mathbb{Z}^d$ and its images under the point-group $[3^{d-2}, 4]$, the translation subgroup $\mathbf{T}(\mathsf{F}_d) < \mathbf{T}(\mathsf{R}_{d+1}{}^{\boldsymbol{\kappa}}) = \mathbb{Z}^d$ is

8B2
$$\mathbf{T}(\mathsf{F}_d) = \begin{cases} \boldsymbol{\Lambda}(1^d), & \text{if } d \text{ is odd}, \\ \boldsymbol{\Lambda}(2, 0^{d-1}), & \text{if } d \text{ is even}. \end{cases}$$

Summarizing this discussion, we have

8B3 Theorem *In each dimension $d \geqslant 3$, there are two discrete regular $(d+1)$-apeirotopes $\{4, 3^{d-2}, 4\}$ and $\{4, 3^{d-2}, 4\}^{\boldsymbol{\kappa}}$ with square 2-faces; for $d = 3$ and $d \geqslant 5$, these are the only such apeirotopes with finite 2-faces. The 3-faces of $\{4, 3^{d-2}, 4\}^{\boldsymbol{\kappa}}$ are Petrie–Coxeter apeirohedra $\{4, \frac{6}{1,3} \mid 4\}$.*

The geometry of the facet F_d of $\{4, 3^{d-2}, 4\}^{\boldsymbol{\kappa}}$ can be explained as follows. Its vertex-figure is the mix $\{3^{d-2}\} \diamond \{2\}$ of a $(d-1)$-simplex and a digon. It has the vertices and some of the edges of the staurotope $\{3^{d-2}, 4\}$; however, the edges of two opposite facets do not occur, and correspond to holes. These holes themselves form the faces of d-cubes; translates of one such cube by the sublattice $\boldsymbol{\Lambda}(1^d)$ (generated by the vectors $(\varepsilon_1, \ldots, \varepsilon_d)$, with $\varepsilon_i = \pm 1$ for $i = 1, \ldots, d$) give all the holes. The 2-faces of F_d are then the remaining square faces of the cubic tiling $\{4, 3^{d-2}, 4\}$. Note that F_d is also the typical d-face of $\{4, 3^{m-2}, 4\}^{\boldsymbol{\kappa}}$, for each $m > d$.

We end this part by saying a little more about the apeirotope $\{4, 3^{d-2}, 4\}^{\kappa}$ as an abstract polytope. By definition, its vertex-figure is $\{3^{d-2}, 4\}^{\varsigma}$. Its (abstract) facet $\mathcal{F}^d$, say, is a face of $\{4, 3^{m-2}, 4\}^{\kappa}$ for each $m \geqslant d$, as we have remarked, and so the general facial structure is specified by this. In fact, it is universal, in the following sense.

8B4 Theorem *If $d \geqslant 3$, then the facet $\mathcal{F}^d$ of $\{4, 3^{d-2}, 4\}^{\kappa}$ is the universal polytope*

$$\mathcal{F}^d = \{\mathcal{F}^{d-1}, \{3^{d-2}\} \# \{2\}\},$$

with starting point $\mathcal{F}^3 = \{4, 6 \mid 4\}$.

Proof. To see this, we apply the circuit criterion Theorem 2D4 to the geometric apeirotope; recall that this says that the group of a regular polytope is given by that of its vertex-figure, together with relations arising from its edge-circuits. The vertex-figure is undoubtedly $\{3^{d-2}\} \diamond \{2\}$. This has the same vertices as the staurotope $\{3^{d-2}, 4\}$; however, the edges of some pair of opposite facets $\{3^{d-2}\}$ correspond to square holes $\{4\}$ (those of 3-faces $\{4, \frac{6}{1,3} \mid 4\}$), while the remaining edges correspond to faces $\{4\}$. Now the concrete realization F_d of $\mathcal{F}^d$ has all the vertices and edges of the cubic tiling $\{4, 3^{d-2}, 4\}$, while the square faces of the tiling are either faces of F_d or holes. It follows at once that any edge-circuit can be contracted over the faces or holes, and so no extra relations arise from edge-circuits, apart from those already given by those of $\{4, \frac{6}{1,3} \mid 4\}$. The claim of the theorem follows at once. □

As an abstract polytope, $\{4, 3^{d-2}, 4\}^{\kappa}$ admits a similar description.

8B5 Theorem *With $\mathcal{F}^d$ the facet of $\{4, 3^{d-2}, 4\}^{\kappa}$ as before, the apeirotope itself is*

$$\{4, 3^{d-2}, 4\}^{\kappa} = \{\mathcal{F}^d, \{3^{d-2}, 4\}^{\varsigma}\},$$

the universal polytope of its type. The group of $\{3^{d-2}, 4\}^{\varsigma}$ is as briefly described in Remark 8A9.

Proof. The argument is just like that for Theorem 8B4; once again, the edge-circuits can be contracted over squares and holes of 3-faces $\{4, \frac{6}{1,3} \mid 4\}$ (in fact, only the squares are needed here), and an appeal to the circuit criterion of Theorem 2D4 finishes the proof. □

8B6 Remark We can rephrase the results of Theorems 8B4 and 8B5 in the following way. The facet $\mathcal{F}^d$ of $\{4, 3^{d-2}, 4\}^{\kappa}$ is universal with 3-face $\{4, 6 \mid 4\}$ and vertex-figure $\{3^{d-2}\} \diamond \{2\}$, while (as the proof of the latter theorem shows) $\{4, 3^{d-2}, 4\}^{\kappa}$ itself is universal with 2-face $\{4\}$ and vertex-figure $\{3^{d-2}, 4\}^{\varsigma}$.

We have one final comment about this apeirotope.

8B7 Remark Since the vertex-figure $\{3^{d-2}, 4\}^{\varsigma}$ of $\mathsf{R}_{d+1}{}^{\kappa}$ is flat with 2^{d-1} facets, it follows that $\mathsf{R}_{d+1}{}^{\kappa}$ itself is flat, also with 2^{d-1} facets. The difference between even and odd dimensions is reflected in the fact that the facets of $\mathsf{R}_{d+1}{}^{\kappa}$ can be 2-coloured if d is even, but not if d is odd.

We now consider the alternative case, when R_0 is a point. Necessarily, then, we are in the situation of Theorem 5E2, since R_0 must be the base vertex of the vertex-figure. The final result of this section is thus the following.

8B8 Theorem *Each of the regular d-polytopes* P *of (8A3) and Theorems 8A4, 8A7 and 8A10 is the vertex-figure of a discrete regular* $(d+1)$*-apeirotope* P^{α}.

Proof. Indeed, each of the potential vertex-figures P is rational (as is extremely easy to check), and so, by Theorem 5E2, yields a regular apeirotope P^{α}. Of course, it is important to note that all these apeirotopes have even edge-circuits, and thus are isomorphic to the corresponding blended apeirotopes apeir P. □

It is of interest to describe the structure of these apeirotopes in a little detail. The easiest way to present $\{3^{d-1}\}^{\alpha}$ and its relative $\{3^{d-1}\}^{\zeta\alpha}$ is – as usual – to work in the symmetric hyperplane $\mathbb{L}^d$ of (1E15), and, for even greater simplicity, to take the vertices to lie in the intersection lattice $\boldsymbol{\Lambda} := \mathbb{Z}^{d+1} \cap \mathbb{L}^d$.

8B9

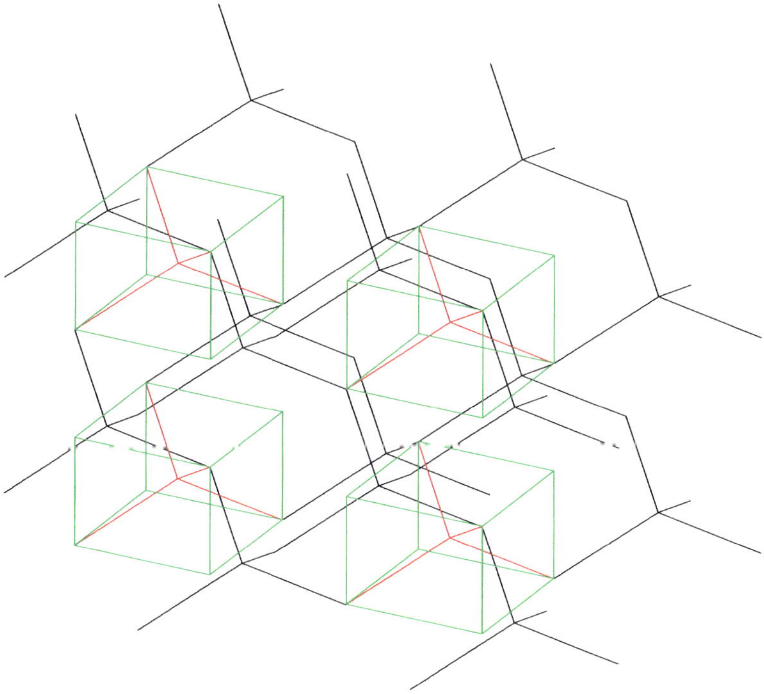

The apeirotope $\{3,3\}^{\alpha} = \{\frac{2}{0,1}, 3, 3 : \frac{4}{0,1}\}$

We may take the $d+1$ vertices of $\{3^{d-1}\}$ to be the points $\frac{1}{2}w_j$, where $w_j := ((d+1)e_j - u)$ for $j = 0, \ldots, d$, with $\{e_0, \ldots, e_d\}$ the standard basis of $\mathbb{E}^{d+1}$ and $u := (1^{d+1}) = e_0 + \cdots + e_d$; the initial vertex of apeir$\{3^{d-1}\}$ is the centre

$v := o$ of $\{3^{d-1}\}$, and that of the broad vertex-figure is w_0. The underlying translation lattice of the apeirotope is thus $(d+1)\mathbf{\Lambda} = (d+1)\mathbb{Z}^{d+1} \cap \mathbb{L}^d$, and the vertex-set of the apeirotope is $(d+1)\mathbf{\Lambda} + \{o, w_0\}$. The case $d = 3$ corresponds to the well-known *diamond net*; see Figure 8B9 for a picture of a small part of it. (We have inserted a skeleton of cubes around some of the vertices, to attempt to make the picture clearer.)

As so often, the question of universality arises. On the abstract level, the starting point is the fairly obvious

$$\operatorname{apeir}\{3\} \cong \{\infty, 3 : 6\} = \{6, 3\}^{\pi}.$$

The general result is

8B10 Theorem *For each $d \geqslant 3$,*

$$\operatorname{apeir}\{3^{d-1}\} \cong \{\operatorname{apeir}\{3^{d-2}\}, \{3^{d-1}\}\}$$

is the universal apeirotope of its kind.

Recall our convention, that apeir Q denotes the class of all apeirotopes obtained from Q by the 'apeir' construction, whereas Q^{α} denotes that particular one with the centre of Q as the initial vertex.

Proof. We proved the theorem in [99, Theorem 7F13] (see also [98]) for the 3-dimensional case, but the proof there was rather special to that case. The following argument is simpler, and at the same time completely general.

We must begin by describing the basic (minimal) edge-circuits in the edge-graph. From any given vertex x, two edges go to the vertices y, z of an edge E of the (broad) vertex-figure at x. Then from y, z two edges $\{y, z'\}$ and $\{z, y'\}$ go in the direction of the outer normal to some facet G of the vertex-figure Q which contains E. Then y', z' are vertices of an edge E' in a facet G' of the vertex-figure at a vertex x', with the hyperplanes aff G and aff G' parallel. Then the edges $\{y', x'\}$ and $\{z', x'\}$ complete a centrally symmetric edge-circuit $\mathcal{D}$. Such hexagonal edge-circuits are clearly visible in Figure 8B9.

It suffices to treat $\mathsf{P} = \{3^{d-1}\}^{\alpha}$, so pick an initial vertex v of it. We can think of the outer normals to the vertex-figure $\mathsf{Q} = \{3^{d-1}\}$ at v to be $e_0, \ldots, e_d$ (strictly speaking, they are scaled projections of these vectors on the hyperplane $\mathbb{L}^d$). Then we can suppose that the vertices of the initial facet F of P which contains v lie in two parallel hyperplanes J_0, J_1 in $\mathbb{L}^d$ with normal e_d. Let $\mathcal{C}$ be an edge-circuit, and suppose that $\mathcal{C}$ does not lie in F. Without loss of generality, we can suppose that $\mathcal{C}$ has a vertex on the positive side of J_0, J_1 with respect to e_d; let x be such a vertex with $\gamma := \langle x, e_d \rangle$ maximal. Then the two edges $\{x, y\}$ and $\{x, z\}$ of $\mathcal{C}$ which contain x must have $\langle y, e_d \rangle = \langle z, e_d \rangle < \gamma$. Complete these two edges to a basic circuit $\mathcal{D}$ as in the previous paragraph, with G the facet of the vertex-figure at x whose outer normal is $-e_d$, and concatenate $\mathcal{C}$ with $\mathcal{D}$. We thus obtain a new edge-circuit $\mathcal{C}'$ (or possibly more than one) with one fewer vertex at height γ.

We now repeat this procedure, dealing with all vertices at a given maximal height before moving lower. We then end with no vertices higher than those in J_1. We deal similarly with the vertices on the negative side of J_0. A little thought shows that the latter reduction process does not produce new vertices on the positive side of the pair J_0, J_1, because parallel facets are joined by sets of edges in the normal direction to the pairs of hyperplanes which contain their vertices.

In conclusion, then, we have reduced our initial edge-circuit to ones lying in the initial facet, and the obvious inductive argument completes the proof. □

Concerning $\{3^{d-1}\}^{\zeta\alpha}$, since its vertex-figure $\{3^{d-1}\}^{\zeta}$ is centrally symmetric, it is immediately clear that the vertex-set of the apeirotope is a lattice, and is in fact generated by the permutations of the initial vertex $(d, (-1)^d)$ of the broad vertex-figure. One basic edge-circuit has successive vertices

$$o,\ (d, (-1)^d),\ (d{-}1, d{-}1, (-2)^{d-1}),\ (-1, d{+}1, (-1)^{d-1}).$$

The fact that the translations in directions $(d, (-1)^d)$ and $(-1, d{+}1, (-1)^{d-1})$ commute can be seen to correspond to the index cycle $(01021212)^2$, and from the known description of $\{3^{d-1}\}^{\zeta} \cong \{3^{d-1}\} \# \{2\}$ in Theorem 8A6, we deduce

8B11 Proposition *As an abstract regular apeirotope,*

$$\{3^{d-1}\}^{\zeta\alpha} \cong \{\infty, 6, 3^{d-2}\}/\langle\langle (01021212)^2, (212123)^2 \rangle\rangle.$$

The characterization of apeir$\{4, 3^{d-2}\}$ is very similar to that of apeir$\{3^{d-1}\}$. Let us begin by describing the (isomorphic) apeirotope $\{4, 3^{d-2}\}^{\alpha}$. If we take the initial vertex to be $v := o$, and the (narrow) vertex-figure to consist of all points $\frac{1}{2}(\pm 1, \ldots, \pm 1) \in \mathbb{E}^d$, then $\{4, 3^{d-2}\}^{\alpha}$ has vertex-set $2\mathbb{Z}^d + \{o, u\}$, with $u := (1, \ldots, 1)$. This is (in the 3-dimensional case) the *body-centred cubic lattice*, and can be described alternatively as the set of all $(\nu_1, \ldots, \nu_d) \in \mathbb{Z}^d$ with $\nu_1 \equiv \cdots \equiv \nu_d \pmod 2$.

8B12 Theorem *For each $d \geqslant 3$,*

$$\text{apeir}\{4, 3^{d-2}\} \cong \{\text{apeir}\{4, 3^{d-3}\}, \{4, 3^{d-2}\}\}$$

is the universal apeirotope of its kind.

Proof. As in the proof of Theorem 8B10, we apply the circuit criterion. A minimal edge-circuit in $\mathsf{P} := \{4, 3^{d-2}\}^{\alpha}$ goes from a vertex z to the two vertices x, y of an edge $\{x, y\}$ of the broad vertex-figure at z, and thence to another vertex w whose broad vertex-figure shares a facet with that at z and also contains the edge $\{x, y\}$. It follows that $w - z = \pm 2e_j$ for some $j = 1, \ldots, d$; note particularly that we have a free choice of j, except for that corresponding to the pair of opposite facets of the vertex-figure which contain x and y.

The vertices of the initial facet $\mathsf{F} \in \text{apeir}\{4, 3^{d-3}\}$ of P lie in the parallel hyperplanes $J_k := \{(\xi_1, \ldots, \xi_d) \in \mathbb{E}^d \mid \xi_d = k\}$ for $k = 0, 1$. If an edge-circuit

$\mathcal{C}$ of P contains vertices other than those in $J_0 \cup J_1$, we may suppose (without losing generality) that there are such vertices $z = (\nu_1, \ldots, \nu_d)$ with $\nu_d > 1$. Choose such a z with ν_d maximal. Then the two edges $\{z, x\}$ and $\{z, y\}$ in $\mathcal{C}$ which contain z lie below z relative to the direction e_d. If x, y do not belong to the same edge of the broad vertex-figure Q of P at z, then take any edge-path $x = x_0, \ldots, x_r = y$ in the common facet G of Q at height $\nu_d - 1$. We then concatenate $\mathcal{C}$ with the minimal edge-circuits $\mathcal{C}_s := \{z, x_{s-1}, z - 2e_d, x_s\}$ for $s = 1, \ldots, r$, observing that the concatenation of $\mathcal{C}_1, \ldots, \mathcal{C}_r$ is just the edge-circuit $\{z, x, z - 2e_d, y\}$.

Just as for Theorem 8B10, we iterate this procedure (and the corresponding one for vertices z with $\nu_d < 0$), until we have reduced our initial edge-circuit to ones lying in F. Finally, we appeal to induction on d to contract edge-circuits in F using minimal circuits, noting that there is nothing to show if $d = 2$. This completes the proof. □

When d is even, the alternative cube $\{4, 3^{d-2}\}^\varsigma$ has the same vertices as the cube, and the same group. Thus $\operatorname{apeir}(\{4, 3^{d-2}\}^\varsigma)$ has the same vertices and the same group as $\operatorname{apeir}\{4, 3^{d-2}\}$. However, when $d = 2m + 1$ is odd, $\{4, 3^{d-2}\}^\varsigma \in \{4, 3^{d-2} : d\}$ is not centrally symmetric. In this case, the translation lattice $\boldsymbol{\Lambda}$ of $\operatorname{apeir}(\{4, 3^{d-2}\}^\varsigma)$ is generated by all $2(e_j \pm e_k)$ with $1 \leqslant j < k \leqslant d$, and the vertex-set is $\boldsymbol{\Lambda} + \{o, u\}$, with $u := (1^d) = e_1 + \cdots + e_d$.

Taking the vertices of the (centrally symmetric) staurotope $\{3^{d-2}, 4\}$ to be all $\pm\frac{1}{2}e_j$ for $j = 1, \ldots, d$, we see that the vertex-set of $\operatorname{apeir}\{3^{d-2}, 4\}$ (and its translation group) is just $\mathbb{Z}^d$. Indeed, its symmetry group $\langle S_0, \ldots, S_d\rangle$ is just the symmetry group $\langle R_0, \ldots, R_d\rangle = [4, 3^{d-2}, 4]$ of the cubic lattice; it is obtained by means of the involutory mixing operation

$$\boldsymbol{\kappa}_{02} \colon (R_0, \ldots, R_d) \mapsto (R_0(R_2 \cdots R_d)^{d-1}, R_1, \ldots, R_d) =: (S_0, \ldots, S_d).$$

In the same way, the group of

$$\operatorname{apeir}(\{3^{d-2}, 4\}^\varsigma) = (\operatorname{apeir}\{3^{d-2}, 4\})^{\boldsymbol{\kappa}} = \{4, 3^{d-2}, 4\}^{\boldsymbol{\kappa}\boldsymbol{\kappa}_{02}}$$

is also $[4, 3^{d-2}, 4]$. Since the vertex-figure of the latter is flat, so is the apeirotope; it has 2^{d-1} facets.

We may notice that the same analysis applied to the case $d = 2$ has nearly completed the enumeration of the regular apeirohedra. The rational polygons are $\{3\}$, $\{4\}$ and $\{6\} = \{3\} \otimes \{2\}$ alone, and so these yield the only discrete apeirohedra $\operatorname{apeir}\{q\}$ (with $q > 2$ rational); they fit into the scheme we have just described. However, if (R_0, R_1, R_2) is the generatrix of any other regular apeirohedron P in $\mathbb{E}^2$, then each of R_0, R_1 and R_2 must be a line. But now the Petrie operation $\boldsymbol{\pi}$ results in the mirror $S_0 = R_0R_2$ being a point, and so we are back in the former case. Thus a regular apeirohedron of Schläfli type $\{p, q\}$ (with p, q finite) must have $q = 3, 4$ or 6, and then $p = 6, 4$ or 3, respectively. Of course, this is a somewhat unusual way to approach the classification problem.

The non-classical regular apeirotopes of full rank in dimensions at least 5 are listed in Table 8B13.

8B13

$$\begin{array}{ccc}
\{3^{d-1}\}^{\alpha} & \overset{\kappa}{\longleftrightarrow} & \{3^{d-1}\}^{\zeta\alpha} \\[2ex]
\{4, 3^{d-2}, 4\} & \overset{\kappa}{\longleftrightarrow} & \{4, \frac{6}{1,3}, 3^{d-3}, 4\} \\[2ex]
\kappa_{02}\Big\updownarrow & & \kappa_{02}\Big\updownarrow \\[2ex]
\{3^{d-2}, 4\}^{\alpha} & \overset{\kappa}{\longleftrightarrow} & \{3^{d-2}, 4\}^{\zeta\alpha} \\[2ex]
\{4, 3^{d-2}\}^{\alpha} & \overset{\kappa}{\longleftrightarrow} & \{4, 3^{d-2}\}^{\zeta\alpha}
\end{array}$$

$(d+1)$-apeirotopes for $d \geqslant 5$

We have given only one indication of type here: recall that $\{4, \frac{6}{1,3}, 3^{d-3}, 4\}$ has 3-face the Petrie–Coxeter sponge $\{4, \frac{6}{1,3} \mid 4\}$ and vertex-figure $\{3^{d-2}, 4\}^{\zeta} = \{3^{d-2}, 4\} \otimes \{2\}$.

8C Apeirohedra and Polyhedra

Each of the three non-classical regular regular apeirohedra in $\mathbb{E}^2$ can be obtained in two ways: first, as a Petrial of a classical apeirohedron and, second, through the free abelian apeirotope construction. This should be as expected, since the product R_0R_2 of two reflexions R_0 and R_2 in perpendicular lines in $\mathbb{E}^2$ is a point-reflexion. Observe that these must be the only possibilities, since Theorem 4B9 tells us that the only allowable mirror vectors are $(1,1,1)$ (for the classical apeirohedra) and $(0,1,1)$.

8C1

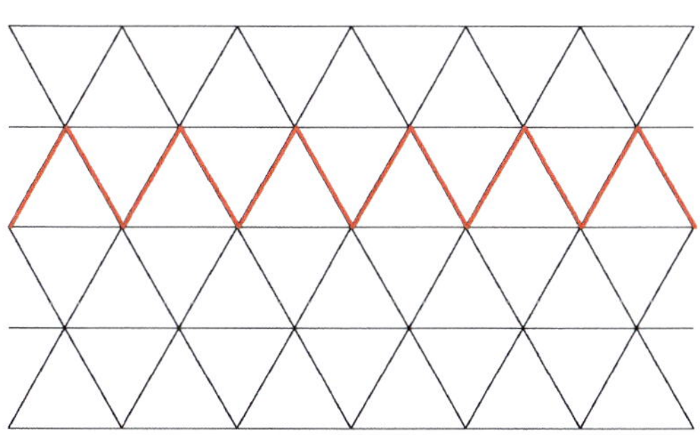

The apeirohedron $\{\frac{2}{0,1}, 6 : 3\} = \{6\}^{\alpha}$

8C2

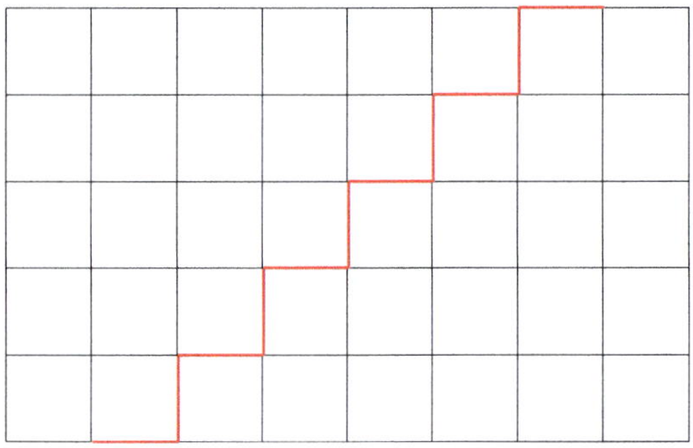

The apeirohedron $\{\frac{2}{0,1}, 4 : 4\} = \{4\}^{\alpha}$

8C3

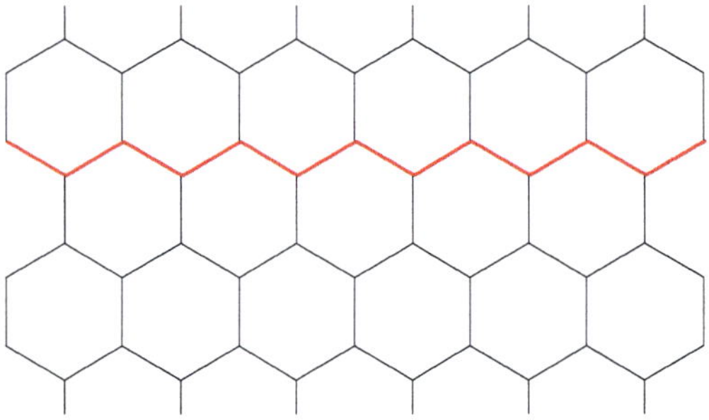

The apeirohedron $\{\frac{2}{0,1}, 3 : 6\} = \{3\}^{\alpha}$

The 2-faces of these apeirohedra are skew apeirogons $\{\frac{2}{0,1}\} = \{\infty\} \# \{2\}$; in Figures 8C1, 8C2 and 8C3, which illustrate portions of the apeirohedra, typical such faces are indicated by red zigzags.

In Table 8C4 we list these apeirohedra, and the various relationships among them.

8C4

$$\{6\}^{\alpha} \quad \overset{\pi}{\longleftrightarrow} \quad \{3,6\} \quad \overset{\delta}{\longleftrightarrow} \quad \{6,3\} \quad \overset{\pi}{\longleftrightarrow} \quad \{3\}^{\alpha}$$

$$\{4,4\} \quad \overset{\pi}{\longleftrightarrow} \quad \{4\}^{\alpha}$$

The 2-dimensional apeirohedra

The (finite) non-classical regular polyhedra in $\mathbb{E}^3$ can be enumerated on the same lines. The mirror vector of such a polyhedron can only be $(1, 2, 2)$, and then either $\boldsymbol{\zeta}$ or $\boldsymbol{\pi}$ will lead to a corresponding classical polyhedron. Naturally, though, in this case the two related polyhedra will generally be different; the only exception is

$$\{\tfrac{6}{1,3}, 4 : 3\}^{\zeta} = \{\tfrac{6}{1,3}, 4 : 3\}^{\pi} = \{3, 4\}.$$

We list the relationships in Tables 8C5 and 8C6; one result of the two forms of relationship is that the crystallographic polyhedra now fall into a single family, because $\boldsymbol{\zeta}$ can halve or double the group order.

8C5

$$\begin{array}{ccccccc}
\{3,3\} & \overset{\pi}{\longleftrightarrow} & \{\frac{4}{1,2},3:3\} & & & & \\
\zeta\updownarrow & & \zeta\updownarrow & & & & \\
\{\frac{6}{1,3},3:4\} & \overset{\pi}{\longleftrightarrow} & \{4,3\} & \overset{\delta}{\longleftrightarrow} & \{3,4\} & \overset{\pi,\zeta}{\longleftrightarrow} & \{\frac{6}{1,3},4:3\}
\end{array}$$

The crystallographic polyhedra

8C6

$$\begin{array}{ccc}
\{\frac{10}{1,5},5:3\} & \overset{\pi}{\longleftrightarrow} & \{3,5\} \\
\zeta\updownarrow & & \zeta\updownarrow \\
\{\frac{5}{2},5\mid 3\} & \overset{\pi}{\longleftrightarrow} & \{\frac{6}{1,3},5:\frac{5}{2},3\} \\
\delta\updownarrow & & \\
\{5,\frac{5}{2}\mid 3\} & \overset{\pi}{\longleftrightarrow} & \{\frac{6}{1,3},\frac{5}{2}:5,3\} \\
\zeta\updownarrow & & \zeta\updownarrow \\
\{\frac{10}{3,5},\frac{5}{2}:3\} & \overset{\pi}{\longleftrightarrow} & \{3,\frac{5}{2}\} \\
 & & \delta\updownarrow \\
\{\frac{10}{3,5},3:\frac{5}{2}\} & \overset{\pi}{\longleftrightarrow} & \{\frac{5}{2},3\} \\
\zeta\updownarrow & & \zeta\updownarrow \\
\{5,3\} & \overset{\pi}{\longleftrightarrow} & \{\frac{10}{1,5},3:5\}
\end{array}$$

The pentagonal polyhedra

8C7 Remark Because the regular polyhedra are each related to two others by $\boldsymbol{\pi}$ and $\boldsymbol{\zeta}$, this yields another easy way of finding their Petrie polygons. For example,

$$\{\tfrac{5}{2},5\mid 3\}^{\pi}=\{3,5\}^{\zeta}=\{\tfrac{6}{1,3},5:\tfrac{5}{2},3\}$$

(we only need its 2-face), showing that the Petrie polygon of $\{\frac{5}{2}, 5 \mid 3\}$ (and its dual $\{5, \frac{5}{2} \mid 3\}$) is $\{\frac{6}{1,3}\}$.

8D Higher-Dimensional Exceptions

In this section we complete the classification, by listing the remaining regular polytopes and apeirotopes of full rank. Of the latter, the exceptional cases only occur in $\mathbb{E}^4$, since we have already dealt with the regular apeirohedra in $\mathbb{E}^2$.

4-Polytopes

We begin with polytopes. For them, our first observation is that the groups of the remaining regular 4-polytopes of full rank are all generated by hyperplane reflexions, even if the mirrors of the generating reflexions are not necessarily hyperplanes; thus products of these with $\pm I$ are not needed. This takes care of the group $\mathbf{G}^v$ of the vertex-figure. Additionally, as we remarked in Section 8A, if $\dim R_0 = 1$ (rather than 3), then we can replace R_0 by $-R_0$ (an application of $\boldsymbol{\zeta}$). Our conclusion is the following.

8D1 Proposition *The symmetry group of a regular 4-polytope of full rank is either a group* $\mathbf{G}$ *generated by hyperplane reflexions, or a product* $\mathbf{G} \times \mathbf{Z}_2$ *of such a group with the group* $\mathbf{Z}_2$ *generated by the central inversion.*

Thus our procedure is straightforward. We can start with a reflexion group $\mathbf{G}$ which is the symmetry group of a classical regular polytope, and try to apply to it operations $\boldsymbol{\zeta}$, $\boldsymbol{\kappa}$ or $\boldsymbol{\pi}$ to obtain another C-group.

Let us deal with the pentagonal family first. Theorem 5B10 has already excluded the possibility of applying $\boldsymbol{\pi}$ to any of them. By the same token, this excludes $\boldsymbol{\kappa}_1$ as well, as the classification of the 3-polytopes shows (any vertex-figure obtainable by $\boldsymbol{\zeta}$ is obtainable – from another polyhedron – by $\boldsymbol{\pi}$). There remain $\boldsymbol{\zeta}$ and $\boldsymbol{\zeta\pi}$. However, it is easy to see that the first always works. Consider any such polytope $\mathsf{P} = \{p, q, r\}$, with generatrix $(R_0, \ldots, R_3)$. Thus, with Z the central inversion as usual, the group of the new polytope $\mathsf{Q} := \mathsf{P}^{\zeta}$ is given by

$$(R_0, \ldots, R_3) \mapsto (R_0 Z, R_1, R_2, R_3) =: (S_0, \ldots, S_3).$$

The new facet will be $\{p, q\} \diamond \{2\}$, with twice as many vertices; the new edges will join vertices u, w, with $u \in \mathsf{F}$ (where F is an original facet of P) and $w \in -\mathsf{F}$ (the opposite facet in P), such that $\{u, -w\}$ is an original edge of P (in F). If v is the base vertex of P, which is also the base vertex of Q, then an element of $\langle S_0, S_1, S_2\rangle$ which fixes v must take F into itself (rather than $-\mathsf{F}$), and hence actually belongs to $\langle R_0, R_1, R_2\rangle$. Therefore,

$$\langle S_0, S_1, S_2\rangle \cap \langle S_1, S_2, S_3\rangle = \langle R_0, R_1, R_2\rangle \cap \langle R_1, R_2, R_3\rangle = \langle R_1, R_2\rangle = \langle S_1, S_2\rangle,$$

which verifies the intersection property. Thus $\boldsymbol{\zeta}$ always yields another polytope; observe that the group remains the same.

On the other hand, as might be expected $\zeta\pi$ fails. The operation here is

$$(R_0, \ldots, R_3) \mapsto (R_0 Z, R_1 R_3, R_2, R_3) =: (S_0, \ldots, S_3),$$

with $Z = -I$ as before. If P is of Schläfli type $\{p, q, r\}$, so that $p, q, r = 3$ or 5 are odd, then the following successively belong to $\mathbf{G}_0 = \langle S_0, S_1, S_2 \rangle$:

$$(S_0 S_1)^p = (R_0 Z \cdot R_1 R_3)^p = Z R_3,$$
$$(S_2 \cdot Z R_3)^{r-1} S_2 = (R_2 R_3)^r R_3 = R_3 = S_3.$$

The last violates the intersection property, as we wanted.

For the simplex $\{3, 3, 3\}$, the same arguments show that neither π nor κ can be applied, but, as in Section 8A, applying ζ gives the mix $\{3, 3, 3\} \diamond \{2\}$.

The analysis for the staurotope $\{3, 3, 4\}$ (which we already saw in Section 8A) and the 24-cell $\{3, 4, 3\}$ is again the same. In particular, π cannot be applied, neither can κ to $\{3, 3, 4\}$, but ζ can in both cases. Moreover, the last argument only used the fact that p and r were odd, so that we can also deduce that $\{3, 4, 3\}^{\zeta\pi}$ is non-polytopal.

We finally come to the most interesting case, that of the cube $\{4, 3, 3\}$. We saw in Section 8A that applying ζ yields the allomorphic cube $\{4, 3, 3\}^{\zeta}$. Its facets are abstractly, of course, 3-cubes $\{4, 3\}$. However, to see what they look like geometrically, we apply Theorem 5D13; thus they occur in the form

$$\{4, 3\} \diamond \{2\} = \{4, 3\}^{\zeta} \# \{2\} = \{\tfrac{4}{1,2}, 3 : 3\} \# \{2\}.$$

To obtain one such, in a facet of the usual 4-cube $\{4, 3, 3\}$ inscribe a tetrahedron, or, rather, its Petrial $\{\tfrac{4}{1,2}, 3 : 3\}$. In the opposite facet of the 4-cube is a translate of $\{\tfrac{4}{1,2}, 3 : 3\}$. The edges of the new facet $\{4, 3\} \diamond \{2\}$ then switch between these two copies.

In contrast to the other cases, π does apply to $\{4, 3, 3\}$; Theorem 5B10 is not relevant here. The new facets are tori $\{4, \tfrac{4}{1,2} \mid 4\} \cong \{4, 4 \mid 4\}$; we thus obtain a polytope of type $\{\{4, 4 \mid 4\}, \{4, 3 : 3\}\}$. In terms of the generatrices, the operation is

$$(R_0, \ldots, R_3) \mapsto (R_0, R_1 R_3, R_2, R_3) =: (S_0, \ldots, S_3).$$

The 'hole' relation $(S_0 S_1 S_2 S_1)^4 = E$ for the torus derives from $(R_0 R_1)^4 = E$. The Petrie operation is involutory, and it is straightforward to check that we recover the relations for the group of the 4-cube from the abstract relations for $\{\{4, 4 \mid 4\}, \{4, 3 : 3\}\}$. In other words, we have

8D2 Theorem *There is a universal regular* 4*-polytope*

$$\{\{4, \tfrac{4}{1,2} \mid 4\}, \{\tfrac{4}{1,2}, 3 : 3\}\} \cong \{\{4, 4 \mid 4\}, \{4, 3 : 3\}\}$$

of full rank. Moreover, the fine Schläfli symbol is rigid.

The rigidity follows from Theorem 6D2. Naturally, if we apply $\boldsymbol{\pi}$ to $\{4,3,3\}^{\zeta}$, or $\boldsymbol{\zeta}$ to $\{\{4,\frac{4}{1,2} \mid 4\},\{\frac{4}{1,2},3:3\}\}$, then we obtain an allomorph of the latter.

Finally, we must consider $\boldsymbol{\kappa}_1$. The analysis in Section 8A does not exclude the case $\{4,3,3\}^{\kappa_1}$ immediately, because we actually obtain a finite group (in fact, it is $[3,4,3]$). However, the resulting 'polytope' would have vertex-figure $\{3,3\}^{\zeta} = \{\frac{6}{1,3},3:4\}$, whose vertices are those of the cube $\{4,3\}$, and at least the edges of $\{4,3,3\}$, and so would have to have the edges of $\{3,4,3\}$. With these edges and vertex-figure, we then see that we have reached $\{3,4,3\}^{\pi}$, which is not permitted by Theorem 5B10.

Observe, by the way, that we can apply $\boldsymbol{\kappa}_1$ to $\{3,4,3\}$ but, in accord with halving the order of the group of the vertex-figure, this results in passing to one of the three copies of $\{\{4,\frac{4}{1,2} \mid 4\},\{\frac{4}{1,2},3:3\}\}$ that are inscribed in it (as three copies of $\{4,3,3\}$ are). However, we can apply $\boldsymbol{\eta}$ to $\{3,4,3\}$ and $\{3,4,3\}^{\zeta}$, with $\{3,4,3\}^{\eta} = \{4,3,3\}$, to link them in with the polytopes derived from $\{4,3,3\}$, as in Table 8D3.

8D3

$$\begin{array}{ccccc}
\{3,3,4\}^{\zeta} & \stackrel{\zeta}{\longleftrightarrow} & \{3,3,4\} & & \\
 & & \delta \updownarrow & & \\
\{3,4,3\} & \stackrel{\eta}{\longrightarrow} & \{4,3,3\} & \stackrel{\pi}{\longleftrightarrow} & \{\{4,\frac{4}{1,2} \mid 4\},\{\frac{4}{1,2},3:3\}\} \\
\zeta \updownarrow & & \zeta \updownarrow & & \zeta \updownarrow \\
\{3,4,3\}^{\zeta} & \stackrel{\eta}{\longrightarrow} & \{\frac{4}{1,2},3,3\} & \stackrel{\pi}{\longleftrightarrow} & \{\{\frac{4}{1,2},\frac{4}{1,2} \mid \frac{4}{1,2}\},\{\frac{4}{1,2},3:3\}\}
\end{array}$$

4-*Apeirotopes*

We now move on to the regular apeirotopes, starting with those in $\mathbb{E}^3$. We saw at the beginning of Section 8B that, apart from the abelian apeirotopes Q^{α}, the only regular apeirotopes of full rank in $\mathbb{E}^3$ are $\{4,3,4\}$ and $\{4,3,4\}^{\kappa}$. However, the Petrie operation $\boldsymbol{\pi}$ is also clearly applicable to $\{4,3,4\}$, from which we conclude the first part of

8D4 Theorem *The two 4-apeirotopes* $\{4,3,4\}^{\pi}$ *and* $\{4,3,4\}^{\kappa}$ *coincide, up to a change of generatrix. Moreover,*

$$\{4,3,4\}^{\pi} = \{\{4,\tfrac{6}{1,3} \mid 4\},\{\tfrac{6}{1,3},4:3\}\} \simeq \mathbf{\{\{4,6 \mid 4\},\{6,4:3\}\}}$$

as a universal polytope.

Proof. The fact that $\{3,4\}^{\pi} = \{3,4\}^{\zeta}$ from Table 8C5 implies the coincidence. For the isomorphism, we first note that $\{4,3,4\} \cong \mathbf{\{4,3,4\}}$ is universal. We then apply π to $\{\{4,6 \mid 4\},\{6,4:3\}\}$, which is necessarily universal in view of the flatness of $\{6,4:3\}$ and Theorem 2F11. Note that π is valid here, since the relators of $\{\{4,6 \mid 4\},\{6,4:3\}\}$ are all 0-even in the terminology of Section 2D. We conclude that

$$\pi\colon \{\{4,6 \mid 4\},\{6,4:3\}\} \longleftrightarrow \{4,3,4\}$$

(more strictly, the result of applying π to one yields a quotient of the other), and the theorem follows. □

5-*Apeirotopes*

Finally, we look at 5-apeirotopes in $\mathbb{E}^4$, bearing in mind the classification of the finite regular 4-polytopes in $\mathbb{E}^4$ which we have just completed. We can discard the pentagonal polytopes as putative vertex-figures, and so there remain just the ten rational ones, namely, the simplex, staurotope, cube and its Petrial, and 24-cell, together with their centriverts. Since the free abelian apeirotope construction $\mathsf{Q} \mapsto \mathsf{Q}^{\alpha}$ applies to each of them, we are left to consider the 5-apeirotopes for which the initial reflexion mirror R_0 is a hyperplane.

8D5 Remark Since $\{3,4,3\}^{\alpha}$ has trigonal edge-circuits, a small check of its polytopality is in order. However, no problem arises, since $\{3,4,3\}$ is centrally symmetric.

We can immediately eliminate $\{3,3,3\}$ and $\{3,3,3\}\diamond\{2\}$ from consideration, since there is no discrete infinite group $[p,3,3,3]$ in $\mathbb{E}^4$ (again, the argument is like that of Theorem 8A4, in that $(R_0R_1)^2$ would be a rotation through an irrational angle). We are thus reduced to considering $\{3,3,4\}$, $\{4,3,3\}$, $\{3,4,3\}$, and the various polytopes derived from them by ζ and π (in the second case only). We shall see that all yield regular apeirotopes.

We have already treated $\{4,3,3,4\}$ in Section 8B, but we recall that the 3-faces of $\{4,3,3,4\}^{\kappa}$ are Petrie–Coxeter apeirohedra $\{4,\frac{6}{1,3} \mid 4\}$, and that the whole apeirotope has eight facets.

The same analysis as in Section 8B shows that the 3-faces of $\{3,3,4,3\}^{\kappa}$ are $\{6,\frac{6}{1,3} \mid 3\}$, while those of $\{3,4,3,3\}^{\kappa}$ are $\{6,\frac{4}{1,2} \mid 4\}$. While we could appeal to Section 5D for their basic properties, nevertheless it is worth describing these two apeirotopes in greater detail.

For the first, we take the generatrix of $[3,3,4,3]$ in the form

$$\begin{aligned}
R_0\colon &\quad x \mapsto (1+\xi_2, -1+\xi_1, \xi_3, \xi_4),\\
R_1\colon &\quad x \mapsto (\xi_1, \xi_3, \xi_2, \xi_4),\\
R_2\colon &\quad x \mapsto (\xi_1, \xi_2, \xi_4, \xi_3),\\
R_3\colon &\quad x \mapsto (\xi_1, \xi_2, \xi_3, -\xi_4),\\
R_4\colon &\quad x \mapsto x(I - \tfrac{1}{2}u^{\mathsf{T}}u),
\end{aligned}$$

where $u := (1,1,1,1)$. (For the last, note that the reflexion in the hyperplane through o with unit normal v is $I - 2v^{\mathsf{T}}v$ in matrix form.) We choose this form for the generators, because then the vertices of $\{3, 3, 4, 3\}$ form the lattice

8D6 $\quad \mathbf{\Lambda} := \{(\zeta_1, \zeta_2, \zeta_3, \zeta_4) \in \mathbb{Z}^4 \mid \zeta_1 + \zeta_2 + \zeta_3 + \zeta_4 \equiv 0 \pmod 2\}$

generated by the vectors $e_j \pm e_k$ for $1 \leqslant j < k \leqslant 4$, and so have integer cartesian coordinates.

The generatrix $(S_0, \ldots, S_4)$ of $\{3, 3, 4, 3\}^\kappa$ is then given by $S_j := R_j$ for $j \neq 1$, and

$$xS_1 = -xR_1 = (-\xi_1, -\xi_3, -\xi_2, -\xi_4).$$

The base edge joins o to $e_1 - e_2 = (1, -1, 0, 0)$. The base 3-face $\mathsf{F}_3 = \{6, \frac{6}{1,3} \mid 3\}$ lies in the hyperplane L_3 through o with normal u. Since $(S_1 S_2)^3 = K$ is the point-reflexion in o, its conjugate by S_0 is the point-reflexion in $e_1 - e_2$, and so their product $((S_1 S_2)^3 S_0)^2$ is the translation (in L_3) by $2(e_1 - e_2)$. The conjugates of this under $\langle S_0, S_1, S_2 \rangle$, namely, all $2(e_j - e_k)$ with $1 \leqslant j < k \leqslant 4$, actually generate the whole translation subgroup of F_3.

Since $uS_3 = (1, 1, 1, -1)$, transforming further by $\langle S_0, \ldots, S_3 \rangle$ shows that the base facet F_4 is composed of copies of F_3 in hyperplanes with normals all $(1, \pm 1, \pm 1 \pm 1)$. Further, we have already noted that $K = (S_1 S_2)^3 \in \langle S_0, \ldots, S_3 \rangle$, so that $\langle R_0, \ldots, R_3 \rangle \leqslant \langle S_0, \ldots, S_3 \rangle$. It follows that the point-reflexion in the centre e_1 of the base facet of $\{3, 3, 4, 3\}$ belongs to $\langle S_0, \ldots, S_3 \rangle$. Thus the product of K with this, which is the translation by $2e_1$, is in the group; this and its conjugates generate the translation subgroup $2\mathbb{Z}^4$ of F_4. Observe that the ridges (images of F_3) lie in parallel hyperplanes spanned by vertices at minimal distance apart.

There is a nice picture of the facet F_4. Its vertex-figure is the mix $\{3, 4\} \diamond \{2\}$ of an octahedron and a digon. The octahedra are opposite facets of the vertex-figure $\{3, 4, 3\}$ of $\{3, 3, 4, 3\}$, and their edges correspond to holes $\{3\}$ of the 3-faces $\{6, \frac{6}{1,3} \mid 6\}$. These holes fit together to form 4-staurotopes $\{3, 3, 4\}$; these staurotopes, in turn, are those of the truncated tiling $\left\{\begin{matrix} 4 \\ 3, 3, 4 \end{matrix}\right\}$ whose vertices are the mid-points of the edges of $\{4, 3, 3, 4\}$. Moreover, the 2-faces of F_4 are hexagonal cross-sections of the cuboctahedra $\left\{\begin{matrix} 4 \\ 3 \end{matrix}\right\}$ of this truncated tiling, and typical 3-faces lie in sections of this tiling spanned by the tetrahedral faces of the cross-polytopal holes.

Finally, the facets of the apeirotope occur in three different orientations, corresponding to the three sets of orthogonal axes spanned by opposite vertices of $\{3, 4, 3\}$, each facet being associated with two of these sets; there are $8 = [\mathbf{\Lambda} : 2\mathbb{Z}^4]$ translates in each set, giving $3 \cdot 8 = 24$ facets in all. The index 8 also counts the number of distinct ridges in each hyperplane which they span.

We next look at $\{3, 4, 3, 3\}^\kappa$. So as to preserve the underlying symmetry that we had before, we take the generators of $[3, 4, 3, 3]$ in a similar form to those of $[3,3,4,3]$, but in the reverse order. We thus take

$$\begin{aligned} R_0&\colon\ x \mapsto u + x(I - \tfrac{1}{2}u^{\mathsf{T}}u),\\ R_1&\colon\ x \mapsto (-\xi_1, \xi_2, \xi_3, \xi_4),\\ R_2&\colon\ x \mapsto (\xi_2, \xi_1, \xi_3, \xi_4),\\ R_3&\colon\ x \mapsto (\xi_1, \xi_3, \xi_2, \xi_4),\\ R_4&\colon\ x \mapsto (\xi_1, \xi_2, \xi_4, \xi_3),\end{aligned}$$

where $u := (1,1,1,1)$ as before. However, we have doubled the previous scale, since we wish the underlying vertex-set of $\{3,4,3,3\}$, namely,

$$\{(\zeta_1,\zeta_2,\zeta_3,\zeta_4) \in \mathbb{Z}^4 \mid \zeta_1 \equiv \zeta_2 \equiv \zeta_3 \equiv \zeta_4 \pmod 2\} \setminus 2(\mathbf{\Lambda} + e_4),$$

with $\mathbf{\Lambda}$ as in (8D6), to consist of integer points. For later reference to the geometry, we note that the base facet $\{3,4,3\}$ has vertices

$$2(e_4 \pm e_j) \quad (j = 1,2,3,4), \qquad \pm e_1 \pm e_2 \pm e_3 + (2 \pm 1)e_4,$$

a standard form shifted by $2e_4$. Further, the translation subgroup of the group is $2\mathbf{\Lambda}$, generated by all $2(e_j \pm e_k)$ for $1 \leqslant j < k \leqslant 4$. With these coordinates, the group $\mathbf{G} = \langle S_0, \ldots, S_4\rangle$ of $\{3,4,3,3\}^{\kappa}$ has $S_j := R_j$ for $j \neq 1$, and

$$S_1\colon\ x \mapsto (\xi_1, -\xi_2, -\xi_3, -\xi_4).$$

8D7

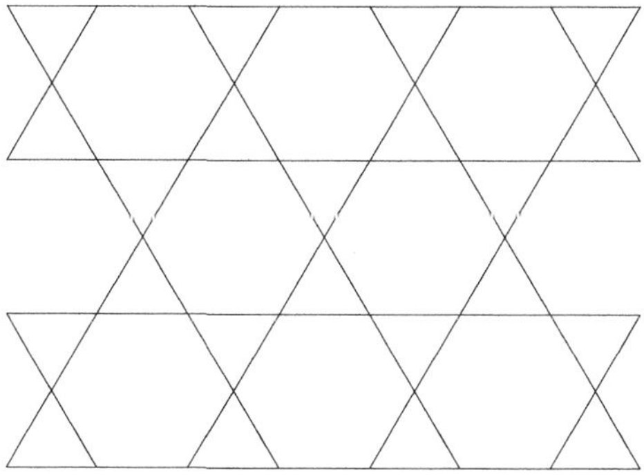

The facet of $\{3,4,3,3\}^{\kappa}$

The base ridge $\mathsf{F}_3 = \{6, \frac{4}{1,2} \mid 4\}$ spans the same hyperplane L_3 through o with normal $e_3 - e_4$ as does the original ridge $\{3,4\}$ of $\{3,4,3,3\}$. We can first check that $(S_0S_1)^3$ is the point-reflexion in $2e_1$; under $\langle S_1, S_2\rangle$ we similarly obtain point-reflexions in $-2e_1$ and $2e_2$, and hence translations in $\langle S_0, S_1, S_2\rangle$ by $4(e_1 \pm e_2)$. The conjugate of that by $4(e_1 + e_2)$ under S_0 is the translation by $4(e_3 + e_4)$; these three translations generate the subgroup of translation symmetries of F_3. However, intrinsically, F_3 has twice as many translations as this, namely, those generated by

$$2((e_1+e_2)\pm(e_1-e_2)\pm(e_3+e_4)) = 4e_j \pm 2(e_3+e_4)$$

for $j = 1, 2$. In fact, in the subgroup $\langle S_0, S_1, S_2\rangle$, these correspond to glide reflexions, consisting of the product of the translation with the reflexion in L_3 (we do not reproduce the details).

An idea of how a 3-face (such as F_3) fits may be useful. The edges of F_3 occur among those of $\{3,4,3,3\}$, and lie in its section by the affine hull of an octahedron $\{3,4\}$. This section is, in fact, a copy of the truncated tiling $\left\{\begin{matrix}4\\3,4\end{matrix}\right\}$; the hexagonal faces $\{6\}$ of $\{6, \frac{4}{1,2} \mid 4\}$ are central sections of the cuboctahedra $\left\{\begin{matrix}4\\3\end{matrix}\right\}$, while the holes $\{4\}$ are central sections of the octahedra $\{3,4\}$. However, these faces and holes form a somewhat sparse subset. The easiest starting point is an octahedron. This is met by eight cuboctahedra at its triangular faces, and then by six further octahedra at its vertices alone. If we delete this initial octahedron and the edges which meet it, we see that it is surrounded by eight diametral hexagons of cuboctahedra, and six diametral squares of cuboctahedra; these form a truncated octahedron. In a similar fashion, we delete the octahedra which we find by going across the hexagonal faces; carrying on like this, we obtain a tiling of a 3-space by truncated octahedra; its hexagons are those of $\{6, \frac{4}{1,2} \mid 4\}$. Note that each diametral square which survives meets eight diametral hexagons of cuboctahedra; these are the hexagons of $\{6, \frac{4}{1,2} \mid 4\}$ which meet the hole. It should be clear that exactly a quarter of the octahedra are used to form the truncated octahedra; thus the translation subgroup of F_3 has index 4 in that of the original section of the tiling $\{3,4,3,3\}$.

We readily see that the base facet F_4, with generatrix $(S_0,\ldots,S_3)$, contains ridges in hyperplanes with normals all $e_j \pm e_k$ for $1 \leqslant j < k \leqslant 4$. The hexagons of F_3 lie in parallel planes, and these are met by images of F_3 alternately at angles $\pm\pi/3$ (corresponding to the glide reflexions mentioned earlier). The translation subgroup of $\langle S_0,\ldots,S_3\rangle$ is now generated by all $4(e_j \pm e_k)$ for $1 \leqslant j < k \leqslant 4$, and so is $4\mathbf{\Lambda}$. Figure 8D7 gives a partial view of a typical facet, projected along the affine hull of a hexagonal face, which illustrates this glide reflexion. Only the facets whose normals lie in a plane are shown; the 3-faces project into lines.

Since the translation subgroup of a facet has index $[2\mathbf{\Lambda} : 4\mathbf{\Lambda}] = 16$ in the whole translation group, we now see fairly easily that all facets are translates of F_4, and that there are 16 of them.

8D8 Remark It is an interesting fact that all three of the 3 dimensional Petrie–Coxeter apeirohedra $\{4, \frac{6}{1,3} \mid 4\}$, $\{6, \frac{4}{1,2} \mid 4\}$ and $\{6, \frac{6}{1,3} \mid 3\}$ occur as 3-faces of regular 5-apeirotopes of full rank. Of course, since $\{4, \frac{6}{1,3} \mid 4\} = \{4,3\}^{\kappa}$ and $\{4,3\}$ is the 3-face of $\{4,3,3,4\}$ (and similarly for the other two instances), this is just to be expected.

We have already remarked that we can also apply $\boldsymbol{\pi}$ to $\{3, 4, 3, 3\}$ and $\{3, 4, 3, 3\}^{\kappa}$. The ridges (3-faces) are preserved, and so they remain octahedra $\{3, 4\}$ or Petrie–Coxeter apeirohedra $\{6, \frac{4}{1,2} \mid 4\}$, respectively. The vertex-figure is $\{\{4, \frac{4}{1,2} \mid 4\}, \{\frac{4}{1,2}, 3 : 3\}\}$ or $\{\{4, \frac{4}{1,2} \mid 4\}, \{\frac{4}{1,2}, 3 : 3\}\}^{\varsigma}$; since the vertex-figures are flat, so are the apeirotopes, and each therefore has three facets. We give a few more details about each.

The (base) facet of the former is a locally toroidal regular apeirotope with vertex-figure the toroid $\{4, \frac{4}{1,2} \mid 4\}$, and so is of type $\{\{3, 4\}, \{4, \frac{4}{1,2} \mid 4\}\}$; in fact, it is the universal regular polytope of its type (see Section 12B or [99, Chapter 10]). There is an interesting further feature of the apeirotope. Ignoring the finer structure, it is of type $\{3, 4, 4, 3\}$. Now clearly it is not self-dual (it has infinitely many vertices, but only three facets). However, the middle reflexion S_2 of its symmetry group has a 2-dimensional mirror, and it turns out that, if we replace this mirror by a suitable orthogonal mirror (of the same dimension), then we are, in effect, constructing the reflexion mirrors for another copy, but in dual form (that is, with the order of the reflexions reversed). Of course, we must distinguish carefully what happens here from the notion of self-quasi-duality; for quasi-duality, see Section 5D.

8D9

The facet of $\{3, 4, 3, 3\}^{\kappa\pi}$.

In $\{3, 4, 3, 3\}^{\kappa\pi}$, the new distinguished generator S_2 is now

$$S_2\colon\ x \mapsto (\xi_2, \xi_1, \xi_4, \xi_3).$$

What were formerly only glide reflexions inducing the intrinsic translation group of the base ridge $\mathsf{F}_3 = \{6, \frac{4}{1,2} \mid 4\}$ now become genuine translations. In contrast to what happens in $\{3, 4, 3, 3\}^{\kappa}$, here the full translation group $2\boldsymbol{\Lambda}$ preserves the base facet F_4, while the hyperplanes spanned by the ridges in it now have normals only $e_1 \pm e_2$, $e_1 \pm e_3$, $e_2 \pm e_4$ and $e_3 \pm e_4$, giving two out of three of the sets of orthogonal axes spanned by vertices of $\{3, 4, 3\}$. Figure 8D9 gives a partial view of this, illustrating those facets whose normals lie in a plane; the projection is to the same scale as that of Figure 8D7, and again is a projection along the affine hull of a hexagonal face.

III

Polytopes of Nearly Full Rank

9

General Families

In this chapter, we describe the families of regular polytopes and apeirotopes of nearly full rank which occur in every dimension. (Note that Remark 7A7 has already yielded examples of these, but naturally there are many more.) The following chapters will then be devoted to the anomalous cases, which are confined to dimensions no greater than 8.

We begin in Section 9A by discussing the rôle played by blended polytopes. In Section 9B we look at the part played by twisting certain diagrams. Then, Section 9C treats the (finite) regular polytopes; here, there are four families, three related (as one would expect) to the simplices, staurotopes and cubes and one to half-cubes. One might suppose that the cubic tiling would lead to just one family; in fact, as we shall see in Section 9D, that is far from being the case. Yet other families are connected with the (non-string) groups $\mathbf{P}_{d+1}$ and $\mathbf{S}_{d+1}$.

It should be understood here that we are, in effect, doing things out of order by making a forward reference to Chapters 11 and 13 for the completeness of the classification in small dimensions. The discussion of Section 5F is central to our arguments, based as they are on the knowledge of potential symmetry groups and vertex-figures. In particular, vertex-figures which do not fall into one of the infinite families which we consider in this chapter cannot, by definition, contribute to any of the families.

There will naturally be some overlap with Chapters 11 and 12, because the classification in $\mathbb{E}^4$ is needed to tackle the 'gateway' dimension five. But there we shall encounter the members of the infinite families as particular instances in larger groupings. In view of this overlap, we shall postpone descriptions of the lower-dimensional faces of the polytopes and apeirotopes until later.

9A Blends

Blends play an important part in the theory. We look in turn at the various ways that they appear.

Blended Polytopes

If a finite regular polytope P of nearly full rank is a blend, then Theorem 4C5 implies that one of its components must be a digon $\{2\}$; the other component is then a regular polytope Q of full rank. The situation for a regular apeirotope P of nearly full rank is very similar; one component is a digon $\{2\}$ or apeirogon $\{\infty\}$, while the other is a regular apeirotope Q of full rank. Actually, at this stage what we are claiming is not fully established, because in theory it is possible for the (necessarily untrivial) pure component Q of full rank not to be polytopal; see the notes at the end of the section. Thus one aim is to prove the relevant Theorems 9A6 and 9A7 below. We consider the two cases in turn.

9A1 Remark The blend of a finite regular polytope of full rank n and an apeirogon is an apeirotope of rank n and dimension $n+1$, and is therefore *not* of nearly full rank.

We have already met one important feature of blends $\mathsf{P} = \mathsf{Q} \# \{2\}$ with Q finite. We saw in Section 5D that finite regular polytopes are paired by the operation $\boldsymbol{\zeta}$, which replaces the initial generating reflexion S_0 of (say) Q (with centre at o) by $-S_0 = S_0^{\perp}$ to give Q^{ζ}, but we observed in Theorem 5D13 that $\boldsymbol{\zeta}$, when applied to blends with digons, lowers the dimension. We can retrieve this situation by forming a blend in an alternative way. If Q has codimension 1, then translate Q in a direction orthogonal to $\operatorname{lin} \mathsf{Q}$ and (in effect) apply the same operation $\boldsymbol{\zeta}$; the resulting polytope is denoted $\mathsf{Q} \diamond \{2\}$. (Another way of looking at this geometry is to replace Q by $\mathsf{Q} \# \{1\}$ before applying $\boldsymbol{\zeta}$.) Then blends are also paired by (5D11), namely,

$$\mathsf{Q} \diamond \{2\} = \mathsf{Q}^{\zeta} \# \{2\}.$$

To be more specific, suppose that Q (in $\mathbb{E}^d$) has generatrix $(S_0, \ldots, S_{d-1})$. In $\mathbb{E}^{d+1} = \mathbb{E}^d \times \mathbb{R}$, the respective generatrices (identified witth their mirrors) are

9A2 $$(S_0, S_1 \times \mathbb{R}, \ldots, S_{d-1} \times \mathbb{R}) \quad \text{for } \mathsf{Q} \# \{2\},$$

9A3 $$(S_0^{\perp}, S_1 \times \mathbb{R}, \ldots, S_{d-1} \times \mathbb{R}) \quad \text{for } \mathsf{Q} \diamond \{2\};$$

the orthogonal complement in the second expression is taken in $\mathbb{E}^d$, of course. We shall usually find it more convenient to write $\mathsf{Q} \diamond \{2\}$ rather than $\mathsf{Q}^{\zeta} \# \{2\}$.

9A4 Remark Bear in mind that Proposition 5D12 implies that $\mathsf{Q} \# \{2\}$ and $\mathsf{Q} \diamond \{2\}$ are isomorphic.

9A5 Remark In the case of the d-simplex $\{3^{d-1}\}$, we have the (combinatorial) isomorphism

$$\{3^{d-1}\} \# \{2\} \cong \{3^{d-1}\} \diamond \{2\} \cong \{3^{d-1}\}^{\zeta},$$

because the last polytope (with twice as many vertices as $\{3^{d-1}\}$) has even edge-circuits. Similarly, for the cube $\{4, 3^{d\;\;2}\}$ we have

$$\{4, 3^{d-2}\} \# \{2\} \cong \{4, 3^{d-2}\}^{\zeta} \# \{2\} \cong \{4, 3^{d-2}\},$$

and so on, in view of Proposition 5D12. Recall from the proof of Theorem 8A10 that $\{4, 3^{d-2}\}^{\zeta}$ has the vertices of the half-cube $\{3, 3^{d-3,1}\}$ when d is odd.

There is a further point. The Petrie operation π only works for one classical regular 4-polytope Q:

$$\{4,3,3\}^{\pi} = \{\{4, \tfrac{4}{1,2} \mid 4\}, \{\tfrac{4}{1,2}, 3 : 3\}\};$$

it fails for all other Q, because their 2-faces are odd. It might be thought that, once we blend with $\{2\}$, the Petrie operation will now be valid for these others, because the 2-faces of such blends are even polygons. However, these examples still fail, because their degeneracy actually runs more deeply than the oddness of their faces; observe that we do not need to consider the operation $\zeta\pi$ separately. We shall discuss a closely related situation in Section 11B, and so we postpone any further considerations until then.

Returning to the main discussion, we conclude that no unexpected examples of blends with digons arise in 5-dimensional space. In higher dimensions, even theoretically there is no way of constructing examples other than the obvious ones; essentially the same argument that eliminated the two pre-polytopes of Theorem 8A16 will show that their blends with a digon are also non-polytopal. We can summarize the analysis so far (which confirms what we claimed in the first paragraph) in

9A6 Theorem *There are* 18 *blended finite regular polytopes of nearly full rank in* $\mathbb{E}^4$, 34 *in* $\mathbb{E}^5$, *and* 6 *in* $\mathbb{E}^d$ *for each* $d \geqslant 6$. *Each is constructible in two ways, related by Proposition 5D12.*

The counts in Theorem 9A6 come from the analysis of Part II, as do those in the following Theorem 9A7.

The case of blended regular apeirotopes of nearly full rank can now be dealt with quickly. Exactly the same considerations as in the finite case prevent us from applying the Petrie operation π, except where we already know that it is allowed for the pure component of full rank. That is, as with blending with $\{2\}$, blending with $\{\infty\}$ cannot undo degeneracies which arise from (mis)application of other operations. We thus arrive at

9A7 Theorem *There* 12 *are blended regular apeirotopes of nearly full rank in* $\mathbb{E}^3$, 16 *in* $\mathbb{E}^4$, 36 *in* $\mathbb{E}^5$, *and* 16 *in* $\mathbb{E}^d$ *for each* $d \geqslant 6$.

There are some interesting special cases in low dimensions. First, observe that eversion applied to the planar apeirohedra has the following effects, in view of Remark 5D20 and Theorem 5D21:

$$\{4,4\}^{\kappa} = \{4,4\} \implies (\{4,4\} \# \{2\})^{\kappa} = \{4,4\} \# \{\infty\},$$

with the analogous results for their Petrials. Moreover,

$$\{3,6\}^{\kappa} = \{6,3\} \implies (\{3,6\} \# \{2\})^{\kappa} = \{6,3\} \# \{\infty\},$$

and for their Petrials; the same results hold with $\{3,6\}$ and $\{6,3\}$ interchanged.

Something even more striking occurs in $\mathbb{E}^4$. Bear in mind from Theorem 8D4 that – up to different generators – $\boldsymbol{\kappa}$ and $\boldsymbol{\pi}$ have the same effect on the cubic tiling $\{4,3,4\}$. Hence the corresponding blends fit into a single family:

$$\begin{array}{ccc}
\{4,3,4\}\,\#\,\{2\} & \xleftrightarrow{\ \kappa\ } & \{\{4,\tfrac{6}{1,3} \mid 4\},\{\tfrac{6}{1,3},4:3\}\}\,\#\,\{\infty\} \\
{\scriptstyle\pi}\updownarrow & & {\scriptstyle\pi}\updownarrow \\
\{\{4,\tfrac{6}{1,3} \mid 4\},\{\tfrac{6}{1,3},4:3\}\}\,\#\,\{2\} & \xleftrightarrow{\ \kappa\ } & \{4,3,4\}\,\#\,\{\infty\}
\end{array}$$

Vertex-Figures of Full Rank

The vertex-figure Q of a regular polytope or apeirotope P of nearly full rank falls into one of three classes:

- Q is of full rank;
- Q is blended;
- Q is pure.

We settle the first class in Theorem 9A8 immediately below, and then the second in Theorems 9A11 and 9A12 (but see also Section 13A). The third case relies ultimately on the first two, as working back through successive vertex-figures easily shows; the key classification is that of the finite regular polyhedra of nearly full rank, which will be accomplished in Chapter 11.

The basic result here is the following.

9A8 Theorem *If $m \geqslant 4$, and* P *is an m-polytope of nearly full rank whose vertex-figure* Q *is an $(m-1)$-polytope of full rank, then* P *is a blend.*

Proof. In order to treat the finite and infinite cases together, we adopt the viewpoint of Section 4B, and take the ambient space $\mathbb{U}$ of P to be a sphere or euclidean space as appropriate.

So, suppose that P is a regular m-polytope in the m-dimensional ambient space $\mathbb{U}$, whose vertex-figure Q is $(m-1)$-dimensional in $\mathbb{U}$, so that Q is of full rank. As usual, the generatrix of P is $(R_0, \ldots, R_{m-1})$, with R_j the canonical generating reflexion. We take v to be the initial vertex of P and w that of Q; we emphasize that here Q is the narrow vertex-figure, so that vert Q consists of the mid-points of the edges of P through v, and thus w is the mid-point of the initial edge. Let $H := \operatorname{span} \mathsf{Q}$ ($= \operatorname{span} \operatorname{vert} \mathsf{Q}$), so that H is a hyperplane. Let J be the hyperplane through w perpendicular to the line $M := \operatorname{span}\{v, w\}$; since reflexion in R_0 interchanges v and the other vertex of P on the initial edge, it is clear that $w \in R_0 \subseteq J$.

Further, write $K := H \cap J$, L for the line in J through w perpendicular to K, and $S_j := H \cap R_j$ for $j = 1, \ldots, m-1$. In fact, because Q is of full rank, from the classification in Chapters 7 and 8 we lose no generality in assuming that $\dim S_j = m-2$ for each $j \geqslant 2$. Then, for each such j, either $R_j = S_j$ or

$R_j = S_j H$, with (as usual) H also standing for the hyperplane reflexion with mirror H; in the latter case, R_j will be a hyperplane reflexion as well.

Since $m \geqslant 4$ by assumption, we have $\dim K \geqslant 2$; this fact will be crucial for our argument. Consider a general point $x \in R_0$; if $x \notin K$, suppose that the line through x perpendicular to K meets K in x'. If $x \notin L$, then the image of x' under $\langle R_2, \ldots, R_{m-1} \rangle$ spans K; in particular, there are some two of these images that do not lie on the same line through w. The corresponding images of x all have the same distance from K; it is now clear that, whether or not these images lie on the same side of K, they span the whole of J (contrast Remark 9A9 below). We infer that the only possibilities for R_0 are $R_0 = \{w\}$, L, K or J.

There are two possibilities. Suppose first that $v \notin H$. Let c be the centre of Q and let A be the line spanned by v and c. Observe that $A \subseteq R_j$ for each $j \geqslant 2$ (thus the images of x under $\langle R_2, \ldots, R_{m-1} \rangle$ must stay on the same side of H). If $R_0 = L$ or J, then there is some $b \in R_0 \cap A$. Since v and vR_0 lie on a common sphere S centred at b, the same holds for all vertices in the broad vertex-figure of P at v, and then for their images under R_0, and so on. We conclude that $\operatorname{vert} \mathsf{P} \subseteq S$, so that P is actually of full rank, contrary to assumption. On the other hand, if $R_0 = \{w\}$ or K, it follows that the vertices of P lie at the same distance either side of H, so that P must be a blend.

Finally, if $v = c$, then all vertices of P lie in H, because each of the four possibilities for R_j preserves H. In this case, P is again of full rank, contrary to the initial supposition. This concludes the proof. □

9A9 Remark It is worth looking into why this result fails for $m = 3$. In this case, $\dim K = 1$; if R_2 is the half-turn about the initial edge, then R_0 can be any line in J through w, and with no further restriction the argument of Theorem 9A8 does not carry over. When $v \notin H$, the argument even goes through in this case.

Blended Vertex-Figures

We begin by observing that three classes of regular polytopes of nearly full rank whose vertex-figures are blends are already known:

- a blend $\mathsf{Q} \# \{\infty\}$, with Q a regular apeirotope of full rank;
- a (finite or infinite) polytope Q^{ϖ}, with Q a regular polytope of full rank;
- an apeirotope Q^{κ}, with Q a crystallographic regular polytope of full rank.

Our task here is to show that these are the only possibilities, and to identify exactly into which case a given instance falls.

9A10 Remark When we recall from Theorem 5E7 that $\boldsymbol{\kappa} = \boldsymbol{\alpha\varpi}$, we see that the third case can be subsumed under the second; this is how the following analysis actually works out.

Suppose that P is a regular polytope or apeirotope of nearly full rank m, whose vertex-figure P^v is a blend. We can write this blend in the alternative forms $\mathsf{Q} \# \{2\}$ or $\mathsf{Q} \diamond \{2\} = \mathsf{Q}^{\varsigma} \# \{2\}$; ultimately one will be preferred over the other. As usual, we take the generatrix of $\mathbf{G} = \mathbf{G}(\mathsf{P})$ to be $(R_0, \ldots, R_{m-1})$, but – just as we did before – initially we allow the ambient space $\mathbb{U}$ to be spherical or euclidean. The finite and infinite cases are a little different, and so we shall treat them in turn. We suppose that m is large enough (we shall be more precise later), so that we can appeal to what we already know about regular blends of nearly full rank. In particular, we may assume that the component Q of P^v is itself polytopal of full rank $m-1$.

Working in $\mathbb{U}$, our analysis proceeds much as does the proof of Theorem 9A8. However, the vertices of P^v now fall into two 'parallel' subspaces, so that H is the one containing the initial vertex w of P^v and $c \in H$ is the centre of the vertices of P^v in H. We then define J, K and L as before. Since $v \notin H$, so that $A = \operatorname{span}\{v, c\} \subseteq R_j$ for each $j \geqslant 2$, we readily see that we have the same four possibilities for R_0. What happens then depends on whether P is finite or infinite; we treat the two cases in turn.

First suppose that P is finite. We set $d := m+1$, so that we work in $\mathbb{E}^d$, and the dimensions of the mirrors are thus increased by 1. Since $\dim \mathsf{Q} = d-2$, it follows that $\dim R_1 = d-3+1 = d-2$ or $1+1 = 2$. In the latter case, we write the blend as $\mathsf{Q} \diamond \{2\}$, and so in both cases we can assume that Q is classical. We thus write the generatrix of Q as $(S_2, \ldots, S_{d-1})$ (as linear hyperplane reflexions, with a shift of indices). If $\dim R_0 = 1$ or $d-2$, corresponding to the cases $R_0 = \{w\}$ or K, then we apply $\boldsymbol{\zeta}$, resulting in $\dim R_0 = d-1$ or 2 respectively, so that we can assume that we have one of the cases $R_0 = L$ or J, which ensures that $R_0 \cap S_2 \cap \cdots \cap S_{d-1} \neq \{o\}$; we then set $S_1 := R_0$. The mirror $R_1 S_2$ corresponding to the component $\{2\}$ of P^v is a line or a hyperplane; we define $S_0 = -R_1 S_2$ or $R_1 S_2$, respectively. Now $(S_0, \ldots, S_{d-1})$ is the generatrix of a regular pre-polytope of full rank; observe that Remark 4B10 does hold. Indeed, we have arranged things so that each S_j is a hyperplane, except possibly for S_1.

As we saw in Section 8D, we do have two instances of polytopes of full rank in $\mathbb{E}^4$ with mirror vectors $(3,2,3,3)$ and $(1,2,3,3)$. However, when $d \geqslant 5$, we showed in Section 8A that no regular d-polytope of full rank has generatrix $(S_0, \ldots, S_{d-1})$ with $\dim S_1 = 2$. The crucial cases were covered in Theorem 8A8, and so we must show that neither excluded case can give rise to a polytope under Petrie contraction $\boldsymbol{\varpi}$.

What we have to show is that the generatrix

$$(R_0, \ldots, R_{d-2}) := (-\bar{1}(2\ 3), \varepsilon(1\ 2)(3\ 4), (4\ 5) \ldots, (d{-}1\ d), \bar{d}),$$

with $\varepsilon = \pm 1$, does not have the intersection property (recall that $\bar{k}$ changes the sign of ξ_k). First, since

$$(R_1 R_2)^3 = \varepsilon(1\ 2) \implies \varepsilon(1\ 2),\ (3\ 4) \in \mathbf{G}_{0,d-2},$$

we clearly have

$$\begin{aligned} \mathbf{G}_0 &= \mathbf{Z}_2 \times \mathbf{C}_{d-2}, \\ \mathbf{G}_{0,d-2} &= \mathbf{Z}_2 \times \mathbf{A}_{d-3}, \end{aligned}$$

where $\mathbf{Z}_2 := \langle \varepsilon(1\ 2) \rangle$, $\mathbf{A}_{d-3} = \langle (3\ 4), \ldots, (d{-}1\ d) \rangle$ and $\mathbf{C}_{d-2}$ adjoins $\bar{d}$ to the latter. It follows that

$$-\bar{1} = (R_0(3\ 4))^3 \in \mathbf{G}_{d-2} \implies (2\ 3) \in \mathbf{G}_{d-2},$$

and then

$$\varepsilon I = (\varepsilon(1\ 2)(2\ 3))^3 \in \mathbf{G}_{d-2} \implies \bar{1},\ (1\ 2) \in \mathbf{G}_{d-2}.$$

We can now conjugate $\bar{1}$ by $(1\ 2)$, $(2\ 3), \ldots,$ $(d{-}1\ d)$ in turn, to show that $\bar{d} \in \mathbf{G}_{d-2}$, which violates the intersection property. Observe that applying ζ changes R_0 to $-R_0 = \bar{1}(2\ 3)$, and so we come to the same conclusion here as well.

This discussion has led to

9A11 Theorem *If $d \geqslant 5$ and P is a finite regular $(d{-}1)$-polytope of nearly full rank whose vertex-figure is blended, then $\mathsf{P} = \mathsf{Q}^{\varpi}$ or $\mathsf{Q}^{\zeta\varpi}$ for some classical regular d-polytope Q.*

We now consider when P is an apeirotope; we take its initial vertex to be the origin o. Once again, we have the same four possibilities for R_0. If $R_0 = \{w\}$ or K, then it is not hard to see that $\operatorname{vert} \mathsf{P}$ falls into a family of hyperplanes parallel to H. Indeed, if we write $H_0 := H$ and H_1 for the parallel hyperplane through o, then (H_0, H_1) is the generatrix for an apeirogon $\mathsf{R} = \{\infty\}$, and $\operatorname{vert} \mathsf{P}$ projects onto $\operatorname{vert} \mathsf{R}$, and we see that P is actually a blend $\mathsf{Q} \# \{\infty\}$ with one component a regular pre-apeirotope Q of full rank and the other R. What remains to be proved is that Q is polytopal. Now $\mathsf{P}^{v} = \mathsf{Q}^{v} \# \{2\}$, and by what we showed previously, Q^{v} is polytopal. But now the initial generator S_0 (say) of $\mathbf{G}(\mathsf{Q})$ is either a point or hyperplane reflexion, and we see at once that Q itself must be polytopal.

In the alternative cases, $R_0 = L$ or J; we express P^{v} in the form $\mathsf{R} \diamond \{2\}$ or $\mathsf{R} \# \{2\}$, with S_0 the generator of $\mathbf{G}(\{2\})$, which is a point or hyperplane respectively. Since $d \geqslant 5$, we see that R is actually classical. If the generatrix of Q is $(S_2, \ldots, S_d)$ with each S_j a hyperplane in $\mathbb{E}^d$, then the line $S_2 \cap \cdots \cap S_d$ through o is perpendicular to H and meets R_0, in v, say. Writing $S_1 := R_0$, we see that $(S_0, \ldots, S_d)$ is the generatrix of a regular apeirotope Q of full rank with initial vertex v, and that $\mathsf{P} = \mathsf{Q}^{\varpi}$. Note that, if S_0 is a point, then we actually have $\mathsf{P} = \mathsf{Q}^{v\kappa}$.

From this discussion, we deduce

9A12 Theorem *For $d \geqslant 5$, a regular d-apeirotope P of nearly full rank whose vertex-figure is blended is one of the following:*

- $\mathsf{Q} \# \{\infty\}$ *for some regular d-apeirotope Q of full rank,*
- Q^{ϖ} *for some regular $(d{+}1)$-apeirotope Q of full rank,*
- Q^{κ} *for some finite regular d-polytope Q of full rank.*

It may help keep track of the various operations if we bear in mind the corresponding mirror vectors; we do this for the three relevant cases.

9A13 Proposition *Under the applications of ϖ to finite polytopes, the mirror vectors behave as*

$$\begin{array}{ccccc}
\big((d{-}1)^d\big) & \xrightarrow{\varpi} & \big(d{-}1, d{-}2, (d{-}1)^{d-3}\big) & \xleftrightarrow{\zeta} & \big(1, d{-}2, (d{-}1)^{d-3}\big) \\
\zeta\updownarrow & & \zeta_1\updownarrow & & \zeta_1\updownarrow \\
\big(1, (d{-}1)^{d-1}\big) & \xrightarrow{\varpi} & \big(d{-}1, 2, (d{-}1)^{d-3}\big) & \xleftrightarrow{\zeta} & \big(1, 2, (d{-}1)^{d-3}\big)
\end{array}$$

The starting point at top left is a classical regular d-polytope.

9A14 Proposition *Under the applications of ϖ to honeycombs, the mirror vectors behave as*

$$\begin{array}{ccc}
\big((d{-}1)^{d+1}\big) & \xrightarrow{\varpi} & \big(d{-}1, d{-}2, (d{-}1)^{d-2}\big) \\
\kappa\updownarrow & & \\
\big(d{-}1, 1, (d{-}1)^{d-1}\big) & \xrightarrow{\varpi} & \big(1, d{-}2, (d{-}1)^{d-3}\big)
\end{array}$$

The starting point at top left is a classical regular honeycomb in $\mathbb{E}^d$.

9A15 Proposition *Under the applications of κ to finite polytopes, the mirror vectors behave as*

$$\begin{array}{ccc}
\big((d{-}1)^{d}\big) & \xleftrightarrow{\kappa} & \big(d{-}1, 1, (d{-}1)^{d-2})\big) \\
\zeta\updownarrow & & \\
\big(1, (d{-}1)^{d-1}\big) & \xleftrightarrow{\kappa} & \big(1, 1, (d{-}1)^{d-2})\big)
\end{array}$$

The starting point at top left is a classical regular d-polytope.

Let us end with a very general result concerning blended vertex-figures; we shall not usually bother to draw attention to each of its occurrences.

9A16 Proposition *If Q is a finite rational regular polytope, then*

$$(\mathsf{Q} \# \{2\})^{\alpha} = \mathsf{Q}^{\alpha} \# \{\infty\}.$$

Proof. Let H_0 be the hyperplane spanned by Q, and H_1 the parallel hyperplane through the centre o of $\mathsf{Q} \# \{2\}$. If $(S_1, S_2, \ldots)$ is the generatrix of Q and $(R_0, R_1, \ldots)$ that of $(\mathsf{Q} \# \{2\})^{\alpha}$, then we think of the point-reflexion R_0 in the initial vertex w of Q as the point-reflexion S_0 about w in H_0 composed with the hyperplane reflexion in H_0; that is, $R_0 = S_0 H_0$ in our familiar convention. We similarly have $R_1 = S_1 H_1$, while $R_j = S_j \times L$ for $j \geqslant 2$, where $L := H_1^{\perp}$. Since $(S_0, S_1, \ldots)$ is the generatrix of Q^{α}, the claim of the proposition follows. □

9A17 Remark Using (5D11), we see that

$$(\mathsf{Q} \diamond \{2\})^{\alpha} = \mathsf{Q}^{\zeta\alpha} \# \{\infty\}.$$

Notes to Section 9A

1. In essence, we need to recover the component $\{2\}$ of the blend $\mathsf{Q} \# \{2\}$ or $\mathsf{Q} \diamond \{2\}$ or, rather, its generating reflexion. On examination, it will be seen that the only cases where this is not possible are when the 2-face of Q is of type $\{4n\}$ for some n, and these will be dealt with by other arguments.

9B Twisting Small Diagrams

We carry on the systematic investigation with a crucial observation, bearing in mind that we are concerned here with polytopes of nearly full rank. This is that, if we do not allow redundant generators (we shall see examples to the contrary later), then the only diagrams which can be twisted to yield unblended regular polytopes of nearly full rank at least 4 are those of the form of Figure 9B1:

9B1 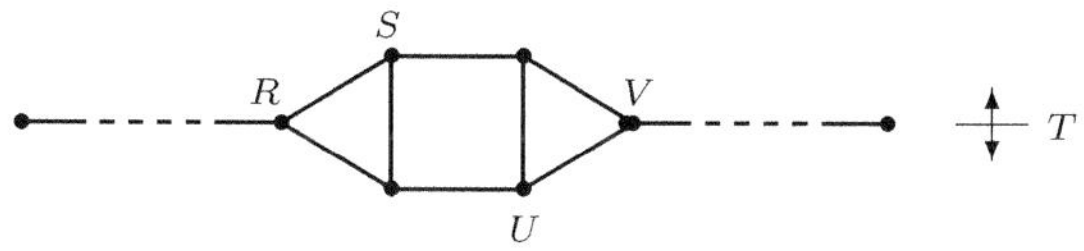

The twist is denoted by T. Branches may carry marks (symmetrically with respect to T, of course), and either vertical branch may be absent. The left and right branches form strings, and either may be absent (including the nodes R or V). We shall not prove this claim formally, but we observe that each twist contributes 1 to the rank, while each node dropped from the generating set (and thus obtained by applying a twist to another node) lowers the rank by 1. With two commuting twists we obtain a square or hexagonal sub-diagram (with diagonals but here we will be obliged to introduce redundant generators if we wish to preserve the symmetry of the whole diagram), whereas two non-commuting twists (which must correspond to adjacent generating reflexions) will act on a single circuit; in neither case can we extend the diagram non-redundantly by further branches. With a single twist, we just have the picture above.

It is worth repeating what we said in Section 5C, that we distinguish four kinds of twist. First, a twist can be *outer* or *inner*, according as the induced automorphism is outer or inner. Second, while a twist must permute the mirrors of the hyperplane reflexions of the diagram, it need not permute the (unit) normal vectors to the mirrors – it may change some signs, while preserving others. If the normals can be chosen to be permuted by the twist, then it is *proper*; otherwise the twist is *improper*.

9B2 Remark Note that basic blends with the segment $\{2\}$ can be obtained by twisting diagrams beginning with the right triangle.

Because any twist of the whole diagram induces one on each appropriate sub-diagram, we begin by describing what twists are allowed on the various small circuit diagrams that can be components of Figure 9B1. These are of two kinds: the group $\mathbf{B}_d$ in the form of a circuit of unmarked branches with the circuit itself marked 2, and an unmarked circuit $\mathbf{P}_{d+1}$ of the infinite group. Those circuits with more than six nodes play no part at all, while those with six nodes make no contribution to infinite families (see Chapter 13, where they do occur). We can thus confine our attention to the circuits that have at most five nodes; as we shall see, those with three or five nodes arise in two ways. We have to bear in mind that our aim is to construct string C-groups; again, refer to Figure 9B1.

9B3 Remark Once the possibility of a twist has been established, the group may then undergo modifications by operations of the kind that we discussed in Chapter 5.

The Group $\mathbf{B}_d$

Recall that the group $\mathbf{B}_d$ consists of all permutations of the coordinates of $\mathbb{E}^d$, together with an even number of changes of sign. Throughout the chapter, we shall freely employ the notation of signed permutations introduced at the end of Section 1D. The choices that follow are made with later applications in mind; indices will then all shift by the same amount. Here, we are concerned with the cases $d \leqslant 6$, with the diagram for $\mathbf{B}_d$ in the form of a single circuit of unmarked edges, which is itself marked 2.

When $d = 3$, we can pick the nodes to correspond to the transpositions $(1\ \overline{2}), (1\ 3), (2\ 3)$, the twist being $\varepsilon(2\ \overline{3})$ with $\varepsilon = \pm 1$. In the case $\varepsilon = 1$, with generatrix $\big(\varepsilon(2\ \overline{3}), ((1\ 3), (2\ 3)\big)$, we appear to obtain the tetrahedron $\{3, 3\}$; however, strictly speaking the trigonal faces are actually doubly-covered by hexagons, and so we shall not allow this. For $\varepsilon = -1$, we have the Petrial $\{\frac{6}{1,3}, 3 : 4\}$ of the 3-cube.

For $d = 4$, the transpositions corresponding to the nodes can be taken to be $(1\ \overline{2}), (2\ 4), (3\ 4), (1\ 3)$. Here, suitable twists are $\varepsilon\overline{1}(2\ 3)$ with $\varepsilon = \pm 1$. With generatrix $\big((1\ \overline{2}), \varepsilon\overline{1}(2\ 3), (3\ 4)\big)$, now it is the choice $\varepsilon = -1$ that leads to degeneracy. We shall describe in detail what results when $\varepsilon = 1$ in Section 11B, although we do encounter this case earlier.

If $d = 5$, then the circuit has transpositions $(1\ \overline{2}), (2\ 3), (3\ 5), (4\ 5), (1\ 4)$, and the twist is $\varepsilon(1\ 2)(3\ 4)$ with $\varepsilon = \pm 1$. As in the previous case, the choice $\varepsilon = -1$ results in degeneracy; with $\varepsilon = 1$, the generatrix $\big((1\ \overline{2}), (2\ 3), (1\ 2)(3\ 4), (4\ 5)\big)$ gives initial vertex $(1, 1, 1, 1, 1)$, from which we obtain the 16 vertices of the half-5-cube.

With the generatrix in the opposite order, the situation is reversed (this case will not occur for the present). For $\varepsilon = 1$, the initial vertex is e_5, and what should be the initial hexagonal 2-face degenerates to a trigon. When $\varepsilon = -1$ the initial vertex is $(1, -1, -1, 1, 0)$, and then we obtain the 80 mid-points of the edges of the 5-cube.

For $d = 6$, the initial generators are $(1\ \overline{2}), (2\ 3), (3\ 6), (5\ 6), (4\ 5), (1\ 4)$ going round the circuit, and the potential twists are $\varepsilon(1\ 2)(3\ 4)(5\ 6)$ with $\varepsilon = \pm 1$. With generatrix $\big((1\ \overline{2}), (2\ 3), \varepsilon(1\ 2)(3\ 4)(5\ 6), (4\ 5), (5\ 6)\big)$, for non-degeneracy we need $\varepsilon = 1$ as before; for reversal of the order we must have $\varepsilon = -1$.

The Group $\mathbf{P}_{d+1}$

As usual in working with $\mathbf{P}_{d+1}$, we often find it convenient to think of it as acting on the symmetric hyperplane $\mathbb{L}^d$ of (1E15) by permutations and changes of sign of the coordinates. Since we need hyperplane reflexions in other than linear hyperplanes here, we use $S^\bullet$ to mean the following: if S is the reflexion in the linear hyperplane with normal u, then $S^\bullet$ will stand for the reflexion in some parallel hyperplane that does not pass through o (it will not matter what). For instance, in $\mathbb{E}^5$, if $S = (0\ 3)$, then we would have

9B4 $$S^\bullet \colon x \mapsto (\xi_3 + 1, \xi_1, \xi_2, \xi_0 - 1, \xi_5).$$

Twisting $\mathbf{P}_3$ by an inner automorphism leads to the finite group $[3]$ (bear in mind that we replace one of its three generators by the automorphism), while an outer twist gives $[3, 6]$ (or $[6, 3]$); we thus need consider it no further.

The group $\mathbf{P}_4$ behaves in a similar way. With an inner automorphism the group reduces to the finite group $[3, 3]$; we shall look into this in more depth in Chapter 10. The outer automorphism is productive; again, Chapter 10 goes into this in more detail.

For $\mathbf{P}_5$, we specify its generators using the notation of (9B4). Going round an unmarked circuit, we have $(0\ 1)^\bullet, (1\ 2), (2\ 4), (3\ 4), (0\ 3)$; the twist is $\varepsilon(0\ 1)(2\ 3)$ with $\varepsilon = \pm 1$, and the new generatrix is $\big((0\ 1)^\bullet, (1\ 2), \varepsilon(0\ 1)(2\ 3), (3\ 4)\big)$. This is where Remark 4A2 comes into play. The initial vertex x satisfies

$$\xi_1 = \xi_2,\ \xi_3 = \xi_4, \begin{cases} \xi_0 = \xi_1,\ \xi_2 = \xi_3, & \text{if } \varepsilon = 1, \\ \xi_0 = -\xi_1,\ \xi_2 = -\xi_3, \xi_4 = 0, & \text{if } \varepsilon = -1; \end{cases}$$

in either case (bearing in mind that we are in $\mathbb{L}^4$) we see that $x - o$. However, when $\varepsilon = 1$, we see that $v(0\ 1)^\bullet = e_0 - e_1 \notin K_2 = \operatorname{lin}\{(3, 3, -2, -2, -2)\}$ (in the notation of the remark). If we take the generators in the reverse order, then we must also have $\varepsilon = -1$; we postpone further details until Section 12C.

The situation for $\mathbf{P}_6$ is very similar. Here, the generators round the circuit are $(0\ 1)^\bullet, (1\ 2), (2\ 4), (4\ 5), (3\ 4), (0\ 3)$; the twist is $\varepsilon(0\ 1)(2\ 3)(4\ 5)$ with $\varepsilon = \pm 1$, and the generatrix is $(0\ 1)^\bullet, (1\ 2), \varepsilon(0\ 1)(2\ 3)(4\ 5), (3\ 4), (4\ 5)$. Once again, Remark 4A2 forces $\varepsilon = -1$. We say little about this group at this point, since it is not involved in the rest of the chapter.

9C Families of Polytopes

We begin by noting that, for each regular $(d-1)$-polytope Q of full rank, we have a blend $\mathsf{Q} \# \{2\}$; remember as well that

$$\mathsf{Q}^\zeta \# \{2\} = \mathsf{Q} \diamond \{2\}.$$

In particular, we may take $\mathsf{Q} = \{3^{d-2}\}, \{3^{d-3}, 4\}, \{4, 3^{d-3}\}$, giving six infinite sequences. Apart from these, we shall group together polytopes that are derived from a given diagram, either directly or by applying one or more operations. We shall see that there is one sub-family with symmetry group $\mathbf{A}_d$ or $\mathbf{A}_d \times \mathbf{Z}_2$, one with group $\mathbf{B}_d$ and two with group $\mathbf{C}_d$.

Before going further, it is appropriate to prove a general result.

9C1 Theorem *If Q is a finite classical regular d-polytope with symmetry group $\mathbf{G}(\mathsf{Q}) = \langle S_0, \ldots, S_{d-1} \rangle$ and the polytope $\mathsf{Q}^{\zeta\varpi}$ is non-degenerate, then its initial vertex can be taken to be a normal u_0 to the hyperplane S_0.*

Proof. The generatrix of $\mathsf{P} := \mathsf{Q}^{\zeta\varpi}$ is $(S_1, -S_0S_2, S_3, \ldots, S_{d-1})$ in terms of the generatrix $(S_0, \ldots, S_{d-1})$ of Q (as usual, we take o to be the centre of Q). The mirror of $-S_0S_2$ is $(S_0 \cap S_2)^{\perp}$, so that the Wythoff space of P (containing its initial vertex) is

$$(S_0S_2)^{\perp} \cap S_3 \cap \cdots \cap S_{d-1} = (S_0^{\perp} + S_2^{\perp}) \cap S_3 \cap \cdots \cap S_{d-1}.$$

Now, if u_j is a (unit) normal to S_j for $j = 0, 2$, then

$$S_0^{\perp} + S_2^{\perp} = \operatorname{lin}\{u_0, u_2\}.$$

Since $u_0 \in S_3 \cap \cdots \cap S_{d-1}$ but $u_2 \notin S_3 \cap \cdots \cap S_{d-1}$, it follows that the Wythoff space of P is actually $\operatorname{lin}\{u_0\}$, as claimed. □

9C2 Remark In fact, Theorem 9C1 gives an easy criterion for degeneracy, when the images of u_0 under $\mathbf{G}(\mathsf{Q})$ yield too few vertices compared to the index $[\mathbf{G}(\mathsf{Q}) : \langle S_3, \ldots, S_{d-1} \rangle]$.

We shall list the families according to the related groups.

Relatives of $\mathbf{A}_d$

Our first family is given by

9C3 Proposition *For each $d \geqslant 4$, there are four regular $(d-1)$-polytopes derived from the diagram*

Proof. By convention, a number j against a node of the diagram refers to the generator R_j. Note that the whole tail to the right is missing if $d = 4$; this will happen in similar cases below. We have similarly indicated the twist R_1 and two other relevant transpositions (we shall not make a habit of the latter); as usual, we work in the symmetric hyperplane $\mathbb{L}^d$ of (1E15). We have already remarked that we allow inner automorphisms as twists, so we begin by illustrating this point. The diagram actually corresponds to the Petrie contraction $\{3^{d-1}\}^{\varpi}$ of the d-simplex $\{3^{d-1}\}$. In terms of the transpositions $(j\ k)$ in the symmetric group $\mathbf{A}_d \cong \boldsymbol{S}_{d+1}$, its generatrix $(R_0, \ldots, R_{d-2})$ is given by

9C4
$$\begin{aligned} R_0 &:= (1\ 2), \\ R_1 &:= (0\ 1)(2\ 3), \\ R_j &:= (j{+}1\ j{+}2) \quad \text{for } j = 2, \ldots, d-2. \end{aligned}$$

In this case, R_1 is a proper inner twist.

On the other hand, if we consider instead $\{3^{d-1}\}^{\zeta\varpi}$, then we change the sign of the reflexion R_1, to obtain $-R_1 = -(0\ 1)(2\ 3)$. Thus $-R_1$ twists the same diagram, but now acting as an outer automorphism; it is an improper outer twist. The symmetry group is now $\mathbf{A}_d \times \mathbf{Z}_2$, with $\mathbf{Z}_2 = \{\pm I\}$ generated by the central reflexion $-I$.

The vertex-sets of these two polytopes are easily found from the general discussion of Petrie contraction in Section 5B. Thus the vertices of $\{3^{d-1}\}^{\varpi}$ are the mid-points of the edges of the simplex $\{3^{d-1}\}$, while those of $\{3^{d-1}\}^{\zeta\varpi}$ are all permutations of $(1, -1, 0^{d-1})$, namely, the vertex-set of the difference body $\Delta\mathsf{T}^d$ of Example 3D5.

We further note that the operation $\boldsymbol{\zeta}$ can be applied to both polytopes, but that no other operations are appropriate. Thus we have found the four polytopes claimed in Proposition 9C3. □

The source of the family is the 4-dimensional regular polyhedron $\mathsf{Q} := \{4, \frac{6}{2,3} \mid 3\} \cong \{4, 6 : 5 \mid 3\}$, for more details of which we refer to Section 11B; note that Theorem 6D5 implies that Q is rigid, so that we can drop mention of the Petrie polygon in its geometric notation. This polytope was described in [94] and briefly discussed in [83], but we did not specify its group until [84]. Since we place considerable emphasis on universality, for future purposes it behoves us to analyse in detail exactly why the group $\mathbf{G}(\mathsf{Q})$ of Q is $[3, 3, 3] = \mathbf{A}_4 \cong \boldsymbol{S}_5$, the symmetry group of the regular 4-simplex.

Because Q slots in at different places in polytopes of various families, initially we adopt a neutral notation $(\boldsymbol{s}, \boldsymbol{t}, \boldsymbol{u})$ for the generatrix of the automorphism group $\boldsymbol{G}$ of the corresponding abstract polytope $\mathcal{Q}$, say. Thus $\boldsymbol{s}, \boldsymbol{t}, \boldsymbol{u}$ correspond to the geometric reflexions S, T, U of Figure 9B1. If we write $(\boldsymbol{x}, \boldsymbol{s}, \boldsymbol{y}, \boldsymbol{u})$ for the natural generators of the Coxeter group $[3, 3, 3]$, then $\boldsymbol{G}$ is obtained by the mixing operation

$$(\boldsymbol{x}, \boldsymbol{s}, \boldsymbol{y}, \boldsymbol{u}) \mapsto (\boldsymbol{s}, \boldsymbol{xy}, \boldsymbol{u}) =: (\boldsymbol{s}, \boldsymbol{t}, \boldsymbol{u}).$$

From this, it is purely routine to verify that $\boldsymbol{G} = \langle \boldsymbol{s}, \boldsymbol{t}, \boldsymbol{u} \rangle$ indeed satisfies the relations implied by the notation $\{4, 6 : 5 \mid 3\}$ for $\mathcal{Q}$. The more difficult task is to establish the relationship in reverse.

In fact, [83] showed how to recover the original group (see also the discussion in Section 11B); however, it did not treat the problem on an abstract level. Therefore, we only assume that we have involutions $\boldsymbol{s}, \boldsymbol{t}, \boldsymbol{u}$ satisfying

$$(\boldsymbol{st})^4 = (\boldsymbol{tu})^6 = (\boldsymbol{stu})^5 = (\boldsymbol{stut})^3 = \boldsymbol{e}, \quad \boldsymbol{s} \rightleftharpoons \boldsymbol{u},$$

where the relation $\rightleftharpoons$ means 'commutes with'. We begin by defining

$$\boldsymbol{x} := (\boldsymbol{tu})^3 = (\boldsymbol{ut})^3, \quad \boldsymbol{y} := \boldsymbol{tx},$$

and see at once that

$$\boldsymbol{x}^2 = \boldsymbol{y}^2 = \boldsymbol{e}, \quad \boldsymbol{x} \rightleftharpoons \boldsymbol{u},\ \boldsymbol{y}, \quad \boldsymbol{y} = \boldsymbol{xt}$$

(thus $\langle \boldsymbol{x}, \boldsymbol{s}, \boldsymbol{y}, \boldsymbol{u} \rangle$ is already an sggi in the terminology of [99] or Section 2C). Moreover,

$$\boldsymbol{yu} = (\boldsymbol{ut})^2 \implies (\boldsymbol{yu})^3 = \boldsymbol{e}.$$

Next,

$$\boldsymbol{sy} = \boldsymbol{sututu} \sim \boldsymbol{stut} \implies (\boldsymbol{sy})^3 = \boldsymbol{e};$$

as usual, $\sim$ denotes conjugacy. For the final relation, we use (twice)

$$(\boldsymbol{stut})^3 = \boldsymbol{e} \implies \boldsymbol{tutstut} = \boldsymbol{stuts}.$$

Not forgetting that $\boldsymbol{s} \rightleftharpoons \boldsymbol{u}$ and $(\boldsymbol{st})^4 = \boldsymbol{e}$ also, we thus have

$$\begin{aligned}
(\boldsymbol{sx})^3 &= \boldsymbol{sutu} \cdot \boldsymbol{tutstut} \cdot \boldsymbol{utusututut} \\
&= \boldsymbol{sutu} \cdot \boldsymbol{stuts} \cdot \boldsymbol{utstutut} \\
&= \boldsymbol{sutustutst} \cdot \boldsymbol{stuts} \cdot \boldsymbol{ut} \\
&= \boldsymbol{sutustu} \cdot \boldsymbol{sts} \cdot \boldsymbol{utsut} \\
&= (\boldsymbol{sut})^5 = \boldsymbol{e}.
\end{aligned}$$

To draw attention to the relations used, we have indicated appropriate separations of blocks of terms. Observe that the Petrie polygon relation $(\boldsymbol{stu})^5 = \boldsymbol{e}$ ultimately played a crucial rôle.

Since we are really looking at the general case $d \geqslant 5$, we must similarly see how to incorporate the abstract generator $\boldsymbol{v}$ corresponding to the reflexion V of Figure 9B1; for the moment, we can ignore R. So, what we have here, as a subgroup of the original group, is

$$\langle \boldsymbol{t}, \boldsymbol{u}, \boldsymbol{v} \rangle = \langle \boldsymbol{xy}, \boldsymbol{u}, \boldsymbol{v} \rangle \cong [6, 3:4] = [6, 3]/\langle\langle (012)^4 \rangle\rangle,$$

as is easy to see. Indeed, since $\boldsymbol{x} \rightleftharpoons \boldsymbol{y}, \boldsymbol{u}, \boldsymbol{v}$, the group is just $[3, 3] \times \boldsymbol{C}_2 \cong [3, 4]$.

We must thus show that we can recover the original group from $\langle \boldsymbol{t}, \boldsymbol{u}, \boldsymbol{v} \rangle \cong [6, 3]/\langle\langle (012)^4 \rangle\rangle$; it is fairly clear what we have to do. We thus define $\boldsymbol{x}, \boldsymbol{y}$ as before, so that $(\boldsymbol{yu})^3 = \boldsymbol{e}$. Next,

$$\boldsymbol{yv} = \boldsymbol{ututuv} \sim \boldsymbol{uvutut} = \boldsymbol{vuvtut} = (\boldsymbol{vut})^2$$

of period 2, so that $\boldsymbol{y} \rightleftharpoons \boldsymbol{v}$. In other words, $\langle \boldsymbol{y}, \boldsymbol{u}, \boldsymbol{v} \rangle \cong [3, 3]$ in the natural way. Finally, appealing to the Petrie period 4 and using $\boldsymbol{t} \rightleftharpoons \boldsymbol{v}$ and $(\boldsymbol{uv})^3 = \boldsymbol{e}$ several times, we have

$$\begin{aligned}
\boldsymbol{xvx} &= (\boldsymbol{ut})^2 \cdot \boldsymbol{utvut} \cdot (\boldsymbol{ut})^2 = \boldsymbol{utu} \cdot \boldsymbol{vuv} \cdot \boldsymbol{t} \cdot \boldsymbol{uvu} \cdot \boldsymbol{tut} \\
&= \boldsymbol{utvu} \cdot \boldsymbol{tvuv} \cdot \boldsymbol{tut} = (\boldsymbol{utv})^4 \boldsymbol{v} \\
&= \boldsymbol{v},
\end{aligned}$$

so that $\boldsymbol{x} \rightleftharpoons \boldsymbol{v}$. We have separated terms in these expressions, in the hope that the calculations will thereby be clarified.

Summarizing this discussion, we see that we have proved

9C5 Theorem *As an abstract regular polytope,*

$$\{3^{d-1}\}^{\varpi} \cong \{4,6,3^{d-4}\}/\langle\langle (012)^5, (0121)^3, (123)^4 \rangle\rangle$$

for each $d \geqslant 5$.

In Example 3D5 we have, in effect, already found the cosine matrices of two of the three; moreover, $\{3^{d-1}\}^{\varsigma\varpi}$ and $\{3^{d-1}\}^{\varsigma\varpi\varsigma}$, being centrally symmetric, have essentially the same realization domains. We can therefore just read off most of the information from the example. So, confining our attention to the top left corner of the cosine matrix of Proposition 3D7, we have first

9C6 Theorem *The cosine matrix of the polytope* $\{3^{d-1}\}^{\varpi}$ *is*

$$\begin{bmatrix} 1 & 1 & 1 \\ 1 & \frac{d-3}{2(d-1)} & -\frac{2}{d-1} \\ 1 & -\frac{1}{d-1} & \frac{2}{(d-2)(d-1)} \end{bmatrix},$$

with layer and dimension vectors

$$\Lambda = \bigl(1, 2(d-1), \tfrac{1}{2}(d-1)(d-2)\bigr), \quad D = \bigl(1, d, \tfrac{1}{2}(d-2)(d+1)\bigr).$$

Next, we have

9C7 Theorem *The cosine matrix of the polytope* $\{3^{d-1}\}^{\varpi\varsigma}$ *is*

$$\begin{bmatrix} 1 & 1 & 1 & 1 & 1 & 1 \\ 1 & \frac{d-3}{2(d-1)} & -\frac{2}{d-1} & -\frac{2}{d-1} & \frac{d-3}{2(d-1)} & 1 \\ 1 & -\frac{1}{d-1} & \frac{2}{(d-2)(d-1)} & \frac{2}{(d-2)(d-1)} & -\frac{1}{d-1} & 1 \\ 1 & -1 & 1 & -1 & 1 & -1 \\ 1 & -\frac{d-3}{2(d-1)} & -\frac{2}{d-1} & \frac{2}{d-1} & \frac{d-3}{2(d-1)} & -1 \\ 1 & \frac{1}{d-1} & \frac{2}{(d-2)(d-1)} & -\frac{2}{(d-2)(d-1)} & -\frac{1}{d-1} & -1 \end{bmatrix},$$

with layer and dimension vectors

$$\Lambda = \bigl(1, 2(d-1), \tfrac{1}{2}(d-1)(d-2), \tfrac{1}{2}(d-1)(d-2), 2(d-1), 1\bigr),$$
$$D = \bigl(1, d, \tfrac{1}{2}(d-2)(d+1), 1, d, \tfrac{1}{2}(d-2)(d+1)\bigr).$$

Proof. Here, we just remark that Γ_3 is the cosine vector of the digon $\{2\}$ (as a realization of $\{3^{d-1}\}^{\varpi\varsigma}$), and that $\Gamma_{j+3} = \Gamma_3\Gamma_j$ for $j = 4, 5$; compare here Theorem 4A16. ◻

Finally, we repeat Proposition 3D7, with the observation that $\{3^{d-1}\}^{\varsigma\varpi}$ has the vertex-set of the difference body of the d-simplex.

9C8 Theorem *The cosine matrix of the polytopes* $\{3^{d-1}\}^{\varsigma\varpi}$ *and* $\{3^{d-1}\}^{\varsigma\varpi\varsigma}$ *is*

$$\begin{bmatrix} 1 & 1 & 1 & 1 & 1 \\ 1 & \frac{d-3}{2(d-1)} & -\frac{2}{d-1} & \frac{d-3}{2(d-1)} & 1 \\ 1 & -\frac{1}{d-1} & \frac{2}{(d-2)(d-1)} & -\frac{1}{d-1} & 1 \\ 1 & \frac{1}{2} & 0 & -\frac{1}{2} & -1 \\ 1 & -\frac{1}{d-1} & 0 & \frac{1}{d-1} & -1 \end{bmatrix},$$

with layer and dimension vectors

$$\Lambda = (1, 2(d-1), (d-1)(d-2), 2(d-1), 1),$$
$$D = (1, d, \tfrac{1}{2}(d-2)(d+1), d, \tfrac{1}{2}d(d-1)).$$

Relatives of $\mathbf{B}_d$

The polytope that lies at the heart of our next family is not regular; it is the half-d-cube.

9C9 Proposition *For each* $d \geqslant 5$, *there are two regular* $(d-1)$*-polytopes derived from the diagram*

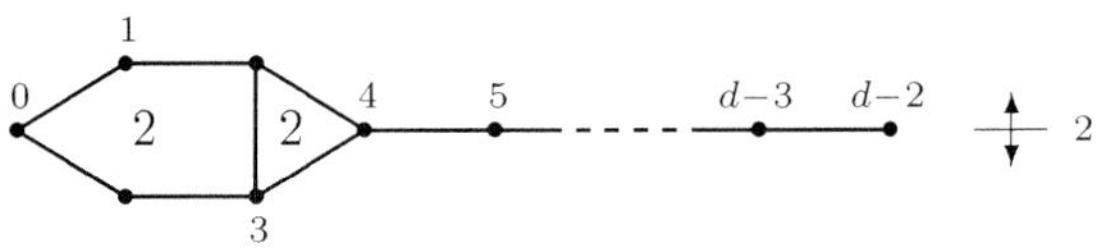

Proof. The initial member of the family arises from a mixing operation which is actually closely related to Petrie contraction. This mixing operation, in fact on the hyperplane reflexion group $\mathbf{B}_d$, can be represented schematically in the spirit of Section 5B by Figure 9C10:

9C10

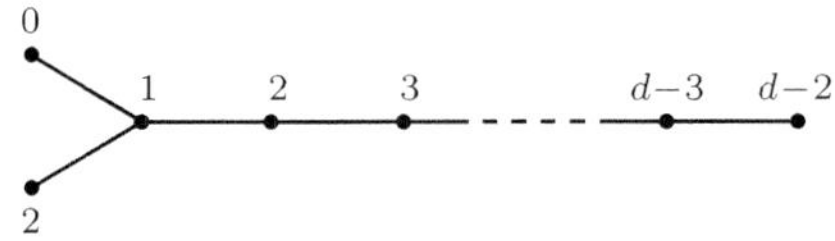

What we are doing here is applying Petrie contraction to the subgroup $\mathbf{A}_{d-1}$ generated by all the reflexions except R_0; thus, when we interpret this picture as resulting from a diagram with a twist, we obtain the diagram of Proposition 9C9.

In terms of elements of the group $\mathbf{B}_d$, we may write the generators of the contracted group as

9C11

$$\begin{aligned} R_0 &:= (1\ \overline{2}), \\ R_1 &:= (2\ 3), \\ R_2 &:= (1\ 2)(3\ 4), \\ R_j &:= (j{+}1\ j{+}2) \quad \text{for } j = 3, \ldots, d-2. \end{aligned}$$

Recall that
$$x(1\,\overline{2}) = (-\xi_2, -\xi_1, \xi_3, \ldots, \xi_d).$$
The generatrix is chosen so that the initial vertex of the resulting polytope $\mathsf{G}_{1,d-3}$ can be taken to be (1^d) (the notation anticipates that of Section 13D); hence, the vertex-set is that of the half-cube $\{3, 3^{d-3,1}\}$, namely,
$$\{(\varepsilon_1, \ldots, \varepsilon_d) \in \{\pm 1\}^d \mid \varepsilon_1 \cdots \varepsilon_d = 1\}.$$
The vertex-figure is clearly $\{3^{d-2}\}^{\varpi}$, whose vertices here are all permutations of $(-1, -1, 1^{d-2})$.

We cannot change the sign of the twist R_2; it is easy to see that the resulting Wythoff space would just be $\{o\}$ (compare also the discussion of Section 9B). However, we can apply ζ to $\mathsf{G}_{1,d-3}$, noting that what happens depends on the parity of d. If d is even, then $\mathsf{G}_{1,d-3}$ is centrally symmetric, and so ${\mathsf{G}_{1,d-3}}^{\zeta}$ has the same 2^{d-1} vertices of the half-cube. On the other hand, if d is odd, then ${\mathsf{G}_{1,d-3}}^{\zeta}$ has all 2^d vertices of the d-cube as its vertex-set. The reader is invited to compare the analogous behaviour of $\{4, 3^{d-2}\}^{\zeta}$, which was treated in Section 8A. □

Since the polytope $\mathsf{G}_{1,d-3}$ is a particular case of a Gosset–Elte polytope of Section 13D, we may denote the corresponding abstract regular polytope by $\mathcal{G}_{1,d-3}$. We then easily deduce the following from Theorem 9C5.

9C12 Theorem *As an abstract regular polytope,*
$$\mathcal{G}_{1,d-3} \cong \{3, 4, 6, 3^{d-5}\}/\langle\!\langle (123)^5, (1232)^3, (234)^4 \rangle\!\rangle$$
for each $d \geqslant 5$, where the last relator is absent if $d = 5$.

9C13 Remark In a similar spirit, we could write $\{3^{d-1}\}^{\varpi} := \mathcal{G}_{0,d-2}$.

We end with a brief comment about realizations. It is not hard to see that the cosine matrix of $\mathcal{G}_{1,d-3}$ can be obtained from that of the d-cube $\{4, 3^{d-2}\}$ by deleting all the odd columns (corresponding to entries $\gamma_1, \gamma_3, \ldots$ of the cosine vectors) and the last $\lfloor \frac{1}{2}d \rfloor$ rows (which now just repeat the first $\lfloor \frac{1}{2}d \rfloor$ in reverse order).

Relatives of $\mathbf{C}_d$

We next have

9C14 Proposition *For each $d \geqslant 5$, there are four regular $(d-1)$-polytopes derived from the diagram*

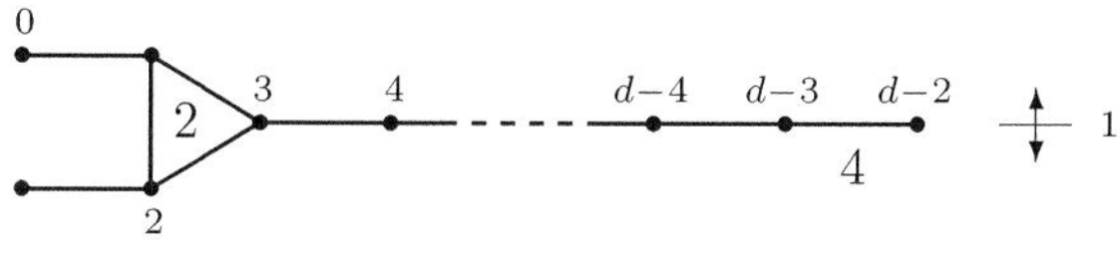

Proof. The diagram now represents the Petrie contractions $\{3^{d-2},4\}^{\varpi}$ and $\{3^{d-2},4\}^{\zeta\varpi}$. Here, the twists are just as for the case of the simplex, except that now both are inner, because the group $\mathbf{C}_d = [3^{d-2},4]$ contains $-I$. When $d=5$, so that the nodes $4,\ldots,d{-}1$ are absent, the mark 4 is transferred to the two slanting edges of the diagram (this already suggests that we cannot allow $d=4$ – we shall see why in Section 11B). The generatrix $(R_0,\ldots,R_{d-2})$ is given by

9C15
$$\begin{aligned} R_0 &:= (2\ 3),\\ R_1 &:= \pm(1\ 2)(3\ 4),\\ R_j &:= (j{+}2\ j{+}3), \quad \text{for } j=2,\ldots,d-3,\\ R_{d-2} &:= \overline{d}. \end{aligned}$$

Recall that $\overline{d}$ just changes the sign of ξ_d.

The two polytopes have the same vertex-set, namely, the mid-points of the edges of the d-staurotope; however, their facets are different, being $\{3^{d-2}\}^{\varpi}$ and $\{3^{d-2}\}^{\zeta\varpi}$ as the sign of R_1 is positive or negative. The symmetry group of both is, of course, $\mathbf{C}_d$. Observe also that we can also apply ζ to each, and we still have the same symmetry group and vertex-set. □

The construction of these polytopes can be carried out on an abstract level, to yield $\{3^{d-2},4\}^{\varpi}$ and $\{3^{d-2},4\}^{\zeta\varpi}$. Since they have the same vertices (even on the abstract level), so far as realizations are concerned we need only look at the former. We therefore show

9C16 Theorem *The common cosine matrix of the polytopes $\{3^{d-2},4\}^{\varpi}$ and $\{3^{d-2},4\}^{\varpi\zeta}$ is*

$$\begin{bmatrix} 1 & 1 & 1 & 1 & 1 & 1\\ 1 & \frac{d-4}{2(d-2)} & -\frac{2}{d-2} & 1 & \frac{d-4}{2(d-2)} & 1\\ 1 & -\frac{1}{d-2} & \frac{2}{(d-2)(d-3)} & 1 & -\frac{1}{d-2} & 1\\ 1 & 0 & 0 & -1 & 0 & 1\\ 1 & \frac{1}{2} & 0 & 0 & -\frac{1}{2} & -1\\ 1 & -\frac{1}{2(d-2)} & 0 & 0 & \frac{1}{2(d-2)} & 1 \end{bmatrix},$$

with layer and dimension vectors

$$\begin{aligned} \Lambda &= (1, 4(d{-}2), 2(d{-}2)(d{-}3), 2, 4(d{-}2), 1),\\ D &= (1, d{-}1, \tfrac{1}{2}d(d{-}3), \tfrac{1}{2}d(d{-}1), d, d(d{-}2)). \end{aligned}$$

Proof. The vertices of the geometric $\mathsf{P}_4 := \{3^{d-2},4\}^{\varpi}$ are the mid-points of the edges of the d-staurotope, and so consist of all $2d(d-1)$ permutations of

$(\pm 1, \pm 1, 0^{d-2})$. Since $\{3^{d-2}, 4\}^{\varpi} \cong \mathsf{P}_4$, we may read off from it the layer vector given above, and cosine vector

$$\Gamma_4 = (1, \tfrac{1}{2}, 0, 0, -\tfrac{1}{2}, -1).$$

Note that layer $\mathcal{L}_3$ consists of the vertices with first two coordinates 0, while layer $\mathcal{L}_4$ contains the two vertices $\pm(1, -1, 0^{d-2})$; clearly, these two layers are distinct.

We next employ the same trick that we used in Example 3D5. If we abolish the minus signs in each vertex, then they coincide in fours at the mid-points of the edges of the $(d-1)$-simplex. Another way of looking at this is that the identifications are induced by the collapse of the d-staurotope $\{3^{d-2}, 4\}$ onto its facet $\{3^{d-2}\}$. We may denote the result of the identification by $\{3^{d-2}, 4\}^{\varpi}/4$. This gives us two pure realizations apart from $\{1\}$, with cosine vectors obtained from the entries Γ_1 and Γ_2 of the matrix of (3D7) on replacing d by $d-1$.

At this stage, we do a count. We have $r = 5$ non-trivial diagonal classes, and so far we have found three pure realizations $\Gamma_1, \Gamma_2, \Gamma_4$ of dimensions $d_1 = d-1$, $d_2 = \frac{1}{2}d(d-3)$ and $d_4 = d$, respectively. Since we need at least one more component of the small simplex realization S (of dimension $d(d-1)$) and one more of the staurotope realization X (also of dimension $d(d-1)$), we see that we must have exactly one of each; moreover, we know their dimensions $d_3 = \frac{1}{2}d(d-1)$ and $d_5 = d(d-2)$ already. We can then use the component equation of Theorem 3C11 to find Γ_3 and Γ_5. □

9C17 Remark We may observe that $\Gamma_2\Gamma_4 = \Gamma_5$, so that $\mathsf{P}_2 \otimes \mathsf{P}_4 = \mathsf{P}_5$. We could also have worked out the realization P_3 directly, since it has dimension $d_3 = \frac{1}{2}d(d-1)$ with twice as many vertices, obviously paired as arising from fours like $(\pm 1, \pm 1, 0^{d-2})$ with opposites identified. In other words, its vertices *must* be those of the staurotope of dimension $\frac{1}{2}d(d-1)$.

Our final family is given by

9C18 Proposition *For each $d \geqslant 4$, there are two regular $(d-1)$-polytopes derived from the diagram*

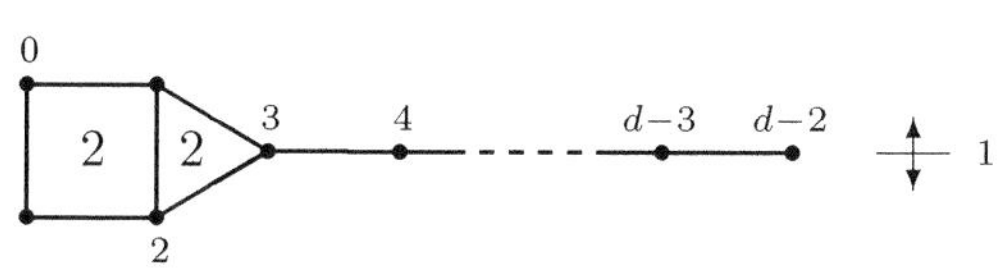

Proof. The diagram now corresponds to the Petrie contraction $\{4, 3^{d-2}\}^{\varpi}$ of the cube. The initial generatrix is given by

9C19
$$\begin{aligned} &R_0 \colon (1\ 2),\\ &R_1 \colon \overline{1}\,(2\ 3),\\ &R_j \colon (j{+}1\ \ j{+}2) \quad \text{for } j = 2, \ldots, d-2. \end{aligned}$$

As usual, we can apply ζ to $\{4, 3^{d-2}\}^{\varpi}$. However, Theorem 9C1 (with the discussion of Section 9B) implies that we cannot change the sign of the twist

R_1, as we were able to do in other cases. The theorem would give a normal $e_1 = (1, 0^{d-1})$ to S_0 as the initial vertex, leading to only $2d$ vertices in all, rather than the $2^d \cdot d!/2 \cdot 2^{d-2} \cdot (d-2)! = 2d(d-1)$ expected from the index of the group of the vertex-figure (this illustrates Remark 9C2). To bolster this argument, we see that the 2-face is a doubly-covered triangle. Moreover, of course, the example cannot be rescued by a further application of $\boldsymbol{\zeta}$ (changing the sign of the new generator R_0); one obtains 3-faces which should be of type $\{6, 6\}$, with 8 vertices and 4 hexagonal faces, an obvious contradiction. Again as we would expect, we cannot obtain genuine polytopes by applying $\boldsymbol{\pi}$ to these non-polytopes (when $d = 5$). □

In analogy to Theorem 9C5 (compare also Proposition 5B21), we can show

9C20 Theorem *As an abstract regular polytope,*

$$\{4, 3^{d-2}\}^{\varpi} \cong \{6, 6, 3^{d-4}\}/\langle\langle(0121)^3, (012121)^4, (123)^4\rangle\rangle$$

for each $d \geqslant 5$.

9C21 Remark Theorem 6D5 shows that the 3-face of $\{4, 3^{d-2}\}^{\varpi}$ actually has fine Schläfli symbol $\{6, \frac{6}{2,3} \mid 3\}$, so that mention of the 3-hole $\{4\}$ is unnecessary.

We shall not describe the realization domain of $\{4, 3^{d-2}\}^{\varpi}$, because it is far too large.

Completeness of the Enumeration

In talking about completeness, we must bear in mind that we are only looking at the general situation here. In other words, forward references to Chapters 11 and 13 will be needed, to make sure that we have caught all possible vertex-figures. However, once the dimension is high enough, the very restricted list of available groups makes the task relatively easy. Nevertheless, it is more sensible to postpone this discussion until we have filled in the low-dimensional gaps.

Naturally, the same considerations will also apply to apeirotopes of nearly full rank.

9D Families of Apeirotopes

Although there is nothing very important to say about them, for the sake of completeness we must list – in principle at least – all the instances obtainable by the free abelian apeirotope construction (recall that we have dealt with the blended apeirotopes in Section 9A). A potential vertex-figure is one of two kinds: first, a blend of a polytope of full rank with a digon, or, second, a pure polytope of nearly full rank. In addition, of course, the polytopes must be rational, but our restriction to infinite families automatically ensures this.

For the first type, the vertex-figure is either $\mathsf{Q} \# \{2\}$ or – using (5D11) – $\mathsf{Q} \diamond \{2\} = \mathsf{Q}^{\varsigma} \# \{2\}$, with Q one of the three polytopes $\{3^{d-2}\}$, $\{3^{d-3}, 4\}$

or $\{4, 3^{d-3}\}$. Using Proposition 9A16 and Remark 9A17, we obtain the six apeirotopes

9D1
$$(\mathsf{Q} \# \{2\})^{\alpha} = \mathsf{Q}^{\alpha} \# \{\infty\},$$
$$(\mathsf{Q} \diamond \{2\})^{\alpha} = \mathsf{Q}^{\varsigma\alpha} \# \{\infty\},$$

with these three Q.

For the second type, we just have all the apeirotopes of the form Q^{α}, with Q one of the regular polytopes enumerated in the previous Section 9C.

9D2 Remark In Section 9A we have already taken account of the blends of apeirotopes of full rank with digons $\{2\}$ or apeirogons $\{\infty\}$. However, it is worth bearing in mind that the general member of the family apeir Q is of the form $\mathsf{Q}^{\alpha} \# \{2\}$; when Q is rational of full rank, then such a general member is of nearly full rank.

In each case, just as we did for the regular polytopes, we shall base a family on a suitable diagram that admits a twist. We take a regular d-apeirotope P to have initial vertex o, and its broad vertex-figure $\mathsf{Q} := \mathsf{P}^{v}$ to have initial vertex v, say. Thus the generatrix $(R_0, \ldots, R_{d-1})$ of P will have all R_j except R_0 a linear reflexion. We can write R_0 in the form

9D3
$$xR_0 = x\Phi + v,$$

with Φ a linear reflexion such that $v\Phi = -v$; as a mirror, $\Phi = R_0 - R_0$, the linear subspace parallel to R_0. Since v can be scaled freely, using the notation introduced before (9B4) we can write $R_0 = \Phi^{\bullet}$ as a shorthand for (9D3).

9D4 Remark We shall usually be able to arrange things so that the vertices of our apeirotopes have integer cartesian coordinates. In this case, we take the term v in (9D3) to be as small an integer vector as possible.

It is also convenient to introduce the operation ς by

9D5
$$\varsigma\colon (R_0, \ldots, R_{d-1}, \Phi) \mapsto ((-\Phi)R_0, R_1, \ldots, R_{d-1}) =: (S_0, \ldots, S_{d-1});$$

in other words, $\mathsf{P}^{\varsigma} = \mathsf{Q}^{\alpha}$. In general, ς may only tie in Q^{α} with a wider family; however, we shall see that occasionally it has further applications.

9D6 Remark It may be the case that $\varsigma = \boldsymbol{\tau}$ $(= \boldsymbol{\kappa}_{02})$, but the former will always exist (as an operation), whereas the latter may not. If the vertex-figure Q is pure, then we do have $\varsigma = \boldsymbol{\tau}$. A core question is whether the reflexion in

$$K_2 = R_2 \cap \cdots \cap R_{d-1}$$

is in $\mathbf{G}^{v} = \mathbf{G}(\mathsf{Q})$ or not; in the former case, P and Q^{α} have the same symmetry group.

We make one further comment before embarking on the classification. The operation κ does not always yield a new apeirotope. For instance, if $\mathsf{P} = \mathsf{Q}^{\kappa}$ with Q a finite polytope, then κ just returns to Q, but it also fails in other cases. However, what is important is the new 3-face, and this will be considered in more depth in Chapter 12. Therefore, here we shall confine ourselves to saying when κ is applicable or otherwise.

Our classification follows the lines of that of the polytopes. We recall that, apart from applications of $\boldsymbol{\alpha}$, there are two basic ways of constructing regular apeirotopes of nearly free rank from regular polytopes or apeirotopes of full rank. First, the discussion of Section 5D shows that, in general, eversion κ applied to a (finite) crystallographic regular polytope results in a regular apeirotope; here, we appeal to the classifications of Sections 7B and 8A. Second, we have Petrie contraction ϖ of a regular apeirotope of Sections 7H and 8B; in the present context, ϖ will always be reversible, and so preserves the group. Both operations preserve dimension, but increase the rank deficit by 1. As with the polytopes, we discuss these families according to the related groups.

Relatives of $\mathbf{P}_{d+1}$

Before we describe the two families of apeirotopes with symmetry groups related to $\mathbf{P}_{d+1}$, we need to make a preliminary comment. We work in the subspace $\mathbb{L}^d$ of $\mathbb{E}^{d+1}$; particularly relevant is the lattice $\mathbb{Z}^{d+1} \cap \mathbb{L}^d$, which is isomorphic to the subgroup of translations in $\mathbf{P}_{d+1}$. The point-group of its symmetry group is $\mathbf{A}_d \times \mathbf{Z}_2$, with $\mathbf{Z}_2 = \{\pm I\}$. As a consequence, we can take a compatible reflexion to be of the form $x \mapsto x\Phi + v$, with $v \in \mathbb{Z}^{d+1} \cap \mathbb{L}^d$ and $\Phi \in \mathbf{A}_d \times \mathbf{Z}_2$ an involution, and hence a product of disjoint transpositions, possibly multiplied by $-I$. This fact places a severe restriction on the putative generatrix of a relevant apeirotope of nearly full rank.

For the first of our two families we have

9D7 Proposition *For each $d \geqslant 3$, there are two regular d-apeirotopes derived from the diagram*

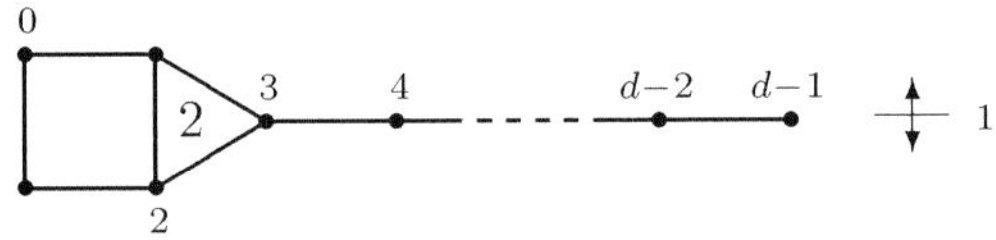

Proof. First, recall our earlier comment about the missing tail in the lowest dimension. The two apeirotopes here are $\{3^{d-1}\}^{\kappa}$ and $\{3^{d-1}\}^{\zeta\kappa}$ in $\mathbb{L}^d$. We can verify the claim by specifying the generatrices $(R_0, \ldots, R_{d-1})$; we write $u_k := (1^k) \in \mathbb{E}^k$ for the vector with all coordinates 1. We have

9D8
$$R_0 := \begin{cases} (0\ 1)^{\bullet}, & \text{for } \{3^{d-1}\}^{\kappa}, \\ -(0\ 1)^{\bullet}, & \text{for } \{3^{d-1}\}^{\zeta\kappa}, \end{cases}$$
$$R_1 := -(1\ 2),$$
$$R_j := (j\ j{+}1) \quad \text{for } j = 2, \ldots, d-1.$$

We employ the usual conventions, together with (9B4). In the present case, which will serve to illustrate the general principle, we can take

$$xR_0 = \begin{cases} x(0\ 1) + e_0 - e_1, & \text{for } \{3^{d-1}\}^{\kappa}, \\ -x(0\ 1) + (d+1)(e_0 + e_1) - 2u_{d+1}, & \text{for } \{3^{d-1}\}^{\varsigma\kappa}; \end{cases}$$

the translational component has been chosen to be an integer vector whose entries are (nearly) as small as possible (for the latter apeirotope, when d is odd this vector can be halved). Note that R_1 is indeed the composition of the transposition $(1\ 2)$ with the reflexion in the initial vertex of the simplex $\{3^{d-1}\}$, when this is translated to o. The 3-face of the first apeirotope is the Petrie–Coxeter apeirohedron $\{6, \frac{6}{1,3} \mid 3\} \cong \{6, 6 \mid 3\}$, which we shall discuss in more detail in the next Chapter 10.

The vertex-figure is $\mathsf{Q} = \{3^{d-2}\} \diamond \{2\}$, and the operation ς induced by $\Phi = (0\ 1)$ leads to the apeirotope Q^{α}. We saw in the discussion of Section 9B that we cannot apply $\boldsymbol{\kappa}$ to change the sign of R_1 (see also Chapter 10). Moreover,

$$K_2 = \{x \in \mathbb{L}^d \mid \xi_2 = \cdots = \xi_d\},$$

and the reflexion in K_2 is not of the form described above. It follows that no further members of the family can be found. □

9D9 Remark In several cases, such as that of Proposition 9D7, we have apeirotopes with proper faces that are already infinite. We shall not comment further on this common phenomenon.

The other family is given by

9D10 Proposition *For each $d \geqslant 4$, there are four regular d-apeirotopes derived from the diagram*

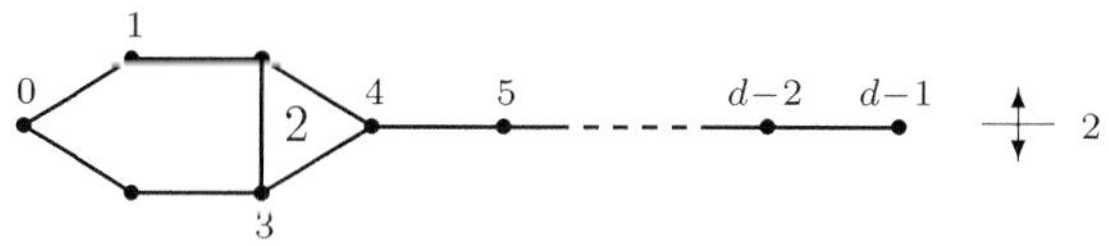

Proof. Here, the generatrix of the initial apeirotope P (say) is given by

$$\begin{aligned} R_0 &:= (0\ 1)^{\bullet}, \\ R_1 &:= (1\ 2), \\ R_2 &:= -(0\ 1)(2\ 3), \\ R_j &:= (j\ j{+}1) \quad \text{for } j = 3, \ldots, d-1. \end{aligned} \tag{9D11}$$

Note that the twist $R_2 = -(0\ 1)(2\ 3)$ is outer but improper. The vertex-figure is $\mathsf{P}^v = \{3^{d-2}\}^{\varpi}$; it is not hard to see that the vertex-set is $\operatorname{vert}\mathsf{P} = \mathbb{L}^d \cap \mathbb{Z}^{d+1}$.

We next observe that the group $\mathbf{G}^v = \mathbf{G}_0$ of the vertex-figure is $\mathbf{A}_d \times \mathbf{Z}_2$. We see this easily from calculations like $S := (R_2R_3)^3 = -(0\ 1)$, $(SR_1)^3 = -I$, and so on. Hence, we can apply $\boldsymbol{\kappa}$ by changing the sign of R_1; observe that $-R_1 \in \mathbf{G}^v$, and so the group remains the same. We also have

$$K_2 = \{x \in \mathbb{L}^d \mid \xi_0 + \xi_1 = \xi_2 = \cdots = \xi_d = 0\},$$

and the reflexion in K_2 inducing the operation ς is $\Phi = -(0\ 1) = (R_2R_3)^3 \in \mathbf{G}^v$, so that in this case $\mathsf{Q}^\alpha = \mathsf{P}^\varsigma$ has the same symmetry group. Further, $-I \in \mathbf{G}^v$ as we have seen, and this induces $\boldsymbol{\kappa}$. The operations $\boldsymbol{\kappa}$ and ς commute, and we thus obtain a fourth apeirotope $\mathsf{P}^{\boldsymbol{\kappa}\varsigma} = \mathsf{Q}^{\varsigma\alpha}$ in the family, again with the same group.

Finally, the discussion of $\mathbf{P}_5$ in Section 9B shows that we cannot change the sign of the twist R_2; our enumeration in this case is thus complete. □

Relatives of $\mathbf{R}_{d+1}$

We begin here with a family that we have encountered already.

9D12 Proposition *For each $d \geqslant 3$, there are two regular d-apeirotopes derived from the diagram*

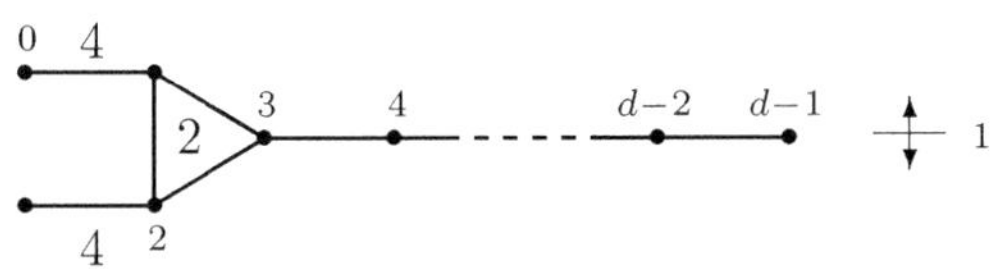

Proof. We described the facet of the regular apeirotope $\{4, 3^{d-2}, 4\}^{\boldsymbol{\kappa}}$ for $d \geqslant 4$ in Section 8B, noting that it is full-dimensional, and hence (in our present terms) of nearly full rank. In fact, this facet must be $\mathsf{P} := \{4, 3^{d-2}\}^{\boldsymbol{\kappa}}$. It is derivable by an improper outer twist of the diagram. We have labelled the triangle rather than one of its edges; either of the two slanting edges can carry the label $\frac{3}{2}$, but the twist changes which is so labelled.

The operation ς leads to Q^α, with $\mathsf{Q} = \{3^{d-2}\} \diamond \{2\}$ the vertex-figure of P; observe that Q has the vertices of the staurotope $\{3^{d-2}, 4\}$. □

Recalling how the apeirotope of Proposition 9D12 was derived from the cubic tiling $\{4, 3^{d-2}, 4\}$, it is easy to write down its generatrix, and check that it does indeed correspond to the diagram (with twist). Thus we have

9D13
$$\begin{aligned} R_0 &:= \overline{1}^{\bullet}, \\ R_1 &:= -(1\ 2), \\ R_j &:= (j\ j{+}1) \quad \text{for } j = 2, \ldots, d-1. \end{aligned}$$

Observe that we can extend the diagram to the right by a further branch labelled 4 (and add a further generator R_d changing the sign of ξ_d) to (9D13), to give the analogous diagram corresponding to $\{4, 3^{d-2}, 4\}^{\boldsymbol{\kappa}}$ itself (of course, this is of full rank, and was treated in Section 8B).

Once we notice that $(R_1R_2)^3 = -I$, it is clear that we obtain $\mathbb{Z}^d$ as the vertex-set of the apeirotope; of course, we really already knew this, but it is always useful to work with any given set of geometric group generators rather than appeal to extraneous information.

The other relatives of $\mathbf{R}_{d+1}$ are given by

9D14 Proposition *For each $d \geqslant 4$, there are four regular d-apeirotopes derived from the diagram*

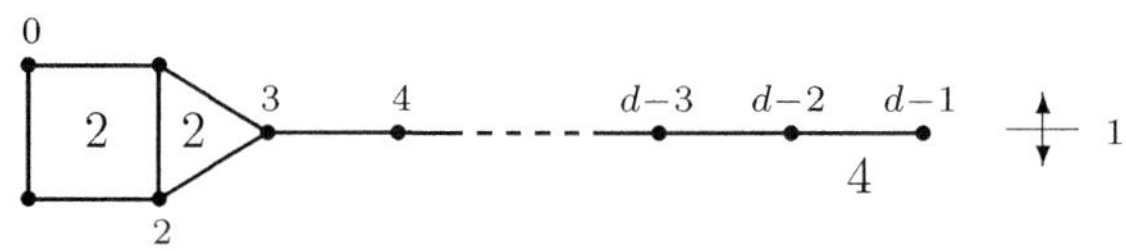

Proof. When $d = 4$, the mark 4 is transferred to the two sloping branches of the diagram, and the rest of the tail disappears. The family starts with the Petrie contraction $\mathsf{P} := \{4, 3^{d-2}, 4\}^{\varpi}$ of the cubic tiling, which corresponds to the diagram. Of course, this is just the diagram of Proposition 9C18, extended to the right by a single node and branch marked 4. However, we choose the generators of the resulting symmetry group (which is still $\mathbf{R}_{d+1}$) to make o the initial vertex, so that they are a little different from those in (9C19):

9D15
$$\begin{aligned} R_0 &:= (1\ 2)^{\bullet}, \\ R_1 &:= \bar{1}\,(2\ 3), \\ R_j &:= (j{+}1\ j{+}2) \quad \text{for } j = 2, \ldots, d-2, \\ R_{d-1} &:= \bar{d}. \end{aligned}$$

Observe that the vertex-figure is $\{3^{d-3}, 4\} \# \{2\}$.

As a reflexion, $K_2 := R_2 \cap \cdots \cap R_{d-1} = -\bar{1}\,\bar{2}$ belongs to the group $\mathbf{G}_0$ of the vertex-figure. Hence the operation $\boldsymbol{\tau} = \boldsymbol{\kappa}_{02}$ leads to another regular apeirotope with the same symmetry group; the new R_0 is therefore

$$x \mapsto (1 - \xi_2, 1 - \xi_1, -\xi_3, \ldots, -\xi_d).$$

The type of the new 2-face is $\{\frac{3}{0,1}\}$.

The operation ς is induced by the reflexion $\Phi := -(1\ 2)$ in the line through the initial vertex o and the other vertex $v := e_1 - e_2$ of the initial edge. From its definition, we have $\boldsymbol{\alpha} = \boldsymbol{\varpi}\varsigma$. Now $\Phi \notin \mathbf{G}_0$, so that the new group is not the same. However, Φ commutes with $S_2, \ldots, S_{d-1}$, and so is compatible with $\mathbf{G}$. Thus ς commutes with $\boldsymbol{\tau}$, and we obtain a fourth apeirotope in the extended family, namely,

$$\{4, 3^{d-2}, 4\}^{\varpi\tau} = (\{3^{d-2}, 4\} \# \{2\})^{\alpha\varsigma} = (\{3^{d-2}, 4\}^{\alpha} \# \{\infty\})^{\varsigma};$$

here, we have appealed to Proposition 9A16. The 2-face of this is of type $\{\frac{4}{0,1}\}$.

Finally, we observe that, just as with the Petrie contraction of the cube, we cannot apply $\boldsymbol{\kappa}$ to change the sign of the twist R_1. Thus the family extends no further. □

9D16 Remark There is an alternative generatrix, directly derived from that of the cubic tiling, in which R_0 and R_1 are replaced by

$$\begin{aligned} &S_0\colon (1\ 2), \\ &S_1\colon x \mapsto (2 - \xi_1, \xi_3, \xi_2, \xi_4, \ldots, \xi_d); \end{aligned}$$

we set $S_j := R_j$ for $j \geqslant 2$.

This form of S_1 avoids fractions, because we know that the vertices are the mid-points of the edges of the cubic tiling. Thus we have taken the vertex-set of the tiling $\{4, 3^{d-2}, 4\}$ to be $2\mathbb{Z}^d$, so that the original edge-length is 2. With this generatrix, the vertex-set of the apeirotope $\{4, 3^{d-1}, 4\}^{\varpi}$ is

$$\{(\nu_0, \ldots, \nu_d) \in \mathbb{Z}^d \mid \text{just one of } \nu_1, \ldots, \nu_d \text{ odd}\}.$$

There are further families which exhibit the same phenomenon of having the diagram of an infinite group as a proper subdiagram.

9D17 Proposition *For each $d \geqslant 4$, there are four regular d-apeirotopes derived from the diagram*

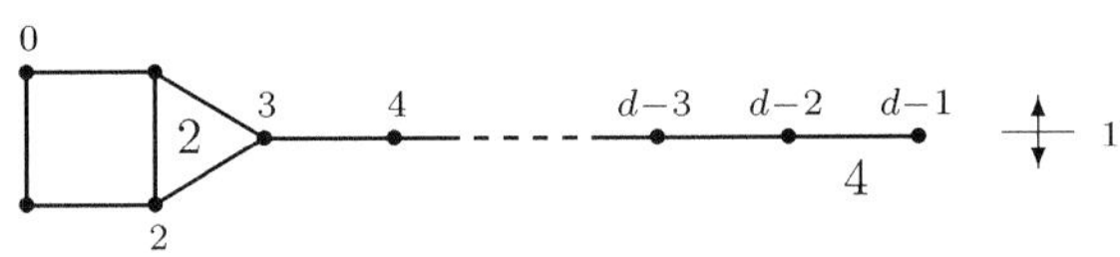

Proof. As in the previous case, for $d = 4$ the mark 4 is attached to the two sloping sides of the triangle in the diagram (we shall discuss this case separately in Section 12B). We have seen the facets of these apeirotopes in Proposition 9D7; the present diagram is that for $\{3^{d-2}, 4\}^{\kappa}$. It is to be hoped that the superficial similarity between the diagrams of Propositions 9D14 and 9D17 will not mislead the reader.

Once again, we can verify everything by listing the generating reflexions. The generatrix $(R_0, \ldots, R_{d-1})$ of $\mathsf{P} = \{3^{d-2}, 4\}^{\kappa}$ is given by

9D18

$$\begin{aligned} R_0 &:= (1\ 2)^{\bullet}, \\ R_1 &:= -(2\ 3), \\ R_j &:= (j{+}1\ j{+}2) \quad \text{for } j = 2, \ldots, d{-}2, \\ R_{d-1} &:= \overline{d}, \end{aligned}$$

with the usual conventions. As before, we have translated so that the initial vertex is o.

We see that

$$\begin{aligned} K_2 &= R_2 \cap \cdots \cap R_{d-1} \\ &= \{x \in \mathbb{E}^d \mid \xi_3 = \cdots = \xi_d = 0\} \\ &= \operatorname{lin}\{e_1, e_2\} = -\overline{1}\,\overline{2} \end{aligned}$$

as a reflexion; observe that $-\overline{1}\,\overline{2} \in \mathbf{G}_0$. Thus $\boldsymbol{\tau}$ is different from ς, given by $\Phi = -(1\ 2)$. As a consequence, P gives rise to three other apeirotopes with the same symmetry group: P^{τ}, $\mathsf{P}^{\varsigma} = \mathsf{Q}^{\alpha}$ (with $\mathsf{Q} = \{3^{d-3}, 4\} \diamond \{2\}$ the vertex-figure of P) and $\mathsf{P}^{\tau\varsigma} = \mathsf{Q}^{\alpha\tau}$.

Once more, $\boldsymbol{\kappa}$ is not applicable; since we cannot change the sign of the twist R_1, there are no more apeirotopes in this family. □

More directly, we can deduce

9D19 Proposition *The apeirotopes of Propositions 9D14 and 9D17 have the same symmetry group* $\mathbf{R}_{d+1}$ *and the same vertex-sets.*

Proof. By construction, the first apeirotope $\{4, 3^{d-2}, 4\}^{\varpi}$ has symmetry group $[4, 3^{d-2}, 4] = \mathbf{R}_{d+1}$. The two generatrices of (9D15) and (9D18) differ only in R_1. Conjugating R_{d-1} in either successively by $R_{d-2}, \ldots, R_1$ shows that each of $\overline{d}, \overline{d-1}, \ldots, \overline{2}$ is in the relevant group. We then observe that $(R_1R_2)^3 = \overline{1}$ or $-I$, respectively, and since $-\overline{2} \cdots \overline{d} = \overline{1}$, it follows that the two groups coincide.

In fact, we have actually shown that the vertex-figures $\{3^{d-3}\} \# \{2\}$ and $\{3^{d-3}\} \diamond \{2\}$ have the same symmetry group (and the same vertices $\pm e_1 \pm e_j$ with $j = 2, \ldots, d$). Since the reflexions R_0 are the same in each group, this shows that the vertex-sets of the two apeirotopes also coincide. □

We describe P in more detail, in particular so as to distinguish it from the next apeirotopes. We say that a vertex $v \in \operatorname{vert} \mathsf{P}$ is of type k if the set of adjacent vertices is $v + V_k$, with $V_k := \{\pm e_j \pm e_i \mid i \neq k\}$; thus the initial vertex o is of type 1. Moreover, how a vertex of one type is transformed under a symmetry to another depends only on the corresponding element of the point-group $\mathbf{C}_d$. In particular, a reflexion $(j\ k)$ or $(j\ \overline{k})$ takes type k into type j; reflexions $(i\ j)$ or $(i\ \overline{j})$ with $i, j \neq k$ leave the type unchanged.

Since we have the (point-)reflexion in one vertex of $\{3^{d-2}, 4\}$ in the group, it must contain the reflexion in every vertex. As one consequence, the lattice of translational symmetries contains all $2(e_1 \pm e_j)$ for $j = 2, \ldots, d$, and hence $2\mathbb{Z}^d_e$, with $\mathbb{Z}^d_e := \{z = (\zeta_1, \ldots, \zeta_d) \in \mathbb{Z}^d \mid \zeta_1 + \cdots + \zeta_d \text{ even}\}$ the *even* sublattice of $\mathbb{Z}^d$. However, the vertex $e_1 + e_2 = (e_1 - e_2)\overline{2}$ of the vertex-figure P^{v} is of type 2, so that $2e_1 = (e_1 + e_2)R_0$ is a vertex that is again of type 1. We deduce that $2\mathbb{Z}^d \subset \operatorname{vert} \mathsf{P}$ consists of vertices of type 1, and therefore that each vector in $2\mathbb{Z}^d$ is a translational symmetry of P.

From this follows a complete description of $\operatorname{vert} \mathsf{P}$. Since each point $v \in 2\mathbb{Z}^d$ is a vertex of type 1, there are no vertices of type $v \pm e_j \pm e_k$ with $2 \leqslant j < k \leqslant d$. Therefore, if $\Lambda := \mathbb{Z}^d_e \cap \{(\zeta_1, \ldots, \zeta_d) \in \mathbb{Z}^d \mid \zeta_1 = 0\}$, then we have

9D20 $$\operatorname{vert} \mathsf{P} = \mathbb{Z}^d_e \setminus \big(2\mathbb{Z}^d + \Lambda\big).$$

Relatives of $\mathbf{S}_{d+1}$

In an exactly similar way, we can extend the diagram of Proposition 9C9 to the right by a single node and branch marked 4.

9D21 Proposition *For each $d \geqslant 5$, eight regular d-apeirotopes can be derived from the diagram*

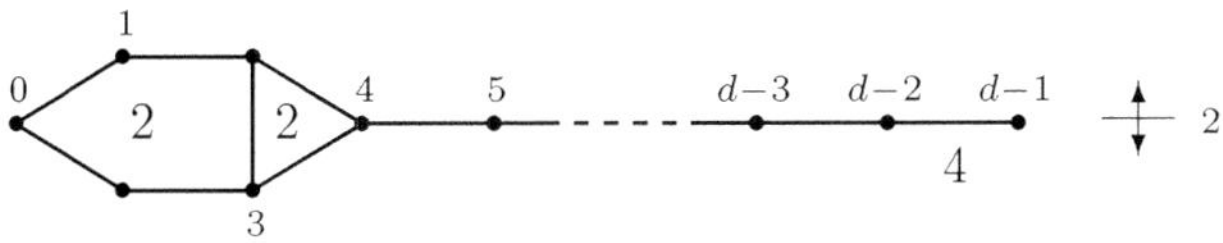

Proof. The symmetry group here is clearly $\mathbf{S}_{d+1}$; indeed, it is derived from it by an invertible mixing operation just like that depicted in Figure 9C10. As before, the mark 4 is attached to the sloping branches of the diagram when $d = 5$. We may write the generatrix $(R_0, \ldots, R_{d-1})$ as

9D22
$$\begin{aligned} R_0 &:= (1\ \overline{2})^{\bullet}, \\ R_1 &:= (2\ 3), \\ R_2 &:= (1\ 2)(3\ 4), \\ R_j &:= (j{+}1\ j{+}2) \quad \text{for } j = 3, \ldots, d-2, \\ R_{d-1} &:= \overline{d}. \end{aligned}$$

We first note that, just as with $\mathsf{G}_{1,d-3}$, we cannot change the sign of the twist R_2. Observe that the vertex-figure of the basic apeirotope P (say) is $\mathsf{Q} = \{3^{d-2}, 4\}^{\varpi}$.

In exactly the same way as we have just seen, there are two commuting operations that can be applied to this apeirotope. First,

$$\begin{aligned} K_2 := R_2 \cap \cdots \cap R_{d-1} &= \{x \in \mathbb{E}^d \mid \xi_1 - \xi_2 = \xi_3 = \cdots = \xi_d = 0\} \\ &= \operatorname{lin}\{e_1 + e_2\}. \end{aligned}$$

As a reflexion, $K_2 = -(1\ \overline{2})$, and this leads in the obvious way to the operation $\boldsymbol{\tau}$. However, $\Phi := -\overline{1}\,\overline{2}$ also commutes with $R_2, \ldots, R_{d-1}$, giving the different operation $\boldsymbol{\varsigma}$ that leads to Q^{α}. (Note that, as reflexions, Φ and K_2 have swapped their rôles from the previous example.) Of course, we can also apply both $\boldsymbol{\tau}$ and $\boldsymbol{\varsigma}$. So far, then we have found four apeirotopes.

What is more, the operation $\boldsymbol{\kappa}$ can be applied to each of these four, giving a total of eight in all. □

In the same way, we can extend the diagram of Figure 9D10 to the right by a single node and branch marked 4.

9D23 Proposition *For each $d \geqslant 5$, eight regular d-apeirotopes can be derived from the diagram*

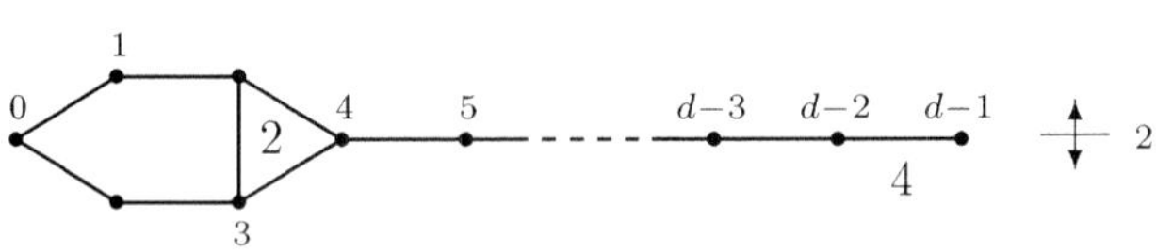

Proof. The case $d = 5$ is as in Proposition 9D21. The symmetry group here is again $\mathbf{S}_{d+1}$, although this may not be immediately obvious; we shall show this in Proposition 9D25 below. We may write the generatrix $(R_0, \ldots, R_{d-1})$ as

9D24
$$\begin{aligned} R_0 &:= (1\ 2)^{\bullet}, \\ R_1 &:= (2\ 3), \\ R_2 &:= -(1\ 2)(3\ 4), \\ R_j &:= (j{+}1\ j{+}2) \quad \text{for } j = 3, \ldots, d-2, \\ R_{d-1} &:= \overline{d}. \end{aligned}$$

We first note that, just as in Proposition 9D10, we cannot change the sign of the twist R_2 (see the notes at the end of the section). Observe that the vertex-figure of the basic apeirotope P (say) is $\mathsf{Q} = \{3^{d-2}, 4\}^{\varsigma\varpi}$, with the full symmetry group $[3^{d-2}, 4]$.

The generatrix seems to make the group look like that of (9D22). However, this is deceptive – we know from its construction that it is different, and we shall see this confirmed below.

In exactly the same way as we have just seen, there are two commuting operations that can be applied to this apeirotope. First,

$$K_2 := R_2 \cap \cdots \cap R_{d-1} = \{x \in \mathbb{E}^d \mid \xi_1 + \xi_2 = \xi_3 = \cdots = \xi_d = 0\}.$$

As a reflexion, $K_2 = -(1\ \overline{2})$, and this leads in the obvious way to the operation $\boldsymbol{\tau}$. However, $\Phi := -(1\ 2)$ also commutes with $R_2, \ldots, R_{d-1}$, giving the different operation ς that leads to Q^{α}. (Note that, as reflexions, Φ and K_2 have swapped their rôles from the previous Proposition 9D21.) Of course, we can also apply both $\boldsymbol{\tau}$ and ς, noting that their composition is $x \mapsto -x\,\overline{1}\,\overline{2}$. So far, then, we have found four apeirotopes.

What is more, the operation $\boldsymbol{\kappa}$ can be applied to each of these four, giving a total of eight in all. □

Let us say a little more about the last two apeirotopes.

9D25 Proposition *The apeirotopes of Propositions 9D21 and 9D23 have the same vertex-set* $\mathbb{Z}_e^d$ *and same symmetry group* $\mathbf{S}_{d+1}$.

Proof. We first treat the groups, observing that their generatrices only differ in R_0 and R_2. In either of them, successively conjugating R_{d-1} by $R_{d-2}, \ldots, R_2, R_1$ and then R_2 again yields all the reflexions $\overline{k}$ in the coordinate hyperplanes, and hence also the central inversion $-I$. Next, multiplying by $-I$ interchanges the twists R_2. Finally, conjugating one of the reflexions R_0 by $\overline{2}$ gives the other.

That both apeirotopes have vertex-set $\mathbb{Z}_e^d$ is now an easy consequence. □

9D26 Remark Proposition 9D25 shows that the mixing operation that gives the generatrix (9D24) from the standard generators of $\mathbf{S}_{d+1}$ is a complicated one. The two apeirotopes of Propositions 9D21 and 9D23 are actually not very closely related.

Notes to Section 9D

1. Strictly speaking, we can change the sign of the twist in (9D24), but then we have to adopt a new R_0 as well.
2. It was mistakenly asserted in [86] that the common groups of Propositions 9D14 and 9D17 and Propositions 9D21 and 9D23 were identical in pairs.

10

Three-Dimensional Apeirohedra

As an illustration of realization theory, in Section 4G we described the realization domain of each finite regular polygon $\{p\}$, and commented on that of $\{\infty\}$. For rank 3, in Chapter 7 we have also classified the regular polyhedra and apeirohedra of full rank. Thus the first non-trivial cases of nearly full rank, that is, corank 1, are apeirohedra in $\mathbb{E}^3$ or (finite) polyhedra in $\mathbb{E}^4$. We shall treat the first case in this chapter, and the second in the next.

We next remark that we have dealt with the blended apeirohedra in $\mathbb{E}^3$ in Section 9A. The core of the chapter is therefore the classification of the twelve pure 3-dimensional regular apeirohedra in Section 10A; we shall expand on the treatment of [99, Chapter 7], which closely followed [98], by displaying new relationships among these apeirohedra. Certain of these relationships are used in Section 10B to describe the automorphism groups of the apeirohedra (as abstract polytopes). In Section 10C we show that the fine Schläfli symbols for nine of the twelve apeirohedra are rigid; the exceptions are the three with finite skew faces.

10A The Classification

We begin the section by establishing the basic classification of the apeirohedra, and then move on to describing the relationships among them. We first need a subsidiary result.

10A1 Lemma *The possible mirror vectors of the* 3*-dimensional discrete pure apeirohedra are* $(2,1,2)$, $(1,1,2)$, $(1,2,1)$ *and* $(1,1,1)$.

Proof. Let P be a pure 3-dimensional apeirohedron in $\mathbb{E}^3$, whose generatrix is (R_0, R_1, R_2). Thus R_0, R_1 and R_2 are involutory isometries of $\mathbb{E}^3$ such that R_0 and R_2 commute, while R_1 does not commute with R_0 or R_2. As usual, we identify a reflexion with its mirror; our classification depends on determining the possible mirror vectors $(\dim R_0, \dim R_1, \dim R_2)$.

We first show that each of R_0, R_1 and R_2 must be a line or a plane; in other words, generating reflexions in points are excluded. Without loss of generality, we may take the initial vertex of P to be o. We then have $o \in R_1 \cap R_2$, and so this latter intersection must be non-empty and strictly contained in both. Hence $\dim R_j \geqslant 1$ for $j = 1, 2$. For these j, we write $S_j = R_j$ or $-R_j$ as R_j is a plane or line (recall that $-R_j = R_j^{\perp}$ is the orthogonal complement of R_j). We are thus employing the same kind of trick as in Section 8C, and we shall see it used again for Theorem 10A3. Let L be the plane through o perpendicular to S_1 and S_2. Now, $o \notin R_0$. If $\dim R_0 = 0$, then it easily follows that $\mathbf{G}(\mathsf{P}) = \langle R_0, R_1, R_2 \rangle$ is reducible, since each R_j permutes the planes parallel to L, which is contrary to the assumption that P is pure. Indeed, we would have $R_0 \subset R_2$, since these reflexions commute. Thus $\dim R_0 \geqslant 1$ also.

We next exclude the possibility that $\dim R_j = 2$ for each $j = 1, 2$. In this case R_1 and R_2 would be planes through o, with some acute angle between them. Then R_0 is a line or plane, whose reflexion commutes with R_2, but not with R_1. If R_0 is a plane, then since $\mathbf{G}(\mathsf{P})$ is irreducible it will follow that $R_0 \cap R_1 \cap R_2 \neq \emptyset$; hence $\mathbf{G}(\mathsf{P})$ will be a discrete orthogonal group, and so finite. Similarly, if R_0 is a line, there are two possibilities. First, R_0 may lie in R_2, giving $R_0 \cap R_1 \cap R_2 \neq \emptyset$ since $\mathbf{G}(\mathsf{P})$ is irreducible, and as before $\mathbf{G}(\mathsf{P})$ is finite. Second, R_0 may be perpendicular to R_2; this makes the group $\mathbf{G}(\mathsf{P})$ reducible, which again is not permitted.

There is now a further case to be excluded; we cannot have $\dim R_0 = 2$ and $\dim R_2 = 1$. If this were so, then the line R_2 would have to be perpendicular to the plane R_0 (since $o \notin R_0$, the possibility $R_2 \subset R_0$ is forbidden). As in the previous case, the group $\mathbf{G}(\mathsf{P})$ would then be reducible, which we do not allow.

In conclusion, then, the mirror vector $(\dim R_0, \dim R_1, \dim R_2)$ can take only the four values of the lemma. □

We list the apeirohedra in Table 10A2 according to their mirror vectors, which are the entries on the left. The columns are indexed by the finite regular polyhedra to which the respective apeirohedra correspond; we have added the Petrials to complete the picture.

10A2

(2,2,2)	$\{3,3\}$	$\{3,4\}$	$\{4,3\}$
(1,2,2)	$\{\frac{6}{1,3}, 3 : 4\}$	$\{\frac{6}{1,3}, 4 : 3\}$	$\{\frac{4}{1,2}, 3 : 3\}$
(2,1,2)	$\{6, \frac{6}{1,3} \mid 3\}$	$\{6, \frac{4}{1,2} \mid 4\}$	$\{4, \frac{6}{1,3} \mid 4\}$
(1,1,2)	$\{\frac{3}{0,1}, \frac{6}{1,3} : 4\}$	$\{\frac{3}{0,1}, \frac{4}{1,2} : 6\}$	$\{\frac{4}{0,1}, \frac{6}{1,3} : 6\}$
(1,2,1)	$\{\frac{6}{1,3}, 6 : \frac{4}{1,2}, \cdot, \frac{1}{0}\}$	$\{\frac{6}{1,3}, 4 : \frac{6}{1,3}, \frac{1}{0}\}$	$\{\frac{4}{1,2}, 6; \frac{6}{1,3}, \cdot, \frac{1}{0}\}$
(1,1,1)	$\{\frac{3}{0,1}, 3 : \frac{4}{0,1}\}$	$\{\frac{3}{0,1}, 4 : \frac{3}{0,1}\}$	$\{\frac{4}{0,1}, 3 : \frac{3}{0,1}\}$

The pure 3-dimensional regular apeirohedra

We can now state the classification result.

10A3 Theorem *The list of twelve discrete pure 3-dimensional apeirohedra in Table 10A2 is complete.*

Proof. From the generatrix (R_0, R_1, R_2) of our apeirohedron P, we derive the planar reflexions S_1 and S_2 as before. We now define a third reflexion S_0, whose mirror is also a plane, as follows. We let R_0' be the translate of R_0 which contains the origin o, and then set $S_0 := R_0'$ or $-R_0'$ as R_0 is a plane or a line. We write $\mathbf{G}' := \langle R_0', R_1, R_2 \rangle$, which is the point-group of $\mathbf{G}(\mathsf{P})$, and set $\mathbf{H} := \langle S_0, S_1, S_2 \rangle$. Then $\mathbf{H}$ is a finite irreducible (plane) reflexion group, namely one of $[3, 3]$, $[3, 4]$ or $[3, 5]$, and $\mathbf{G}'$ is either again one of these reflexion groups, or its rotation subgroup (this can happen only when $\dim R_j = 1$ for each j). Since $\mathbf{G}(\mathsf{P})$ has to be discrete, Proposition 1D22 excludes 5-fold rotations, and hence $\mathbf{G}'$ cannot be $[3, 5]$ or its rotation subgroup. In other words, $\mathbf{H}$ must be $[3, 3]$ or $[3, 4]$.

With four possibilities for the mirror vector $(\dim R_0, \dim R_1, \dim R_2)$, and three for the group $\mathbf{H}$ (which can also be taken as $[4, 3]$, of course), we see that we have just twelve possibilities. These twelve all occur; we may reverse the method of the proof, and observe that different positions of R_0 not containing o, but meeting R_2, lead to similar apeirohedra. □

The notation here takes into account the rigidity properties which we will prove in Section 10C. It is important to bear in mind that their expressions as abstract regular apeirohedra may differ considerably from their geometric descriptions. We shall discuss their groups in the following Section 10B; this will show that, in a sense, the apeirohedra fall into a single family, derived from the regular honeycomb $\{4, 3, 4\}$.

It is appropriate to make one further comment here. The symmetry groups of the three apeirohedra associated with the mirror vector $(1, 1, 1)$ are generated by rotations (half-turns) in $\mathbb{E}^3$; thus the whole groups contain only direct isometries. This implies that the three apeirohedra are *handed*, and occur in enantiomorphic (mirror-image) pairs, with their facets consisting of either all left-hand helices, or all right-hand helices. The other six apeirohedra with helical facets (three blended and three pure) contain both left- and right-handed helices, since there is a plane or point-reflexion among the generators of each of their symmetry groups.

In Figure 10A4, we sketch part of the Petrie pair of apeirohedra $\{\frac{3}{0,1}, 3 : \frac{4}{0,1}\}$ and $\{\frac{4}{0,1}, 3 : \frac{3}{0,1}\}$. We have inserted green scaffolding around certain 4-helices, which we hope makes the structure more clear. The red lines to the right represent another vertical 4-helix, whose scaffolding has been omitted; the others complete some basic edge-circuits, for the significance of which see the following Section 10B.

We shall discuss relationships which lead to descriptions of the automorphism groups of the apeirohedra in the next section. In this, we look at the various ways in which the apeirohedra can be constructed in the first place. As we have seen in Sections 5B and 5D, there are two basic constructions that can be applied to regular apeirotopes or polytopes of full rank to yield regular apeirotopes of nearly full rank. While we have already considered these in Section 9D, it is worth briefly revisiting them here; we shall also observe that, in some respects, the 3-dimensional case does not quite fit into the general pattern. The context of this discussion is Section 5D, where we modify mirrors rather than perform mixing operations.

10A4

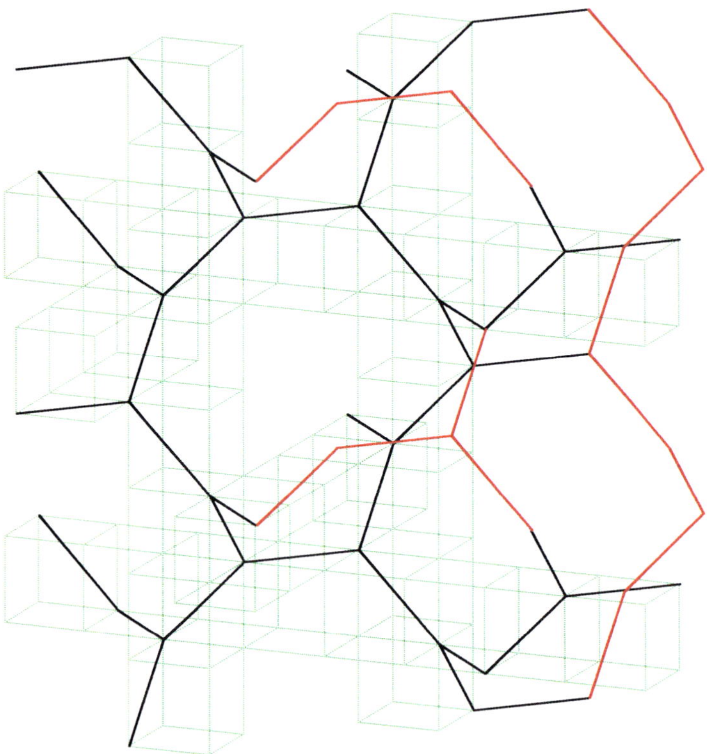

The apeirohedra $\{\frac{3}{0,1}, 3 : \frac{4}{0,1}\}$ and $\{\frac{4}{0,1}, 3 : \frac{3}{0,1}\}$

The first construction is Petrie contraction ϖ. There are only two regular 4-apeirotopes in $\mathbb{E}^3$ (apart from those of the form Q^α), namely, the cubic tiling $\{4,3,4\}$ and its Petrial $\{4,3,4\}^\pi = \{4,3,4\}^\kappa$ (it is the latter expression which is appropriate here). In its second guise, Section 8B shows that the facet of $\{4,3,4\}^\kappa$ is $\{4,3\}^\kappa$, but it is the first which leads to the automorphism group. Under ϖ, though, we have

10A5 $$\{4,3,4\}^\varpi = \{6, \tfrac{4}{1,2} \mid 4\}, \quad \{4,3,4\}^{\kappa\varpi} = \{\tfrac{3}{0,1}, \tfrac{4}{1,2} : 6\};$$

we shall use the fact that $\pi\varpi = \varpi\pi$ in the next section.

The second operation is κ applied to the three crystallographic polyhedra and their Petrials; of course, κ and π commute. Recall from Section 8C that the Petrials can also be derived by applications of ζ, possibly to different polyhedra.

In this way, we obtain

10A6
$$\{3,3\}^{\boldsymbol{\kappa}} = \{6, \tfrac{6}{1,3} \mid 3\}, \qquad \{3,3\}^{\zeta\boldsymbol{\kappa}} = \{\tfrac{3}{0,1}, \tfrac{6}{1,3} : 4\},$$
$$\{3,4\}^{\boldsymbol{\kappa}} = \{6, \tfrac{4}{1,2} \mid 4\}, \qquad \{3,4\}^{\zeta\boldsymbol{\kappa}} = \{\tfrac{3}{0,1}, \tfrac{4}{1,2} : 6\},$$
$$\{4,3\}^{\boldsymbol{\kappa}} = \{4, \tfrac{6}{1,3} \mid 4\}, \qquad \{4,3\}^{\zeta\boldsymbol{\kappa}} = \{\tfrac{4}{0,1}, \tfrac{6}{1,3} : 6\}.$$

However, it is worth noting that $\boldsymbol{\kappa}$ is also applicable to the apeirohedra with planar vertex-figures, and so will link them in pairs:

10A7
$$\{\tfrac{6}{1,3}, 6 : \tfrac{4}{1,2}, \cdot, \tfrac{1}{0}\} = \{\tfrac{3}{0,1}, 3 : \tfrac{4}{0,1}\}^{\boldsymbol{\kappa}},$$
$$\{\tfrac{6}{1,3}, 4 : \tfrac{6}{1,3}, \tfrac{1}{0}\} = \{\tfrac{3}{0,1}, 4 : \tfrac{3}{0,1}\}^{\boldsymbol{\kappa}},$$
$$\{\tfrac{4}{1,2}, 6; \tfrac{6}{1,3}, \cdot, \tfrac{1}{0}\} = \{\tfrac{4}{0,1}, 3 : \tfrac{3}{0,1}\}^{\boldsymbol{\kappa}}.$$

Notes to Section 10A

1. The basic ideas behind the classification Theorem 10A3 are due to McMullen; they were presented in [98] as part of a common programme with Schulte. The treatment in [99, Section 7E] followed that of the paper; we have somewhat varied it here, particularly in the use of $\boldsymbol{\kappa}$.
2. Coxeter [24] uses the term *skew polyhedra* for what are now generally named after him and Petrie.

10B Groups of the Apeirohedra

We now describe the automorphism groups of the apeirohedra of Section 10A; once again, we must recall that the fine Schläfli symbols sometimes bear little resemblance to the abstract descriptions. Since we are discussing the groups, we work on the abstract rather than the geometric level.

Our starting point for the present purposes is $\{4, \tfrac{6}{1,3} \mid 4\} \cong \mathbf{\{}4, 6 \mid 4\mathbf{\}}$. We can verify this isomorphism in various ways. First, directly, we can derive it from twisting the Coxeter group $\mathbf{R}_4$ with diagram

10B1

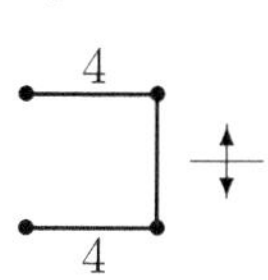

Of course, this also gives the dual $\{6, \tfrac{4}{1,2} \mid 4\} \cong \mathbf{\{}6, 4 \mid 4\mathbf{\}}$ and, though we will find an alternative derivation, we can similarly obtain $\{6, \tfrac{6}{1,3} \mid 3\} \cong \mathbf{\{}6, 6 \mid 3\mathbf{\}}$ by twisting $\mathbf{P}_4$, namely,

10B2

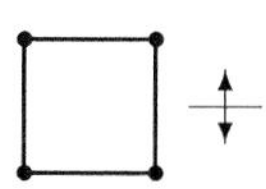

A different approach is to think of $\{4, \frac{6}{1,3} \mid 4\}$ as the facet of the regular 4-apeirotope $\{4,3,4\}^{\pi} = \{\{4, \frac{6}{1,3} \mid 4\}, \{\frac{6}{1,3}, 4 : 3\}\}$, which we already met in Section 8B and mentioned in the previous section. From either notion, we see that we can take the generatrix (S_0, S_1, S_2) of $\{4, \frac{6}{1,3} \mid 4\}$ to be

10B3

$$\begin{aligned} S_0 &= \overline{1}^{\bullet}\colon\ x \mapsto (1-\xi_1, \xi_2, \xi_3), \\ S_1 &= (1\ 2)\overline{3}\colon\ x \mapsto (\xi_2, \xi_1, -\xi_3), \\ S_2 &= (2\ 3)\colon\ x \mapsto (\xi_1, \xi_3, \xi_2). \end{aligned}$$

Table 10B4 gives the relationships among the apeirohedra, including those that are implicit in the proof of Theorem 10A3.

10B4

$$\begin{array}{ccccc}
\{\frac{3}{0,1}, \frac{4}{1,2} : 6\} & \overset{\pi^*}{\longleftrightarrow} & \{\frac{3}{0,1}, \frac{6}{1,3} : 4\} & & \\
\updownarrow \pi & & \updownarrow \pi & & \\
\{6, \frac{4}{1,2} \mid 4\} & \overset{\delta}{\longleftrightarrow} & \{4, \frac{6}{1,3} \mid 4\} & & \\
& & \downarrow \eta & & \\
& & \{\frac{6}{1,3}, 6 : \frac{4}{1,2}, \cdot, \frac{1}{0}\} & \overset{\kappa}{\longleftrightarrow} & \{\frac{3}{0,1}, 3 : \frac{4}{0,1}\} \\
& & \updownarrow \pi & & \updownarrow \pi \\
\{\frac{6}{1,3}, 4 : \frac{6}{1,3}, \frac{1}{0}\} & \overset{\delta}{\longleftrightarrow} & \{\frac{4}{1,2}, 6 : \frac{6}{1,3}, \cdot, \frac{1}{0}\} & \overset{\kappa}{\longleftrightarrow} & \{\frac{4}{0,1}, 3 : \frac{3}{0,1}\} \\
\updownarrow \kappa & & \downarrow \eta & & \\
\{\frac{3}{0,1}, 4 : \frac{3}{0,1}\} & & \{6, \frac{6}{1,3} \mid 3\} & \overset{\pi}{\longleftrightarrow} & \{\frac{4}{0,1}, \frac{6}{1,3} : 6\}
\end{array}$$

We are left to prove that ten of the apeirohedra have the automorphism groups that their notations signify, and to determine the groups of the other two. In doing this, we also verify the relationships of Table 10B4. (We shall not treat the blended apeirohedra here, since they are described by their geometric structures.)

For the moment, we shall leave aside $\{4, \frac{6}{1,3} \mid 4\}$ and its dual; we shall thus take their groups as given. In fact, Theorem 8B4 tells us that their groups are as indicated by the notation. (The same assumption could apply to $\{6, \frac{6}{1,3} \mid 3\}$, but we shall actually obtain its group here, as it fits into our general scheme.) As we said, though, we work on the abstract level, since we are investigating automorphism groups; the geometric correspondences will follow automatically.

Since the Petrie operation interchanges k-holes and k-zigzags, we see that

$$\{4,6 \mid 4\} \overset{\pi}{\longmapsto} \{\infty,6:4,4\},$$
$$\{6,4 \mid 4\} \overset{\pi}{\longmapsto} \{\infty,4:6,4\},$$

as claimed. (Strictly speaking, perhaps we ought to replace '∞' by '$\cdot$'!) We next appeal to Theorem 5A22, to obtain

$$\{4,6 \mid 4\} \overset{\eta}{\longmapsto} \{6,6:4\}.$$

The Petrie operation and duality then yield $\{4,6:6\}$ and $\{6,4:6\}$. Another appeal to Theorem 5A22 then yields

$$\{4,6:6\} \overset{\eta}{\longmapsto} \{6,6 \mid 3\}.$$

From this last, just as above we obtain

$$\{6,6 \mid 3\} \overset{\pi}{\longmapsto} \{\infty,6:6,3\}.$$

We have three apeirohedra remaining. These we must obtain by operations different from those of Table 10B4, since the effects of $\boldsymbol{\kappa}$ do not readily translate to changes of generators. One needs an application of $\boldsymbol{\sigma}$ $(= \boldsymbol{\pi}^*\boldsymbol{\eta}\boldsymbol{\pi}^*)$, and the other two applications of φ_2.

The generatrix of the group $\mathbf{S}_4 = [4,3,4]$ of the cubic tiling is (T_0,T_1,T_2,T_3) $= (\overline{1}^{\bullet},(1\,2),(2\,3),\overline{3})$ in the abbreviated notation used above. Because $\{4,\frac{6}{1,3} \mid 4\}$ is the facet of $\{4,3,4\}^{\pi}$, it follows that its generatrix (S_0,S_1,S_2) is given by $S_0 = T_0$, $S_1 = T_1T_3$ and $S_2 = T_2$, as in (10B3). As usual, the initial vertex is o.

From this, we can verify directly that $\{\frac{3}{0,1},4:\frac{3}{0,1}\} = \{4,\frac{6}{1,3} \mid 4\}^{\sigma}$. So, if we follow through the operations at the abstract level, we obtain

$$\begin{aligned}\{4,6 \mid 4\} &\overset{\pi^*}{\longmapsto} \{4,\infty:6,*4\}\\ &\overset{\eta}{\longmapsto} \{\infty,\infty:4 \mid 3\}\\ &\overset{\pi^*}{\longmapsto} \{\infty,4:\cdot,*3\}.\end{aligned}$$

The $*$ prefix to a suffix was explained in Section 2D. We have replaced the ∞ in the final suffix by $\cdot$, to indicate that the corresponding entry is unspecified. The only step left unexplained is the application of $\boldsymbol{\eta}$ to $\{4,\infty:6,*4\}$. The operation is

$$\boldsymbol{\eta}\colon (\boldsymbol{s}_0,\boldsymbol{s}_1,\boldsymbol{s}_2) \mapsto (\boldsymbol{s}_0\boldsymbol{s}_1\boldsymbol{s}_0,\boldsymbol{s}_2,\boldsymbol{s}_1) =: (\boldsymbol{r}_0,\boldsymbol{r}_1,\boldsymbol{r}_2).$$

The first suffix 6 is dealt with by Theorem 5A22; observe that the graph of $\{4,\infty:6,*4\}$ is indeed bipartite. For the second suffix $*4$, the relation gives the period of

$$\boldsymbol{s}_2(\boldsymbol{s}_1\boldsymbol{s}_0)^2 = \boldsymbol{s}_2\boldsymbol{s}_1\boldsymbol{s}_0\boldsymbol{s}_1\boldsymbol{s}_0 = \boldsymbol{r}_1\boldsymbol{r}_2\boldsymbol{r}_0 \sim \boldsymbol{r}_0\boldsymbol{r}_1\boldsymbol{r}_2,$$

namely, that of the Petrie polygon of the second apeirohedron.

The last two apeirohedra, those of type $\{\infty, 3\}$, must be characterized by direct methods. We shall work with $\{\frac{4}{0,1}, 3 : \frac{3}{0,1}\}$ rather than with its Petrial $\{\frac{3}{0,1}, 3 : \frac{4}{1,2}\}$, because its structure is a little easier to describe. For this, we can check that φ_2 has the same effect as κ on $\{\frac{4}{1,2}, 6 : \frac{6}{1,3}, \cdot, \frac{1}{0}\}$. So, if we start from $\{4, \frac{6}{1,3} \mid 4\}$ and trace through the three mixing operations which lead to $\{\frac{4}{0,1}, 3 : \frac{3}{0,1}\}$ (namely η, π and φ_2), we find that the generatrix (R_0, R_1, R_2) of the latter is given by

10B5
$$\begin{aligned} R_0 &= (S_0S_1)^2\colon\ x \mapsto (1-\xi_1, 1-\xi_2, \xi_3),\\ R_1 &= S_2S_1S_2\colon\ x \mapsto (\xi_3, -\xi_2, \xi_1),\\ R_2 &= S_1\colon\ x \mapsto (\xi_2, \xi_1, -\xi_3); \end{aligned}$$

these are indeed reflexions in lines.

We may picture $\{\frac{4}{0,1}, 3 : \frac{3}{0,1}\}$ in the following way. As we have already remarked, its facets are all helices with the same sense. The initial vertex is still o, and hence all the vertices are points of $\mathbb{E}^3$ with integer cartesian coordinates. We easily see from the generators that, in fact, the sum of the coordinates of each vertex is even.

The initial edge has vertices $o = (0,0,0)$ and $oR_0 = (1,1,0)$, which is a diagonal of a 2-face of $\{4,3,4\}$; hence all edges are such diagonals. Next, R_1R_0, which preserves the initial facet and takes o into $(1,1,0)$, is

$$R_1R_0\colon\ x \mapsto (1-\xi_3, 1+\xi_2, \xi_1),$$

which is a translation by $(0,1,0)$ together with a right-hand (or negative) twist of $\pi/2$ about the axis through $(\frac{1}{2}, 0, \frac{1}{2})$ in direction $(0,1,0)$. Hence the facets are helices of type $\{\infty\}\#\{4\}$. Finally, R_1 takes $(1,1,0)$ into $(1,1,0)R_1 = (0,-1,1)$, and R_2 takes $(0,-1,1)$ into $(0,-1,1)R_2 = (-1,0,-1)$; indeed, $R_2R_1 : x \mapsto (-\xi_3, -\xi_1, \xi_2)$ is a cyclic permutation of the signed basis vectors $e_1, -e_2, -e_3$.

It follows from this that $(R_1R_0)^4$ is a translation, by $(0,4,0)$. However, such translations in the directions of the three coordinate axes do not generate the whole translation group. Instead, we observe that

$$xR_2R_1R_0 = (1+\xi_3, 1+\xi_1, \xi_2),$$

so that $(R_2R_1R_0)^3$ is the translation by $(2,2,2)$. Since the images of $(2,2,2)$ under R_1R_2 and its inverse are $(-2,2,-2)$ and $(-2,-2,2)$, we see that the translation subgroup is actually the lattice $\Lambda := \Lambda_{(2,2,2)}$, which is generated by $(2,2,2)$ and its transforms under changes of signs of the coordinates. Moreover, since $R_2R_1R_0$ is conjugate to $R_1 \cdot R_0R_2$, which takes one vertex of a facet of the Petrial $\{\frac{3}{0,1}, 3 : \frac{4}{0,1}\}$ into the next (with the same initial edge as that of the original face), we see that these facets are of type $\{\infty\} \# \{3\}$; this time, they are helices with a left-hand (positive) twist.

The axes of the helical faces of $\{\frac{4}{0,1}, 3 : \frac{3}{0,1}\}$ are parallel to the coordinate axes; as we have seen, these three axes are permuted by R_2R_1. To visualize the way in which the facets fit together, it is more convenient to concentrate on the vertical ones. The cubes in $\{4,3,4\}$ fall into vertical stacks or (infinite) towers. Just an eighth of these towers are associated with facets; they are all the images of one fixed tower under the translation lattice Λ. A typical facet winds upwards (or downwards) in a right-hand spiral around the tower, crossing its square faces diagonally; we may envisage it as a staircase. (In Figure 10B6, we are looking at the vertical towers from above. As we go around a tower in the clockwise direction, we rise by a floor each time we traverse an edge.)

10B6

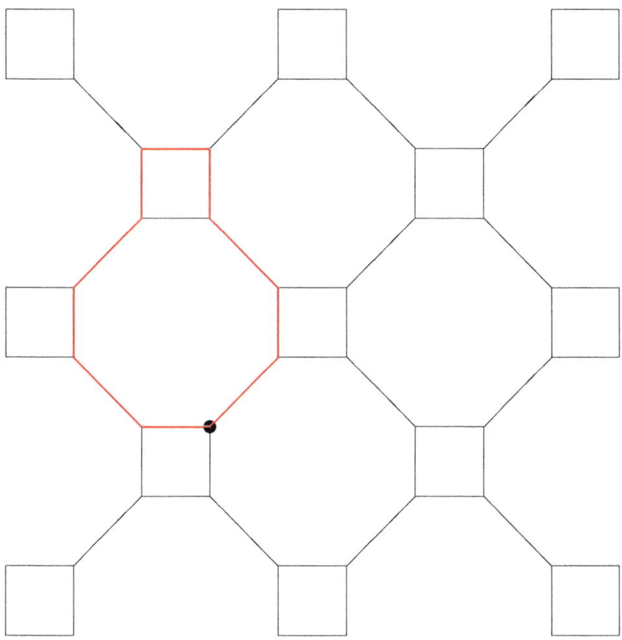

The apeirohedra $\{\frac{3}{0,1}, 3 : \frac{4}{0,1}\}$ and $\{\frac{4}{0,1}, 3 : \frac{3}{0,1}\}$ in projection

The origin o is a vertex of a vertical tower; we think of it as lying at ground level. Ascending four flights of stairs brings us to $(0,0,4)$ on the fourth floor, immediately above our starting point. At each floor is a single horizontal bridge, leading away from one tower to an *adjacent* tower, across the diagonal of a horizontal square of $\{4,3,4\}$. If we ascend one flight to the first floor, cross the bridge, descend one flight in the adjacent tower to the ground floor, and then cross the next bridge, we shall have gone four edges along a face of $\{\frac{4}{0,1}, 3 : \frac{3}{0,1}\}$ whose axis is horizontal. Each bridge belongs to two such horizontal facets, of course, according as it was reached by an ascending or descending flight.

Theorem 2D4 shows that we can find a presentation of the automorphism group of $\{\frac{4}{0,1}, 3 : \frac{3}{0,1}\}$ by considering its edge-circuits (the vertex-figure is, of course, known). A minimal or *basic* circuit is constructed as follows. Ascend four floors of a tower by the staircase, cross the bridge to the adjacent tower, descend four floors by its staircase, and then cross back over the bridge to the starting point. The other basic circuits are then the images of this one under the symmetry group $\mathbf{G}$ of $\{\frac{4}{0,1}, 3 : \frac{3}{0,1}\}$. A typical basic circuit using horizontal towers is formed similarly, although the description appears a little different. From a starting vertex, cross a bridge and ascend one floor of a staircase. Repeat this twice, then cross a final bridge and descend three floors. Of course, we may interchange 'ascend' and 'descend' in this description. Such a circuit uses four towers in a square formation; the circuit goes round the inside of three towers, and the outside of the fourth. (In Figure 10B6, such a horizontal basic circuit is indicated by red edges.)

If $\mathcal{C}$ and $\mathcal{D}$ are two edge-circuits, then we *concatenate* them by taking their symmetric difference $\mathcal{C}\triangle\mathcal{D}$. (In taking the symmetric difference, we are of course only considering the edges, not their vertices; isolated vertices can obviously be eliminated.) Observe that $(\mathcal{C}\triangle\mathcal{D})\triangle\mathcal{D} = \mathcal{C}$, so that concatenating twice with a fixed circuit has no effect. The key result for categorizing $\mathbf{G}$ is

10B7 Theorem *An arbitrary edge-circuit in* $\{\frac{4}{0,1}, 3 : \frac{3}{0,1}\}$ *is a concatenation of basic circuits.*

Proof. It is clear that, at any stage, we may confine our attention to a single connected circuit $\mathcal{C}$; if, after any concatenation, a circuit becomes disconnected, then we simply consider the resulting components.

We now reduce the circuit $\mathcal{C}$ to a vertex by means of two kinds of operation. First, if $\mathcal{C}$ uses two or more bridges between the same towers, then we can concatenate with vertical basic circuits to eliminate these bridges in pairs. Thus we may assume that $\mathcal{C}$ contains no more than one bridge between any two towers.

We now look down on $\mathcal{C}$ from a vertical direction, as in Figure 10B6. Since the plane is simply-connected, we may contract the projection of $\mathcal{C}$ to a single vertex. For this purpose, we can safely identify the vertices of $\mathcal{C}$ in any one tower, since there is now no more than one bridge between any two towers. A contraction over a diamond formed by four towers is achieved by concatenating with a horizontal basic circuit (like that in red in Figure 10B6) which uses these four towers and shares one of the bridges of $\mathcal{C}$. Of course, further reductions of the first kind will then also generally be needed, since horizontal bridges along the other three sides of the diamond may be introduced (and some may disappear). It is clear that systematic application of these two kinds of operation will eventually reduce $\mathcal{C}$ to a single vertex of $\{\frac{4}{0,1}, 3 : \frac{3}{0,1}\}$. In other words, if we reverse the successive concatenations, then we shall recover the original circuit, as was claimed. □

By Theorem 2D4, a basic edge-circuit in $\{\frac{4}{0,1}, 3 : \frac{3}{0,1}\}$ corresponds to a relation in $\mathbf{G}$ between its distinguished generators R_0, R_1 and R_2. Let us consider the following horizontal basic circuit. It starts from the initial vertex o, and contains the first four successive edges of the initial face F_2. This sequence of four edges is continued at each end by the two edges (corresponding to bridges) joining F_2 to the face in an adjacent (horizontal) stack of cubes, and is completed by the four intermediate edges of that face. The symmetry group of this basic circuit has two generators. The first is the conjugate $U_1 := (R_0R_1)^3R_0$ of R_1 by $(R_1R_0)^2$, which fixes F_2 and interchanges the two bridging edges. The second is the conjugate $U_2 := R_2R_1R_0R_1R_2$ of R_0 by R_1R_2, which interchanges F_2 and the adjacent face, and fixes the two bridging edges. The relation which imposes this basic circuit is then $U_1U_2 = U_2U_1$, or $(U_1U_2)^2 = I_3$, the identity in $\mathbb{E}^3$. When expressed in terms of the generators R_0, R_1 and R_2, the relation $(U_1U_2)^2 = I_3$ involves R_0 ten times, in keeping with the fact that a basic circuit in $\{\frac{4}{0,1}, 3 : \frac{3}{0,1}\}$ has ten edges.

In order to state the main theorem, we provide an alternative interpretation of the group relation given by this basic circuit. Define

$$S := (R_0R_1)^4,$$
$$T := (R_0R_1R_2)^3.$$

Thus S and T are the translational symmetries of a face of $\{\frac{4}{0,1}, 3 : \frac{3}{0,1}\}$ and a Petrie polygon. Then, freely using $R_0R_2 = R_2R_0$ and $R_1R_2R_1 = R_2R_1R_2$, but not the fact that S and T are actually translations, and hence commute, it is straightforward but tedious to verify that

$$(U_1U_2)^2 = ST^{-1}S^{-1}T.$$

Rearranging, we see that $(U_1U_2)^2 = I_3$ and $ST = TS$ are equivalent relations; it is the latter which we shall employ.

10B8 Theorem *The automorphism groups of the non-planar pure apeirohedra of type $\{\infty, 3\}$ in $\mathbb{E}^3$ are the Coxeter group $[\infty, 3] = \langle \boldsymbol{r}_0, \boldsymbol{r}_1, \boldsymbol{r}_2 \rangle$, subject to the single extra relation*

$$\boldsymbol{st} = \boldsymbol{ts},$$

where $\boldsymbol{s} := (\boldsymbol{r}_0\boldsymbol{r}_1)^3$ and $\boldsymbol{t} := (\boldsymbol{r}_0\boldsymbol{r}_1\boldsymbol{r}_2)^4$ for $\{\frac{3}{0,1}, 3 : \frac{4}{0,1}\}$, or $\boldsymbol{s} := (\boldsymbol{r}_0\boldsymbol{r}_1)^4$ and $\boldsymbol{t} := (\boldsymbol{r}_0\boldsymbol{r}_1\boldsymbol{r}_2)^3$ for $\{\frac{4}{0,1}, 3 : \frac{3}{0,1}\}$.

Proof. The given relation for $\{\frac{4}{0,1}, 3 : \frac{3}{0,1}\}$ is equivalent to the one given above, since $\boldsymbol{s}$ corresponds to S and $\boldsymbol{t}$ corresponds to T. By Theorem 2D4, any relation on the automorphism group $\mathbf{G}$ of $\{\frac{4}{0,1}, 3 : \frac{3}{0,1}\}$ corresponds to an edge-circuit, and we have shown in Theorem 10B7 that these are formed by concatenating basic circuits, each of which is obtained by conjugating the extra relation. Thus the automorphism group of $\{\frac{4}{0,1}, 3 : \frac{3}{0,1}\}$ is as claimed.

The corresponding relation for the other apeirohedron is obtained from that for $\{\frac{4}{0,1}, 3 : \frac{3}{0,1}\}$ by means of the Petrie operation substitution of $\boldsymbol{r}_0\boldsymbol{r}_2$ for $\boldsymbol{r}_0$. Indeed, in terms of the generators of $\mathbf{G}$, and with $\boldsymbol{s}$ and $\boldsymbol{t}$ retaining (for the moment) their original definitions, we have

$$(\boldsymbol{r}_0\boldsymbol{r}_2\boldsymbol{r}_1)^3 = \boldsymbol{r}_2\boldsymbol{t}\boldsymbol{r}_2,$$
$$(\boldsymbol{r}_0\boldsymbol{r}_2\boldsymbol{r}_1\boldsymbol{r}_2)^4 = \boldsymbol{r}_2\boldsymbol{s}\boldsymbol{r}_2.$$

Thus the relations between the new $\boldsymbol{s}$ and $\boldsymbol{t}$ are just the old ones (conjugated by $\boldsymbol{r}_2$) with $\boldsymbol{s}$ and $\boldsymbol{t}$ interchanged, again as asserted. □

Let us make a further comment on this group. We have $\{\frac{4}{0,1}, 3 : \frac{3}{0,1}\} \cong \{4, 6 : 6\}^{\varphi_2}$, by means of the operation

$$(\boldsymbol{s}_0, \boldsymbol{s}_1, \boldsymbol{s}_2) \mapsto (\boldsymbol{s}_0, \boldsymbol{s}_1\boldsymbol{s}_2\boldsymbol{s}_1, \boldsymbol{s}_2) =: (\boldsymbol{r}_0, \boldsymbol{r}_1, \boldsymbol{r}_2)$$

on the (larger) automorphism group $[4, 6 : 6] = \boldsymbol{G}(\{4, 6 : 6\}) = \langle \boldsymbol{s}_0, \boldsymbol{s}_1, \boldsymbol{s}_2 \rangle$. If we substitute for $\boldsymbol{r}_0$, $\boldsymbol{r}_1$ and $\boldsymbol{r}_2$ in this way, it is easy to check that (as we must) we do obtain a valid relation in $[4, 6 : 6]$.

10B9 Remark We used the fact that $\boldsymbol{\kappa}$ has the same effect as φ_2 in obtaining the two apeirohedra $\{\frac{3}{0,1}, 3 : \frac{4}{0,1}\}$ and $\{\frac{4}{0,1}, 3 : \frac{3}{0,1}\}$. Since $\boldsymbol{\kappa}$ is involutory, we see that φ_2 is inverted by $\boldsymbol{\kappa}$ in this situation. Bear in mind, though, that $\boldsymbol{\pi}$ operates on the abstract level as well, whereas $\boldsymbol{\kappa}$ is purely geometric.

We end with a remark on the pure apeirohedra with finite faces. We put them in Table 10B10, together with the pairs of finite regular polyhedra to which they correspond; as before, the entry in the first column is the mirror vector $(\dim R_0, \dim R_1, \dim R_2)$.

10B10

(2,2,2)	$\{3, 3\}$	$\{3, 4\}$	$\{4, 3\}$
(1,2,2)	$\{\frac{6}{1,3}, 3 : 4\}$	$\{\frac{6}{1,3}, 4 : 3\}$	$\{\frac{4}{1,2}, 3 : 3\}$
(2,1,2)	$\{6, \frac{6}{1,3} \mid 3\}$	$\{6, \frac{4}{1,2} \mid 4\}$	$\{4, \frac{6}{1,3} \mid 4\}$
(1,2,1)	$\{\frac{6}{1,3}, 6 : \frac{4}{1,2}, \cdot, \frac{1}{0}\}$	$\{\frac{6}{1,3}, 4 : \frac{6}{1,3}, \frac{1}{0}\}$	$\{\frac{4}{1,2}, 6 : \frac{6}{1,3}, \cdot, \frac{1}{0}\}$

A convex regular polyhedron of type $\{3, q\}$ (or $\{q, 3\}$) has holes $\{h\}$ with $h = q$, while its Petrie polygon is a skew r-gon with

$$r = \frac{2q + 10}{7 - q}$$

(this is derived from [27, 4.91], and is just one of many possible expressions; we write r here for Coxeter's h, for obvious reasons). Recall that, for $p \geqslant 2$, we define p'' by

$$\frac{1}{p} + \frac{1}{p''} = \frac{1}{2}.$$

The corresponding polyhedra in each column are then

$$\{p, q\}, \quad \{p'', q : r''\}, \quad \{p'', q'' \mid h\}, \quad \{p'', q'' : r\},$$

with the omission in the last of the linear apeirogon needed to ensure rigidity (see the next section). Of course, the fact that the holes or Petrie polygons of the derived polyhedra are those given follows from the relationship between the generating reflexions R_0, R_1 and R_2 of their groups, and the corresponding plane reflexions S_0, S_1 and S_2 which generate the group of the convex polyhedron.

Notes to Section 10B

1. The circuit criterion of Theorem 2D4 was actually devised by McMullen in [98] to treat the two apeirohedra of Theorem 10B8 whose groups are not immediately derivable from those of other apeirohedra.

10C Rigidity of the Apeirohedra

In this section, we consider the 3-dimensional pure regular apeirohedra, with a view to determining which of them are rigid. We treat them according to their mirror vectors. As we shall see, three of the four classes consist of rigid apeirohedra, while those of the fourth need an extra condition to ensure rigidity. This section follows [89] fairly closely, except that certain background material in the paper has been dealt with earlier here.

Mirror Vector $(2,1,2)$

The apeirohedra under discussion here are those discovered by Petrie and Coxeter; see [24]. The appropriate fine Schläfli symbols for these are

$$\begin{aligned}\{6, \tfrac{6}{1,3} \mid 3\} &\cong \{6,6 \mid 3\},\\ \{4, \tfrac{6}{1,3} \mid 4\} &\cong \{4,6 \mid 4\},\\ \{6, \tfrac{4}{1,2} \mid 4\} &\cong \{6,4 \mid 4\}.\end{aligned}$$

The only difference between the fine Schläfli symbols and the abstract ones is that the geometry of the faces, vertex-figures and holes is specified; it is worth emphasizing that (for example) an entry '6' refers to a *planar* hexagon, whereas an entry '$\frac{6}{1,3}$' means a *skew* hexagon.

The result here is a special case of Theorem 6D5. In the present context, this says that a regular polyhedron (or apeirohedron) with planar faces and holes and vertex-figures that are blends of planar polygons with digons is rigid.

10C1 Theorem *The three regular polyhedra*

$$\{6, \tfrac{6}{1,3} \mid 3\}, \quad \{4, \tfrac{6}{1,3} \mid 4\}, \quad \{6, \tfrac{4}{1,2} \mid 4\}$$

are rigid, and hence are the corresponding Petrie–Coxeter apeirohedra.

Thus this class consists of rigid apeirohedra. We shall see further applications of Theorem 6D5 in what follows.

Mirror Vector $(1,1,2)$

Theorem 10C1 was, perhaps, exactly what should have been expected. In this section, though, we show that the situation for the Petrials of the Petrie–Coxeter apeirohedra is a little different. In [98], we specified these Petrials by their 1- and 2-zigzags, the 1-zigzag being the Petrie polygon. However, the notation there gave abstract descriptions of the apeirohedra; in particular, the faces were merely indicated by $\{\infty\}$. If we restore the geometric information about the faces, then things change. In fine, we shall show that describing the faces by generalized fractions enables us to drop mention of the 2-zigzags.

So, assuming the result of Theorem 10C2, we shall have

$$\begin{aligned}\{\tfrac{4}{0,1}, \tfrac{6}{1,3} : 6\} &\cong \{\infty, 6 : 6, 3\},\\ \{\tfrac{3}{0,1}, \tfrac{6}{1,3} : 4\} &\cong \{\infty, 6 : 4, 4\},\\ \{\tfrac{3}{0,1}, \tfrac{4}{1,2} : 6\} &\cong \{\infty, 4 : 6, 4\}.\end{aligned}$$

The key observation in this case is that the Petrie polygons, being planar, determine the geometry of the helical faces completely. For example, the face $\{\frac{3}{0,1}\}$ of $\{\frac{3}{0,1}, \frac{6}{1,3} : 4\}$ has the same angle $\pi/2$ at a vertex as that of the Petrie polygon $\{4\}$, and this shows that it is congruent to the Petrie polygon of the tiling of $\mathbb{E}^3$ by cubes (with the same edge-length) or – more importantly – that of $\{4, \frac{6}{1,3} \mid 4\}$. In general, let $\ldots, a_{-1}, a_0, a_1, a_2, \ldots$ be the successive vertices of the initial face F, say, of the given apeirohedron P (we know from its faces that P must be infinite). For each j, let G_j be the Petrie polygon of P of which a_{j-1}, a_j, a_{j+1} are successive vertices; since $G_j \subset \operatorname{aff}\{a_{j-1}, a_j, a_{j+1}\}$, it lies in the 3-dimensional space $\mathbb{E}$ spanned by F. If $\{a_0, c_j\}$ is the other edge of G_j through a_0 for $j = -1, 1$, then c_{-1}, a_{-1}, a_1, c_1 are four successive vertices of the vertex-figure Q of P at a_0. Since the skew polygon $Q \subset \operatorname{aff}\{c_{-1}, a_{-1}, a_1, c_1\}$, we deduce that Q lies in the same 3-space $\mathbb{E}$, and that a_0 must be the centre of Q. Then every face of P through a_0 (and similarly though each a_j) also lies in $\mathbb{E}$; connectedness implies that P lies in $\mathbb{E}$, and so is 3-dimensional. Since it is clear that P will coincide locally with the Petrial of the corresponding (geometric) Petrie–Coxeter apeirohedron, it will therefore coincide globally, which is what we want.

We conclude that we have established

10C2 Theorem *The three regular apeirohedra*

$$\{\tfrac{4}{0,1}, \tfrac{6}{1,3} : 6\}, \quad \{\tfrac{3}{0,1}, \tfrac{6}{1,3} : 4\}, \quad \{\tfrac{3}{0,1}, \tfrac{4}{1,2} : 6\}$$

are rigid; they are the Petrials of the corresponding Petrie–Coxeter apeirohedra.

There is an interesting consequence of this characterization.

10C3 Corollary *The three regular apeirohedra*

$$\{6, \tfrac{6}{1,3} : \tfrac{4}{0,1}\}, \quad \{4, \tfrac{6}{1,3} : \tfrac{3}{0,1}\}, \quad \{6, \tfrac{4}{1,2} : \tfrac{3}{0,1}\}$$

are rigid; they are the corresponding Petrie–Coxeter apeirohedra.

Proof. This is clear from Theorem 10C2, since we are just representing these apeirohedra as the Petrials of those characterized by that theorem. □

In other words, we have alternative expressions for the (geometric) Petrie–Coxeter apeirohedra:

$$\begin{aligned}\{6, \tfrac{6}{1,3} : \tfrac{4}{0,1}\} &\cong \{6, 6 \mid 3\},\\ \{4, \tfrac{6}{1,3} : \tfrac{3}{0,1}\} &\cong \{4, 6 \mid 4\},\\ \{6, \tfrac{4}{1,2} : \tfrac{3}{0,1}\} &\cong \{6, 4 \mid 4\}.\end{aligned}$$

It is now the case that the geometric and abstract descriptions are significantly different. Note also that we can write $\{6, \frac{6}{1,3} : \frac{4}{0,1}\} \approx \{6, \frac{6}{1,3} \mid 3\}$, in the notation introduced in Chapter 6, and so on.

Mirror Vector $(1, 1, 1)$

The three regular apeirohedra in this class are

$$\begin{aligned}\{\tfrac{3}{0,1}, 3 : \tfrac{4}{0,1}\} &\cong \{\infty, 3\}^{(a)},\\ \{\tfrac{4}{0,1}, 3 : \tfrac{3}{0,1}\} &\cong \{\infty, 3\}^{(b)},\\ \{\tfrac{3}{0,1}, 4 : \tfrac{3}{0,1}\} &\cong \{\infty, 4 : \cdot, *3\}.\end{aligned}$$

The notation for the first two is, as in [98] (see also [99, Section 7B]) a shorthand for the combinatorial descriptions given in the previous Section 10B; as we have said, these are two of the regular apeirohedra found by Grünbaum [58], and form a Petrie pair. The third (the only one of the twelve under discussion missed by Grünbaum – see the notes at the end of the section) was due to Dress [41, 42]; its notation as in Section 2D was also justified in the previous section.

More precisely, we showed in Theorem 10B8 that the two apeirohedra of type $\{\infty, 3\}$ are determined abstractly by the fact that the translations induced by the appropriate number of steps along a certain helical face and a certain Petrie apeirogon commute. In one sense, this is an obvious condition arising from the geometry. However, there are hidden assumptions here, which need to be brought out into the open and properly addressed. The basic one is that the translational symmetries of a helix (whether face or Petrie apeirogon) extend to ones of the whole apeirotope; this will be true if the apeirotope can be shown to be 3-dimensional, but will not necessarily hold otherwise. In fact, while the helical symmetries will be appealed to, the core of the proofs of the two main theorems of the section will depend on purely local properties.

We first have

10C4 Theorem *The regular apeirohedra* $\{\frac{3}{0,1}, 3 : \frac{4}{0,1}\}$ *and* $\{\frac{4}{0,1}, 3 : \frac{3}{0,1}\}$ *are rigid, and are isomorphic to* $\{\infty, 3\}^{(a)}$ *and* $\{\infty, 3\}^{(b)}$*, respectively.*

Proof. We begin the proof by observing that, while the notation demands that the faces and Petrie apeirogons be 3-dimensional, there is no initial requirement that this be true of the apeirohedron itself. As in Section 10B, we shall work with the latter apeirohedron; to deal with the former, we merely swap the rôles of face and Petrie apeirogon. The key fact employed is that, as already mentioned in Section 6B, k steps along a k-helix (here, $k = 3$ or 4) is a translational symmetry of the helix, though not necessarily of the whole apeirohedron.

10C5

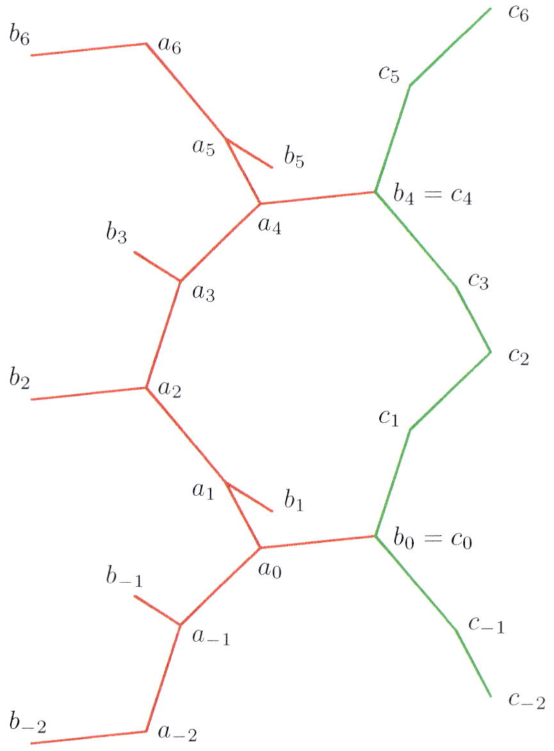

Local structure of $\{\frac{4}{0,1}, 3 : \frac{3}{0,1}\}$

Let $\ldots, a_{-2}, a_{-1}, a_0, a_1, a_2, a_3, \ldots$ be successive vertices of an initial 4-helical face F_1 of $\mathsf{P} \in \{\frac{4}{0,1}, 3 : \frac{3}{0,1}\}$ (as a realization cone, of course). For each such j, there is a third edge $\{a_j, b_j\}$ (that is, other than $\{a_{j-1}, a_j\}$ and $\{a_j, a_{j+1}\}$) containing a_j. For each j again, there is a Petrie apeirogon $\mathsf{G}_j = \{\frac{3}{0,1}\}$ of P, with successive vertices $\ldots, b_{j-1}, a_{j-1}, a_j, a_{j+1}, b_{j+1}, \ldots$; thus $a_{j-1} - b_{j-1} = b_{j+1} - a_{j+1}$. Repeating this shows that

$$b_{j+4} - a_{j+4} = b_j - a_j$$

for each $j \in \mathbb{Z}$. Observe that we already have a strong suggestion here of a global translational symmetry of P.

Let F' be the face of P which contains b_0, but not the edge $\mathsf{E} := \{a_0, b_0\}$, and suppose that it has successive vertices $\ldots, c_{-1}, c_0, c_1, c_2, \ldots$, with $c_0 = b_0$. The Petrie apeirogons G_{-1} and G_1 of P which share the edge E can be written

$$\begin{aligned} \mathsf{G}_{-1}:&\quad \ldots, b_{-2}, a_{-2}, a_{-1}, a_0, b_0 = c_0, c_1, c_2, \ldots \\ \mathsf{G}_1:&\quad \ldots, b_2, a_2, a_1, a_0, b_0 = c_0, c_{-1}, c_{-2}, \ldots \end{aligned}$$

up to changing signs of the indices in F'.

Again using the local translational symmetries of G_{-1}, G_1 and F_1 (refer to Figure 10C5), we have

$$\begin{aligned} c_{-1} - c_{-2} &= a_1 - a_0, \\ c_0 - c_{-1} &= a_2 - a_1 \\ c_1 - c_0 &= a_{-1} - a_0 = a_3 - a_2 \\ c_2 - c_1 &= a_0 - a_{-1} = a_4 - a_3. \end{aligned}$$

We deduce three things at once. First, $\mathsf{F}' = \mathsf{F}_1 + s_{-1} = \mathsf{F}_1 - s_1$, where

$$s_j := a_{j+1} - b_{j-1} = b_{j+1} - a_{j-1}$$

is the basic translational symmetry of G_j for $j \in \mathbb{Z}$. Second, the translational symmetry t_1 of F_1 such that $a_{j+4} = a_j + t_1$ is

$$t_1 = s_{j+1} - s_{j-1}$$

for each j. Third, it follows that $c_{4j} = b_{4j}$ for each j; that is, the sets of parallel edges $\{a_{4j}, b_{4j}\}$ all lead from F_1 to F'.

The same then holds for the other sets of parallel edges from F_1, and we easily conclude that the vertices of P fall into the vertex-sets of disjoint faces of P parallel to F_1. The same will then be true of the other two faces F_2 and F_3 of P which contain a_0. Finally, it follows that we have (commuting) translations S_j by s_j of P (with $S_{j+4} = S_j$) and T_k by t_k which satisfy

$$S_{j-1}^{-1} S_{j+1} = T_1$$

and similar relations, showing that the translation group has rank 3; moreover, all this shows that P itself is just 3-dimensional. We now really need no more description than this to recognize $\{\frac{4}{0,1}, 3 : \frac{3}{0,1}\}$ as the apeirohedron $\{\infty, 3\}^{(b)}$ of Theorem 10B8. □

10C6 Remark Since a k-helix induces a k-fold rotation in the point-group, we deduce that the point-group of $\mathsf{P} = \{\frac{4}{0,1}, 3 : \frac{3}{0,1}\}$ is the rotation group $[4, 3]^+ \cong \mathcal{S}_4$ of the cube. Moreover, the edges of P are translates of the six edges of the regular tetrahedron $\{3, 3\}$, perhaps better regarded in the present context as those of its Petrial $\{\frac{4}{1,2}, 3 : 3\}$.

The remaining case is the apeirohedron $\{\frac{3}{0,1}, 4 : \frac{3}{0,1}\} \cong \{\infty, 4 : \cdot, *3\}$. As the notation indicates, this apeirohedron is self-Petrie; its faces and Petrie polygons are 3-helices. We shall prove

10C7 Theorem *The regular apeirohedron $\{\frac{3}{0,1}, 4 : \frac{3}{0,1}\}$ is 3-dimensional, and is isomorphic to the abstract apeirohedron $\{\infty, 4 : \cdot, *3\}$.*

10C8

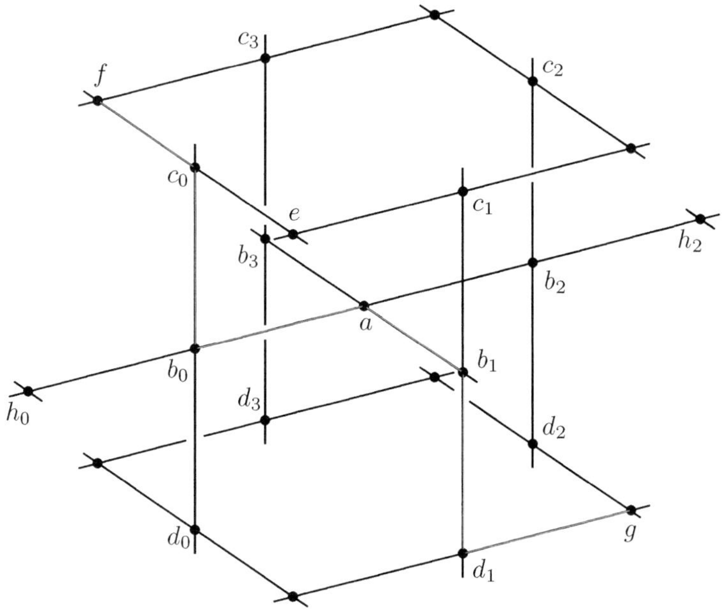

The regular apeirohedron $\{\frac{3}{0,1}, 4 : \frac{3}{0,1}\}$

Proof. Our proof is somewhat different from that of Theorem 10C4, but again – apart from appealing to the intrinsic translational symmetries of the faces or Petrie apeirogons, which we do not need to distinguish – we use purely local properties. Let $\mathsf{P} \in \{\frac{3}{0,1}, 4 : \frac{3}{0,1}\}$, and let $a \in \operatorname{vert} \mathsf{P}$ have adjacent vertices b_0, b_1, b_2, b_3 in this order around the square vertex-figure. For $j = 0, 1, 2, 3$, let the remaining vertices adjacent to b_j be c_j, h_j, d_j, so that a, c_j, h_j, d_j are again the vertices (in this order) of its vertex-figure (see Figure 10C8, where we have suppressed h_1 and h_3 for clarity). For each $j = 0, \ldots, 3$ (with indices j modulo 4) there is a (unique) face F_j of P containing b_j, a, b_{j+1}; let this have vertices $\ldots, c_j, b_j, a, b_{j+1}, d_{j+1}, \ldots$ (Figure 10C8 illustrates the case $j = 0$ in green). Then the Petrie apeirogon G_j of P which also contains b_j, a, b_{j+1} has vertices $\ldots, d_j, b_j, a, b_{j+1}, c_{j+1}, \ldots$.

Bearing in mind that the faces are 3-helices, for each j we obtain

10C9 $$d_{j+1} - b_j = b_{j+1} - c_j, \quad c_{j+1} - b_j = b_{j+1} - d_j.$$

Subtracting one equation from the other yields

10C10 $$c_{j+1} - d_{j+1} = c_j - d_j =: t_3,$$

say.

By assumption, the vertices $w_0, \ldots, w_3$ of P (in this order) adjacent to the vertex v of P are coplanar, and so the *displacement vector* $d(v)$ of v from the centre of its broad vertex-figure is

$$d(v) := v - \tfrac{1}{2}(w_0 + w_s) = v - \tfrac{1}{2}(w_1 + w_3).$$

From (10C9), we deduce that

$$d(b_{j+1}) = -d(b_j), \qquad d(b_{j+2}) = d(b_j)$$

for each j. The second equation shows that the hole $\ldots, h_0, b_0, a, b_2, h_2, \ldots$ must actually be a zigzag $\{\frac{2}{0,1}\}$, so that

$$d(a) = -d(b_0).$$

But from the hole $\ldots, h_1, b_1, a, b_3, h_3, \ldots$ we now conclude that $d(a) = -d(a)$, so that $d(a) = o$.

It follows that each vertex lies at the centre of its vertex-figure. Hence the holes are linear apeirogons $\{\frac{0}{1}\}$, the edges fall into three mutually orthogonal families, and thus P itself is 3-dimensional. Consequently, the holes and helices link up to form skew hexagons, as in the picture, and we have $\mathsf{P} \cong \{\infty, 4 : \cdot, *3\}$, as claimed. As in the previous case, the point-group is $[3,4]^+$. Once more, of course, the translational symmetries of the 3-helices extend to those of P itself. □

10C11 Remark The abstract notation for $\{\frac{3}{0,1}, 4 : \frac{3}{0,1}\}$ in Section 10B results from following through the operations which derive it from one of the Petrie–Coxeter apeirohedra; we could equally obtain it by an application of the circuit criterion Theorem 2D4. The extra relation corresponds to the skew hexagon with vertices a, b_0, c_0, e, c_1, b_1; in a slightly different context, we shall meet this again in the proof of Theorem 10C14.

We used the planarity of the vertex-figures in an essential way in the proof of Theorem 10C7. It may be of interest to see what happens if we relax this condition.

10C12 Theorem *The general apeirohedron in* $\{\frac{3}{0,1}, \frac{4}{1,2} : \frac{3}{0,1}\}$ *is 6-dimensional, and is a blend*

$$\{\tfrac{3}{0,1}, \tfrac{4}{1,2} : \tfrac{3}{0,1}\} = \{\tfrac{3}{0,1}, 4 : \tfrac{3}{0,1}\} \# \{\infty\} \# \{3\}.$$

Proof. We begin by observing that

$$\{\tfrac{3}{0,1}, 4 : \tfrac{3}{0,1}\},\ \{\infty\},\ \{3\} \preccurlyeq \{\tfrac{3}{0,1}, \tfrac{4}{1,2} : \tfrac{3}{0,1}\},$$

and that their blend is 6-dimensional. Therefore, we only have to show that $\{\frac{3}{0,1}, \frac{4}{1,2} : \frac{3}{0,1}\}$ is at most 6-dimensional. At this stage, we may also notice that

$$\{\tfrac{3}{0,1}, 4 : \tfrac{3}{0,1}\} \# \{3\} \cong \{\infty, 4 : \cdot, *3\},$$

but that this isomorphism fails if we further blend with $\{\infty\}$; indeed, the typical hexagon which provided the defining relation for $\{\infty, 4 : \cdot, *3\}$ opens up into an apeirogon $\{\frac{6}{0,1,3}\}$.

Our proof carries on from the point in that of Theorem 10C7 when we obtained the vector t_3 in (10C10). We now proceed along the holes of P though a, and then along those though the h_j, and so on; that is, we move to any new hole which diverges after an even number of edges from a – we can think of these as 'horizontal' holes. We make no assumption about how these different holes might subsequently meet. However, what we do see is that the vertices of P adjacent to those at odd edge distance from a in the horizontal holes are paired up by the translation vector t_3.

It follows from this that t_3 actually induces a translational symmetry T_3 of P, and that the 'vertical' holes (like those through d_j, b_j, c_j) are actually zigzags. In a similar way, we see that the 'horizontal' holes give rise to translational symmetries T_1 and T_2 (say). Crucially, it follows that there are at most six translation classes of edges, so that P is, as claimed, at most 6-dimensional. □

10C13 Remark The translation subgroup here has rank 4, as one would expect. Notice that, even in this case, we have appealed to very little of the abstract structure of the apeirohedron.

Mirror Vector (1, 2, 1)

We have left this class until last, because there is a contrast between it and the others.

The apeirohedra here have (finite) skew polygonal faces, planar vertex-figures and (finite) skew polygonal Petrie polygons. They are

$$\{\tfrac{6}{1,3}, 6 : \tfrac{4}{1,2}\} \cong \{6, 6 : 4\},$$
$$\{\tfrac{4}{1,2}, 6 : \tfrac{6}{1,3}\} \cong \{6, 6 : 6\},$$
$$\{\tfrac{6}{1,3}, 4 : \tfrac{6}{1,3}\} \cong \{6, 4 : 6\}.$$

The fine Schläfli symbols imply the corresponding isomorphisms. However, it is obvious that they cannot ensure that the apeirohedra be 3-dimensional. The reason is simple: the faces and Petrie polygons of each are blends $\{6\} \# \{2\}$ or $\{4\} \# \{2\}$, and so blending the apeirohedra with digons will not change their types. Observe that the edge-graphs of these apeirohedra are bipartite. Thus the best that we can achieve is

10C14 Theorem *In general, the three apeirohedra with fine Schläfli symbols* $\{\tfrac{6}{1,3}, 6 : \tfrac{4}{1,2}\}$, $\{\tfrac{4}{1,2}, 6 : \tfrac{6}{1,3}\}$ *and* $\{\tfrac{6}{1,3}, 4 : \tfrac{6}{1,3}\}$ *are 4-dimensional, being blends of a 3-dimensional pure component and a digon.*

Proof. We deal with $\mathsf{P} \in \{\tfrac{6}{1,3}, 4 : \tfrac{6}{1,3}\}$ first, while the treatment of $\{\tfrac{3}{0,1}, 4 : \tfrac{3}{0,1}\}$ is still (perhaps) fresh in the mind. The two apeirohedra have isomorphic edge-graphs, and so we shall adopt the same initial notation. Now, a typical skew hexagonal face (or Petrie polygon – we shall not bother to distinguish them) F of type $\{\tfrac{6}{1,3}\}$ has vertices a, b_0, c_0, e, c_1, b_1; since F is centrally symmetric, we see that $c_j - b_j = w_0$ and $d_j - b_j = w_1$ for each j. Then $t_1 := c_j - d_j$ (for each j)

is a translation vector of P as in the proof of Theorem 10C12; we clearly have analogous translations t_2, t_3 as well, but now there is no translation s_1.

The same calculations of centres of vertex-figures yields

$$v_0 + v_2 = v_1 + v_3 = -(w_0 + w_1),$$

but no contradictions of signs. Indeed, we see that the displacements of vertices from the centres of their vertex-figures are equal but alternating in sign at adjacent vertices. It follows immediately that, if this displacement is non-zero, then P is a blend.

Similar considerations apply to the other two cases, which we can treat together; we take $\mathsf{P} \in \{\frac{6}{1,3}, 6 : \frac{4}{1,2}\}$. Let a be the initial vertex, whose vertex-figure has vertices $b_0, \ldots, b_5$ (in cyclic order). Let a, b_0, c, b_1 be a skew tetragonal Petrie polygon G of type $\{\frac{4}{1,2}\}$ of P, and let b_5, a, b_0, c, d_5, e_5 and b_2, a, b_1, c, d_1, d_2 be the two skew hexagonal faces $\mathsf{F}_5, F_2 = \{\frac{6}{1,3}\}$ of P which share two edges of G. Then $v_j = b_j - a = d_j - c$ for $j = 2, 5$ (again, parallel sides of centrally symmetric skew hexagons); consequently, the vectors from a and c to the centres of their vertex-figures are the same. Tracing edge-paths shows that the same holds for alternate vertices; once again, then, P will generally be a blend. Note that the displacement from the remaining vertices must be equal and opposite in sign. □

10C15 Remark In view of the following Theorem 10C16, this actually suffices for a proof. However, the interested reader might like to show directly that alternate vertices genuinely form 3-dimensional configurations. For example, from $\{\frac{4}{1,2}, 6 : \frac{6}{1,3}\}$ we obtain $\{6, \frac{6}{1,3} \mid 3\}$, as comparison with the abstract case demonstrates (see also [98]). Actually, here it is even easier to show that the mid-points of the edges of $\{\frac{6}{1,3}, 6 : \frac{4}{1,2}\}$ yield $\{6, \frac{4}{1,2} \mid 4\}$, with the faces and vertex-figures providing the planar hexagons, and the Petrie polygons providing the planar holes. There is no associated *regular* figure in the third case, but there is enough symmetry to establish the 3-dimensionality of the alternate vertex configuration.

It is natural to ask whether we can impose rigidity on these apeirohedra. In fact, we can.

10C16 Theorem *Three of the Grünbaum apeirohedra are determined by the fine Schläfli symbols* $\{\frac{6}{1,3}, 6 : \frac{4}{1,2}, \cdot, \frac{1}{0}\}$, $\{\frac{4}{1,2}, 6 : \frac{6}{1,3}, \cdot, \frac{1}{0}\}$ *and* $\{\frac{6}{1,3}, 4 : \frac{6}{1,3}, \frac{1}{0}\}$ *as rigid regular apeirohedra.*

Proof. We deliberately write $\{\frac{1}{0}\}$ rather than $\{\infty\}$ here, to emphasize that we are talking about the linear apeirogon. The reason is straightforward. Specifying these apeirogons as linear ensures that a vertex is the centre of its vertex figure (compare the previous proof), and hence that the corresponding apeirohedron is 3-dimensional. □

10C17 Remark We observed in Section 2B that, if the vertex-figure of a regular apeirohedron $\mathcal{P}$ is a $2k$-gon and the k-holes or k-zigzags of $\mathcal{P}$ are infinite, then they coincide. Hence we can represent these rigid apeirohedra alternatively as

$$\{\tfrac{6}{1,3}, 6 : \tfrac{4}{1,2} \mid \cdot, \tfrac{1}{0}\}, \quad \{\tfrac{4}{1,2}, 6 : \tfrac{6}{1,3} \mid \cdot, \tfrac{1}{0}\}, \quad \{\tfrac{6}{1,3}, 4 : \tfrac{6}{1,3} \mid \tfrac{1}{0}\}.$$

Notes to Section 10C

1. The concept of rigidity arose from the wish to find a more geometric description than that of [98] or [99, Section 7E] of the apeirohedra with mirror vector $(1, 1, 1)$. That they could be described as we have seen here was a pleasant surprise.
2. As we noted, the sole regular apeirohedron in $\mathbb{E}^3$ which was missed by Grünbaum in [58] is $\{\frac{3}{0,1}, 4 : \frac{3}{0,1}\}$. In a way, this is surprising: first, since $\{\frac{3}{0,1}, 4 : \frac{3}{0,1}\}$ has the same edge-graph as the apeirohedron $\{\frac{6}{1,3}, 4 : \frac{6}{1,3}\}$ and, second, because eleven would be a strange number of such apeirohedra in a complete classification (at least, so it seems to us). On the other hand, $\{\frac{3}{0,1}, 4 : \frac{3}{0,1}\}$ cannot be derived from any of the other regular apeirohedra in $\mathbb{E}^3$ by immediately obvious operations such as duality, Petriality, and the like. As we have seen, the two apeirohedra are actually related by $\boldsymbol{\kappa}$.

11

Four-Dimensional Polyhedra

As in the case of the 3-dimensional regular apeirohedra described in the previous chapter, the mirror vector plays an important rôle in the classification of the 4-dimensional regular polyhedra. Thus in the first Section 11A of this chapter we determine the possible mirror vectors of such polyhedra. Then Section 11B is devoted to the polyhedra with mirror vector $(3, 2, 3)$ and their relatives under standard operations such as Petriality. Section 11C deals with one particular family of these polyhedra, where we describe their realization domains in detail. Section 11D concerns the mirror vector $(2, 3, 2)$; here, most of the standard operations lead to polyhedra in the same class.

However, though there is a close analogy between the infinite and finite cases, there is one family whose symmetry groups need not be related to reflexion groups. In Section 11E, the treatment therefore departs from the previous pattern by recalling quaternions from Section 1K, and then treats the class of polyhedra with mirror vector $(2, 2, 2)$.

Finally, in Section 11F, we describe various connexions among these regular polyhedra. The most interesting of these is the way that the skewing operation $\boldsymbol{\sigma}$ takes certain polyhedra in class $(3, 2, 3)$ into ones in class $(2, 2, 2)$.

11A Mirror Vectors

As a first step in describing the geometrically realized regular polyhedra in $\mathbb{E}^4$, we need to determine each possible mirror vector $(\dim R_0, \dim R_1, \dim R_2)$, with (R_0, R_1, R_2) the generatrix of the corresponding symmetry group $\mathbf{G}(P)$. Theorem 4B8 provides a starting point; the mirror vector must satisfy

$$(1, 2, 2) \leqq (\dim R_0, \dim R_1, \dim R_2) \leqq (3, 3, 3),$$

where $\leqq$ indicates inequality in each component individually.

We now proceed as follows. Using essentially the same trick as in earlier chapters (see also [82, 98], for example), if the mirror R_0 satisfies $\dim R_0 = 1$, then we can replace it by

$$-R_0 = R_0^{\perp},$$

its orthogonal complement, which (as an isometry) is its product with the central inversion $-I$; as we recall from Section 5D, this operation is centriversion, and is denoted by $\boldsymbol{\zeta}$. (We shall employ more general operations $\boldsymbol{\zeta}_j$ in Section 11F, but they are not needed here.) We always obtain another finite group $\mathbf{G}'$; in fact,

$$|\mathbf{G}'| = \tfrac{1}{2}|\mathbf{G}|, \ |\mathbf{G}| \text{ or } 2|\mathbf{G}|.$$

It will turn out that we always obtain another regular polyhedron as well.

Next, if $\dim R_0 = 2$ and $\dim R_2 = 3$ (or vice versa, but we shall shortly see that this case will have to be excluded), then we can replace R_0 by R_0R_2, that is, apply (or reverse) the Petriality $\boldsymbol{\pi}$; bearing in mind (1D17), the new R_0 has $\dim R_0 = 1$ or 3, and in the former case we can proceed as previously.

Finally, so long as our (possibly new) group contains a hyperplane reflexion (that is, $\dim R_j = 3$ for some j), we can regard $\mathbf{G}$ as a reflexion (Coxeter) group, on which certain involutions with 2-dimensional mirrors act as automorphisms (more precisely, $\mathbf{G}$ is the corresponding semi-direct product). When we have carried out the foregoing procedures, it will later become clear that we need analyse in detail only those polyhedra whose groups have mirror vectors $(3, 2, 3)$ or $(2, 3, 2)$. For classification purposes, we then reverse the procedure: the starting point is a Coxeter (hyperplane reflexion) group, not necessarily with standard generators, which is represented by a diagram that permits twists (corresponding to the automorphisms).

The one case which is not covered by the previous analysis is mirror vector $(2, 2, 2)$. This class is closed under $\boldsymbol{\pi}$ and $\boldsymbol{\zeta}$; in theory, it is also closed under duality $\boldsymbol{\delta}$ but, as we shall see, this is not applicable here. In fact, there are polyhedra in this class whose symmetry groups are not subgroups of reflexion groups. The approach here will be through quaternions (see Section 1K), which will enable us to relate such a polyhedron to a pair of (sometimes degenerate) polyhedra in $\mathbb{E}^3$ or, rather, the corresponding regular projective polyhedra.

We now briefly analyse the possibilities in general terms, leaving the details and the enumeration problems for each class until subsequent sections. In the listing of the classes, we group together those which are related by $\boldsymbol{\pi}$ or $\boldsymbol{\zeta}$; as we have just remarked, we cannot always apply $\boldsymbol{\delta}$.

- $(3, 3, 3)$ Here, the intersection $R_0 \cap R_1 \cap R_2$ of three linear hyperplanes will be a line, and so a corresponding polyhedron will only be 3-dimensional. It must therefore be excluded (but only on these grounds).
- $(1, 3, 3)$ This case is permitted; however, these polyhedra are blends of polyhedra of full rank with a digon $\{2\}$, since applying $\boldsymbol{\zeta}$ leads to the case $(3, 3, 3)$.
- $(2, 3, 3)$ An appeal to (1D17) shows that applying $\boldsymbol{\pi}$ to this case yields either case $(3, 3, 3)$ or case $(1, 3, 3)$. The first possibility must be excluded on the same grounds as before, but the second will occur.

11A1 Remark The discussion so far shows that we do not need to consider these three mirror vectors any further; the enumeration of the corresponding polyhedra has already been carried out in Chapter 7 or in Sections 8C and 9A.

- (3, 3, 2) This would be the dual class to (2, 3, 3). However, even in the permitted case (the Petrials of class (1, 3, 3)), the faces of the original polyhedron are centred at o, and so the dual must be excluded.
- (1, 3, 2) If this class occurred, it would be obtained from the class (3, 3, 2) by applying $\boldsymbol{\zeta}$; we must therefore disallow it.
- (3, 2, 3) Let the group be $\mathbf{G} = \langle R_0, R_1, R_2 \rangle$. Then R_1 acts as a twist of the hyperplane reflexion group $\mathbf{H} = \langle S_0, \ldots, S_3 \rangle$, given by

$$(S_0, \ldots, S_3) := (R_0, R_1R_2R_1, R_2, R_1R_0R_1).$$

The underlying diagram of $\mathbf{H}$ is denoted $\mathcal{D}_1(p, q; r)$ as in Figure 11A2 (the notation indicates that p and q play a similar rôle, while that of r is different). The labels on the nodes of the diagram indicate the reflexions R_0 and R_2, while R_1 is indicated by the implied flip of the diagram. Moreover, we must have $p, q > 1$, where p is not a fraction with an even denominator (we shall explain why in Section 11B) and $r > 2$; the resulting polyhedron is of type $\{2p, 2q \mid r\}$, where we must interpret these entries as integers indicating periods rather than fractions. In case q is also not a fraction with an even denominator, then $\boldsymbol{\delta}$ interchanges p and q. There is an alternative approach when q is a fraction with an even denominator; in this event, the automorphism R_1 of the diagram is inner, and the group of the polyhedron can be obtained from that of some regular 4-polytope by Petrie contraction ϖ (see Section 5B).

11A2

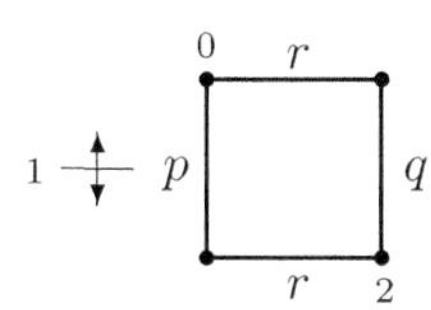

The diagram $\mathcal{D}_1(p, q; r)$

- (1, 2, 3) This case arises from the case (3, 2, 3) by applying $\boldsymbol{\zeta}$.
- (2, 2, 3) This case is obtained from either case (3, 2, 3) or case (1, 2, 3) by Petriality $\boldsymbol{\pi}$; naturally, the two possibilities have to be distinguished. These three classes of polyhedra (or four if we make the distinction implied in the last sentence) fall into two families with the same vertices.
- (2, 3, 2) This class gives rise to rich families. In this case, R_0 and R_2 are the automorphisms, which act on a diagram given by

$$(S_0, \ldots, S_3) := (R_1, R_0R_1R_0, R_0R_2R_1R_2R_0, R_2R_1R_2).$$

The general case is therefore derived from a diagram $\mathcal{D}_2(p, q, r)$, as in Figure 11A3. Once again, the labels on the nodes and automorphisms correspond to the original reflexions. The resulting polyhedron is of type

$\{2p, 2q : 2r\}$ (with the same convention about the entries as previously), from which up to five others are obtained by $\boldsymbol{\delta}$ and $\boldsymbol{\pi}$; observe that these two operations preserve the class $(2, 3, 2)$. We have $p, q, r > 1$, but q must not be a fraction with even denominator; we shall see why in Section 11D.

11A3

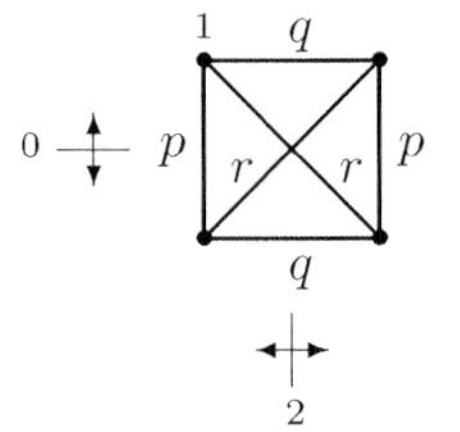

The diagram $\mathcal{D}_2(p, q, r)$

- $(3, 2, 2)$ This case would arise from $(2, 2, 3)$ by duality. However, it may be seen that, whichever way the original group of type $(2, 2, 3)$ arises, the product R_0R_1 of its corresponding reflexions R_0 and R_1 is a double rotation (in two orthogonal planes), since $R_0 \cap R_1 = \{o\}$; it follows that the class cannot occur.
- $(1, 2, 2)$ This case would be obtained from $(3, 2, 2)$ by applying $\boldsymbol{\zeta}$, and so it too must be excluded.
- $(2, 2, 2)$ This is the anomalous case, to which the notion of a Coxeter group with outer automorphisms is inapplicable. Indeed, some examples of this kind cannot be related to Coxeter groups in any meaningful way. The approach here is through quaternions; see Section 1K for the appropriate background.

11A4 Remark Lowering each entry of these mirror vectors by 1 yields a potential mirror vector for polyhedra or apeirohedra in $\mathbb{E}^3$. If we reinstate $(3, 3, 3)$ (which yields only 3-dimensional polyhedra), then exactly the analogues of those which give polyhedra in $\mathbb{E}^4$ are valid mirror vectors in $\mathbb{E}^3$.

It is notable that only the symmetry groups $[3, 3, 3]$ and $[3, 4, 3]$ of the regular 4-polytopes give rise to polyhedra in the classes $(3, 2, 3)$ and $(2, 3, 2)$ and those derived from them (we leave aside the groups $[s] \times [s]$). Even though other finite reflexion groups in $\mathbb{E}^4$ permit diagram automorphisms (for suitably chosen generators), these are inner, and then the putative polyhedra degenerate.

11B Mirror Vector $(3, 2, 3)$ and Its Relatives

In this section, we consider the regular polyhedra with mirror vector $(3, 2, 3)$; except in one particularly interesting case (see the next Section 11C), we shall say rather less about those which are derived from them using the operations $\boldsymbol{\zeta}$, $\boldsymbol{\pi}$ or both. Where appropriate, though, we shall look at the effect of the operations φ_k; however, we leave discussion of η and σ (which change the mirror vector) until Section 11F.

Let the group of a polyhedron in this class be $\mathbf{G} = \langle R_0, R_1, R_2 \rangle$, and consider the operation

11B1 $$(R_0, R_1, R_2) \mapsto (R_0, R_1R_0R_1, R_2, R_1R_2R_1) =: (S_0, S_1, S_2, S_3);$$

each of the S_j is now a hyperplane reflexion. Since R_0 and R_2 commute, the picture here – as we saw in Section 11A – is of a (generalized) Coxeter diagram of the form

11B2

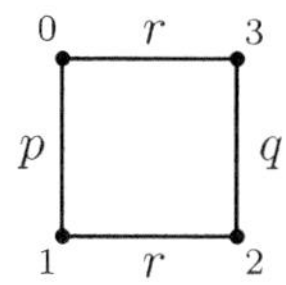

The label j on a node in Figure 11B2 now corresponds to the generator S_j of the reflexion subgroup $\mathbf{H} := \langle S_0, S_1, S_2, S_3 \rangle$. For geometric reasons, we cannot have $r = 2$; otherwise, $\langle S_0, S_1 \rangle$ and $\langle S_2, S_3 \rangle$ act on orthogonally complementary planes of $\mathbb{E}^4$, and the corresponding 'polyhedron' degenerates to a polygon (or, more strictly perhaps, a dihedron). However, $p = 2$ or $q = 2$ (or both) are allowed. Observe that any of the marks p, q or r may be fractional (with a restriction on p which will be explained shortly); in other words, the mark p indicates the angle π/p between the mirrors S_0 and S_1, and similarly for q or r, and so we only demand that the marks p and q be greater than 1.

Let us go a little deeper here into the geometric detail. The mirrors R_0 and R_1 whose reflexions generate the group of the initial 2-face are of dimensions 3 and 2, respectively, and so (generically) their intersection has dimension 1. The initial vertex lies in R_1 but not in R_0, and so its images under $\mathbf{G}_2 = \langle R_0, R_1 \rangle$ will only span a plane; that is, the 2-faces of polyhedra in this class are planar. However, for the vertex-figure, the situation is reversed: R_1 has dimension 2 while R_2 has dimension 3, and so the vertex-figure will be (in general) a skew (3-dimensional) polygon. With the notation of the diagram, the faces will thus be planar of type $\{2p\}$, with the appropriate interpretation when p is fractional. We thus see at once that $p = s/t$ cannot be a fraction with an even denominator t, because then the resulting polygon doubly covers $\{\frac{s}{t/2}\}$; this restriction is that foreshadowed in Section 11A.

We can employ Wythoff's construction of Section 4A to show us what the polyhedron looks like. In general, its (planar) faces (which thus include its vertices and edges) are the

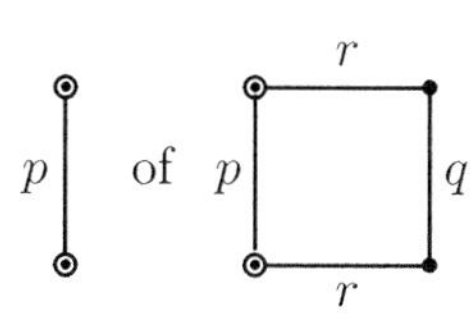

Moreover, the remaining faces

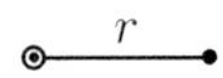

of the 4-polytope form the (again planar) holes in the polyhedron. Finally, the vertex-figure is a skew polygon; if $q = \frac{s}{t}$, then it is in fact $\{\frac{2s}{t,s}\}$.

In view of the rigidity criterion of Theorem 6D5, the polyhedron is therefore denoted unambiguously by its fine Schläfli symbol

11B3 $$\{p, \tfrac{2s}{t,s} \mid r\},$$

with $p, q = \frac{s}{t}$ and r as in Figure 11A2.

For the enumeration, we begin with the case $p = q = 2$; any $r > 2$ is allowed, and in particular r can be a fraction. The corresponding diagram is

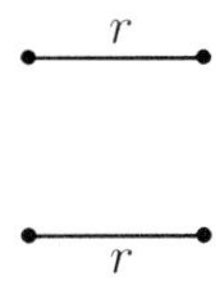

with the horizontal flip understood. From (11B3), we obtain $\{4, \frac{4}{1,2} \mid r\}$, which is isomorphic to the universal toroid $\{4, 4 \mid s\}$ ($= \{4, 4\}_{(s,0)}$ in the appropriate adaptation of the notation of [99, Section 1D]) when $r = \frac{s}{t}$ in its lowest terms. Its group is $(\mathbf{D}_2^s \times \mathbf{D}_2^s) \rtimes \mathbf{Z}_2$, an extension of the product of two copies of the geometric dihedral group $\mathbf{D}_2^s$ by an outer automorphism of order 2, and so has order $2 \cdot (2s)^2 = 8s^2$. The geometric description is straightforward: take the cartesian product $\{r\} \times \{r\}$ of two congruent polygons $\{r\}$, bearing in mind that r can be a fraction (so that $\{r\}$ would then be a star-polygon). The vertices and edges are those of the product, while the square faces $\{4\}$ are the products of edges, one from each component; the vertex-figure is a skew square $\{\frac{4}{1,2}\}$. Note that this polyhedron is self-dual.

Henceforth, then, we can assume that at least one of p or q exceeds 2, so that the diagram is connected. It thus follows that the corresponding Coxeter group will be irreducible. One approach is to look through suitable presentations of such Coxeter groups (that is, Coxeter diagrams with, possibly, fractional marks – *Goursat tetrahedra* in the terms of [27, Section 14.8]) which admit appropriate automorphisms.

The Group $\mathbf{A}_4 \rtimes \mathbf{Z}_2$

The next examples derive from the diagram of the extended group $\mathbf{A}_4 \rtimes \mathbf{Z}_2 \cong [3, 3, 3] \times \boldsymbol{C}_2$ of the regular 4-simplex; we have seen some of them already, a part of the general discussion of Section 9C. Again, the horizontal flip is understood, with the extension of the group $[3, 3, 3]$ by an outer automorphism of order 2 as before (in fact, the group is actually a direct product):

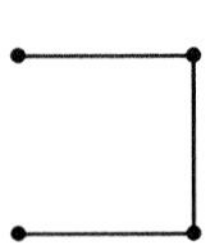

There result two polyhedra which are isomorphic to the universal $\{4, 6 \mid 3\}$ and its dual $\{6, 4 \mid 3\}$; these polyhedra were first described by Coxeter in [24]. In our more geometrical notation, their fine Schläfli symbols are $\{4, \frac{6}{1,3} \mid 6\}$ and $\{6, \frac{4}{1,2} \mid 4\}$. We shall say more about these two polyhedra and their relatives in Sections 11C and 11F; for the time being, we just give a picture in Figure 11B4 of the latter projected into the plane of a decagonal component of its Petrie polygon $\{\frac{10}{1,3}\}$.

11B4

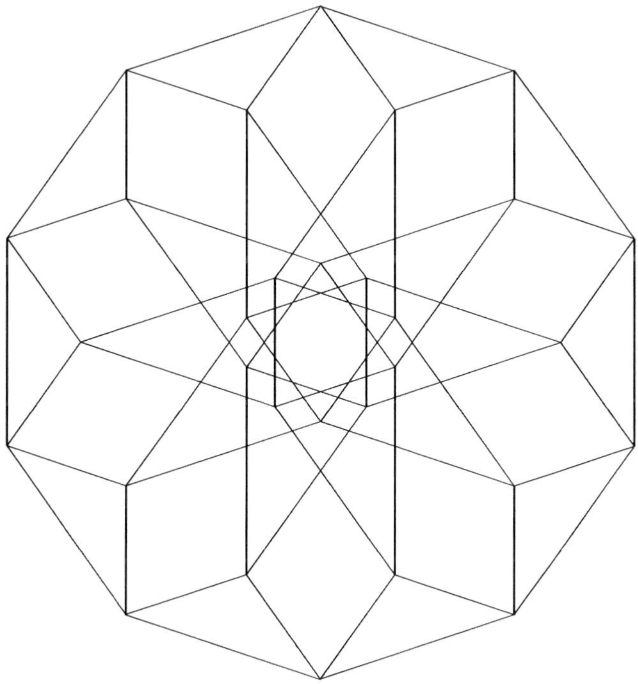

The polyhedron $\{6, \frac{4}{1,2} \mid 4\}$

The Group $\mathbf{F}_4 \rtimes \mathbf{Z}_2$

We now move on to the extended group $\mathbf{F}_4 \rtimes \mathbf{Z}_2 \cong (\mathbf{O}/\mathbf{O}; \mathbf{O}/\mathbf{O})^*$ of the regular 24-cell, from which we derive the diagrams

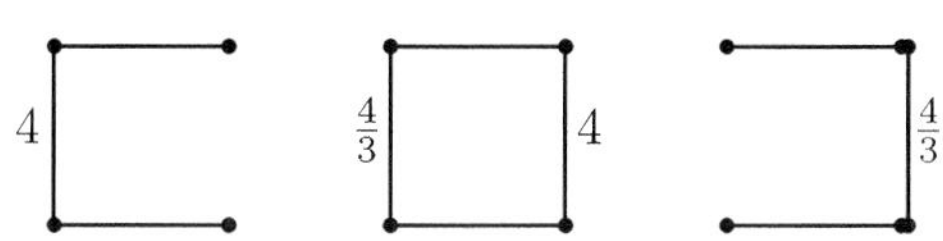

Each permits a top-to-bottom flip, and so gives two dual regular polyhedra with mirror vectors $(3, 2, 3)$. From the first diagram, we obtain polyhedra isomorphic to the universal dual regular polyhedra $\{4, 8 \mid 3\}$ and $\{8, 4 \mid 3\}$ of [24]. Following

the notation of (11B3), we can arrange the six resulting polyhedra (given by their fine Schläfli symbols) in a pattern similar to that of Table 7G7:

11B5

$\{8, \frac{4}{1,2} \mid 3\}$	$\{4, \frac{8}{1,4} \mid 3\}$		
	$\{8, \frac{8}{3,4} \mid 3\}$	$\{\frac{8}{3}, \frac{8}{1,4} \mid 3\}$	
		$\{4, \frac{8}{3,4} \mid 3\}$	$\{\frac{8}{3}, \frac{4}{1,2} \mid 3\}$

Polyhedra in the same row of Table 11B5 are related by duality $\boldsymbol{\delta}$, while those in the same column are related by the faceting operation φ_3 (because the index $k = 3$ in φ_k is odd, the mirror vector $(3, 2, 3)$ is preserved). For those in the middle row, bear in mind here that, while the notation only indicates the geometry of the faces, vertex-figures and holes, it nevertheless ensures rigidity by Theorem 6D5.

11B6 Remark Because $\{8, \frac{8}{3,4} \mid 3\} = \{4, \frac{8}{1,4} \mid 3\}^{\varphi_3}$, it follows that (abstractly) we have

$$\{8, \tfrac{8}{3,4} \mid 3\} \cong \{8, 8 \mid 3, 4\};$$

combinatorially, the polyhedron is self-dual.

Let us say a little about the geometry of these polyhedra. Their symmetry group is $[3, 4, 3] \rtimes \mathbf{Z}_2 = (\mathbf{O}/\mathbf{O}; \mathbf{O}/\mathbf{O})^*$ in the notation of Section 1K. The vertex-set of those with octagonal vertex-figures is that of the (Minkowski) sum of congruent polar copies of the 24-cell, and consists of all 144 permutations with all changes of sign of

$$(2 + \sqrt{2}, \sqrt{2}, 0, 0),\ (1 + \sqrt{2}, 1 + \sqrt{2}, 1, 1).$$

The sum has 48 octahedral facets; two arising from adjacent facets of the same copy of $\{3, 4, 3\}$ are joined by a triangular prism, whose square faces $\{4\}$ are those of $\mathsf{P} := \{4, \frac{8}{1,4} \mid 3\}$. The triangular faces of the octahedra now become the holes $\{3\}$ of P.

The vertex-set of the two other polyhedra is that of the intersection of the copies of the 24-cell, and consists of all permutations with all changes of sign of

$$(\sqrt{2} + 1, 1, 1, \sqrt{2} - 1), \quad (2, \sqrt{2}, \sqrt{2}, 0).$$

The 48 facets of the intersection are truncated cubes, whose octagonal faces are those of $\mathsf{Q} := \{8, \frac{4}{1,2} \mid 3\}$. Two truncates arising from adjacent facets of the same copy of $\{3, 4, 3\}$ meet in a triangular face, which becomes a hole of Q.

We can express these vertex-sets in a different way employing quaternions. Analogously to what we saw for the pentagonal 4-polytopes, they form cosets of $\mathbf{O}$ in the whole quaternion group $\mathbf{Q}$. For example, for vert$\{4, \frac{8}{1,4} \mid 3\}$, the (left or right) cosets representatives can be taken to be

$$2+\sqrt{2}+\sqrt{2}\mathbf{i}, \quad 2+\sqrt{2}+\sqrt{2}\mathbf{j}, \quad 2+\sqrt{2}+\sqrt{2}\mathbf{k},$$

which form a typical hole $\{3\}$.

We must also bear in mind the polyhedra obtained from these by applying π, ζ or both. First, the Petrie polygons are of type $\{\frac{24}{1,7}\}$ for $\{8, \frac{4}{1,2} \mid 3\}$ and its dual, $\{\frac{12}{1,5}\}$ for $\{8, \frac{8}{3,4} \mid 3\}$ and its dual, and $\{\frac{24}{5,11}\}$ for $\{4, \frac{8}{3,4} \mid 3\}$ and its dual. So, for instance,

$$\{8, \tfrac{4}{1,2} \mid 3\}^{\pi} = \{\tfrac{24}{1,7}, 4 : 8, 3\},$$

which specifies it abstractly and geometrically; compare here Proposition 5A15. If we further apply ζ, we obtain

$$\{\tfrac{24}{1,7}, 4 : 8, 3\}^{\zeta} = \{\tfrac{24}{5,11}, 4 : \tfrac{8}{3,4}, \tfrac{6}{1,3}\};$$

the fine Schläfli symbol does specify it, because applying ζ again returns us to what we had before. Observe how ζ affects the zigzags as well as the face, and note as well that this polyhedron is distinct from

$$\{\tfrac{8}{3}, \tfrac{4}{1,2} \mid 3\}^{\pi} = \{\tfrac{24}{5,11}, 4 : \tfrac{8}{3}, 3\}.$$

Consequently, from each of the six polyhedra of Table 11B5 we obtain three others, leading to 24 related polyhedra in all in this section.

The Group $\mathbf{G}_4$

We now need to comment on the remaining possibilities. As we have said, we can exclude the possibility that p is a fraction with an even denominator. Leaving aside for now the case when q is a fraction with an even denominator, this denies any further examples arising from the two groups already considered, or from the remaining Coxeter groups apart from $\mathbf{G}_4 \cong [3,3,5]$. However, among the Coursat tetrahedra in this last group is

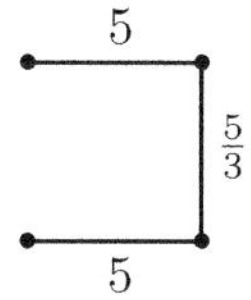

(see the notes at the end of the section). This diagram apparently fulfils all the required conditions. In fact, the automorphism here is inner, because $\mathbf{G}_4$ is not a proper subgroup of any finite orthogonal group in $\mathbf{O}_4$ (see, for example, [43]); therefore, it does not lead to a new regular polyhedron, because the realization is degenerate. More precisely, if the polyhedron were to exist, then its vertices and edges would be (as we have observed) those of

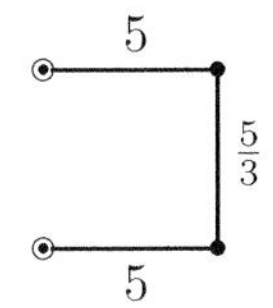

From the symmetry, the $14400/10 = 1440$ vertices would lie in pairs in the (possibly extended) edges of the 600-cell $\{3,3,5\}$. However, direct calculation shows that the initial vertex is actually one of $\{3,3,5\}$, and so the vertices coincide in twelves with those of $\{3,3,5\}$. Similar considerations apply to other putative examples.

Polyhedra Obtained by ϖ

There remain the cases when q is a fraction with an even denominator (these were overlooked in [100]). We observe, as in [94, Section 6.1 (10)], that Petrie contraction ϖ of Section 5D, namely,

$$(T_0,\ldots,T_3) \mapsto (T_1, T_0T_2, T_3) =: (R_0, R_1, R_2),$$

applied to the group $\langle T_0,\ldots,T_3\rangle$ of a regular 4-polytope Q yields that of a regular polyhedron P. With one exception, P has the following description. The edges of its faces join successive mid-points of edges of Petrie polygons of the facets of Q; the faces can then be thought of as central sections of these facets (this is actually the case when Q is convex). If $\mathsf{Q} = \{q_1, q_2, q_3\}$, with $q_3 = \frac{s}{t} \neq 4$ (this is the exception), then the vertex-figure is $\{\frac{2s}{t,s}\}$; note that each edge of a facet of Q belongs to two of its Petrie polygons. In case $\mathsf{Q} = \{3,3,4\}$, the resulting 'polyhedron' splits into three copies of the torus $\{4, \frac{4}{1,2} \mid 4\}$.

We thus obtain 15 regular polyhedra, whose generatrices have mirror vector $(3,2,3)$. Applying ϖ and (11B1) in turn yields

$$(T_0,\ldots,T_3) \mapsto (T_1, T_0T_2T_1T_0T_2, T_3, T_2T_3T_2),$$

which corresponds to the diagram $\mathcal{D}_1(p,q;r)$, where

- $p = \frac{1}{2}h$, with $\{h\}$ the Petrie polygon of the facet $\{q_1,q_2\}$ of Q (strictly speaking, the Petrie polygon is skew, and p corresponds to the planar polygon whose basic rotation is $(T_1T_0T_2)^2$),
- $q = \frac{1}{2}q_3$,
- $\{r\}$ is the hole of the vertex-figure $\{q_2,q_3\}$ of Q, namely, the face of $\{q_2,q_3\}^{\varphi_2}$.

Summarizing this discussion, we see that we have

11B7 Proposition *If* $\mathsf{Q} = \{q_1,q_2,q_3\}$ *is a classical regular 4-polytope with* $q_3 \neq 4$, *then*

$$\mathsf{Q}^{\varpi} = \{h, \tfrac{2s}{t,s} \mid r\}$$

is a 4-dimensional regular polyhedron with mirror vector $(3,2,3)$. *Here,* $\{h\}$ *is the planar component of the Petrie polygon of* $\{q_1,q_2\}$, $q_3 = \frac{s}{t}$ *and* $\{r\}$ *is the hole of* $\{q_2,q_3\}$.

We may also observe that we can recover the original 4-polytope Q from the diagram $\mathcal{D}_1(p, q; r)$:

- $q_3 = 2q$;
- q_3 and r yield q_2 – indeed, $r = t$ for $\{3, t\}$ and $\{t, 3\}$, while $r = 3$ for $\{5, \frac{5}{2}\}$ and $\{\frac{5}{2}, 5\}$;
- q_2 and $h = 2p$ then give q_1 by

$$\cos^2 \frac{\pi}{q_1} = \cos^2 \frac{\pi}{h} - \cos^2 \frac{\pi}{q_2}$$

(compare [27, 2.33]).

Of course, because q is a fraction with an even denominator (q_3 is odd), none of these 15 polyhedra has a geometric dual in $\mathbb{E}^4$. We should observe that reversing the generatrix of such a polyhedron does yield a string C-group. However, when we look at the geometry, we see that a reversed generatrix fails to yield a non-degenerate polyhedron. Because there is an original face for each Petrie polygon of each facet of the initial 4-polytope, it follows that corresponding vertices of the putative geometric dual fall together (in threes, fours or sixes); thus the dual degenerates. Moreover, this degeneracy clearly survives applications of $\boldsymbol{\zeta}$ (and $\boldsymbol{\pi}$ as well).

However, with a few exceptions, these polyhedra all have quasi-duals, in the sense introduced at the end of Section 5D. Changing the sign of the twist (here, T_0T_2) results in replacing p, q in the diagram $\mathcal{D}_1(p, q; r)$ by p', q', with

$$\frac{1}{p} + \frac{1}{p'} = 1,$$

and similarly for q. Interchanging the rôles of p' and q' then yields

$$\mathcal{D}_1(p, q; r) \longleftrightarrow \mathcal{D}_1(q', p'; r).$$

When p and q have odd numerators, but the denominator of p is odd and that of q is even, we see that the regular polyhedron associated with the new diagram $\mathcal{D}_1(q', p'; r)$ will be a quasi-dual of that associated with $\mathcal{D}_1(p, q; r)$. Indeed, there are certain cases where $p' = q$ (and $q' = p$), so that the associated polyhedra are self-quasi-dual.

The polyhedra derived from the pentagonal 4-polytopes are of particular interest; we list the polytopes, their Petrie contractions, and the corresponding quasi-dualities. Except in the last block, isomorphic polytopes occur on the same line.

The first two polyhedra do not even have quasi-duals.

11B8 $\quad \{3, 3, 5\} \xrightarrow{\varpi} \{4, \frac{10}{2,5} \mid 5\} \qquad \{3, 3, \frac{5}{2}\} \xrightarrow{\varpi} \{4, \frac{10}{4,5} \mid \frac{5}{2}\}$

The next grouping consists of those polyhedra which are self-quasi-dual.

$\{3, 5, \frac{5}{2}\} \xrightarrow{\varpi} \{10, \frac{10}{4,5} \mid 3\} \qquad \{3, \frac{5}{2}, 5\} \xrightarrow{\varpi} \{\frac{10}{3}, \frac{10}{2,5} \mid 3\}$

11B9 $\quad \{\frac{5}{2}, 5, 3\} \xrightarrow{\varpi} \{6, \frac{6}{2,3} \mid 5\} \qquad \{5, \frac{5}{2}, 3\} \xrightarrow{\varpi} \{6, \frac{6}{2,3} \mid \frac{5}{2}\}$

$\{5, 3, \frac{5}{2}\} \xrightarrow{\varpi} \{10, \frac{10}{4,5} \mid \frac{5}{2}\} \qquad \{\frac{5}{2}, 3, 5\} \xrightarrow{\varpi} \{\frac{10}{3}, \frac{10}{2,5} \mid 5\}$

Finally, we have examples of pairs linked by quasi-duality $\boldsymbol{\delta}^*$ (say).

11B10

$$\{5,3,3\} \xrightarrow{\varpi} \{10, \tfrac{6}{2,3} \mid 3\} \xleftrightarrow{\delta^*} \{6, \tfrac{10}{4,5} \mid 3\} \xleftarrow{\varpi} \{\tfrac{5}{2}, 5, \tfrac{5}{2}\}$$

$$\{\tfrac{5}{2},3,3\} \xrightarrow{\varpi} \{\tfrac{10}{3}, \tfrac{6}{2,3} \mid 3\} \xleftrightarrow{\delta^*} \{6, \tfrac{10}{2,5} \mid 3\} \xleftarrow{\varpi} \{5, \tfrac{5}{2}, 5\}$$

It is worth repeating that these fine Schläfli symbols are all rigid; while they may not determine the abstract combinatorial type, they do determine the geometry.

There are further connexions among these polyhedra, induced by the faceting operation φ_3. (This corresponds to the operation φ_2 on the original classical 4-polytopes, but we know that φ_2 and φ_3 are equivalent operations there.) The specific cases are

11B11

$$\{4, \tfrac{10}{2,5} \mid 5\} \xleftrightarrow{\varphi_3} \{\tfrac{10}{3}, \tfrac{10}{2,5} \mid 3\} \qquad \{4, \tfrac{10}{4,5} \mid \tfrac{5}{2}\} \xleftrightarrow{\varphi_3} \{10, \tfrac{10}{4,5} \mid 3\}$$

$$\{6, \tfrac{10}{2,5} \mid 3\} \xleftrightarrow{\varphi_3} \{10, \tfrac{10}{4,5} \mid \tfrac{5}{2}\} \qquad \{6, \tfrac{10}{4,5} \mid 3\} \xleftrightarrow{\varphi_3} \{\tfrac{10}{3}, \tfrac{10}{2,5} \mid 5\}$$

Applying $\zeta\varpi$

We have seen in Section 9C that, if Q is a (finite) classical regular polytope, then $\mathsf{Q}^{\zeta\varpi} = \mathsf{Q}^{\varpi\zeta_1}$ usually degenerates. Here, there is a single exception, which we have already noted:

$$\{3,3,3\}^{\zeta\varpi} = \{4, \tfrac{6}{2,3} \mid 3\}^{\zeta_1} = \{4, \tfrac{6}{1,3} \mid 3\}.$$

It is appropriate to explain what is happening here, particularly since we passed over these matters rather lightly in [82, 83]. The relevant cases here are those with mirror vectors $(3,2,3)$ or $(1,2,3)$, the latter being related to the former by $\boldsymbol{\zeta}$. As we have seen, these are derived by twisting diagrams $\mathcal{D}_1(p,q;r)$, as in Figure 11A2. Now, if we apply $\boldsymbol{\zeta}_1$, changing the sign of the second generator R_1 of the symmetry group, then we replace p and q by their complements p' and q', with

$$\frac{1}{p} + \frac{1}{p'} = 1, \qquad \frac{1}{q} + \frac{1}{q'} = 1.$$

For the previously excluded polyhedra, p' now has odd denominator (included here is the case when it is an integer). If q' is a fraction with even denominator, the previous discussion shows that they are obtained from classical regular 4-polytopes by Petrie contraction $\boldsymbol{\varpi}$. This then exhibits the excluded cases as applications of $\boldsymbol{\zeta\varpi}$ to the 4-polytopes; using Theorem 9C1 shows that fewer vertices are obtained than the theoretical quotient of the order of the whole group by that of the vertex-figure (compare Remark 9C2). This is not surprising, because we know that vertices are identified by the degeneracy of the faces (at least, with mirror vector $(3,2,3)$). We then see that applying $\boldsymbol{\zeta}$ cannot rescue this situation.

In the two remaining cases, with $(p,q,r) = (\tfrac{5}{3},2,5)$ or $(5,\tfrac{5}{2},2)$, to which the previous considerations do not apply, direct calculations show that the $1800 =$

14400/8 expected vertices coincide in threes with the 600 vertices of $\{5,3,3\}$. Thus applying ζ will not produce a genuine polyhedron in either case.

For the 4-polytopes of [82] (with the exception of $\{4,3,3\}$), if we apply π, then we obtain facets which are polyhedra of the type we have just excluded; in fact, we are applying the operation (7.2) of [83] in the dual situation (see also [95]). Thus – even apart from the degeneracy of the faces – the new facets degenerate, again in a way that applying ζ will not remedy.

With two exceptions, for which see the following Section 11C, we shall say little about the related classes (1, 2, 3) and (2, 2, 3), which are derived from the class (3, 2, 3) by ζ, π, or both, except to remark that they contribute to the final classification. It should be observed (as the analysis of Section 11A shows) that none of these derived polyhedra can have geometrically realizable duals in $\mathbb{E}^4$. For example, the faces of the Petrial of $\{4, \frac{4}{1,2} \mid r\}$ are helices of type $\{\frac{2s}{t,s-t}\}$ (if $r = \frac{s}{t}$ as before); all have the same vertex-figure $\{\frac{4}{1,2}\}$.

Note that ζ interchanges polyhedra in the class (2, 2, 3) derived from the class (3, 2, 3) with those derived from $(1,2,3) = (3,2,3)^{\pi}$ (in an obvious informal notation). In summary, we thus derive a further $3 \cdot 23 = 69$ regular polyhedra, apart from infinitely many derived from the toroids, to add to the $8 + 15 = 23$ listed previously.

Notes to Section 11B

1. The diagrams that we have used correspond to spherical tetrahedra giving rise to finite reflexion groups, whose enumeration was proposed in [54] by Goursat, after whom they are named.

11C A Family of Petrials

In this section, we take a closer look at one of the families of polyhedra we described in Section 11B or, rather, a closely related family. At its heart lies a subfamily of six (of which $\{4, 6 : 5\}$ can be taken as a starting point), which are connected by duality and Petriality. Two of the family are non-orientable, and their orientable double covers then connect them with a closely related family. One reason for choosing to look at this family is that the realizations of its members exhibit a range of phenomena which cover several anomalous behaviours:

- there are polytopes whose centriverts are not polytopal;
- there are abstract regular polytopes with faithful realizations, none of which is pure;
- there are realizations with Wythoff spaces which are not lines.

There are also interesting connexions with the regular polyhedra of index 2 in $\mathbb{E}^3$ as introduced by Wills in [126, 127, 128] and classified by Cutler and Schulte in [35, 36]; indeed, these provide a useful picture for investigating various 6-dimensional realizations.

We remarked in Section 11B that applications of $\boldsymbol{\zeta}$ or $\boldsymbol{\pi}$ to polyhedra with mirror vector $(3,2,3)$ did not, with two exceptions, lead to anything notable. One exception is $\{4, \frac{6}{1,3} \mid 3\} \cong \{4,6 \mid 3\}$, with generatrix (R_0, R_1, R_2), say (the other exception is its dual). Under $\boldsymbol{\zeta}$: $(R_0, R_1, R_2) \mapsto (R_0 Z, R_1, R_2) =: (S_0, S_1, S_2)$, with Z the central inversion, we have

11C1 $$\{4, \tfrac{6}{1,3} \mid 3\}^{\zeta} = \{\tfrac{4}{1,2}, \tfrac{6}{1,3} : \tfrac{5}{1,2}\} \cong \{4,6:5\}.$$

This isomorphism may initially seem somewhat surprising. However, the two polyhedra have the same vertex-set and the same group $\mathbf{A}_4 \times \mathbf{Z}_2$; since the period of $S_0 S_1 S_2$ is now 5, the new polyhedron is a quotient of the universal $\{4,6:5\}$ with isomorphic automorphism group (as one sees from [33, Table 8]), and so is isomorphic to it. (Note, by contrast, that the new hole is $\{\frac{6}{1,3}\}$.)

11C2 Remark The interchange

$$\{4,6 \mid 3\} \overset{\zeta}{\longleftrightarrow} \{4,6:5\}$$

as abstract regular polyhedra is of interest in its own right. With their respective generatrices $(\boldsymbol{r}_0, \boldsymbol{r}_1, \boldsymbol{r}_2)$ and $(\boldsymbol{s}_0, \boldsymbol{s}_1, \boldsymbol{s}_2)$, we see that $(\boldsymbol{s}_0\boldsymbol{s}_1\boldsymbol{s}_2\boldsymbol{s}_1)^3 = \boldsymbol{z}$ is a central involution, and that the abstract change of generators is effected by

$$\boldsymbol{s}_0 = (\boldsymbol{r}_0\boldsymbol{r}_1\boldsymbol{r}_2)^5\boldsymbol{r}_0, \qquad \boldsymbol{r}_0 = (\boldsymbol{s}_0\boldsymbol{s}_1\boldsymbol{s}_2\boldsymbol{s}_1)^3\boldsymbol{s}_0.$$

This is easily checked. Since $\boldsymbol{r}_j = \boldsymbol{s}_j$ for $j = 2,3$, we have

$$\begin{aligned} \boldsymbol{s}_0\boldsymbol{s}_1\boldsymbol{s}_2 = (\boldsymbol{r}_0\boldsymbol{r}_1\boldsymbol{r}_2)^6 &\implies (\boldsymbol{s}_0\boldsymbol{s}_1\boldsymbol{s}_2)^5 = (\boldsymbol{r}_0\boldsymbol{r}_1\boldsymbol{r}_2)^{30} = \boldsymbol{z}, \\ \boldsymbol{r}_0\boldsymbol{r}_1\boldsymbol{r}_2\boldsymbol{r}_1 = (\boldsymbol{s}_0\boldsymbol{s}_1\boldsymbol{s}_2\boldsymbol{s}_1)^4 &\implies (\boldsymbol{r}_0\boldsymbol{r}_1\boldsymbol{r}_2\boldsymbol{r}_1)^3 = (\boldsymbol{s}_0\boldsymbol{s}_1\boldsymbol{s}_2\boldsymbol{s}_1)^{12} = \boldsymbol{z}, \end{aligned}$$

so that the defining relations for the group of one imply those of the other. Observe that we actually use the isomorphism of the abstract groups with the geometric ones in this argument.

In a similar way, we have

11C3 $$\{6, \tfrac{4}{1,2} \mid 3\}^{\zeta} = \{\tfrac{6}{2,3}, \tfrac{4}{1,2} : \tfrac{5}{1,2}\} \cong \{6,4:5\}.$$

Note that the face of $\{6, \frac{4}{1,2} \mid 3\}^{\zeta}$ is $\{6\} \diamond \{2\} = \{3\} \# \{2\} = \{\frac{6}{2,3}\}$.

11C4 Remark In view of the actual duality $\{4, \frac{6}{1,3} \mid 3\}^{\delta} = \{6, \frac{4}{1,2} \mid 3\}$, it should not be surprising that we have a quasi-duality

$$\{\tfrac{4}{1,2}, \tfrac{6}{1,3} : \tfrac{5}{1,2}\} \overset{\delta^*}{\longleftrightarrow} \{\tfrac{6}{2,3}, \tfrac{4}{1,2} : \tfrac{5}{1,2}\};$$

in terms of the generatrices, this is given by $(S_0, S_1, S_2) \mapsto (-S_2, S_1, -S_0)$. (The fine Schläfli symbol of the latter shows that we cannot have genuine geometric duality.)

Quasi-Regular Polyhedra

We begin, not with the actual family of regular polyhedra, but instead with some closely related quasi-regular polyhedra and their duals in $\mathbb{E}^3$. For a more general discussion of uniform polyhedra, we refer the reader to [32]. In the present context, we consider certain quasi-regular polyhedra that are regular of index 2 in the sense of [126, 127, 128]. In other words, these polyhedra are combinatorially regular, but have only half the symmetries that would be needed to make them regular. Our descriptions will be self-contained, but we refer the reader to [35, 36] for pictures of these polyhedra, as well as many others.

In the notation of [32], the two basic polyhedra are $2\,|\,5, \frac{5}{2}$ and $3\,|\,5, \frac{5}{2}$. The first is the well-known dodecadodecahedron, whose 30 vertices are the mid-points of the edges of the dual polyhedra $\{5, \frac{5}{2}\}$ and $\{\frac{5}{2}, 5\}$, and whose faces are pentagons and pentagrams alternating around each vertex. It is thus of Schläfli type $\{5, 4\}$ (we shall be more specific shortly); for a picture, see the left column of [35, Figure 4]. The second is probably less familiar; its 20 vertices are those of the regular dodecahedron, and its faces are pentagrams inscribed in the pentagonal faces of the dodecahedron $\{5, 3\}$ and pentagons escribed to the pentagrammal faces of the great stellated dodecahedron $\{\frac{5}{2}, 3\}$. In other words, its edges are those of the five cubes inscribed in the dodecahedron. This polyhedron is of type $\{5, 6\}$; for a picture, see the left column of [35, Figure 3].

In fact, these two polyhedra are isomorphic to $\{5, 4 : 6\}$ and $\{5, 6 : 4\}$, respectively. Notice, though, that their Petrie polygons are skew. To complete the family, we need (say) the dual of the former; Petriality will then give the dual of the latter. Things now get interesting, with important implications for the future. The strict (geometric) dual of type $\{4, 5 : 6\}$ has planar rhombic faces, whose diagonals are the edges of $\{5, \frac{5}{2}\}$ and $\{\frac{5}{2}, 5\}$ placed in dual position (with corresponding edges meeting at their mid-points). The vertices are thus those of two copies of the icosahedron, one inside the other (in this special case, the ratio of their circumradii is τ). Edges are just like those of the icosahedron, except that they pass between the outer icosahedron and the inner. Now we clearly lose none of the symmetry if we vary the relative radii of the two icosahedra; indeed, if we multiply the outer by λ and the inner by μ, we may even allow λ and μ to have opposite signs, with the edges changing accordingly. For the moment, we may suppose that λ and μ are non-zero; later, we shall allow one or other to proceed to the limit.

11C5 Remark In the discussion of [36], the cases $\lambda\mu > 0$ and $\lambda\mu < 0$ are regarded as forming two different families; see [36, Figure 5]. What we do here obliges us to identify these families.

Let us give more details, which will feed into our subsequent discussions. Though we ultimately need to work in $\mathbb{E}^3$, it helps if we first work over the field $\mathbb{F} := \mathbb{Q}[\sqrt{5}]$. We have to be rather careful, because we want to employ the automorphism ‡ of $\mathbb{F}$ which changes the sign of $\sqrt{5}$; of course, this has no counterpart in $\mathbb{R}$. We also let Ξ be the involution on $\mathbb{F}^3$ defined by

$$(\xi_1, \xi_2, \xi_3)\Xi := (\xi_1{}^\ddagger, \xi_3{}^\ddagger, \xi_2{}^\ddagger).$$

First, we define V to consist of all even permutations with all changes of sign of $(2,0,0)$ and $(\tau,\tau^{-1},1)$, with $\tau=\frac{1}{2}(1+\sqrt{5})$ the usual golden section. Thus V is the vertex-set of the icosidodecahedron; moreover, $\frac{1}{2}V$ is the set of unit normals to the mirror planes of the group $\mathbf{G}_3 \cong [3,5]$. Note further that $V\Xi = V$.

We now take the polyhedra in order. For the model of $\{5,4:6\}$ of index 2, we define $u_0 := \frac{1}{2}(1,\tau,-\tau^{-1})$ and $u_1 := (0,-1,0)$, and for $j=0,1$, let R_j be the reflexion in the plane whose unit normal is u_j. Further, we set $R_2 := \Xi$, observing that $u_0\Xi = u_0$, so that R_0 and R_2 commute. With $(2,0,0)$ as the initial vertex, we see that V itself is the resulting vertex-set.

Even though $\{5,6:4\}$ is obtained abstractly from the generatrix $(\boldsymbol{r}_0,\boldsymbol{r}_1,\boldsymbol{r}_2)$ of $\{5,4:6\}$ by the mixing operation

$$\boldsymbol{\pi}^* := \boldsymbol{\delta\pi\delta}\colon (\boldsymbol{r}_0,\boldsymbol{r}_1,\boldsymbol{r}_2) \mapsto (\boldsymbol{r}_0,\boldsymbol{r}_1,\boldsymbol{r}_0\boldsymbol{r}_2) =: (\boldsymbol{s}_0,\boldsymbol{s}_1,\boldsymbol{s}_2),$$

replacing R_2 by R_0R_2 here will not serve; again bear in mind that we only have half the symmetry group. Instead, set $v_0 := (-1,0,0)$ and $v_1 := \frac{1}{2}(\tau,-\tau^{-1},-1)$, and for $j=0,1$ let S_j be the reflexion in the plane whose unit normal is v_j. Further, set $S_2 := \Xi$. With $(1,1,1)$ the initial vertex, the vertex-set is now that of the dodecahedron, namely, all even permutations of $(1,1,1)$ and $(\tau,0,\tau^{-1})$ with all changes of sign.

Finally, for $\{4,5:6\}$, we can directly dualize $\{5,4:6\}$, so obtaining the group generators $(T_0,T_1,T_2) := (R_2,R_1,R_0)$. However, we should say a little more about this case. We can pick $(1,0,\tau)$ as the initial vertex; then the vertex-set is that of two concentric icosahedra, the ratio of whose circumradii is τ^{-1}; the initial edge goes to $(1,-\tau^{-1},0)$. However, if we multiply the initial vertex by $\alpha+\beta\tau$, say, with $\alpha,\beta\in\mathbb{Q}$, then we easily see that the new ratio of circumradii is $\tau^{-1}(\alpha-\beta\tau^{-1})/(\alpha+\beta\tau)$; such numbers are dense in $\mathbb{R}$. In other words, in this case we actually have a whole family of essentially different models. This is not quite what we shall do later, but it does point us in the right direction.

$\{4,6:5\}$ *and* $\{4,6\mid 3\}$

We begin the analysis of the realization domains with $\{4,6\mid 3\}$, noting that we already have its faithful realization $\{4,\frac{6}{1,3}\mid 3\}$ in $\mathbb{E}^4$. (We pick this, rather than the closely related realization $\{\frac{4}{1,2},\frac{6}{1,3}:\frac{5}{1,2}\}$ of $\{4,6:5\}$, because its geometry is initially easier to describe.) Its symmetry group has generators

$$\begin{aligned} S_0 &= (0\ 1),\\ S_1 &= -(0\ 4)(1\ 3),\\ S_2 &= (2\ 3). \end{aligned}$$

As we commonly do when the symmetry group involved is closely related to that of a simplex, we work in the hyperplane

$$\mathbb{L}^4 := \{x=(\xi_0,\ldots,\xi_4)\in\mathbb{E}^5 \mid \xi_0+\cdots+\xi_4=0\}.$$

Thus the reflexions S_j are permutations of the coordinates (in the natural way), combined with changing all signs for S_1. We can take $04 := (1, 0, 0, 0, -1) = e_0 - e_4$ to be the initial vertex, where $jk := e_j - e_k$ for $j, k = 0, \ldots, 4$ with $j \neq k$, and $\{e_0, \ldots, e_4\}$ is the usual orthonormal basis of coordinate vectors of $\mathbb{E}^5$. It is clear that all 20 points jk are vertices of $\{4, \frac{6}{1,3} \mid 3\}$. Since inner products between vectors of the form jk take the values $\pm 2, \pm 1$ and 0, we see that the cosine vector of this realization is

$$\Gamma_3' := (1, \tfrac{1}{2}, 0, -\tfrac{1}{2}, -1).$$

(The index is chosen with the future in mind.) Moreover, the layer vector is

11C6 $$\Lambda = (1, 6, 6, 6, 1).$$

Since the vertices, edges and faces of $\{4, \frac{6}{1,3} \mid 3\}$ are among those of the Minkowski difference $\Delta\mathsf{T} := \mathsf{T} - \mathsf{T}$ of a regular 4-simplex T, we can see that the first and second entries of Γ_3' correspond to edges and face-diagonals (remember that we always have zeroth entry 1).

Now among the realizations of $\{4, 6 \mid 3\}$ must also be those of its quotient $\{4, 6 : 5 \mid 3\}$, which it doubly covers. This quotient also has a 4-dimensional pure realization $\{4, \frac{6}{2,3} \mid 3\}$ (as we pointed out previously, the rigidity criterion of Theorem 6D5 enables us omit reference to the Petrie polygon $\{\frac{5}{1,2}\}$), whose group $\langle T_0, T_1, T_2\rangle$ is given by $T_j = S_j$ for $j = 0, 2$ and $T_1 = (0\ 4)(1\ 3)$; that is, we just drop the sign change in S_1. Its ten vertices are all permutations of $(3, -2, -2, -2, 3)$, with this vertex the initial one. The cosine vector (as a realization of $\{4, 6 \mid 3\}$) is then

$$\Gamma_1 := (1, \tfrac{1}{6}, -\tfrac{2}{3}, \tfrac{1}{6}, 1).$$

Further recall from Section 11B (see also [86], for example) that $\{4, \frac{6}{2,3} \mid 3\} = \{3, 3, 3\}^{\varpi}$ (ϖ is Petrie contraction), so that its vertices are mid-points of the edges of the 4-simplex T, and its faces are tetragonal central sections of the tetrahedral facets of T.

11C7 Remark We recall from similar situations earlier that a fruitful way to think about the relationship between $\{4, \frac{6}{1,3} \mid 3\}$ and $\{4, \frac{6}{2,3} \mid 3\}$ is that we can obtain the vertices of the latter from those of the former just by dropping the minus signs. For example, $(0, -1, 0, 1, 0) \mapsto (0, 1, 0, 1, 0) \sim (-2, 3, -2, 3, -2)$ when we centre the new vertices at o and rescale.

We can now appeal to Theorem 3C11 to find the rest of the cosine matrix of $\{4, 6 \mid 3\}$; as a check, bear in mind the layer vector Λ of (11C6). Thus there is a 5-dimensional realization of $\{4, 6 : 5 \mid 3\}$, with cosine vector Γ_2 given by

$$\Gamma_2 = \tfrac{1}{5}\big(10(1, 0^4) - \Gamma_0 - 4\Gamma_1\big) = (1, -\tfrac{1}{3}, \tfrac{1}{3}, -\tfrac{1}{3}, 1);$$

the usual conventions prevail, so that $\Gamma_0 = (1^5)$ is the cosine vector of the henogon. Finally, there is a 6-dimensional faithful realization, with cosine vector Γ_4' given by

$$\Gamma_4' = \tfrac{1}{6}\big(10(1, 0^3, -1) - 4\Gamma_3\big) = (1, -\tfrac{1}{3}, 0, \tfrac{1}{3}, -1).$$

11C8 Remark It is relatively easy to work with the given coordinate system, and show that

$$\mathsf{P}_4 = \{4, \tfrac{6}{2,3} \mid 3\} \wedge \{4, \tfrac{6}{1,3} \mid 3\} = \mathsf{P}_1 \wedge \mathsf{P}_3,$$

say, as an alternating product in the sense of Section 4E. We need the analogues of the non-symmetric mixed cosine vectors $\varGamma_{13}$ and $\varGamma_{31}$, with

$$\begin{aligned} \varGamma_{13} &= \big(\langle v_1 \varPhi_{1s}, v_3 \rangle \bigm| s = 0, \ldots, 4\big), \\ \varGamma_{31} &= \big(\langle v_3 \varPhi_{3s}, v_1 \rangle \bigm| s = 0, \ldots, 4\big); \end{aligned}$$

here, v_k is the initial vertex of P_k for $k = 1, 3$, and $\varPhi_{ks}$ is the corresponding symmetry. We find that

$$\begin{aligned} \varGamma_{13} &= \sqrt{\tfrac{5}{12}}(0, 1, 0, 1, 0)), \\ \varGamma_{31} &= \sqrt{\tfrac{5}{12}}(0, 1, 0, -1, 0)), \end{aligned}$$

giving

$$\begin{aligned} \varGamma(\mathsf{P}_1 \wedge \mathsf{P}_3) &= \varGamma_1 \varGamma_3 - \varGamma_{13} \varGamma_{31} \\ &= (1, \tfrac{1}{12}, 0, -\tfrac{1}{12}, -1) - \tfrac{5}{12}(0, 1, 0, -1, 0) \\ &= (1, -\tfrac{1}{3}, 0, \tfrac{1}{3}, -1) = \varGamma_4, \end{aligned}$$

exactly as required. Observe, by the way, that $\langle \varGamma_{13}, \varGamma_{31} \rangle_\varLambda = 0$, even though orthogonality here does not fit into the context of (3H16).

Now we should be concentrating on $\{4, 6 : 5\}$. However, as we know,

$$\{\tfrac{4}{1,2}, \tfrac{6}{1,3} : \tfrac{5}{1,2}\} = \{4, \tfrac{6}{1,3} \mid 3\}^\zeta = \{4, \tfrac{6}{1,3} \mid 3\} \otimes \{2\}.$$

Indeed, we have seen that this relationship is combinatorial in nature. What this implies is that $\{4, 6 : 5\}$ has faithful realizations of dimensions 4 and 6, which are just the products of those of $\{4, 6 \mid 3\}$ with $\{2\}$. Moreover, since $\{4, 6 : 5\}$ is another double cover of $\{4, 6 : 5 \mid 3\}$, the realizations of the latter are also realizations of the former. In summary, therefore, we have proved

11C9 Theorem *The cosine matrix of* $\{4, 6 : 5\}$ *is*

$$\begin{bmatrix} 1 & 1 & 1 & 1 & 1 \\ 1 & \frac{1}{6} & -\frac{2}{3} & \frac{1}{6} & 1 \\ 1 & -\frac{1}{3} & \frac{1}{3} & -\frac{1}{3} & 1 \\ 1 & -\frac{1}{2} & 0 & \frac{1}{2} & -1 \\ 1 & \frac{1}{3} & 0 & -\frac{1}{3} & -1 \end{bmatrix},$$

with corresponding layer and dimension vectors

$$\varLambda = (1, 6, 6, 6, 1), \quad D = (1, 4, 5, 4, 6).$$

The last two rows are just the earlier Γ_3' and Γ_4', with appropriate sign changes to take account of $\boldsymbol{\zeta}$ (multiplication by $\{2\}$).

There is a somewhat surprising consequence of Theorem 11C9.

11C10 Proposition *The fine Schläfli symbol* $\{\frac{4}{1,2}, \frac{6}{1,3} : \frac{5}{1,2}\}$ *is rigid.*

Proof. Inspection of the faithful 4-dimensional realization $\mathsf{P}_3 = \{\frac{4}{1,2}, \frac{6}{1,3} : \frac{5}{1,2}\}$ of $\mathcal{P} = \{4, 6 : 5\}$ shows that the induced cosine vector of the vertex-figure $\mathcal{P}^v$ is $\Gamma^v = (\gamma_0, \gamma_2, \gamma_3, \gamma_1)$. It is straightforward to check that the vertex-figure of each pure realization of $\mathcal{P}$ except for $\mathsf{P}_0 = \{1\}$ and P_3 has a trigonal component $\{3\}$. However, the fine Schläfli symbol of P_3 excludes such a component, which implies that P_3 must therefore be rigid. □

There is a natural way to extend this family further. Because the period 5 of its Petrie polygon is odd, the abstract polyhedron $\{4, 6 : 5\}$ is non-orientable, and so has an orientable double cover $\mathcal{P}$, say, which will be a quotient of $\{4, 6 : 10 \mid 6\}$; the pentagonal Petrie polygons will expand to decagons, but the hexagonal holes will remain. Now $\mathcal{P}$ will also be a double cover of $\{4, 6 \mid 3\}$, and so among its realizations are all those of both $\{4, 6 \mid 3\}$ and $\{4, 6 : 5\}$. In particular, among these must be $\{2\}$ itself, because $\{4, \frac{6}{1,3} \mid 3\} \otimes \{4, \frac{6}{1,3} \mid 3\}$ has a component $\{1\}$, and so

$$\{4, \tfrac{6}{1,3} \mid 3\} \otimes \{\tfrac{4}{1,2}, \tfrac{6}{1,3} : \tfrac{5}{1,2}\} = \{4, \tfrac{6}{1,3} \mid 3\} \otimes \{4, \tfrac{6}{1,3} \mid 3\} \otimes \{2\}$$

has a component $\{2\}$. It follows that the products of the pure realizations of $\{4, 6 : 5 \mid 3\}$ with $\{2\}$, which must also be pure, are also realizations of $\mathcal{P}$. Since $\mathcal{P}$ has 40 vertices, a count of dimensions shows that we have found all the pure realizations of $\mathcal{P}$. In particular, $\mathcal{P}$ has no pure faithful realizations. At first sight this may seem surprising; however, since the group of $\mathcal{P}$ is $\boldsymbol{S}_5 \times \boldsymbol{C}_2 \times \boldsymbol{C}_2$, which has no irreducible faithful representations, in retrospect it is not.

The geometric description leads to a combinatorial picture of our polyhedra. We may identify the vertices of the abstract polyhedra with those of their faithful 4-dimensional realizations. The edges of $\{4, 6 \mid 3\}$ are of the form $\{ij, ik\}$ or $\{ik, jk\}$, with distinct indices i, j, k. More importantly, those of $\{4, 6 : 5\}$ are of the form $\{ij, jk\}$. We see at once that the 24 Petrie polygons $\{5\}$ of $\{4, 6 : 5\}$ can be identified with cyclic orderings of $\{0, \ldots, 4\}$; for example, with 02341 we associate the vertices $02, 23, 34, 41, 10$, in that order. (Indeed, we shall further associate these Petrie polygons with cyclic permutations of $\{0, \ldots, 4\}$, our example corresponding to $(0\ 2\ 3\ 4\ 1)$.) Moreover, since an edge involves three of the five indices, we see that the two Petrie polygons containing that edge are obtained by switching the two remaining (adjacent) indices in their cycles. Similarly, the 30 faces $\{4\}$ of $\{4, 6 : 5\}$ are associated with 4-cycles; however, there is no simple rule for writing down the new face found by crossing an edge.

In a sense, we are now done. However, it will be useful to see what the other realizations look like geometrically, and here the pictures of the previous section come in handy. For the moment, we defer the 5-dimensional realization of $\{4,6:5 \mid 3\}$, which will need a different idea.

Our picture in $\mathbb{E}^3$ was of $\{4,6:5\}$, with vertices those of the dodecahedron $\{5,3\}$. The obvious way to lift quasi-regularity into regularity is to blend with the conjugate, under the change of sign of $\sqrt{5}$. Thus the lifted vertices are of the form $(x, x^{\ddagger})$ with x a vertex of the dodecahedron, namely, those obtained from

$$\big((1,1,1),(1,1,1)\big), \qquad \big((\tau,\tau^{-1},0),(-\tau^{-1},-\tau,0)\big),$$

by applying the same even permutations and the same changes of sign in each of the two blocks. For example, we have $(-\tau^{-1},0,\tau,\tau,0,-\tau^{-1})$, but not $(1,-1,1,-1,-1,1)$. An easy check shows that we obtain the cosine vector $\Gamma_4 = (1,\frac{1}{3},0,-\frac{1}{3},-1)$; as we have already observed, the vector $\Gamma_4' = (1,-\frac{1}{3},0,\frac{1}{3},-1)$ found above yields this on application of $\boldsymbol{\zeta}$.

We can even write down the generating reflexions, by lifting them (in a sense) from $\mathbb{E}^3$; here, the S_j are as defined there. Thus, for $j = 0, 1$, S_j lifts to $\overline{S}_j := S_j \times S_j{}^{\ddagger}$ on $\mathbb{E}^6 = \mathbb{E}^3 \times \mathbb{E}^3$. On the other hand, for $S_2 = \Xi$, we have

$$\overline{S}_2 = (1\ 4)(2\ 6)(3\ 5)\colon (\xi_1,\ldots,\xi_6) \mapsto (\xi_4,\xi_6,\xi_5,\xi_1,\xi_3,\xi_2);$$

bear in mind the connexion between the two blocks of three coordinates.

For $\{4,6:5 \mid 3\}$, we proceed as follows. First, let us recall the skewing operation

$$\boldsymbol{\sigma} := \boldsymbol{\pi}^*\boldsymbol{\eta}\boldsymbol{\pi}^* : (\boldsymbol{r}_0,\boldsymbol{r}_1,\boldsymbol{r}_2) \mapsto (\boldsymbol{r}_0\boldsymbol{r}_1\boldsymbol{r}_0, \boldsymbol{r}_0\boldsymbol{r}_2, (\boldsymbol{r}_0\boldsymbol{r}_1)^2) =: (\boldsymbol{s}_0,\boldsymbol{s}_1,\boldsymbol{s}_2)$$

of (5A25), where $\boldsymbol{\pi\delta\pi} = \boldsymbol{\delta\pi\delta} = \boldsymbol{\pi}^*$ as in the previous section. When we apply $\boldsymbol{\sigma}$ to $\{4,6:5 \mid 3\}$, we obtain

$$\{4,6:5 \mid 3\}^{\boldsymbol{\sigma}} = \{5,3:5\},$$

as is easy to verify. Even more to the point, when we trace the operation $\boldsymbol{\sigma}$ on a geometric level in $\mathbb{E}^4$, we find the generatrix (S_0,S_1,S_2) of $\{4,\frac{6}{2,3} \mid 3\}^{\boldsymbol{\sigma}}$ to be

$$\begin{aligned} S_0 &= R_0R_1R_0 = (0\ 3)(1\ 4),\\ S_1 &= R_0R_2 = (0\ 1)(2\ 3),\\ S_2 &= (R_0R_1)^2 = (0\ 1)(3\ 4). \end{aligned}$$

Noting that their groups are the same (the edge graph of $\{4,6:5 \mid 3\}^{\boldsymbol{\pi}^*}$ is not bipartite), we deduce that the two polyhedra $\{4,6:5 \mid 3\}$ and $\{4,6:5 \mid 3\}^{\boldsymbol{\sigma}} = \{5,3:5\}$ have the same vertex-sets, though with different initial vertices.

However, it is easy to find the realization of $\{5,3:5\}$ in $\mathbb{E}^5$; we may take the generating set (T_0,T_1,T_2) of its symmetry group to be the permutations

$$\begin{aligned} T_0 &:= (0\ 1)(2\ 3),\\ T_1 &:= (1\ 2)(3\ 4),\\ T_2 &:= (0\ 1)(4\ 5). \end{aligned}$$

Hence its vertices – and thus those of a realization of $\{4,6:5 \mid 3\}$ also – are among the permutations of $(1,1,1,-1,-1,-1)$ (which is the vertex-set of 0_{22}), and yield the cosine vector Γ_2.

To be more specific, because $\{4,6:5 \mid 3\}^{\pi^*}$ has odd edge-circuits we can invert $\boldsymbol{\sigma}$. If we do this, and then conjugate by the transposition (1 3) (for sake of tidiness), we obtain the putative generators

$$\begin{aligned} R_0 &:= \pm(0\ 5)(1\ 2)(3\ 4), \\ R_1 &:= (1\ 2)(4\ 5), \\ R_2 &:= \pm(0\ 5)(1\ 3)(2\ 4), \end{aligned}$$

where the signs $\pm$ must go together. In fact, the positive sign yields $\{o\}$ as the Wythoff space, and so we must take the negative sign.

As we have chosen things, the initial vertex is $(1,1,1,-1,-1,-1)$. Now, from an initial vertex, the vertices of 0_{22} fall into layers of $1,9,9,1$. The first nine form the vertex-figure $\{3\}\times\{3\}$; in this, we take some three vertices forming a regular trigon which is symmetric with respect to the two components $\{3\}$. In the opposite vertex-figure, we take the complementary – in an obvious sense – set of six vertices; these form the vertex-figure of our realized polyhedron. Thus we have $1+3+6=10$ vertices in all, as we require.

11C11 Remark The reader will not be surprised to learn that all 20 vertices of 0_{22} are obtained on multiplication of this realization by $\{2\}$, that is, on replacing R_0 by $-R_0 = (0\ 5)(1\ 2)(3\ 4)$.

$\{6,4:5\}$ *and* $\{6,4 \mid 3\}$

In exact analogy to what we did with the polyhedra with tetragonal faces, we begin our investigation here with $\{6,4 \mid 3\}$. (Note that, just as for the dual, we have $\{6,4:5\} = \{6,4 \mid 3\}^{\varsigma}$, with ς as before indicating replacing the initial generator r_0 by its product with the central element $(r_0r_1r_2)^5$.) This is the dual of $\{4,6 \mid 3\}$, and so we can take as generators of its 4-dimensional faithful realization $\{6, \frac{4}{1,2} \mid 3\}$

$$\begin{aligned} R_0 &= (2\ 3), \\ R_1 &= -(0\ 4)(1\ 3), \\ R_2 &= (0\ 1). \end{aligned}$$

We can take $(1,1,0,-1,-1)$ as initial vertex, and with little difficulty find that the vertices of $\{6, \frac{4}{1,2} \mid 3\}$ are all permutations of the initial vertex, and that the cosine vector is

$$\Gamma_4' := (1, \tfrac{3}{4}, \tfrac{1}{4}, 0, -\tfrac{1}{4}, -\tfrac{3}{4}, -1).$$

Once again, the index is chosen with forethought. Observe that $\{6,4 \mid 3\}$ has layer vector

11C12 $$\Lambda = (1,4,8,4,8,4,1).$$

In just the same way as we realized $\{4, 6 : 5 \mid 3\}$ by dropping the minus sign before S_1, we can similarly realize the quotient $\{6, 4 : 5 \mid 3\}$ of $\{4, 6 \mid 3\}$. However, we very quickly see that the realization is degenerate; indeed, as was pointed out in the previous Section 11B (see also [83]), $\{4, \frac{6}{2,3} \mid 3\}$ does not have a faithful geometric dual in $\mathbb{E}^4$. However, again as with its dual $\{4, \frac{6}{1,3} \mid 3\}$, if we drop the minus signs in the vertices of $\{6, \frac{4}{1,2} \mid 3\}$, we obtain (in a concealed form) the vertex-set of a 4-simplex. Here, the vertices of $\{6, 4 : 5 \mid 3\}$ coincide in threes at the vertices of the simplex, while the original faces $\{6\}$ collapse onto trigons. It is clear that (as a realization of $\{6, 4 \mid 3\}$) the cosine vector of this realization C_1, say, is

$$\Gamma_1 := (1, -\tfrac{1}{4}, -\tfrac{1}{4}, 1, -\tfrac{1}{4}, -\tfrac{1}{4}, 1).$$

Now we need two more pure realizations of $\{6, 4 : 5 \mid 3\}$; we see that their blend has cosine vector

$$\tfrac{1}{10}\big(15(1, 0^5, 1) - \Gamma_0 - 4\Gamma_1\big) = (1, 0, 0, -\tfrac{1}{2}, 0, 0, 1).$$

Note that the vertices of the corresponding realization are those of five trigons in mutually orthogonal 2-planes in $\mathbb{E}^{10}$. Applying the layer equation and Λ-orthogonality with respect to Γ_1 shows that we must consider cosine vectors of the form $\Gamma = (1, \gamma, -\frac{1}{2}\gamma, -\frac{1}{2}, -\frac{1}{2}\gamma, \gamma, 1)$, and with the given blend the two solutions have equal and opposite values of γ. Knowing this, we can proceed in at least two different ways.

First, consider $\mathsf{Q} \otimes \mathsf{Q}$, with $\mathsf{Q} := \{6, \frac{4}{1,2} \mid 3\}$. We calculate

$$(\Gamma_4')^2 = (1, \tfrac{9}{16}, \tfrac{1}{16}, 0, \tfrac{1}{16}, \tfrac{9}{16}, 1) = \tfrac{1}{4}\Gamma_0 + \tfrac{1}{12}\Gamma_1 + \tfrac{2}{3}\Gamma_2,$$

with

$$\Gamma_2 := (1, \tfrac{1}{2}, -\tfrac{1}{4}, -\tfrac{1}{2}, -\tfrac{1}{4}, \tfrac{1}{2}, 1).$$

The coefficient $\frac{1}{4}$ of Γ_0 comes from Theorem 3F5, and then the other two scalars are forced by the fact that the middle entry of Γ_2 is $-\frac{1}{2}$. Applying Theorem 3F5 to Γ_2 gives the correct dimension $d_2 = 5$, so that C_2 is pure. We see immediately that the third cosine vector of realizations arising from those of $\{6, 4 : 5 \mid 3\}$ must be

$$\Gamma_3 := (1, -\tfrac{1}{2}, \tfrac{1}{4}, -\tfrac{1}{2}, \tfrac{1}{4}, -\tfrac{1}{2}, 1);$$

again, the dimension is $d_3 = 5$.

In the alternative approach, we just ask whether we can geometrically dualize the realization of $\{4, 6 : 5 \mid 3\}$ in $\mathbb{E}^5$. In fact, we can. We thus take the generators in the form

$$\begin{aligned} R_0 &:= \pm(0\ 5)(1\ 3)(2\ 4), \\ R_1 &:= (1\ 2)(4\ 5), \\ R_2 &:= \pm(0\ 5)(1\ 2)(3\ 4), \end{aligned}$$

Since we are wanting two 5-dimensional realizations, we may hope that both choices of sign will work.

Taking the negative signs first (as the strict dual), the initial vertex is $(1, 0, 0, 1, -1, -1)$, and applying $R_0R_1 = -(0\ 4\ 1\ 3\ 2\ 5)$, we find as vertices of the initial hexagonal face

$$(1, 0, 0, 1, -1, -1),\ (1, 1, -1, 0, -1, 0),\ (0, 1, 0, -1, -1, 1),\ (-1, 1, 1, -1, 0, 0),\ \ldots$$

These are enough to show that this realization C_2 has cosine vector Γ_2.

For C_3, say, we can again look at things in two ways: either directly, or by dropping the minus signs in the vertices of C_2, we see that the vertices of C_3 are the mid-points of the edges of the 5-simplex, and direct calculation leads to the cosine vector Γ_3.

We now move on to the remaining faithful realizations, bearing in mind that we already know one in $\mathbb{E}^4$. These have dimensions 5 and 6.

We begin with the latter. Once again, we can appeal to our picture of regular polyhedra of index 2 in $\mathbb{E}^3$; in this case, we work directly with $\{5, 4 : 6\}$. Thus, for $j = 0, 1$, we lift the ordinary reflexion R_j that we had earlier to $\overline{R}_j := R_j \times R_j^{\ddagger}$ on $\mathbb{E}^6 = \mathbb{E}^3 \times \mathbb{E}^3$ and, as before, $R_2 = \Xi$ lifts to

$$\overline{R}_2 = (1\ 4)(2\ 6)(3\ 5)\colon (\xi_1, \ldots, \xi_6) \mapsto (\xi_4, \xi_6, \xi_5, \xi_1, \xi_3, \xi_2).$$

The initial vertex is $(2, 0, 0, 2, 0, 0)$, and the vertex-set is obtained from

$$\big((2, 0, 0), (2, 0, 0)\big), \quad \big((\tau, \tau^{-1}, 1), (-\tau^{-1}, -\tau, 1)\big),$$

by applying the same even permutations and the same changes of sign in each of the two blocks. Strictly speaking, we should apply the Petrie operation here, because (as we shall see) we really want to look at $\{6, 4 : 5\}$. After some tedious calculation, we find that the realization has cosine vector

$$\Gamma_6 := (1, \tfrac{1}{2}, \tfrac{1}{4}, 0, -\tfrac{1}{4}, -\tfrac{1}{2}, -1).$$

With

$$\Gamma_4 := (1, -\tfrac{3}{4}, \tfrac{1}{4}, 0, -\tfrac{1}{4}, \tfrac{3}{4}, -1)$$

the cosine vector of the 4-dimensional (faithful) realization of $\{6, 4 : 5\}$ (derived from Γ_4' by applying ζ), we easily calculate the missing cosine vector Γ_5 by

$$\Gamma_5 = \tfrac{1}{5}\big(15(1, 0^5, -1) - \Gamma_0 - 4\Gamma_4 - 6\Gamma_6\big) = (1, 0, -\tfrac{1}{2}, 0, \tfrac{1}{2}, 0, -1).$$

It remains for us to identify this realization geometrically. It is reasonable to expect that its group be related in some fashion to those of the two 5-dimensional realizations of $\{6, 4 : 5 \mid 3\}$; an analogy between this situation and the 4-dimensional realizations of $\{4, 6 : 5\}$ and $\{4, 6 : 5 \mid 3\}$ suggests that we might try changing the sign of R_1, and so take generating reflexions of the form

$$\begin{aligned} R_0 &:= \pm(0\ 5)(1\ 3)(2\ 4), \\ R_1 &:= -(1\ 2)(4\ 5), \\ R_2 &:= \pm(0\ 5)(1\ 2)(3\ 4). \end{aligned}$$

We quickly find that, with the plus sign for R_2, the Wythoff space is $\{o\}$; with the minus sign, the initial vertex is (say) $(0, 1, -1, 0, 0, 0)$. Both signs for R_0 lead to valid polyhedra. Looking at the resulting holes and Petrie polygons immediately shows that the plus sign yields a realization of $\{6, 4 \mid 3\}$, while the minus sign gives $\{6, 4 : 5\}$; again recall that the two polyhedra are related by ζ. The vertices are all permutations of $(1, -1, 0, 0, 0, 0)$, and are those of the Minkowski difference body $\Delta\mathsf{T} = \mathsf{T} - \mathsf{T}$ of the 5-simplex T; both polyhedra have cosine vector Γ_5.

In summary, then, we see that we have shown

11C13 Theorem *The cosine matrix of* $\{6, 4 : 5\}$ *is*

$$\begin{bmatrix} 1 & 1 & 1 & 1 & 1 & 1 & 1 \\ 1 & -\frac{1}{4} & -\frac{1}{4} & 1 & -\frac{1}{4} & -\frac{1}{4} & 1 \\ 1 & \frac{1}{2} & -\frac{1}{4} & -\frac{1}{2} & -\frac{1}{4} & \frac{1}{2} & 1 \\ 1 & -\frac{1}{2} & \frac{1}{4} & -\frac{1}{2} & \frac{1}{4} & -\frac{1}{2} & 1 \\ 1 & -\frac{3}{4} & \frac{1}{4} & 0 & -\frac{1}{4} & \frac{3}{4} & -1 \\ 1 & 0 & -\frac{1}{2} & 0 & \frac{1}{2} & 0 & -1 \\ 1 & \frac{1}{2} & \frac{1}{4} & 0 & -\frac{1}{4} & -\frac{1}{2} & -1 \end{bmatrix},$$

with layer and dimension vectors

$$\Lambda = (1, 4, 8, 4, 8, 4, 1), \quad D = (1, 4, 5, 5, 4, 5, 6).$$

In principle, we can use this matrix to work out the multiplication table for the pure realizations. However, we shall leave this to the interested reader.

We end this part by remarking that, just as for their duals, the polyhedra $\{6, 4 \mid 3\}$ and $\{6, 4 : 5\}$ have a common double cover $\mathcal{P}$. This cover $\mathcal{P}$ has no pure faithful realizations, because the group of $\mathcal{P}$ is $\mathcal{S}_5 \times \mathcal{S}_2 \times \mathcal{S}_2$; its pure realizations are those of $\{6, 4 : 5\}$, together with those obtained from them on multiplication by $\{2\}$.

$\{4, 5 : 6\}$ *and* $\{6, 5 : 4\}$

We begin our discussion by describing the two polyhedra combinatorially, and to do this we appeal to our picture of $\{\frac{5}{1,2}, \frac{6}{1,3} : \frac{4}{1,2}\}$ above or, rather, that of its Petrial $\mathsf{P} := \{\frac{4}{1,2}, \frac{6}{1,3} : \frac{5}{1,2}\}$. Now recall that we identified faces $\{\frac{4}{1,2}\}$ of P with 4-cycles, and Petrie polygons $\{\frac{5}{1,2}\}$ with 5-cycles. The latter shows that we can label the vertices of the abstract polyhedron $\mathcal{P} := \{4, 5 : 6\}$ by 5-cycles on $\{0, \ldots, 4\}$; moreover, two vertices are joined by an edge when one cycle-label is obtained from the other by a transposition of adjacent entries. From this, it is straightforward to see that to a 4-cycle with missing entry k, say, corresponds a tetragonal face $\{4\}$ of $\mathcal{P}$ whose vertices are obtained by inserting k into the four possible places in that cycle. This can then be used to show that the six vertices of a typical hexagonal Petrie polygon $\{6\}$ of $\mathcal{P}$ are those for which two adjacent entries, j and k, say, are fixed. So, for example, $01234, 02341, 03412, 04123$ are the successive vertices of a face (obtained from the cycle 1234, with edges given by transpositions with 0), while $01234, 01324, 01342, 01432, 01423, 01243$ are the successive vertices of a Petrie polygon (with edges given by transposing entries other than the fixed pair 01).

It is clear from this description that we can immediately find another copy of $\{4, 5 : 6\}$ with the same vertices, now obtaining edges by transposing *non-adjacent* entries in the vertex-labels. We call this operation *squaring*, because we can think of it as obtained by squaring the vertex-labels when we regard them as cyclic permutations, but keeping the existing edges. Thus, for example, $01234 \mapsto 02413$, and so on. Thought of in this way, squaring is not involutory, but instead, when applied twice, yields the central involution of $\mathcal{P}$.

This picture also tells us that the layer vector of $\mathcal{P}$ is $\Lambda = (1, 5, 5, 2^*, 5, 5, 1)$; the first and second entries correspond to edges and face-diagonals, respectively. The middle entry 2 lists the two vertices obtained by taking the square and cube of the initial one. Observe, by the way, that these two diagonals are asymmetric as indicated by the asterisk; they correspond to crossing a face by a diagonal, followed by tracing the edge opposite (see Figure 11C14). We deduce that we have a realization with a Wythoff space which is at least 2-dimensional.

11C14

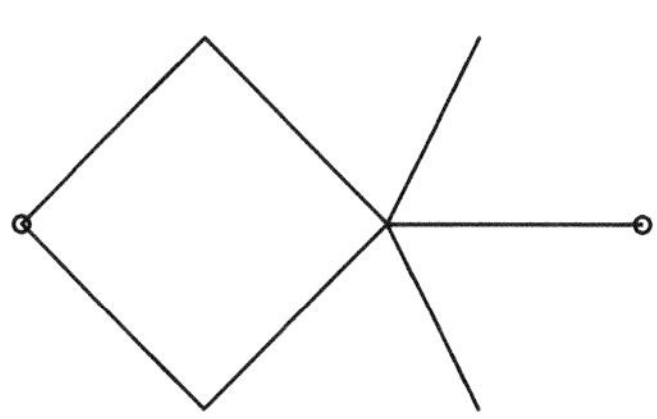

An asymmetric diagonal of $\{4, 5 : 6\}$

Another operation which is applicable here is the faceting operation φ_2. For our purposes, we combine it with the Petrie operation π to give $\pi_2 = \pi\varphi_2 = \varphi_2\pi$. A typical new edge-path is $01234, 02134, 02143, 01243$, of length 4. We conclude that we have yet another copy of $\{4, 5 : 6\}$ with the same vertices and edges as before. Just as a check, note that φ_2 alone yields a typical face $01234, 01324, 04132, 02413, 02431, 02341$.

A third operation is the abstract operation corresponding to ζ; in other words, if we write $\langle \boldsymbol{r}_0, \boldsymbol{r}_1, \boldsymbol{r}_2 \rangle$ for the automorphism group of $\{4, 5 : 6\}$, then $\boldsymbol{r}_0$ is combined with the central involution $\boldsymbol{z}$, which we deduce from the earlier discussion (tracing the various changes of generators) to be $\boldsymbol{z} = (\boldsymbol{r}_0\boldsymbol{r}_1\boldsymbol{r}_0\boldsymbol{r}_1\boldsymbol{r}_2)^3$. Now both faces $\{4\}$ and Petrie polygons $\{6\}$ become polygons of the same kind under ζ, and so ζ yields yet another copy of $\mathcal{P}$ with the same vertices, but now with a different set of edges.

Something else to bear in mind is that the edge-graph of $\mathcal{P}$ is bipartite (so that $\{2\} \in \mathcal{P}$), and that

$$\{4, 5 : 6\}^\eta = \{5, 5 \mid 3\}.$$

Hence, in each realization of $\{4, 5 : 6\}$ is inscribed (potentially at least) two copies of realizations of $\{5, 5 \mid 3\}$.

What we have just seen shows that identifying opposite vertices of $\mathcal{P}$ under the central involution yields

$$\{4, 5 : 6\}/2 = \{4, 5 : 6, *3\}.$$

Thus the edge-graph of $\mathcal{P}/2$ is still bipartite, and therefore, in fact, $\{2\} \in \mathcal{P}/2$. There is a nice picture of this identification:

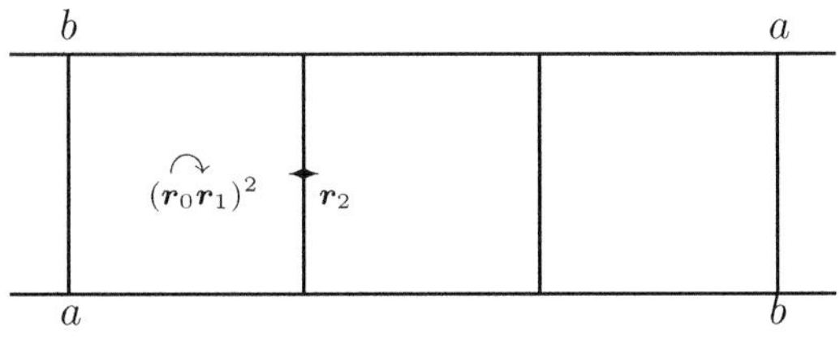

Thus $\mathcal{P}/2$ has short Möbius strips running through it; naturally, it is non-orientable.

We can say a little more about $\mathcal{P}/2$, using the identification of vertices of $\mathcal{P} = \{4, 5 : 6\}$ with 5-cycles regarded as permutations. Thus, if v is a vertex of $\mathcal{P}$, then we can write v^k for its kth power, with $k = \pm 1, \pm 2$; the central symmetry of $\mathcal{P}$ is therefore $v \mapsto v^{-1}$. Now we can easily see that, if u and v are adjacent vertices of $\mathcal{P}$, then so are u^2 and v^{-2}; for example, the edge $\{01234, 01324\}$ corresponds to $\{02413, 02143\}$ (this particular choice makes the transposes obvious). Hence, to each face (u, v, w, x) of $\mathcal{P}$ corresponds another regular tetragon $(u^2, v^{-2}, w^2, x^{-2})$; this is a 2-zigzag of $\mathcal{P}$. The quotient mapping $\mathcal{P} \mapsto \mathcal{P}/2$, given by the identification $v \leftrightarrow v^{-1}$, preserves the distinction between faces and 2-zigzags.

We now move on to the realizations themselves. It is here that the regular polyhedra of index 2 which we discussed at the beginning of the section really contribute essentially to our understanding. We follow the outline presented there, but now we put flesh on the bones. As we said, the standard picture has the vertices of the polyhedra of type $\{4, 5 : 6\}$ as all even permutations (with all changes of sign) of vectors of the form $\lambda(\pm\tau, \pm 1, 0)$ and $\mu(\pm 1, \pm\tau^{-1}, 0)$ with $\lambda = \mu$. We now allow λ and μ to vary, and we pair these vertices up with those obtained by changing the sign of $\sqrt{5}$ and, to maintain the same circumradius, interchanging the rôles of λ and μ. Thus the vectors of the 6-dimensional realizations are obtained from

$$\big(\lambda(\tau, 1, 0), \mu(-\tau^{-1}, 1, 0)\big), \qquad \big(\mu(1, \tau^{-1}, 0), \lambda(1, -\tau, 0)\big)$$

by applying the same even permutations and same sign changes in each of the two blocks of three coordinates. Observe, in particular, that the correspondence reverses the sign relationship between λ and μ in the two blocks; as we said earlier, this identifies two families regarded as different in [36].

Since we wish to express everything in terms of cosine vectors, we must choose λ and μ so that these vectors have length 1. That is, we must have

$$\lambda^2(\tau^2+1)+\mu^2(\tau^{-2}+1)=1;$$

for reasons that will become clear, we define ϑ with $0 \leqslant \vartheta < 2\pi$ by

$$\cos\vartheta = \mu\sqrt{\tau^{-2}+1}, \quad \sin\vartheta = \lambda\sqrt{\tau^2+1}.$$

As should be familiar by now, recall that changing the signs of both λ and μ produces a congruent realization, and so effectively we can take $0 \leqslant \vartheta < \pi$ (up to such a simultaneous change of sign).

Calling this realization P_ϑ, we may calculate its cosine vector to be

$$\Gamma_\vartheta = \bigl(1, \tfrac{1}{\sqrt{5}}\sin 2\vartheta, \tfrac{1}{\sqrt{5}}\cos 2\vartheta, 0, -\tfrac{1}{\sqrt{5}}\cos 2\vartheta, -\tfrac{1}{\sqrt{5}}\sin 2\vartheta, -1\bigr).$$

Observe that we have planar (square) faces $\{4\}$ for the single value $\vartheta = \arctan\tau$; this case, of course, is given by $\lambda = \mu$.

We have shown that various changes of generators of the abstract group of $\mathcal{P} = \{4, 5 : 6\}$ yield different copies of $\mathcal{P}$ with the same vertices (and sometimes with the same edges as well). It is straightforward to find the corresponding effects on the parameter angle ϑ, where we write $\boldsymbol{\beta}$ for the squaring operation:

$$\begin{aligned} \boldsymbol{\zeta}&: \vartheta \mapsto -\vartheta, \\ \boldsymbol{\pi}_2&: \vartheta \mapsto \tfrac{1}{2}\pi - \vartheta, \\ \boldsymbol{\beta}&: \vartheta \mapsto \vartheta + \tfrac{1}{2}\pi. \end{aligned}$$

Putting these three operations together gives

11C15 Proposition *The operations which yield different copies of* $\{4, 5 : 6\}$ *with the same vertices are related by*

$$\boldsymbol{\beta} = \boldsymbol{\zeta}\boldsymbol{\pi}_2.$$

11C16 Remark Note the two special values $\vartheta = 0$ and $\frac{1}{2}\pi$, for which $\mathsf{P}_\vartheta^{\boldsymbol{\zeta}} = \mathsf{P}_\vartheta$. Similarly, $\mathsf{P}_\vartheta^{\boldsymbol{\pi}_2} = \mathsf{P}_\vartheta$ for $\vartheta = \frac{1}{4}\pi$ and $\frac{3}{4}\pi$.

These 6-dimensional realizations of $\{4, 5 : 6\}$ account for all those that are faithful; note that the cross-polytope realization X is such that

$$\mathsf{X} = \tfrac{1}{\sqrt{2}}\bigl(\mathsf{P}_\vartheta \mathbin{\#} \mathsf{P}_{\vartheta+\pi/2}\bigr)$$

for each ϑ. Hence all the remaining pure realizations of $\mathcal{P} = \{4, 5 : 6\}$ must be those of $\mathcal{Q} := \mathcal{P}/2$. In fact, we can dispose of this problem very quickly. We have already noticed that $\{2\} \in \mathcal{Q}$. Thus we see that there is an obvious quotient with vertices those of the 5-simplex Q, say; indeed, we just identify each vertex-label with those of its powers, as before thinking of the label as a cyclic permutation. So, for example, $01234 \sim 02413 \sim 03142 \sim 04321$. With

$\{1\}$, $\{2\}$ and Q we must also have $\mathsf{Q} \otimes \{2\}$ as a fourth pure component, and counting dimensions $(1 + 1 + 5 + 5 = 12)$ shows that we now have all of them. Observe that $\mathsf{Q} \otimes \{2\}$ must actually be a faithful realization of $\mathcal{Q}$.

For reference, we single this out as an example.

11C17 Example It is important to observe that Q itself cannot be polytopal. It has six vertices and so, if it were to be a genuine polyhedron, its group would have order $6 \cdot 10 = 60$. However, the order 8 of the group of a tetragon is not a divisor of 60; hence degeneracy must occur. Note particularly that we have an example of a non-polytope Q for which $\mathsf{Q}^\varsigma = \mathsf{Q} \otimes \{2\}$ is polytopal; turning this around, we have a regular polytope P $(= \mathsf{Q} \otimes \{2\})$ for which P^ς is not polytopal. Furthermore, since $\mathsf{Q} \# \{2\} = \mathsf{Q}^\varsigma \diamond \{2\} \cong \mathsf{Q}^\varsigma \# \{2\}$ is polytopal, we have a finite example where a blend with a segment is a polytope, but the main component is not (in [84], we already had an apeirotope with this property, with a 4-dimensional main component).

By the way, note that the group of T does actually have order 120. It is straightforward to see that we can take $\mathbf{G}(\mathsf{Q} \otimes \{2\})$ to have generators

$$\begin{aligned} R_0 &:= -(0\ 1)(2\ 4)(3\ 5), \\ R_1 &:= (1\ 2)(3\ 5), \\ R_2 &:= (2\ 5)(3\ 4). \end{aligned}$$

We think of the vertices of Q labelled $0, \ldots, 5$, with initial vertex 0, initial edge $\{0, 1\}$, and vertices of the vertex-figure $1, \ldots, 5$ in cyclic order. The vertices of $\mathsf{Q} \otimes \{2\}$ are then these six, together with their opposites, which we write as $\overline{k}$ for $k = 0, \ldots, 5$. In moving from $\mathbf{G}(\mathsf{Q} \otimes \{2\})$ to $\mathbf{G}(\mathsf{Q})$ we change the sign of R_0; when we observe that a minus sign is attached just to the odd permutations, we see that this yields an isomorphic group. Using these generators and the geometry, it is easy to check that $\mathsf{Q} \otimes \{2\}$ is indeed polytopal.

Observe also that, while the edge-graph of $\mathsf{Q} \otimes \{2\}$ is centrally symmetric, the polyhedron itself is not. One symmetry which interchanges 0 and $\overline{0}$ is

$$X := -(2\ 3\ 5\ 4) = (R_2R_1)^{-2} \cdot R_0R_1R_0 \cdot (R_2R_1)^2 \cdot R_0,$$

which we express in this way to draw attention to the underlying geometry (note that $(0, \ldots, 5)X = (\overline{0}, \overline{1}, \overline{3}, \overline{5}, \overline{2}, \overline{4})$, and that $X^2 = R_2$). We then see that conjugating (R_0, R_1, R_2) by X leads to new generators

$$\begin{aligned} S_0 &= -(0\ 1)(2\ 3)(4\ 5) = R_0R_2, \\ S_1 &= (1\ 3)(4\ 5) = R_1R_2R_1, \\ S_2 &= (2\ 5)(3\ 4) = R_2, \end{aligned}$$

which are obtained from the originals by the operation $\pi_2 = \varphi_2\pi$. In other words, we have shown that

$$-(\mathsf{Q} \otimes \{2\}) = (\mathsf{Q} \otimes \{2\})^{\pi_2}.$$

This gives another, very obvious, explanation as to why Q itself cannot be polytopal.

In conclusion, we have shown

11C18 Theorem *The cosine matrix of* $\{4, 5 : 6\}$ *is*

$$\begin{bmatrix} 1 & 1 & 1 & 1 & 1 & 1 & 1 \\ 1 & -1 & 1 & -1 & 1 & -1 & 1 \\ 1 & -\frac{1}{5} & -\frac{1}{5} & 1 & -\frac{1}{5} & -\frac{1}{5} & 1 \\ 1 & \frac{1}{5} & -\frac{1}{5} & -1 & -\frac{1}{5} & \frac{1}{5} & 1 \\ 1 & 0 & \frac{1}{\sqrt{5}} & 0 & -\frac{1}{\sqrt{5}} & 0 & -1 \\ 0 & \frac{1}{\sqrt{5}} & 0 & 0 & 0 & -\frac{1}{\sqrt{5}} & 0 \\ 1 & 0 & -\frac{1}{\sqrt{5}} & 0 & \frac{1}{\sqrt{5}} & 0 & -1 \end{bmatrix},$$

with layer and dimension vectors

$$\Lambda = (1, 5, 5, 2^*, 5, 5, 1), \quad D = (1, 1, 5, 5, 6, 12^\dagger, 6).$$

Following our conventions, we use an asterisk to denote an asymmetric diagonal class and (as in Theorems 7K36 and 7K37) an obelisk for a notional dimension.

As usual, we shall say little about the multiplication table for realizations of $\{4, 5 : 6\}$, since we have not needed products in our treatment. It should be clear in any event that we would only need to consider $\mathsf{P}_\vartheta \otimes \mathsf{Q}$ and $\mathsf{P}_\vartheta \otimes \mathsf{P}_\psi$, once we recall that $\mathsf{P}_\vartheta \otimes \{2\} = \mathsf{P}^\varsigma = \mathsf{P}_{-\vartheta}$ from above. As an example, writing $\Gamma_0, \ldots, \Gamma_3$ for the first four rows of the cosine matrix (so that $\Gamma_2 := \Gamma(Q)$) and Γ_ϑ for the last, we have

$$\begin{aligned} \Gamma_\vartheta \Gamma_2 &= \big(1, -\tfrac{1}{5\sqrt{5}} \sin 2\vartheta, -\tfrac{1}{5\sqrt{5}} \cos 2\vartheta, 0, \tfrac{1}{5\sqrt{5}} \cos 2\vartheta, \tfrac{1}{5\sqrt{5}} \sin 2\vartheta, -1\big) \\ &\quad - \tfrac{2}{5}\Gamma_\vartheta + \tfrac{3}{5}\Gamma_{\vartheta+\pi/2}, \end{aligned}$$

irrespective of the value of ϑ. For the latter, all we remark is that, since each P_ϑ is a component of the staurotope realization X, it follows that the cosine vector of $\mathsf{P}_\vartheta \otimes \mathsf{P}_\psi$ is a convex combination of $\Gamma_0, \ldots, \Gamma_3$ alone.

11D Mirror Vector $(2, 3, 2)$

As we saw in Section 11A, a polyhedron in the class $(2, 3, 2)$ is derived from a diagram $\mathcal{D}_2(p, q, r)$, as in Figure 11A3. Thus, if its generatrix is (R_0, R_1, R_2), then R_0 and R_2 are automorphisms which act on the diagram given by

11D1 $\qquad (S_0, \ldots, S_3) := (R_1, R_0R_1R_0, R_0R_2R_1R_2R_0, R_2R_1R_2).$

Again, as we have already indicated, the resulting polyhedron is of type $\{2p, 2q : 2r\}$ (we shall explain how to interpret these numbers below), from which are obtained up to five others by $\boldsymbol{\delta}$ and $\boldsymbol{\pi}$. Note that the three operations $\boldsymbol{\delta}$, $\boldsymbol{\pi}$ and $\boldsymbol{\zeta}$ preserve the class $(2,3,2)$; indeed, we have

11D2 $\qquad \boldsymbol{\delta}\colon\, p \leftrightarrow q, \qquad \boldsymbol{\pi}\colon\, p \leftrightarrow r, \qquad \boldsymbol{\zeta}\colon\, p \leftrightarrow p',\ r \leftrightarrow r',$

where as before

$$\frac{1}{p} + \frac{1}{p'} = 1, \qquad \frac{1}{r} + \frac{1}{r'} = 1.$$

Applying the same analysis as in Section 11B, we see that the 2-faces of the polyhedron are skew $2p$-gons of the form $\{p\} \# \{2\}$. Here, it does not matter if p is a fraction with even denominator; the skew nature of the face prevents a collapse. However, the vertex-figure is a planar polygon so that, if q is a fraction with even denominator, then collapse occurs and the polyhedron degenerates. This is the restriction on q which we mentioned in Section 11A.

Observe that R_0R_2 is also an automorphism of the diagram. To ensure non-degeneracy, at least one of R_0, R_2, R_0R_2 must be an outer automorphism; in general, then, two of them will be outer and the third inner.

Bearing these considerations in mind, we readily see that the only diagrams of the above form which admit suitable automorphisms are given by the following unordered triples $\{p,q,r\}$ (these should not be confused with Schläfli symbols of regular 4-polytopes – none actually is such a symbol). First, $\{p,2,2\}$ (with an arbitrary fractional $p > 2$, except that such a fraction with even denominator cannot play the rôle of q); second, $\{3,3,\frac{3}{2}\}$, $\{4,4,\frac{3}{2}\}$, $\{\frac{4}{3},\frac{4}{3},\frac{3}{2}\}$ (in these three we cannot have $q = \frac{3}{2}$, and so each gives only two polyhedra); third, $\{3,4,\frac{4}{3}\}$ (here, any permutation is allowed, and so we obtain a full family of six). The group $[3,3,5]$ makes no contribution, even though (for instance) we do have a Goursat tetrahedron corresponding to $\mathcal{D}_2(3,5,\frac{5}{3})$. As we observed in Section 11B, $[3,3,5]$ has no outer automorphisms as a subgroup of $\mathbf{O}_4$, and we shall see in Section 11F why the remaining possiblilities must also be eliminated.

We briefly discuss these polyhedra. In the first case, the two automorphisms (symmetries) R_2 and R_0R_2 of the diagram

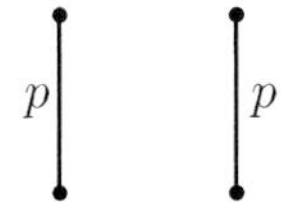

are both outer; their composition R_0 is also an outer automorphism, except when p is a fraction with even denominator. (Note that R_2 and R_0R_2 give the two involutory ways of interchanging the two dihedral subdiagrams.) In any event, if $p = s/t$ (in lowest terms), then the resulting group is of order

$$\begin{cases} 2\cdot 2\cdot (2s)^2 = 16s^2, & \text{if } t \text{ is odd,} \\ 2\cdot (2s)^2 = 8s^2, & \text{if } t \text{ is even.} \end{cases}$$

Whatever t is, we obtain a polyhedron with (planar) square vertex-figures and skew $2s$-gonal faces, namely, one of type

$$\{\tfrac{2s}{t,s}, 4 : \tfrac{4}{1,2}\}.$$

The Petrial $\{\frac{4}{1,2}, 4 : \frac{2s}{t,s}\}$ is a toroid of type $\{4, 4 : 2s\}$ when t is odd, but of type $\{4, 4 \mid s\}$ if t is even. Note that the group order gives the type; we shall discuss the relationship between these toroids and those of the Section 11B in Section 11F. If t is odd, then there is a polyhedron of type $\{\frac{4}{1,2}, \frac{2s}{t} : \frac{4}{1,2}\}$ dual to the first; however, if t is even, then the dual is not realizable in $\mathbb{E}^4$.

The two polyhedra in the family $\{3, 3, \frac{3}{2}\}$ have the same symmetry group $\mathbf{A}_4 \rtimes \mathbf{Z}_2$ of order $2 \cdot 120 = 240$ (and the same 20 vertices) as $\{4, 6 \mid 3\}$; we shall see why in Section 11F. They form a Petrial pair, and are also interchanged by the operation $\boldsymbol{\zeta}$, so that we have

11D3
$$\{\tfrac{6}{1,3}, 6 : \tfrac{6}{2,3}\} \stackrel{\pi,\zeta}{\longleftrightarrow} \{\tfrac{6}{2,3}, 6 : \tfrac{6}{1,3}\}.$$

Further, the first polyhedron is self-dual; the second has no dual in $\mathbb{E}^4$.

The remaining polyhedra in the class $(2, 3, 2)$ have symmetry group $\mathbf{F}_4 \rtimes \mathbf{Z}_2 = (\mathbf{O}/\mathbf{O}; \mathbf{O}/\mathbf{O})^*$of order $2 \cdot 1152 = 2304$. Indeed, the hyperplane reflexions in the three diagrams are the same, and it is only the choice of outer normal vectors to them which gives rise to different diagrams; it is then the automorphisms which differ. In the last case $\{p, q, r\} = \{4, \frac{4}{3}, 3\}$, we have the whole family of six related by duality (interchange R_0 and R_2) and Petriality (replace R_0 by R_0R_2). For the other two (as we have said), $\frac{3}{2}$ cannot play the rôle of q.

11D4
$$\begin{array}{ccc}
\{\frac{8}{1,4}, 8 : \frac{6}{2,3}\} & \stackrel{\pi}{\longleftrightarrow} & \{\frac{6}{2,3}, 8 : \frac{8}{1,4}\} \\
\updownarrow \zeta & & \updownarrow \zeta \\
\{\frac{8}{3,4}, 8 : \frac{6}{1,3}\} & \stackrel{\pi}{\longleftrightarrow} & \{\frac{6}{1,3}, 8 : \frac{8}{3,4}\} \\
\updownarrow \delta & & \\
\{\frac{8}{1,4}, \frac{8}{3} : \frac{6}{1,3}\} & \stackrel{\pi}{\longleftrightarrow} & \{\frac{6}{1,3}, \frac{8}{3} : \frac{8}{1,4}\} \\
\updownarrow \zeta & & \updownarrow \zeta \\
\{\frac{8}{3,4}, \frac{8}{3} : \frac{6}{2,3}\} & \stackrel{\pi}{\longleftrightarrow} & \{\frac{6}{2,3}, \frac{8}{3} : \frac{8}{3,4}\}
\end{array}$$

We shall see in Section 11F that the polyhedra for which $q = 4$ or $\frac{4}{3}$ have the same 144 vertices as $\{4, 8 \mid 3\}$, namely, those of the (Minkowski) sum of

$\{3,4,3\}$ and its dual (of the same size). When $q = 3$ (recall that $q = \frac{3}{2}$ is not permitted), we have $192 = 2 \cdot 96$ vertices; these are the mid-points of the edges of one copy of $\{3,4,3\}$ and those of its dual copy.

We list the polyhedra in the class in Table 11D4 according to their Schläfli symbols and the connecting relationships. As we have already observed, it is worth noting that the various operations $\boldsymbol{\delta}, \boldsymbol{\pi}$ and $\boldsymbol{\zeta}$ all preserve the mirror vector $(2,3,2)$, and thus the whole family can be linked together. Recall also that centriversion $\boldsymbol{\zeta}$ preserves vertex-figures; here, it interchanges the pairs of marks $\{\frac{8}{1,4}, \frac{8}{3,4}\}$ and $\{\frac{6}{1,3}, \frac{6}{2,3}\}$ in both face and Petrie polygon.

11D5 Remark On the face of it, the polyhedra $\{\frac{6}{1,3}, 8 : \frac{8}{3,4}\}$ and $\{\frac{6}{1,3}, \frac{8}{3} : \frac{8}{1,4}\}$ should have (geometric) duals, to complete a family of six related alternately by $\boldsymbol{\delta}$ and $\boldsymbol{\pi}$. In fact, these duals (which would have been a Petrie pair $\{\frac{8}{1,4}, 6 : \frac{8}{3,4}\}$ and $\{\frac{8}{3,4}, 6 : \frac{8}{1,4}\}$ – see the notes at the end of the section) degenerate; their vertices fall together in fours at those of a copy of a 24-cell and its dual. Part of the reason is that the twist R_1 is now inner; we shall provide a complete justification in Section 11F.

Finally, note that φ_3 also links pairs of polyhedra whose vertex-figures are octagons. Namely, we first have

$$\{\tfrac{8}{1,4}, 8 : \tfrac{6}{2,3}\} \overset{\varphi_3}{\longleftrightarrow} \{\tfrac{8}{1,4}, \tfrac{8}{3} : \tfrac{6}{1,3}\}, \qquad \{\tfrac{8}{3,4}, \tfrac{8}{3} : \tfrac{6}{2,3}\} \overset{\varphi_3}{\longleftrightarrow} \{\tfrac{8}{3,4}, 8 : \tfrac{6}{1,3}\}.$$

In other words, we actually have $\varphi_3 = \boldsymbol{\zeta\delta}$ here. Of course, since $\boldsymbol{\pi}$ and φ_3 commute, we can similarly link their Petrials:

$$\{\tfrac{6}{2,3}, 8 : \tfrac{8}{1,4}\} \overset{\varphi_3}{\longleftrightarrow} \{\tfrac{6}{1,3}, \tfrac{8}{3} : \tfrac{8}{1,4}\}, \qquad \{\tfrac{6}{2,3}, \tfrac{8}{3} : \tfrac{8}{3,4}\} \overset{\varphi_3}{\longleftrightarrow} \{\tfrac{6}{1,3}, 8 : \tfrac{8}{3,4}\}.$$

Notes to Section 11D

1. The existence of $\{\frac{8}{1,4}, 6 : \frac{8}{3,4}\}$ and $\{\frac{8}{3,4}, 6 : \frac{8}{1,4}\}$ was mistakenly claimed in [83]. As we said, their 192 vertices coincide in fours. The vertex-figures at a given vertex are the four diametral hexagons in the central stratum of the dual 24-cell.

11E Mirror Vector $(2,2,2)$

Let P be a regular polyhedron of Schläfli type $\{p,q\}$ in the class $(2,2,2)$ (with p,q generally fractional), and let $(R_0,R_1,R_2) \subset \mathbf{SO}_4$ be its generatrix, so that $\dim R_j = 2$ for each j. As in Section 1K, we regard the symmetry group $\mathbf{G} = \langle R_0,R_1,R_2\rangle$ as a group of elements $\mathbf{g}(\mathbf{a},\mathbf{b})$, with $\mathbf{a},\mathbf{b} \in \mathbf{Q}$, the group of unit quaternions. Our approach will be to determine which rotation groups $\widetilde{\mathbf{G}}_L, \widetilde{\mathbf{G}}_R \leqslant \mathbf{SO}_3$ are involved, and how they lift to groups $\mathbf{G}_L, \mathbf{G}_R \leqslant \mathbf{Q}$.

We can therefore suppose that the generating reflexions are $R_j = \mathbf{g}(\mathbf{a}_j,\mathbf{b}_j)$ for $j = 0,1,2$, with each $\mathbf{a}_j, \mathbf{b}_j \in \mathbf{Q}$ pure imaginary, as we observed after (1K9). Interpreted as unit vectors in $\mathbb{E}^3$, we can think of $\mathbf{a}_0,\mathbf{a}_1,\mathbf{a}_2$ as outer normals to the cone generated by some Schwarz triangle on the unit sphere (see [27, Section 6.8]). If this triangle is $(r_1\ q\ 2)$, it means that

$$\langle \mathbf{a}_0, \mathbf{a}_1 \rangle = -\cos(\pi/r_1), \quad \langle \mathbf{a}_1, \mathbf{a}_2 \rangle = -\cos(\pi/q), \quad \langle \mathbf{a}_0, \mathbf{a}_2 \rangle = 0.$$

Similarly, $\mathbf{b}_0, \mathbf{b}_1, \mathbf{b}_2$ give a Schwarz triangle $(r_2\ q\ 2)$. To some extent, this already fixes the possible liftings into quaternions; however, there is still some ambiguity as to the choice of signs of (say) the $\mathbf{a}_j$, bearing in mind that $\mathbf{g}(-\mathbf{a}, -\mathbf{b}) = \mathbf{g}(\mathbf{a}, \mathbf{b})$.

Since $R_1 R_2$ is an ordinary rotation through $2\pi/q$ (and we can take $q > 2$ without loss of generality), we see that we must have

$$\langle \mathbf{a}_1, \mathbf{a}_2 \rangle = \langle \mathbf{b}_1, \mathbf{b}_2 \rangle = \pm\cos(\pi/q).$$

By changing signs of both $\mathbf{a}_1$ and $\mathbf{b}_1$ if necessary, we can assume that the sign is negative, fitting in with the convention of Section 1K. Further, we have $r_1, r_2 > 1$ such that

$$\langle \mathbf{a}_0, \mathbf{a}_1 \rangle = -\cos(\pi/r_1), \quad \langle \mathbf{b}_0, \mathbf{b}_1 \rangle = -\cos(\pi/r_2);$$

we shall see that it is necessary to allow $r_j \leqslant 2$.

We first look at just one of the groups $\mathbf{G}_L$ and $\mathbf{G}_R$, say, the former. Here, we make the choice of coordinates

$$\begin{aligned} \mathbf{a}_0 &= \mathbf{i}, \\ \mathbf{a}_1 &= \alpha_1 \mathbf{i} + \alpha_2 \mathbf{j} + \alpha_3 \mathbf{k}, \\ \mathbf{a}_2 &= \mathbf{j}. \end{aligned}$$

We therefore have

$$\alpha_1 = -\cos(\pi/r_1), \quad \alpha_2 = -\cos(\pi/q).$$

Since

$$\mathbf{a}_0 \mathbf{a}_2 = \mathbf{ij} = \mathbf{k},$$

it follows that

$$\alpha_3 = -\cos(\pi/h_1),$$

where $\widetilde{\mathbf{G}}_L$ is the group of a (possibly degenerate) polyhedron $\{r_1, q : h_1\}$, with the notation h_1 indicating its Petrie type (this may perhaps better be thought of as a polyhedron lying in the real projective plane). Note that

$$\cos^2(\pi/r_1) + \cos^2(\pi/q) + \cos^2(\pi/h_1) = \alpha_1^2 + \alpha_2^2 + \alpha_3^2 = 1;$$

this gives a different way of looking at Coxeter's formula [27, 2.33], which is easily deducible from it.

We gain even more information (enough to distinguish the possibilities) if we look further into the geometry. Recalling the free choice of coordinates we have for our generating quaternions $\mathbf{a}_i$ and $\mathbf{b}_j$, we now pick them to be

$$\begin{aligned}\mathbf{a}_0 &= \alpha_1\mathbf{i} + \alpha_2\mathbf{j}, \qquad \mathbf{b}_0 = \beta_1\mathbf{i} + \beta_2\mathbf{j},\\ \mathbf{a}_1 &= \mathbf{b}_1 = \sin(\pi/q)\mathbf{j} - \cos(\pi/q)\mathbf{k},\\ \mathbf{a}_2 &= \mathbf{b}_2 = \mathbf{k}.\end{aligned}$$

(We remark that these coordinates are only chosen to elucidate the geometry; they will not usually be best in other circumstances.) In view of our previous discussion, we have

$$\alpha_2 \sin(\pi/q) = -\cos(\pi/r_1) \implies \alpha_2 = -\frac{\cos(\pi/r_1)}{\sin(\pi/q)},$$

and similarly for β_2. (Since we only insist that $r_1 > 1$, we allow $\cos(\pi/r_1) < 0$.) Finally, applying the Petrie operation, and so replacing $\mathbf{a}_0$ by

$$\mathbf{a}_0\mathbf{a}_2 = (\alpha_1\mathbf{i} + \alpha_2\mathbf{j})\mathbf{k} = \alpha_2\mathbf{i} - \alpha_1\mathbf{j},$$

we see that

$$-\alpha_1 \sin(\pi/q) = -\cos(\pi/h_1) \implies \alpha_1 = \frac{\cos(\pi/h_1)}{\sin(\pi/q)},$$

and similarly for β_1.

At this stage, we begin to see the rôle that different liftings of the groups $\widetilde{\mathbf{G}}_L$ and $\widetilde{\mathbf{G}}_R$ can play. Since we may transfer a change of sign freely between $\mathbf{a}_j$ and $\mathbf{b}_j$, we need only impose such sign changes on the $\mathbf{a}_j$ (in practice, we do not always do this, particularly when we wish to emphasize some symmetry). Moreover, we must ensure that $\langle \mathbf{a}_1, \mathbf{a}_2 \rangle = \langle \mathbf{b}_1, \mathbf{b}_2 \rangle = -\cos(\pi/q)$, and so $\mathbf{a}_1$ and $\mathbf{a}_2$ must change sign together. Thus our freedom of choice is to change the sign of $\mathbf{a}_0$, change signs of both $\mathbf{a}_1$ and $\mathbf{a}_2$, or change all three signs. The first is $\boldsymbol{\zeta}$, and replaces r_1 and h_1 by r_1' and h_1', given by

$$\frac{1}{r_1} + \frac{1}{r_1'} = 1, \qquad \frac{1}{h_1} + \frac{1}{h_1'} = 1.$$

The second replaces r_1 by r_1', while the third replaces h_1 by h_1'. Potentially, then, this gives four distinct liftings, giving rise to four different polyhedra. Bearing in mind what we said about congruency of these groups in Section 1K, we can see that equivalence of the sets of vectors $(\mathbf{a}_0, \mathbf{a}_1, \mathbf{a}_2)$ and $(\mathbf{b}_0, \mathbf{b}_1, \mathbf{b}_2)$ under $\mathbf{SO}_3$ is what matters, subject to possible exchanges of sign of an $\mathbf{a}_j$ and corresponding $\mathbf{b}_j$. It should be noticed that changing the signs of all three $\mathbf{a}_j$ (which is the easiest way of changing relative orientation) will, in general, not yield the same polyhedron (if it exists), because h_1 is replaced by h_1' while r_1 is preserved.

Our notation keeps track of these sign changes. As we have said, the triple r_j, q, h_j is associated with the regular projective polyhedron $\{r_j, q : h_j\}$ (which indicates the type of the polyhedron and its Petrial), with the understanding that replacing r_j by r'_j or h_j by h'_j gives the same set of generators of a finite subgroup of $\mathbf{SO}_3$, but that different choices may give different liftings of the combined pair of groups.

From the discussion of Section 1K, we already know the type of face of our polyhedron. It is of the form $\{p_1\}\#\{p_2\}$, with the p_j given by

$$\frac{1}{p_j} = \pm\frac{1}{2}\left(\frac{1}{r_1} \pm \frac{1}{r_2}\right)$$

for some choice of signs (different inside the bracket); here, we replace r_i by r'_i for both i if necessary, to ensure that suitable signs can be chosen so that $p_j > 2$ for $j = 1, 2$ (it is easy to see that this is possible). This choice, by the way, may result in changing signs of some $\mathbf{b}_j$ (as well as, of course, the corresponding $\mathbf{a}_j$). As usual, it is convenient to indicate the face by a generalized fraction, as

$$\left\{\frac{p}{d_1, d_2}\right\}, \qquad \text{with} \quad p_j = \frac{p}{d_j}$$

(when reduced to lowest terms) for $j = 1, 2$. Thus we have $1 \leqslant d_1 < d_2 < \frac{1}{2}p$; the faces are helices, so that $d_2 = \frac{1}{2}p$ is not allowed.

Observe that, with the correct choice of signs above, we can recover r_1 and r_2 from

$$\frac{1}{r_j} = \frac{d_2 \mp d_1}{p},$$

with reduction of the fractions to their lowest terms. These are the r_1 and r_2 which we usually employ in the notation for the polyhedra.

The type of the Petrie polygon is given in exactly the same way, with h_1, h_2 instead of r_1, r_2. Using the conventions just introduced, we denote the resulting polyhedron (when it exists) by

11E1 $$\{r_1, q : h_1\} \bowtie \{r_2, q : h_2\}.$$

With these coordinates, 1 is an initial vertex of the putative polyhedron. Then the vertex $1R_0$ adjacent to 1 is

$$\begin{aligned}\overline{\mathbf{a}_0}1\mathbf{b}_0 = -\mathbf{a}_0\mathbf{b}_0 &= -(\alpha_1\mathbf{i} + \alpha_2\mathbf{j})(\beta_1\mathbf{i} + \beta_2\mathbf{j}) \\ &= (\alpha_1\beta_1 + \alpha_2\beta_2) + (\alpha_2\beta_1 - \alpha_1\beta_2)\mathbf{k},\end{aligned}$$

so that, if the edge-length is 2λ, then

$$\begin{aligned}\lambda^2 &= \tfrac{1}{4}\left((1 - (\alpha_1\beta_1 + \alpha_2\beta_2))^2 + (\alpha_2\beta_1 - \alpha_1\beta_2)^2\right) \\ &= \tfrac{1}{2}(1 - \alpha_1\beta_1 - \alpha_2\beta_2) \\ &= \frac{1}{2}\left(1 - \frac{\cos(\pi/r_1)\cos(\pi/r_2) + \cos(\pi/h_1)\cos(\pi/h_2)}{\sin^2(\pi/q)}\right),\end{aligned}$$

after a little calculation (note that $\alpha_1^2 + \alpha_2^2 = \beta_1^2 + \beta_2^2 = 1$).

11E2 Remark Since the symmetries of a polyhedron in this class consist only of rotations, such polyhedra occur in enantiomorphic pairs, where the 2-faces are always left or always right helices. The situation is exactly analogous to that of the regular apeirohedra in $\mathbb{E}^3$ whose symmetry groups have mirror vector $(1,1,1)$; compare Remark 11A4. We discussed these apeirohedra in Chapter 10; see also [98] or [99, Section 7E]. Any opposite isometry (that is, in $\mathbf{O}_4 \setminus \mathbf{SO}_4$) interchanges the enantiomorphism classes.

11E3 Remark The whole symmetry groups $[3,4]$ and $[3,5]$ of the octahedron and icosahedron are not rotation groups. Notice, by the way, that this accords with the well-known facts that $[3,4] \cong \boldsymbol{S}_4 \times \boldsymbol{Z}_2$ and $[3,5] \cong \boldsymbol{A}_5 \times \boldsymbol{Z}_2$; in the geometric groups, $\boldsymbol{Z}_2$ is generated by the central inversion in the origin o.

One problem which we encounter here is that the corresponding polyhedron may degenerate, although there are pointers to when degeneracy occurs. As a specific example, ignoring the rôle of the h_j for the moment, consider the case $r_1 = 3$, $r_2 = \frac{5}{2}$ and $q = 5$, when we should obtain a polyhedron

$$\{3,5\} \bowtie \{\tfrac{5}{2},5\} = \{\tfrac{30}{1,11},5\},$$

with the latter symbol indicating its type. If we replace $\frac{5}{2}$ by $\frac{5}{3}$ (or 3 by $\frac{3}{2}$), giving the other choice of lifting for R_0, then we should obtain type

$$\{3,5\} \bowtie \{\tfrac{5}{3},5\} = \{\tfrac{15}{2,7},5\}.$$

(Up to enantiomorphism, these two choices can be Petrials of each other.) Now reference to [43] shows that the only possible group these polyhedra can have is the rotation group $[3,3,5]^+$ of the regular 600-cell $\{3,3,5\}$, of order 7200. (There are limited possibilities which ensure that $\mathbf{G}_L/\mathbf{N}_L \cong \mathbf{G}_R/\mathbf{N}_R$, because the 120 icosians $\mathbf{I}$ of (1K13) form a double cover of the simple group $[3,5]^+ \cong \boldsymbol{A}_5$.) Thus the polyhedra themselves should have $7200/10 = 720$ vertices. However, if we choose (for the first polyhedron) $\mathbf{G}_L = \langle \mathbf{a}_0, \mathbf{a}_1, \mathbf{a}_2 \rangle$ and $\mathbf{G}_R = \langle \mathbf{b}_0, \mathbf{b}_1, \mathbf{b}_2 \rangle$, with generators

$$\begin{aligned} \mathbf{a}_0 &= \mathbf{k}, \quad \mathbf{b}_0 = \mathbf{j}, \\ \mathbf{a}_1 &= \mathbf{b}_1 = -\tfrac{1}{2}(\tau\mathbf{i} + \tau^{-1}\mathbf{j} + \mathbf{k}), \\ \mathbf{a}_2 &= \mathbf{b}_2 = \mathbf{i}, \end{aligned}$$

where $\tau = \frac{1}{2}(1+\sqrt{5})$, in which case (up to scaling) 1 is the initial vertex, it is obvious that the vertices obtained form a subset of $\mathbf{I}$, that is, the vertex set of $\{3,3,5\}$; in fact, we get all of $\mathbf{I}$. It follows that the 720 putative vertices of the two polyhedra must coincide in sixes at the vertices of the 600-cell, and so the polyhedra degenerate. (We shall return to this example later, and see that it can be redeemed by – for example – changing signs of all the $\mathbf{b}_j$.)

We must thus look at the possible component rotation subgroups of $\mathbf{SO}_3$ systematically. We first list those groups which contribute to the enumeration;

these are really groups which act on the corresponding projective space and, as we have seen, we can specify them by (rational) triples $\{p,q,r\}$ with

11E4 $$\cos^2(\pi/p)+\cos^2(\pi/q)+\cos^2(\pi/r)=1.$$

This is *Gordan's equation* [52]. Here, p,q,r stand indiscriminately for the type of the face, vertex-figure and Petrie polygon. In lifting them, we can replace (for example) p by p', except that the mark q, say, for the vertex-figure always satisfies $q>2$. In the listing, we also need to recall that q'' is defined by

$$\frac{1}{q}+\frac{1}{q''}=\frac{1}{2}.$$

We then have

11E5 $$\{2,q,q''\}\ (q>2\text{ arbitrary}),\quad \{3,3,4\},\quad \{3,5,\tfrac{5}{2}\}.$$

11E6 Remark We have not used the fact that these triples are (effectively) the only rational solutions of Gordan's equation (11E4), because our listing is derived from the rotation groups.

For the binary dihedral group, let $q=\frac{s}{t}>2$. The half-turns R_1' and R_2' in $\mathbb{E}^3$ corresponding to R_1 and R_2 are about axes subtending the angle $\pi(1-1/q)$, ensuring that R_1R_2 is a single rotation through π/q; we can take the corresponding quaternions to be

$$\begin{aligned}\mathbf{a}_1=\mathbf{b}_1&=-(\cos(\pi/q)\mathbf{j}+\sin(\pi/q)\mathbf{k}),\\ \mathbf{a}_2=\mathbf{b}_2&=\mathbf{j}.\end{aligned}$$

The half-turn R_0' corresponding to R_0 must have an axis orthogonal to that of R_2'; it can be in the plane of R_1' and R_2' (regarded as lines), or orthogonal to it. The different choices for the corresponding quaternions are thus

$$\mathbf{a}_0=\pm\mathbf{i},\qquad \mathbf{b}_0=\pm\mathbf{k}.$$

We list them like this, but note that $a_0=\mathbf{k}$ (and so on) is equally possible. This gives the triple $\{2,q,q''\}$.

Which dihedral group $\langle R_0',R_1',R_2'\rangle$ results depends on whether or not s is even. In fact, we obtain

$$\begin{cases}\mathbf{D}_s, & \text{if } s \text{ is even,}\\ \mathbf{D}_{2s}, & \text{if } s \text{ is odd.}\end{cases}$$

To see why, note that $\langle R_1',R_2'\rangle\cong \boldsymbol{D}_s$, and $R_0'\in\langle R_1',R_2'\rangle$ if s is even, while $R_0'\notin\langle R_1',R_2'\rangle$ if s is odd. Thus the quaternion group is $\mathbf{D}_s$ if s is even, but $\mathbf{D}_{2s}$ if s is odd (see (1K11)).

After only a little work, we find that $\{2,q:q''\}\bowtie\{q'',q:2\}$ has $2s$ vertices

$$\cos\frac{2k\pi}{q}+\sin\frac{2k\pi}{q}\mathbf{i},\qquad \cos\frac{2k\pi}{q}\mathbf{j}+\sin\frac{2k\pi}{q}\mathbf{k},$$

for $k = 0, \ldots, s-1$, where $q = s/t$ (in lowest terms) as before. In our abbreviated notation,

$$\{2, q : q''\} \bowtie \{q'', q : 2\} = \{\tfrac{2s}{t,s-t}, q : \tfrac{2s}{t,s-t}\};$$

the faces are therefore $2s$-gons, so that the polyhedron is flat (meaning that every vertex belongs to every face – see Section 2F). Moreover, it is self-Petrie (strictly, its Petrial is enantiomorphic), and is taken into a congruent copy by $\boldsymbol{\zeta}$. Our four potential possibilities here reduce to one; note that changing signs of $\mathbf{a}_1$ and $\mathbf{a}_2$ leads to a directly congruent set in $\mathbb{E}^3$, as does changing signs of all three $\mathbf{b}_j$ for the second set.

The symmetry group has order $4s^2$ in either case. It is worth remarking why this order does not depend on s being even or odd: if s is even, the normal subgroups $\mathbf{N}_L$ and $\mathbf{N}_R$ are both $\mathbf{D}_{s/2}$, while if s is odd they are $\mathbf{Z}_s$, a cyclic group of order s. (It should be remarked that this case was omitted in the brief sketch in [100].)

Next, we consider the case when one of the component quaternion groups, $\mathbf{G}_L$ say, is binary dihedral, while $\mathbf{G}_R$ is not. Then $q = 3,\ 4,\ 5$ or $\frac{5}{2}$, and the other component $\mathbf{G}_R$ is the binary octahedral group $\mathbf{O}$ of (1K12) or $\mathbf{I}$. Here, different liftings will generally give different polyhedra. For example,

$$\{2, 3 : 6\} \bowtie \{5, 3 : \tfrac{5}{2}\} = \{\tfrac{20}{3,7}, 3 : \tfrac{60}{7,17}\},$$
$$\{2, 3 : 6\} \bowtie \{5, 3 : \tfrac{5}{3}\} = \{\tfrac{20}{3,7}, 3 : \tfrac{60}{13,23}\}.$$

However, we get nothing new if we change 5 to $\frac{5}{4}$; the variation is in the Petrie polygon alone. Of course, if we take the Petrial, the opposite is true.

The enumeration is straightforward; we list Petrial pairs together. For $q = 3$, we have two pairs with $\mathbf{G}_R = \mathbf{O}$ and group order 96 (see the notes at the end of the section), and four pairs with $\mathbf{G}_R = \mathbf{I}$ and group order $24 \cdot 120/2 = 1440$. For $q = 4,\ 5$ or $\frac{5}{2}$, we have two pairs each; for the first, $\mathbf{G}_R = \mathbf{O}$ and the group has order $16 \cdot 48/2 = 384$, and for the other two $\mathbf{G}_R = \mathbf{I}$ and the group order is $40 \cdot 120/2 = 2400$. In total, we thus obtain 24 polyhedra of this kind.

11E7 Remark The polyhedra $\{2, 3 : 6\} \bowtie \{4, 3 : 3\}$ and $\{2, 3 : 6\} \bowtie \{\frac{4}{3}, 3 : 3\}$ do not fit into the general pattern; their common group is $(\mathbf{D}_6/\mathbf{C}_4; \mathbf{O}/\mathbf{V})$ of order 96. Observe that this corrects a mistaken claim in [83, Section 10]. The reason is connected to the quotient relationship $[4, 3] \searrow [2, 3]$.

Indeed, we may choose generators of the symmetry group to be

$$\mathbf{a}_0 = \pm\tfrac{1}{\sqrt{3}}(\mathbf{i} + \mathbf{j} + \mathbf{k}),\ \ \mathbf{b}_0 = \mathbf{i},$$
$$\mathbf{a}_1 = \mathbf{b}_1 = \tfrac{1}{\sqrt{2}}(\mathbf{j} - \mathbf{i}),$$
$$\mathbf{a}_2 = \mathbf{b}_2 = \tfrac{1}{\sqrt{2}}(\mathbf{k} - \mathbf{j}).$$

With initial vertex $v = \sqrt{3} = (\sqrt{3}, 0, 0, 0)$, we obtain 16 vertices

$$\pm\sqrt{3}e_j\ (j = 1, \ldots, 4), \quad \pm(0, 1, 1, 1), \quad \pm(1, 1, 0, -1),$$

and those derived from the last by cyclic permutation of the last three coordinates. The common edge-graph of the polyhedra is the Petersen type graph $G(8, 3)$ of [48] (the polyhedra themselves are of type $\{p, 3\}$ with $2p$ vertices – see also below).

Particularly note the handedness of the polyhedra, which are related by ζ (and thus have the same vertices); projections of these two polyhedra on the plane are illustrated in Figure 11E8. Reflexion in the hyperplane with normal e_1 yields enantiomorphic copies.

11E8

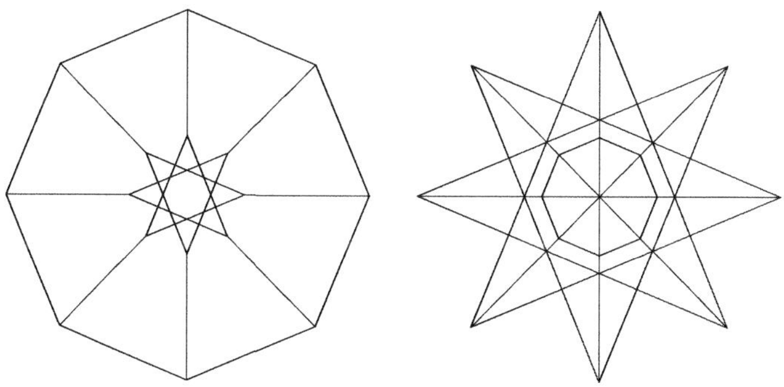

$\{2, 3 : 6\} \bowtie \{4, 3 : 3\}$ and $\{2, 3 : 6\} \bowtie \{\frac{4}{3}, 3 : 3\}$

These polyhedra are the sole pure faithful realizations of the corresponding abstract regular polyhedron $\mathcal{P}$, say. The remainder arise from the quotient of $\mathcal{P}$ onto the 3-cube $\{4, 3\}$, which folds each octagon onto a tetragon (the first picture in Figure 11E8 should make this identification obvious).

11E9 Remark It is also worth noting the two polyhedra $\{4, 4 : 2\} \bowtie \{3, 4 : 3\}$ and $\{4, 4 : 2\} \bowtie \{\frac{3}{2}, 4 : 3\}$ and their duals, with common group $(\mathbf{D}_4/\mathbf{V}; \mathbf{O}/\mathbf{T})$ of order 192, since (as we shall see in the final Section 11F) they fall into a larger family.

For the rest, the component groups $\mathbf{G}_L$ and $\mathbf{G}_R$ are binary octahedral or binary icosahedral. The case where we have one of each – say $\mathbf{G}_L = \mathbf{O}$ and $\mathbf{G}_R = \mathbf{I}$ – is the easiest, and so we treat that first. Here, we have $q = 3$; the group order is $48 \cdot 120/2 = 2880$, and so the polyhedra all have $2880/6 = 480$ vertices (note that these comprise four left cosets of a conjugate of $\mathbf{I}$ or ten right cosets of one of $\mathbf{O}$). For example, the normalized polyhedra derived from $\{4, 3 : 3\}$ and $\{5, 3 : \frac{5}{2}\}$, with possible replacement of 4 by $\frac{4}{3}$ or $\frac{5}{2}$ by $\frac{5}{3}$, have edge-lengths 2λ given by

$$\begin{aligned}\lambda^2 &= \frac{1}{2}\left(1 - \frac{\pm\frac{1}{\sqrt{2}} \cdot \frac{1}{2}\tau \pm \frac{1}{2} \cdot \frac{1}{2}\tau^{-1}}{3/4}\right) \\ &= \tfrac{1}{6}(3 \mp \tau\sqrt{2} \mp \tau^{-1}).\end{aligned}$$

Thus the different liftings – corresponding to the choices of signs – give four distinct polyhedra, whose types are

$$\{\tfrac{40}{1,9}, 3 : \tfrac{30}{1,11}\},\ \{\tfrac{40}{1,9}, 3 : \tfrac{15}{2,7}\},\ \{\tfrac{40}{11,19}, 3 : \tfrac{30}{1,11}\},\ \{\tfrac{40}{11,19}, 3 : \tfrac{15}{2,7}\};$$

naturally, we also have their Petrials. Note particularly the way that the Petrie polygons differ in length. We obtain another four Petrie pairs by interchanging 5 and $\frac{5}{2}$ (and thus τ by τ^{-1}); thus we have sixteen polyhedra in all. It is somewhat tedious to verify that these polyhedra actually exist, because there are no convenient coordinates (though they can be taken to have the same vertex-set).

Finally, we take the two component groups to be the same. We begin with the binary octahedral group $\mathbf{O}$. Here, we find that some liftings give degenerate polyhedra. In order to see why some choices lead to degeneracy, we must specify suitable generating reflexions. We thus first take

$$\begin{aligned} \mathbf{a}_0 &= \tfrac{1}{\sqrt{2}}(\mathbf{j}+\mathbf{k}), \quad \mathbf{b}_0 = -\mathbf{i}, \\ \mathbf{a}_1 &= \mathbf{b}_1 = \tfrac{1}{\sqrt{2}}(\mathbf{i}-\mathbf{j}), \\ \mathbf{a}_2 &= \mathbf{b}_2 = \tfrac{1}{\sqrt{2}}(\mathbf{j}-\mathbf{k}). \end{aligned}$$

The Petrie operation $\boldsymbol{\pi}$ replaces $\mathbf{a}_0$ and $\mathbf{b}_0$ by

$$\mathbf{a}_0' = \mathbf{a}_0\mathbf{a}_2 = -\mathbf{i}, \quad \mathbf{b}_0' = \mathbf{b}_0\mathbf{b}_2 = -\tfrac{1}{\sqrt{2}}(\mathbf{j}+\mathbf{k}),$$

so that the (putative) polyhedron would be

11E10 $$\{\tfrac{24}{1,7}, 3 : \tfrac{24}{5,11}\} = \{3,3:4\} \bowtie \{4,3:\tfrac{3}{2}\}.$$

Note that this would *not* be self-Petrie. Since the initial vertex can clearly be taken to be 1 (as in our earlier general discussion), we see that the vertex-set is a subset of $\mathbf{O}$ (regarded as a set of points in $\mathbb{E}^4$); it is not hard to see that we obtain all these 48 points. However, the group has order $48^2/2 = 1152$ (it is not the group $[3,4,3]$ of the 24-cell but, instead, the extension of its rotation subgroup $[3,4,3]^+$ by an involutory outer automorphism), and so we should expect (if the polyhedron were non-degenerate) $1152/6 = 192$ vertices. It follows that the vertices actually collapse in sets of four.

The other non-self-Petrie cases degenerate in the same way. It is not clear immediately that the self-Petrie liftings do not degenerate, but we shall see why (by means of a direct construction) in Section 11F. Given that they exist, we then obtain two polyhedra:

$$\begin{aligned} \{3,3:4\} \bowtie \{4,3:3\} &= \{\tfrac{24}{1,7}, 3 : \tfrac{24}{1,7}\}, \\ \{\tfrac{3}{2},3:4\} \bowtie \{4,3:\tfrac{3}{2}\} &= \{\tfrac{24}{5,11}, 3 : \tfrac{24}{5,11}\}. \end{aligned}$$

Observe that these two polyhedra are related by $\boldsymbol{\zeta}$.

We now move on to the binary icosahedral group $\mathbf{I}$. We first choose

$$\begin{aligned} \mathbf{a}_0 &= \mathbf{i}, \quad \mathbf{b}_0 = \mathbf{j}, \\ \mathbf{a}_1 &= \mathbf{b}_1 = -\tfrac{1}{2}(\tau\mathbf{i} - \tau^{-1}\mathbf{j} + \mathbf{k}), \\ \mathbf{a}_2 &= \mathbf{b}_2 = \mathbf{k}. \end{aligned}$$

Ignoring the Petrie polygon, we obtain the type $\{5,3\} \bowtie \{\frac{5}{3},3\} = \{\frac{5}{1,2},3\}$. If we apply the Petrie operation $\boldsymbol{\pi}$, then we replace $\mathbf{a}_0$ and $\mathbf{b}_0$ by

$$\mathbf{a}_0' = \mathbf{ik} = -\mathbf{j}, \qquad \mathbf{b}_0' = \mathbf{jk} = \mathbf{i},$$

and the type is $\{\frac{5}{2},3\} \bowtie \{5,3\} = \{\frac{10}{1,3},3\}$. In other words, what we actually have for the first is

$$\{5,3:\tfrac{5}{2}\} \bowtie \{\tfrac{5}{3},3:5\} = \{\tfrac{5}{1,2},3:\tfrac{10}{1,3}\}.$$

Reference to Section 7D identifies this polyhedron as the faithful realization in $\mathbb{E}^4$ of the dodecahedron $\{5,3\}$, while the second polyhedron is that of its Petrial $\{10,3:5\}$. The group order is 120, and the two polyhedra are related by $\boldsymbol{\zeta}$ as well as $\boldsymbol{\pi}$.

If, however, we change the sign of each $\mathbf{b}_j$ (rather than the $\mathbf{a}_j$ for convenience here), then the Petrie operation $\boldsymbol{\pi}$ yields

$$\mathbf{a}_0' = \mathbf{ik} = -\mathbf{j}, \qquad \mathbf{b}_0' = \mathbf{j}(-\mathbf{k}) = -\mathbf{i}$$

as before, but the type is now

$$\{5,3:\tfrac{5}{3}\} \bowtie \{\tfrac{5}{3},3:5\} = \{\tfrac{5}{1,2},3:\tfrac{5}{1,2}\}.$$

In this case, we therefore have the faithful realization in $\mathbb{E}^4$ of the self-Petrie hemi-dodecahedron $\{5,3:5\}$ (in view of our previous remark, the two Petrials are actually enantiomorphic). Finally, changing the sign of $\mathbf{b}_0$ again, we get

$$\{5,3:\tfrac{5}{2}\} \bowtie \{\tfrac{5}{2},3:5\} = \{\tfrac{10}{1,3},3:\tfrac{10}{1,3}\},$$

for the polyhedron and its Petrial, which identifies it as the polyhedron of type $\{10,3:10\}$ of [78] whose graph is $G(10,3)$. The order of the group of the first polyhedron is 60, and that of the second is 120. The two polyhedra, whose projections on the plane are illustrated in Figure 11E12, are related by $\boldsymbol{\zeta}$, which here changes the group order.

11E11 Remark This case is in contrast to that of the binary octahedral group, in that all four polyhedra exist (that is, are non-degenerate). Note, however, that the group orders are much smaller than the $120 \cdot 120/2 = 7200$ that can occur when the component groups are both $\mathbf{I}$.

11E12

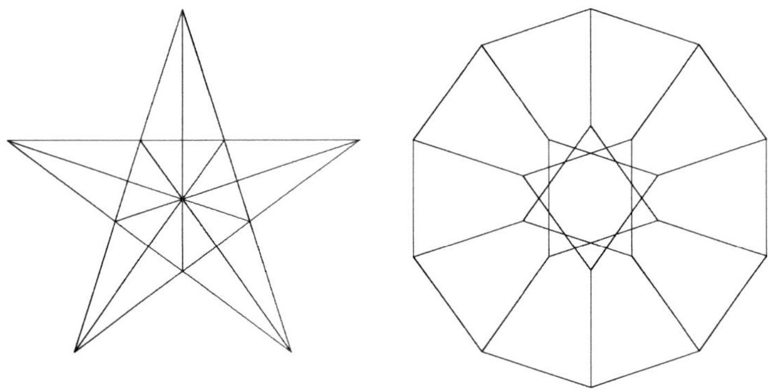

The 4-dimensional realizations of $\{5, 3 : 5\}$ and $\{10, 3 : 10, *3\}$

As an abstract regular polyhedron, $\{\frac{10}{1,3}, 3 : \frac{10}{1,3}\} \cong \{10, 3 : 10, *3\}$, with the last relation implied by the entry $*3$ corresponding to a certain hexagon in the edge-graph, which is actually regular when an automorphism $\boldsymbol{t}$ which interchanges the polyhedron and its Petrial is adjoined (in other words, $\boldsymbol{t}\boldsymbol{r}_0\boldsymbol{t} = \boldsymbol{r}_0\boldsymbol{r}_2$, and $\boldsymbol{t}\boldsymbol{r}_j\boldsymbol{t} = \boldsymbol{r}_j$ for $j = 1, 2$ on the generatrix $(\boldsymbol{r}_0, \boldsymbol{r}_1, \boldsymbol{r}_2)$ of $\{10, 3 : 10, *3\}$). This hexagon shows up quite clearly in Figure 11E12. Moreover, again on the abstract level, $\{10, 3 : 10, *3\} = \{5, 3 : 5\} \otimes \{2\}$, which implies that all its pure realizations are of the form P or $\mathsf{P} \otimes \{2\}$, with P a pure realization of $\{5, 3 : 5\}$.

11E13

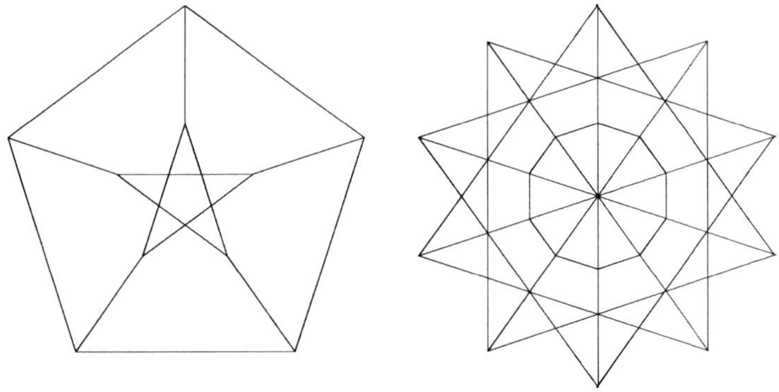

The 5-dimensional realizations of $\{5, 3 : 5\}$ and $\{10, 3 : 10, *3\}$

While it is irrelevant here, but to complete the picture, Figure 11E13 depicts symmetric projections of the 5-dimensional pure realizations of $\{5, 3 : 5\}$ and $\{10, 3 : 10, *3\}$. Finding the relative radii of the circles of (projected) vertices uses calculations like those for $\{3, 4, 3\}$ in Section 7C. Note that the cosine matrix of $\{5, 3 : 5\}$ is given – in effect – by Theorem 7D8.

The Petrie polygons of the pure realizations are $\{\frac{5}{1,2}\}$ in each case, with induced cosine vector $(1, \alpha, \beta)$. Thus, for the radii ρ_1, ρ_2 of the components $\{5\}, \{\frac{5}{2}\}$ of the Petrie polygon, we have

$$\begin{aligned}\rho_1^2 &= \tfrac{2}{5}(1 + \tau^{-1}\alpha - \tau\beta),\\ \rho_2^2 &= \tfrac{2}{5}(1 - \tau\alpha + \tau^{-1}\beta),\end{aligned}$$

where we have suppressed some intermediate calculations. Substituting for α, β from the cosine matrix surprisingly shows that $\rho_1 : \rho_2 = \tau : \tau^{-1}$ in both cases. In dimension 4, the Petrie polygon is centred (as it must be, since it is full dimensional); in dimension 5, on the other hand, it is not.

The remaining examples (in the binary icosahedral family) display exactly the same behaviour as those for which $\mathbf{G}_L$ and $\mathbf{G}_R$ are the binary octahedral group. That is, if the putative polyhedron is self-Petrie then it exists, otherwise it degenerates (once again, we make a forward reference to Section 11F). For example, the degenerate case which we looked at earlier is

$$\{3, 5 : \tfrac{5}{3}\} \bowtie \{\tfrac{5}{2}, 5 : 3\} = \{\tfrac{30}{1,11}, 5 : \tfrac{15}{2,7}\}.$$

There are just four edge-lengths 2λ of the polyhedra in the non-degenerate family, given by

$$\lambda^2 = \tfrac{1}{2}(1 \pm 2/\sqrt{5}) = \frac{\tau^{\pm 3}}{2\sqrt{5}}.$$

However, the effect of the faceting operation $\varphi := \varphi_2$ should be noticed; this ties the two families of two (related by $\boldsymbol{\zeta}$ – they are already self-Petrie) into a single family of four. We then have

$$\begin{aligned}\{3, 5 : \tfrac{5}{2}\} \bowtie \{\tfrac{5}{2}, 5 : 3\} &= \{\tfrac{30}{1,11}, 5 : \tfrac{30}{1,11}\},\\ \{5, \tfrac{5}{2} : 3\} \bowtie \{3, \tfrac{5}{2} : 5\} &= \{\tfrac{15}{1,4}, \tfrac{5}{2} : \tfrac{15}{1,4}\},\\ \{3, 5 : \tfrac{5}{3}\} \bowtie \{\tfrac{5}{3}, 5 : 3\} &= \{\tfrac{15}{2,7}, 5 : \tfrac{15}{2,7}\},\\ \{5, \tfrac{5}{2} : \tfrac{3}{2}\} \bowtie \{\tfrac{3}{2}, \tfrac{5}{2} : 5\} &= \{\tfrac{30}{7,13}, \tfrac{5}{2} : \tfrac{30}{7,13}\};\end{aligned}$$

these polyhedra are related by

$$\begin{array}{ccc}\{\tfrac{30}{1,11}, 5 : \tfrac{30}{1,11}\} & \stackrel{\varphi}{\longleftrightarrow} & \{\tfrac{15}{1,4}, \tfrac{5}{2} : \tfrac{15}{1,4}\}\\ \zeta \updownarrow & & \zeta \updownarrow\\ \{\tfrac{15}{2,7}, 5 : \tfrac{15}{2,7}\} & \stackrel{\varphi}{\longleftrightarrow} & \{\tfrac{30}{7,13}, \tfrac{5}{2} : \tfrac{30}{7,13}\}\end{array}$$

Notes to Section 11E

1. Since $\{3, 5 : \frac{5}{3}\} \bowtie \{\frac{5}{2}, 5 : 3\} = \{\frac{30}{1,11}, 5 : \frac{15}{2,7}\}$ is degenerate, it was unfortunate that we chose this as an illustrative example in [100].
2. The relevant entry 18 in the first table of [43, p. 57] is incorrect; it should read (in that notation) $(\mathbf{D}_{3m}/\mathbf{C}_{2m}; \mathbf{O}/\mathbf{V})$, with the common quotient having order 6.
3. It is also worth pointing out an omission in [43, p. 56]; for each r, there is a normal subgroup $\mathbf{C}_{2n}$ in $\mathbf{D}_{nr}$ with quotient $\mathfrak{D}_r$.

11F Further Connexions

In this final section, we discuss some further connexions among the various classes which we have considered previously. Some of the operations which we introduced in Chapter 5 automatically lead from one polyhedron to another in the same class; thus, what we shall do in this section is look at the remaining possibilities. We have already dealt with $\boldsymbol{\delta}$ and $\boldsymbol{\zeta}$ in the general discussion. We have also observed the effect of $\boldsymbol{\pi}$ in the way that we have grouped the classes. The only faceting operations $\boldsymbol{\varphi}_k$ which preserve these classes are $\boldsymbol{\varphi}_2$ for those polyhedra with pentagonal vertex-figures and $\boldsymbol{\varphi}_3$ for polyhedra with octagonal or decagonal vertex-figures; we have already listed their effects. However, the polyhedra with planar vertex-figures $\{6\}$, $\{8\}$ and $\{\frac{8}{3}\}$ also admit the singular operation φ_2 (it degenerates on those polyhedra whose vertex-figures are skew polygons). Finally, we are left with halving $\boldsymbol{\eta}$ and skewing $\boldsymbol{\sigma}$ of Section 5A.

Faceting

If we try to apply the (singular) faceting operation $\varphi = \varphi_2$ to polyhedra with skew vertex-figures, then we obtain 3-dimensional polyhedra. However, when the vertex-figure is planar, and so a hexagon $\{6\}$, an octagon $\{8\}$ or octagram $\{\frac{8}{3}\}$, the mirror vector is $(2, 3, 2)$, and there results a polyhedron with mirror vector $(2, 2, 2)$ (see the notes at the end of the section). The specific examples are

11F1
$$\begin{aligned}
\{\tfrac{6}{1,3}, 6 : \tfrac{6}{2,3}\} &\mapsto \{\tfrac{5}{1,2}, 3 : \tfrac{10}{1,3}\} = \{5, 3 : \tfrac{5}{2}\} \bowtie \{\tfrac{5}{3}, 3 : 5\},\\
\{\tfrac{8}{1,4}, 8 : \tfrac{6}{2,3}\} &\mapsto \{\tfrac{12}{1,5}, 4 : \tfrac{24}{1,7}\} = \{3, 4 : 3\} \bowtie \{2, 4 : 4\},\\
\{\tfrac{8}{3,4}, \tfrac{8}{3} : \tfrac{6}{2,3}\} &\mapsto \{\tfrac{12}{1,5}, 4 : \tfrac{24}{5,11}\} = \{3, 4 : \tfrac{3}{2}\} \bowtie \{2, 4 : 4\}.
\end{aligned}$$

The Petrie operation $\boldsymbol{\pi}$ commutes with $\boldsymbol{\varphi}$, and hence there are corresponding relationships between the Petrials. Moreover, since $\boldsymbol{\varphi}_3\boldsymbol{\varphi}_2 = \boldsymbol{\varphi}_2$ (in effect) for the latter two polyhedra, we similarly have

11F2
$$\begin{aligned}
\{\tfrac{8}{1,4}, \tfrac{8}{3} : \tfrac{6}{1,3}\} &\mapsto \{\tfrac{12}{1,5}, 4 : \tfrac{24}{1,7}\},\\
\{\tfrac{8}{3,4}, 8 : \tfrac{6}{1,3}\} &\mapsto \{\tfrac{12}{1,5}, 4 : \tfrac{24}{5,11}\}.
\end{aligned}$$

As promised, we justify what we said in Remark 11D5 about the degeneracy of $\mathsf{P} := \{\frac{8}{1,4}, 6 : \frac{8}{3,4}\}$ and its Petrial. Applying $\boldsymbol{\varphi}$ yields

$$\mathsf{P} \mapsto \{\tfrac{24}{1,7}, 3; \tfrac{24}{5,11}\} = \{3, 3 : 4\} \bowtie \{4, 3 : \tfrac{3}{2}\}.$$

The discussion around (11E10) showed that this latter polyhedron degenerates, with 192 vertices coinciding in fours in $\mathbf{O}$ (as a set of quaternions). Since φ preserves initial vertices, it follows that at least one vertex of P also lies in $\mathbf{O}$. However, the symmetry group of P is the extended group $(\mathbf{O}/\mathbf{O};\mathbf{O}/\mathbf{O})^*$, and so this forces all vertices of P to lie in $\mathbf{O}$, which is what we needed to show.

Halving

For η, if the original mirror vector is (d_0, d_1, d_2), then the new one is (d_1, d_2, d_1); in our cases, we have $(3,2,3) \mapsto (2,3,2)$ or $(2,3,2) \mapsto (3,2,3)$. Note, however, that the polyhedra in the other classes related to $(3,2,3)$ do not permit the application of η, and that only the toroidal polyhedra in class $(2,3,2)$ allow it.

It is straightforward to keep track of the generating (hyperplane) reflexions of the original diagrams of Figures 11A2 and 11A3, to show that

11F3 $\quad \mathcal{D}_1(2,q;r) \overset{\eta}{\longmapsto} \mathcal{D}_2(q,q,\tfrac{r}{2}), \qquad \mathcal{D}_2(2,q,r) \overset{\eta}{\longmapsto} \mathcal{D}_1(q,q;r);$

that is, corresponding to the diagram $\mathcal{D}_1(2,q;r)$ in class $(3,2,3)$ with $p=2$ (and so with the relevant branch of the diagram missing) is the diagram $\mathcal{D}_2(q,q,\frac{r}{2})$ in class $(2,3,2)$ with $p=q$ and r replaced by $\frac{r}{2}$, and similarly for $\mathcal{D}_2(2,q,r)$. We look at the individual cases.

We begin with the toroidal polyhedra and their relatives. For those of Schläfli type $\{4,4\}$, we have

$$\{4, \tfrac{4}{1,2} \mid \tfrac{s}{t}\} \overset{\eta}{\longmapsto} \{\tfrac{4}{1,2}, 4 : \tfrac{2s}{t,s}\}, \qquad \{\tfrac{4}{1,2}, 4 : \tfrac{2s}{t,s}\} \overset{\eta}{\longmapsto} \{4, \tfrac{4}{1,2} \mid \tfrac{s}{t}\}.$$

Note that, if t is even, then η does not halve the group order in the latter case. If $r = \frac{s}{t}$ with t odd, then the dual of its Petrial is

$$\{\tfrac{4}{1,2}, \tfrac{2s}{t} : \tfrac{4}{1,2}\}.$$

It might appear that we could apply η to it, however, as (11F3) shows, the result degenerates to a polyhedron with digonal holes.

There are no more polyhedra with square faces in the class $(2,3,2)$, and so our only further applications are to the class $(3,2,3)$. Thus we next have

$$\{4, \tfrac{6}{1,3} \mid 3\} \overset{\eta}{\longmapsto} \{\tfrac{6}{1,3}, 6 : \tfrac{6}{2,3}\},$$

with the same group $[3,3,3] \rtimes \mathbf{Z}_2$.

Next, η links up the two families of polyhedra derived from the extension $[3,4,3] \rtimes \mathbf{Z}_2 = (\mathbf{O}/\mathbf{O};\mathbf{O}/\mathbf{O})^*$ of the group $[3,4,3]$. Here, we have

$$\{4, \tfrac{8}{1,4} \mid 3\} \overset{\eta}{\longmapsto} \{\tfrac{8}{1,4}, 8 : \tfrac{6}{2,3}\}, \qquad \{4, \tfrac{8}{3,4} \mid 3\} \overset{\eta}{\longmapsto} \{\tfrac{8}{3,4}, \tfrac{8}{3} : \tfrac{6}{2,3}\},$$

derived from the diagrams $\mathcal{D}_2(4,4,\frac{3}{2})$ and $\mathcal{D}_2(\frac{4}{3},\frac{4}{3},\frac{3}{2})$ in class $(2,3,2)$. Observe that these polyhedra all have the same vertices and group $[3,4,3] \rtimes \mathbf{Z}_2$; this follows from the existence of odd edge-circuits (the holes) in the graphs of the first polyhedra in each pair. Naturally, the same then remains true when we further apply operations such as $\boldsymbol{\pi}$, $\boldsymbol{\zeta}$ and $\boldsymbol{\varphi}_3$, which explains why we did not discuss the geometric structure of the polyhedra in the class $(2,3,2)$ in more detail in Section 11D.

Last, consider polyhedra with square faces derived by Petrie contraction $\boldsymbol{\varpi}$ from 4-polytopes; these come from $\{3,3,q_3\}$ with $q_3 = 3, 5$ or $\frac{5}{2}$. In such a case, q is a fraction with an odd denominator, and the vertex-figure of the corresponding polyhedron of class $(2,3,2)$ will then collapse. A similar line of argument shows that the other possible polyhedra in class $(2,3,2)$ derived from diagrams with marks 5 or its fractions must also degenerate, because they can be obtained from such degenerate polyhedra in class $(3,2,3)$ by $\boldsymbol{\eta}$. In particular, note that $\boldsymbol{\eta}$ degenerates on $\{4, \frac{6}{2,3} \mid 3\}$.

Skewing

We now come to the skewing operation $\boldsymbol{\sigma} = \boldsymbol{\pi}^*\boldsymbol{\eta}\boldsymbol{\pi}^*$ with $\boldsymbol{\pi}^* = \boldsymbol{\delta}\boldsymbol{\pi}\boldsymbol{\delta} = \boldsymbol{\pi}\boldsymbol{\delta}\boldsymbol{\pi}$; this applies to a regular polyhedron of Schläfli type $\{4,q\}$. The result degenerates for toroidal polyhedra in the class $(2,3,2)$, and there are no others in that class to which it applies (in any event, since both $\boldsymbol{\pi}$ and $\boldsymbol{\delta}$ preserve this class, one is – in effect – reduced to considering $\boldsymbol{\eta}$). It also degenerates for the few polyhedra in class $(2,2,2)$ to which it might be applicable; what goes wrong here is that the component regular projective polyhedra end up having different values of q.

However, $\boldsymbol{\sigma}$ does work on class $(3,2,3)$; in fact, it yields polyhedra in class $(2,2,2)$ (see the notes at the end of the section). Now

$$\boldsymbol{\sigma}\colon (R_0, R_1, R_2) \mapsto (R_1, R_0R_2, (R_0R_1)^2) =: (S_0, S_1, S_2).$$

Thus, in terms of the diagram $\mathcal{D}_1(2,q;r)$, the new generators (S_0, S_1, S_2) are given by

11F4

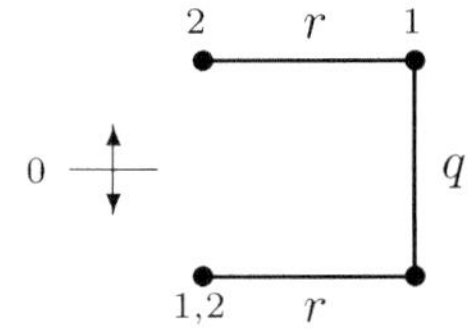

The labels on the diagram of Figure 11F4 index the reflexions S_j; thus S_0 is the flip. Note that S_1 and S_2 are both products of two (commuting) hyperplane reflexions, one of which is shared between them; observe that the new vertex-figure corresponds to the old hole. Further, $S_0 = R_1$ is the flip in the original diagram; hence, subsequently applying $\boldsymbol{\zeta}$ is equivalent to replacing the second mark q in the diagram by q'.

For the toroids, it should not occasion any surprise that

$$\{4, \tfrac{4}{1,2} \mid r\}^{\boldsymbol{\sigma}} = \{\tfrac{2s}{t,s-t}, r; \tfrac{2s}{t,s-t}\} = \{2, r : r''\} \bowtie \{r'', r : 2\},$$

with $r = s/t$.

Next, we have

$$\{4, \tfrac{6}{1,3} \mid 3\}^{\sigma} = \{\tfrac{10}{1,3}, 3; \tfrac{10}{1,3}\} = \{5, 3 : \tfrac{5}{2}\} \bowtie \{\tfrac{5}{2}, 3 : 5\}.$$

As we said in Section 11E, the edge-graph of this polyhedron is the Petersen-type graph $G(10, 3)$; it doubly covers the ordinary Petersen graph $G(5, 2)$. Indeed, the latter is the graph of the hemi-dodecahedron, which is obtained as

$$\{4, \tfrac{6}{2,3} \mid 3\}^{\sigma} = \{\tfrac{5}{1,2}, 3; \tfrac{5}{1,2}\} = \{5, 3 : \tfrac{5}{3}\} \bowtie \{\tfrac{5}{3}, 3 : 5\}.$$

This is an interesting example, in that the whole group is isomorphic to each of the component rotation groups in $\mathbf{SO}_3$.

11F5 Remark If we chase through the various changes of generators, we find that this 4-dimensional realization of $\{5, 3 : 5\}$ has the same vertices as the initial polyhedron $\{4, \frac{6}{2,3} \mid 3\}$, but with edges and diagonals interchanged. Indeed, this carries over to the abstract level: $\{5, 3 : 5\}$ has the same vertices as $\{4, 6 : 5 \mid 3\}$, and hence as well those of $\{3^{2,1}\}$ (the abstract version of the truncate 0_{21} of the 4-simplex), but with edges and diagonals swapped. Moreover, if we adjoin the automorphism which takes $\{5, 3 : 5\}$ into its Petrial, then the groups also coincide.

Moving on, we have

$$\{4, \tfrac{8}{1,4} \mid 3\}^{\sigma} = \{\tfrac{24}{1,7}, 3 : \tfrac{24}{1,7}\} = \{3, 3 : 4\} \bowtie \{4, 3 : 3\}.$$

The geometry of this polyhedron is interesting and, incidentally, shows that it exists. It may be seen that its 192 vertices lie at the mid-points of the edges of the 24-cell $\{3, 4, 3\}$ and its congruent polar, and are thus all permutations with all changes of sign of

$$(3, 1, 1, 1), \ (2, 2, 2, 0), \ (2\sqrt{2}, \sqrt{2}, \sqrt{2}, 0).$$

A given vertex is joined to the mid-points of the edges of the corresponding triangular face of the polar. Given two adjacent edges, there are two choices to continue to a third edge; these two choices correspond to the faces and the Petrie polygons (the polyhedron is self-Petrie). In view of our earlier comment about $\boldsymbol{\zeta}$, we have directly

$$\{4, \tfrac{8}{3,4} \mid 3\}^{\sigma} = \{\tfrac{24}{5,11}, 3 : \tfrac{24}{5,11}\} = \{\tfrac{3}{2}, 3 : 4\} \bowtie \{4, 3 : \tfrac{3}{2}\}.$$

Recall that the pentagonal diagrams, corresponding to regular polytopes such as $\{5, \frac{5}{2}, 5\}$, give rise to degenerate polyhedra in class $(3, 2, 3)$. What is surprising is that $\boldsymbol{\sigma}$ does apply to these diagrams, yielding genuine polyhedra in class $(2, 2, 2)$; we can see this as follows.

We begin with $\mathsf{Q} = \{3, 3, r\}$ for $r = 3, 5, \frac{5}{2}$ (in fact, we could even include the case $r = 4$, though this degenerates), and apply $\boldsymbol{\varpi\sigma}$; the combined operation on the generatrix is

$$\boldsymbol{\varpi\sigma}\colon (R_0, \ldots, R_3) \mapsto (R_0R_2, R_1R_3, (R_1R_0R_2)^2) =: (S_0, S_1, S_2).$$

We can see that

$$R_1R_3(R_1R_0R_2)^2 \sim R_2R_3$$

after a little calculation; we thus end up with a self-Petrie polyhedron P of type $\{h, r : h\}$, where $\{h\}$ is the Petrie polygon of Q. In particular,

$$\begin{aligned}\{3,3,5\} &\mapsto \{\tfrac{30}{1,11}, 5 : \tfrac{30}{1,11}\} = \{\tfrac{5}{2}, 5 : 3\} \bowtie \{3, 5 : \tfrac{5}{2}\},\\ \{3,3,\tfrac{5}{2}\} &\mapsto \{\tfrac{30}{7,13}, \tfrac{5}{2} : \tfrac{30}{7,13}\} = \{5, \tfrac{5}{2} : \tfrac{3}{2}\} \bowtie \{\tfrac{3}{2}, \tfrac{5}{2} : 5\}.\end{aligned}$$

As we saw in Section 11E, we can apply $\boldsymbol{\zeta}$ to each of these polyhedra P, to obtain

$$\begin{aligned}\{\tfrac{30}{1,11}, 5 : \tfrac{30}{1,11}\}^{\zeta} &= \{\tfrac{15}{2,7}, 5 : \tfrac{15}{2,7}\} = \{\tfrac{5}{3}, 5 : 3\} \bowtie \{3, 5 : \tfrac{5}{3}\},\\ \{\tfrac{30}{7,13}, \tfrac{5}{2} : \tfrac{30}{7,13}\}^{\zeta} &= \{\tfrac{15}{1,4}, 5 : \tfrac{15}{1,4}\} = \{5, \tfrac{5}{2} : 3\} \bowtie \{3, \tfrac{5}{2} : 5\}.\end{aligned}$$

On the other hand, this change of sign of S_0 is equivalent to changing the sign of the twist in Q^{ϖ}, which when $r = 5$ or $\frac{5}{2}$ leads to the degenerate cases discussed in Section 11B. (In contrast, there was no problem when $r = 3$.)

This confirms the listings in Section 11E, but the alternative approach now enables us to describe their geometry, and thus establish their existence. So, for example, the 720 vertices of $\{\frac{30}{1,11}, 5 : \frac{30}{1,11}\}$ are the mid-points of the edges of the 600-cell $\{3, 3, 5\}$ (or $\{5, \frac{5}{2}, 5\}$), and each vertex is joined by an edge to the mid-points of the edges of the pentagonal link of the original edge in the boundary complex of $\{3, 3, 5\}$. It is not hard to see that the other three polyhedra have the same vertex-set.

Notes to Section 11F

1. In the original paper [83], skewing – as defined there – applied mainly to degenerate polyhedra. Employing the redefined $\boldsymbol{\sigma}$ of Section 5A instead means that we usually operate on genuine polyhedra, which is much to be preferred.
2. The connexions in (11F1) and (11F2) were not made in [83]; all but the first were in notes made by McMullen in the mid-1980s.

12

Four-Dimensional Apeirotopes

In the discussion of Chapter 9, we have already found several regular apeirotopes of nearly full rank in $\mathbb{E}^4$. Unlike in the general case, various operations applied to these lead to many more apeirotopes. In addition, other symmetry groups give rise to families unrelated to these. We describe all of them in this chapter which, for various reasons, probably differs more than most from the original paper [84]. Indeed, the necessary corrections to that paper are too numerous to mention individually.

However, before we go further, we should remark that there is little to add to the discussion of Section 9A about the part played by blending. As we shall see in Section 12B, there are connexions that tie together the two basic ways of constructing apeirotopes of nearly full rank from polytopes or apeirotopes of full rank which are not available in other dimensions. Otherwise, there are no surprises here.

There are three sections in this chapter. First, this is the only dimension to which imprimitive symmetry groups can make a contribution to apeirotopes of nearly full rank; we consider these in Section 12A. The largest family of apeirotopes is derived from the group $\mathbf{U}_5 = [3,3,4,3]$; we describe this family in Section 12B, and include here those related to $\mathbf{R}_5 = [4,3,3,4]$. The final family consists of the apeirotopes with group related to $\mathbf{P}_5$; these are treated in Section 12C.

12A Imprimitive Groups

Among the symmetry groups of regular 4-apeirotopes of rank 4 are three that are imprimitive; their vertex-figures are correspondingly imprimitive. This section will be devoted to two of these; the treatment of the third falls more naturally into Section 12B.

For the vertex-figure, the starting point in this case is the torus

$$\{4, \tfrac{4}{1,2} \mid 6\} \cong \{4, 4 \mid 6\},$$

whose vertices are those of the (orthogonal direct) product $\{6\} \times \{6\}$ of two congruent regular hexagons. It is convenient here to use complex coordinates $z = (x, y)$, with $x, y \in \mathbb{C}$ further regarded as pairs of real numbers. Writing $\omega := \exp(2i\pi/3) = (-1+i\sqrt{3})/2$, so that $\varphi := -\overline{\omega} = -\omega^2 = \exp(i\pi/3)$ is a sixth root of one, the vertex-set of $\{4, \frac{4}{1,2} \mid 6\}$ is $D_1 := \{(\varphi^m, \varphi^n) \mid 0 \leqslant m, n < 6\}$. The generatrix (R_1, R_2, R_3) of its group $\mathbf{G}_0$ is given by

12A1
$$\begin{aligned} R_1 &\colon z \mapsto (\varphi\overline{x}, y), \\ R_2 &\colon z \mapsto (y, x), \\ R_3 &\colon z \mapsto (x, \overline{y}). \end{aligned}$$

The initial vertex is $(1, 1)$, and the initial edge joins $(1, 1)$ to $(\varphi, 1)$.

12A2 Remark It is worth recalling that the isomorphism class of a regular torus of (Schläfli) type $\{4, 4\}$ is determined by the number of its vertices. If this is r^2, then the torus is $\{4, 4 \mid r\}$; if it is $2r^2$, then the torus is $\{4, 4 : 2r\}$.

Successive applications of halving $\boldsymbol{\eta}$ lead to

$$\{\tfrac{4}{1,2}, 4 : \tfrac{6}{1,3}\}, \quad \{4, \tfrac{4}{1,2} \mid 3\}, \quad \{\tfrac{4}{1,2}, 4 \mid 3\};$$

the first has vertex-set $D_2 := \{(\varphi^m, \varphi^n) \mid 0 \leqslant m, n < 6 \text{ and } m + n \text{ even}\}$, while the last two have the same vertex-set $D_3 := \{(\omega^m, \omega^n) \mid 0 \leqslant m, n < 3\}$ (and hence the same group – indeed, $\{\frac{4}{1,2}, 4 \mid 3\}^{\boldsymbol{\eta}} = \{4, \frac{4}{1,2} \mid 3\}$ again). We retain the initial vertex $(1, 1)$, which is now joined by the initial edge to (φ, φ), $(\omega, 1)$ or (ω, ω), respectively (after passing to suitable conjugate sets of generators).

We now adjoin a suitable reflexion R_0; this must interchange the initial vertex $o = (0, 0)$ and the initial vertex $(1, 1)$ of the vertex-figure. Such reflexions are linked in pairs by the operation $\boldsymbol{\tau} = \boldsymbol{\kappa}_{02}$; this is induced by the reflexion T with mirror $R_2 \cap R_3 = \{(x, y) \in \mathbb{C}^2 \mid x, y \in \mathbb{R}\}$, so that $T = (R_2R_3)^2 \colon z \mapsto (\overline{x}, \overline{y})$. The point-reflexion for the free abelian apeirotope $\mathsf{Q}^{\boldsymbol{\alpha}}$ is $z \mapsto (1-x, 1-y)$, so that we also have the choice

12A3
$$R_0 \colon z \mapsto (1 - \overline{x}, 1 - \overline{y}).$$

In fact, these two are the only possibilities.

12A4 Lemma *A reflexion R_0 with 1- or 3-dimensional mirror is incompatible with any of the vertex-figures above.*

Proof. The two cases are again linked by $\boldsymbol{\tau}$, and so we need only address the latter. Moreover, since $\boldsymbol{\eta}$ links the different vertex-figures (so inducing subgroup relationships), proving the incompatibility for the vertex-figure $\{4, \frac{4}{1,2} \mid 3\}$ will suffice. We therefore have the interaction of two hyperplane reflexions: $R_0 \colon z \mapsto (1 - \eta_1, \xi_2, 1 - \xi_1, \eta_2)$ and $R_1 \colon z \mapsto (\omega\overline{x}, y)$. Here, we write $x = (\xi_1, \xi_2)$ and $y = (\eta_1, \eta_2)$ as real vectors; moreover, the R_1 is that appropriate for the given vertex-figure. The two unit normals to the corresponding mirrors are thus

$$u_0 = \tfrac{1}{\sqrt{2}}(1, 0, 1, 0), \quad u_1 = (i\overline{\omega}, 0) = \tfrac{1}{2}(\sqrt{3}, -1, 0, 0);$$

since the angle between them, namely, $\arccos\langle u_0, u_1\rangle = \arccos\sqrt{\frac{3}{8}}$, is not a multiple of $\pi/6$ or $\pi/4$, such a choice of reflexion cannot give rise to a discrete group. □

12A5

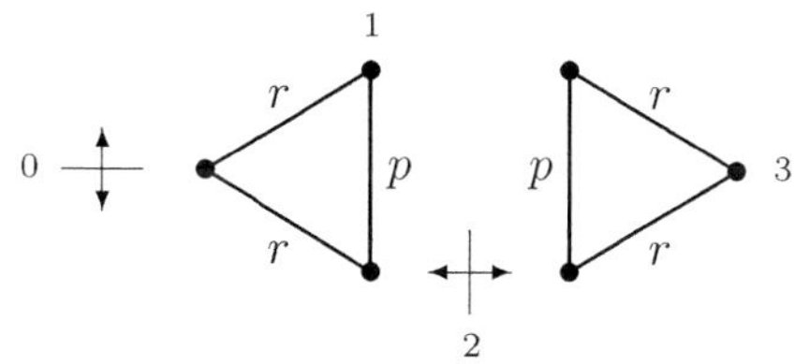

The diagram $\mathcal{A}(p, r)$

With the original reflexions $R_0, \ldots, R_3$ of (12A1) and (12A3), we actually have a twisted Coxeter group $(\mathbf{V}_3 \times \mathbf{V}_3) \rtimes \mathbf{Z}_2$ with diagram $\mathcal{A}(\frac{3}{2}, 6)$, where the general diagram $\mathcal{A}(p, r)$ is as in Figure 12A5. Thus the 3-dimensional R_1 and R_3 correspond to the marked nodes of the diagram, while the 2-dimensional R_0 and R_2 correspond to the indicated diagram automorphisms. Writing P for the resulting (at this stage putative) apeirotope, we find it more illuminating to look at the Petrial $\mathsf{P}^{f\pi}$ of its facet P^f rather than at the facet itself. For this,

$$zR_0R_2 = (1 - \overline{y}, 1 - \overline{x}),$$

from which we have

$$zR_1R_0R_2 = (1 - \overline{y}, 1 - \overline{\varphi}x) = (1 - \overline{y}, 1 + \omega x).$$

We can verify that this has period 4; indeed, we have successive vertices $o, (1, 1), (0, -\overline{\omega}), (-\overline{\omega}, 1)$, giving a tetragon $\{\frac{4}{1,2}\}$. Hence $\mathsf{P}^{f\pi}$ is a torus.

For the hole, we have

$$zR_1R_2R_1 \cdot R_0R_2 = (1 - \overline{\varphi}x, 1 - \overline{\varphi}y) = (1 + \omega x, 1 + \omega y);$$

this has period 3, and gives a trigon $\{3\}$. In other words,

$$\mathsf{P}^{f\pi} = \{\tfrac{4}{1,2}, 4 \mid 3\} \implies \mathsf{P}^f = \{\tfrac{6}{2,3}, 4 : \tfrac{4}{1,2}, 3\}.$$

We can check the face $\{\frac{6}{2,3}\}$ directly, since $zR_1R_0 = (1+\omega x, 1-\overline{y})$. In conclusion, then,

$$\mathsf{P} = \{\{\tfrac{6}{2,3}, 4 : \tfrac{4}{1,2}, 3\}, \{4, \tfrac{4}{1,2} \mid 6\}\};$$

we can appeal to the geometry in $\mathbb{E}^4$ to see that P genuinely is an apeirotope.

We can say a little more about P. First, as a fine Schläfli symbol, it is rigid, because its facet and vertex-figure both are (the facet is rigid because its Petrial is). Second, it does not appear to be universal (of type $\{\{6, 4 : 4, 3\}, \{4, 4 \mid 6\}\}$). Third, it has no geometric dual, since its facet does not; refer to Section 11D for this.

We can now apply the operations $\boldsymbol{\pi}, \boldsymbol{\kappa}$ and $\boldsymbol{\pi\kappa}$ to P (see the notes at the end of the section). Under $\boldsymbol{\pi}$, according to Theorem 5B14 the new vertex-figure is $\{\frac{12}{1,5}, \frac{4}{1,2} : 4, 6\}$, while the 2-face remains $\{\frac{6}{2,3}\}$. We know little about the new facet of type $\{\frac{6}{2,3}, \frac{12}{1,5}\}$, other than that it is infinite because its vertex-figure is centred. Further applying $\boldsymbol{\kappa}$ to these, we obtain a new vertex-figure of the same type $\{12, 4 : 4, 6\}$ but distinct from the original; the new 2-face is now an apeirogon $\{\frac{6}{0,1}\}$, as Theorem 5D19 indicates. Once again, none of these apeirotopes has a geometric dual.

We next see what can be derived from the alternative diagram $\mathcal{A}(3,3)$; the group is now $(\mathbf{P}_3 \times \mathbf{P}_3) \rtimes \mathbf{Z}_2$. We start from vertex-figure $\{4, \frac{4}{1,2} \mid 3\}$, and then apply $\boldsymbol{\pi}, \boldsymbol{\kappa}$ and $\boldsymbol{\pi\kappa}$; recall that (abstractly) $\{4,4 : 6\} = \{4,4 \mid 3\} \otimes \{2\}$. (We postpone discussion of the effect of $\boldsymbol{\eta}$.) The generators R_0, R_2, R_3 of (12A1) and (12A3) remain; for the last, the replacement is

12A6 $$R_1 \colon z \mapsto (\omega\overline{x}, y).$$

Calculations like those previously show that we now have

$$\{\{\tfrac{6}{1,3}, 4 : \tfrac{4}{1,2}\}, \{4, \tfrac{4}{1,2} \mid 3\}\} \overset{\boldsymbol{\delta}}{\longleftrightarrow} \{\{4, \tfrac{4}{1,2} \mid 3\}, \{\tfrac{4}{1,2}, 6 : \tfrac{4}{1,2}\}\}.$$

Here, as indicated, we have a geometric dual. We saw in Section 11F that $\{\frac{4}{1,2}, 6 : \frac{4}{1,2}\}^{\boldsymbol{\eta}}$ degenerates, and so we cannot apply $\boldsymbol{\eta}$ to this dual. Indeed, if we try, then we obtain facets with triangular faces and planar hexagonal vertex-figures, in other words, planar tilings $\{3, 6\}$. Moreover, the vertex-figure $\{\frac{4}{1,2}, 6 : \frac{4}{1,2}\}$ is preserved by $\boldsymbol{\pi}$ and $\boldsymbol{\zeta}$, so that applying $\boldsymbol{\pi}, \boldsymbol{\kappa}$ or $\boldsymbol{\pi\kappa}$ leads to nothing new; note that $\{4\}^{\boldsymbol{\kappa}} = \{4\}$.

Of course, we can apply these operations to $\mathsf{P} = \{\{\frac{6}{1,3}, 4 : \frac{4}{1,2}\}, \{4, \frac{4}{1,2} \mid 3\}\}$ itself.

Finally, let us address universality and rigidity; we can only comment on the original P here. From the generatrix of $\{\{6, 4 : 4\}, \{4, 4 \mid 3\}\}$, we obtain

12A7 $$\begin{aligned}(\boldsymbol{r}_0, \ldots, \boldsymbol{r}_3) &\mapsto (\boldsymbol{r}_1, \boldsymbol{r}_0\boldsymbol{r}_1\boldsymbol{r}_0, \boldsymbol{r}_2\boldsymbol{r}_3\boldsymbol{r}_2; \boldsymbol{r}_2\boldsymbol{r}_1\boldsymbol{r}_2, \boldsymbol{r}_2\boldsymbol{r}_0\boldsymbol{r}_1\boldsymbol{r}_0\boldsymbol{r}_2, \boldsymbol{r}_3) \\ &=: (\boldsymbol{s}_1, \ldots, \boldsymbol{s}_6),\end{aligned}$$

giving (abstractly) two copies of an unmarked triangle $\mathbf{P}_3$ (with a semicolon separating the two copies); the universality of P follows (see the notes at the end of the section). As it stands, the fine Schläfli symbol for P is not rigid; indeed, it permits blending with the digon $\{2\}$. However, we can make it rigid by specifying the 2-zigzag $\{6\}$ of the facet, which corresponds to the hole of its Petrial $\{\frac{4}{1,2}, 4 : \frac{6}{1,3}\}$; that is,

$$\{\{\tfrac{6}{1,3}, 4 : \tfrac{4}{1,2}, 6\}, \{4, \tfrac{4}{1,2} \mid 3\}\}$$

is rigid (see the notes at the end of the section).

Applying $\boldsymbol{\eta}$ to $\{\{\frac{6}{2,3}, 4 : \frac{4}{1,2}\}, \{4, \frac{4}{1,2} \mid 6\}\}$ yields a new generatrix $(S_0, \ldots, S_3)$ for which

$$zS_1S_0 = (1 + \omega y, 1 + \omega x);$$

hence $z(S_1S_0)^3 = (y,x)$. But this says that $(S_1S_0)^3 = R_2 = S_3$, which violates the intersection property, so that the corresponding apeirotope degenerates. In fact, we know for a different reason that something goes wrong here. Tracing $(1,1)$ under S_1S_0 yields a triangle, and with triangular faces, we would obtain facets of type $\{3, \frac{4}{1,2}\}$, which is impossible (regular octahedra in $\mathbb{E}^4$ can only have planar vertex-figures – see Example 3D4 in the case $d = 3$). In fact, this triangle is actually a doubly-covered hexagon.

However, $\boldsymbol{\eta}$ does work in the case not yet covered, and we obtain

$$\{\{\tfrac{6}{1,3}, 4 : \tfrac{4}{1,2}\}, \{4, \tfrac{4}{1,2} \mid 3\}\} \xrightarrow{\boldsymbol{\eta}} \{\{6, \tfrac{4}{1,2} \mid 6\}, \{\tfrac{4}{1,2}, 4 \mid 3\}\}.$$

The fine Schläfli symbol of the facet is rigid, and so describes it geometrically; however, we lack an expression for its combinatorics. Because the facet is infinite (since its Petrie polygon is), the apeirotope can have no (discrete geometric) dual.

We now attempt to apply $\boldsymbol{\pi}$ and axiversion $\boldsymbol{\kappa}$ to these examples. Bearing in mind that $\boldsymbol{\kappa\eta} = \boldsymbol{\eta}$, we see that no more cases will arise to which we can apply $\boldsymbol{\eta}$. Of course, $\boldsymbol{\pi}$ and $\boldsymbol{\kappa}$ commute, and both operations commute with $\boldsymbol{\tau}$ (when it is applicable). As it happens, none of the new examples will be dualizable. Since there is no problem with apeirotopes of the form Q^{α}, we need only consider those cases with R_0 as in (12A3).

Two of the four vertex-figures derived from the tori are self-Petrie, as their notation indicates; thus $\boldsymbol{\pi}$ yields no new apeirotope in these cases. For the other two, we have

$$\{4, \tfrac{4}{1,2} \mid 6\}^{\boldsymbol{\pi}} = \{\tfrac{12}{1,5}, \tfrac{4}{1,2} : 4, 6\}, \quad \{4, \tfrac{4}{1,2} \mid 3\}^{\boldsymbol{\pi}} = \{\tfrac{6}{1,2}, \tfrac{4}{1,2} : 4, 3\};$$

the Petrials are indicated by the notation. We shall say no more about the effects on the corresponding apeirotopes, except that their facets must be infinite since their vertex-figures are 4-dimensional; Theorem 5B14 tells us that the ridges are unaltered.

We now move on to $\boldsymbol{\kappa}$ which, as we recall, changes the sign of R_1. First, it fails on $\{\{6, \tfrac{4}{1,2} \mid \tfrac{6}{1,3}\}, \{\tfrac{4}{1,2}, 4 \mid 3\}\}$; since $\{\tfrac{4}{1,2}, 4 \mid 3\}^{\boldsymbol{\zeta}} = \{\tfrac{4}{1,2}, 4 : \tfrac{6}{1,3}\}$, it produces the degenerate example that we have just encountered (this provides another picture of why its face is a doubly-covered triangle). In the remaining cases, we have

$$\begin{aligned} \{4, \tfrac{4}{1,2} \mid 6\}^{\boldsymbol{\zeta}} &= \{\tfrac{4}{1,2}, \tfrac{4}{1,2} \mid \tfrac{6}{1,3}\} \cong \{4, 4 \mid 6\}, \\ \{4, \tfrac{4}{1,2} \mid 3\}^{\boldsymbol{\zeta}} &= \{\tfrac{4}{1,2}, \tfrac{4}{1,2} : \tfrac{6}{1,3}\} \cong \{4, 4 : 6\}; \end{aligned}$$

note that $D_2 = \pm D_3$. By Theorem 5D19, $\{\{\tfrac{6}{2,3}, 4 : \tfrac{4}{1,2}\}, \{4, \tfrac{4}{1,2} \mid 6\}\}^{\boldsymbol{\kappa}}$ has facets of type $\{\tfrac{6}{0,1}, \tfrac{4}{1,2} : \tfrac{4}{0,1}\}$, while $\{\{\tfrac{6}{1,3}, 4 : \tfrac{4}{1,2}\}, \{4, \tfrac{4}{1,2} \mid 3\}\}^{\boldsymbol{\kappa}}$ has facets of type $\{\tfrac{3}{0,1}, \tfrac{4}{1,2} : \tfrac{4}{0,1}\}$.

Next, we see that $\boldsymbol{\kappa}$ has the same effect as $\boldsymbol{\pi}$ on $\{\tfrac{4}{1,2}, 6 : \tfrac{4}{1,2}\}$, which is self-Petrie; thus $\boldsymbol{\kappa}$ preserves $\{\{4, \tfrac{4}{1,2} \mid 3\}, \{\tfrac{4}{1,2}, 6 : \tfrac{4}{1,2}\}\}$. This does not contradict Theorem 5B10, since the hole $\{3\}$ of the facet is unaffected by Theorem 5D19.

Finally, under $\boldsymbol{\pi\kappa} = \boldsymbol{\kappa\pi}$, the only cases not yet effectively considered are

$$\begin{aligned}\{4, \tfrac{4}{1,2} \mid 6\}^{\pi\kappa} &= \{\tfrac{12}{1,5}, \tfrac{4}{1,2} : \tfrac{4}{1,2}, \tfrac{6}{1,3}\},\\ \{4, \tfrac{4}{1,2} \mid 3\}^{\pi\kappa} &= \{\tfrac{6}{2,3}, \tfrac{4}{1,2} : \tfrac{4}{1,2}, \tfrac{6}{2,3}\}.\end{aligned}$$

The faces of the resulting apeirotopes can be determined using Proposition 5A15 and Theorem 5D19; we leave this to the interested reader.

Under $\boldsymbol{\sigma}$, we have

$$\begin{aligned}\{4, \tfrac{4}{1,2} \mid 3\} &\mapsto \{\tfrac{6}{1,2}, 3 : \tfrac{6}{1,2}\},\\ \{4, \tfrac{4}{1,2} \mid 4\} &\mapsto \{\tfrac{8}{1,3}, 4 : \tfrac{8}{1,3}\},\\ \{4, \tfrac{4}{1,2} \mid 6\} &\mapsto \{\tfrac{12}{1,5}, 6 : \tfrac{12}{1,5}\}.\end{aligned}$$

These polyhedra were described in Section 11E; as pointed out in Section 11F, $\boldsymbol{\sigma}$ degenerates on tori of type $\{\tfrac{4}{1,2}, 4\}$. We have included the additional case here, because they all exhibit the same behaviour. (Note that $\{\tfrac{6}{1,2}, 3 : \tfrac{6}{1,2}\} \cong \{6,3\}/\langle\langle (01012)^2 \rangle\rangle$ is also a torus; see Figure 5A18.)

We can take the new generators to be

$$\begin{aligned}R_1 &\colon z \mapsto (x, y),\\ R_2 &\colon z \mapsto (\overline{x}, \varphi\overline{y}),\\ R_3 &\colon z \mapsto (\overline{x}, \overline{y}),\end{aligned}$$

where $\varphi = \omega$, i or $-\overline{\omega} = -\omega^2$ according as $r = 3$, 4 or 6, indicating the holes of the original tori or the vertex-figures of the new polyhedra. The rth polyhedron has vertex-set $\{(\varphi^n, 0), (0, \varphi^n) \mid 0 \leqslant n < r\}$; the initial vertex is $(1, 0)$, which is joined to $(0, 1)$ by the initial edge. Further, define the reflexion T by

$$T \colon z \mapsto (x, -y).$$

For $r = 4$ or 6, we see that $T = (R_2R_3)^{r/2}$ is as we have generally defined it; it is convenient to extend the notation to the case $r = 3$ as well, since $T \rightleftarrows R_2, R_3$ there also.

For each of these r, let s be given by

$$\frac{1}{s} + \frac{1}{r} = \frac{1}{2},$$

so that $\{s, r\}$ is the tiling of the plane $\mathbb{E}^2$ by regular s-gons. It turns out that the vertex- and edge-set of each apeirotope for r must be those of the product $\{s, r\} \times \{s, r\}$. Indeed, for each r we have

$$z \mapsto \begin{cases}(1 - x, -y),\\ (1 - \overline{x}, -y),\\ (1 - x, y),\\ (1 - \overline{x}, y),\end{cases}$$

according as $\dim R_0 = 0,\ 1,\ 2$ or 3. Clearly, the first and third, and second and fourth, are related by $\boldsymbol{\tau}\colon R_0 \mapsto R_0 T$. Now it is obvious that each of these reflexions is compatible with the decomposition $\mathbb{E}^4 = \mathbb{E}^2 \oplus \mathbb{E}^2$, and only a little thought is needed to show that the resulting vertex- and edge-sets are as claimed. For the first and third, the face is of type $\{\frac{2}{0,1}\}$, while for the second and fourth it is $\{4\}$ (such a square is the product of an edge of each of the components M_r, and all such products are faces). The vertex-figures are self-Petrie; the Petrie operation $\boldsymbol{\pi}$ in any case does not change the type of the face. Finally, $\boldsymbol{\kappa}$ preserves $\{\frac{12}{1,5}, 6 : \frac{12}{1,5}\}$ and $\{\frac{6}{1,2}, 3 : \frac{6}{1,2}\}$ (actually, it sends the latter into a congruent copy).

However, two of the resulting apeirotopes degenerate, namely, those with $r = 6$ and $\dim R_0$ odd. To see this, note that R_0 will take the initial vertex-figure into an enantiomorphic copy, reversing the twist of its helical faces. Hence, going along an odd edge-circuit $\{3\}$ (a face of the section $\{3,6\}$) will take a vertex-figure into its enantiomorph, which is not allowed.

Notes to Section 12A

1. We say little about the apeirotopes derived from $\mathsf{P} = \{\{\frac{6}{2,3}, 4 : \frac{4}{1,2}, 3\}, \{4, \frac{4}{1,2} \mid 6\}\}$, because we know even less about their combinatorics than we do about those of P.
2. It is interesting that the fine Schläfli symbol $\{\{\frac{6}{1,3}, 4 : \frac{4}{1,2}\}, \{4, \frac{4}{1,2} \mid 3\}\}$ is universal but not rigid. If we replace $\boldsymbol{r}_0$ in the generatrix $(\boldsymbol{r}_0, \ldots, \boldsymbol{r}_3)$ of $\boldsymbol{\{\{}6, 4 : 4\boldsymbol{\}}, \boldsymbol{\{}4, 4 \mid 3\boldsymbol{\}\}}$ by $\boldsymbol{r}_0\boldsymbol{t}$, with $\boldsymbol{t}$ an involution commuting with all $\boldsymbol{r}_j$, then $\boldsymbol{t}$ disappears in the mixing operation of (12A7).

12B Group $\mathbf{U}_5$ and Relatives

As we saw in Section 1F, the reflexion groups $\mathbf{R}_5 = [4,3,3,4]$ and $\mathbf{U}_5 = [3,3,4,3]$ are closely related. It should thus not be surprising that the derived families of 4-apeirotopes that fall into the patterns of Proposition 12B1 should also be somewhat intertwined. Indeed, this must be the case: there is one regular 5-honeycomb with group $\mathbf{R}_5$ and two with group $\mathbf{U}_5$, whereas (corresponding to the point groups) there are two regular 4-polytopes with group $\mathbf{C}_4$ and one with group $\mathbf{F}_4$.

We treat these three cases in order; it turns out that all the other apeirotopes whose groups are subgroups of $[3,3,4,3]$ fall into families related to one of the three, including some mentioned in Section 12A.

As we did in Section 9A, it may help keep track of the various operations if we bear in mind the corresponding mirror vectors; the picture here joins together the diagrams of Propositions 9A14 and 9A15.

12B1 Proposition *Under applications of $\boldsymbol{\varpi}$ and $\boldsymbol{\kappa}$ in $\mathbb{E}^4$, mirror vectors* *behave as follows:*

$$\begin{array}{ccccccc}
(3,3,3,3,3) & \xrightarrow{\varpi} & (3,2,3,3) & \xleftrightarrow{\pi} & (3,1,3,3) & \xleftrightarrow{\kappa} & (3,3,3,3) \\
\kappa \big\updownarrow & & & & & & \zeta \big\updownarrow \\
(3,1,3,3,3) & \xrightarrow{\varpi} & (1,2,3,3) & \xleftrightarrow{\pi} & (1,1,3,3) & \xleftrightarrow{\kappa} & (1,3,3,3)
\end{array}$$

The starting point at top left is a classical regular honeycomb in $\mathbb{E}^4$, *while that at top right is a crystallographic classical regular* 4*-polytope.*

12B2 Remark The way that the two diagrams join depends on the fact that π is actually applicable. Observe that the condition of Theorem 5B10 cannot occur since, if P or P^{κ} is a classical honeycomb in $\mathbb{E}^4$, then the 2-faces of P^{ϖ} are even. However, the example pictured in Figure 5A16 shows that this alone may not be enough to ensure polytopality, and so we must check each individual case.

12B3 Remark Since $\boldsymbol{\kappa} = \boldsymbol{\alpha\varpi}$, we can rewrite the diagram of Proposition 12B1 as follows: change $\boldsymbol{\kappa}$ to $\boldsymbol{\varpi}$, add 0 to the beginning of each of the right-most mirror vectors, and change $\boldsymbol{\zeta}$ to $\boldsymbol{\kappa}$.

We begin by applying Proposition 12B1 to the three individual cases.

12B4

$$\begin{array}{ccc}
\{4,3,3,4\} & \xleftrightarrow{\kappa} & \{4,\frac{6}{1,3},3,4\} \\
\varpi \big\downarrow & & \varpi \big\downarrow \\
\{\{6,\frac{6}{2,3} \mid 3\},\{\frac{6}{2,3},4:\frac{6}{1,3}\}\} & & \{\{\frac{3}{0,1},\frac{6}{2,3} \mid \frac{6}{2,3}\},\{\frac{6}{2,3},4:\frac{6}{1,3}\}\} \\
\pi \big\updownarrow & & \pi \big\updownarrow \\
\{\{6,\frac{6}{1,3} \mid 3\},\{\frac{6}{1,3},4:\frac{6}{2,3}\}\} & & \{\{\frac{3}{0,1},\frac{6}{1,3} \mid \frac{6}{2,3}\},\{\frac{6}{1,3},4:\frac{6}{2,3}\}\} \\
\kappa \big\updownarrow & & \kappa \big\updownarrow \\
\{3,3,4\} & \xleftrightarrow{\zeta} & \{\frac{6}{1,3},3,4\}
\end{array}$$

The facet of the apeirotope $\{\{6,\frac{6}{1,3} \mid 3\},\{\frac{6}{1,3},4:\frac{6}{2,3}\}\}$ at the left of the third row is isomorphic to the universal $\{6,6 \mid 3\}$. The vertex-figures in the second and third rows are $\{3,4\} \# \{2\}$ and $\{3,4\} \diamond \{2\}$, respectively.

12B5

$$\begin{array}{ccc}
\{3,4,3,3\} & \overset{\kappa}{\longleftrightarrow} & \{6,\frac{4}{1,2},3,3\} \\
\varpi\big\downarrow & & \varpi\big\downarrow \\
\{\{6,\frac{6}{2,3}\mid 4\},\{\frac{6}{2,3},3:\frac{4}{1,2}\}\} & & \{\{\frac{3}{0,1},\frac{6}{2,3}\mid\frac{4}{1,2}\},\{\frac{6}{2,3},3:\frac{4}{1,2}\}\} \\
\pi\big\updownarrow & & \pi\big\updownarrow \\
\{\{6,\frac{4}{1,2}\mid 4\},\{\frac{4}{1,2},3:\frac{6}{2,3}\}\} & & \{\{\frac{3}{0,1},\frac{4}{1,2}\mid\frac{4}{1,2}\},\{\frac{4}{1,2},3:\frac{6}{2,3}\}\} \\
\kappa\big\updownarrow & & \kappa\big\updownarrow \\
\{3,4,3\} & \overset{\zeta}{\longleftrightarrow} & \{\frac{6}{1,3},4,3\}
\end{array}$$

The facet of the apeirotope $\{\{6,\frac{4}{1,2}\mid 4\},\{\frac{4}{1,2},3\}\}$ at the left of the third row is isomorphic to the universal $\{6,4\mid 4\}$. The vertex-figures in the second and third rows are $\{3,3\}\#\{2\}$ and $\{4,3\}\diamond\{2\}$, respectively.

12B6

$$\begin{array}{ccc}
\{3,3,4,3\} & \overset{\kappa}{\longleftrightarrow} & \{6,\frac{6}{1,3},4,3\} \\
\varpi\big\downarrow & & \varpi\big\downarrow \\
\{\{4,\frac{4}{1,2}\mid 4\},\{\frac{4}{1,2},3:\frac{6}{1,3}\}\} & & \{\{\frac{4}{0,1},\frac{4}{1,2}:\frac{8}{1,3}\},\{\frac{4}{1,2},3:\frac{6}{1,3}\}\} \\
\pi\big\updownarrow & & \pi\big\updownarrow \\
\{\{4,\frac{6}{1,3}\mid 3\},\{\frac{6}{1,3},3:\frac{4}{1,2}\}\} & & \{\{\frac{4}{0,1},\frac{6}{1,3}\},\{\frac{6}{1,3},3:\frac{4}{1,2}\}\} \\
\kappa\big\updownarrow & & \kappa\big\updownarrow \\
\{4,3,3\} & \overset{\zeta}{\longleftrightarrow} & \{\frac{4}{1,2},3,3\}
\end{array}$$

The facets at the left of the second and third rows are isomorphic to the universal $\{4,4 \mid 4\}$ and $\{4,6 \mid 3\}$, respectively. The vertex-figures in the second and third rows are $\{4,3\} \# \{2\}$ and $\{3,3\} \diamond \{2\}$, respectively.

From the last Table 12B6, we obtain a dual:

$$\{\{3,4\},\{4,\tfrac{4}{1,2} \mid 4\}\} = \{\{4,\tfrac{4}{1,2} \mid 4\},\{\tfrac{4}{1,2},3 : \tfrac{6}{1,3}\}\}^{\delta}.$$

This is the starting point for a rich family of apeirotopes. A partial picture of the relationships among them is illustrated in Table 12B7. What is particularly notable is that each of the two tori $\{4,\frac{4}{1,2} \mid 4\}$ and $\{\frac{4}{1,2},4 : \frac{4}{1,2}\}$ occurs as vertex-figure with $\dim R_0$ taking each possible value 0, 1, 2 or 3 (we have not listed the two free abelian apeirotopes).

12B7

$$\begin{array}{ccc}
\{\{\tfrac{6}{1,3},4 : \tfrac{6}{1,3}\},\{4,\tfrac{4}{1,2} \mid 4\}\} & \stackrel{\eta}{\longrightarrow} & \{\{\tfrac{2}{0,1},\tfrac{4}{1,2} : \tfrac{4}{1,2}\},\{\tfrac{4}{1,2},4 : \tfrac{4}{1,2}\}\} \\
\tau \updownarrow & & \tau \updownarrow \\
\{\{3,4\},\{4,\tfrac{4}{1,2} \mid 4\}\} & \stackrel{\eta}{\longrightarrow} & \{\{4,\tfrac{4}{1,2} \mid 4\},\{\tfrac{4}{1,2},4 : \tfrac{4}{1,2}\}\} \\
 & & \delta \updownarrow \\
\{\{4,\tfrac{4}{1,2}\},\{\tfrac{4}{1,2},4 : \tfrac{4}{1,2}\}\} & \stackrel{\eta}{\longleftarrow} & \{\{\tfrac{4}{1,2},4 : \tfrac{4}{1,2}\},\{4,\tfrac{4}{1,2} \mid 4\}\}
\end{array}$$

Recall that $\{\frac{2}{0,1}\}$ is the zigzag apeirogon.

To keep track of the changes of generators implied by the diagram, it is helpful to choose the initial ones as simply as possible (and so we make a change from the previous ones implied by the duality). A particularly suitable choice, which respects the decomposition $\mathbb{E}^4 = \mathbb{E}^2 \oplus \mathbb{E}^2$ implicit in the tori, is those of the generatrix $(R_0,\ldots,R_3)$ of $\{\{3,4\},\{4,\frac{4}{1,2} \mid 4\}\}$:

12B8

$$\begin{aligned}
R_0 &\colon x \mapsto (1-\xi_3,\xi_2,1-\xi_1,\xi_4),\\
R_1 &\colon x \mapsto (\xi_2,\xi_1,\xi_3,\xi_4),\\
R_2 &\colon x \mapsto (\xi_3,\xi_4,\xi_1,\xi_2),\\
R_3 &\colon x \mapsto (\xi_1,\xi_2,\xi_3,-\xi_4).
\end{aligned}$$

It is easily checked that these do indeed correspond to those of $\mathcal{A}_1(2,4,2,3)$ in the reverse order.

Before we move on, we identify some of the facets that occur in Table 12B7, and show that some at least of the apeirotopes are universal. Beginning at the top, $\{\frac{6}{1,3},4 : \frac{6}{1,3}\} \cong \{6,4 : 6\}$, which is self-Petrie; see Chapter 10 for more details. In fact, it appears in the form $\{\frac{6}{1,3},4 : \frac{6}{1,3}\} \# \{2\}$; compare Theorem 10C14. The skew faces and Petrie polygons of the original are Petrie polygons of 3-cubes (recall that $\{\frac{6}{1,3},4 : \frac{6}{1,3}\}$ in $\mathbb{E}^3$ has the same edge-graph as $\{\frac{3}{0,1}, 4 : \frac{3}{0,1}\}$ – the picture is thus the same as Figure 10C8), and are lifted in $\mathbb{E}^4$ to Petrie polygons of octahedra. Next, still at the top, $\{\frac{2}{0,1}, \frac{4}{1,2} : \frac{4}{1,2}\}$ stands for the Petrial of the square tessellation $\{4, 4\}$, which appears in its own right as $\{4, \frac{4}{1,2}\}$ near the bottom. The apeirotope $\{\{3, 4\}, \{4, \frac{4}{1,2} \mid 4\}\}$ (which we prove in Theorem 12B9 to be universal) is the facet of the 5-apeirotope $\{3, 4, 3, 3\}^{\pi}$; see Section 8D. Finally, at bottom left, $\{4, \frac{6}{1,3} \mid 4\}$ is the universal Petrie–Coxeter apeirohedron $\{4, 6 \mid 4\}$, and the apeirotope itself is the facet of the regular 5-apeirotope $\{4, 3, 3, 4\}^{\kappa}$; see Section 8A.

Moreover, since they are derived from diagrams $\mathcal{A}_1(2, 4, 2, 3)$ or $\mathcal{A}_2(4, 4)$, and these can be recovered from the groups of the apeirotopes, we have

12B9 Theorem *The following apeirotopes are universal:*

$$\{\{\tfrac{4}{1,2}, 4 : \tfrac{4}{1,2}\}, \{4, \tfrac{4}{1,2} \mid 4\}\} \cong \{\{4, 4 : 4\}, \{4, 4 \mid 4\}\},$$

$$\{\{4, \tfrac{4}{1,2} \mid 4\}, \{\tfrac{4}{1,2}, 4 : \tfrac{4}{1,2}\}\} \cong \{\{4, 4 \mid 4\}, \{4, 4 : 4\}\},$$

$$\{\{3, 4\}, \{4, \tfrac{4}{1,2} \mid 4\}\} \cong \{\{3, 4\}, \{4, 4 \mid 4\}\},$$

$$\{\{4, \tfrac{4}{1,2} \mid 4\}, \{\tfrac{4}{1,2}, 3\}\} \cong \{\{4, 4 \mid 4\}, \{4, 3\}\}.$$

Of course, we must also consider the effects of $\boldsymbol{\kappa}$ and $\boldsymbol{\pi}$. We note first that the vertex-figure $\{\frac{4}{1,2}, 4 : \frac{4}{1,2}\}$ is preserved by $\boldsymbol{\kappa}$, with a minor change of generators; as the notation indicates, the torus is also self-Petrie, but the isometry that takes the original into its Petrial does not respect the decomposition $\mathbb{E}^4 = \mathbb{E}^2 \oplus \mathbb{E}^2$. Thus applying $\boldsymbol{\kappa}$ or $\boldsymbol{\pi}$ in these cases cannot change the type of the apeirotope, although it will produce a different congruent copy.

However, under $\boldsymbol{\kappa}$ the vertex-figure $\{4, \frac{4}{1,2} \mid 4\}$ becomes $\{\frac{4}{1,2}, \frac{4}{1,2} \mid \frac{4}{1,2}\}$, an isomorphic but not congruent copy, while $\{4, \frac{4}{1,2} \mid 4\}^{\pi} = \{\frac{8}{1,3}, \frac{4}{1,2} : 4, 4\}$ (these polygons $\{\frac{8}{1,3}\}$ are Petrie polygons of 4-cubes), and further applying $\boldsymbol{\kappa}$ changes $\{\frac{8}{1,3}\}$ into a non-congruent polygon of the same kind (and each zigzag $\{4\}$ into $\{\frac{4}{1,2}\}$). Theorem 5D19 tells us how the various data of the corresponding apeirotopes change under $\boldsymbol{\kappa}$. Ignoring the free abelian apeirotope, for the initial effects of $\boldsymbol{\kappa}$, we have

$$\{\{3, 4\}, \{4, \tfrac{4}{1,2} \mid 4\}\} \stackrel{\kappa}{\longleftrightarrow} \{\{6, \tfrac{4}{1,2} \mid 4\}, \{\tfrac{4}{1,2}, \tfrac{4}{1,2} \mid \tfrac{4}{1,2}\}\},$$

12B10 $$\{\{\tfrac{6}{1,3}, 4 : \tfrac{6}{1,3}\}, \{4, \tfrac{4}{1,2} \mid 4\}\} \stackrel{\kappa}{\longleftrightarrow} \{\{\tfrac{3}{0,1}, \tfrac{4}{1,2} : \tfrac{3}{0,1}\}, \{\tfrac{4}{1,2}, \tfrac{4}{1,2} \mid \tfrac{4}{1,2}\}\},$$

$$\{\{\tfrac{4}{1,2}, 4 : \tfrac{4}{1,2}\}, \{4, \tfrac{4}{1,2} \mid 4\}\} \stackrel{\kappa}{\longleftrightarrow} \{\{\tfrac{4}{0,1}, \tfrac{4}{1,2} : \tfrac{4}{0,1} \mid 4\}, \{\tfrac{4}{1,2}, \tfrac{4}{1,2} \mid \tfrac{4}{1,2}\}\}.$$

The first facet on the right in (12B10) is the Petrie–Coxeter apeirohedron; the apeirotope itself is the facet of the 5-apeirotope $\{3, 4, 3, 3\}^{\pi\kappa}$ (in [82, p. 33] it was denoted $\{6, 4, 4, 3\}^{(s)}$). That of the second is $\{\frac{3}{0,1}, 4 : \frac{3}{0,1}\} \# \{\infty\}$. We have additionally noted the hole in the third; other than that, we know little about this apeirohedron. We shall not list the effects of $\boldsymbol{\pi}$ (as usual, the data can be found using Theorem 5B14), because we have not identified any facets of interest among the resulting apeirotopes; we merely recall that π is not valid for $\{\{3,4\},\{4,\frac{4}{1,2} \mid 4\}\}$, but will work on the other five apeirotopes in the diagram.

We conclude the section by seeing how the apeirotopes in case $q = 4$ discussed at the end of Section 12A fit into the present context. Translated into coordinate terms, the generating reflexions of the vertex-figure $\{\frac{8}{1,3}, 4 : \frac{8}{1,3}\}$ are

$$\begin{aligned} R_1 &: x \mapsto (\xi_3, \xi_4, \xi_1, \xi_2), \\ R_2 &: x \mapsto (\xi_1, -\xi_2, \xi_4, \xi_3), \\ R_3 &: x \mapsto (\xi_1, -\xi_2, \xi_3, -\xi_4). \end{aligned}$$

The four reflexions R_0 take the common form

$$x \mapsto (1 - \xi_1, \pm\xi_2, \pm(\xi_3, \xi_4)),$$

with an obvious meaning. The plus sign in the second coordinate gives $\dim R_0 = 1$ or 3, while that in the third and fourth coordinates (together) gives $\dim R_0 = 2$ or 3 (and, in case there is any doubt, all minuses gives the case $\dim R_0 = 0$ of the free abelian apeirotope). In view of what we said before, the vertices and edges are those of the cubic tiling $\{4,3,3,4\}$, which also contains the faces $\{4\}$ when $\dim R_0 = 2$ or 3 (the faces in the other two cases are $\{\frac{2}{0,1}\} = \{\infty\} \# \{2\}$); it is abundantly clear that each group is a subgroup of $[4,3,3,4]$.

12C Twisting $\mathbf{P}_5$

We finally look at apeirotopes whose symmetry groups are subgroups of $\mathbf{P}_5 \ltimes \mathbf{D}_2^5$; we recall that $\mathbf{P}_5$ has Coxeter diagram

12C1

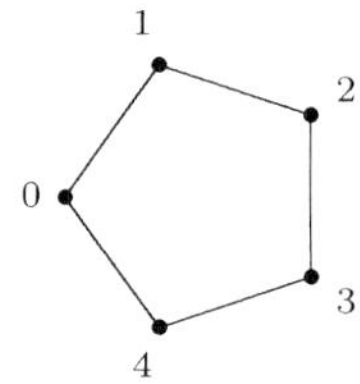

Remember as well that we find it convenient here to work in the hyperplane

$$\mathbb{L}^4 := \{x = (\xi_0, \ldots, \xi_4) \in \mathbb{E}^5 \mid \xi_0 + \cdots + \xi_4 = 0\}.$$

In particular, all our polytopes will have vertices in the lattice $\Lambda := \mathbb{L}^4 \cap \mathbb{Z}^5$, whose generators are the $e_0 - e_j$ for $j = 1, \ldots, 4$, where $\{e_0, \ldots, e_4\}$ is the standard basis of $\mathbb{E}^5$. The generators $(R_0, \ldots, R_4)$ of $\mathbf{P}_5$ can be taken to be

12C2 $$R_0 \colon x \mapsto (\xi_1 + 1, \xi_0 - 1, \xi_2, \xi_3, \xi_4),$$

with $R_1, \ldots, R_4$ obtained by cyclic permutation of the coordinates modulo 5. Further, we can identify the symmetry group $[3,3,3]$ of the 4-simplex $\{3,3,3\}$ with the permutation group $\boldsymbol{S}_5$ acting on the standard basis; it is useful to note that the (linear) reflexions in $\boldsymbol{S}_5 \ltimes \boldsymbol{C}_2 = \boldsymbol{S}_5 \times \boldsymbol{C}_2$ are composed of one or two (disjoint) transpositions, together with $\pm I$.

12C3 Remark As an alternative choice of generators of $\mathbf{P}_5$, we keep R_0 as in (12C2), but take $R_j := (j\ j{+}1)$ for $j = 1, \ldots, 4$ (with indices modulo 5); in effect, this corresponds to the description in Section 1E. The outer automorphisms are now not so nice, but one advantage is that the images of o under the group comprise the whole of the lattice Λ (which is actually the underlying translation subgroup).

Blended Vertex-Figures

As in Section 12B, we enter the family through applications of $\boldsymbol{\kappa}$ – here to $\{3,3,3\}$ and $\{3,3,3\}^{\varsigma}$. Since there is no classical regular honeycomb whose symmetry group is related to $\mathbf{P}_5$, it is clear that the pattern of Proposition 12B1 cannot go through in its entirety. What survives is

$$\begin{array}{lccc}
 & \{3,3,3\} & \stackrel{\varsigma}{\longleftrightarrow} & \{\frac{6}{1,3},3,3\} \\
 & \boldsymbol{\kappa}\big\downarrow & & \boldsymbol{\kappa}\big\updownarrow \\
\textbf{12C4} & \{\{6,\frac{6}{1,3} \mid 3\},\{\frac{6}{1,3},3:\frac{4}{1,2}\}\} & & \{\{\frac{3}{0,1},\frac{6}{1,3}:\frac{4}{0,1}\},\{\frac{6}{1,3},3:\frac{4}{1,2}\}\} \\
 & \pi\big\updownarrow & & \pi\big\updownarrow \\
 & \{\{6,\frac{4}{1,2} \mid 3\},\{\frac{4}{1,2},3:\frac{6}{1,3}\}\} & & \{\{\frac{3}{0,1},\frac{4}{1,2}:\frac{5}{1,2}\},\{\frac{4}{1,2},3:\frac{6}{1,3}\}\}
\end{array}$$

We next observe

12C5 Theorem *The following isomorphisms hold:*

$$\{\{6,\tfrac{4}{1,2} \mid 3\},\{\tfrac{4}{1,2},3\}\} \cong \boldsymbol{\{\{6,4 \mid 3\},\{4,3\}\}},$$

$$\{\{6,\tfrac{6}{1,3} \mid 3\},\{\tfrac{6}{1,3},3:\tfrac{4}{1,2}\}\} \cong \boldsymbol{\{\{6,6 \mid 3\},\{6,3:4\}\}}.$$

Both fine Schläfli symbols are rigid.

Proof. The universality of the first is guaranteed by the fact that it is the dual of the universal apeirotope $\{\{3,4\},\{4,\frac{6}{1,3} \mid 3\}\} \cong \boldsymbol{\{\{3,4\},\{4,6 \mid 3\}\}}$, for which we make a forward reference to Theorem 12C13. For the second, we merely observe that its faces and holes of facets are the same as those of the first (these hexagons and triangles form quasi-regular planar tessellations). Any contraction of an edge-circuit in the first therefore gives a corresponding contraction in the second, which is all that we need.

The rigidity is a consequence of standard arguments. The facets are rigid, and the vertex-figures are blends of rigid polyhedra with segments. □

Vertex-Sets

It is helpful to list from the outset certain point-sets that play a rôle in our analysis. We define them by means of a typical point of each, bearing in mind that they are acted upon by the symmetric group of permutations on the coordinates $\xi_0, \ldots, \xi_4$.

12C6
$$\begin{aligned} V_1 &: \quad (1, 0, 0, 0, -1), \\ V_2 &: \quad (1, 1, 0, -1, -1), \\ V_3 &: \quad (3, -2, -2, -2, 3), \\ V_4 &: \quad (4, -1, -1, -1, -1, -1). \end{aligned}$$

In fact, we need $\pm V_3$ as well as V_3; V_4 and $\pm V_4$ are of less importance, but are useful as references. We begin by observing

12C7 Lemma *The point-sets V_2 and $\pm V_3$ violate the nearest vertices criterion of Proposition 5F1.*

Of course, this does not imply that they cannot occur as vertex-sets of vertex-figures – indeed, each of them does, as we shall soon see.

12C8 Remark Note the close connexion between V_1 and V_2, in that $V_2 \subset V_1 + V_1$ and $V_1 \subset V_2 + V_2$. We shall exploit this connexion in the following.

The V_1 Family

The apeirotope $\{\{6, \frac{4}{1,2} \mid 3\}, \{\frac{4}{1,2}, 3 : \frac{6}{1,3}\}\}$ that we met in Table 12C4 has a geometric dual. However, while that is a new starting point, we approach it instead through a natural twist on the group $\mathbf{P}_5$. In terms of the generators of Remark 12C3, we define

12C9
$$\begin{aligned} R_0 &:= (0\ 4)^\bullet, \\ R_1 &:= (0\ 1), \\ R_2 &:= -(0\ 4)(1\ 3), \\ R_3 &:= (2\ 3); \end{aligned}$$

the vector term in R_0 is $e_0 - e_4$. This is represented by the diagram

12C10

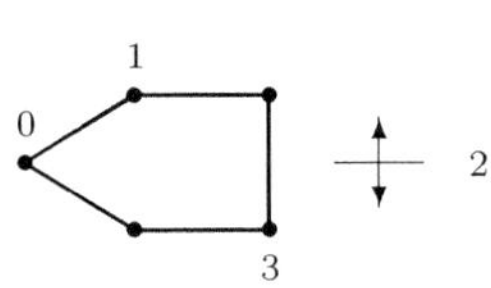

As usual, the labels correspond to the generators; observe that we just have the case $d = 4$ of Proposition 9D10. Referring to Section 11B for the vertex-figure, we see that the resulting apeirotope is $\{\{3, 4\}, \{4, \frac{6}{1,3} \mid 3\}\}$. The initial vertex is o, and the vertex-set of the vertex-figure is V_1. Indeed, very little work is required to show

12C11 Proposition *The vertex-set of* $\{\{3,4\},\{4,\frac{6}{1,3} \mid 3\}\}$ *is the lattice* $\Lambda = \mathbb{L}^4 \cap \mathbb{Z}^5$.

From this, we obtain by the standard operations $\boldsymbol{\kappa}$, $\boldsymbol{\pi}$ and $\boldsymbol{\eta}$ various other apeirotopes whose vertex-figures have vertex-set V_1; we postpone for the moment a discussion of skewing $\boldsymbol{\sigma}$. The effects of these operations have already been described in general terms in earlier chapters; however, we shall make some additional comments below.

$$\begin{array}{lccc}
 & \{\{3,4\},\{4,\frac{6}{1,3} \mid 3\}\} & \overset{\kappa}{\longleftrightarrow} & \{\{6,\frac{4}{1,2} \mid 4\},\{\frac{4}{1,2},\frac{6}{1,3} : \frac{5}{1,2}\}\} \\
 & \eta \big\downarrow & & \pi \big\updownarrow \\
\textbf{12C12} & \{\{4,\frac{6}{1,3} \mid 3\},\{\frac{6}{1,3},6 : \frac{6}{2,3}\}\} & & \{\{6,\frac{5}{1,2} \mid 4\},\{\frac{5}{1,2},\frac{6}{1,3} : \frac{4}{1,2}\}\} \\
 & \kappa \big\updownarrow \pi & & \\
 & \{\{4,\frac{6}{2,3} \mid 3\},\{\frac{6}{2,3},6 : \frac{6}{1,3}\}\} & &
\end{array}$$

Referring to (12C9), we see that, as a reflexion,

$$K_2 = -(0\ 4) = (R_2R_3)^3 \in \mathbf{G}^v,$$

and so the operation $\boldsymbol{\tau}$ coincides with ς, yielding the generatrix of $\{4,\frac{6}{1,3} \mid 3\}^{\boldsymbol{\alpha}}$. Indeed, for each apeirotope P of Table 12C12, we have $\mathsf{P}^{\boldsymbol{\tau}} = \mathsf{P}^{\boldsymbol{v\alpha}}$. This ties in free abelian apeirotopes to the family of Table 12C12.

At this stage, we pause to address universality and rigidity.

12C13 Theorem *The following isomorphisms hold:*

$$\begin{aligned}
\{\{3,4\},\{4,\tfrac{6}{1,3} \mid 3\}\} &\cong \boldsymbol{\{}\{3,4\},\{4,6 \mid 3\}\boldsymbol{\}}, \\
\{\{6,\tfrac{4}{1,2} \mid 3\},\{\tfrac{4}{1,2},3\}\} &\cong \boldsymbol{\{}\{6,4 \mid 3\},\{4,3\}\boldsymbol{\}}, \\
\{\{6,\tfrac{4}{1,2} \mid 4\},\{\tfrac{4}{1,2},\tfrac{6}{1,3} : \tfrac{5}{1,2}\}\} &\cong \boldsymbol{\{}\{6,4 \mid 4\},\{4,6 : 5\} \mid 3\boldsymbol{\}}, \\
\{\{6,\tfrac{5}{1,2} \mid 4\},\{\tfrac{5}{1,2},\tfrac{6}{1,3} : \tfrac{4}{1,2}\}\} &\cong \boldsymbol{\{}\{6,5 \mid 4\},\{5,6 : 4\} \mid 3\boldsymbol{\}}.
\end{aligned}$$

Moreover, each of the four fine Schläfli symbols is rigid.

Proof. If we write $(\boldsymbol{r}_0,\ldots,\boldsymbol{r}_3)$ for the generatrix of $\boldsymbol{\{}\{3,4\},\{4,6 \mid 3\}\boldsymbol{\}}$, then we see that the mixing operation

$$(\boldsymbol{r}_0,\ldots,\boldsymbol{r}_3) \mapsto (\boldsymbol{r}_0,\boldsymbol{r}_1,\boldsymbol{r}_2\boldsymbol{r}_3\boldsymbol{r}_2,\boldsymbol{r}_3,\boldsymbol{r}_2\boldsymbol{r}_1\boldsymbol{r}_2) =: (\boldsymbol{s}_0,\ldots,\boldsymbol{s}_4)$$

yields the generators of a quotient of the abstract group $\mathbf{P}_5$; the isomorphism of the theorem is a direct consequence. The second case follows from the first by duality.

For the third and fourth, we first note that

$$\{\{6,4 \mid 4\},\{4,6:5\} \mid 3\} \stackrel{\pi}{\longleftrightarrow} \{\{6,5 \mid 4\},\{5,6:4\} \mid 3\};$$

more exactly, π applied to one gives a quotient of the other, from which the claimed relationship follows. It thus suffices to prove our claim for the third. Now we certainly need some extra relation on $\{\{6,4 \mid 4\},\{4,6:5\}\}$, since its edge-graph is bipartite, while $\{\{6,\frac{4}{1,2} \mid 4\},\{\frac{4}{1,2},\frac{6}{1,3}:\frac{5}{1,2}\}\}$ has the same edges as $\{\{3,4\},\{4,\frac{6}{1,3} \mid 3\}\}$, and so has trigonal edge-circuits (see the notes at the end of the section). Taking the generatrix of $\{\{6,4 \mid 4\},\{4,6:5\}\}$ to be $(\boldsymbol{t}_0,\ldots,\boldsymbol{t}_3)$ and comparing with Remark 11C2 shows that the mixing operation

$$(\boldsymbol{t}_0,\ldots,\boldsymbol{t}_3) \mapsto (\boldsymbol{t}_0,\boldsymbol{t}_1(\boldsymbol{t}_1\boldsymbol{t}_2\boldsymbol{t}_3\boldsymbol{t}_2)^3,\boldsymbol{t}_2,\boldsymbol{t}_3) =: (\boldsymbol{r}_0,\ldots,\boldsymbol{r}_3)$$

must capture the generatrix of $\{\{3,4\},\{4,6 \mid 3\}\}$. This then requires that

$$\boldsymbol{t}_0\boldsymbol{t}_1(\boldsymbol{t}_1\boldsymbol{t}_2\boldsymbol{t}_3\boldsymbol{t}_2)^3 \sim \boldsymbol{t}_0\boldsymbol{t}_1\boldsymbol{t}_2\boldsymbol{t}_3\boldsymbol{t}_2\boldsymbol{t}_1$$

have period 3, which is just the deep hole relation of the theorem.

To show rigidity, for all except the second we can appeal to Theorem 6D2: the 2-faces are planar, and the vertex-figures are themselves rigid (for the first we need Theorem 6D5, and for the third and fourth we need Proposition 11C10). But observe also that Theorem 6D5 implies that the facets of the second, third and fourth are rigid, and it is then easy to proceed directly from there. □

12C14 Remark It is worth noting the universality of the facet $\{6,\frac{5}{1,2} \mid 4\} \cong \{6,5 \mid 4\}$ itself. Indeed, if its generatrix is $(\boldsymbol{r}_0,\boldsymbol{r}_1,\boldsymbol{r}_2)$, then $\boldsymbol{r}_0$ and its four conjugates under $\langle \boldsymbol{r}_1,\boldsymbol{r}_2\rangle$ generate a quotient of $\mathbf{P}_5$ (as an abstract group), from which the desired isomorphism follows. In addition, the fine Schläfli symbols $\{6,\frac{5}{1,2} \mid 4\}$ and $\{4,\frac{5}{1,2} \mid 6\}$ are rigid, since the faces and holes determine the shape of the vertex-figure, though we did not need to use that fact in proving Theorem 12C13.

The V_2 Family

Another geometrically dualizable apeirotope in the V_1 family leads us into the V_2 family:

$$\{\{4,\tfrac{6}{1,3} \mid 3\},\{\tfrac{6}{1,3},6:\tfrac{6}{2,3}\}\}^{\delta} = \{\{\tfrac{6}{1,3},6:\tfrac{6}{2,3}\},\{6,\tfrac{4}{1,2} \mid 3\}\}.$$

For this case (with the previous generators in reverse order), $T = (R_2R_3)^2$ is

$$T\colon x \mapsto (\xi_1,\xi_0,\xi_2,\xi_4,\xi_3).$$

Applying the corresponding operation $\boldsymbol{\tau}$ to the point-reflexion for $\{6,\frac{4}{1,2} \mid 3\}^{\boldsymbol{\alpha}}$, we arrive at the generators

$$\begin{aligned}
R_0 &\colon x \mapsto (1-\xi_1,1-\xi_0,-\xi_2,-1-\xi_4,-1-\xi_3),\\
R_1 &\colon x \mapsto (\xi_0,\xi_1,\xi_3,\xi_2,\xi_4),\\
R_2 &\colon x \mapsto -(\xi_4,\ldots,\xi_0),\\
R_3 &\colon x \mapsto (\xi_1,\xi_0,\xi_2,\xi_3,\xi_4).
\end{aligned}$$

Since $\dim T = 2$, we see that $\dim R_0 = 2$ also. Observe (see also [83, p. 272]) that the facet $\{\frac{6}{1,3}, 6 : \frac{6}{2,3}\}$ is self-dual.

This is now the jumping-off point for applying $\boldsymbol{\kappa}$ and $\boldsymbol{\pi}$. Without going into details (but see the notes at the end of the section), there results

12C15 Theorem *There is a family of* 4*-apeirotopes in* $\mathbb{E}^4$ *whose vertex-figures have vertex-set* V_2*, related as follows:*

$$\begin{array}{ccc}
\{\{\frac{6}{1,3}, 6 : \frac{6}{2,3}\}, \{6, \frac{4}{1,2} \mid 3\}\} & \xleftrightarrow{\ \kappa\ } & \{\{\frac{3}{0,1}, \frac{6}{2,3} : \frac{6}{0,1}\}, \{\frac{6}{2,3}, \frac{4}{1,2} : \frac{5}{1,2}\}\} \\
\pi \updownarrow & & \pi \updownarrow \\
\{\{\frac{6}{1,3}, \frac{10}{1,3} : \frac{5}{1,2}\}, \{\frac{10}{1,3}, \frac{4}{1,2} : 6, 3\}\} & \xleftrightarrow{\ \kappa\ } & \{\{\frac{3}{0,1}, \frac{5}{1,2} : \frac{10}{1,3}\}, \{\frac{5}{1,2}, \frac{4}{1,2} : \frac{6}{2,3}\}\}
\end{array}$$

Apart from apeirotopes Q^α*, this list is complete.*

The vertex-figures of the two apeirotopes on the right are of interest; they are (abstractly) the Petrie pair $\{6, 4 : 5\}$ and $\{5, 4 : 6\}$. As with the previous isomorphisms $\{\frac{4}{1,2}, \frac{6}{1,3} : \frac{5}{1,2}\} \cong \{4, 6 : 5\}$ and $\{\frac{5}{1,2}, \frac{6}{1,3} : \frac{4}{1,2}\} \cong \{5, 6 : 4\}$, these isomorphisms follow from the fact that the geometric polyhedra have the same group $\mathcal{S}_5 \times \mathcal{C}_2$ satisfying the same relations as the abstract ones.

The V_3 *Family*

The final basic vertex-figure in this family is $\{4, \frac{6}{2,3} \mid 3\} = \{3, 3\}^{\varpi}$, whose group is just $\mathbf{A}_4$. We therefore change the sign of R_2 in the generatrix of $\{4, \frac{6}{1,3} \mid 3\}$, to obtain the new generators

$$\begin{aligned}
R_0 &: x \mapsto (3 - \xi_4, -2 - \xi_1, -2 - \xi_2, -2 - \xi_3, 3 - \xi_0), \\
R_1 &: x \mapsto (\xi_0, \xi_1, \xi_3, \xi_2, \xi_4), \\
R_2 &: x \mapsto (\xi_4, \text{[illegible]}, \xi_0), \\
R_3 &: x \mapsto (\xi_1, \xi_0, \xi_2, \xi_3, \xi_4).
\end{aligned}$$

It is clear that the vertices of the vertex-figure consist of the set V_3 of the ten permutations of $(3, -2, -2, -2, 3)$.

In this case,

$$T = (R_2 R_3)^3 : x \mapsto (\xi_4, \xi_1, \xi_2, \xi_3, \xi_0);$$

applying the operation $\boldsymbol{\tau}$ to the point-reflexion for $\{4, \frac{6}{2,3} \mid 3\}^\alpha$ leads to the line-reflexion R_0. The corresponding apeirotope is $\{\{\frac{6}{1,3}, 4 : \frac{6}{1,3}\}, \{4, \frac{6}{2,3} : \frac{5}{1,2} \mid 3\}\}$, with facet isomorphic to the universal $\{6, 4 : 6\}$; it is, in fact, $\{\frac{6}{1,3}, 4 : \frac{6}{1,3}\} \# \{2\}$ (note that the hyperplane $\xi_2 = -1$ is fixed by each of R_0, R_1 and R_2, and that the vertices of the facet lie in $\xi_2 = 0$ or -2). Of course, this apeirotope is not dualizable – its facet is infinite, and its vertex-figure has no dual.

As we saw in Section 11F, we cannot apply $\boldsymbol{\eta}$ to $\{4, \frac{6}{2,3} : \frac{5}{1,2} \mid 3\}$, since the result degenerates. However, we can apply both $\boldsymbol{\pi}$ and $\boldsymbol{\kappa}$; note that $\boldsymbol{\kappa}$ doubles the number of vertices of the vertex-figures, because $\{4, \frac{6}{2,3} : \frac{5}{1,2} \mid 3\}$ is not centrally symmetric. What we obtain is then depicted in the following diagram; we have not given details of the facets of the last two, since we know little about them other than that they are infinite.

$$\begin{array}{ccc} \{\{\frac{6}{1,3}, 4 : \frac{6}{1,3}\}, \{4, \frac{6}{2,3} \mid 3\}\} & \overset{\boldsymbol{\kappa}}{\longleftrightarrow} & \{\{\frac{3}{0,1}, \frac{4}{1,2} : \frac{3}{0,1}\}, \{\frac{4}{1,2}, \frac{6}{2,3} : \frac{10}{1,3} \mid 3\}\} \\ \boldsymbol{\pi}\updownarrow & & \boldsymbol{\pi}\updownarrow \\ \{\{\frac{6}{1,3}, \frac{5}{1,2}\}, \{\frac{5}{1,2}, \frac{6}{2,3} : 4, 3\}\} & \overset{\boldsymbol{\kappa}}{\longleftrightarrow} & \{\{\frac{3}{0,1}, \frac{10}{1,3}\}, \{\frac{10}{1,3}, \frac{6}{2,3} : \frac{4}{1,2}, \frac{6}{1,3}\}\} \end{array}$$

Returning to the vertex-figures, since $\{4, \frac{6}{2,3} : \frac{5}{1,2} \mid 3\}$ and $\{\frac{5}{1,2}, 3 : \frac{5}{1,2}\}$ have the same vertex-set V_3, it follows that we can regard the latter as inscribed in the former. In fact, it is so inscribed in two ways, which are the Petrials of each other. Similarly, $\{4, \frac{6}{1,3} \mid 3\}$ and $\{\frac{5}{1,2}, 3 : \frac{10}{1,3}\}$ have the same vertex-set V_1, and so the latter is inscribed in the former. Again, this is in two ways; the other can be thought of as $\{\frac{5}{1,2}, 3 : \frac{10}{1,3}\}^{\boldsymbol{\pi\kappa}}$, since the double operation $\boldsymbol{\pi\kappa} = \boldsymbol{\kappa\pi}$ produces an enantiomorphic copy of the original with the same vertices. Naturally, we then have corresponding apeirotopes inscribed in one another, but only those resulting from the 'apeir' construction.

Applications of Sskewing

We finally come (in this section) to possible applications of $\boldsymbol{\sigma}$; since they yield the dodecahedron and some relatives, these are of potential interest. However, we shall not go into too many details, for reasons that will emerge.

The pattern of relationships of vertex-figures is depicted in (12C16); for convenience we have repeated what we said in Section 11F, to which we refer for more details.

12C16

$$\begin{array}{ccccc} \{4, \frac{6}{1,3} \mid 3\} & \overset{\boldsymbol{\sigma}}{\longrightarrow} & \{\frac{10}{1,3}, 3 : \frac{10}{1,3}\} & \overset{\boldsymbol{\kappa}_2\boldsymbol{\kappa}_3}{\longrightarrow} & \{\frac{5}{1,2}, 3 : \frac{10}{1,3}\} \\ \boldsymbol{\kappa}_2\updownarrow & & \boldsymbol{\kappa}\updownarrow & & \boldsymbol{\kappa}\updownarrow\boldsymbol{\pi} \\ \{4, \frac{6}{2,3} : \frac{5}{1,2} \mid 3\} & \overset{\boldsymbol{\sigma}}{\longrightarrow} & \{\frac{5}{1,2}, 3 : \frac{5}{1,2}\} & \overset{\boldsymbol{\kappa}_2\boldsymbol{\kappa}_3}{\longrightarrow} & \{\frac{10}{1,3}, 3 : \frac{5}{1,2}\} \end{array}$$

The polyhedron $\{\frac{5}{1,2}, 3 : \frac{5}{1,2}\} \cong \{5, 3 : 5\}$ is the hemi-dodecahedron, whose edge-graph is the famous Petersen graph $G(5, 2)$; we can take its vertex-set to be V_3, with initial vertex (if we follow the effect of $\boldsymbol{\sigma}$ on the original generators) $(3, 3, -2, -2, -2)$. (See [48] for the complete classification of the cubic Petersen graphs $G(p, k)$ with $2p$ vertices, and [78] for the fact that each is the edge-graph of a regular polyhedron of type $\{p, 3\}$.) Next, $\{\frac{10}{1,3}, 3 : \frac{10}{1,3}\}$ is a double cover of it (different from the dodecahedron), whose edge-graph is the Petersen-type graph $G(10,3)$, with the same initial vertex $(3,3,-2,-2,-2)$, and whole vertex-set $\pm V_3$. Both these polyhedra are self-Petrie or, rather, the Petrial pairs are enantiomorphs, with helical faces of opposite twists.

Since the vertex-figure of these two polyhedra is $\{3\}$, there is no corresponding reflexion T. Nevertheless, we can try to replace the point-reflexion of the free abelian apeirotope by

$$R_0\colon x \mapsto (3-\xi_1, 3-\xi_0, -2-\xi_2, -2-\xi_3, -2-\xi_4),$$

which does commute with R_2 and R_3, just as if T did exist. However, there are odd edge-circuits: go successively in cyclic permutations of the direction $(3,3,-2,-2,-2)$. Because $\dim R_0 = 1$ is odd, such a circuit will change the initial vertex-figure into its enantiomorph; we thus have a contradiction, and the resulting apeirotope degenerates.

If we now apply $\boldsymbol{\kappa}_2\boldsymbol{\kappa}_3$ (that is, change the signs of both R_2 and R_3), then we obtain the dodecahedron $\{\frac{5}{1,2}, 3 : \frac{10}{1,3}\} \cong \{5,3\}$ and its Petrial $\{\frac{10}{1,3}, 3 : \frac{5}{1,2}\}$. Each has initial vertex $(1,-1,0,0,0)$ and vertex-set V_1; of course, they have the same edges as well. Once again, we do not have a reflexion T, but we can act as if we did, and introduce the hyperplane reflexion

$$R_0\colon x \mapsto (1+\xi_1, -1+\xi_0, \xi_2, \xi_3, \xi_4).$$

As before, though, there are odd edge-circuits (here they are triangles), and since $\dim R_0 = 3$ is odd, we have a clash of enantiomorphic vertex-figures, and so the apeirotope degenerates.

In a way, this is unfortunate, since the facet of the first of these latter two would be $\{4, \frac{5}{1,2} \mid 6\} \cong \{4,5 \mid 6\}$, the universal apeirotope, with group $\mathbf{P}_5 \rtimes \mathrm{D}_5$. The faceting operation $\varphi = \varphi_2$ of Section 5A interchanges face and hole here, and results in $\{6, \frac{5}{1,2} \mid 4\} \cong \{6,5 \mid 4\}$ (which we have already encountered as the facet of the apeirotope $\{\{6, \frac{5}{1,2} \mid 4\}, \{\frac{5}{1,2}, \frac{6}{1,3} : \frac{4}{1,2}\}\}$).

12C17 Remark It is of interest to note that, if we blend these degenerate examples with a segment or an apeirogon, then they become polytopal, that is, non-degenerate.

In conclusion, then, in $\mathbb{E}^4$ we only have the free abelian apeirotopes Q^α for these pentagonal vertex-figures Q.

12C18 Remark We cannot change the sign of R_2 (or R_3) alone; the group $\langle -R_2, R_3 \rangle$ has o as its only fixed point. In Section 11E (see also [83]) we analysed these polyhedra using quaternions, and the reason for the exclusion of this change of sign emerged from the general theory.

Notes to Section 12C

1. The proof of Theorem 12C13 shows that the second assertion of [84, Theorem 8.2] is false.
2. Note that Theorem 12C15 corrects the table of [84, p. 251].

13

Higher-Dimensional Cases

In the last chapter of Part III, which is largely based on [86], we treat the remaining regular polytopes and apeirotopes of nearly full rank. The 'gateway' dimension 5 is crucial to the investigation, since there is a severe restriction on the possible symmetry groups, and hence on the corresponding (finite) regular polytopes. In Section 13A, we look at this dimension only in general terms, since the polytopes not previously described fall naturally into families that are considered in later sections. However, one case is dealt with in full detail in Section 13B; there is a sole regular polytope in $\mathbb{E}^5$ (and none in higher dimensions) whose symmetry group consists only of rotations. The facet of this polytope is a polyhedron of independent interest, but it is more appropriate to consider it in Chapter 16, and so we postpone discussion of it (or, rather, its Petrial) until Section 16A.

Since the new families of regular polytopes of nearly full rank are closely related to the corresponding Gosset–Elte polytopes r_{s1}, it is appropriate that we briefly describe some of these polytopes in Section 13C. (A more comprehensive account, including finding their realization domains, is relegated to Chapter 14.) The largest of the families is then considered in Section 13D. Another Gosset–Elte polytope 1_{22} gives rise to a second family; we consider this in Section 13E.

The Petrie operation turns out to be very productive. In Section 13F we consider a third family, whose facets are Petrials of certain members of the first class; these polytopes were overlooked in [86], and were first described in [87]. In [86], we claimed the existence of another family, whose starting point would be the 5-dimensional Petrial of the Petrie contraction of the 5-cube; however, we show in Section 13G that the putative 6-dimensional member of the family actually degenerates.

In all these families, although the groups are Coxeter groups (possibly with automorphisms), the best approach to them is by twisting of diagrams; indeed, we have already seen in Chapter 9 how effective this idea can be.

Recall that we have already dealt with blended polytopes (and apeirotopes) in Section 9A, and the infinite families in every dimension in Sections 9C and 9D. Therefore, the classification of the polytopes here completes the list of those which can be thought of as exceptional. Just as the exceptional regular polytopes of full rank are only of dimension at most 4, so those of nearly full rank have dimension at most 8.

13A The Gateway

The first step in the classification problem in higher dimensions is to tackle the 'gateway' dimension $d = 5$. We follow the general procedure that we outlined in Section 5F, and so begin by identifying those potential symmetry groups which admit the known symmetry groups of the 4-dimensional polyhedra as suitable subgroups.

Blended Vertex-Figures

Noting that the blends $\{4,3\} \# \{2\}$ and $\{4,3\} \diamond \{2\}$ with a 3-cube $\{4,3\}$ actually have group $[3,3] \times \mathbf{Z}_2$, we see that this and $[3,4] \times \mathbf{Z}_2$ are the only possible groups $\mathbf{H}$. In effect, we can recover the symmetry giving the blend. In the case of cubes, we pass to the Petrial first (and then return afterwards). If the vertex-figure Q has group $\mathbf{H} = \mathbf{H}(\mathsf{Q}) = \langle R_1, R_2, R_3 \rangle$, then $(R_1R_2)^6 = I$ and $T := (R_1R_2)^3$ is the blending element. Although, strictly speaking, they are isomorphic groups, we find it convenient to distinguish $[3,3] \times \mathbf{Z}_2$ and $[3,3] \rtimes \mathbf{Z}_2$. In the latter case, we can have $\mathbf{G} = [3,3,3,3] \rtimes \mathbf{Z}_2$, with vertex-set that of $\Delta\mathrm{T}^5$ (the difference body – see Example 3D5). For the remaining cases, the vertices are the mid-points of the edges of a regular 5-polytope; see Section 9C.

Pure Vertex-Figures

For these, we first look at potential vertex-figures Q with mirror vectors of types $(r,2,3)$ for $r = 1,2,3$ or $(2,3,2)$, so that the group $\mathbf{H}$ of Q contains hyperplane reflexions, and therefore is a hyperplane reflexion group, possibly with outer automorphisms. We postpone until the next Section 13B the case $(2,2,2)$. Now we must exclude groups which contain $[r] \times [r]$ with $r \geqslant 5$, $[3,4,3]$ or $[3,3,5]$ from the outset, so reference to Chapter 11 shows that the only groups which survive are subgroups of $([r] \times [r]) \rtimes \mathbf{Z}_2$ (with $r = 3,4$), $[3,3,3] \rtimes \mathbf{Z}_2$ and $[3,3,4]$. We consider these in turn.

The group $\mathbf{H} = ([3] \times [3]) \rtimes \mathbf{Z}_2$ is a subgroup of $\mathbf{G} = [3,3,3,3] \rtimes \mathbf{Z}_2$, but not of $[3,3,3,4]$. Moreover, the corresponding polytope P must have the vertices of the truncate 0_{22} of the 5-simplex $\{3^4\}$; these are the centres of the triangular 2-faces of $\{3^4\}$. Observe that the vertex-figure of 0_{22} is the product $\{3\} \times \{3\}$.

The group $([4] \times [4]) \rtimes \mathbf{Z}_2$ is a subgroup of $[3,3,3,4]$, actually in two distinct ways. However, the putative polytope P would then have to have the vertices of the 5-staurotope $\{3,3,3,4\}$. This immediately excludes the torus $\{4, \frac{4}{1,2} \mid 4\}$ with 16 vertices as a vertex-figure. It might seem to permit $\{4, \frac{4}{1,2} \mid 4\}^\eta = \{\frac{4}{1,2}, 4 : \frac{4}{1,2}\}$ with the 8 vertices of the half-cube $\{3, 3^{1,1}\}$, which are the same as those of $\{3,3,4\}$. This case does not occur, but the reason put forward in [86, Section 8] makes no sense. So, let us do it directly. We take e_1 as the initial vertex of our putative polytope P, and e_2 as that of the vertex-figure $\mathsf{Q} = \{\frac{4}{1,2}, 4 : \frac{4}{1,2}\}$. Then the generatrix $(R_0, \dots, R_3)$ of P may be assumed to satisfy

$$xR_j = \begin{cases} (\xi_1, \xi_3, \xi_2, \xi_5, \xi_4), & \text{if } j = 1, \\ (\xi_1, \xi_2, \xi_4, \xi_3, \xi_5), & \text{if } j = 2, \\ (\xi_1, \xi_2, \xi_3, -\xi_4, -\xi_5), & \text{if } j = 3. \end{cases}$$

(This choice is unique, up to permuting the last three coordinates.) Now R_0 must do two things: it must interchange e_1 and e_2, and it must commute with R_2 and R_3. A little thought then shows that there are only two possibilities, namely,

$$xR_0 = \begin{cases} (\xi_2, \xi_1, \xi_3, \xi_4, \xi_5), \\ (\xi_2, \xi_1, -\xi_3, -\xi_4, -\xi_5). \end{cases}$$

In both cases, R_0R_1 has period 6. The first is thus ruled out, because the initial 2-face is a trigon with vertices e_1, e_2, e_3. The other needs more work, but once it is observed that $(R_0R_1)^3 = -I$ it is then not too hard to see that $\langle R_0, R_1, R_2\rangle = [3, 3, 3, 4]$, the whole symmetry group of the 5-staurotope (actually, a larger group than we thought that we had started with). This contains R_3, of course, and so the intersection property does not hold.

The group $\mathbf{H} = [3, 3, 3]$ is a subgroup of both $[3, 3, 3, 3]$ and $\mathbf{G} = [3, 3, 3, 4]$. In the former case, a putative polytope P would have to have the vertices of $\mathsf{T} = \{3^4\}$, possibly together with the copy $-\mathsf{T}$ of T obtained by reflexion in its centre; the vertex-figure would then have to have the vertices of $\{3, 3, 3\}$, which is not possible. In the latter case, the vertices of P would then be a subset of those of the 5-cube $\{4, 3, 3, 3\}$, so that we should write $\mathbf{G} = [4, 3, 3, 3]$; consequently, the vertex-figure must then have the vertices of the truncate 0_{21} of $\{3, 3, 3\}$ (again, $\{3, 3, 3\}$ itself cannot occur), which are the mid-points of the edges of $\{3, 3, 3\}$. Also covered here is the case of the half-cube $\{3, 3^{2,1}\}$, with $\mathbf{G} = [3^{2,1,1}]$. We have already seen instances of this kind in Section 9C, and more will turn up later in this chapter.

Finally, though $[3, 3, 3] \rtimes \mathbf{Z}_2$ is a subgroup of $[3, 3, 3, 3] \rtimes \mathbf{Z}_2$, we come back to the first case in the previous paragraph, which we have already excluded.

13A1 Remark Of course, the same kind of analysis applies in higher rank, with an appropriately restricted list of (vertex-sets of) potential vertex-figures. It should be noted that in Section 13B a regular polytope derived from the basic one does have opposite symmetries in its group, and so contains hyperplane reflexions. However, since none of its generators is a hyperplane reflexion, it is easier to approach it through the techniques that we employ.

13B Rotational Symmetry Groups

In dimensions 3 and 4, we saw that there are many examples of regular polytopes (or apeirotopes) of nearly full rank whose symmetry groups consist of direct (orientation preserving) isometries only. We shall call such polytopes *handed* (suitable words of latin or greek origin have been pre-empted for other purposes). To be more exact, in $\mathbb{E}^3$ we have three handed apeirotopes, while in $\mathbb{E}^4$ we have a family of handed polytopes which we treated using quaternions, and some sporadic apeirotopes related to them. Note that, as we saw in Section 4B, a polytope of full rank m cannot be handed, since the mirror of the last of its generating reflexions R_{m-1} must be a hyperplane (see also [82]).

A natural question is whether handed examples exist in $\mathbb{E}^d$ for $d \geqslant 5$. In fact, there is one solitary example of a handed regular 4-polytope in $\mathbb{E}^5$, and while there is a closely related polytope in $\mathbb{E}^5$ (actually, a double cover), it is not handed. There are no handed regular polytopes or apeirotopes of higher rank.

Before we describe our example, we make some remarks about the general situation.

13B1 Proposition *Let* P *be a regular polytope (or apeirotope) of nearly full rank, whose vertex-figure* Q *is handed. If* P *is not itself handed, then the edge-circuits of* P *are all even.*

Proof. This is clear. If P is not handed, then its symmetry R_0 interchanging the two vertices of the initial edge must be opposite; an odd edge-circuit would then reverse the local orientation of the vertex-figure at the initial vertex. □

Naturally, we must begin with the (finite) regular 4-polytopes in $\mathbb{E}^5$. The potential vertex-figures Q here are the regular polyhedra in $\mathbb{E}^4$ with dimension vector $(2, 2, 2)$; these fall into two classes. The first are those which can be derived by the skewing operation

$$\sigma\colon (S_1, S_2, S_3) \mapsto (S_2, S_1S_3, (S_1S_2)^2) =: (R_1, R_2, R_3)$$

of Section 5A on the generatrix of some other regular polyhedron with tetragonal faces. (The indices emphasize that we are thinking of Q as a vertex-figure.) For our purposes, it is crucial to recall from Section 5A (see also [98] – as applied to the dual polyhedron) that these are self-Petrie, and that the Petrie operation π is induced by conjugating by S_1. (In this class we also include two closely related polyhedra.) The second class consists of the remainder; we begin by dismissing these.

Recall from Section 11E how we represented the symmetry group of a regular polyhedron Q in $\mathbb{E}^4$ with mirror vector $(2, 2, 2)$ by quaternions. Thus we have a left group $\mathbf{G}_L$ and right group $\mathbf{G}_R$, and a symmetry of Q is of the form $\mathbf{x} \mapsto \overline{\mathbf{a}}\mathbf{x}\mathbf{b}$, where $\mathbf{a} \in \mathbf{G}_L$ and $\mathbf{b} \in \mathbf{G}_R$. Moreover, $\mathbf{G}_L$ $\mathbf{G}_R$ are liftings of subgroups of $\mathbf{SO}_3$ generated by half-turns; the resulting permissible quaternion groups are the binary dihedral group $\mathbf{D}_n$ of order $4n$ (for some even n), the binary octahedral group $\mathbf{O}$ of order 48, and the binary icosahedral group $\mathbf{G}$ of order 120. (The underlying rotation groups of $\mathbf{D}_n$ for odd n and the binary tetrahedral group $\mathbf{A}$ are not generated by appropriate half-turns.) As point-sets in $\mathbb{E}^4$, $\mathbf{D}_n$ consists of two regular $2n$-gons in orthogonal planes, $\mathbf{O}$ consists of the vertices of a 24-cell $\{3, 4, 3\}$ and its dual, while $\mathbf{G}$ is the vertex-set of a 600-cell $\{3, 3, 5\}$.

In the second class of polyhedra, we can take $\mathbf{G}_R$ to be $\mathbf{O}$ or $\mathbf{I}$. More to the point, because the only identifications between elements of $\mathbf{G}_L$ and $\mathbf{G}_R$ are $(\mathbf{a}, \mathbf{b}) \leftrightarrow (-\mathbf{a}, -\mathbf{b})$, it follows that any initial point $\mathbf{x} \neq \mathbf{o}$ will give rise to at least one left coset of $\mathbf{O}$ or $\mathbf{I}$; regarded as a point-set in $\mathbb{E}^4$, such a coset violates the nearest points criterion of Proposition 5F1. In other words, no such polyhedron can be a vertex-figure of a regular 4-polytope in $\mathbb{E}^5$.

For the first class, we can employ a trick. As we have just observed, the Petrie opertion $\boldsymbol{\pi}$ on one of these polyhedra Q is induced by conjugation by the hyperplane reflexion S_1 in the group with generators (S_1, S_2, S_3) to which $\boldsymbol{\sigma}$ was applied. Indeed, this is nearly the case even for the dodecahedron $\mathsf{Q} = \{\frac{5}{1,2}, 3 : \frac{10}{1,3}\}$ and its Petrial; here, S_1 transforms Q into an enantiomorphic copy with the same vertices. Thus, if we reintroduce S_1 into the new group generators (R_0, R_1, R_2, R_3) of our putative regular polytope P, then we obtain a group generated by hyperplane reflexions which acts on the vertices and edges of P (and 2-faces as well, for all but the dodecahedron and its Petrial). Our argument has thus led to

13B2 Theorem *The group of a* 5*-dimensional regular* 4*-polytope with handed vertex-figure is a subgroup of a hyperplane reflexion group, possibly with outer automorphisms.*

The corresponding groups of the potential vertex-figures are $([n] \times [n]) \rtimes \mathbf{Z}_2$ (for some $n \geqslant 3$), $[3, 3, 3]$, $[3, 3, 3] \rtimes \mathbf{Z}_2$ and $[3, 4, 3] \rtimes \mathbf{Z}_2$. In the second group, $[3, 3, 3]$ is acted on by an inner automorphism, while the other automorphisms are outer; we shall see that all but the second can be eliminated.

In fact, the last fails for the same reason as did those in the second class, namely, each $\mathbf{x} \neq \mathbf{o}$ gives rise to at least one left coset of $\mathbf{O}$; the polyhedra cannot even be vertex-figures of apeirotopes in $\mathbb{E}^4$, since they are excluded by the nearest vertices criterion of Proposition 5F1 (see also [84]). In any event, there is no suitable reflexion group of which $[3, 4, 3]$ is a subgroup, let alone one permitting the outer automorphism. For a similar reason to this last, $[3, 3, 3] \rtimes \mathbf{Z}_2$ is not possible; $[3, 3, 3]$ is a subgroup of each of the finite Coxeter groups $[3^4]$, $[3^{2,1,1}]$ and $[3^3, 4]$, but none permits a suitable automorphism. For the first group, only the case $n = 3$ could be possible. Indeed, we do have a subgroup $[3] \times [3]$ of the group $[3^4]$ of the 5-simplex, and there is an appropriate outer automorphism. However, after we have obtained (R_1, R_2, R_3) using $\boldsymbol{\sigma}$, it is not too hard to see that there is no possible choice for R_0 within $[3^4] \rtimes \mathbf{Z}_2$.

We now move on to the sole example of a handed regular 4-polytope P in $\mathbb{E}^5$. In view of the foregoing discussion, the vertex-figure can only be the self-Petrie hemi-dodecahedron $\mathsf{Q} = \{\frac{5}{1,2}, 3 : \frac{5}{1,2}\} \cong \{5, 3 : 5\}$, whose ten vertices are the mid-points of the edges of the regular 4-simplex, that is, the vertex-set of the truncated 4-simplex 0_{21}. The latter is the vertex-figure of the half-cube $\{3, 3^{2,1}\} = 1_{21}$. We know the group of $\{\frac{5}{1,2}, 3 : \frac{5}{1,2}\}$ from Section 11E (see also [83]); if we are to have this as vertex-figure with the vertices of 1_{21} (resulting in the edges of 1_{21} being those of the putative polytope P as well), then Proposition 13B1 tells us that R_0 must be a direct isometry as well as R_1, R_2, R_3.

A little experimentation then yields (essentially) the only possibility:

13B3
$$\begin{aligned} R_0 &:= \overline{1}\,\overline{2}, \\ R_1 &:= (1\ 5)(2\ 4), \\ R_2 &:= (1\ 2)(3\ 4), \\ R_3 &:= (1\ 2)(4\ 5). \end{aligned}$$

We have taken the generators R_1, R_2, R_3 (which just indicate permutations of coordinates) from Section 11E; R_0 changes the signs of the first and second coordinates. It is worth recalling from the discussion and Section 11E that the Petrie operation $\boldsymbol{\pi}$ on Q (replacing R_1 by $R_1R_3 = (1\ 4)(2\ 5)$) is induced by conjugation under the hyperplane reflexion $T = (1\ 2)$; we have concluded that we are obliged to allow T to be incorporated into the appropriately enlarged group.

We suppress the details of the working, but it is routine to calculate that the (initial) facet F has planar tetragon faces $\{4\}$, Petrie polygons and 2-zigzags $\{\frac{5}{1,2}\}$ (these are not congruent) and holes $\{\frac{4}{1,2}\}$. It has 16 vertices – that is, all of those of $\{3, 3^{2,1}\}$, so that P is flat – and 20 faces. Looking in the tables of [33] shows that we must have $\mathsf{F} = \{4, \frac{5}{1,2} : \frac{5}{1,2}\} \cong \{4, 5 : 5\}$, with symmetry group $\mathbf{G}(F)$ of order 160; see Section 16A, where we discuss its Petrial $\{5, 5 : 4\}$ in detail. The whole symmetry group thus consists of all even permutations of the coordinates with all even numbers of changes of sign. Either directly, or as a consequence, there then follows

13B4 Lemma *There is a regular polytope* $\mathsf{P} = \{\{4, \frac{5}{1,2} : \frac{5}{1,2}\}, \{\frac{5}{1,2}, 3 : \frac{5}{1,2}\}\}$ *with vertices those of* $1_{21} = \{3, 3^{2,1}\}$, *and symmetry group* $\mathbf{G}(\mathsf{P}) = [3^{2,1,1}]^+$, *the rotation subgroup.*

We observe that it is straightforward to establish the intersection property of Theorem 2B7 for $\mathbf{G}(\mathsf{P})$ with respect to the given generatrix.

Let us add a little to the description of F and P. A typical face of F is a diametral tetragon in a staurotopal facet $\{3, 3, 4\}$ of the half-cube $\{3, 3^{2,1}\} = 1_{21}$; thus we should really regard this facet as $\{3, 3^{1,1}\} = 1_{11}$ rather than as $\{3, 3, 4\}$. With each diametral tetragon in a given facet 1_{11} is associated a disjoint tetragon determined by the remaining four vertices; the 20 faces of F fall into ten pairs of such disjoint diametral tetragons. Two different copies of F (under $\mathbf{G}(\mathsf{P})$) then meet in the two pairs of disjoint tetragons in opposite facets of $\{3, 3^{2,1}\}$, and so in four faces in all.

The self-Petriality of Q (induced, as we have said, under conjugation by T) then yields an enantiomorphic copy of P with the same vertices, edges and 2-faces. In other words, P itself is self-Petrie.

It is natural to ask whether our polytope is universal with the given facet and vertex-figure (see the notes at the end of the section). In fact, it is, as an application of the circuit criterion of Theorem 2D4 shows. The basic edge-circuits correspond to the triangular faces of the half-cube $\{3, 3^{2,1}\}$, and it is clear that any edge-circuit can be obtained by concatenating such triangles.

What we have to show is that this carries over to the abstract level. In other words,

13B5 Lemma *The edge-graph $\mathcal{E}$ of $\mathcal{P} := \{\{4,5:5\},\{5,3:5\}\}$ coincides with that of the abstract half-cube $\{3,3^{2,1}\}$.*

Proof. We first show that some triangle can be obtained by concatenating edge-circuits arising from copies of the facet $\{4, \frac{5}{1,2} : \frac{5}{1,2}\}$ of P. We adopt a temporary shorthand:

- the initial vertex $(1,1,1,1,1)$ is denoted 0;
- for $1 \leqslant j < k \leqslant 5$, the vertex $(\varepsilon_1,\ldots,\varepsilon_5)$ with $\varepsilon_i = -1$ for $i = j,k$ and $\varepsilon_i = 1$ otherwise is denoted jk;
- for $j = 1,\ldots,5$, the vertex $(\varepsilon_1,\ldots,\varepsilon_5)$ with $\varepsilon_j = 1$ and $\varepsilon_i = -1$ for $i \neq j$ is denoted $\overline{\jmath}$.

The initial tetragonal face $\{4\}$ with rotational symmetry R_1R_0 has successive vertices $0, 12, \overline{3}, 45$, while the Petrie polygon $\{\frac{5}{1,2}\}$ with rotational symmetry $R_2R_1R_0$ has successive vertices $0, 12, \overline{3}, \overline{1}, 35$; the two polygons share the vertices $0, 12, \overline{3}$ and corresponding edges. We then apply the subgroup [3] generated by the two symmetries $T_0 := (1\ 3)(2\ 5)$ and $T_1 := (1\ 4)(2\ 5)$ to this pair of polygons. It is easy to check that, after concatenation, we are left with the trigon $\overline{3}, \overline{1}, \overline{4}$; Figure 13B6 (with the trigon in red) illustrates how this works schematically.

13B6

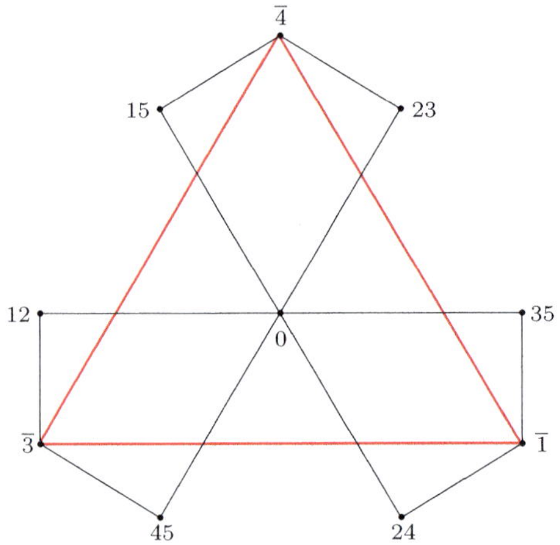

13B7 Remark Perhaps surprisingly, we do not have the expected $T_1 = R_1$, but instead $T_1 = R_1R_3$.

More importantly, it can be seen that this construction of a trigon in the edge-graph of P carries over to $\mathcal{E}$ at the abstract level, since the concatenation of the copies of the face and Petrie polygon is afforded by automorphisms of the vertex-figure alone. Of course, by symmetry the initial vertex of the abstract polytope $\mathcal{P}$ must also be a vertex of such a trigon in $\mathcal{E}$.

To complete the proof, we observe that, since the third edge of a trigon through the initial vertex of the universal polytope $\mathcal{P}$ is a diagonal of the vertex-figure $\{5,3:5\}$, it follows that all such diagonals are edges of $\mathcal{P}$. Bearing in mind Remark 11F5, we thus see that the diagonal-graph of the vertex-figure is the edge-graph of $\{3^{2,1}\}$, with full symmetry if we adjoin the operation which interchanges the vertex-figure and its Petrial. Since $\{3,3^{2,1}\}$ is clearly universal with trigonal 2-faces $\{3\}$ and vertex-figure $\{3^{2,1}\}$, we conclude that $\mathcal{P}$ itself must have the same 16 vertices and 80 edges as the abstract half-cube $\{3,3^{2,1}\}$. □

Putting Lemma 13B5 together with the previous discussion, we have thus shown

13B8 Theorem *There is a single handed* 5*-dimensional* 4*-polytope, namely,*

$$\{\{4,\tfrac{5}{1,2}:\tfrac{5}{1,2}\},\{\tfrac{5}{1,2},3:\tfrac{5}{1,2}\}\} \cong \{\{4,5:5\},\{5,3:5\}\},$$

the latter being the universal regular 4*-polytope.*

13B9 Remark The five vertices in layer 2 of $\{\{4,5:5\},\{5,3:5\}\}$ from the initial vertex are those of a complete subgraph of the edge-graph $\mathcal{G}$, as should be clear from the proof of Lemma 13B5.

Applying ζ to $\mathsf{P} = \{\{4,\frac{5}{1,2}:\frac{5}{1,2}\},\{\frac{5}{1,2},3:\frac{5}{1,2}\}\}$ yields a regular polytope $\mathsf{Q} := \mathsf{P}^{\zeta} = \mathsf{P} \otimes \{2\}$ with the 32 vertices of the 5-cube $\{4,3^3\}$; this is no longer handed. It is again flat; the Petrie polygons and 2-zigzags $\{\frac{5}{1,2}\}$ of the original facet F are now expanded to decagons $\{\frac{10}{1,3}\}$. Its facets are actually of type $\{\frac{4}{1,2},\frac{5}{1,2}\mid\frac{4}{1,2}\} \cong \{4,5\mid 4\}$; however, Q is not isomorphic to the universal $\{\{4,5\mid 4\},\{5,3:5\}\}$, as we shall see. The group $\mathbf{G}(\mathsf{Q})$ has order $32\cdot 60 = 1920$.

13B10 Remark The group $\mathbf{G}(\mathsf{Q})$ does contain hyperplane reflexions, but only the five in the coordinate hyperplanes. Thus Q escapes the restrictions underlying the general treatment, which rely on the symmetry group containing a large hyperplane reflexion subgroup.

13B11 Remark It is worth noting that the facet $\{\frac{4}{1,2},\frac{5}{1,2}\mid\frac{4}{1,2}\}$ of Q is not the standard $2^{\{5\}}$; this latter has planar faces and, indeed, is obtained from the generatrix of the group of the 5-cube by the mixing operation $(S_0,\ldots,S_4)\mapsto(S_0,S_1S_3,S_2S_4) =: (R_0,R_1,R_2)$.

To complete the list in $\mathbb{E}^5$, note that both these polytopes P and Q are vertex-figures of apeirotopes P^{α} and Q^{α}; since a point reflexion in $\mathbb{E}^5$ is opposite, neither of these apeirotopes is handed. Their symmetry groups $\mathbf{G}(\mathsf{P})$ and $\mathbf{G}(\mathsf{Q})$ preclude their being vertex-figures of any other apeirotopes; bear in mind our ability to adjoin suitable hyperplane reflexions to these groups.

Proposition 13B1 tells us that the initial hyperplane reflexion $R_0'\colon x\mapsto(\xi_2,\xi_1,\xi_3,\xi_4,\xi_5)$ of $\{3,3^{2,1}\}$ cannot replace R_0; however, it might be thought that $S_0 := -R_0'$ could (with $S_j := R_j$ for $j=1,2,3$), yielding another – this time handed – regular 4-polytope with the 32 vertices of $\{4,3^3\}$ and group

$[4,3,3,3]^+$, the rotation group of $\{4,3^3\}$. In fact, this is not the case. A little work shows that the resulting polytope would be flat, but that its facets would have (regular) hexagonal edge-circuits; since 12 does not divide the putative group order $32 \cdot 10 = 320$ of the facet, it is clear that the polytope degenerates.

There is a different way of looking at this example, which usefully generalizes. The group $\mathbf{G}$ above is actually isomorphic to $[3^{2,1,1}]$; all we are doing is replacing each opposite element $G \in [3^{2,1,1}] \cap \mathbf{SO}_5$ by $-G$. In fact, just thinking about the putative initial facet, we are introducing an element $-R$ for each reflexion R taking the initial vertex of $1_{21} = \{3, 3^{2,1}\}$ into some five vertices of its vertex-figure. It is then not too hard to see that we obtain enough such reflexions $-R$ from the facet alone to generate the whole group $\mathbf{G}$; the group of the facet thus contains S_3, contrary to the intersection property.

13B12 Remark The dual polytope $\{\{3,5:5\},\{5,4:5\}\}$ is also flat, and thus has a single pure realization (other than $\{1\}$) of dimension 5, with the vertices of the regular 5-simplex. However, this realization cannot be faithful; in fact, not only do the facets $\{3, \frac{5}{1,2} : \frac{5}{1,2}\}$ have the same vertices, they even coincide. Note that there are two copies of a hemi-icosahedron $\{3, \frac{5}{1,2} : \frac{5}{1,2}\}$ with the same 6 vertices, each of which is determined by an initial choice of a trigonal face. The coincidences result from adjacent facets sharing the same face $\{3\}$.

We now move on to 5-polytopes in $\mathbb{E}^6$, still in the context of handedness (bear in mind that we have already classified the remainder). We thus naturally ask whether $\mathsf{Q} = \{\{4, \frac{5}{1,2} : \frac{5}{1,2}\}, \{\frac{5}{1,2}, 3 : \frac{5}{1,2}\}\}$ could be the vertex-figure of such a polytope. Once again, we can introduce the hyperplane reflexion which interchanges the vertex-figure of Q with its Petrial, and conclude that the only possibility essentially arises from the Gosset polytope $2_{21} = \{3^2, 3^{2,1}\}$, whose vertex-figure is $\{3, 3^{2,1}\}$. Since we shall see in Section 13G that there is a regular 5-polytope with the same 27 vertices, we might hope that our 4-polytope Q could be the vertex-figure of another regular 5-polytope with these vertices. However, this is not to be. Proposition 13B1 again shows that the initial reflexion S_0 switching the vertices of an initial edge E of 2_{21} cannot be adjoined to the group of the putative vertex-figure, because 2_{21} has odd edge-circuits. Shifting the previous indices by 1, it follows that the reflexion R_0 would have to be the product of the S_0 with a half-turn preserving E and commuting with R_2, R_3, R_4; since there is no such half-turn, this possibility must be excluded.

Once again, though, it could be thought feasible to replace S_0 by $R_0 := -S_0$, yielding a non-handed 6-dimensional regular 5-polytope with the 54 vertices of 2_{21} and its reflexion in o. However, exactly the same argument we deployed previously shows that, just from the initial facet, we would already obtain a group $\mathbf{G} \cong [3^{2,2,1}]$, with opposite elements G replaced by $-G$ as before (note that they remain opposite).

Let us finally remark that the double cover of $\{\{4, \frac{5}{1,2} : \frac{5}{1,2}\}, \{\frac{5}{1,2}, 3 : \frac{5}{1,2}\}\}$ of type $\{\{\frac{4}{1,2}, \frac{5}{1,2} \mid \frac{4}{1,2}\}, \{\frac{5}{1,2}, 3 : \frac{5}{1,2}\}\}$ is even more obviously to be excluded as a possible vertex-figure of a regular 5-polytope in $\mathbb{E}^6$.

Recall that the focus of this section is on handed regular polytopes. The conclusion which we gain from the present discussion is that handedness – for polytopes of nearly full rank – is a phenomenon of low dimensions; indeed, we have a solitary instance of rank 4 and dimension 5 to add to those in $\mathbb{E}^3$ and $\mathbb{E}^4$.

Realizations of $\{\{4,5:5\},\{5,3:5\}\}$

The realization domain of $\mathcal{Q} := \{\{4,5:5\},\{5,3:5\}\}$ is easily determined. Indeed, since $\mathcal{Q}$ has only two (non-trivial) diagonal classes, it is clear that it has just two non-trivial pure realizations, namely, the 5-dimensional one Q_1 (say) we started with, and a 10-dimensional complement Q_2. Moreover, the vertices of Q_2 have to be those of the pure 10-dimensional realization of the hemi-cube $\{4,3,3,3\}/2$. Since Q_1 has cosine vector $\Gamma_1 = (1, \frac{1}{5}, -\frac{3}{5})$, it follows that Q_2 has cosine vector

13B13 $$\Gamma_2 = \tfrac{1}{10}\big(16(1,0,0) - \Gamma_0 - 5\Gamma_1\big) = \big(1, -\tfrac{1}{5}, \tfrac{1}{5}\big),$$

by the layer equation of Theorem 3C7. In other words,

13B14 Theorem *The cosine matrix of the polytope* $\{\{4,5:5\},\{5,3:5\}\}$ *is*

$$\begin{bmatrix} 1 & 1 & 1 \\ 1 & \frac{1}{5} & -\frac{3}{5} \\ 1 & -\frac{1}{5} & \frac{1}{5} \end{bmatrix},$$

with layer and dimension vectors

$$\Lambda = (1,10,5), \quad D = (1,5,10).$$

Notes to Section 13B

1. The fact that $\{\{4, \frac{5}{1,2} : \frac{5}{1,2}\}, \{\frac{5}{1,2}, 3 : \frac{5}{1,2}\}\} \cong \{\{4,5:5\},\{5,3:5\}\}$ was settled in [86, Section 12] by an appeal to the computer algebra program GAP [50]; this information was kindly provided by Barry Monson. The proof in Lemma 13B5 using the circuit criterion is new.
2. Egon Schulte has made the following observations, which we repeat from [86]. Consider the polytope $\mathbf{2}^{\mathcal{Q}}$ defined in Section 5C (see also [99, Section 8C]), with $\mathcal{Q} = \{5,3:5\}$. Its facets are isomorphic to $\mathbf{2}^{\{5\}} = \{4,5 \mid 4\}$; this latter polyhedron has group order $320 = 2^5 \cdot 10$, and doubly covers $\{4,5:5\}$ (its Petrie polygons are of length 10). Further, by [99, Theorem 8E6], the universal polytope $\{\mathbf{2}^{\{5\}}, \mathcal{Q}\}$ is simply $\mathbf{2}^{\mathcal{Q}}$, since $\mathcal{Q}$ is weakly neighbourly (any two vertices of $\mathcal{Q}$ lie in a common facet of $\mathcal{Q}$). Hence, our polytope is a quotient of $\mathbf{2}^{\mathcal{Q}}$, whose group is $\mathbf{Z}_2^{10} \rtimes [5,3]_5$ of order $2^{10} \cdot 60$, by a factor 2^5. The extra relations that must be imposed on the group of $\mathbf{2}^{\mathcal{Q}}$ to yield the polytope are as yet unknown; we do know that the group relation arising from the lift of the triangle of $\mathcal{P}$ (namely, $(\boldsymbol{r}_0\boldsymbol{r}_3 \cdot \boldsymbol{r}_1\boldsymbol{r}_2\boldsymbol{r}_1)^6 = \boldsymbol{e}$) is not needed, since it already holds in $\mathbf{2}^{\mathcal{Q}}$.

13C The Gosset–Elte Polytopes

The geometric polytopes r_{21} were found by Gosset; the remaining polytopes r_{st} (with the exception of 1_{41}) were discovered by Elte. For further details see the notes at the end of the section, and also the historical remarks in [27, Section 11.x].

We use the notation $\{3^r, 3^{s,t}\}$ for the abstract polytope isomorphic to the Gosset–Elte polytope r_{st}. This adapts the notation of Coxeter [27, (8.91)] (for example), so that

13C1 $$\{3^r, 3^{s,t}\} := \left\{3^r, \begin{matrix} 3^s \\ 3^t \end{matrix}\right\} = \left\{3, \ldots, 3, \begin{matrix} 3, \ldots, 3 \\ 3, \ldots, 3 \end{matrix}\right\},$$

with the strings consisting of r, s or t 3s, as appropriate; observe the use of heavy braces to denote abstract polytopes. Recall that, rather than duality, the Gosset–Elte polytopes exhibit a triality, in which the vertices of $\{3^r, 3^{s,t}\}$ correspond to the facets $\{3^s, 3^{r-1,t}\}$ of $\{3^s, 3^{r,t}\}$ and $\{3^t, 3^{r-1,s}\}$ of $\{3^t, 3^{r,s}\}$.

As an abstract polytope, the original Gosset polytope $\{3^r, 3^{2,1}\} \cong r_{21}$ of rank $r+4$ is semi-regular with two kinds of facet, namely, $(r+3)$-staurotopes and $(r+3)$-simplices. These meet two staurotopes and one simplex around each $(r+1)$-face. We can think of the facets of the staurotopes 2-coloured black and white and those of the simplices coloured white, with two staurotopes abutting on black facets, and staurotopes and simplices abutting on white ones. We shall discuss these polytopes – in particular, their realizations – in more detail in the following Chapter 14.

It is natural to extend the Gosset–Elte family to include the various truncates $\{3^{s,t}\}$ of the $(s+t+1)$-simplex (the case $s = t = 2$ is important here); we have already implicitly included the half-d-cube $\{3, 3^{d-3,1}\}$. More generally, if $r, s, t \geqslant 1$, then $\{3^r, 3^{s,t}\}$ has vertex-figure $\{3^{r-1}, 3^{s,t}\}$ and facets of two types $\{3^r, 3^{s-1,t}\}$ and $\{3^r, 3^{s,t-1}\}$.

The two polytopes $\{3, 3^{3,2}\}$ and $\{3, 3^{4,2}\}$ are of less interest, since they are not involved in this chapter. In any event, they have too many vertices for our realization techiques to be applicable, and so we do not tackle them in Chapter 14. As we shall see, $\{3^2, 3^{4,1}\}$ and a kind of dual of $\{3, 3^{2,2}\}$ (the abstract polytope corresponding to the difference polytope $\Delta 1_{22}$) are also rather large; in spite of that, we have been able to determine their cosine matrices, although both required a lot of work. In fact, this work was worth doing, in that it illustrates many of the techniques of Chapters 3 and 4.

Notes to Section 13C

1. The (convex) polytopes r_{21} were found by Gosset [53] in his enumeration of those that are *semi-regular*, meaning that their facets are regular (though maybe without full symmetry), and their symmetry groups are transitive on their vertices. The exceptional polytope 1_{42} (curiously missed by Elte [46]) does not contribute to any of our families. See also the historical remarks in [27, Section 11.x].

13D The First Gosset Class

This section is devoted to a description of a family of regular polytopes and apeirotopes of nearly full rank, which are closely related to the Gosset–Elte polytopes. We naturally also include apeirotopes here, even though two of these families are not, strictly speaking, derived from Gosset polytopes.

As we have already pointed out in Section 9C, where we considered its group in detail, the source of the family is the 4-dimensional regular polyhedron $Q := \{4, \frac{6}{2,3} \mid 3\} \cong \{4, 6 : 5 \mid 3\}$, which we discussed in Section 11B. For the general case, in Figure 13D1 we have the Coxeter diagram of the group $[3^{r,s,1}]$ (in Chapter 9 we extended such a diagram by a single branch marked q when $r = 1$); as usual, the nodes represent the involutory generators, and two nodes are joined by a branch or not according as the product of the corresponding pair of generators has period 3 or 2. The labelling X, R, Y, T of the geometric hyperplane reflexions corresponds to the abstract $\boldsymbol{x}, \boldsymbol{r}, \boldsymbol{y}, \boldsymbol{t}$ that we worked with in Section 9C. We have treated the cases $r = 0, 1$ in Section 9C, so that we may take $r, s \geqslant 2$; however, the previous cases are covered by our arguments and so, in particular, we may allow one or other of U and V to be absent.

13D1

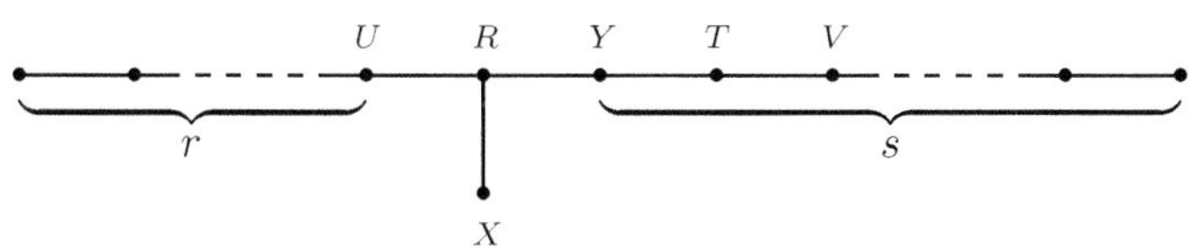

Recall that, if $r, s \geqslant 1$, then $[3^{r,s,1}]$ is a finite or euclidean reflexion group just when

$$\frac{1}{r+1} + \frac{1}{s+1} \geqslant \frac{1}{2} \iff (r-1)(s-1) \leqslant 4;$$

see [31, p. 31] for a neat proof of this.

We first need to see how U and V interact with the new generator $S = XY$. Since $U \rightleftharpoons X, Y$, we see that $U \rightleftharpoons S$, and it follows that $\langle U, R, S\rangle \cong [3, 4]$ in the natural way. Next, since $V \rightleftharpoons X, Y$, we see that $V \rightleftharpoons S$. However,

$$VTS = VTY \cdot X \implies (VTS)^4 = I,$$

and since (as we already know) ST has period 6, we see that $\langle S, T, V\rangle$ is the group of a polyhedron of type $\{6, 3 : 4\}$. In fact (comparing [83]), it is $\{3, 3\} \# \{2\} = \{\frac{6}{2,3}, 3 : \frac{4}{1,2}\} \cong \{6, 3 : 4\}$.

In the spirit of Chapter 9 (see also [84]), we can also represent the group by a diagram, acted upon by an (inner) automorphism S, as in Figure 13D2:

13D2

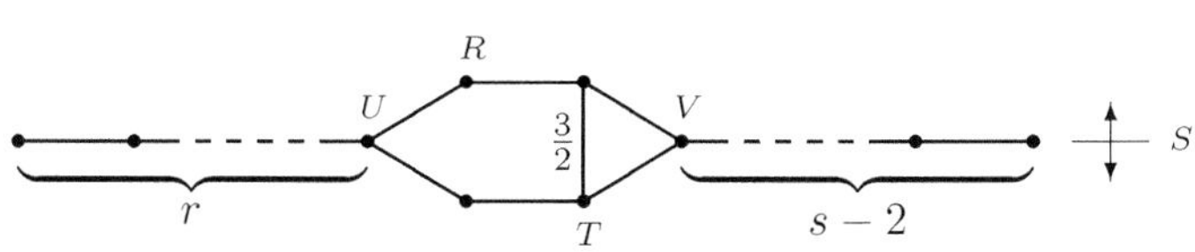

The left horizontal branch (including U) would be missing if $r = 0$; the right one (including V) is missing if $s = 2$. Another way of drawing the diagram is to replace the mark $\frac{3}{2}$ on the branch by marks 2 in the pentagon and triangle.

We are now in a position to describe the polytopes in this family.

13D3 Proposition *For each $r \geqslant 0$ and $s \geqslant 2$ with $(r-1)(s-1) \leqslant 4$ there is a regular polytope (or apeirotope) $\mathsf{G}_{r,s}$ of nearly full rank with the following properties:*

- *it has rank $r+s+1$,*
- *its group $\mathbf{G}_{r,s}$ is $[3^{r,s,1}]$,*
- *it has the same j-faces as r_{s1} for $j = 0, \ldots, r+1$ (namely, j-simplices),*
- $\mathsf{G}_{0,2} = \{4, \frac{6}{2,3} \mid 3\} \cong \{4,6\}/\langle\langle (012)^5, (0121)^3 \rangle\rangle$,
- *if $s = 2$, then its facet is the $(r+2)$-staurotope $\{3^r, 4\}$,*
- *if $s > 2$, then its facet is $\mathsf{G}_{r,s-1}$,*
- *if $r = 0$, then its vertex-figure is the blend $\{3^{s-1}\} \# \{2\}$ of an s-simplex and a segment,*
- *if $r > 0$, then its vertex-figure is $\mathsf{G}_{r-1,s}$.*

More generally, as an abstract regular polytope,

$$\mathsf{G}_{r,s} \cong \{3^r, 4, 6, 3^{s-2}\}/\langle\langle \mathsf{A}, \mathsf{B}, \mathsf{C} \rangle\rangle,$$

with $\mathsf{A}, \mathsf{B}, \mathsf{C}$ the relators

$$\begin{aligned} \mathsf{A} &:= (r(r{+}1)(r{+}2))^5, \\ \mathsf{B} &:= (r(r{+}1)(r{+}2)(r{+}1))^3, \\ \mathsf{C} &:= ((r{+}1)(r{+}2)(r{+}3))^4; \end{aligned}$$

the last relator is absent if $s = 2$,

Let us say a little more about the case $s = 2$. Recall that the Gosset polytope r_{21} has two kinds of facet, namely, $(r{+}3)$-dimensional staurotopes and simplices. The facets of $\mathsf{G}_{r,2}$ are then the diametral $(r{+}2)$-staurotopes of these staurotopal facets of r_{21}, and we see from this description that they do indeed meet in pairs on each of the simplicial $(r+1)$-faces of r_{21}.

We can now apply the operation $\boldsymbol{\zeta}$ to all the finite polytopes in the family; the results are not particularly interesting, and so we shall say no more about this. On the other hand, applications of $\boldsymbol{\pi}$ (where appropriate) lead to some interesting polytopes. It is clear that $\boldsymbol{\pi}$ can only apply to $\mathsf{G}_{r,s}$ when $s = 2$ or 3. In fact, if $s = 2$, the only case is $r = 0$, so we have a polyhedron; we have dealt with this in Chapter 11 (see also [83]). Thus $s = 3$ is the only productive case.

We first treat the sub-case $r = 0$. Going back to the way we constructed the group from $[3^{0,3,1}]$ (that is, in the first diagram, with no left branch, and only V in the right branch), we see that the new group has generators $(R, SV, T, V) = (R, XYV, T, V)$. A little work shows that the facet of the resulting polytope is $\{4, \frac{4}{1,2} \mid 3\} \cong \{4,4 \mid 3\}$, while the vertex-figure is $\{\frac{4}{1,2}, 3 : \frac{6}{2,3}\} \cong \{4,3\}$.

However, this is not universal of type $\{\{4,4 \mid 3\},\{4,3\}\}$, as we shall shortly see. Indeed, if we anticipate a little, we find that we actually have

$$\mathsf{G}_{0,3}{}^{\pi} = \{\{4, \tfrac{4}{1,2} \mid 3\}, \{\tfrac{4}{1,2}, 3 : \tfrac{6}{2,3}\} : \tfrac{5}{1,2}\} \cong \{4,4,3\}/\langle\langle (0121)^3, (0123)^5 \rangle\rangle.$$

In turn, we can use these as successive vertex-figures, to construct $\mathsf{G}_{r,3}{}^{\pi}$ for $r = 1, 2, 3$, the last being an apeirotope in $\mathbb{E}^7$. The case $r = 1$ is worth describing in some detail, particularly in view of the fact that it can be derived from the hexagonal diagram of Figure 13D4 by twisting using an inner automorphism; this is an example of an improper twist, since just one – or an odd number – of its branches can be marked $\frac{3}{2}$.

13D4

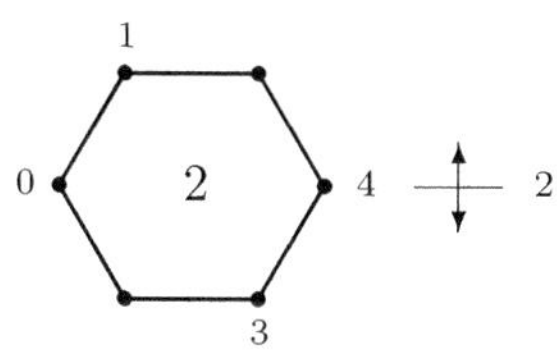

The mark 2 at the centre of the diagram indicates the period of a certain product of the generating reflexions corresponding to the nodes of the hexagon; this notation was introduced in Section 1E. Note that the situation here is not the same as that in the case $\ell = 2$ of [95, (39)], where an *outer* automorphism was applied to this hexagonal diagram.

Writing $(R_0, \ldots, R_4) := (U, R, XYV, T, V)$ with the notation of Figure 13D1, we can take

13D5

$$\begin{aligned} R_0 &:= (1\ \overline{2}),\\ R_1 &:= (2\ 3),\\ R_2 &:= (1\ 2)(3\ 4)(5\ 6),\\ R_3 &:= (4\ 5),\\ R_4 &:= (5\ 6). \end{aligned}$$

As usual, the permutations are those of the standard basis vectors, with $(1\ \overline{2})$ indicating the interchange of e_1 and $-e_2$.

The vertices are those of the half-6-cube, which we write as

$$c_{jk\ldots} = (\varepsilon_1, \varepsilon_2, \varepsilon_3, \varepsilon_4, \varepsilon_5, \varepsilon_6),$$

with each $\varepsilon_i = \pm 1$, to indicate that $\varepsilon_i = -1$ precisely for $i = j, k, \ldots$ (with $c := c_{\emptyset}$); thus, for example $c_{1246} = (-1, -1, 1, -1, 1, -1)$. The initial vertex is c, which is joined to c_{12} by the initial edge; the initial 2-face then has the additional vertex c_{13}. A little work shows that the initial 3-face (which is an octahedron) has opposite pairs of vertices $c, c_{1234} : c_{12}, c_{34} : c_{13}, c_{24}$. Finally, the vertex-figure of the initial facet consists of all c_{jk} with $j = 1, 4, 5$ and $k = 2, 3, 6$, whose square faces are given by two out of three of the choices for j and k. The facet itself is centrally symmetric about the origin o. We thus see that the facet

$\{\{3,4\},\{4,\frac{4}{1,2} \mid 3\}\}$ is, in fact, the universal $\mathbf{\{\{3,4\},\{4,4 \mid 3\}\}}$ with 20 vertices and 30 octahedral 3-faces.

However, the vertex-figure cannot be the dual of this, in spite of the apparent symmetry of the diagram, because it only has 15 vertices and 10 toric 3-faces. Since the Petrie polygon of $\mathbf{\{\{4,4 \mid 3\},\{4,3\}\}}$ has period 10, while that of the vertex-figure has period 5, we see that it is indeed obtained by imposing the extra relator $(0123)^5$ on the group of the former. In other words, we conclude that

$$\mathsf{G}_{13}{}^{\pi} \cong \{3,4,4,3\}/\langle\langle (1232)^3, (1234)^5 \rangle\rangle,$$

when we bear in mind that we can recover the original group $[3^{1,3,1}]$ by reversing the various operations which led to $\mathsf{G}_{13}{}^{\pi}$.

Our polytope $\mathsf{G}_{13}{}^{\pi}$ has no geometric dual, because its vertex-figure does not. However, if we change the sign of R_2, and so replace it by $-R_2 = R_2^{\perp}$, and reverse the order of the generators, namely,

$$(R_0,\ldots,R_4) \mapsto (R_4,R_3,-R_2,R_1,R_0) =: (S_0,\ldots,S_4),$$

then we have the group $\langle S_0,\ldots,S_4\rangle$ of another copy of $\mathsf{G}_{13}{}^{\pi}$. In other words, the polytope is self-quasi-dual, in the sense of Section 5D. Its vertices consist of those of the 'complementary' copy of the half-6-cube in the 6-cube $\{4,3^4\}$; the initial vertex c_{145} (or c_{236}) corresponds in an obvious way to the initial facet of the original.

We shall say little about $\mathsf{G}_{r3}{}^{\pi}$ for $r = 2$ or 3. The former has group $[3^{2,3,1}]$ and the vertices of 2_{31}. The latter has group $[3^{3,3,1}]$ and the vertices of the apeirotope 3_{31}; one feature is that its facets are also apeirotopes with group $[3^{2,2,2}] \rtimes \mathbf{Z}_2$; we shall see in Section 13F that each $\mathsf{G}_{r3}{}^{\pi}$ is itself the facet of another regular polytope or apeirotope.

13E The Second Gosset Class

We next have the Coxeter diagram of the group $[3^{r,2,2}]$ acted on by a proper outer twist, as depicted in Figure 13E1:

13E1

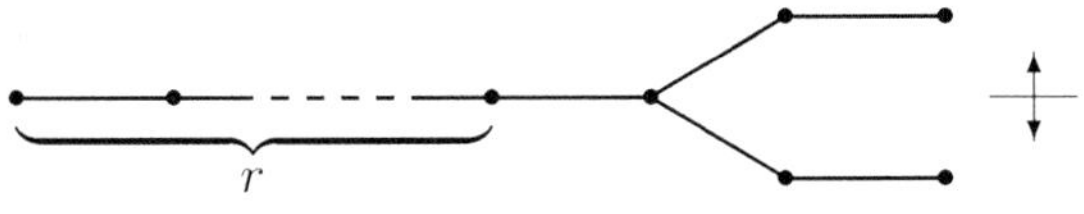

The Coxeter diagram for $[3^{r,2,2}]$

The case $r = 0$ is the group $[3,3,3,3] = \mathbf{A}_5 \cong \boldsymbol{S}_6$ of the 5-simplex, which we have already encountered. Since the group is infinite for $r = 2$, the only relevant cases are $r = 0,1,2$; the corresponding polytope is denoted J_{r+4}, where the suffix is the rank (see the notes at the end of the section). The twist acts as indicated in the picture. The case $r = 0$ is the universal polytope

$$\{\{3,4\},\{4,\tfrac{4}{1,2} \mid 3\}\} \cong \mathbf{\{\{3,4\},\{4,4 \mid 3\}\}} = \{4,4,3\}/\langle\langle (1232)^3 \rangle\rangle.$$

In general, the facet is the $(r+3)$-staurotope in the form $\{3^r, 3^{1,1}\}$, for $r = 1, 2$ the vertex-figure is the case $r - 1$, and the whole polytope (or apeirotope) J_{r+4} (say – here, $r + 4$ is the rank) is then universal with this facet and vertex-figure. Its vertex-set is the same as that of the Gosset polytope or honeycomb r_{22}, and each of these polytopes has a dual.

13E2 Remark The universality follows straightforwardly from the fact that we can recover the original Coxeter diagram of $[3^{r,2,2}]$ from the symmetry group of the polytope. Compare the discussion of the 4-dimensional polyhedra in Section 11B.

The case $r = 2$ is of some interest because it is a sponge in $\mathbb{E}^6$, analogous to the Petrie–Coxeter apeirohedra in $\mathbb{E}^3$. The (convex) Gosset polytope $T = 2_{21}$ and its negative $-T$ tile $\mathbb{E}^6$ by translates. On a simplicial facet $\{3, 3, 3, 3\}$, a translate of T meets another translate of T (and similarly for $-T$), while on a staurotopal facet $\{3, 3, 3, 4\}$ it meets a translate of $-T$. If we delete the copies of the simplices $\{3, 3, 3, 3\}$ and the tiles T and $-T$ themselves, then the staurotopes $\{3, 3, 3, 4\}$ are the facets of J_6; they meet four around each 3-face $\{3, 3\}$, and form a surface which splits $\mathbb{E}^6$ into two congruent parts.

When $r = 2$, the dual is an apeirotope with blended vertex-figure. Hence, potentially at least, we can apply the operation $\boldsymbol{\kappa}$. However, this does not yield a genuine regular 6-polytope of full rank; instead, we obtain the first of the two examples which were excluded by Theorem 8A16 (see also [82, Section 10]) as non-polytopes. Indeed, this viewpoint can be made to provide an alternative explanation for its non-polytopality.

Theorem 5B10 prevents us from applying the Petrie operation $\boldsymbol{\pi}$ to the polytopes J_n, but the halving operation $\boldsymbol{\eta}$ will work. In the present case, $\boldsymbol{\eta}$ is actually invertible (a double application of $\boldsymbol{\eta}$ recovers the original polytope, with a change of generators). When $n = 4$ (that is, $r = 0$), as in [99, Section 10E] we obtain the universal polytope

$$\{\{4, \tfrac{4}{1,2} \mid 3\}, \{\tfrac{4}{1,2}, 4 \mid 3\}\} \cong \{\{4, 4 \mid 3\}, \{4, 4 \mid 3\}\};$$

it has the same 20 vertices as before (those of 0_{22}). Abstractly, this polytope is self-dual; however, geometrically it is not, the dual actually being

$$\{\{\tfrac{4}{1,2}, 4 \mid 3\}, \{4, \tfrac{4}{1,2} \mid 3\}\}.$$

The symmetry betwen these two polytopes suggests that we may be able to apply $\boldsymbol{\eta}$ to the latter. Indeed we can; perhaps it is not surprising that the result is

$$\{\{\tfrac{4}{1,2}, 4 \mid 3\}, \{4, \tfrac{4}{1,2} \mid 3\}\}^{\boldsymbol{\eta}} = \{\{3, \tfrac{4}{1,2}\}, \{\tfrac{4}{1,2}, 4 \mid 3\}\},$$

isomorphic to J_4, but geometrically distinct. In contrast to J_4, this polytope does not have a geometric dual, as its mirror vector $(3, 3, 4, 3)$ may indicate.

This correspondence is underlined by realization theory. Since J_4 has the same vertices as 0_{22} and the same symmetries, we can really work with the latter. Since 0_{22} is the central section of the 6-cube, its vertices fall into four equally spaced parallel sections, giving its layer vector $\Lambda = (1, 9, 9, 1)$, and initial cosine vector $(1, \frac{1}{3}, -\frac{1}{3}, -1)$. Now the abstract polytope $\{3^{2,2}\}/2 \cong 0_{22}$ has a single non-trivial diagonal, and so its vertices must be those of the 9-simplex, with cosine vector $(1, -\frac{1}{9}, -\frac{1}{9}, 1)$. Either directly, or by considering the components of the staurotope realization X, we find that we must have another pure 5-dimensional realization, with cosine vector $(1, -\frac{1}{3}, \frac{1}{3}, -1)$.

Let us adopt a convenient shorthand for the polytopes of this section. Instead of $\{4, \frac{4}{1,2} \mid 3\}$ we write just $\{4, \frac{4}{1,2}\}$, and instead of $\{\frac{4}{1,2}, 4 \mid 3\}$ we write $\{\frac{4}{1,2}, 4\}$; we then extend the notation to infaces. For instance, we have

$$\begin{aligned}\{\{4, \tfrac{4}{1,2} \mid 3\}, \{\tfrac{4}{1,2}, 4 \mid 3\}\} &=: \{4, \tfrac{4}{1,2}, 4\},\\ \{\{3, \tfrac{4}{1,2}\}, \{\tfrac{4}{1,2}, 4 \mid 3\}\} &=: \{3, \tfrac{4}{1,2}, 4\}.\end{aligned}$$

Of course, we could employ the $\langle\langle \ldots \rangle\rangle$ notation, as

$$\{\{4, \tfrac{4}{1,2} \mid 3\}, \{\tfrac{4}{1,2}, 4 \mid 3\}\} = \{4, \tfrac{4}{1,2}, 4\}/\langle\langle (0121)^3, (1232)^3 \rangle\rangle$$

(strictly, we should write $(0 \cdot 121)^3$ rather than $(0121)^3$, but there can be no ambiguity here); however, this does not lead to more concise expressions.

So, since we have Petrie pairs with the same vertices, our 5-dimensional 4-polytopes are

$$\begin{aligned}\{3, 4, \tfrac{4}{1,2}\},\ \{4, \tfrac{4}{1,2}, 4\}: \quad &\Gamma = (1, \tfrac{1}{3}, -\tfrac{1}{3}, -1),\\ \{3, \tfrac{4}{1,2}, 4\},\ \{\tfrac{4}{1,2}, 4, \tfrac{4}{1,2}\}: \quad &\Gamma = (1, -\tfrac{1}{3}, \tfrac{1}{3}, -1).\end{aligned}$$

Observe as well that this emphasizes that there must also be two distinct 5-dimensional realizations of $\{\{4, 4 \mid 3\}, \{4, 4 \mid 3\}\}$; the asymmetric mirror vector $(4, 3, 4, 3)$ of $\{4, \frac{4}{1,2}, 4\}$ (see the following Remark 13E3) shows that it cannot be geometrically self-dual.

13E3 Remark The symmetry group of $\{4, \frac{4}{1,2}, 4\}$ has generators

$$\begin{aligned}R_0 &= (2\ 3),\\ R_1 &= -(0\ 5)(1\ 3)(2\ 4),\\ R_2 &= (4\ 5),\\ R_3 &= -(0\ 5)(1\ 4)(2\ 3),\end{aligned}$$

working in the hyperplane $\mathbb{H}^5 \leqslant \mathbb{E}^6$, with the same conventions as before.

We may now apply $\boldsymbol{\pi}$ to both these (isomorphic) polytopes, to obtain

$$\begin{aligned}\{\{4, \tfrac{4}{1,2} \mid 3\}, \{\tfrac{4}{1,2}, 4 \mid 3\}\}^{\pi} &= \{\{4, \tfrac{6}{2,3} : \tfrac{5}{1,2} \mid 3\}, \{\tfrac{6}{2,3}, 4 : \tfrac{4}{1,2} \mid 3\}\},\\ \{\{\tfrac{4}{1,2}, 4 \mid 3\}, \{4, \tfrac{4}{1,2} \mid 3\}\}^{\pi} &= \{\{\tfrac{4}{1,2}, \tfrac{6}{1,3} : \tfrac{5}{1,2} \mid 3\}, \{\tfrac{6}{1,2}, \tfrac{4}{1,2} : 4 \mid 3\}\}.\end{aligned}$$

These polytopes can have no geometric duals, because their facets do not (at least in $\mathbb{E}^4$, which would be needed for them to act as vertex-figures of 5-dimensional polytopes). The Petrials have facets isomorphic to $\{4, 6 : 5 \mid 3\}$, with dimensions 4 and 5, respectively; see Theorem 11C9 for these realizations.

The polytope $\{4, \frac{4}{1,2}, 4\}$ and its Petrial are distinguished (among finite 4-polytopes of dimension 5) in having vertex-figures with mirror vector $(2, 3, 2)$. As we have already pointed out, the only way to incorporate two automorphisms is to have a diagram with a redundant generator; for the first polytope, we thus have the diagram of Figure 13E4 (see the notes at the end of the section):

13E4

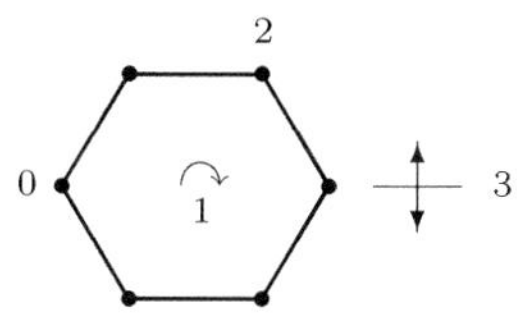

Going clockwise around the hexagonal diagram from $R_0 = (2\ 3)$, the other nodes correspond to $(3\ 5)$, $(4\ 5)$, $(1\ 4)$, $(0\ 1)$ and $(0\ 2)$; any of these generators can be taken to be the redundant one. After applying π, the new R_1 is $(1\ 2)(3\ 4)$, which reflects the diagram in a vertical line.

In (5B17) we introduced the operation μ_k on a regular m-polytope of type $\{3^{m-3}, 4, q\}$ for some q, which – potentially at least – produces a new regular polytope of type $\{3^{k-2}, 4, q, q, 4, 3^{m-k-3}\}$. We now apply the μ_k to the J_m. With J_4, the sequence is completed to

$$\{3, 4, \tfrac{4}{1,2}\}, \quad \{4, \tfrac{4}{1,2}, 4\}, \quad \{\tfrac{4}{1,2}, 4, \tfrac{4}{1,2}\}, \quad \{4, \tfrac{4}{1,2}, 3\};$$

the first and last are geometric duals. For J_5 we obtain the sequence

$$\{3, 3, 4, \tfrac{4}{1,2}\}, \quad \{3, 4, \tfrac{4}{1,2}, 4\}, \quad \{4, \tfrac{4}{1,2}, 4, \tfrac{4}{1,2}\}, \quad \{\tfrac{4}{1,2}, 4, \tfrac{4}{1,2}, 3\}, \quad \{4, \tfrac{4}{1,2}, 3, 3\},$$

the first two being J_5 and $\mathsf{J}_5{}^{\eta}$. From J_6 we obtain

$$\{3, 3, 3, 4, \tfrac{4}{1,2}\}, \quad \{3, 3, 4, \tfrac{4}{1,2}, 4\}, \quad \{3, 4, \tfrac{4}{1,2}, 4, \tfrac{4}{1,2}\},$$
$$\{4, \tfrac{4}{1,2}, 4, \tfrac{4}{1,2}, 3\}, \quad \{\tfrac{4}{1,2}, 4, \tfrac{4}{1,2}, 3, 3\}, \quad \{4, \tfrac{4}{1,2}, 3, 3, 3\};$$

the first two are J_6 and $\mathsf{J}_6{}^{\eta}$. In contrast to the case of J_4, symmetry about the middle of the two sequences yields geometrically dual polytopes or apeirotopes.

We can now apply η to those two polytopes with 4-coface $\{\frac{4}{1,2}, 4, \frac{4}{1,2}\}$. We then obtain

$$\{4, \tfrac{4}{1,2}, 4, \tfrac{4}{1,2}\}^{\eta} = \{4, 3, \tfrac{4}{1,2}, 4\},$$
$$\{3, 4, \tfrac{4}{1,2}, 4, \tfrac{4}{1,2}\}^{\eta} = \{3, 4, 3, \tfrac{4}{1,2}, 4\}.$$

In these, $\{4, 3, \frac{4}{1,2}\} = \{4, 3, \frac{4}{1,2} \mid 3\} \cong \{4, 3, 4 \mid 3\}$ is a torus; in the obvious adaptation of the notation of [97] or [99, Section 6D], this would be written $\{4, 3, 4\}_{(3,0,0)}$. The facet of the second (an apeirotope) is isomorphic to the universal locally toroidal polytope

$$\mathcal{T}^5_{(3,0,0)} = \{\{3, 4, 3\}, \{4, 3, 4 \mid 3\}\}$$

of [97] or [99, Section 12B]; its facet is the 24-cell. Neither of these has a geometric dual.

We next look at $\{3, 3^{2,2}\}$, whose 4-staurotopal faces are the facets of the locally toroidal 5-polytope $\{3,3,4,4\}/\langle\langle (2343)^3 \rangle\rangle$. In the 'standard' realization with the vertices of 1_{22}, this is $\{3,3,4,\frac{4}{1,2}\}/\langle\langle (2343)^3 \rangle\rangle$, with 3-coface the torus $\{4,\frac{4}{1,2} \mid 3\}$.

13E5 Remark Since $\{4,3,\frac{4}{1,2},4\}$ has the same vertex-set as $\{3,3,4,\frac{4}{1,2}\}$, namely, that of 1_{22}, this polytope illustrates the fact that $[3^2, 3^{2,1}]$, being the point-group of $[3^2, 3^{2,2}]$, has $[3] \times [3] \times [3]$ as a subgroup. (In these polytopes, we understand that each $\{4,\frac{4}{1,2}\}$ or $\{\frac{4}{1,2},4\}$ is (abstractly) a torus $\{4,4 \mid 3\}$.) Moreover, vert 1_{22} contains the vertices of the torus $\{4,3,\frac{4}{1,2} \mid 3\} \cong \{4,3,4 \mid 3\}$; its edges and trigonal holes occur among the faces of 1_{22}, while its tetragonal faces $\{4\}$ are diametral squares of certain of the octahedral 3-faces of 1_{22}.

We introduced the idea of vertex-figure replacement in the classification of the classical regular star-polytopes in Section 7G. For other than the classical polytopes, there is – in general – no equivalent notion. However, with the J_n families, we can clearly see parallel ways of listing the polytopes. Thus, for $n = 4$, we have

13E6

$\{4,\frac{4}{1,2},3\}$	$\{3,4,\frac{4}{1,2}\}$	
	$\{4,\frac{4}{1,2},4\}$	$\{\frac{4}{1,2},4,\frac{4}{1,2}\}$
		$\{3,\frac{4}{1,2},4\}$

For $n = 5$, our family looks like

13E7

$\{4,\frac{4}{1,2},3,3\}$	$\{3,3,4,\frac{4}{1,2}\}$	
	$\{3,4,\frac{4}{1,2},4\}$	$\{\frac{4}{1,2},4,\frac{4}{1,2},3\}$
	$\{4,\frac{4}{1,2},4,\frac{4}{1,2}\}$	
	$\{4,3,\frac{4}{1,2},4\}$	

Finally, for $n = 6$, we have

13E8

$\{4, \frac{4}{1,2}, 3, 3, 3\}$	$\{3, 3, 3, 4, \frac{4}{1,2}\}$	
	$\{3, 3, 4, \frac{4}{1,2}, 4\}$	$\{\frac{4}{1,2}, 4, \frac{4}{1,2}, 3, 3\}$
	$\{3, 4, \frac{4}{1,2}, 4, \frac{4}{1,2}\}$	
	$\{3, 4, 3, \frac{4}{1,2}, 4\}$	
	$\{4, \frac{4}{1,2}, 4, \frac{4}{1,2}, 3\}$	

The conventions of Section 7G hold, namely, that vertex-figure replacement operates on columns, while rows contain duals. Observe that $\{4, \frac{4}{1,2}, 4, \frac{4}{1,2}, 3\}$ could also be listed in Table 13E8 as the dual of $\{3, 4, \frac{4}{1,2}, 4, \frac{4}{1,2}\}$, but we have suppressed this to save space.

These tables do not exhibit as much symmetry as do those of Section 7G, mainly because several of the polytopes in the lists do not have geometric duals.

We cannot apply $\boldsymbol{\pi}$ to any of the polytopes in the sequences in the cases $r = 1, 2$; we fall foul of Theorem 5B10, because of holes or infaces $\{3\}$. We can always apply $\boldsymbol{\zeta}$ to the polytopes in the J_4 and J_5 families, and $\boldsymbol{\kappa}$ to the apeirotopes in the J_6 family. However, as is frequently the case we obtain nothing of any striking interest; note that applying $\boldsymbol{\kappa}$ will result in infinite facets (or even infinite 2-faces $\{\frac{4}{0,1}\}$ from original faces $\{\frac{4}{1,2}\}$).

We might think of applying $\boldsymbol{\eta}$ to those polytopes with 3-cofaces $\{\frac{4}{1,2}, 3\}$. However, we actually obtain

$$\begin{aligned} \{4, \tfrac{4}{1,2}, 3\}^{\eta} &= \{3, 3, 3\}, \\ \{\tfrac{4}{1,2}, 4, \tfrac{4}{1,2}, 3\}^{\eta} &= \{\tfrac{4}{1,2}, 3, 3, 3\}, \\ \{4, \tfrac{4}{1,2}, 4, \tfrac{4}{1,2}, 3\}^{\eta} &= \{4, \tfrac{4}{1,2}, 3, 3, 3\}. \end{aligned}$$

The first of these shows that, in each case, we end up with a proper subgroup of the original. In the second, we have

$$\{\tfrac{4}{1,2}, 3, 3, 3\} = \{4, 3, 3, 3\} \# \{2\} = \{4, 3, 3, 3\}^{\zeta} \diamond \{2\},$$

the latter more exactly indicating the algebra as well as the geometry, in view of the fact that $\{4, 3, 3, 3\}^{\zeta}$ has the vertices of the half-cube 1_{31}. However, the last case is of some interest, because it shows that the vertices of a copy of $\mathsf{J}_6{}^{\delta}$ can be chosen to lie among those of J_6 itself. (There are implied subgroup relationships, with indices 10, 27 and 64, respectively, the last illustrating the expected index $64 = 2^6$ of $[3^{2,2,2}]$ in itself.) In any event, we have added nothing new to our previous lists.

Notes to Section 13E

1. We have kept (roughly) to the notation of the original papers [86, 87], even though it results in the polytope classes appearing out of alphabetical order.
2. Figure 13E4 looks like [99, Figure 10C1], which was unfortunately mislabelled – the labels $0, \ldots, 5$ on the hexagon should be shifted one step anti-clockwise, but the latter represents the infinite group, where one of the hyperplane mirrors is displaced from the origin o.

13F The Third Gosset Class

13F1

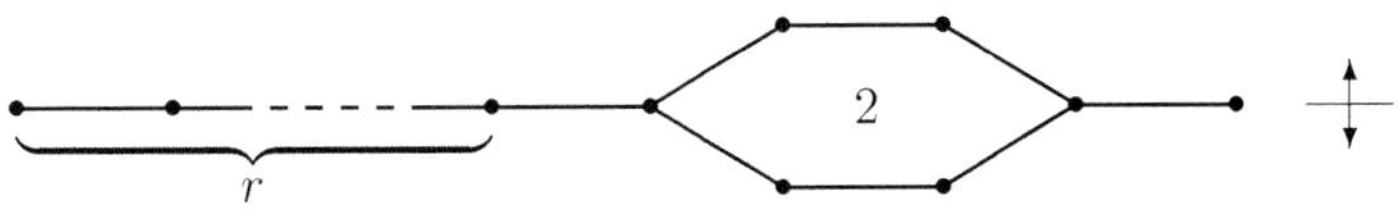

The final family of regular polytopes (or apeirotopes) related to the Gosset polytopes is derived by applying an improper inner twist to the diagram of Figure 13F1 (see the notes at the end of the section). If we omit the right-most node (and corresponding branch), then we obtain the diagram for the Petrial ${\mathsf{G}_{r+1,3}}^{\pi}$ of the polytope $\mathsf{G}_{r+1,3}$ of the first Gosset class, which we considered in Section 13D.

It is clear that we cannot extend the diagram by any further nodes to the right, because we would then obtain the diagram of the infinite group $[3^{2,2,2}]$ as a subdiagram. However, as we shall shortly see, we can allow any $r \leqslant 2$, in spite of the fact that there is again an 'infinite' subdiagram; we encountered such a situation several times in Section 9D.

So as to treat the members of the family together, we list the generators R_j of $\mathbf{G}$ in the extreme case $r = 2$ in (13F2) by giving the equations of their mirrors, in terms of coordinate vectors $(\xi_1, \ldots, \xi_8) \in \mathbb{E}^8$. They are

13F2

$$R_j\colon \begin{cases} \xi_1 + \xi_8 = 2, & \text{if } j = 0, \\ \xi_1 + \cdots + \xi_8 = 0, & \text{if } j = 1, \\ \xi_2 + \xi_7 = 0, & \text{if } j = 2, \\ \xi_2 = \xi_3, & \text{if } j = 3, \\ \xi_1 = \xi_8,\ \xi_2 = \xi_7,\ \xi_3 = \xi_6,\ \xi_4 = \xi_5, & \text{if } j = 4, \\ \xi_5 = \xi_6, & \text{if } j = 5, \\ \xi_4 = \xi_5, & \text{if } j = 6, \\ \xi_1 + \cdots + \xi_4 = \xi_5 + \cdots + \xi_8, & \text{if } j = 7. \end{cases}$$

The situation is depicted in Figure 13F3, whose labels indicate the corresponding generators of the symmetry group $\mathbf{G}$. Thus R_4 is the diagram twist, which just reverses the order of the coordinates $\xi_1, \ldots, \xi_8$.

13F3 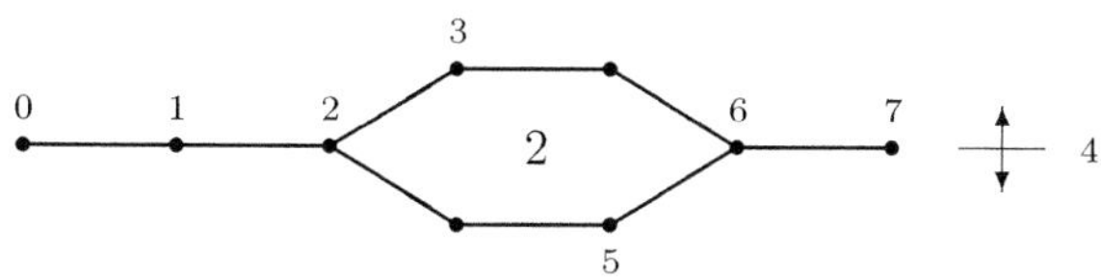

Of course, as we know, to show that we do obtain genuine polytopes, we must verify the intersection property of Theorem 2B7. However, the geometric picture given by the generators makes this straightforward, if a little tedious.

For $r = 5, \ldots, 8$, the symmetry group of the general member K_m of the family is

$$\mathbf{K}_m := \langle R_{8-m}, \ldots, R_7 \rangle.$$

For $m = 5, 6, 7$, K_m is a (finite) $(m+1)$-dimensional regular polytope of rank r, and thus of nearly full rank; similarly, K_8 is an 8-dimensional apeirotope of nearly full rank.

The initial vertex of K_8 is the origin o; which R_0 takes into the initial vertex $(2, 0^6, 2)$ of the vertex-figure K_7; as usual in this context, α^k denotes a string $\alpha, \ldots, \alpha$ of length k. Under the group $\mathbf{G}^v = \mathbf{K}_7$ of the vertex-figure, we obtain all permutations with an even number of changes of sign of $(2, 2, 0^6)$ and (1^8), namely, the vertex-set of the Gosset polytope 4_{21}. Thus K_8 has the vertices of the semi-regular tiling 5_{21} of $\mathbb{E}^8$.

More generally, the k-coface K_k has the same vertices as $(k-3)_{21}$. Moreover, as we said above, the facet of K_k is the Petrial $\mathsf{G}_{k-5,3}{}^\pi$ of the regular polytope $\mathsf{G}_{k-5,3}$ of the first Gosset class. We pointed out in Section 13D that $\mathsf{G}_{33}{}^\pi$ is an apeirotope, whose facets are themselves apeirotopes of type J_6 of Section 13E; see the notes at the end of the section. Hence K_8 even has ridges which are apeirotopes.

So far as K_5 is concerned, its group is obtained from that of $\mathsf{J}_5{}^\delta$ by changing the sign of the diagram twist T (that is, replace the mirror T by its orthogonal complement $T^\perp$); this changes a proper outer automorphism to an improper inner one.

13F4 Remark In spite of the apparent symmetry of the diagram, K_7 is not self-dual; indeed, like each of the polytopes K_r, it has no geometric dual. However, just as with other cases, if we reverse the order of the generators $R_1, \ldots, R_7$ and change the sign of the twist R_4, then we obtain the symmetry group of another copy of K_7 (or, rather, the same copy, but with different initial vertex and so on). In other words, in the sense of Section 5D, K_7 is self-quasi-dual.

Notes to Section 13F

1. At the end of [86], we expressed the hope that our enumeration of the regular polytopes of nearly full rank was then complete. In retrospect, it is fortunate that we did not make an absolute claim for completeness, since we overlooked the family of this section.

2. Since such changes of sign of reflexions were used quite often in [86, Section 12], this makes the fact that K_5 was overlooked even less excusable.
3. It was mistakenly asserted in [86, Section 12] that the facet of K_k was of type A_6.

13G A Degenerate Gosset Class

We have already observed that the polytopes G_{rs} do not have (geometric) duals. Nevertheless, we can remove the right branch from the diagram (beginning with the node V) and reverse it, to obtain the diagram of Figure 13G1. The mark 2 on the circuit indicates as before that one of the branches – but not the vertical one – is to be thought of as marked $\frac{3}{2}$.

13G1

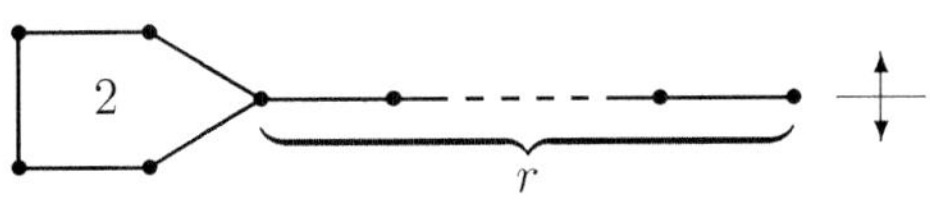

The group specified by the diagram is, again as before, $[3^{r,2,1}]$, so that we must have $r \leqslant 5$, with $r = 5$ the infinite case. We can then – potentially at least – apply an improper twist, to obtain a regular polytope H_{r+3}, where (as before) the suffix denotes the rank.

However, something rather unexpected happens, in that we do not obtain the full family as might be expected. The starting point for $r = 1$ is $\mathsf{H}_4 = \mathsf{C}_5{}^{\varpi\pi}$, the Petrial of the Petrie contraction of the 5-cube $\mathsf{C}_5 = \{4, 3, 3, 3\}$, whose 80 vertices are the mid-points of the edges of C_5. Its facets $\{6, \frac{4}{1,2} \mid 3\}$ (with 30 vertices) lie in central sections of C_5; Proposition 4D15 then implies that H_{m-1} must be a central section of H_n for $n \geqslant 5$, if the latter polytope exists. We immediately conclude – if for no other reason – that we cannot actually have an apeirotope H_8 derived from $[3^{5,2,1}]$.

13G2 Remark The diagram for H_4 is the same as that for $\mathsf{G}_{1,2}$. Indeed, the generators of the symmetry group are the same (in reverse order), except that the sign of the twist is changed. The relationship between their symmetry groups illustrates Proposition 1F1(b): that of $\mathsf{G}_{1,2}$ is $\mathbf{B}_5$, while that of $\mathbf{H}_4$ is $\mathbf{C}_5 = \mathbf{B}_5 \times \mathbf{Z}_2$.

In fact, the family has no more members, in that the putative 6-dimensional polytope H_5 actually degenerates. To see this, we choose suitable coordinates, working within the group $[3^{4,2,1}]$ as we described it in Section 1E, which ensures that everything works properly. So our diagram, now labelled, is

13G3

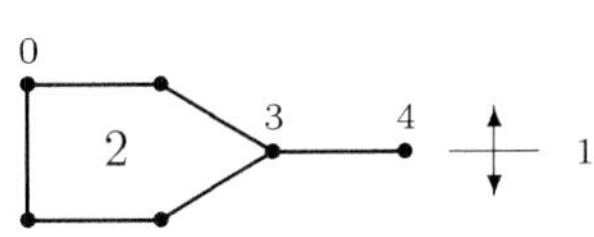

The generators of the corresponding group **G** are

13G4
$$\begin{aligned} R_0 &:= (2\ 3),\\ R_1 &:= (3\ \overline{6})(4\ \overline{5}),\\ R_2 &:= (5\ 6),\\ R_3 &:= (4\ 5),\\ R_4 &:= v^{\perp} \end{aligned}$$

with $v := (1^4, (-1)^4)$, where (as in Section 13F) R_4 is represented by its mirror. (Recall that $(3\ \overline{6})$ is the transposition that interchanges e_3 and $-e_6$.) We observe that $R_0, \ldots, R_3$ effectively act on the subspace spanned by $e_2, \ldots, e_6$.

Since the vector $e_1 + e_8$ is fixed by the whole group, we may take our points to lie in its orthogonal complement $\{(\xi_1, \ldots, \xi_8) \in \mathbb{E}^8 \mid \xi_1 + \xi_8 = 0\}$. We readily find that the Wythoff space is

$$W = \{(\alpha, -\beta, 0^4, \gamma, -\alpha) \mid \beta + \gamma = 2\alpha\}.$$

However, under the subgroup $\mathbf{H} = \langle R_0, \ldots, R_3 \rangle$ of the facet, we only obtain the ten points

$$\big(\alpha, (\pm\beta, 0^4), \gamma, -\alpha\big),$$

where we freely permute the coordinates $\xi_2, \ldots, \xi_6$. Of course, we should get the 80 vertices of H_4; since we do not, we see that H_5 degenerates, as we said.

13G5 Remark A closer inspection shows that the group is actually $[3^{3,2,1}]$; with initial vertex $(1, 2, 0^5, -1)$ we obtain 3_{21}, while starting from $(2, 0^6, -2)$ we get 2_{31}. The choice $\alpha = 0$ (and $\beta \neq 0$) gives 2_{21}.

13G6 Remark We should point out that H_7 was encountered in Section 8A (see also [82, Section 10]) as the supposed facet of the second non-existent regular polytope of full rank (its group coinciding, as we have seen, with the whole group $[3^{4,2,1}]$). In fact, we have seen that even the '5-face' H_5 was degenerate, let alone the whole 'polytope'.

IV

Miscellaneous Polytopes

14

Gosset–Elte Polytopes

We saw in the previous Chapter 13 that the Gosset–Elte polytopes play an important rôle in the theory of regular polytopes of nearly full rank; it therefore seems appropriate to collect here some facts about them. Though we make no use of them, their realization domains are also of interest, since they provide good examples of how the general theory of Chapters 3 and 4 works. In addition, some simple projections of the Gosset–Elte polytopes into the plane can reveal a lot about their structure. It must be emphasized that these projections do not display the large amount of symmetry of, for example, those into 18- or 14-gons of 3_{21}, or the projection of 4_{21} into a triacontagon. Their purpose is quite different, namely, to illustrate suitable sections, to show how components of the polytopes fit together.

We briefly discussed the Gosset–Elte polytopes $\{3^r, 3^{s,t}\}$ in general terms in Section 13C. The following sections will augment this discussion, as well as describing their realization domains, with the exception of $\{3, 3^{s,2}\}$ for $s = 3, 4$. These latter two polytopes have too many vertices to be amenable to our treatment, but in any case they do not underlie regular polytopes of nearly full rank. In Section 14C we also have the abstract difference body of $\{3^2, 3^{2,1}\}$ which occurs in the second Gosset class of Section 13E; as we shall see, this and $\{3^2, 2^{4,1}\}$ in Section 14G both pose considerable problems.

A preliminary comment is in order. We have taken the basic combinatorics of these polytopes from their 'standard' realizations as Gosset–Elte polytopes, which we briefly described in Section 13C. As abstract polytopes, of course, these combinatorics could be found from their automorphism groups.

14A Rank 6: $\{3^2, 3^{2,1}\}$

Figure 14A1 illustrates the notion of simple projections with 2_{21}; this results from projection on a plane that takes a typical ridge (which is a 4-simplex) to a point; those points marked 5 (giving the numbers of original vertices that coincide in the projection) correspond to 4-simplices $\{3, 3, 3\}$. The vertical lines faithfully represent edges. The top and bottom horizontal lines represent 5-

simplices $\{3^4\}$, while the bounding slanted lines represent 5-staurotopes in the form 3_{11}. Note that opposite ridges in a staurotope meet a simplex and another staurotope. Thus the angles of the pentagonal outline at the vertices labelled 5 faithfully represent the corresponding dihedral angles at ridges of 2_{21}; we shall see similar features in other projections. The central horizontal line represents a truncated 5-simplex 0_{31}, while the central slanted lines represent half-5-cubes 1_{21}, which are the vertex-figures at the vertices labelled 1.

14A1

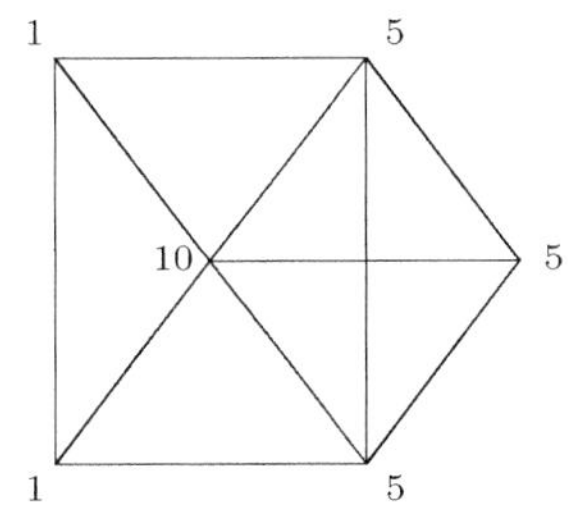

The Gosset polytope 2_{21}

Another projection of 2_{21}, more symmetric, but slightly less informative, is Figure 14A2. Here, the labels 8 indicate 4-staurotopes in the form 1_{11}, so that the three edges of the picture represent 5-staurotopes 2_{11}, while the other three edges give vertex-figures 1_{21}.

14A2

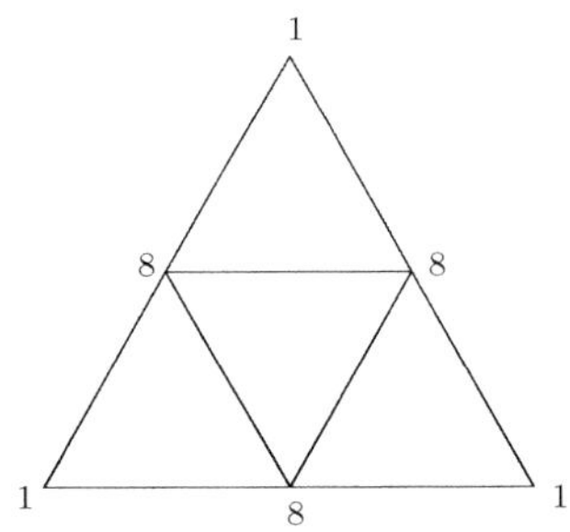

Another projection of 2_{21}

Let us remark that the neatest coordinates for the vertices of 2_{21} are complex in $\mathbb{C}^3$, namely, all even permutations of $(\omega^r, -\omega^s, 0)$, where $\omega := \frac{1}{2}(-1 + i\sqrt{3})$ and $r, s = 0, 1, 2$. Figure 14A2 (see the notes at the end of the section), if thought of as given by the projection $(\xi, \eta, \zeta) \mapsto \xi - \eta$ from $\mathbb{C}^3$ to the complex line $\mathbb{C}$ (this is orthogonal, up to a scaling factor), actually illustrates this.

The abstract polytope $\{3^2, 3^{2,1}\}$, with 27 vertices and layer vector $\Lambda = (1, 16, 10)$, has a very simple realization space. The vertices of the standard realizations of all the Gosset–Elte polytopes fall into equally spaced parallel hyperplanes (refer again to Figure 14A1 or 14A2). Hence, the cosine vector of the 6-dimensional realization $\mathsf{P}_1 = 2_{21}$ of $\{3^2, 3^{2,1}\}$ is of the form $\Gamma_1 = (1, \gamma, 2\gamma - 1)$, and from the layer equation $\langle \Gamma_1, \Lambda \rangle = 0$ there follows immediately

$\gamma = \frac{1}{4}$, so that $\Gamma_1 = (1, \frac{1}{4}, -\frac{1}{2})$. Of course, since P_1 is pure, we know from Corollary 3F6 that

$$\|\Gamma_1\|_\Lambda^2 = \tfrac{1}{27}\big(1 + 16(\tfrac{1}{4})^2 + 10(-\tfrac{1}{2})^2\big) = \tfrac{1}{6}.$$

There will be just one other non-trivial pure realization P_2, whose cosine vector Γ_2 can be calculated in two ways. First, since P_2 has dimension $d_2 = 27 - 1 - 6 = 20$, we find from the component equation Theorem 3C11 that

$$\Gamma_2 = \tfrac{1}{20}\big(27(1,0,0) - (1,1,1) - 6(1, \tfrac{1}{4}, -\tfrac{1}{2})\big) = (1, -\tfrac{1}{8}, \tfrac{1}{10}).$$

Second, $\mathsf{P}_1 \otimes \mathsf{P}_1$ will have a non-trivial component P_2 of dimension at most $\frac{1}{2}(6-1)(6+2) = 20$, which must be pure; it cannot have a component P_1, since otherwise we would need another pure component of dimension at most 14, and then we would not have enough pure components whose dimensions sum to 27. So, we calculate that

$$\Gamma_1{}^2 = (1, \tfrac{1}{16}, \tfrac{1}{4}) = \tfrac{1}{6}\Gamma_0 + \tfrac{5}{6}\Gamma_2,$$

with Γ_2 as before. Of course, we already know from Corollary 3F6 that $\|\Gamma_1\|_\Lambda^2 = \frac{1}{6}$ gives the coefficient of Γ_0.

The latter calculations also show that $d_2 = 20$, which we can verify in yet a third way from

$$\|\Gamma_2\|_\Lambda^2 = \tfrac{1}{27}\big(1 + 16(-\tfrac{1}{8})^2 + 10(\tfrac{1}{10})^2\big) = \tfrac{1}{20}.$$

Observe as well the Λ-orthogonality of Theorem 3F5, namely,

$$\langle \Gamma_1, \Gamma_2 \rangle_\Lambda = \tfrac{1}{27}\big(1 + 16(\tfrac{1}{4})(-\tfrac{1}{8}) + 10(-\tfrac{1}{2})(\tfrac{1}{10})\big) = 0,$$

so that (as we know) $\mathsf{P}_1 \otimes \mathsf{P}_2$ is centred. We also remark that

$$\Gamma_1\Gamma_2 = \tfrac{1}{4}\Gamma_1 + \tfrac{3}{4}\Gamma_2.$$

Finally, we can make an appeal to Theorem 3C14, since the diagonals of $\{3^2, 3^{2,1}\}$ are symmetric. The vertex-figure $\{3, 3^{2,1}\}$, which is a half-5-cube with layer vector $(1, 10, 5)$, has induced cosine vector $(1, \alpha, \beta)$ in terms of the general cosine vector $(1, \alpha, \beta)$ of $\{3^2, 3^{2,1}\}$. From that theorem, we solve

$$\begin{aligned} 16\alpha^2 &- 1 + 10\alpha + 5\beta \\ 0 &= 1 + 16\alpha + 10\beta, \end{aligned}$$

and eliminating β leads at once to $\alpha = \frac{1}{4}$ or $-\frac{1}{8}$. We then find the cosine vectors Γ_1 and Γ_2 as before, but now we can calculate the dimensions $d_1 = 6$ and $d_2 = 20$ from Theorem 3F5; that is, we do not need to assume that we know d_1.

14A3 Remark We have done these easy calculations in detail to illustrate the general theory; we shall employ the last method for the smaller polytopes, since it reduces the number of calculations. Subsequently, we shall often take for granted that comparable calculations can be performed as verifications of our assertions. While we often use the component equation of Theorem 3C11 we shall not otherwise verify it, nor the Λ-orthogonality properties of Theorem 3F5.

In conclusion, we have proved

14A4 Theorem *The cosine matrix of* $\{3^2, 3^{2,1}\}$ *and its associated regular polytopes of nearly full rank is*

$$\begin{bmatrix} 1 & 1 & 1 \\ 1 & \frac{1}{4} & -\frac{1}{2} \\ 1 & -\frac{1}{8} & \frac{1}{10} \end{bmatrix},$$

with layer and dimension vectors

$$\Lambda = (1, 16, 10), \quad D = (1, 6, 20).$$

14A5 Remark One kind of facet of $\{3^2, 3^{2,1}\}$ is the 5-simplex $\{3^4\}$; its layer vector and induced cosine vector are $\Lambda^{\boldsymbol{f}} = (1, 5)$ and $\Gamma^{\boldsymbol{f}} = (1, \gamma_1)$. The two pure realizations P_1 and P_2 satisfy $\eta_{\boldsymbol{f}} = \frac{1}{6}(1 + 5\gamma_1) = \frac{3}{8}$ and $\frac{1}{16}$, respectively, which tells us that the trial $\{3, 3^{2,2}\}$ must also have pure realizations of dimensions 6 and 20.

14A6 Remark We said in the preamble to the chapter that we were not particularly interested in the more highly symmetric projections of, for example, 2_{21}. However, it is, perhaps, of interest to see how we might say certain things about them, just as we did for $\{3, 4, 3\}$ and $\{3, 3, 5\}$ in Chapter 7. If we are told that the edge-graph of 2_{21} contains a regular enneagon, then we might reasonably assume that its induced cosine vector is $(1, \alpha, \alpha, \beta, \beta)$, since three steps along an enneagon gives a trigon. The three pure faithful realizations $\mathsf{Q}_k := \{\frac{9}{k}\}$ of the abstract enneagon $\{9\}$ have cosine vectors

$$\left(1, \cos\tfrac{2k\pi}{9}, \cos\tfrac{4k\pi}{9}, \cos\tfrac{6k\pi}{9}, \cos\tfrac{8k\pi}{9}\right)$$

for $k = 1, 2, 4$. The coefficient λ_k of the cosine vector Γ_k of Q_k is therefore given by

$$\begin{aligned} \lambda_k &= \tfrac{2}{9}\left(1 + 2 \cdot \tfrac{1}{4}(\cos\tfrac{2k\pi}{9} + \cos\tfrac{4k\pi}{9}) + 2 \cdot (-\tfrac{1}{2})(\cos\tfrac{6k\pi}{9} + \cos\tfrac{8k\pi}{9})\right) \\ &= \tfrac{1}{3}(1 - \cos\tfrac{8k\pi}{9}), \\ &= \tfrac{2}{3}\sin^2\left(\tfrac{4\pi}{9}\right), \end{aligned}$$

after applying $\cos\frac{2k\pi}{9} + \cos\frac{4k\pi}{9} + \cos\frac{8k\pi}{9} = 0$ (which actually follows from the layer equation for $\{9\}$). Since $\lambda_1 + \lambda_2 + \lambda_4 = 1$, these are the only three components of the enneagon (in other words, there are no components $\{1\}$ or $\{3\}$); of course, with these three components, the enneagon is already 6-dimensional.

Since $\lambda_k = \rho_k^2$, with ρ_k the radius of the corrresponding component, the three circles on which the vertices of a projection on a planar enneagon lie have radii ρ_1, ρ_4, ρ_2 in order of decreasing size. Various relationships such as

$$\sin\tfrac{8\pi}{9} = 2\sin\tfrac{4\pi}{9}\cos\tfrac{4\pi}{9} \implies \cos\tfrac{4\pi}{9}\rho_1 = \tfrac{1}{2}\rho_2$$

suggest how to construct the projection. We shall not pursue this idea any further; in any event, the projection – attractive as it may appear – gives only a vague notion of the geometry of 2_{21}.

Notes to Section 14A

1. All the drawings are only approximations. In the cases where we have trigonal or hexagonal symmetry, we have made use of the fact that $15^2 + 26^2 = 901$, so that the right-angled triangle with perpendicular sides 15 and 26 is very close to half an equilateral triangle (the difference is imperceptible at the scale of our pictures).

14B Rank 6: $\{3, 3^{2,2}\}$

We next look at the trial $\{3, 3^{2,2}\}$, whose 4-staurotopal faces are the facets of the locally toroidal 5-polytope $\{3, 3, 4, \frac{4}{1,2}\}/\langle\langle (2343)^3 \rangle\rangle$, with 3-coface the torus $\{4, \frac{4}{1,2} \mid 3\}$, which we discussed in Section 13E. In Remark 13E5, we noted that its group contains the product $[3] \times [3] \times [3]$ as a subgroup.

14B1

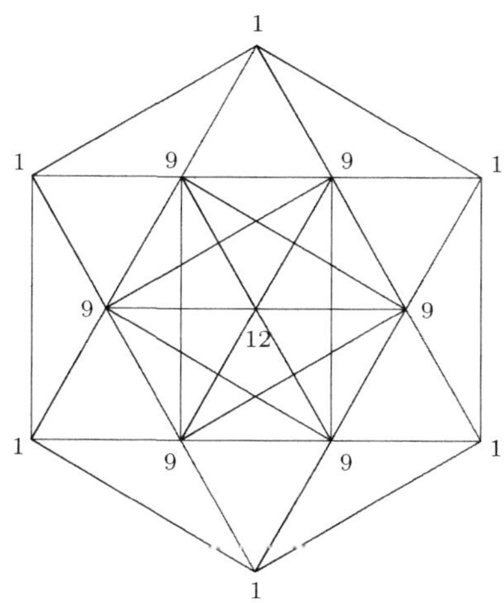

The Gosset–Elte polytope 1_{22}

Figure 14B1 (which corresponds to the triangular projection of 2_{21}) may help in visualizing part of what is happening here. This is the orthogonal projection of 1_{22} on the plane of one of its diametral hexagons. The points labelled '9' are the images of the products $\{3\} \times \{3\}$ of two trigons, while the label '12' corresponds to two more mutually orthogonal hexagons. Thus diametral hexagons of 1_{22} fall into mutually orthogonal triples. In fact, this projection points to a neat expression for the vertices of 1_{22} using complex coordinates in $\mathbb{C}^3$, analogous to those given above for vert 2_{21}. These are

$$\begin{aligned} &(\omega^r - \omega^s, 0, 0), && \text{all permutations, } r, s = 0, 1, 2, \ r \neq s, \\ &\pm(\omega^r, \omega^s, \omega^t), && r, s, t = 0, 1, 2, \end{aligned}$$

with $\omega := \frac{1}{2}(-1 + i\sqrt{3})$ as before. In particular, the vertices $(\omega^r, \omega^s, \omega^t)$ are clearly those of an inscribed torus $\{4, 3, \frac{4}{1,2} \mid 3\}$; these correspond (in the picture) to one of the two sets of three 9s forming an equilateral triangle.

There is an alternative projection of 1_{22}, analogous to the first one of 2_{21}. In Figure 14B2, the labels '5' and '10' stand for 4-simplices and truncated 4-simplices, as before. The central label '20' indicates the difference body $\Delta\mathsf{T}^4 = \mathsf{T}^4 - \mathsf{T}^4$ of a 4-simplex; the central horizontal section is similarly the difference body $\Delta\mathsf{T}^5$ of a 5-simplex. The other two horizontal sections are copies of 0_{22}. The slanting edges depict facets 1_{21}, and the slanting central sections are truncated 5-staurotopes (that is, the vertices are the mid-points of the edges of $\{3^3, 4\}$ or, perhaps better, of 2_{11}).

14B2

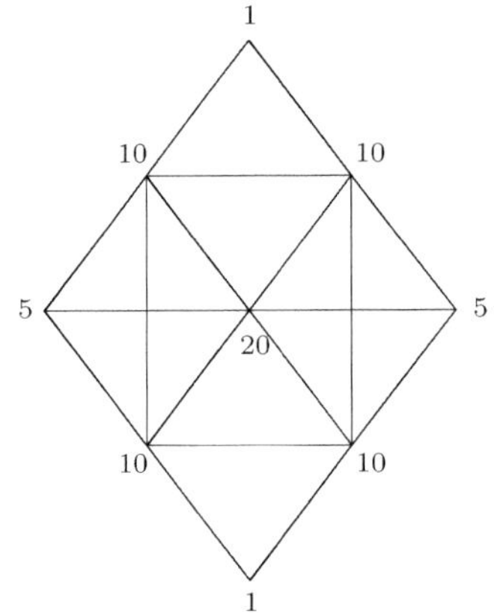

Another projection of 1_{22}

The realization space of $\{3, 3^{2,2}\}$ has a straightforward structure. The polytope is centrally symmetric, with symmetric diagonals and layer vector $\Lambda = (1, 20, 30, 20, 1)$, giving 72 vertices in all. The vertex-figure is $\{3^{2,2}\}$, with layer vector $\Lambda^v = (1, 9, 9, 1)$ and induced cosine vector $(1, \gamma_1, \gamma_2, \gamma_3)$ in terms of that of $\{3, 3^{2,2}\}$ itself.

Thus, for the components of the staurotope realization X with cosine vectors of the form $(1, \alpha, 0, -\alpha, -1)$, Theorem 3C14 implies that

$$20\alpha^2 = 1 + 9\alpha + 9 \cdot 0 + (-\alpha) = 1 + 8\alpha,$$

from which follows $\alpha = \frac{1}{2}$ or $-\frac{1}{10}$. Theorem 3F5 then yields the corresponding dimensions $d_3 = 6$ and $d_4 = 30$. Of course, the polytope corresponding to the former is the 'standard' 6-dimensional realization $\mathsf{P}_3 = 1_{22}$, whose cosine vector $\Gamma_3 = (1, \frac{1}{2}, 0, -\frac{1}{2}, -1)$ we could have worked out by hand.

For the components of the small simplex realization S, with cosine vectors of the form $(1, \alpha, \beta, \alpha, 1)$, Theorem 3C14 similarly tells us that

$$\begin{aligned} 20\alpha^2 &= 1 + 9\alpha + 9\beta + \alpha = 1 + 10\alpha + 9\beta, \\ 0 &= 1 + 20\alpha + 15\beta, \end{aligned}$$

the latter being the layer equation of Theorem 3C7. Eliminating β leads to $\alpha = -\frac{1}{5}$ or $\frac{1}{10}$, with corresponding dimensions $d_1 = 15$ and $d_2 = 20$.

Summarizing this discussion, we have shown

14B3 Theorem *The polytope* $\{3, 3^{2,2}\}$ *and associated locally toroidal regular polytopes of nearly full rank have cosine matrix*

$$\begin{bmatrix} 1 & 1 & 1 & 1 & 1 \\ 1 & -\frac{1}{5} & \frac{1}{5} & -\frac{1}{5} & 1 \\ 1 & \frac{1}{10} & -\frac{1}{5} & \frac{1}{10} & 1 \\ 1 & \frac{1}{2} & 0 & -\frac{1}{2} & -1 \\ 1 & -\frac{1}{10} & 0 & \frac{1}{10} & -1 \end{bmatrix},$$

with layer and dimension vectors

$$\Lambda = (1, 20, 30, 20, 1), \quad D = (1, 15, 20, 6, 30).$$

14B4 Remark The induced cosine vectors show that both the ridges and facets of P_1 of type $\{3, 3^{k,1}\}$ (for $k = 1, 2$) are central; the corresponding facets of P_4 are also central. This is just as Theorem 4D6 implies, bearing in mind triality: $\{3^2, 3^{2,1}\}$ does not have pure realizations of dimensions 15 or 30.

14B5 Remark We have not used the fact that ${\Gamma_3}^2 = \frac{1}{6}\Gamma_0 + \frac{5}{6}\Gamma_2$ would have to yield a pure realization P_2 of dimension at most $\frac{1}{2}(6-1)(6+2) = 20$. Note also that $\Gamma_4 = \Gamma_1\Gamma_3$, so that $\mathsf{P}_4 = \mathsf{P}_1 \otimes \mathsf{P}_3$.

14C Rank 6: $\{3, 3^{2,2}\}^*$

The notation of the section heading is designed to indicate that the polytope $\mathcal{P} := \{3, 3^{2,2}\}^*$ under discussion here is a kind of dual to $\{3, 3^{2,2}\}$ (this notation $\mathcal{P}$ is retained for the rest of the section). More exactly, just as $\{3, 3^{2,2}\}$ underlies the regular polytope $\{3, 3, 4, 4\}/\langle\langle (2343)^3 \rangle\rangle$ of rank 5, so $\{3, 3^{2,2}\}^*$ underlies its dual $\{4, 4, 3, 3\}/\langle\langle (0121)^3 \rangle\rangle$. It is also useful to recall that one view of realization theory was that of a representation of a group acting on the cosets of a subgroup; in this case, we have $[3^{2,2,1}]$ acting on $[3^{1,1,1}]$.

We could get by with relatively little information about the automorphism group of the polytopes in the previous two sections. However, here we do need to go into it in some detail. As we did in Section 1E, we actually embed our group in a larger one; we refer here to the diagram of Figure 1E17 for the groups $\mathbf{E}_d$ and $\mathbf{T}_{d+1}$. We thus work in the subspace $\{x \in \mathbb{L}^7 \mid \xi_1 + \xi_8 = 0\}$ of the familiar subspace $\mathbb{L}^7$ of (1E15). In terms of permutations of coordinates, the generators of the symmetry group $\mathbf{G}$ of the convex polytope P_5 which underlies

$\{4, \frac{4}{1,2}, 3, 3\}/\langle\langle(0121)^3\rangle\rangle$ are

$$\begin{aligned} R_0 &:= (2\ 3), \\ R_1 &:= -(1\ 8)(2\ 7)(3\ 6)(4\ 5), \\ R_2 &:= (5\ 6), \\ R_3 &:= (4\ 5), \\ R_4 &:= v^\perp, \end{aligned}$$

where we represent R_4 by its mirror with $v := (1^4, (-1)^4)$ (as a unit vector normal to the mirror, we should divide v by $\sqrt{8}$). As we have done before, the minus sign attached to R_1 indicates changing signs of all coordinates, in addition to the given permutation. Note that $R_1 R_0 R_1 = (6\ 7)$ and $R_1 R_3 R_1 = (3\ 4)$, so that we do have the whole symmetry group of P_5 here. The suffix of P_5 is chosen with the ultimate result in mind; the corresponding dimension is, of course, $d_5 = 6$.

14C1 Remark Since $\mathcal{P}$ is just the abstract polytope corresponding to P_5, so far as the combinatorics of $\mathcal{P}$ are concerned we may safely work with P_5.

A little work, some of which we leave to the interested reader, shows that we obtain 270 vertices, which are all allowable permutations in their coordinates of vectors of the form $\big((\pm 2)^4, 0^4\big)$ and $\big(\pm 3, (\pm 1)^7\big)$ (that is, the vectors must lie in $\mathbb{L}^7$ and satisfy $\xi_1 + \xi_8 = 0$).

14C2 Remark All the vectors of this form, with the restriction that there is an odd number of minus signs in the vectors $\big(\pm 3, (\pm 1)^7\big)$, and with the addition of the scaled basis vectors $\pm 4e_j$ for $j = 1, \ldots, 8$, comprise the vertex-set of 2_{41}, which we shall treat in the last section of the chapter.

In fact, it may be seen that $\mathsf{P}_5 = \Delta\mathsf{Q}$, the difference body of the Gosset polytope $\mathsf{Q} := 2_{21}$; the vertices of $\mathsf{P}_5 = \mathsf{Q} - \mathsf{Q}$ are (uniquely) of the form $x - y$, where $x \in \operatorname{vert}\mathsf{Q}$ and $y \in \operatorname{vert}\mathsf{G}$, with G the facet of Q opposite x, giving the required $27 \cdot 10 = 270$.

However, we do need the layer vector Λ, and for this we must look at the effect of $\mathbf{G}$ in some detail.

14C3 Lemma *The layer vector of the polytope $\mathcal{P}$ is*

$$\Lambda = (1, 16, 32, 2, 64, 32^*, 8, 64, 2, 32, 16, 1),$$

where as usual the asterisk denotes an asymmetric diagonal class.

Proof. Assuming, for the moment, that this is the case, then the cosine vector of the (standard) realization $\mathsf{P}_5 = \Delta 2_{21}$ is

$$\Gamma_5 = (1, \tfrac{3}{4}, \tfrac{1}{2}, \tfrac{1}{2}, \tfrac{1}{4}, 0, 0, -\tfrac{1}{4}, -\tfrac{1}{2}, -\tfrac{1}{2}, -\tfrac{3}{4}, -1).$$

Figure 14C4 depicts a simple projection of P_5; this illustrates much of what we require, though not quite all. The twist of the diagram $[3^{2,2,1}]$ of $\mathbf{E}_6$ is the reflexion R_1, which flips the picture from top to bottom, if we take the left-most point (labelled '1') as the initial vertex $(2, -2, 0^4, 2, -2)$. There are two things to explain: how the split of the 34 vertices at height $\frac{1}{2}$ (and similarly $-\frac{1}{2}$) into $32 + 2$ arises, and why the 40 vertices at height 0 split into an asymmetric class of 32 and a symmetric class of 8. An important fact to bear in mind in the whole discusion is that the subgroup $\mathbf{G}_0$ fixing the initial vertex contains all permutations of coordinates $\xi_3, \ldots, \xi_6$; our listings will assume this without further comment.

14C4

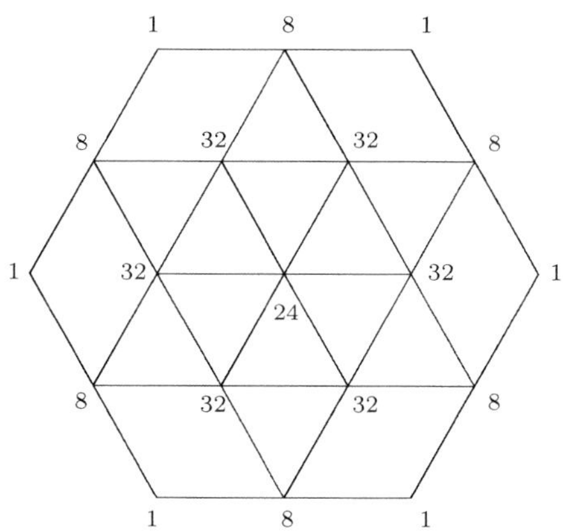

The polytope $\Delta 2_{21}$

The first problem is straightforward. The two vertices $(1, -3, 1^4, -1, -1)$ and $(1, 1, (-1)^4, 3, -1)$ at height $\frac{1}{2}$ are interchanged by the twist R_1 and fixed by each of R_2, R_3, R_4. The remaining 32 vertices at height $\frac{1}{2}$ are those of a (combinatorial) truncated 4-cube.

At height 0, we may first check that the set of eight vertices

$$\perp(2, 2, 0^4, -2, -2), \ (0, 0, (2, 2, -2, -2), 0, 0)$$

is closed under $\mathbf{G}_0$. The remaining 32 vertices fall into two subsets of 16 each under $\mathbf{G}_0$:

$$\pm (1, 1, \pm(-3, 1^3), -1, -1),$$
$$\pm(0, 2, (-2, -2, 0, 0), 2, 0), \ \pm (-1, -3, 1^4, -1, 1), \ \pm(1, -1, 1^4, -3, -1).$$

However, we see that

$$\begin{aligned}(2, \quad 2, 0^4, 2, -2) &\mapsto (2, 2, 0^4, -2, -2)\\ &\mapsto (0, 0, 2, 2, -2, -2, 0, 0)\\ &\mapsto (0, 2, 0, 0, -2, -2, 2, 0)\end{aligned}$$

in the second set of 16, where we successively apply $(2\ 7)$, then R_4 and end with

(2 3)(4 7). If we apply these operations in the reverse order, we have

$$\begin{aligned}(2,-2,0^4,2,-2) &\mapsto (2,0,-2,2,0,0,0,-2)\\ &\mapsto (1,-1,-3,1,1,1,1,-1)\\ &\mapsto (1,1,-3,1,1,1,-1,-1)\end{aligned}$$

in the first set of 16; that is, these 32 vertices fall into a single asymmetric layer, as claimed. □

Because cosine vectors here have 12 coordinates, we shall usually write them in abbreviated form. So, for instance, we only need $\gamma_0,\ldots,\gamma_4$ for components of the staurotope realization X, and only $\gamma_0,\ldots,\gamma_6$ for those of the small simplex realization S; moreover, for the latter, we have layer vector $(1,16,32,2,64,16,4)$, to avoid double counting of γ_5 and γ_6.

The first component of our realization domain $\mathcal{N}$ is somewhat degenerate. Suppose that we drop the minus sign in the reflexion R_1; that is, we apply $\boldsymbol{\zeta}_1$ in the notation of Section 5D. We thus set $S_1 := (1\ 8)(2\ 7)(3\ 6)(4\ 5) = -R_1$, $S_j = R_j$ for $j = 0,2,3,4$; in other words, if $Z = -I$ is the central inversion, then we perform the mixing operation

$$(R_0,\ldots,R_4,Z) \mapsto (R_0,R_1Z,R_2,R_3,R_4) =: (S_0,\ldots,S_4).$$

This has the effect of taking the whole group $[3^{2,2,1}] \rtimes \mathbf{Z}_2$ onto its subgroup $[3^{2,2,1}]$. We see that the initial vertex can be taken to be $(3,5,(-1)^5,-3)$, and we obtain all permutations in coordinates $\xi_2,\ldots,\xi_7$ of

$$(\pm 3,5,(-1)^5,\mp 3),\quad (0,4^2,(-2)^4,0),$$

which we can recognize as the 27 vertices of the Gosset polytope 2_{21}. We label this realization P_1, and calculate that its cosine vector is

$$\Gamma_1 = (1,\tfrac14,-\tfrac12,-\tfrac12,\tfrac14,-\tfrac12,1)$$

as a component of the small simplex realization S.

The next observation is crucial to the description.

14C5 Lemma *The vertices of* P_5 *fall into* 45 *disjoint vertex-sets of diametral regular hexagons, which are permuted by* $\mathbf{G}$. *The edges of these hexagons are diameters of staurotopal facets of* P_5. *The stabilizer of the hexagon containing the initial vertex has* $\mathbf{G}_0$ *as a subgroup.*

Note particularly that each vertex of P_5 belongs to a unique such hexagon. For instance, that containing the initial vertex has vertices

$$\pm(2,-2,0^4,2,-2),\ \pm(1,-3,1^4,-1,-1),\ \pm(1,1,(-1)^4,3,-1).$$

As a consequence, we can identify opposite vertices of each of these hexagons to obtain $\mathcal{P}/2$, alternate ones to obtain $\mathcal{P}/3$, and all of them to obtain $\mathcal{P}/6$. These identifications are key to breaking into the realization domain.

Therefore, we tackle $\mathcal{P}/6$ first. This has just two non-trivial diagonal classes, both necessarily symmetric; its intrinsic layer vector is $\Lambda_6 = (1, 12, 32)$, and the embedding of a general cosine vector is

$$(1, \alpha, \beta) \hookrightarrow (1, \alpha, \beta, 1, \beta, \alpha, \alpha)$$

as a component of S. This may be seen from Figure 14C4. As we have said, an edge of a hexagon is a diameter of a staurotopal facet $\{3, 3, 3, 4\}$ (actually in the form $\{3, 3, 3^{1,1}\}$), so that one hexagon – the initial one – is the peripheral one in the picture. Four others use two opposite sets of eight vertices on the boundary and eight of the 24 vertices at the centre; in the picture, these project into line segments. Finally, there are 32 using one vertex from each of the six sets of 32 vertices in the figure. Moreover, since edges and non-edges are inherited from $\mathcal{P}$ itself, we see that the initial hexagon is joined by an edge to each of the 12 which project to segments, and is not joined to the other 32.

We intend to apply Theorem 3C14 to the vertex-figure. The vertices of this consist of the 12 hexagons which project to segments; each of these is joined by an edge to the three others which project to the same segment, and is not joined to the other eight. In other words, the induced layer vector is $(1, 3, 8)$ and the induced cosine vector of the vertex-figure is $(1, \alpha, \beta)$. Theorem 3C14 now takes the form

$$12\alpha^2 = 1 + 3\alpha + 8\beta,$$

and eliminating β using the layer equation

$$1 + 12\alpha + 32\beta = 0$$

results in

$$48\alpha^2 = 3 \implies \alpha = \pm\tfrac{1}{4}.$$

This gives the two solutions

$$(1, \alpha, \beta) = (1, \tfrac{1}{4}, -\tfrac{1}{8}) \text{ or } (1, -\tfrac{1}{4}, \tfrac{1}{16}).$$

In conclusion, the cosine vectors of pure realizations of $\mathcal{P}/6$ (as components of S) are

$$\begin{aligned} \Gamma_u &= (1, \tfrac{1}{4}, -\tfrac{1}{8}, 1, -\tfrac{1}{8}, \tfrac{1}{4}, \tfrac{1}{4}), \\ \Gamma_3 &= (1, -\tfrac{1}{4}, \tfrac{1}{16}, 1, \tfrac{1}{16}, -\tfrac{1}{4}, -\tfrac{1}{4}); \end{aligned}$$

further, we can calculate the corresponding dimensions using Theorem 3F5 as $d_u = 20$ and $d_3 = 24$, and verify that Γ_u and Γ_3 are Λ-orthogonal. The notation Γ_u is strictly temporary.

We next consider the realizations $\mathsf{P} \otimes \mathsf{P}$, with $\mathsf{P} = \mathsf{P}_1$ and P_5. Their cosine vectors are

$$\begin{aligned} \Gamma_1^2 &= \tfrac{1}{6}\Gamma_0 + \tfrac{5}{6}\Gamma_v, \\ \Gamma_5^2 &= \tfrac{1}{6}\Gamma_0 + \tfrac{5}{6}\Gamma_w, \end{aligned}$$

with

$$\begin{aligned}\Gamma_v &= (1, -\tfrac{1}{8}, \tfrac{1}{10}, \tfrac{1}{10}, -\tfrac{1}{8}, \tfrac{1}{10}, 1),\\ \Gamma_w &= (1, \tfrac{19}{40}, \tfrac{1}{10}, \tfrac{1}{10}, -\tfrac{1}{8}, -\tfrac{1}{5}, -\tfrac{1}{5});\end{aligned}$$

again, the notation is temporary. We calculate their dimensions as 20, which is what we should expect. However, neither Γ_v nor Γ_w is Λ-orthogonal to Γ_u or to each other; indeed, we have

$$\langle \Gamma_u, \Gamma_v \rangle_\Lambda = \langle \Gamma_u, \Gamma_w \rangle_\Lambda = \tfrac{1}{50},$$

a fact which will prove very useful.

We see from this that we must have here an essential Wythoff space W^* of dimension at least 2. Actually, it cannot be more than 2-dimensional, because it would then contribute at least $6 = \binom{4}{2}$ to the diagonal count $r = 6$, and we already have 2. Assuming that each of Γ_v and Γ_w is also of the form $\Gamma(\vartheta)$ for some angle ϑ in the notation of Remark 3H12, we use that remark as a guide to finding the corresponding $\Gamma_m, \Gamma_c, \Gamma_s$. We want Γ_u to play the rôle of $\Gamma_m = \Gamma_{2,11}$ in Theorem 3H3, so that the cosine vector $\Gamma_{2,22}$ will be an affine combination of $\Gamma_u, \Gamma_v, \Gamma_w$ with equal coefficients for Γ_v and Γ_w in view of $\langle \Gamma_u, \Gamma_v \rangle_\Lambda = \langle \Gamma_u, \Gamma_v \rangle_\Lambda$, and so of the form $\Gamma = (1 - 2\lambda)\Gamma_u + \lambda(\Gamma_v + \Gamma_w)$ for some λ. We thus require that

$$0 = \langle \Gamma_u, \Gamma \rangle_\Lambda = (1 - 2\lambda)\tfrac{1}{20} + 2\lambda\tfrac{1}{50} \implies \lambda = \tfrac{5}{6}.$$

A little calculation, whose details we suppress, then leads to

$$\Gamma_{2,22} = \bigl(1, \tfrac{1}{8}, \tfrac{1}{4}, -\tfrac{1}{2}, -\tfrac{1}{8}, -\tfrac{1}{4}, \tfrac{1}{2}\bigr),$$

from which we deduce that

$$\begin{aligned}\Gamma_m &:= \tfrac{1}{2}\bigl(\Gamma_{2,11} + \Gamma_{2,22}\bigr) = (1, \tfrac{3}{16}, \tfrac{1}{16}, \tfrac{1}{4}, -\tfrac{1}{8}, 0, \tfrac{3}{8}),\\ \Gamma_c &:= \tfrac{1}{2}\bigl(\Gamma_{2,11} - \Gamma_{2,22}\bigr) = (0, \tfrac{1}{16}, -\tfrac{3}{16}, \tfrac{3}{4}, 0, \tfrac{1}{4}, -\tfrac{1}{8}).\end{aligned}$$

Finally, Γ_s must be a multiple of $\Gamma_v - \Gamma_w$ with $\|\Gamma_s\|_\Lambda^2 = \frac{1}{40}$, and a little work then yields

$$\Gamma_s := \bigl(0, -\tfrac{\sqrt{6}}{8}, 0, 0, 0, \tfrac{\sqrt{6}}{16}, \tfrac{\sqrt{6}}{4}\bigr).$$

The general cosine vector for this subfamily of pure realizations with dimension $d_2 = 20$ is thus

$$\Gamma(\vartheta) = \Gamma_m + \cos 2\vartheta\, \Gamma_c + \sin 2\vartheta\, \Gamma_s,$$

with $0 \leqslant \vartheta < \pi$ and $\Gamma_m, \Gamma_c, \Gamma_s$ as above. Of course, $\Gamma_{2,12} = \Gamma_s$ in Remark 3H12; naturally, we could also write the general cosine vector as

$$\Gamma_2(x) = \xi_1^2\Gamma_{2,11} + 2\xi_1\xi_2\Gamma_{2,12} + \xi_2^2\Gamma_{2,22},$$

with $\Gamma_{2,11} = \Gamma_u$ and $x = (\xi_1, \xi_2)$ such that $\xi_1^2 + \xi_2^2 = 1$.

14C6 Remark We can calculate that

$$\Gamma_v, \Gamma_w = \tfrac{2}{5}\Gamma_{2,11} \pm \tfrac{2\sqrt{6}}{5}\Gamma_{2,12} + \tfrac{3}{5}\Gamma_{2,22},$$

which illustrates Theorem 3H8 in the case rank $A = 1$.

There are many ways of completing the description of the realization domain of $\mathcal{P}/2$; since the number of diagonal classes still to account for is $6-1-3-1 = 1$, with corresponding dimension

$$135 - 1 - 6 - 2 \cdot 20 - 24 = 64,$$

the easiest is to use the component equation of Theorem 3C11. Thus we find that our last cosine vector Γ_4 is given by

$$\Gamma_0 + 6\Gamma_1 + 20\Gamma_{2,11} + 20\Gamma_{2,22} + 24\Gamma_3 + 64\Gamma_4 = 135(1, 0^6)$$
$$\implies \Gamma_4 = \big(1, -\tfrac{1}{16}, -\tfrac{1}{32}, -\tfrac{1}{2}, \tfrac{1}{64}, \tfrac{1}{8}, -\tfrac{1}{4}\big).$$

It is straightforward, but tedious, to check various properties that Γ_4 must have, for example, Λ-orthogonality with respect to the previous cosine vectors $\Gamma_1, \Gamma_{2,jk}$ and Γ_3, dimension $d_4 = 64$ given by $\|\Gamma_4\|_\Lambda^2 = \frac{1}{64}$, and so on. As a last confirmation that we have a genuine cosine vector of $\mathcal{P}/2$, observe that $\Gamma_4 = \Gamma_1\Gamma_3$.

We now move on to the components of the staurotope realization X. We begin by looking at the Λ-orthogonal complement of $\mathcal{P}/6$ in $\mathcal{P}/3$. So, let $\Gamma = (1, \gamma_1, \gamma_2, \gamma_3, \gamma_4)$ be the (truncated) cosine vector of a centrally symmetric $\mathsf{P} \in \mathcal{P}/3$. Then the following hold.

- Alternate vertices of diametral hexagons become antipodal in $\mathcal{P}/3$; hence $\gamma_3 = -1$.
- Vertices of vertex-figures are equidistant from antipodal vertices; hence $\gamma_1 = 0$.
- From $\langle \Gamma_5, \Gamma \rangle_\Lambda = 0$, we have

  $$0 = 1 + 32 \cdot \tfrac{1}{2} \cdot \gamma_2 + 2 \cdot \tfrac{1}{2} \cdot (-1) + 64 \cdot \tfrac{1}{4} \cdot \gamma_4 = 16(\gamma_2 + \gamma_4),$$

  and so $\gamma_2 + \gamma_4 = 0$.

In other words, $\Gamma = (1, 0, \alpha, -1, -\alpha)$ for some α. In particular, the contribution of $\mathcal{P}/3$ to $135(1, 0^4)$ for X will be $45(1, 0, 0, -1, 0)$, with (of course) contribution 45 to the dimension.

Next,

$$\Gamma_9 := \Gamma_3\Gamma_5 = \big(1, -\tfrac{3}{16}, \tfrac{1}{32}, \tfrac{1}{2}, \tfrac{1}{64}\big)$$

can be checked to be Λ-orthogonal to Γ_5 and $\Gamma_* := (1, 0, 0, -1, 0)$. Then we have $\dim \mathsf{P}_9 \geqslant \|\Gamma_9\|_\Lambda^{-2} = 64$ from Theorem 3F5, giving dimension $d_9 \geqslant 64$.

However, with $135 - 6 - 45 - 64 = 20$ in mind, we then find Γ_7 with

$$6\Gamma_5 + 20\Gamma_7 + 45\Gamma_* + 64\Gamma_9 = 135(1, 0^4) \implies \Gamma_7 = \big(1, \tfrac{3}{8}, -\tfrac{1}{4}, \tfrac{1}{2}, -\tfrac{1}{8}\big),$$

and verify that Γ_7 is Λ-orthogonal to each of Γ_5, Γ_9 and Γ_*. Since these last are all cosine vectors of genuine realizations, Theorem 3F10 implies that Γ_7 also corresponds to a realization P_7. Moreover, we obtain $d_7 = 20$ as required, and this tells us that $\dim \mathsf{P}_9 = 64 = d_9$.

Finally, $\mathsf{P}_1 \otimes \mathsf{P}_5$, with cosine vector $\Gamma_1\Gamma_5$, has dimension at most $6 \cdot 6 = 36$; we can verify that $\Gamma_1\Gamma_5$ is Λ-orthogonal to Γ_5 and Γ_9, but that

$$20\langle \Gamma_1\Gamma_5, \Gamma_7\rangle_\Lambda = \tfrac{1}{2}.$$

In other words, there is a $\mathsf{P}_6 \in \mathcal{P}$ of dimension at most $36 - 20 = 16$, whose cosine vector Γ_6 satisfies

$$\Gamma_1\Gamma_5 = \tfrac{1}{2}\Gamma_6 + \tfrac{1}{2}\Gamma_7 \implies \Gamma_6 = \bigl(1, 0, -\tfrac{1}{4}, -1, \tfrac{1}{4}\bigr).$$

We know that $\mathsf{P}_6 \in \mathcal{P}/3$, if only because $\langle \Gamma_1\Gamma_5, \Gamma_6\rangle \neq 0$; the form of the cosine vector Γ_6 confirms this. Now P_6 must be pure, since $\mathcal{P}/3$ has only two pure centrally symmetric components; we calculate its dimension to be $d_6 = 15$. The last pure component P_8 thus has dimension $d_8 = 45 - 15 = 30$, with cosine vector Γ_8 given by

$$\tfrac{1}{3}\Gamma_6 + \tfrac{2}{3}\Gamma_8 = (1, 0, 0, -1, 0) \implies \Gamma_8 = \bigl(1, 0, \tfrac{1}{8}, -1, -\tfrac{1}{8}\bigr).$$

This concludes the description. We shall not write out the 12×12 cosine matrix of $\mathcal{P}$ since it is rather large, instead leaving the task to the interested reader.

14C7 Remark It is worth mentioning here that we could have found the realization P_6 of $\mathcal{P}/3$ using the alternating product of Section 4E, since $\mathsf{P}_6 = \mathsf{P}_1 \wedge \mathsf{P}_5$; note that $d_6 = 15 = \binom{6}{2}$ is as to be expected. This is just the kind of case discussed in Section 4E; the symmetry groups of P_1 and P_5 only differ in the sign of the twist R_1 or $S_1 = -R_1$ (see the notes at the end of the section). Indeed, it is tedious but by no means very hard to carry out the calculations in practice, even though we start in a 6-dimensional subspace of $\mathbb{E}^8$, and so have to end up working in a 15-dimensional subspace of $\mathbb{E}^{\binom{8}{2}} = \mathbb{E}^{28}$. For example, we have typical pairs of vertices (the first are the initial ones)

$$\begin{gathered}(0, 4, -2, -2, -2, -2, 4, 0) \wedge (2, -2, 0, 0, 0, 0, 2, -2)\\ (3, 1, 1, 1, -5, 1, 1, -3) \wedge (1, -1, -1, -1, 1, -1, 3, -1)\\ (0, 4, -2, -2, -2, -2, 4, 0) \wedge (0, 0, 2, 2, -2, -2, 0, 0)\end{gathered}$$

which can be expressed as antisymmetric matrices with row sums 0, and first and last rows equal but opposite in sign; once again, we leave the details to the interested reader (see the notes at the end of the section). These will also illustrate the fact that P_1 has only 27 vertices, rather than the 270 of P_5. Finally, observe that there are 60 vertices of the first (or second) kind, and 30 of the third, giving the required 90 of $\mathcal{P}/3$ in all.

Notes to Section 14C

1. Further to Remark 14C7, we see that the general tensor product of the symmetry groups of P_1 and P_5 (with opposite signs of the twist) splits into an antisymmetric component of degree 15, a symmetric component of degree 20, and a 1-dimensional component (representing the central symmetry $-I$).
2. The coordinates of the alternating product above are multiples of 4, and so can be simplified by multiplying by $\frac{1}{4}$.

14D Rank 7: $\{3^3, 3^{2,1}\}$

We now move on to rank 7, starting with $\mathcal{P} = \{3^3, 3^{2,1}\}$. This is centrally symmetric, with symmetric diagonals and layer vector $\Lambda = (1, 27, 27, 1)$, and so with 56 vertices. The vertex-figure is $\{3^2, 2^{2,1}\}$ with layer vector $\Gamma^v = (1, 16, 10)$ and induced cosine vector $\Gamma^v = (1, \gamma_1, \gamma_2)$ in terms of that of $\mathcal{P}$ itself. Our starting point, the standard realization $\mathsf{P}_2 = 3_{21}$ of dimension $d_2 = 7$, has cosine vector $\Gamma_2 = (1, \frac{1}{3}, -\frac{1}{3}, -1)$, which follows from the even spacing mentioned previously, as well as from the next calculations.

For the components of the staurotope realization X, whose cosine vectors are of the form $(1, \alpha, -\alpha, -1)$, we appeal once more to Theorem 3C14. We therefore solve

$$27\alpha^2 = 1 + 16\alpha + 10(-\alpha) = 1 + 6\alpha,$$

whose solutions are $\alpha = \frac{1}{3}$ and $-\frac{1}{9}$. The corresponding dimensions are the already known $d_2 = 7$, and $d_3 = 21$.

Since $\mathcal{P}$ is centrally symmetric, it has the small simplex realization S, with antipodal vertices identified in pairs. But every vertex of $\mathcal{P}/2$ is joined to every other by an edge; thus P_1 with $d_1 = 27$ and cosine vector $\Gamma_1 = (1, -\frac{1}{27}, -\frac{1}{27}, 1)$ is the sole non-trivial component of S.

In conclusion, then, we have

14D1 Theorem *The cosine matrix of* $\{3^3, 3^{2,1}\}$ *and its associated regular polytopes of nearly full rank is*

$$\begin{bmatrix} 1 & 1 & 1 & 1 \\ 1 & -\frac{1}{27} & -\frac{1}{27} & 1 \\ 1 & \frac{1}{3} & -\frac{1}{3} & -1 \\ 1 & -\frac{1}{9} & \frac{1}{9} & -1 \end{bmatrix},$$

with layer and dimension vectors

$$\Lambda = (1, 27, 27, 1), \quad D - (1, 27, 7, 21).$$

14D2 Remark We can obtain P_1 in another way, as the non-trivial component of $\mathsf{P}_2 \otimes \mathsf{P}_2$, whose dimension is $\frac{1}{2}(7-1)(7+2) = 27$, as required. Observe that

$$\Gamma_2{}^2 = (1, \tfrac{1}{9}, \tfrac{1}{9}, 1) = \tfrac{1}{7}\Gamma_0 + \tfrac{6}{7}\Gamma_1,$$

with calculations similar to those above; the coefficient $\frac{1}{7}$ is, of course, $\|\Gamma_2\|_\Lambda^2$. Moreover, $\Gamma_1 = \Gamma_2\Gamma_3$, so that $\mathsf{P}_1 = \mathsf{P}_2 \otimes \mathsf{P}_3$.

The first projection in Figure 14D3 is that of the standard 7-dimensional $3_{21} \in \{3^3, 3^{2,1}\}$. The points labelled '10' are images of 5-staurotopes, strictly speaking in the form 2_{11}, while those labelled '16' are half-5-cubes 1_{21}. Thus the top and bottom sets of 12 points correspond to the vertices of 6-staurotopes 3_{11}, which are facets of 3_{21}. The slanted lines, each containing 27 points, give the vertex-figures 2_{21} at the vertices labelled '1'; observe the layer vector $\Lambda(2_{21}) = (1, 16, 10)$. Incidentally, the ratio between the horizontal and vertical dimensions of the projection is $\sqrt{2} : 1$.

14D3

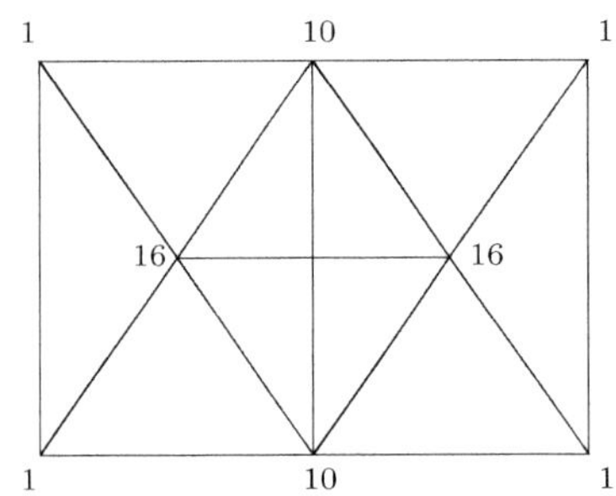

A projection of 3_{21}

In Figure 14D4, we next have an alternative projection of 3_{21}; this is drawn to the same scale. It shows less, and we only put it in because it corresponds to the projection of 2_{31} in Figure 14E3. The points labelled '6' are ridges, which are 5-simplices $\mathsf{T}^5 = \{3^4\}$, while the point marked '20' at the centre is a copy of $0_{22} = \{3^{2,2}\}$.

14D4

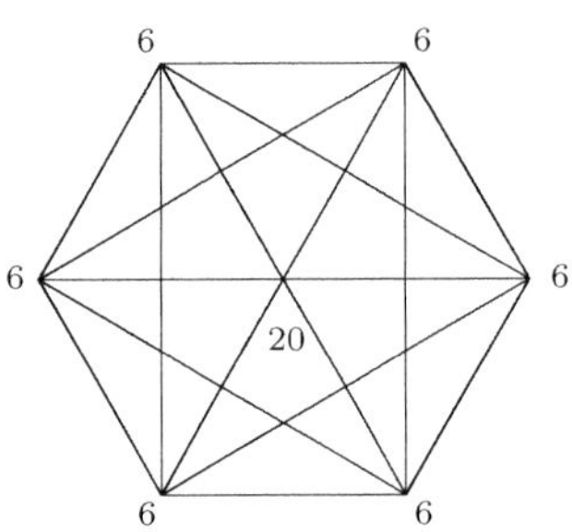

Another projection of 3_{21}

Even though we have no coordinates for the vertices of the 21-dimensional realization P (say) of $\{3^3, 3^{2,1}\}$, nevertheless we have enough information to draw a simple planar projection of it. We can argue as follows. First, from the induced cosine vector, its vertex-figure P^v is the 20-dimensional realization of $\{3^2, 3^{2,1}\}$. It follows that we will have a planar projection in which P^v projects into a line-segment. Moreover, we know both the cosine vectors $(1, -\frac{1}{8}, \frac{1}{10})$ of

P^v and $(1, -\frac{1}{9}, \frac{1}{9}, -1)$ of P itself. And, finally, the centroid of each vertex-figure must lie on the line joining the initial vertex to the antipodal one. Since (in the projection) the initial vertices of P^v lie on the circle with diameter joining these latter opposite vertices, the whole projection is fixed; it is given in Figure 14D5. The ratio between the horizontal and vertical dimensions of this projection is $2 : \sqrt{5}$.

14D5

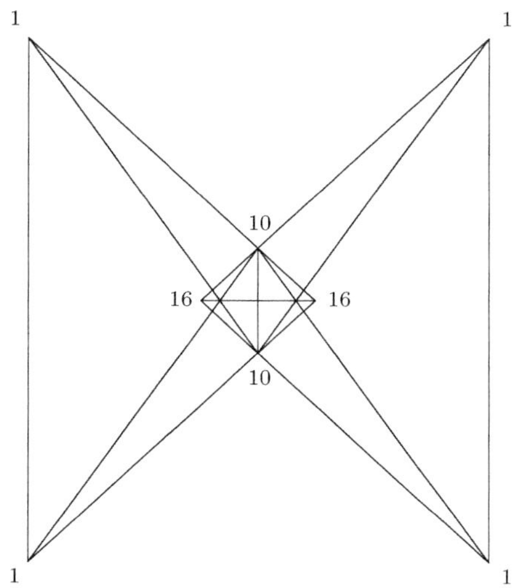

The 21-dimensional realization of $\{3^3, 3^{2,1}\}$

We identified two staurotopal facets in Figure 14D3. In Figure 14D5, one of them has two vertices labelled '1' at top left and right, joined to the lower 5-staurotope labelled '10'. Thus the 6-staurotope is not pure (nor should we expect it to be); instead, it is a blend $\{3^4, 4\} \# \{3^4\}$ of the usual staurotope with its facet.

14E Rank 7: $\{3^2, 3^{3,1}\}$

We next consider $\{3^2, 3^{3,1}\}$. This is centrally symmetric, with all diagonals symmetric and layer vector $\Lambda = (1, 32, 60, 32, 1)$, and so with 126 vertices in all. The vertex-figure $\{3, 3^{3,1}\}$ is the half-6-cube, with layer vector $\Lambda^v = (1, 15, 15, 1)$ and induced cosine vector $\Gamma^v = (1, \gamma_1, \gamma_2, \gamma_3)$. Once again, we shall appeal to Theorem 3C14 to find the pure realizations.

For the components of the small simplex realization S, with cosine vectors of the form $(1, \alpha, \beta, \alpha, 1)$, we solve

$$\begin{aligned} 32\alpha^2 &= 1 + 15\alpha + 15\beta + \alpha = 1 + 16\alpha + 15\beta, \\ 0 &= 1 + 32\alpha + 30\beta; \end{aligned}$$

eliminating β leads to $\alpha = \pm\frac{1}{8}$, with corresponding dimensions $d_1 = 27$ and $d_2 = 35$.

For the components of the staurotope realization X, whose cosine vectors are of the form $(1, \alpha, 0, -\alpha, -1)$, we have

$$32\alpha^2 = 1 + 15\alpha + 15 \cdot 0 + (-\alpha) = 1 + 14\alpha,$$

with solutions $\alpha = \frac{1}{2}$ or $-\frac{1}{16}$ and dimensions $d_3 = 7$ and $d_4 = 56$. The 'standard' realization 2_{31} is P_3.

In summary, then, we have

14E1 Theorem *The cosine matrix of* $\{3^2, 3^{3,1}\}$ *and its associated regular polytopes of nearly full rank is*

$$\begin{bmatrix} 1 & 1 & 1 & 1 & 1 \\ 1 & \frac{1}{8} & -\frac{1}{6} & \frac{1}{8} & 1 \\ 1 & -\frac{1}{8} & \frac{1}{10} & -\frac{1}{8} & 1 \\ 1 & \frac{1}{2} & 0 & -\frac{1}{2} & -1 \\ 1 & -\frac{1}{16} & 0 & \frac{1}{16} & -1 \end{bmatrix},$$

with layer and dimension vectors

$$\Lambda = (1, 32, 60, 32, 1), \quad D = (1, 27, 35, 7, 56).$$

14E2 Remark Just as we have seen before, we have

$$\Gamma_3{}^2 = (1, \tfrac{1}{4}, 0, \tfrac{1}{4}, 1) = \tfrac{1}{7}\Gamma_0 + \tfrac{6}{7}\Gamma_1 .$$

Observe as well that $\Gamma_2 \Gamma_3 = \Gamma_4$, so that $\mathsf{P}_2 \otimes \mathsf{P}_3 = \mathsf{P}_4$.

14E3

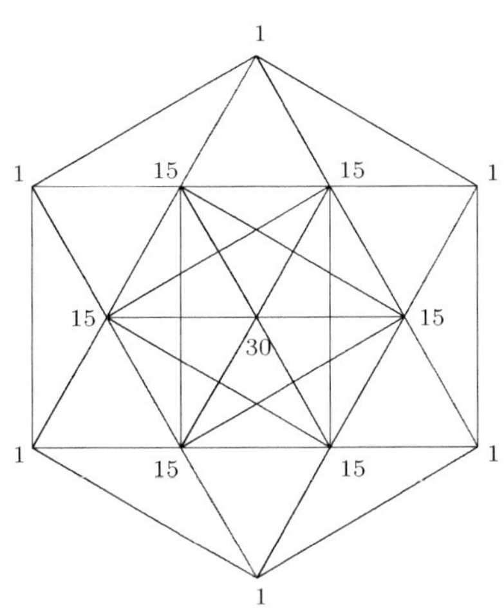

The Gosset–Elte polytope 2_{31}

14E4 Remark Though we do not need to deploy the argument here, because we know for other reasons that P_2 is pure, it is clear that its purity is ensured by the fact that its facet of type $\{3^2, 3^{2,1}\}$ with induced cosine vector $(1, -\frac{1}{8}, \frac{1}{10})$ is pure 20-dimensional and centred (and so with not even a trivial component), and so can have no other component than itself. We can appeal to similar reasoning later.

There are two useful projections of 2_{31}. The first, in Figure 14E3, is more symmetric; the points marked '15' are projections of truncated 5-simplices 0_{31}, while that marked '30' represents a difference body $D\mathsf{T}^5$ of a 5-simplex. The horizontal sections from a vertex are a half-cube 1_{31}, a truncated 6-staurotope, then another half-cube before the antipodal vertex.

14E5

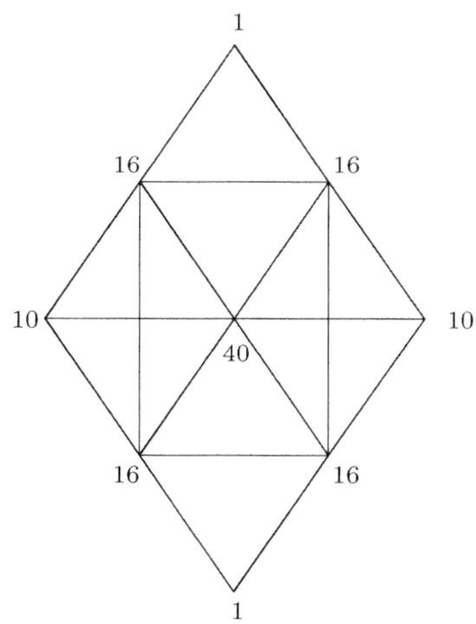

Another projection of 2_{31}

The other projection in Figure 14E5, drawn to the same scale, illustrates rather more of the structure of 2_{31}; this corresponds to the projection of 3_{21} in Figure 14D3. The labels '16' correspond to half-cubes 1_{21}, those marked '10' are 5-staurotopes, while the mark '40' is a truncated 5-staurotope. The horizontal sections are as previously, but now the slanted sections are Gosset–Elte polytopes 2_{21}, 1_{22} and 2_{21} again (in the opposite orientation).

14F Rank 8: $\{3^4, 3^{2,1}\}$

The 'standard' realization 4_{21} of $\{3^4, 3^{2,1}\}$ has diametral hexagons; Figure 14F1 shows the projection of this polytope on the plane spanned by one of them. If it looks just like the projections of 1_{22} and 2_{31}, this should not occasion any surprise. As we shall see, 2_{31} is a central section of 4_{21}, and we know that 1_{22} is a central section of 2_{31}. In Figure 14F1, the points labelled '27' indicate copies of 2_{21} (adjacent ones have opposite orientations), while that at the centre labelled '72' is 1_{22}. Thus the parallel sections (ignoring the antipodal vertices) are 3_{21},

2_{31} and 3_{21} again, respectively. It follows that the layer vector of $\{3^4, 3^{2,1}\}$ is $\Lambda = (1, 56, 126, 56, 1)$; moreover, its diagonals are all symmetric.

14F1

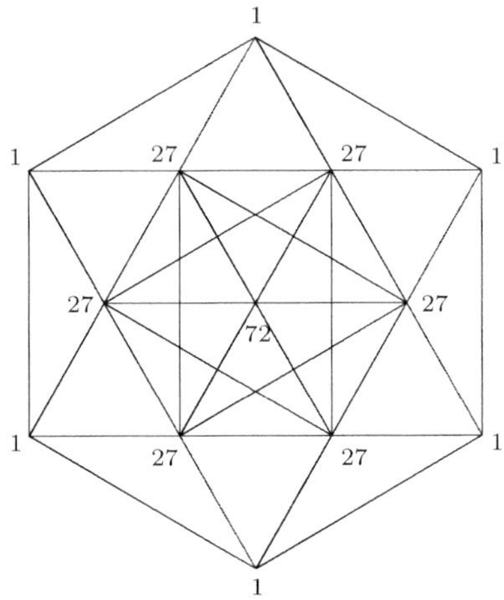

The Gosset polytope 4_{21}

The vertex-figure of $\mathcal{P} = \{3^4, 3^{2,1}\}$ is $\mathcal{P}^v = \{3^3, 3^{2,1}\}$, with layer vector $\Lambda^v = (1, 27, 27, 1)$ and induced cosine vector $\Gamma^v = (1, \gamma_1, \gamma_2, \gamma_3)$ in terms of that of $\mathcal{P}$. For a pure component (other than $\{1\}$) of the small simplex realization S, with cosine vector $(1, \alpha, \beta, \alpha, 1)$, Theorem 3C14 tells us that

$$\begin{aligned} 56\alpha^2 &= 1 + 27\alpha + 27\beta + \alpha = 1 + 28\alpha + 27\beta, \\ 0 &= 1 + 56\alpha + 63\beta; \end{aligned}$$

eliminating β gives $\alpha = \frac{1}{7}$ or $-\frac{1}{14}$. Similarly, for a pure component of the staurotope realization X, we have

$$56\alpha^2 = 1 + 27\alpha + 27 \cdot 0 + (-\alpha) = 1 + 26\alpha,$$

with solutions $\alpha = \frac{1}{2}$ or $-\frac{1}{28}$. Substituting for the respective values of β (where appropriate), and using Theorem 3F5 to calculate the dimensions, we then have

14F2 Theorem *The cosine matrix of $\{3^4, 3^{2,1}\}$ and its associated regular polytopes of nearly full rank is*

$$\begin{bmatrix} 1 & 1 & 1 & 1 & 1 \\ 1 & \frac{1}{7} & -\frac{1}{7} & \frac{1}{7} & 1 \\ 1 & -\frac{1}{14} & \frac{1}{21} & -\frac{1}{14} & 1 \\ 1 & \frac{1}{2} & 0 & -\frac{1}{2} & -1 \\ 1 & -\frac{1}{28} & 0 & \frac{1}{28} & -1 \end{bmatrix},$$

with layer and dimension vectors

$$\Lambda = (1, 56, 126, 56, 1), \quad D = (1, 35, 84, 8, 112).$$

14F3 Remark Note that $\Gamma_2\Gamma_3 = \Gamma_4$, so that $\mathsf{P}_4 = \mathsf{P}_2 \otimes \mathsf{P}_3$, the point being that we can find P_4 using the product structure. Observe that $d_4 = 112$ falls a long way short of $d_2 \cdot d_3 = 84 \cdot 8 = 672$.

There are alternative calculations for P_4, which we want to appeal to when we look at the realizations of $\{3^2, 3^{4,1}\}$. In terms of cosine vectors, these are

14F4 $$\Gamma_3{}^3 = \tfrac{3}{10}\Gamma_3 + \tfrac{7}{10}\Gamma_4,$$

14F5 $$\Gamma_1\Gamma_3 = \tfrac{5}{3}\Gamma_3 + \tfrac{4}{5}\Gamma_4.$$

There is actually an analogous expression for P_2, namely,

14F6 $$\Gamma_3\Gamma_4 = \tfrac{1}{4}\Gamma_1 + \tfrac{3}{4}\Gamma_2;$$

however, we shall see in the next section that we cannot make any use of it.

14F7 Remark For the calculation of (14F4), we can use

$$\langle \Gamma_3{}^3, \Gamma_3\rangle_\Lambda = \langle \Gamma_3{}^2, \Gamma_3{}^2\rangle_\Lambda = (\tfrac{1}{8})^2\|\Gamma_0\|_\Lambda^2 + (\tfrac{7}{8})^2\|\Gamma_1\|_\Lambda^2 = \tfrac{1}{64} + \tfrac{49}{64}\cdot\tfrac{1}{35} = \tfrac{3}{80}$$

from $\Gamma_3{}^2 = \frac{1}{8}\Gamma_0 + \frac{7}{8}\Gamma_1$, whence the coefficient of Γ_3 in $\Gamma_3{}^3$ is $8 \cdot \frac{3}{80} = \frac{3}{10}$, as claimed.

14F8 Remark One might wonder why multiples of 7 play such a prominent rôle in the cosine matrix of $\{3^4, 3^{2,1}\}$. The obvious explanation is that the corresponding pure realizations have dimensions which are multiples of 7, so that sevens are needed in the norm calculations.

14F9

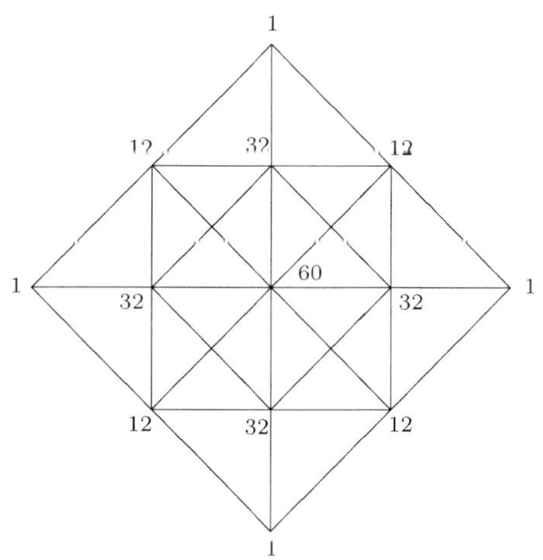

Another projection of 4_{21}

As with other Gosset–Elte polytopes than 4_{21}, an alternative projection can sometimes reveal more about the structure. That of Figure 14F9 is on the same scale as Figure 14F1, so that successive horizontal sections from the top vertex are again 3_{21}, the central section 2_{31}, another copy of 3_{21}, and finally the antipodal vertex. Now, however, we have the slanted sections, which are a staurotopal facet 4_{11}, a half-cube 1_{41}, a central truncated 7-staurotope, and then the half-cube and staurotope again. In fact, these sections are easily picked out from the given vertex coordinates. The points marked '12' represent 6-staurotopes, those marked '32' are half-cubes 1_{31}, and that marked '60' is a truncated 6-staurotope.

14G Rank 8: $\{3^2, 3^{4,1}\}$

We begin by formally stating the main result.

14G1 Theorem *The cosine matrix of* $\{3^2, 3^{4,1}\}$ *and its associated regular polytopes of nearly full rank is*

$$\begin{bmatrix}
1 & 1 & 1 & 1 & 1 & 1 & 1 & 1 & 1 & 1 \\
1 & -\frac{1}{8} & \frac{1}{10} & -\frac{1}{8} & 1 & \frac{1}{10} & -\frac{1}{8} & \frac{1}{10} & -\frac{1}{8} & 1 \\
1 & \frac{1}{16} & -\frac{1}{14} & \frac{1}{16} & 1 & -\frac{1}{14} & \frac{1}{16} & -\frac{1}{14} & \frac{1}{16} & 1 \\
1 & \frac{1}{2} & \frac{1}{7} & -\frac{1}{14} & -\frac{1}{7} & -\frac{1}{7} & -\frac{1}{14} & \frac{1}{7} & \frac{1}{2} & 1 \\
1 & \frac{1}{8} & -\frac{1}{14} & -\frac{1}{56} & -\frac{1}{7} & \frac{1}{14} & -\frac{1}{56} & -\frac{1}{14} & \frac{1}{8} & 1 \\
1 & -\frac{1}{16} & \frac{1}{70} & \frac{1}{112} & -\frac{1}{7} & -\frac{1}{70} & \frac{1}{112} & \frac{1}{70} & -\frac{1}{16} & 1 \\
1 & \frac{3}{4} & \frac{1}{2} & \frac{1}{4} & 0 & 0 & -\frac{1}{4} & -\frac{1}{2} & -\frac{3}{4} & -1 \\
1 & \frac{9}{32} & -\frac{1}{28} & -\frac{19}{224} & 0 & 0 & \frac{19}{224} & \frac{1}{28} & -\frac{9}{32} & -1 \\
1 & -\frac{3}{32} & \frac{1}{20} & -\frac{1}{32} & 0 & 0 & \frac{1}{32} & -\frac{1}{20} & \frac{3}{32} & -1 \\
1 & 0 & -\frac{1}{28} & \frac{1}{28} & 0 & 0 & -\frac{1}{28} & \frac{1}{28} & 0 & -1
\end{bmatrix},$$

with layer and dimension vectors

$$\begin{aligned}
\Lambda &= (1, 64, 280, 448, 14, 560, 448, 280, 64, 1), \\
D &= (1, 50, 84, 35, 210, 700, 8, 112, 400, 560).
\end{aligned}$$

The realization domain of $\mathcal{P} := \{3^2, 3^{4,1}\}$, which has 2160 vertices, displays some curious features. The vertex-set of its standard realization as $\mathsf{P}_6 = 2_{41}$ (with the same symmetries as those given for 4_{21}, and so with dimension $d_6 = 8$) consists of all permutations of $(\pm 4, 0^7)$, $(\pm 3, (\pm 1)^7)$ with an odd number of minus signs, and $((\pm 2)^4, 0^4)$. It is not too hard to see (using these coordinates, since by definition we have a faithful realization) that the layer vector is as claimed in Theorem 14G1. Moreover, P_6 then has cosine vector

$$\Gamma_6 = (1, \tfrac{3}{4}, \tfrac{1}{2}, \tfrac{1}{4}, 0, 0, -\tfrac{1}{4}, -\tfrac{1}{2}, -\tfrac{3}{4}, -1).$$

As usual, the pure realizations are numbered with the final result in mind.

14G2 Remark Observe that the ordering of the entries in the layer vector Λ of $\mathcal{P}$ is such that its first 5 are those of the induced cosine vector of its facet of type $\mathcal{F} := \{3^2, 3^{3,1}\}$. To avoid constant repetition when we come to describing the realization space, we shall adopt the convention that, for each $j = 0, \ldots, 9$, the facet in $\mathcal{F}$ of the realization $\mathsf{P}_j \in \mathcal{P}$ is denoted F_j.

The stabilizer $\mathbf{G}_0$ of the initial vertex $(4, 0^7)$ of 2_{41} is $[3^{4,1,1}]$, consisting of all permutations of the last 7 coordinates, together with all even numbers of changes of their signs. It easily follows from this description that $\mathbf{G}_0$ is transitive on the vertices in each layer, and the isomorphism $2_{41} \cong \{3^2, 3^{4,1}\}$ hence shows that all its diagonal classes are symmetric.

Before we go into further details, it is worth pointing out a consequence of the obvious variant of Corollary 4D10 and the triality relationship with $\{3^4, 3^{2,1}\}$.

14G3 Lemma *A pure realization of* $\mathcal{P}$ *of dimension* $d \neq 1, 8, 35, 84, 112$ *must have centred facets of type* $\mathcal{F}$.

14G4 Remark In fact, we could use Theorem 4D6 (or, rather, the obvious generalization) and appeal to the equality of the values of $\eta_{\boldsymbol{f}}$ (with staurotopal facets of $\{3^4, 3^{2,1}\}$) in the remaining cases, but we do not need to. The induced cosine vector of a staurotopal facet is $(1, \gamma_1, \gamma_2)$ in terms of that of $\{3^4, 3^{2,1}\}$, and so it is easy to check equality. Indeed, since $\eta_{\boldsymbol{f}} > 0$ in each case, triality automatically gives rise to a realization of $\mathcal{P}$ of the same dimension.

It is important to recognize that the vertices in the central section of 2_{41} are *not* all equivalent; from initial vertex $(4, 0^7)$ they fall into 14 points of the form $(0, \pm 4, 0^6)$ and 560 of the form $(0, (\pm 2)^4, 0^3)$ (in each case, the first coordinate is 0, while the remaining seven coordinates are freely permuted). Indeed, any given vertex forms with 15 others a uniquely determined set of 16 vertices of an 8-staurotope, of which 14 vertices are permuted by the group $[3, 3^{4,1}]$ of the vertex-figure; while other sets of 15 similarly form staurotopes, they are not fixed by this subgroup. In fact, these special inscribed staurotopes illustrate the subgroup relationship $[3^{5,1,1}] < [3^{4,2,1}]$ that we drew attention to in Remark 1F4.

It follows that we can identify the vertices of $\{3^2, 3^{4,1}\}$ in sixteens, to yield a realization with the $2160/16 = 135$ vertices of a 134-simplex. But this realization cannot be pure, since two non-trivial diagonal classes survive the identification (the initial vertex is not joined by edges to any vertex of the form $((\pm 2)^4, 0^4)$). The general cosine vector of the quotient $\mathcal{P}/16 := \{3^2, 3^{4,1}\}/16$ (as we can designate it) is therefore of the form

14G5 $$\Gamma = (1, \alpha, \beta, \alpha, 1, \beta, \alpha, \beta, \alpha, 1);$$

applying the layer equation (and dividing by 16) gives $64\alpha + 70\beta = -1$.

Let us look at $\mathcal{P}/16$ in its own right. Formally, we have

14G6 Proposition *The group of* $\{3^2, 3^{4,1}\}/16$ *is* $[3^{4,2,1}]$*, acting on the* 135 *right cosets of its subgroup* $[3^{5,1,1}]$.

We can write the layer vector of $\{3^2, 3^{4,1}\}/16$ in the form $(1, 64, 70)$, and its general cosine vector in the form $(1, \alpha, \beta)$. We need to say a few more things about its group and geometry. We can think about the two sets of 64 and 70 vertices in several ways, but perhaps the best is to consider all the vertices of $\{3^2, 3^{4,1}\}$ which go into the identifications. A fact worth noting to start with is that antipodal vertices of the facets $\{3^2, 3^{3,1}\}$ are identified, so that in $\{3^2, 3^{4,1}\}/16$ they take the form $\{3^2, 3^{3,1}\}/2$. Similarly, two of the diagonal classes of the vertex-figure $\{3, 3^{4,1}\}$ (the half-7-cube) of $\{3^2, 3^{4,1}\}/16$ collapse to one.

At this point, we can state our result; to illustrate different techniques, we shall give two proofs.

14G7 Theorem *The cosine matrix of* $\{3^2, 3^{4,1}\}/16$ *is*

$$\begin{bmatrix} 1 & 1 & 1 \\ 1 & -\frac{1}{8} & \frac{1}{10} \\ 1 & \frac{1}{16} & -\frac{1}{14} \end{bmatrix},$$

with layer and dimension vectors

$$\Lambda = (1, 64, 70), \qquad D = (1, 50, 84).$$

First proof. Since all diagonal classes of $\{3^2, 3^{4,1}\}/16$ are symmetric, we can apply Theorem 3C14 to the induced cosine vector of layer 1, namely, $\{3, 3^{5,1}\}/2$, which is also $(1, \alpha, \beta)$. Thus we have

$$\begin{aligned} 64\alpha^2 &= 1 + 21\alpha + 35\beta + 7\alpha = 1 + 28\alpha + 35\beta, \\ 0 &= 1 + 64\alpha + 70\beta; \end{aligned}$$

this yields $\alpha = -\frac{1}{8}$ or $\frac{1}{16}$, and hence the two cosine vectors as claimed.

We now use Theorem 3F5 to determine the dimension d of each realization:

14G8 $$\frac{1}{d} = \frac{1}{135}(1 + 64\alpha^2 + 70\beta^2).$$

Calling the corresponding realizations P_1 and P_2 (to fit in with the subsequent listing), we easily check that $d_1 = 50$ if $\alpha = -\frac{1}{8}$ and $d_2 = 84$ if $\alpha = \frac{1}{16}$. □

14G9 Remark As a further check of the results of Theorem 3F5, observe that

$$\langle \Gamma_1, \Gamma_2 \rangle_\Lambda = 0.$$

Further note that the columns of the cosine matrix are D-orthogonal. (For further comments on this realization, see the notes at the end of the section.)

Second proof. An alternative approach appeals to Corollary 4D10. Since 135 is not a sum of excluded dimensions in Lemma 14G3, at least one of the pure components of $\mathcal{P}/16$ must have centred facets of type $\mathcal{F}$. Putting together the induced layer equation for $\mathcal{F}$ and the layer equation for $\mathcal{P}/16$, we thus solve

$$\begin{aligned} 1 + 32\alpha + 30\beta &= 0, \\ 1 + 64\alpha + 70\beta &= 0, \end{aligned}$$

to obtain $\Gamma = (1, -\frac{1}{8}, \frac{1}{10})$ with corresponding dimension $d_1 = 50$, as we have seen.

For the complementary (non-trivial) pure realization, using Theorem 3F5 for Λ-orthogonality with respect to $(1, -\frac{1}{8}, \frac{1}{10})$ we solve

$$\begin{aligned} 1 + 64\alpha + 70\beta &= 0, \\ 1 - 8\alpha + 7\beta &= 0, \end{aligned}$$

to find $\Gamma = (1, \frac{1}{16}, -\frac{1}{14})$, again as before. Alternatively, we could appeal to the component equation of Theorem 3C11 as applied to the quotient. □

14G10 Remark Observe that the facets F_2 are *not* centred; of course, 84 is a permitted dimension under Lemma 14G3. This may seem improbable at first, even though we have a realization of $\mathcal{P}/16$ rather than simply $\mathcal{P}/2$.

Now that we have found the pure realizations P_1 and P_2 of $\{3^2, 3^{4,1}\}/16$, we next make the useful observation that the vertices of $\mathsf{P}_6 = 2_{41}$ itself contain copies of the vertices of 4_{21}; the easiest way to see this is to apply to them the similarity $x \mapsto y$ on $\mathbb{E}^8$ given by

$$\begin{aligned} \eta_{2j-1} &:= \xi_{2j-1} + \xi_{2j}, \\ \eta_{2j} &:= \xi_{2j-1} - \xi_{2j}, \end{aligned}$$

for $j = 1, 2, 3, 4$. This implies that we have a 35-dimensional realization P_3 of $\{3^2, 3^{4,1}\}/2$ (the non-trivial component of $\mathsf{P}_6 \otimes \mathsf{P}_6$) and a 112-dimensional centrally symmetric pure realization given by the same calculations as those in (14F4) or (14F5), with Γ_6 instead of Γ_3 and Γ_7 instead of Γ_4; all this follows from simple dimension counts. However, although we obtained an expression for the 84-dimensional realization of $\{3^4, 3^{2,1}\}/2$ in (14F6), the analogous calculation in $\{3^2, 3^{4,1}\}/2$ does not yield its pure 84-dimensional realization (but we know what it is anyway).

14G11 Remark The fact that 4_{21} is inscribed in 2_{41} in a way that does not respect most of its symmetry is irrelevant. All we are considering is tensor products of vectors with themselves; thus prescribed subsets will behave in the appropriate way, however they are embedded.

So, without going into the details here (see the end of the section), we find the following cosine vectors and corresponding dimensions:

$$\begin{aligned} \Gamma_3 &= \left(1, \tfrac{1}{2}, \tfrac{1}{7}, -\tfrac{1}{14}, -\tfrac{1}{7}, -\tfrac{1}{7}, -\tfrac{1}{14}, \tfrac{1}{7}, \tfrac{1}{2}, 1\right), \quad d_3 = 35, \\ \Gamma_7 &= \left(1, \tfrac{9}{32}, -\tfrac{1}{28}, -\tfrac{19}{224}, 0, 0, \tfrac{19}{224}, \tfrac{1}{28}, -\tfrac{9}{32}, -1\right), \quad d_7 = 112. \end{aligned}$$

Counting diagonal classes shows that we need two more pure realizations of $\{3^2, 3^{4,1}\}/2$, whose dimensions sum to $1080 - 1 - 35 - 50 - 84 = 910$. We might guess that these arise from $\mathsf{P}_1 \otimes \mathsf{P}_3$ and $\mathsf{P}_2 \otimes \mathsf{P}_3$. From what we know about $\{3^2, 3^{4,1}\}/16$, we have

14G12 $$\Gamma_0 + 50\Gamma_1 + 84\Gamma_2 = 135(1, 0^3, 1, 0^4, 1),$$

counting the pure realizations (compare (14G5) for the central entry '1'). If we multiply by Γ_3 we obtain

14G13 $$\Gamma_3 + 50\Gamma_1\Gamma_3 + 84\Gamma_2\Gamma_3 = 135(1, 0^3, -\tfrac{1}{7}, 0^4, 1);$$

multiplying (14G13) by 7 and adding to (14G12) then gives

14G14 $$\Gamma_0 + 50\Gamma_1 + 84\Gamma_2 + 7\Gamma_3 + 350\Gamma_1\Gamma_3 + 588\Gamma_2\Gamma_3 = 1080(1, 0^8, 1),$$

the appropriate multiple of the cosine vector of the small simplex realization S, that is, the simplex realization of $\{3^2, 3^{4,1}\}/2$.

However, (14G14) is incorrect as a dimension count, since Γ_3 should occur with multiple 35. This means that one of $\mathsf{P}_1 \otimes \mathsf{P}_3$ and $\mathsf{P}_2 \otimes \mathsf{P}_3$ is blended, with P_3 itself as a component. Aesthetic considerations suggest that the blend is the latter ($588 - 28 = 560$ is clearly preferable to $350 - 28 = 322$, and in any case the latter choice leads to nastier numbers).

In fact, there are several approaches to the problem. First, looking at induced cosine vectors of facets of type $\mathcal{F}$ with layer vector $\Lambda = (1, 32, 60, 32, 1)$, which are given by Remark 14G2, we see that this facet of $\mathsf{P}_1 \otimes \mathsf{P}_3$ is centred, while that of P_3 is not; thus P_3 cannot be a component of $\mathsf{P}_1 \otimes \mathsf{P}_3$.

Better still is to appeal to Theorem 3J1 for the coefficients of Γ_3 in $\Gamma_1\Gamma_3$ and $\Gamma_2\Gamma_3$. These yield

$$35\langle \Gamma_j\Gamma_3, \Gamma_3\rangle_\Lambda = \begin{cases} 0, & \text{if } j = 1,\\ \frac{1}{21}, & \text{if } j = 2. \end{cases}$$

Note that $\frac{1}{21} = \frac{28}{588}$, which exactly accounts for the component of P_3.

These calculations might suggest that $\mathsf{P}_5 = \mathsf{P}_1 \otimes \mathsf{P}_3$ is pure with $\dim P_5 = 350$. However, they are deceptive; as usual, the actual dimension d_5 can be found from Theorem 3F5, namely,

$$\begin{aligned}\|\Gamma_5\|_\Lambda^2 &= \tfrac{1}{2160}\Big(1 + 64\big(-\tfrac{1}{16}\big)^2 + 280\big(\tfrac{1}{70}\big)^2 + 448\big(\tfrac{1}{112}\big)^2 + 14\big(-\tfrac{1}{7}\big)^2 + 560\big(-\tfrac{1}{70}\big)^2\\ &\quad + 448\big(\tfrac{1}{112}\big)^2 + 280\big(\tfrac{1}{70}\big)^2 + 64\big(-\tfrac{1}{16}\big)^2 + 1\Big)\\ &= \tfrac{1}{700},\end{aligned}$$

so that $d_5 = 700$, twice the value one perhaps expected. This, in turn, tells us that $\mathsf{P}_2 \otimes \mathsf{P}_3$ also has a component P_5, and that the final pure component P_4

has dimension $d_4 = 588 - 28 - 350 = 210$. It is easier to calculate its cosine vector Γ_4 directly from the component equation of Theorem 3C11 as

$$\begin{aligned}\Gamma_4 &= \tfrac{1}{210}(1080(1, 0^8, 1) - \Gamma_0 - 35\Gamma_1 - 50\Gamma_2 - 84\Gamma_3 - 700\Gamma_5)\\ &= \left(1, \tfrac{1}{8}, -\tfrac{1}{14}, -\tfrac{1}{56}, -\tfrac{1}{7}, \tfrac{1}{14}, -\tfrac{1}{56}, -\tfrac{1}{14}, \tfrac{1}{8}, 1\right).\end{aligned}$$

As a check of Theorem 3F5, we can verify that

$$\|\Gamma_4\|_\Lambda^2 = \tfrac{1}{210};$$

furthermore

$$\langle \Gamma_4, \Gamma_5 \rangle_\Lambda = 0,$$

which is the appropriate orthogonality property.

14G15 Remark From Lemma 14G3, we know that the facet F_4 of P_4 of type $\mathcal{F}$ is also centred.

It remains for us to find two further pure components of the staurotope realization X. We employ the same idea as before. We have

14G16
$$\begin{aligned}8\Gamma_6 + 400\Gamma_1\Gamma_6 + 672\Gamma_2\Gamma_6 &= 8(\Gamma_0 + 50\Gamma_1 + 84\Gamma_2)\Gamma_6\\ &= 1080(1, 0^8, -1),\end{aligned}$$

the appropriate multiple of the cosine vector of X. An appeal to (3E12) shows that

$$\begin{aligned}\dim(\mathsf{P}_1 \otimes \mathsf{P}_6) &\leqslant 50 \cdot 8 = 400,\\ \dim(\mathsf{P}_2 \otimes \mathsf{P}_6) &\leqslant 84 \cdot 8 = 672;\end{aligned}$$

since these are the correct dimensions in (14G16), we do not need Lemma 3F3 to see that we must have equality.

However, we have to recover a term $112\Gamma_7$ in (14G16) from somewhere. Using Remark 14G2, we calculate that the facet of $\mathsf{P}_8 := \mathsf{P}_1 \otimes \mathsf{P}_6$ of type $\{3^2, 3^{3,1}\}$ is centred. Since, by a similar calculation, this type of facet is not centred in P_7, it follows that the latter cannot be a component of P_8, and so must be a component of $\mathsf{P}_2 \otimes \mathsf{P}_6$. This leads us to P_9 with $d_9 = 560$ and cosine vector Γ_9 given by $\Gamma_2\Gamma_6 = \frac{1}{6}\Gamma_7 + \frac{5}{6}\Gamma_9$. Alternatively, we have

$$112\langle \Gamma_j\Gamma_6, \Gamma_7 \rangle_\Lambda = \begin{cases} 0, & \text{if } j - 1,\\ \frac{1}{6}, & \text{if } j = 2,\end{cases}$$

leading to the same conclusion. Note that this expression for $\Gamma_2\Gamma_6$ illustrates the equality case in Proposition 3F8.

14G17 Remark We have omitted many of the details of checking dimensions using Theorem 3F5, leaving this to the interested reader. The strangest fractions occur in Γ_7 (for instance, $\pm\frac{19}{224}$), and so it is particularly worth looking at this case. Observe as well that, corresponding to (14F6) for $\{3^4, 3^{2,1}\}$, we can show that

$$\Gamma_6\Gamma_7 = \tfrac{1}{8}\Gamma_2 + \tfrac{1}{4}\Gamma_3 + \tfrac{5}{8}\Gamma_4;$$

the coefficient of Γ_3 does agree with that for (14F6), but the analogous term $\frac{3}{4}\Gamma_2$ (for dimension 84) does not carry over. In fact, $\mathsf{P}_6 \otimes \mathsf{P}_7$ now has dimension $84 + 35 + 210 = 329$ rather than $35 + 84 = 119$. (In each case, of course, the dimension is less than the product $8 \cdot 112 = 896$.)

14G18 Remark The simple form of Γ_9 would also suggest that $\mathsf{P}_2 \otimes \mathsf{P}_6$ is the blended component. Indeed, observe that the facet F_9 of P_9 is also centred; while this alone does not ensure that P_9 is pure, it does strongly suggest it.

Although we do not repeat various calculations which have led to this cosine matrix, it may help the reader if we list some of its ingredients. Recall that $\mathsf{P}_6 = 2_{41}$ itself, so that Γ_6 is its cosine vector:

$$\begin{aligned}
\Gamma_1\Gamma_3 &= \Gamma_5,\\
\Gamma_2\Gamma_3 &= \tfrac{1}{21}\Gamma_3 + \tfrac{5}{14}\Gamma_4 + \tfrac{25}{42}\Gamma_5,\\
\Gamma_6{}^2 &= \tfrac{1}{8}\Gamma_0 + \tfrac{7}{8}\Gamma_3,\\
\Gamma_6{}^3 &= \tfrac{3}{10}\Gamma_6 + \tfrac{7}{10}\Gamma_7,\\
\Gamma_3\Gamma_6 &= \tfrac{1}{5}\Gamma_6 + \tfrac{4}{5}\Gamma_7,\\
\Gamma_1\Gamma_6 &= \Gamma_8,\\
\Gamma_2\Gamma_6 &= \tfrac{1}{6}\Gamma_7 + \tfrac{5}{6}\Gamma_9.
\end{aligned}$$

Observe that Γ_7 (with the most complicated entries) occurs as a component in three of these relations, while Γ_5 (with the next most complicated) is a simple product.

14G19 Remark Rather large numbers turn up in the denominators of the fractions in some of the entries of the cosine vectors. In fact, this should occasion less surprise, when one considers (taking P_5 as an example) that 1080 points are being distributed symmetrically in $\mathbb{E}^{700}$, but without central symmetry. Thus the number of points is less than twice the dimension, and one would expect that many of them must be nearly orthogonal to the initial vertex.

Notes to Section 14G

1. We began to investigate the realization domain of $\{3^2, 3^{4,1}\}/16$ before we had fully developed the current theory. First, we guessed the dimensions d_1 and d_2 of the pure realizations P_1 and P_2 ; apart from 64, 70, the most natural choice (summing to 134, of course) is 50, 84. Unfortunately, picking the dimensions does not determine corresponding cosine vectors; the converse is true, of course. Hence we tried to

find plausible cosine vectors whose fractional entries had small denominators. A little experimentation came up with $\Gamma_1 = (1, -\frac{1}{8}, \frac{1}{10})$ and $\Gamma_2 = (1, \frac{1}{16}, -\frac{1}{14})$; these (putative) cosine vectors do yield the dimensions $d_1 = 50$ and $d_2 = 84$.

2. Before we went on to establish Theorem 14G7 formally (not originally by the method presented here), we adduced some more evidence in its favour. First, the induced cosine vector of the facet $\{3^2, 3^{3,1}\}/2$ is $(1, -\frac{1}{8}, \frac{1}{10})$, which is that of the pure realization of dimension 35; moreover, these facets are centred. It follows that, if there is a realization P_1 of $\{3^2, 3^{4,1}\}/16$ with cosine vector $\Gamma_1 := (1, -\frac{1}{8}, \frac{1}{10})$, then it is pure.
3. In fact, we see that P_1 will have ridges which are 20-dimensional pure realizations of $\{3^2, 3^{2,1}\}$, and that these will be centred. Even this suffices to show that P_1 would have to be pure.
4. The geometry of its second layer (using induced cosine vectors, of course) implies that P_1 has to have dimension at least 50. At an early stage, we all but found coordinates for the vertices of P_1, showing that its dimension is indeed 50; we resile from giving the very messy details.
5. We have only appealed to Theorem 4D6 in our treatment in a minor way (see Lemma 14G3), because it does not materially shorten the proof of Theorem 14G1. Taking, for example, the pure realizations of $\mathcal{P}/2$, those with non-centred facets $\mathcal{F}$ are P_0, P_2 and P_3. Since the corresponding facets of the remaining pure realizations are centred, one might expect that this fact is implied by the orthogonality relations of Theorem 3F5. However, this is not the case, as one may fairly easily verify.

15

Locally Toroidal Polytopes

A regular polytope is said to be *locally toroidal* if its minimal infaces which are not spherical are toroidal. Thus we follow the precedent of [76]; see also [99, Section 14A].

This chapter treats those locally toroidal regular polytopes of types $\{3,4,4\}$, $\{4,4,3\}$ and $\{4,4,4\}$ with not too many vertices, and includes some geometric descriptions. Also considered are certain others involving the toroids of type $\{3,6\}$ and $\{6,3\}$.

There are relatively few finite universal regular toroidal polytopes of type $\{4,4,4\}$. Those which have vertex-figure $\{4,4 \mid 2\} = \{4,4\}_{(2,0)}$ are flat, and so their realization domains coincide with those of their facets. Those with facet $\{4,4 \mid 2\}$ are degenerate, and have trivial realization domains. Two more, namely $\{\{4,4 \mid 3\},\{4,4 \mid 4\}\}$ and its dual, are probably too large to be tackled by our techniques; certainly $\{\{4,4 \mid 3\},\{4,4 \mid 5\}\}$ and its dual are hopelessly out of reach. The aim here is to describe the realization domains of the rest.

Some of these locally toroidal polytopes are closely related to the Gosset–Elte polytopes, in which event we will have noted them earlier; indeed, these relationships often provide proofs of universality.

In a similar way, many of the locally toroidal polytopes with infaces of type $\{3,6\}$ or $\{6,3\}$ are either flat (or nearly so) or too large to be tractable. In Sections 15F and 15G we shall discuss those of the rest which are amenable to our treatment, in the course of which we shall find hitherto undescribed families of universal regular polytopes.

15A $\{\{4,4:4\},\{4,3\}\}$ and its Dual

The locally toroidal regular polytope $\mathcal{P} := \{\{4,4:4\},\{4,3\}\}$ is the smallest non-flat polytope of [99, Table 10B1]; it has 16 vertices, 12 facets and automorphism group of order 768. It has a realization P_4 of dimension $d_4 = 6$ as usual, (the labelling looks towards the final classification), with generatrix (R_0, R_1, R_2, R_3)

given by

$$\begin{aligned} R_0 &:= (1\ 4)(2\ 5)(3\ 6), \\ R_1 &:= \overline{4}, \\ R_2 &:= (1\ 2)(4\ 5), \\ R_3 &:= (2\ 3)(5\ 6); \end{aligned}$$

as previously, this means that R_1 changes the sign of the fourth coordinate, while R_0, R_2, R_3 act as the given permutations on the coordinates. We quickly find that P_4 has vertices $((\pm 1)^3, 0^3)$ and $(0^3, (\pm 1)^3)$, where as usual r^k denotes a string $r, \ldots, r$ of length k; these are the vertices of two cubes $\{4,3\}$ in orthogonal 3-dimensional subspaces of $\mathbb{E}^6$.

In view of this, we see that we can write the layer vector as $\Lambda = (1,3,8,3,1)$; to bring out the symmetry, we label the class of an edge $\mathcal{D}_2$. For P_5 itself, we thus have dimension $d_4 = 6$ and cosine vector $\Gamma_4 = (1, \frac{1}{3}, 0, -\frac{1}{3}, -1)$.

The combinatorial structure of $\mathcal{P}$ is so simple that virtually no work is needed to write down all its pure realizations. We observe that $\mathcal{P}$ is both 1- and 2-collapsible (refer to Section 2F for the definition), so that $\mathcal{P}$ has realizations $\mathsf{P}_1 := \{2\}$ with dimension $d_1 = 1$ and cosine vector $\Gamma_1 = (1,1,-1,1,1)$, and $\mathsf{P}_3 := \{4\}$ with $d_2 = 2$ and $\Gamma_3 = (1,-1,0,1,-1)$. With $\mathsf{P}_0 := \{1\}$ with $d_0 = 1$ and $\Gamma_0 = (1^5)$, we see that we need one more pure realization P_2 with $d_4 = 16 - 1 - 1 - 2 - 6 = 6$ also. This is found by collapsing the vertex-figure $\{4,3\}$ onto the Petrial $\{\frac{4}{1,2}, 3 : 3\}$ of the tetrahedron; the cosine vector is therefore $\Gamma_2 = (1, -\frac{1}{3}, 0, -\frac{1}{3}, 1)$.

15A1 Remark From Theorem 2F11, the central quotient $\{\{4,4:4\},\{4,3\}\}/2 = \{\{4,4:4\},\{4,3:3\}\}$ must be the universal polytope of its kind, because its vertex-figure is flat. This is of mixed locally toroidal-projective type. Observe as well that $\{\{4,4:4\},\{4,3\}\}^\eta = \{4,3,3:4\} = \{4,3,3\}/2$.

In summary, then, we have

15A2 Theorem *The cosine matrix of* $\{\{4,4:4\},\{4,3\}\}$ *is*

$$\begin{bmatrix} 1 & 1 & 1 & 1 & 1 \\ 1 & 1 & -1 & 1 & 1 \\ 1 & -\frac{1}{3} & 0 & -\frac{1}{3} & 1 \\ 1 & -1 & 0 & 1 & -1 \\ 1 & \frac{1}{3} & 0 & -\frac{1}{3} & -1 \end{bmatrix},$$

with layer and dimension vectors

$$\Lambda = (1,3,8,3,1), \quad D = (1,1,6,2,6).$$

15A3 Remark It is striking that a polytope with so few vertices should have so complicated a realization domain.

Now let $\mathcal{P} := \{\{3,4\},\{4,4:4\}\}$. By and large, $\mathcal{P}$ will stand for whatever polytope happens to be in the current frame; this saves having to devise new notation on each occasion. So, we take the previous generatrix in the reverse order to find a realization P_3 with dimension $d_3 = 6$. From the initial vertex $e_3 + e_6$, we quickly find that the vertex-set consists of all $\pm e_j \pm e_{j+3}$ with $j = 1,2,3$; these form three tetragons $\{4\}$ in mutually orthogonal planes in $\mathbb{E}^6$. Writing the layer vector in the form $\Lambda = (1,2,1,8)$, with the last entry corresponding to the class of an edge, we have cosine vector $\Gamma_3 = (1,0,-1,0)$.

We observe that $\mathcal{P}$ is 2- and 3-collapsible, so that among its realizations are the trigon $\mathsf{P}_1 = \{3\}$ with $d_1 = 2$ and $\Gamma_1 = (1,1,1,-\frac{1}{2})$, and octahedron $\mathsf{P}_2 = \{3,4\}$ with $d_2 = 3$ and $\Gamma_2 = (1,-1,1,0)$. With $\mathsf{P}_0 = \{1\}$ as usual, we see that we have found pure realizations with dimensions summing to $1+2+3+6 = 12$, and so we must have all of them. In conclusion, we have

15A4 Theorem *The cosine matrix of* $\{\{3,4\},\{4,4:4\}\}$ *is*

$$\begin{bmatrix} 1 & 1 & 1 & 1 \\ 1 & 1 & 1 & -\frac{1}{2} \\ 1 & -1 & 1 & 0 \\ 1 & 0 & -1 & 0 \end{bmatrix},$$

with layer and dimension vectors

$$\Lambda = (1,2,1,8), \quad D = (1,2,3,6).$$

15B $\{\{3,4\},\{4,4 \mid 3\}\}$ and $\{\{4,4 \mid 3\},\{4,4 \mid 3\}\}$

We have already met a realization of $\{\{3,4\},\{4,4 \mid 3\}\}$ as the polytope J_4 in the second Gosset class of Section 13E; recall that J_{r+4} is associated with the Coxeter diagram of $[3^{r,2,2}]$ in Figure 13E1. Moreover, we described the realization domain of the abstract analogue of J_5 in Section 14B.

Since

15B1 $$\{\{3,4\},\{4,4 \mid 3\}\}^{\eta} = \{\{4,4 \mid 3\},\{4,4 \mid 3\}\},$$

with η invertible because of the odd hole $\{3\}$, the polytopes $\{\{3,4\},\{4,4 \mid 3\}\}$ and $\{\{4,4 \mid 3\},\{4,4 \mid 3\}\}$ have the same vertices and group, and so their realizations spaces are essentially the same. Thus we need only consider the former and its dual.

The starting point is the faithful realization $\mathsf{P}_2 = \mathsf{J}_4$, namely,

$$\mathsf{P}_2 := \{\{3,4\},\{4,\tfrac{4}{1,2} \mid 3\}\} \in \mathcal{P} := \{\{3,4\},\{4,4 \mid 3\}\},$$

with the 20 vertices of 0_{22} and group $\mathbf{A}_5 \rtimes \mathbf{Z}_2 \cong \boldsymbol{S}_6 \times \boldsymbol{C}_2$. The generatrix of P_2

is given by

15B2
$$\begin{aligned} R_0 &= (2\ 3), \\ R_1 &= (1\ 2), \\ R_2 &= -(0\ 5)(1\ 4)(2\ 3), \\ R_3 &= (4\ 5), \end{aligned}$$

working in the symmetric hyperplane $\mathbb{L}^5$, with the usual conventions. It follows that the vertex-set of P_2 consists of all 20 permutations of $(1,1,1,-1,-1,-1)$, with this itself as the initial vertex. Thus the facets of P_2 are the octahedral faces 0_{11} of a central section 0_{22} of the 6-cube $\{4,3^4\}$.) We therefore find that the layer vector of the abstract polytope $\mathcal{P}$ is $\Lambda = (1,9,9,1)$, and that the cosine vector of P_2 itself is $\Gamma_2 = (1,\frac{1}{3},-\frac{1}{3},-1)$, with dimension $d_2 = 5$.

Now $\mathcal{P}$ is centrally symmetric; clearly, $-I \in \langle R_0,\ldots,R_3\rangle$. Counting possible components allows only one more pure faithful realization P_3 of $\mathcal{P}$, and a single (non-trivial) pure realization P_1 of the central quotient $\mathcal{P}/2$. Since P_1 has the vertices of the 9-simplex, its cosine vector is $\Gamma_1 = (1,-\frac{1}{9},-\frac{1}{9},1)$ and $d_1 = 9$. Using the component equation of Theorem 3C11 shows that P_3 also has dimension $d_3 = 10-5 = 5$, and cosine vector $\Gamma_3 = (1,-\frac{1}{3},\frac{1}{3},-1)$. We conclude this stage of the discussion by stating

15B3 Theorem *The cosine matrix of* $\{\{3,4\},\{4,4\mid 3\}\}$ *is*

$$\begin{bmatrix} 1 & 1 & 1 & 1 \\ 1 & -\frac{1}{9} & -\frac{1}{9} & 1 \\ 1 & \frac{1}{3} & -\frac{1}{3} & -1 \\ 1 & -\frac{1}{3} & \frac{1}{3} & -1 \end{bmatrix},$$

with layer and dimension vectors

$$\Lambda = (1,9,9,1), \quad D = (1,9,5,5).$$

15B4 Remark At this point, we can observe that $\Gamma_1 = \Gamma_2\Gamma_3$, so that $\mathsf{P}_1 = \mathsf{P}_2 \otimes \mathsf{P}_3$.

Since P_2 and P_3 are both 5-dimensional, it is an obvious guess that they must be closely related. At first sight, one might think that $\mathsf{P}_3 = \mathsf{P}_2{}^{\varsigma}$, obtained by changing the sign of R_0 in (15B2). However, this cannot be the case; the 2-faces of $\mathsf{P}_2{}^{\varsigma}$ will be skew hexagons $\{\frac{6}{1,3}\}$. Nevertheless, this is not too far from the truth.

To see what really happens, we go back to (15B1). Geometrically, we have

$$\begin{aligned} \pi &\colon \{\{3,4\},\{4,\tfrac{4}{1,2}\mid 3\}\} \mapsto \{\{4,\tfrac{4}{1,2}\mid 3\},\{\tfrac{4}{1,2},4\mid 3\}\}, \\ \delta &\colon \{\{4,\tfrac{4}{1,2}\mid 3\},\{\tfrac{4}{1,2},4\mid 3\}\} \mapsto \{\{\tfrac{4}{1,2},4\mid 3\},\{4,\tfrac{4}{1,2}\mid 3\}\}, \\ \pi &\colon \{\{\tfrac{4}{1,2},4\mid 3\},\{4,\tfrac{4}{1,2}\mid 3\}\} \mapsto \{\{3,\tfrac{4}{1,2}\},\{\tfrac{4}{1,2},4\mid 3\}\}. \end{aligned}$$

To anticipate what we consider later, we remark that the last polytope is P_3, and that it has no geometric dual. Indeed, the facet of P_3 is $\{3, \frac{4}{1,2}\} = \{3,4\} \# \{3\}$, the blend of an octahedron and its face (the fine Schläfli symbol here suffices for a geometric description), which is full-dimensional.

An interesting feature is that – as we have seen – $\{\{4, \frac{4}{1,2} \mid 3\}, \{\frac{4}{1,2}, 4 \mid 3\}\}$ is abstractly but not geometrically self-dual; indeed, we easily see that its mirror vector is $(4,3,4,3)$. As a final comment, all these fine Schläfli symbols are rigid.

For the dual polytope $\mathcal{P} := \{\{4,4 \mid 3\},\{4,3\}\}$ (as we said before, we use $\mathcal{P}$ rather than $\mathcal{P}^\delta$), we take the generators $R_0, \ldots, R_3$ of (15B2) in the reverse order, yielding a 5-dimensional realization $\mathsf{P}_3 = \{\{4, \frac{4}{1,2} \mid 3\}, \{\frac{4}{1,2}, 3 : \frac{6}{1,3}\}$ with vertex-set all 30 permutations of $(1, 0^4, -1)$, with the latter as initial vertex. These are the vertices of the Minkowski difference polytope $\Delta\mathsf{T}^5 := \mathsf{T}^5 - \mathsf{T}^5$ of a regular 5-simplex T^5. But we already solved this problem in Example 3D5 (here in the case $d = 5$), and so we need merely state

15B5 Theorem *The cosine matrix of* $\{\{3,4\},\{4,4 \mid 3\}\}$ *is*

$$\begin{bmatrix} 1 & 1 & 1 & 1 & 1 \\ 1 & \frac{1}{4} & -\frac{1}{2} & \frac{1}{4} & 1 \\ 1 & -\frac{1}{4} & \frac{1}{6} & -\frac{1}{4} & 1 \\ 1 & \frac{1}{2} & 0 & -\frac{1}{2} & -1 \\ 1 & -\frac{1}{4} & 0 & \frac{1}{4} & -1 \end{bmatrix},$$

with layer and dimension vectors

$$\Lambda = (1,8,12,8,1), \quad D = (1,5,9,5,10).$$

15C $\{\{4,4:4\},\{4,4 \mid 3\}\}$ and its Dual

In this section, we are looking at $\mathcal{Q} := \{\{4,4 : 4\},\{4,4 \mid 3\}\}$ and its dual (exceptionally, we use $\mathcal{Q}$ instead of $\mathcal{P}$, so as to tie this section in with the next). In [99, Section 10C], a construction is described for universal regular toroidal polytopes of the form $\{\{4,4 : 2r\},\{4,4 \mid s\}\}$. From our present point of view, only one of the finite instances of these is of interest, namely, that under discussion. The basic polytopes and their group sit in $\mathbb{E}^6 = \mathbb{E}^3 \times \mathbb{E}^3$, where we can define the generatrix $(R_0, \ldots, R_3)$ to consist of

$$\begin{aligned} R_0 &= \overline{1}\,\overline{4}, \\ R_1 &= (1\ 2), \\ R_2 &= (1\ 4)(2\ 5)(3\ 6), \\ R_3 &= (5\ 6), \end{aligned}$$

with the usual conventions. In other words, the group is derived from the diagram with twists of Figure 15C1:

15C1

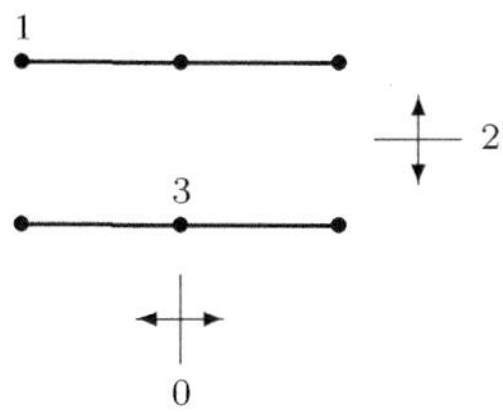

Again as usual, the generator R_j is indicated by the corresponding label j at a node or twist.

It follows easily that we have a realization $\mathsf{Q}_4 := \{\{\frac{4}{1,2}, 4 : \frac{4}{1,2}\}, \{4, \frac{4}{1,2} \mid 3\}\}$ of $\mathcal{Q}$, with initial vertex $(1^6) = (1,\dots,1)$, and general vertex $(\pm 1,\dots,\pm 1)$ with an even number of minus signs; in other words, the vertex-set is that of the 6-dimensional half-cube $1_{31} = \{3, 3^{3,1}\}$. The layer vector is $\Lambda = (1,9,6,6,9,1)$; the cosine vector of this realization of dimension $d_4 = 6$ is $\Gamma_4 = (1, \frac{1}{3}, \frac{1}{3}, -\frac{1}{3}, -\frac{1}{3}, -1)$.

It is worth noting at this stage that $\mathcal{Q}$ is 1-collapsible (that is, has bipartite edge-graph), so that $\{2\} \in \mathcal{Q}$; indeed, $\mathsf{Q}_3 := \{2\}$ is actually centrally symmetric in $\mathcal{Q}$, with dimension $d_3 = 1$ and cosine vector $\Gamma_3 = (1,-1,1,-1,1,-1)$. Hence we have another realization $\mathsf{Q}_1 := \mathsf{Q}_4 \otimes \{2\}$ (actually of $\mathcal{Q}/2$, and so with only 16 vertices) also with dimension $d_1 = 6$, whose cosine vector is $\Gamma_1 = \Gamma_3\Gamma_4 = (1, -\frac{1}{3}, \frac{1}{3}, \frac{1}{3}, -\frac{1}{3}, 1)$. The vertices of Q_1 are those of the product $\mathsf{T}^3 \times \mathsf{T}^3$ of two tetrahedra; the edges of Q_1 are the non-edges of $\mathsf{T}^3 \times \mathsf{T}^3$, and conversely.

To complete the picture of the realization domain of $\mathcal{Q}$, we utilize the fact that $\operatorname{vert}\mathsf{Q}_4 = \operatorname{vert} 1_{31}$, and that the symmetry group of Q_4 is a subgroup of $[3^{1,3,1}]$. Bear in mind that (pure) realizations are paired under multiplication by $\{2\}$, so that we are looking for two 9-dimensional realizations to complete the pattern. However, we also know that we must have a realization whose vertices are a subset of those of the 20-dimensional realization of the 6-cube $\{4,3^4\}$. Moreover, we know from Remark 7B10 that this realization consists of two sets of 32 vertices lying in mutually orthogonal 10-dimensional subspaces, each subset corresponding to an abstract half-cube $\{3,3^{3,1}\} \cong 1_{31}$.

In fact, it is relatively straightforward to lift our given group generators into $\mathbb{E}^{20}$ which, in our context, can be thought of as spanned by the orthonormal vectors e_{ijk} with $1 \leqslant i < j < k \leqslant 6$. We know that our vertex-set will lie in one of the 10-dimensional subsets. A little more calculation shows that e_{123} is invariant under the lifted group, and this reduces the dimension to 9, as was to be expected.

All this aside, we do not need to work out the details of the 9-dimensional realizations (at least, not for the moment). Bear in mind that $\mathcal{Q}/2$ only has three diagonal classes, including the trivial one, and so the cosine vector Γ_2 of its final pure realization Q_2 with $d_2 = 9$ is given from Theorem 3C11 by

$$\Gamma_2 = \tfrac{1}{9}\big(16(1,0^4,1) - (1^6) - 6(1,-\tfrac{1}{3},\tfrac{1}{3},\tfrac{1}{3},-\tfrac{1}{3},1)\big) = (1,\tfrac{1}{9},-\tfrac{1}{3},-\tfrac{1}{3},\tfrac{1}{9},1).$$

The vertices of Q_2 are of the form $x_j \otimes y_k$, where the x_j and y_k are the vertices of two tetrahedra; this is exactly as to be expected from the realization Q_1.

15C2 Remark The realization domain of the abstract symmetric product $\mathcal{T}^m \times \mathcal{T}^m$ of two m-simplices has two non-trivial components: the ordinary direct product $\mathrm{T}^m \times \mathrm{T}^m$ of geometric m-simplices T^m, and the tensor product $\mathrm{T}^m \otimes \mathrm{T}^m$.

Multiplying Q_2 by $\{2\}$ yields the last pure realization $\mathrm{Q}_5 = \mathrm{Q}_2 \otimes \{2\}$ of $\mathcal{Q}$ itself, also with $d_5 = 9$, whose cosine vector is

$$\Gamma_5 = (1, -\tfrac{1}{9}, -\tfrac{1}{3}, \tfrac{1}{3}, \tfrac{1}{9}, -1).$$

In conclusion, then, we have proved

15C3 Theorem *The cosine matrix of* $\{\{4, 4 : 4\}, \{4, 4 \mid 3\}\}$ *is*

$$\begin{bmatrix} 1 & 1 & 1 & 1 & 1 & 1 \\ 1 & -\frac{1}{3} & \frac{1}{3} & \frac{1}{3} & -\frac{1}{3} & 1 \\ 1 & \frac{1}{9} & -\frac{1}{3} & -\frac{1}{3} & \frac{1}{9} & 1 \\ 1 & -1 & 1 & -1 & 1 & -1 \\ 1 & \frac{1}{3} & \frac{1}{3} & -\frac{1}{3} & -\frac{1}{3} & -1 \\ 1 & -\frac{1}{9} & -\frac{1}{3} & \frac{1}{3} & \frac{1}{9} & -1 \end{bmatrix},$$

with layer and dimension vectors

$$\Lambda = (1, 9, 6, 6, 9, 1), \quad D = (1, 6, 9, 1, 6, 9).$$

15C4 Remark By rearranging the rows in order $\Gamma_0, \Gamma_5, \Gamma_1, \Gamma_4, \Gamma_2, \Gamma_3$, we could make the cosine matrix of Theorem 15C3 symmetric.

Moving on to the dual $\mathcal{Q} = \{\{4, 4 \mid 3\}, \{4, 4 : 4\}\}$, we take the previous generatrix in reverse order. From the initial vertex $(0, 0, 1, 0, 0, 1) = e_3 + e_6$, we find that the resulting realization $\mathrm{Q}_4 = \{\{4, \frac{4}{1,2} \mid 3\}, \{\frac{4}{1,2}, 4 : \frac{4}{1,2}\}$ of dimension $d_4 = 6$ has vertices $\pm e_j \pm e_k$, with $j = 1, 2, 3$ and $k = 4, 5, 6$; that is, the vertex-set is that of the (direct) product of two octahedra. We take the layer vector in the form $\Lambda = (1, 2, 1, 8, 8, 16)$, which accords with how much overlap (in terms of basis vectors) a typical vertex has with the initial one $e_3 + e_6$, and then the different cosines obtained by changing signs. So, for example, the two entries '8' correspond to diagonals specified by vertices $\pm e_j \pm e_k$ with just one of j or k equal to 3 or 6, with the positive sign preceding the negative. In other words, the cosine vector of this realization Q_4 is $\Gamma_4 = (1, 0, -1, \frac{1}{2}, -\frac{1}{2}, 0)$.

From this starting point, we can proceed in two different ways. First, we can employ our earlier trick, replacing coordinate entries by their absolute values. This leads to a vertex-set consisting of the $e_j + e_k$ with $j = 1, 2, 3$ and $k = 4, 5, 6$; these nine vertices are clearly those of the facet $\mathrm{Q}_1 = \{4, \frac{4}{1,2} \mid 3\}$ of dimension

$d_1 = 4$, and confirms the observation that $\mathcal{Q}$ is 3-collapsible. Of course, the 3-collapsibility implies that every realization of $\{4,3\mid 3\}$ is one of $\mathcal{Q}$ also; hence, we also have $\mathsf{Q}_2 = \{\frac{4}{1,2},4\mid 3\}$, also of dimension $d_2 = 4$. The corresponding cosine vectors here are $\Gamma_1 = (1,1,1,\frac{1}{4},\frac{1}{4},-\frac{1}{2})$ and $\Gamma_2 = (1,1,1,-\frac{1}{2},-\frac{1}{2},\frac{1}{4})$.

Alternatively, we can consider $\mathsf{Q}_4 \otimes \mathsf{Q}_4$, whose vertices in $\mathbb{E}^{36} = \mathbb{E}^6 \otimes \mathbb{E}^6$ will consist of all those of the form

$$(e_j \otimes e_j + e_k \otimes e_k) \pm (e_j \otimes e_k + e_k \otimes e_j),$$

with (as before) $j = 1,2,3$ and $k = 4,5,6$. Now we should recognize the vertices $e_j \otimes e_j + e_k \otimes e_k$ as just those of Q_1 (with an additional component $\{1\}$). The vertices $\pm(e_j \otimes e_k + e_k \otimes e_j)$, on the other hand, are those of a realization Q_3 of dimension $d_3 = 9$, with cosine vector $\Gamma_3 = (1,-1,1,0,0,0)$.

15C5 Remark The vertex-set of Q_3 is centrally symmetric; however, Q_3 itself is not centrally symmetric (as a realization of $\mathcal{Q}$).

As a check, we verify that

$$\Gamma_0 + 4\Gamma_1 + 4\Gamma_2 + 9\Gamma_3 = 18(1,0,1,0,0,0),$$

the appropriate multiple of the cosine vector of the small simplex realization S.

Since the realization domain of $\mathcal{Q}$ is mostly filled out with realizations of $\mathcal{Q}/2$, we see that we only have one pure faithful realization P_5 to find, of dimension $d_5 = 18 - 6 = 12$, with cosine vector

$$\Gamma_5 = \tfrac{1}{12}\big(18(1,0,-1,0,0,0) - 6(1,0,-1,\tfrac{1}{2},-\tfrac{1}{2},0)\big) = (1,0,-1,-\tfrac{1}{4},\tfrac{1}{4},0).$$

We can now conclude the discussion by stating

15C6 Theorem *The cosine matrix of* $\{\{4,4\mid 3\},\{4,4:4\}\}$ *is*

$$\begin{bmatrix} 1 & 1 & 1 & 1 & 1 & 1 \\ 1 & 1 & 1 & \frac{1}{4} & \frac{1}{4} & -\frac{1}{2} \\ 1 & 1 & 1 & -\frac{1}{2} & -\frac{1}{2} & \frac{1}{4} \\ 1 & -1 & 1 & 0 & 0 & 0 \\ 1 & 0 & -1 & \frac{1}{2} & -\frac{1}{2} & 0 \\ 1 & 0 & -1 & -\frac{1}{4} & \frac{1}{4} & 0 \end{bmatrix},$$

with layer and dimension vectors

$$\Lambda = (1,2,1,8,8,16), \quad D = (1,4,4,9,6,12).$$

Note also that $\Gamma_2\Gamma_4 = \Gamma_5$, implying that $\mathsf{P}_5 = \mathsf{P}_2 \otimes \mathsf{P}_4$.

15D $\{\{4,4:4\},\{4,4:6\}\}$ and its Dual

We begin the investigation of $\mathcal{P} := \{\{4,4:4\},\{4,4:6\}\}$ and its dual with the observation that we have an abstract blend

$$\mathcal{P} = \mathcal{Q} \# \{4\},$$

with $\mathcal{Q} := \{\{4,4:4\},\{4,4 \mid 3\}\}$ as in the previous section. We can do this geometrically, taking $\{\{\frac{4}{1,2},4 \mid \frac{4}{1,2}\},\{4,\frac{4}{1,2} \mid 3\}\} \# \lambda\{4\}$ to identify exactly how the components of the blend interact. The inner product vector (which it is more useful to consider initially) takes the form

$$(6+2\lambda^2, 6-2\lambda^2; 2; 2+2\lambda^2, 2-2\lambda^2; -2; -2+2\lambda^2, -2-2\lambda^2; -6),$$

with layer vector $\Lambda = (1,1;18;6,6;12;9,9;2)$. We use semicolons for some of the dividers, to emphasize the relationship with a cosine vector $(\gamma_0,\ldots,\gamma_5)$ of $\mathcal{Q}$, which becomes $(\gamma_0,\gamma_0;\gamma_1;\gamma_2,\gamma_2;\gamma_3;\gamma_4,\gamma_4;\gamma_5)$ as a cosine vector of $\mathcal{P}$. On the other hand, $\{4\} \in \mathcal{P}$ has cosine vector $(1,-1;0;1,-1;0;1,-1;0)$, while $\{2\}$ has cosine vector $(1,1;-1;1,1;-1;1,1;-1)$.

We have seen that $\mathcal{Q}$ has two pure realizations of each of the dimensions 1, 6 and 9, which are paired under multiplication by $\{2\}$. Because $\{2\} \otimes \{4\} = \{4\}$, we see immediately that each pair gives rise to a single new pure realization of $\mathcal{P}$ of twice the dimension under multiplication by $\{4\}$, namely, $\mathsf{P}_6 := \{4\}$ itself of dimension $d_6 = 2$, P_7 with $d_7 = 12$ and P_8 with $d_8 = 18$. In view of the observations of the previous paragraph, we have thus – in principle – determined the realization domain of $\mathcal{P}$. We shall not write out the cosine matrix in this case; these two paragraphs provide enough information for the interested reader to construct it.

It turns out that the 12-dimensional realization P_7 has no geometric dual; we shall not give the details. In contrast, the 18-dimensional realization P_8 is geometrically dualizable. Working directly from the 9-dimensional realizations of $\mathcal{Q}$ produces a coordinate system which does not seem readily to display the structure of $\mathcal{P}$. However, a suitable relabelling of the basis vectors of $\mathbb{E}^{18}$ makes everything transparent. So, we let the vectors of an orthonormal basis of $\mathbb{E}^{18}$ be e_{rs}, with r,s integers modulo 6 and $r+s$ even; that is, we identify the basis vectors with the vertices of $\{4,4:6\}$.

We need some auxiliary notation first. For a given pair (m,n), we define the reflexion S_{mn} by

$$e_{rs}S_{mn} := \begin{cases} -e_{rs}, & \text{if } (r,s)-(m,n) \in \{(0,0),(\pm1,\pm1),(\pm2,0),(0,\pm2)\}, \\ e_{rs}, & \text{otherwise.} \end{cases}$$

Observe that each S_{mn} changes the signs of exactly half the basis vectors; moreover, $S_{mn}S_{m+3,n+3} = -I$, the central inversion.

The symmetry group $\mathbf{G}$ of P_8 in $\mathbb{E}^{18}$ then has generators R_j for $j = 0,1,2,3$ given by

$$e_{rs}R_j := \begin{cases} e_{rs}S_{33}, & \text{if } j = 0,\\ e_{1-s,1-r}, & \text{if } j = 1,\\ e_{r,-s}, & \text{if } j = 2,\\ e_{sr}, & \text{if } j = 3. \end{cases}$$

Thus the group $\mathbf{G}_0 = \langle R_1, R_2, R_3\rangle$ is just that of $\{4,4:6\}$ with initial vertex e_{00}. There are 64 vertices in all.

15D1

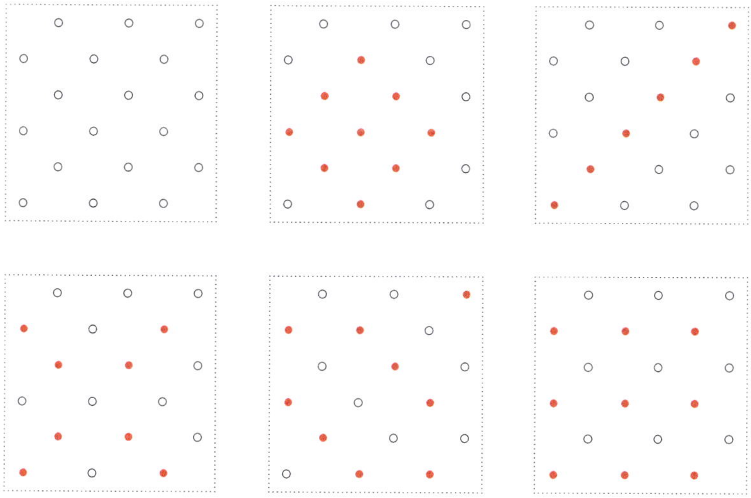

The initial vertex of P_8 is $\sum e_{rs}$, and the general vertex is of the form $\sum \pm e_{rs}$, with the negative sign patterns illustrated in Figure 15D1 by red dots. (We have not illustrated the complements of the first, third and fourth, obtained by multiplying by $-I$; those of the second, fifth and sixth are of the same kind.) For example, the first picture gives the initial vertex, while the second shows the effect on it of S_{00} (say). For the rest, the third is obtained by (for example) $S_{11}S_{22}$, the fourth by $S_{20}S_{40}$, the fifth by $S_{00}S_{11}S_{22}$, and the last by $S_{00}S_{20}S_{40}$. Adjacency of vertices is given by multiplication by any S_{mn}.

We can also look at the dual polytope $\{\{4,4:6\},\{4,4:4\}\}$ with this coordinate system. It is important that we use this opportunity to correct a mistake in [99, Section 10C]. It is claimed there (on p. 373) that

$$\{\{4,4:6\},\{4,4:4\}\} = \{\{4,4 \mid 3\},\{4,4:4\}\} \# \{2\};$$

consequently, it is actually stated in effect (though not in numbers) that $9216 = 2 \cdot 2304$. In fact, as the group orders make clear, the polytope on the right is a quotient of $\{\{4,4:6\},\{4,4:4\}\}$ of index 2.

Taking the generators $R_0, \ldots, R_3$ above in the reverse order, we see the initial vertex is $e_{00} + e_{11} + e_{20} + e_{15}$. Moreover, the general vertex is of the form $\pm e_{r-1,s} \pm e_{r,s+1} \pm e_{r+1,s} \pm e_{r,s-1}$, with r, s integers modulo 6 such that $r + s$ is odd, and with an even number of minus signs. That is, with each face of the original copy of $\{4,4:6\}$ (with vertices the e_{rs}), or with each vertex of its dual, is associated eight vertices of our realization, thus giving $8 \cdot 18 = 144$ vertices in all. We see that two such vertices are adjacent precisely when they have in common two basis vectors e_{rs} with the same signs. Of course, this means that the vertices come from adjacent faces of the torus $\{4,4:6\}$, with an additional condition of compatibility of signs.

We can now describe the quotient mapping taking $\{\{4,4:6\},\{4,4:4\}\}$ to $\{\{4,4 \mid 3\},\{4,4:4\}\}$. Indeed, this mapping is given by

$$\varepsilon_{-1,0}e_{r-1,s} + \varepsilon_{0,1}e_{r,s+1} + \varepsilon_{1,0}e_{r+1,s} + \varepsilon_{0,-1}e_{r,s-1} \mapsto \varepsilon_{1,0}\varepsilon_{0,-1}e_{r+s-1} + e_{r-s},$$

with $\varepsilon_{ij} \in \{\pm 1\}$ having product 1 (thus $\varepsilon_{1,0}\varepsilon_{0,1} = \varepsilon_{-1,0}\varepsilon_{0,-1}$, and so on). Note particularly how adjacency of vertices is preserved.

The layer vector of $\{\{4,4:6\},\{4,4:4\}\}$ is

$$\Lambda = (1,2,4,1;8,16,8;16,16;32;16,16;2,2,4),$$

where the semicolons partition Λ according to the layer vector $(1,4,8,4,1)$ of its facet $\{4,4:6\}$. To a cosine vector $(\gamma_0,\ldots,\gamma_5)$ of $\{\{4,4 \mid 3\},\{4,4:4\}\}$ corresponds

$$(\gamma_0,\gamma_5,\gamma_3,\gamma_0;\gamma_1,\gamma_4,\gamma_1;\gamma_2,\gamma_2;\gamma_2;\gamma_1,\gamma_4;\gamma_0,\gamma_5,\gamma_3).$$

Multiplication by $\{2\}$, which is centrally symmetric as a realization, changes signs in alternate blocks. The 18-dimensional realization which we have just constructed has cosine vector

$$(1,0,0,-1;\tfrac{1}{2},0,-\tfrac{1}{2};\tfrac{1}{4},-\tfrac{1}{4};0;0,0;0,0,0);$$

we thus have another 18-dimensional realization with cosine vector

$$(1,0,0,-1;-\tfrac{1}{2},0,\tfrac{1}{2};\tfrac{1}{4},-\tfrac{1}{4};0;0,0;0,0,0).$$

The list is completed by a 36-dimensional realization, with cosine vector

$$(1,0,0,-1;0,0,0;-\tfrac{1}{4},\tfrac{1}{4};0;0,0;0,0,0).$$

We remark that the rest of the cosine matrix is filled out by the 6 cosine vectors of the pure realizations Q of $\{\{4,4 \mid 3\},\{4,4:4\}\}$ (with dimensions summing to 36), together with 6 more for the realizations $\mathsf{Q}^{\zeta} = \mathsf{Q} \otimes \{2\}$ (and thus adding another 36 to the dimension count); recall that $\{2\}$ is not a realization of $\{\{4,4 \mid 3\},\{4,4:4\}\}$. Thus the calculation of the last cosine vector is purely mechanical, and gives no insight into the geometry of this 36-dimensional realization. Once again, we shall not write out the whole cosine matrix.

15E $\{\{4,4\mid 4\},\{4,4\mid 3\}\}$ and its Dual

While we are not in a position to describe their realization domains completely, nevertheless we can say quite a lot about them. Even though it has more vertices (namely, 512 as opposed to 288), we begin with $\{\{4,4\mid 4\},\{4,4\mid 3\}\}$ rather than its dual.

From [99, (8E7)], we know that

$$\mathcal{P} := \{\{4,4\mid 4\},\{4,4\mid 3\}\} = \mathbf{2}^{\{4,4\,|\,3\}},$$

the finiteness and universality being ensured by the weak neighbourliness of the torus $\{4,4\mid 3\}$. Since $\{4,4\mid 3\}$ has 9 vertices, we directly have a realization of $\mathcal{P}$ in $\mathbb{E}^9$, with the vertices of the 9-cube $\{4,3^7\}$. The symmetry group of $\{4,4\mid 3\}$ suggests that we label the standard orthonormal basis vectors of $\mathbb{E}$ by e_{jk}, with $j,k \in \{0,1,2\}$; subsequent calculations involving the basis vectors assume that the indices are taken modulo 3. Thus the generatrix (R_0,R_1,R_2,R_3) of the first realization $\mathsf{P} \in \mathcal{P}$ is given by

$$R_0\colon e_{jk} \mapsto \begin{cases} -e_{00}, & \text{if } (j,k)=(0,0),\\ e_{jk}, & \text{otherwise,}\end{cases}$$
$$R_1\colon e_{jk} \mapsto e_{1-j,k},$$
$$R_2\colon e_{jk} \mapsto e_{kj},$$
$$R_3\colon e_{jk} \mapsto e_{j,-k}.$$

15E1

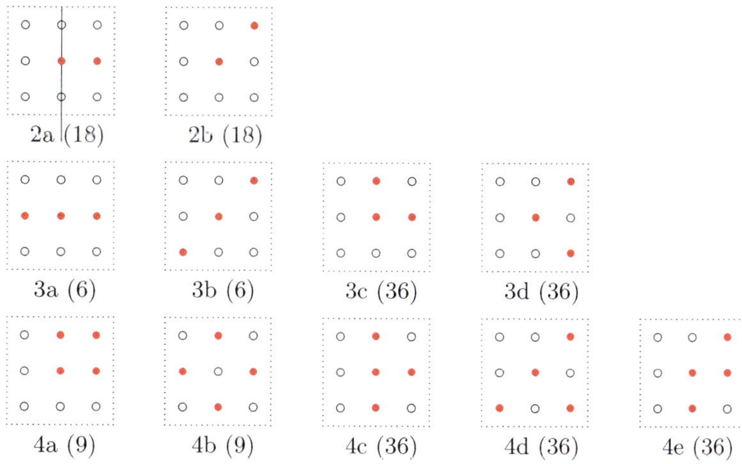

Even if we were unfamiliar with the geometric construction, it is clear that the initial vertex is $u = e_{00}+\cdots+e_{22}$, the initial vertex of the cube. The general vertex is of the form $\pm\, e_{00} \pm \cdots \pm e_{22}$. Since the edge-graph of P is the same as that of the cube, the vertices fall into layers $r = 0,\ldots,9$, according to their (edge-path) distance from u. However, not all vertices in a given layer need be equivalent under the action of $\mathbf{H} := \langle R_1, R_2, R_3 \rangle$, and so they are further divided into sublayers. These can be specified by their *sign patterns*, that is, the specification of negative signs up to symmetry under $\mathbf{H}$. We label these sublayers $ra, rb, \ldots$, with $r = 0, \ldots, 9$. Layers 0 and 1 have just one sublayer, as do the antipodal layers 8 and 9. Layers 2, 3 and 4 are pictured in Figure 15E1; the number in brackets after the label is the number of different copies of that pattern. There is one copy of pattern 0 and there are nine copies of pattern 1.

We label the remaining sublayers according to their complements, so that, for example, the sign pattern of sublayer 5d is

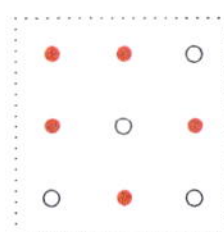

Moreover, bear in mind that each sign pattern actually represents a family of sets equivalent under $\mathbf{H}$, so that, again for example,

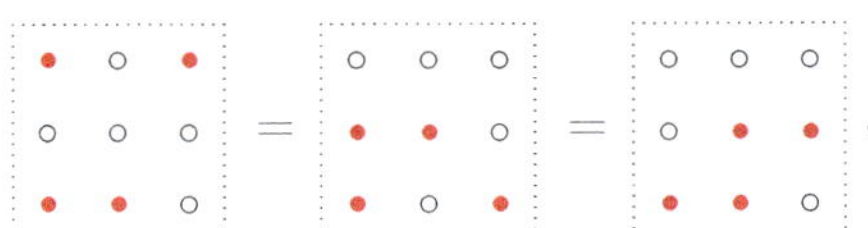

which we recognize as pattern 4e.

We can now write down the layer vector of $\mathcal{P}$; we use semicolons to separate the different layers in the 9-cube, split into sublayers by commas. Thus we have

$$\Lambda = (1; 9; 18, 18; 6, 6, 36, 36; 9, 9, 36, 36, 36; \ldots),$$

the second half repeating the first in reverse order. In other words, we follow sublayer 4e by 5e, then 5d, and so on. In all, we have 26 sublayers.

15E2 Remark All diagonals of $\mathcal{P}$ are symmetric, as is easily seen from our faithful realization P; we just observe that we can interchange any given vertex v (associated with some sign pattern) with u just by changing signs in the pattern. Thus the realization domain has 26 pure components, each with a 1-dimensional Wythoff space (this applies to the realizations $\{1\}$ and $\{2\}$ as well).

15E3 Remark It is obvious that $\mathcal{P}$ is centrally symmetric. Since it has edge-paths of odd length which lead from one vertex of $\mathcal{P}$ to an antipodal one, we see that $\{2\}$ is centrally symmetric in $\mathcal{P}$. In particular, we obtain from P a 9-dimensional realization $\mathsf{P}^{\varsigma} = \mathsf{P} \otimes \{2\}$ of $\mathcal{P}/2$. Note that multiplication by $\{2\}$ changes signs of alternate blocks in cosine vectors of realizations of $\mathcal{P}$.

We know that $\mathcal{Q} := \{\{4, 4 : 4\}, \{4, 4 \mid 3\}\}$ is a quotient of $\mathcal{P}$ of index 16. The quotient is obtained by identifying antipodal vertices of the facets $\{4, 4 \mid 4\}$ of $\mathcal{P}$; it is, perhaps, of interest to see how this works with our realization P. There are nine vertices antipodal to u in facets containing u. We can easily see that the initial facet permits sign changes of those e_{jk} with $j,k = 0,1$, so that these antipodal vertices have sign pattern 4a. Applying two successive sign changes of pattern 4a results either in another of type 4a, or in one of type 6b. Indeed, u and any two vertices with sign patterns 4a or 6b have mutual (edge-path) distances 4 or 6.

Now this set of $1+9+6 = 16$ vertices itself has cosine vector $(1, \frac{1}{9}, -\frac{1}{3})$, which we recognize from the previous discussion as that of the 9-dimensional realization Q_2 of $\{\{4,4 : 4\},\{4,4 \mid 3\}\}$, which is actually one of $\{\{4,4 : 4\},\{4,4 \mid 3\}\}/2$. Applying common sign changes to this set of 16 vertices then results in 32 sets of 16 vertices of P, and hence also of the abstract polytope $\mathcal{P}$, whose identification produces $\mathcal{Q}$.

If we apply the halving operation η to $\{4,4 \mid 3\}$, then we produce another copy with the same vertices, but with different edges and faces. We can similarly apply η to $\{\{4,4 : 4\},\{4,4 \mid 3\}\}$; unsurprisingly, this results in an isomorphic copy, this time with the same vertices and edges. However, if we apply η to our 9-dimensional realization $\{\{4, \frac{4}{1,2}|4\},\{\frac{4}{1,2},4 : 3\}\}$, then the new copy is congruent to the original; that is, no different realization is obtained. What η does do is interchange the rôles of sign patterns 4a and 4b, and patterns 6b and 6a, for the inscribed sets of 16 vertices to be identified under the quotient map from $\mathcal{P}$ to $\mathcal{Q}$.

We can argue as before that, since P has the vertices of the 9-cube, other pure realizations of $\mathcal{P}$ will arise from the remaining realizations of the cube; however, it is not at all obvious how these other realizations split, or even into how many pure components.

We now move on to the dual $\{\{4,4 \mid 3\},\{4,4 \mid 4\}\}$, which we shall denote by $\mathcal{P}$ rather than $\mathcal{P}^{\delta}$. We have a 9-dimensional representation of its group, which is just $\mathbf{G}$ with its generatrix reversed. Looking at its Wythoff space occasions a bit of a surprise, since it is spanned by two vectors $e_{02}+e_{20}+e_{12}+e_{21}$ and e_{22}. Hence, for some choice of λ and μ, the general vertex of the realization, P say, is

$$\lambda(\pm e_{r-1,s} \pm e_{r,s-1} \pm e_{r+1,s} \pm e_{r,s+1}) \pm \mu e_{rs};$$

with 9 choices for (r,s) and 32 possible signs, we obtain the required $32 \cdot 9 = 288$ vertices.

Since we have a 2-dimensional Wythoff space, we must have asymmetric diagonals. The possible inner products of vertices with the initial one, taken to be $v := \lambda(e_{02}+e_{20}+e_{12}+e_{21}) + \mu e_{22}$, are

$$\begin{array}{lll} \pm 4\lambda^2 \pm \mu^2, & \pm 2\lambda^2 \pm \mu^2, & \pm \mu^2; \\ \pm \lambda^2 \pm 2\lambda\mu, & \pm \lambda^2; & \\ \pm 2\lambda^2, & 0. & \end{array}$$

These are grouped according to the displacement of the 'centre' (r,s) from the initial position $(2,2)$; different choices of signs then yield the different inner products. If we normalize by

$$\lambda = \tfrac{1}{2}\cos\vartheta, \quad \mu = \sin\vartheta,$$

then these inner products become entries in the cosine vector.

The asymmetric diagonals then correspond to the values $\pm\lambda^2$. It is enough to consider the value λ^2, which typically arises from (say) $w = \lambda(e_{02} + e_{10} + e_{22} + e_{11}) - \mu e_{12}$. Note that the signs of e_{02} and e_{12}, where the basis vectors overlap, must differ; the signs of e_{10} and e_{11} are actually irrelevant. Applying the reflexion $T\colon e_{jk} \mapsto e_{-j,k}$ brings w to $wT = \lambda(e_{02} + e_{20} + e_{12} + e_{21}) - \mu e_{22}$. Changing the sign of e_{22} to bring wT into line with v simultaneously changes $vT = \lambda(e_{02}+e_{10}+e_{22}+e_{11})+\mu e_{12}$ into $w' = \lambda(e_{02}+e_{10}-e_{22}+e_{11})+\mu e_{12}$, which is not w. We conclude that two of the diagonal classes of $\mathcal{P}$ are asymmetric.

In view of this, we see that we must have another pure realization of $\mathcal{P}$ with a Wythoff space which is not a line. Fortunately, this is comparatively easy to find. We know that $\{4,4 \mid 4\}$ has two 4-dimensional realizations, which are related by $\boldsymbol{\zeta}$. We apply the same operation to P which, in the geometric context, is $\boldsymbol{\kappa}_1$. We saw already in effect (when we looked at the quotient map from $\mathcal{P}$ to $\mathcal{Q}$) that the central symmetry of the initial facet is

$$K_1 : e_{jk} \mapsto \begin{cases} -e_{jk}, & \text{if } j,k = 0,1, \\ e_{jk}, & \text{otherwise.} \end{cases}$$

Thus we have the operation

$$\boldsymbol{\kappa}_1\colon (R_3,\ldots,R_0) \mapsto (R_3, R_2K_1, R_1, R_0) =: (S_0,\ldots,S_3),$$

which produces a geometrically distinct realization of $\mathcal{P}$, with the same 2-dimensional Wythoff space as before. Note that the mirror of

$$S_1 : e_{jk} \mapsto \begin{cases} -e_{kj}, & \text{if } j,k = 0,1, \\ e_{kj}, & \text{otherwise,} \end{cases}$$

has dimension $\dim S_1 = 4$ as opposed to $\dim R_2 = 6$.

15F Polytopes of Type $\{3^{m-2}, 6\}$

Central to the approach of the next Section 15G will be the geometry of the polytope $\mathcal{Q}_m = \{3^{m-2}, 6\}$ with 3-coface of type $\{3, 6 : 6\}$. This has a natural faithful realization Q_m in $\mathbb{C}^m$, which is obtained by twisting the unitary reflexion group with diagram

15F1 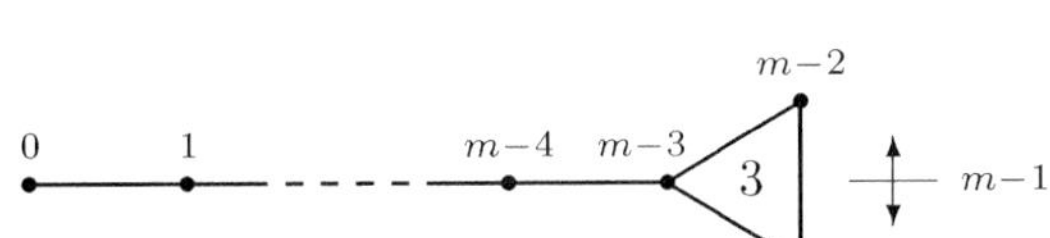

The $3m$ vertices are all $\omega^j e_k$, with $\omega := \frac{1}{2}(-1+\sqrt{3})$ a cube root of 1, $j = 0, 1, 2$, and e_k the kth vector of the standard orthonormal basis of $\mathbb{C}^m$ for $k = 1, \ldots, m$. We shall always write $\overline{\omega} := \omega^2$.

The generatrix $(S_0, \ldots, S_{m-1})$ of the symmetry group of order $2 \cdot 3^{m-1} \cdot m!$ is given by

15F2
$$\begin{aligned} S_j &= (j{+}1 \ j{+}2) \quad \text{for } j = 0, \ldots, m-3, \\ S_{m-2} &= (m{-}1 \ \omega m), \\ S_{m-1} &: z \mapsto \overline{z}. \end{aligned}$$

The transpositions interchange (linearly) the basis vectors with those labels, while S_{m-2} similarly interchanges e_{m-1} and ωe_m, and so is the linear mapping

$$z \mapsto (\zeta_1, \ldots, \zeta_{m-2}, \overline{\omega}\zeta_m, \omega\zeta_{m-1}).$$

We need a little more of the geometric picture for what follows. For the $9\binom{m}{2}$ edges, we easily see that each triple of vertices $\omega^j e_k$ (with $j = 0, 1, 2$) is joined to each other triple with a different index k, giving nine such edges in all. With three different indices k are thus associated 27 trigonal edge-circuits. However, when $m = 3$ (that is, for $\{3, 6 : 6\}$ itself), only two-thirds of these circuits correspond to faces; if the product of the multipliers ω^j is 1, then instead we have a 3-hole.

15F3 Remark In this context, it is worth recalling from Theorem 2E6 that

$$\{3, q : 2r\} = \{3, q \mid \cdot, r\}.$$

At this stage, we could show that our realization is that of the universal polytope, although this is really implicit in the construction. However, since the same ideas will be used in the proof of Theorem 15G2, it is unnecessary to do that here.

15F4

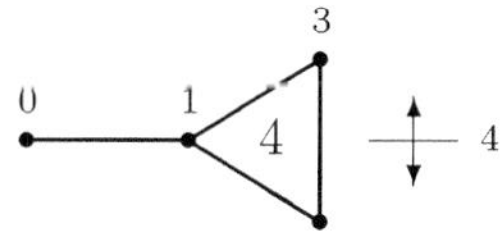

The only other locally toroidal polytope of type $\{3^{m-2}, 6\}$ which is of interest here is $\mathcal{P} := \{\{3, 3\}, \{3, 6 : 8\}\}$, which is derived geometrically by twisting the diagram of Figure 15F4, in the manner shown. Note that, if we extend the diagram by any further branch to the left (even an unmarked one), then the corresponding group is infinite.

The generatrix $(R_0, \ldots, R_3)$ of the realized 4-polytope P_5 in $\mathbb{C}^4$ is given by

15F5
$$\begin{aligned} zR_0 &= z - \tfrac{1}{2}\langle z, v\rangle v, \quad \text{where } v = (-1, 1, 1, 1), \\ zR_1 &= (\zeta_1, -i\zeta_3, i\zeta_2, \zeta_4), \\ zR_2 &= (\zeta_1, \zeta_2, \zeta_4, \zeta_3), \\ zR_3 &= (\overline{\zeta}_1, \overline{\zeta}_3, \overline{\zeta}_2, \overline{\zeta}_4); \end{aligned}$$

see the notes at the end of the section. Observe that $zR_3R_2R_3 = (\zeta_1, \zeta_4, \zeta_3, \zeta_2)$. The initial vertex can be taken to be $(2, 0, 0, 0)$; then the 80 vertices of P_5, with dimension $d_5 = 8$, are all

15F6
$$\begin{array}{ll} 2i^\lambda e_j, & \lambda = 0,1,2,3,\ j = 1,2,3,4, \\ (i^\kappa, i^\lambda, i^\mu, i^\nu), & \kappa,\lambda,\mu,\nu = 0,1,2,3,\ \kappa+\lambda+\mu+\nu \equiv 0 \pmod 4. \end{array}$$

We conclude from this that the abstract polytope $\mathcal{P}$ has automorphism group $\boldsymbol{G}$ of order $80 \cdot 4^2 \cdot 3! = 15360$.

The layer vector Λ is much more complicated than these coordinates for the vertices might suggest. The group $\mathbf{H}$ of the vertex-figure consists of all permutations of the last three coordinates, together with mappings of the form $z \mapsto (\zeta_1, i^\lambda \zeta_2, i^\mu \zeta_3, i^\nu \zeta_4)$ with $\lambda+\mu+\nu \equiv 0 \pmod 4$, and complex conjugation. We find that we have seven layers of which three are central in P_5; they contain the following vertices:

$$\begin{array}{ll} \mathcal{L}_0\colon & (2,0,0,0), \\ \mathcal{L}_1\colon & (1, i^\lambda, i^\mu, i^\nu), \text{with } \lambda+\mu+\nu \equiv 0 \pmod 4, \\ \mathcal{L}_2\colon & (i^\kappa, i^\lambda, i^\mu, i^\nu), \text{with } \kappa \text{ odd}, \\ \mathcal{L}_3\colon & (0, 2i^\lambda, 0, 0), \text{with appropriate permutations and } \lambda = 0,1,2,3, \\ \mathcal{L}_4\colon & (2i^\kappa, 0, 0, 0), \text{with } \kappa \text{ odd}, \\ \mathcal{L}_5\colon & (-1, i^\lambda, i^\mu, i^\nu), \text{with } \lambda+\mu+\nu \equiv 2 \pmod 4, \\ \mathcal{L}_6\colon & (-2,0,0,0). \end{array}$$

In other words, $\Lambda = (1, 16, 32, 12, 2, 16, 1)$. We have arranged things so that the induced cosine vector of the vertex-figure is $\Gamma^{\boldsymbol{v}} = (1, \gamma_1, \gamma_2, \gamma_3)$, where the induced layer vector is $\Lambda^{\boldsymbol{v}} = (1, 6, 6, 3)$.

We notice at once that we can scale the vertices by powers of i. Of course, $\mathcal{P}$ is centrally symmetric, so that we can identify opposite vertices to give $\mathcal{P}/2$. However, $\mathcal{P}/2$ itself is still centrally symmetric, under the action induced by multiplication by $\pm i$, and this yields $\mathcal{P}/4$ with the vertices identified in fours. To be more specific, in $\mathcal{P}/4$ the three layers $\mathcal{L}_0, \mathcal{L}_4$ and $\mathcal{L}_6$ are identified, as are layers $\mathcal{L}_1, \mathcal{L}_2$ and $\mathcal{L}_5$; furthermore, the twelve vertices in $\mathcal{L}_3$ are identified in fours.

At this stage, we could begin the analysis of the realization domain $\mathcal{N}$, using Theorem 3C14 as applied to the vertex-figure of $\mathcal{P}/4$, and then work our way up through $\mathcal{P}/2$ to $\mathcal{P}$ itself. We have seen how such ideas enabled us to tackle some polytopes in Section 4G. However, here we have one more trick up our sleeve. Let $(\boldsymbol{r}_0, \ldots, \boldsymbol{r}_3)$ be the generatrix of $\mathcal{P}$, and let $\boldsymbol{N}$ be the normal closure of $(\boldsymbol{r}_2\boldsymbol{r}_3)^3$. Then $\boldsymbol{G}/\boldsymbol{N} \cong [3,3,3]$; that is, we have a quotient map $\mathcal{P} \searrow \{3,3,3\}$. Indeed, the mapping actually induces a quotient of $\mathcal{P}/4$.

The intrinsic layer vector of $\mathcal{P}/4$ is $(1, 16, 3)$, and then $\mathsf{P}_1 = \{3,3,3\}$ has cosine vector

$$\Gamma_1 = (1, -\tfrac{1}{4}, 1),$$

with dimension $d_1 = 4$. We need one more component P_2 in $\mathcal{P}/4$ (apart from $\{1\}$); for this, we can use the component equation of Theorem 3C11 to find that $d_2 = 20 - 1 - 4 = 15$ and

$$\Gamma_2 = \tfrac{1}{15}\big(20(1,0,0) - (1,1,1) - 4(1,-\tfrac{1}{4},1)\big) = (1,0,-\tfrac{1}{3}).$$

It is also easy to see what this realization looks like: its 20 vertices are those of five tetrahedra in mutually orthogonal subspaces of $\mathbb{E}^{15}$, with edges joining all pairs of vertices in different tetrahedra.

Now we can look at $\mathcal{P}/2$, with intrinsic layer vector $(1, 16, 6, 16, 1)$. The general cosine vector of $\mathcal{P}/4$ embeds as one of $\mathcal{P}/2$ by

$$(1, \alpha, \beta) \hookrightarrow (1, \alpha, \beta, \alpha, 1),$$

as we should expect. However, since $\mathcal{P}/2$ is centrally symmetric, and $\mathcal{P}/4$ takes care of the contributions from the small simplex, here we must consider cosine vectors of the form $(1, \alpha, 0, -\alpha, -1)$. Observe that this follows in any case from Λ-orthogonality with respect to Γ_j for $j = 0, 1, 2$.

The induced cosine vector of the vertex-figure is now of the form $\Gamma^v = (1, \alpha, -\alpha, 0)$, so that Theorem 3C14 yields

$$16\alpha^2 = 1 + 6\alpha + 6(-\alpha) + 3 \cdot 0 = 1,$$

with solutions $\alpha = \pm\frac{1}{4}$. We thus obtain

$$\begin{aligned} \Gamma_3 &= (1, \tfrac{1}{4}, 0, -\tfrac{1}{4}, -1), \\ \Gamma_4 &= (1, -\tfrac{1}{4}, 0, \tfrac{1}{4}, -1). \end{aligned}$$

We shall say a little more about these two realizations later.

We finally treat the remaining pure components of $\mathcal{P}$, noting that the general cosine vector of $\mathcal{P}/2$ embeds in that of $\mathcal{P}$ by

$$(1, \alpha, \beta, \gamma, \delta) \hookrightarrow (1, \alpha, \gamma, \beta, \delta, \beta, \gamma, \alpha, 1).$$

In fact, we already have P_5 with dimension $d_5 = 8$ and cosine vector

$$\Gamma_5 = (1, \tfrac{1}{2}, 0, 0, 0, -\tfrac{1}{2}, -1),$$

and a dimension count shows that we only need one more component P_6 (of the staurotope realization X) of dimension $d_6 = 32$. Then Remark 3C12 in the centrally symmetric case says that

$$\begin{aligned} \Gamma_6 &= \tfrac{1}{32}\big(40(1,0,0,0,0,0,-1) - 8(1,\tfrac{1}{2},0,0,0,-\tfrac{1}{2},-1)\big) \\ &= (1, -\tfrac{1}{8}, 0, 0, 0, \tfrac{1}{8}, -1). \end{aligned}$$

In summary, then, we have shown

15F7 Theorem *The cosine matrix of* $\{\{3,3\},\{3,6:8\}\}$ *is*

$$\begin{bmatrix} 1 & 1 & 1 & 1 & 1 & 1 & 1 \\ 1 & -\frac{1}{4} & -\frac{1}{4} & 1 & 1 & -\frac{1}{4} & 1 \\ 1 & 0 & 0 & -\frac{1}{3} & 1 & 0 & 1 \\ 1 & \frac{1}{4} & -\frac{1}{4} & 0 & -1 & \frac{1}{4} & 1 \\ 1 & -\frac{1}{4} & \frac{1}{4} & 0 & -1 & -\frac{1}{4} & 1 \\ 1 & \frac{1}{2} & 0 & 0 & 0 & -\frac{1}{2} & -1 \\ 1 & -\frac{1}{8} & 0 & 0 & 0 & \frac{1}{8} & -1 \end{bmatrix},$$

with layer and dimension vectors

$$A = (1,16,32,12,2,16,1), \quad D = (1,4,15,10,10,8,32).$$

As we said earlier, we can say a little more about the two 10-dimensional realizations P_3 and P_4. From the forms of their cosine vectors, we might suspect that they have the same vertices; we shall see that this is the case. The point to note is that, given a realization of the vertex-figure $\{3,6:8\}$ with cosine vector $(1,\gamma_1,\gamma_2,\gamma_3)$, there is another with the same vertices in which γ_1 and γ_2 are swapped. This is easy to see by inspection of the picture of the torus in Figure 15F8:

15F8

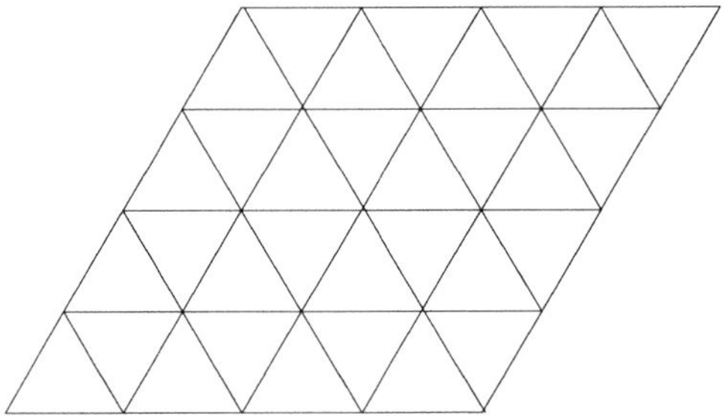

In this, of course, opposite sides of the parallelogram of triangles are identified. Diagonals in class $\mathcal{D}_2$ join opposite vertices of triangles which share an edge.

Now in P_3 the induced cosine vector of the vertex-figure is $(1,\frac{1}{4},-\frac{1}{4},0)$; swapping diagonal classes $\mathcal{D}_1$ and $\mathcal{D}_2$ clearly gives the induced cosine vector $(1,-\frac{1}{4},\frac{1}{4},0)$ of the vertex-figure of P_4. Moreover, under this same change of sign from $\frac{1}{4}$ to $-\frac{1}{4}$, the edges now go from a given vertex to the vertex-figure at the antipodal vertex. The claim that we have made is therefore justified.

While we do not know coordinates for the vertices of P_3 or P_4, the data that we have enable us to draw simple projections of these realizations.

15F9

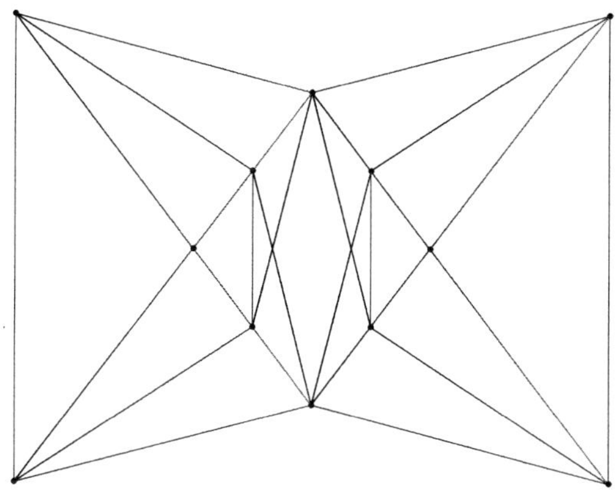

A projection of the realization P_3

The vertex-figure at the (single) vertex at the top left corner (the immediately adjacent vertices) going from the bottom left corner consists of 1 (at the corner), 6, 3 and 6 vertices. The picture has up-down and left-right symmetry.

15F10

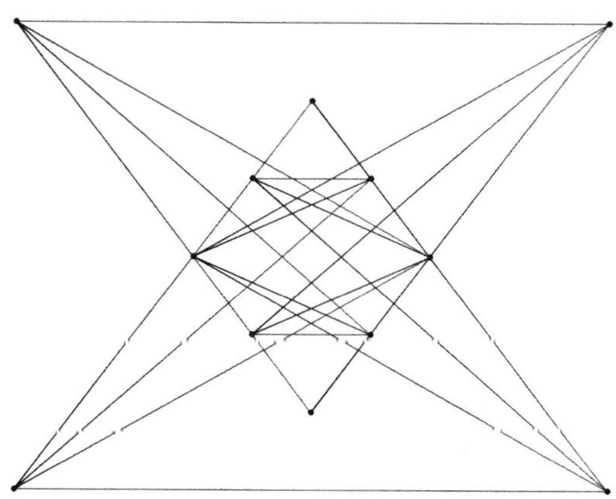

A projection of the realization P_4

15F11 Remark If we adjoin to the vertex-set of P_5 suitably scaled multiples of the normal vectors to the reflexion hyperplanes of its symmetry group, then we obtain another highly symmetric set V. More specifically, we have all 240 permutations of

15F12
$$\begin{aligned} &(2i^{\lambda}, 0, 0, 0), \quad \text{all } \lambda, \\ &(i^{\kappa}, i^{\lambda}, i^{\mu}, i^{\nu}), \quad \kappa+\lambda+\mu+\nu \equiv 0 \pmod 2, \\ &(\pm 1 \pm i, \pm 1 \pm i, 0, 0). \end{aligned}$$

The complex symmetry group $\mathbf{G}$ of V is generated by the involutory reflexions in the hyperplanes whose normals are in V; as for P_5, we can also add in complex conjugation. As a real set of points in $\mathbb{E}^8$, we just have the vertex-set of the Gosset polytope 4_{21}.

It is worth noting that, apart from the groups $G(m, p, n)$, $\mathbf{G}$ is the only finite irreducible reflexion group acting on a space $\mathbb{C}^n$ for some n which is not generated by n reflexions. For further details, see [117, 121].

Notes to Section 15F

1. The generatrix for the underlying unitary group of Figure 15F4 given by Coxeter in [30, p. 172] is incorrect. In fact, those reflexions only generate a subgroup $\mathbf{B}_4$.
2. On the facing page 173 of [30] is a picture of the realization P_5 of $\{\{3,3\},\{3,6:8\}\}$ (drawn by McMullen), with its Petrie polygon projected into a planar icosagon.

15G Polytopes of Type $\{3^{m-1}, 6, 3^{n-1}\}$

The indices of the section heading have been chosen to simplify the total rank $m+n$ though, in fact, there is only one example where we need the implied generality; moreover, in a sense we have already met it. The $(m+n)$-simplex $\{3^{m+n-1}\}$ satisfies the conditions of Proposition 4A15 with respect to its m-face $\{3^{m-1}\}$, which in turn come from those of Proposition 2F7. Thus we write $\mathcal{P} := \{3^{m-1}\} \otimes \{3^{m+n-1}\}$ instead of $\{3^{m-1}\} \# \{3^{m+n-1}\}$ in this case, to point towards the realization domain. That is, $\mathcal{P}$ has four pure realizations, which are just the tensor products in pairs of the pure realizations of the component simplices (and thus of dimensions $1, m, n$ and mn). The reason for mentioning this polytope here is that (for $m, n > 1$) its section of type $\{3,6,3\}$ is actually $\{\{3,6\}_{(2,2)}, \{6,3\}_{(2,0)}\}$. In other words, we have a doubly infinite family of locally toroidal regular polytopes whose realization domains are easily identified.

15G1 Remark Observe that the dual of such a polytope cannot be faithfully realized, because its $(n+1)$-face isomorphic to the dual of $\{2\} \# \{3^n\}$ cannot.

The real origins of this section lay in finding the realization domain of the self-dual locally toroidal regular 4-polytope $\{\{3,6:6\},\{6,3:6\}\}$ with 27 vertices and facets, and automorphism group of order 2916; see [99, Section 11E] for more details of the other known finite members of the family $\{3,6,3\}$. It proved relatively straightforward to find explicit generators for the symmetry group of an 18-dimensional faithful realization in $\mathbb{C}^9$ (see below), and this then suggested the possibility of extending to a family of which each was the vertex-figure of the next. Perhaps a little surprisingly, it turned out that these regular polytopes were all universal of their type.

We begin by stating the main result of the section.

15G2 Theorem *For each $m \geqslant 4$, the universal locally toroidal abstract regular m-polytope $\mathcal{P}_m$ of type $\{3^{m-3}, 6, 3\}$ with 4-coface $\mathcal{P}_4 = \{\{3,6:6\},\{6,3:6\}\}$ is finite. It has $9(m-1)$ vertices, 3^{m-1} facets and automorphism group of order $g_m = 2 \cdot 3^{2m-3} \cdot (m-1)!$.*

We could even initiate the family with the toroid $\mathcal{P}_3 = \{6, 3 : 3\}$; indeed, this is the starting point for a later induction argument.

For reasons that will soon become apparent, we shall actually change rank, and describe the polytope $\mathcal{P}_{m+1}$. Its facet is the polytope of type $\mathcal{Q}_m = \{3^{m-2}, 6\}$ which we described in the previous Section 15F. Our next task is to lift this facet into $\mathbb{C}^{3m}$; its lifted vertices will just be the standard basis vectors $e_1, \ldots, e_{3m}$. For each $j = 1, \ldots, m$, we map

15G3
$$\begin{aligned} e_j &\mapsto e_{3j}, \\ \omega e_j &\mapsto e_{3m-2}, \\ \overline{\omega} e_j &\mapsto e_{3m-1}. \end{aligned}$$

We note here for future reference

15G4 Lemma *A trigonal edge-circuit has vertices of the form $3a_j - b_j$ with distinct indices $a_j = 1, \ldots, m$ and $b_j \in \{0, 1, 2\}$ for $j = 1, 2, 3$. It is a hole if $b_1 + b_2 + b_3 \equiv 0 \pmod 3$, and a face otherwise.*

The generatrix $(S_0, \ldots, S_{m-1})$ similarly lifts to $(R_0, \ldots, R_{m-1})$, where

15G5
$$\begin{aligned} R_j &= (3j{+}1\ 3j{+}4)(3j{+}2\ 3j{+}5)(3j{+}3\ 3j{+}6) \quad \text{if } j = 0, \ldots, m-3, \\ R_{m-2} &= (3m{-}5\ 3m{-}1)(3m{-}4\ 3m)(3m{-}3\ 3m{-}2), \\ R_{m-1} &= (1\ 2)(4\ 5) \cdots (3m{-}2\ 3m{-}1). \end{aligned}$$

We complete the generatrix $(R_0, \ldots, R_m)$ of the realization of the whole polytope $\mathcal{P}_{m+1}$ by

15G6
$$R_m = (1\ \omega 2)(4\ \omega 5) \cdots (3m{-}5\ \omega(3m{-}4))(3m{-}1\ \omega(3m)),$$

with the same convention as that introduced in Section 15F. It is fairly routine, if tedious, to verify the remaining group relations. We conclude that we have a realization P_{m+1} of $\mathcal{P}_{m+1}$ in $\mathbb{C}^{3m} \cong \mathbb{E}^{6m}$ with $9m$ vertices.

15G7 Remark The Wythoff space is 2-dimensional, but there is an obvious centralizing action by the complex numbers of absolute value 1, and hence $w^* = 1$ in the notation of Section 3G.

Observe from this that the $3m$ vertices of each facet of P_{m+1} are of the form $\omega^j e_k$ with distinct indices k. Indeed, the mapping $\omega^j e_k \mapsto e_k$ takes a general facet onto the initial one, preserving the distinction between trigonal faces and holes.

Universality

We begin by remarking that the nine vertices $\omega^j e_k$ for $j = 0, 1, 2$ and $k = 1, 2, 3$ have the same vertex-figure P_m. Moreover, any edge from one of these nine vertices goes to the vertex-figure. For universality, we appeal to the circuit criterion Theorem 2D4, which states (as we recall) that the group of a regular polytope is determined by the group of its vertex-figure and its edge-circuits.

So, suppose that we have an edge-circuit $\mathcal{C}$ in P_{m+1}. If $\mathcal{C}$ is confined to the vertex-figure P_m, then we can appeal to the obvious inductive assumption on m. Note that the initial case here is $\mathsf{P}_3 \cong \mathcal{P}_3 = \{6, 3 : 6\}$, and so our argument actually applies to P_4 as well.

On the other hand, suppose that $\mathcal{C}$ contains one of the nine initial vertices, say u. Then u is joined to two vertices v, w of P_m. Within P_m, we can join v and w by an edge-path, say $v = v_0, \ldots, v_k = w$. As we saw above, for $j = 1, \ldots, k$, each trigonal circuit u, v_{j-1}, v_j is either a face or a hole. Hence, by contracting that part of $\mathcal{C}$ between v and w over faces or holes, we can replace the two edges v, u and u, w by that edge-path in P_m. We repeat this procedure for each relevant initial vertex, and end the proof by appealing to the previous paragraph.

Realizations

We conclude by making a brief remark about the realization domain; we revert to discussing $\mathcal{P}_m$. Let us write $(\boldsymbol{r}_0, \ldots, \boldsymbol{r}_{m-1})$ for the generatrix of the abstract automorphism group of $\mathcal{P}_m$. First, we see that $\mathcal{P}_m$ admits an $(m-2)$-collapse onto its ridge $\{3^{m-3}\}$, given by $\boldsymbol{r}_{m-2} \mapsto \boldsymbol{e}$; it is easy to see this, by going back from P_{m+1} to Q_m, and then applying the same mapping $\omega^j e_k \mapsto e_k$ as we used above.

Next, we have seen that we had an apparent collapse of $\mathcal{P}_m$ onto its facet $\mathcal{Q}_{m-1}$. In fact, this is not quite the case; the identification arises from the quotient $\{6, 3 : 6\} \searrow \{6, 3 \mid 2\}$ of the 3-coface. Since the latter polyhedron is flat, the corresponding quotient of $\mathcal{P}_m$ is also flat, and so its vertices coincide with those of its facet $\mathcal{Q}_{m-1}$.

15G8 Remark The first collapse arises alternatively from the quotient

$$\{\{3, 6 : 6\}, \{6, 3 : 6\}\} \searrow \{\{3, 6 \mid 2\}, \{6, 3 : 6\}\}$$

applied to the 4-coface $\mathcal{P}_4$.

When we do a dimension count, including the henogon realization $\{1\}$, we find that we have a total

$$1 + (m-2) + 2(m-1) + 6(m-1) = 9(m-1),$$

and so these – with P_m itself, of course – must be all the pure realizations of $\mathcal{P}_m$. We can then state

15G9 Theorem *For each $m \geqslant 4$, the cosine matrix of $\mathcal{P}_m$ is*

$$\begin{bmatrix} 1 & 1 & 1 & 1 \\ 1 & -\frac{1}{m-2} & 1 & 1 \\ 1 & 0 & -\frac{1}{2} & 1 \\ 1 & 0 & 0 & -\frac{1}{2} \end{bmatrix},$$

with layer and dimension vectors

$$\Lambda = \big(1, 9(m-2), 6, 2\big), \quad D = \big(1, m-2, 2(m-1), 6(m-1)\big).$$

Layers $0, 2, 3$ (in P_m) consist of the initial vertex e_3, the six vertices $\omega^j e_k$ with $j = 0, 1, 2$ and $k = 1, 2$, and the two vertices $\omega^j e_3$ with $j = 1, 2$, respectively.

15G10 Remark Theorem 15G9 does not carry over to $\mathcal{P}_3 = \{6, 3 : 6\}$, whose layer vector is $(1, 3, 6, 3, 3, 2^*)$. Petriality interchanges the diagonal classes $\mathcal{D}_3$ and $\mathcal{D}_4$, and $\mathcal{D}_5$ is – as indicated – an asymmetric class. The generatrix (R_0, R_1, R_2) of the 6-dimensional pure faithful realizations is given by

$$zR_j = \begin{cases} (\overline{\zeta}_1, \overline{\zeta}_2, \overline{\zeta}_3), & \text{if } j = 0, \\ (\zeta_1, \overline{\omega}\zeta_3, \omega\zeta_2), & \text{if } j = 1, \\ (\zeta_2, \zeta_1, \zeta_3), & \text{if } j = 2, \end{cases}$$

with $z = (\zeta_1, \zeta_2, \zeta_3) \in \mathbb{C}^3$. The initial vector of a normalized realization is thus $\lambda(1, 1, \omega)$ for any λ with $|\lambda| = 1/\sqrt{3}$, and this leads to a family whose general cosine vector is

$$\Gamma = \big(1, -(\alpha + \beta), 0, \alpha, \beta, -\tfrac{1}{2}\big),$$

where $\alpha^2 + \alpha\beta + \beta^2 = \frac{1}{4}$. The collapses or quotients $\mathcal{P}_3 \searrow \{2\}$ or $\{6, 3 \mid 2\}$ still give realizations of dimensions 1 or 4, which do fit into the previous pattern.

16

A Family of 4-Polytopes

There is a certain quotient of the regular hyperbolic honeycomb $\{3,3,3,5\}$ whose automorphism group is the simple group $S_4(4)$; we shall discuss this and its relatives in the following Chapter 17. Two of the maximal subgroups of $S_4(4)$ are $[3,3,5]/2 = (A_5 \times A_5) \rtimes C_2$, the automorphism group of the 300-cell $\{3,3,5\}/2 = \{3,3,5 : 15\}$, and $L_2(16).2 = O(4,2^2,-1)$ of order 8160 (the former notation follows [16] (which we shall henceforth cite as $\mathbb{ATLAS}$), while the latter is that used in [101]). These groups turn up naturally as those of facets or vertex-figures of 5-polytopes in the family.

This chapter is devoted to those regular 4-polytopes with automorphism group $L_2(16).2$. One of them is a quotient of the hyperbolic honeycomb $\{3,5,3\}$; another quotient with the same group is of type $\{5,3,5\}$. We shall see that there are interesting parallels between this family and that of the pentagonal 4-polytopes described in Sections 7E and 7F; more specifically, the parallels are rather with $\{3,3,5\}/2$ and $\{5,3,3\}/8$. Indeed, there are relationships much like those among the regular star-polytopes, as can be seen by comparing Tables 16F1 and 16F2.

The polyhedron $\{5,5 : 4\}$ plays a vital part in the family, and so we look at it in detail in Section 16A. In the following Section 16B we give a permutation representation of the group $L_2(16).2$. In Sections 16C and 16D we describe various features of the geometry of $\{\{5,5 : 4\},\{5,3\}\}$, including its realization domain. In Section 16E we consider the dual polytope $\{\{3,5\},\{5,5 : 4\}\}$. From these two polytopes can be derived the extended family just mentioned; this is treated in Section 16F. Finally, some interesting relationships between the two polytopes of Schläfli types $\{3,5,3\}$ and $\{5,3,5\}$ are exhibited in Section 16G.

16A The Polyhedron $\{5,5 : 4\}$

In this chapter and the next, the regular polyhedron $\{5,5 : 4\}$ plays a central rôle. We have already met its Petrial $\{4,5 : 5\}$; as we saw in Section 13B, this is isomorphic to the facet of the sole regular 4-polytope in $\mathbb{E}^5$ whose symmetry group has direct isometries only (see the notes at the end of the section). There are several different ways of introducing $\{5,5:4\}$. One is by specifying its automorphism group as $[5,5]/\langle\langle(012)^4\rangle\rangle$, although this does not immediately yield the order of the group. Another entails the construction of the polyhedron as a map on a surface; this is fairly straightforward, but a little tedious. Yet a third – the one that we follow – starts from a closely related polyhedron, which is easy to describe. This in turn gives a realization of our polyhedron, and a representation of its group. Though it may seem unnecessary to go through these details, the insight gained will enable us to construct the polytope $\{\{5,5:4\},\{5,3\}\}$ of Section 16C in a similar direct way.

As somewhat of a diversion, we also discuss the wider family containing $\{5,5:4\}$.

The Construction

So, we begin with the related polyhedron. This is one of a family found by Coxeter [24], and is constructed in a very general context in [99, Section 8C] as the polyhedron $\mathbf{2}^{\{5\}}$. It has a standard realization in $\mathbb{E}^5$, with generatrix (S_0,S_1,S_2) given in our usual abbreviated notation by

16A1
$$\begin{aligned} S_0 &:= \overline{1}, \\ S_1 &:= (1\ 5)(2\ 4), \\ S_2 &:= (2\ 5)(3\ 4). \end{aligned}$$

We easily verify that its vertices and edges are those of the 5-cube $\{4,3,3,3\}$, while its faces consist of 40 of the 80 tetragonal faces of the cube; the other 40 tetragonal faces of the cube are the holes of the polyhedron. In view of this, an easy application of the circuit criterion of Theorem 2D4 proves

16A2 Lemma *The polyhedron* $\mathbf{2}^{\{5\}}$ *is isomorphic to* $\{4,5\mid 4\}$.

Proof. Indeed, any edge-circuit in the 5-cube can be contracted over its 2-faces, which are faces or holes of the polyhedron. □

16A3 Remark As a geometric polyhedron, what we have constructed here is $\mathsf{Q} := \{4,\frac{5}{1,2}\mid 4\}$, which is rigid. As we saw, the faces $\{4\}$ of Q form exactly half of those of the 5-cube $\{4,3^3\}$. Moreover, the Petrie operation π interchanges faces and holes.

The geometric description shows that $\{4,5\mid 4\}$ has a bipartite edge-graph. By Theorem 5A22, we can halve this polyhedron to obtain

$$\{4,5\mid 4\}^\eta = \{5,5:4\},$$

the polyhedron under consideration. As a geometric polyhedron P, its generatrix is therefore $(R_0,R_1,R_2) := (S_0S_1S_0,S_2,S_1)$, namely,

16A4
$$\begin{aligned} R_0 &:= (\overline{1}\ 5)(2\ 4), \\ R_1 &:= (2\ 5)(3\ 4), \\ R_2 &:= (1\ 5)(2\ 4). \end{aligned}$$

We know that S_0 induces the self-duality (by conjugation), but it is useful to observe that we can replace S_0 by $-S_0 = -\bar{1}$, which changes the signs of all coordinates except the first; this belongs to the smaller group of (16A4). Indeed, its 'rotation' group $\langle R_0R_1, R_1R_2\rangle$ consists of cyclic permutations of the coordinates, together with an even number of changes of sign.

Operations

The polyhedron $\{5, 5 : 4\}$ has a remarkable property, which we can show using the faithful realization P. We write a vertex of the 5-cube $\{4, 3, 3, 3\}$ as a string of 1s and $\bar{1}$s, where $\bar{1} = -1$; for instance, $\bar{1}111\bar{1} = (-1, 1, 1, 1, -1)$. The initial vertex of P is 11111, forming layer $\mathcal{L}_0$. For the rest, $\mathcal{L}_1$ corresponds to edges, $\mathcal{L}_2$ to face-diagonals and $\mathcal{L}_3$ to diagonals of holes $\{5\}$. Then $\mathcal{L}_1$ consists of cyclic permutations of $\bar{1}111\bar{1}$, $\mathcal{L}_2$ of $\bar{1}11\bar{1}1$ and $\mathcal{L}_3$ of $\bar{1}\bar{1}\bar{1}\bar{1}1$.

There are two operations from Section 5A that will play a useful rôle in the future. The first is faceting

$$\varphi\colon (\boldsymbol{r}_0, \boldsymbol{r}_1, \boldsymbol{r}_2) \mapsto (\boldsymbol{r}_0, \boldsymbol{r}_1\boldsymbol{r}_2\boldsymbol{r}_1, \boldsymbol{r}_2),$$

so that $\varphi = \varphi_2$ in the notation of (5A1) (no other φ_k occurs here, and so as usual we shorten the notation). The other is quartering of (5A27):

$$\psi\colon (\boldsymbol{r}_0, \boldsymbol{r}_1, \boldsymbol{r}_2) \mapsto (\boldsymbol{r}_0\boldsymbol{r}_1\boldsymbol{r}_0, \boldsymbol{r}_1, \boldsymbol{r}_2\boldsymbol{r}_1\boldsymbol{r}_2);$$

recall that this is valid only because $\boldsymbol{r}_0\boldsymbol{r}_1\boldsymbol{r}_0 \cdot \boldsymbol{r}_2\boldsymbol{r}_1\boldsymbol{r}_2 = (\boldsymbol{r}_0\boldsymbol{r}_1\boldsymbol{r}_2)^2$ has period 2. It is an easy exercise (using the generatrix of (16A4)) to show that φ fixes $\mathcal{L}_1$ and interchanges $\mathcal{L}_2$ and $\mathcal{L}_3$, while ψ fixes $\mathcal{L}_3$ and interchanges $\mathcal{L}_1$ and $\mathcal{L}_2$. There follows

16A5 Proposition *There are operations that freely permute the non-trivial diagonal classes of* $\{5, 5 : 4\}$ *while preserving the combinatorial type.*

There is an immediate consequence of Proposition 16A5.

16A6 Proposition *The cosine matrix of the polyhedron* $\{5, 5 : 4\}$ *is*

$$\begin{bmatrix} 1 & 1 & 1 & 1 \\ 1 & -\frac{3}{5} & \frac{1}{5} & \frac{1}{5} \\ 1 & \frac{1}{5} & -\frac{3}{5} & \frac{1}{5} \\ 1 & \frac{1}{5} & \frac{1}{5} & -\frac{3}{5} \end{bmatrix},$$

with layer and dimension vectors

$$\Lambda = (1, 5, 5, 5), \quad D = (1, 5, 5, 5).$$

Proof. The realization with generatrix (16A4) corresponds to the last row Γ_3 of the matrix; the other two are then given by Proposition 16A5. □

16A7 Remark It is not too hard to see that the cosine vector of a general centred realization of $\{5, 5 : 4\}$ (that is, with no trivial component) is of the form $\Gamma = (1, \gamma_1, \gamma_2, \gamma_3)$ with $\gamma_j \leqslant \frac{1}{5}$ for each j and $\gamma_1 + \gamma_2 + \gamma_3 = -\frac{1}{5}$. The relations imply that each $\gamma_j \geqslant -\frac{3}{5}$. The fact that our operations freely permute $\gamma_1, \gamma_2, \gamma_3$ verifies the assertion made at the beginning of the section. What we have shown is that the vertices of such a general realization are those of five others as well.

The operations φ and ψ are only involutory up to an inner automorphism, so that composing them need not lead to an expected result. However, there is another mixing operation, which does work unambiguously.

16A8 Proposition *The mixing operation*

$$(\boldsymbol{r}_0, \boldsymbol{r}_1, \boldsymbol{r}_2) \mapsto (\boldsymbol{r}_1\boldsymbol{r}_2\boldsymbol{r}_1\boldsymbol{r}_0\boldsymbol{r}_1\boldsymbol{r}_0\boldsymbol{r}_1\boldsymbol{r}_2\boldsymbol{r}_1, \boldsymbol{r}_1, \boldsymbol{r}_2) =: (\boldsymbol{s}_0, \boldsymbol{s}_1, \boldsymbol{s}_2)$$

on a generatrix of $\{5, 5 : 4\}$ *is cyclic of period* 3, *and permutes layers* $\mathcal{L}_1, \mathcal{L}_2, \mathcal{L}_3$.

Proof. We sketch the proof, which has three stages; all these involve messy calculations using the group presentation, which for the most part we suppress.

We first verify that $(\boldsymbol{s}_0, \boldsymbol{s}_1, \boldsymbol{s}_2)$ is a generatrix of $\{5, 5 : 4\}$. One curiosity is that

$$\boldsymbol{s}_0\boldsymbol{s}_1\boldsymbol{s}_2 = \boldsymbol{r}_1\boldsymbol{r}_2\boldsymbol{r}_1\boldsymbol{r}_0\boldsymbol{r}_1\boldsymbol{r}_0\boldsymbol{r}_1 \sim \boldsymbol{r}_2\boldsymbol{r}_1\boldsymbol{r}_0\boldsymbol{r}_1\boldsymbol{r}_0,$$

which is the basic 'rotation' of the 2-zigzag of the dual.

We next check how the layers change. A typical involution taking the initial vertex v (say) into $\mathcal{L}_2$ is $\boldsymbol{r}_0\boldsymbol{r}_1\boldsymbol{r}_0$; since $\boldsymbol{s}_0$ is the conjugate of this by $\boldsymbol{r}_1\boldsymbol{r}_2\boldsymbol{r}_1$ which fixes v, it follows that $\boldsymbol{s}_0$ also takes v into $\mathcal{L}_2$. It is really tedious to show that

$$\boldsymbol{s}_0\boldsymbol{s}_1\boldsymbol{s}_2\boldsymbol{s}_1\boldsymbol{s}_0 = \boldsymbol{r}_2\boldsymbol{r}_1\boldsymbol{r}_0\boldsymbol{r}_1\boldsymbol{r}_2;$$

as in the previous case, this shows that layer 3 with respect to the new generatrix is the original $\mathcal{L}_1$. Since it is the only one left, we do not need to show directly that the new layer 2 is $\mathcal{L}_3$.

Finally, since $\boldsymbol{r}_1$ and $\boldsymbol{r}_2$ remain under the operation, we need only look at what happens to $\boldsymbol{r}_0$ when we apply the operation twice and thrice. If, say, $(\boldsymbol{s}_0, \boldsymbol{s}_1, \boldsymbol{s}_2) \mapsto (\boldsymbol{t}_0, \boldsymbol{t}_1, \boldsymbol{t}_2)$, then

$$\boldsymbol{t}_0 = \boldsymbol{r}_0\boldsymbol{r}_1\boldsymbol{r}_0\boldsymbol{r}_2\boldsymbol{r}_1\boldsymbol{r}_2\boldsymbol{r}_1\boldsymbol{r}_2\boldsymbol{r}_0\boldsymbol{r}_1\boldsymbol{r}_0 = \boldsymbol{r}_2\boldsymbol{r}_1\boldsymbol{r}_0\boldsymbol{r}_2\boldsymbol{r}_1\boldsymbol{r}_2\boldsymbol{r}_1\boldsymbol{r}_2\boldsymbol{r}_0\boldsymbol{r}_1\boldsymbol{r}_2$$

(this equation is not trivial), and then showing that $(\boldsymbol{t}_0, \boldsymbol{t}_1, \boldsymbol{t}_2) \mapsto (\boldsymbol{r}_0, \boldsymbol{r}_1, \boldsymbol{r}_2)$ is (relatively) straightforward. □

We need one more feature of the polyhedron. As mirrors, we have

$$R_0 \cap R_2 = \{(0, \lambda, \mu, \lambda, 0) \in \mathbb{E}^5 \mid \lambda, \mu \in \mathbb{R}\}.$$

With a general initial vertex in $R_0 \cap R_2 = S_0 \cap S_1$ in (16A1), we recover the polyhedron $\{5, 4 \mid 4\}$ dual to that we began with; observe that this Wythoff space is 2-dimensional. The particular values $\lambda = 0$ or $\mu = 0$ yield degenerate realizations with 10 and 20 vertices, respectively. We shall meet the latter again subsequently; as a symmetric set in its own right, it has layer vector $(1, 8, 8, 2, 1)$, so that each vertex is now adjacent to eight others.

Combinatorial Description

Another way to see what happens to $\{5, 5 : 4\}$ under φ and ψ is to describe its combinatorics directly. There are 16 vertices, labelled a and b_j, c_j, d_j with $j = 0, \ldots, 4$ and indices taken modulo 5. We label the edge, face-diagonal and hole-diagonal classes $\mathcal{D}_1, \mathcal{D}_2, \mathcal{D}_3$, respectively, so that (again with $j = 0, \ldots, 4$ (mod 5)) we have

$$\begin{aligned}\mathcal{D}_1\colon\ & ab_j,\ b_jc_{j\pm 2},\ b_jd_{j\pm 1},\ c_jc_{j+2},\ c_jd_j,\ d_jd_{j+1};\\ \mathcal{D}_2\colon\ & ac_j,\ b_jb_{j+1},\ b_jc_{j\pm 1},\ b_jd_j,\ c_jd_{j\pm 2},\ d_jd_{j+2};\\ \mathcal{D}_3\colon\ & ad_j,\ b_jb_{j+2},\ b_jc_j,\ b_jd_{j\pm 1},\ c_jc_{j+1},\ c_jd_{j\pm 1}.\end{aligned}$$

Up to equivalence, we have

$$\begin{aligned}\varphi\colon (a, b_j, c_j, d_j) &\mapsto (a, b_{2j}, d_{2j}, c_{2j}),\\ \psi\colon (a, b_j, c_j, d_j) &\mapsto (a, c_{2j}, b_{2j}, d_{2j}).\end{aligned}$$

As usual, these operations are involutory up to inner automorphism. Moreover,

$$\varphi^{-1}\psi\colon (a, b_j, c_j, d_j) \mapsto (a, c_j, d_j, b_j).$$

Halving

In Section 13B, the facet was the Petrial $\{4, 5 : 5\}$ of $\{5, 5 : 4\}$, and so we now return to that. The dual polyhedron has an interesting realization domain which we also discuss. We also briefly consider the double cover of our polyhedron.

From $\{4, 5 : 5\}$ the halving operation η of (5A21) yields a polyhedron of type $\{5, 5\}$; since $\mathcal{P}$ has odd edge-circuits, we see that $\mathcal{P}^{\eta}$ has the same vertices as $\mathcal{P}$ and the same automorphism group $\boldsymbol{G}$, so that we must have $\{4, 5 : 5\}^{\eta} = \{5, 5 : 4\}$ again. If we follow η by the Petrie operation π of (5A8), we find that this again has the effect of a cyclic permutation of the diagonal classes; we suppress the details.

16A9 Remark Only the pure realization P_2 of $\{4, 5 : 5\}$ corresponding to Γ_2 has planar faces $\{4\}$ (as opposed to skew faces $\{\frac{4}{1,2}\}$). It follows that the fine Schläfli symbol $\{4, \frac{5}{1,2} : \frac{5}{1,2}\}$, which indicates the geometry of the faces (and so on – see Chapter 6), is rigid.

To prepare for consideration of the dual, we note that from φ we obtain the generatrix given by

16A10

$$\begin{aligned}S_0 &:= \overline{1}\,\overline{2},\\ S_1 &:= (2\ 3)(4\ 5),\\ S_2 &:= (1\ 2)(3\ 4),\end{aligned}$$

which corresponds to P_3. Similarly, $\eta\pi$ leads to

16A11
$$\begin{aligned} T_0 &:= \bar{1}\,\bar{2}\,\bar{4}\,\bar{5} = -\bar{3}, \\ T_1 &:= (1\ 2)(3\ 4), \\ T_2 &:= (1\ 5)(2\ 4), \end{aligned}$$

which yields P_1.

The Dual

We next look at realizations of the self-Petrie dual $\mathcal{Q} := \mathcal{P}^\delta = \{5,4:5\}$. To avoid elaborate relabelling, we just use the (abstract and geometric) generators of the group of $\mathcal{P}$ in the reverse order.

We first note that $\{5,4:5\}$ is 2-collapsible in the sense of Section 2F; that is, the quotient given by

$$(\boldsymbol{r}_2, \boldsymbol{r}_2, \boldsymbol{r}_0) \mapsto (\boldsymbol{r}_2, \boldsymbol{r}_1, \boldsymbol{e})$$

induces a covering of the pentagon $\{5\}$ by $\{5,4:5\}$, so that setting $\boldsymbol{r}_0 = \boldsymbol{e}$ is compatible with the original group relations. Thus among the realizations of $\mathcal{Q}$ are those of $\{5\}$, namely, the pentagon $\{5\}$ and pentagram $\{\frac{5}{2}\}$.

We now find the Wythoff spaces of the various (putative) realizations given by our generating reflexions. First, $x \in R_0 \cap R_1$ satisfies $\xi_1 = \xi_2 = \xi_3 = \xi_4 = 0$, so that

$$R_0 \cap R_0 = \operatorname{lin}\{e_3\},$$

with $(e_1, \ldots, e_4)$ the standard orthonormal basis of $\mathbb{E}^5$. We shall return to this realization Q_3, say, in a moment.

For the second case, if $x \in S_0 \cap S_1$, then $\xi_1 = \xi_2 = \xi_3 = 0$ and $\xi_3 = \xi_4$, so that

$$S_0 \cap S_1 = \operatorname{lin}(e_3 + e_4).$$

We treat this case in detail first, because the realization Q_4, say, is faithful. The vertices are $\pm\, e_j \pm e_k$, with $\{j,k\} = \{4,5\}, \{1,2\}, \{3,5\}, \{2,4\}, \{1,3\}$. It follows that there are five non-trivial diagonal classes; we list these in an unusual order, so that the layer vector is $(1,2,1,4,8,4)$ (the first 4 here corresponds to the class of an edge). The cosine vector is then $\Gamma_4 = (1,0,-1,\frac{1}{\sqrt{2}},0,-\frac{1}{\sqrt{2}})$. Observe that the edge-graph is centrally symmetric; however, the central inversion $-I$ is not a symmetry of the polyhedron; instead, it acts like the Petrie operation π, interchanging Q_4 and its Petrial.

We now see that Q_3 is obtained by identifying diametrically opposite vertices of Q_4. Now this means that Q_3 actually coincides with its Petrial, so that each of its pentagonal faces is also a Petrie polygon. It follows that Q_3 is not polytopal. By the way, observe further that the vertex-set of Q_3 is also centrally symmetric, and that identifying opposite vertices induces the coverings of the pentagon and pentagram. The cosine vector of Q_3 is $\Gamma_3 = (1,-1,1,0,0,0)$; this accounts for the ordering of the layer vector.

With this ordering, the cosine vectors of the pentagon and pentagram are

$$\Gamma_1 = (1, 1, 1, \tfrac{1}{2}\tau^{-1}, -\tfrac{1}{2}\tau, \tfrac{1}{2}\tau^{-1}),$$
$$\Gamma_2 = (1, 1, 1, -\tfrac{1}{2}\tau, \tfrac{1}{2}\tau^{-1}, -\tfrac{1}{2}\tau),$$

respectively.

We have one more pure realization Q_5 to find. Theorem 3C11 tells us that its dimension must be $20 - 1 - 2 - 2 - 5 - 5 = 5$, and (with $\mathsf{Q}_0 = \{1\}$ as before) its cosine vector is given by

$$\Gamma_5 = \tfrac{1}{5}\big(20(1, 0^5) - \Gamma_0 - 2\Gamma_1 - 2\Gamma_2 - 5\Gamma_3 - 5\Gamma_4\big) = \big(1, 0, -1, -\tfrac{1}{\sqrt{2}}, 0, \tfrac{1}{\sqrt{2}}\big),$$

which perhaps we ought to have expected. Note that Q_5 is not derived from (T_2, T_1, T_0), whose Wythoff space is just $\{o\}$. Instead, it is obtained by changing the signs of S_2 and S_1, giving new generators

$$S_0' = \overline{1}\,\overline{2},$$
$$S_1' = -(2\ 3)(4\ 5),$$
$$S_2' = -(1\ 2)(3\ 4)),$$

which do not correspond to a realization of $\{4, 5 : 5\}$; thus we have a quasi-dual of the realization Q_3 of $\mathcal{P}$ in the sense of Section 5D.

Again, we can summarize our discussion in

16A12 Theorem *The cosine matrix of the regular polyhedron* $\{5, 4 : 5\}$ *is*

$$\begin{bmatrix} 1 & 1 & 1 & 1 & 1 & 1 \\ 1 & 1 & 1 & \tfrac{1}{2}\tau^{-1} & -\tfrac{1}{2}\tau & \tfrac{1}{2}\tau^{-1} \\ 1 & 1 & 1 & -\tfrac{1}{2}\tau & \tfrac{1}{2}\tau^{-1} & -\tfrac{1}{2}\tau \\ 1 & -1 & 1 & 0 & 0 & 0 \\ 1 & 0 & -1 & \tfrac{1}{\sqrt{2}} & 0 & -\tfrac{1}{\sqrt{2}} \\ 1 & 0 & -1 & -\tfrac{1}{\sqrt{2}} & 0 & \tfrac{1}{\sqrt{2}} \end{bmatrix},$$

with layer and dimension vectors

$$\Lambda = (1, 2, 1, 4, 8, 4), \quad D = (1, 2, 2, 5, 5, 5).$$

The Double Cover

As we have already observed, the polyhedron $\{4, 5 : 5\}$ is non-orientable; its orientable double cover is $\{4, 5 \mid 4\} = \mathbf{2}^{\{5\}}$, which was the starting point of the section. Moreover, we had a realization whose vertices are those of the regular 5-cube. Indeed, there are three faithful pure 5-dimensional realizations, one corresponding to each of those of $\{4, 5 : 5\}$. If the latter are denoted $\mathsf{P}_1, \mathsf{P}_2, \mathsf{P}_3$ as previously, then the realization corresponding to P_j is $\mathsf{Q}_j := \mathsf{P}_j{}^{\zeta} = \mathsf{P}_j \otimes \{2\}$, obtained by changing the sign of the first generator of the symmetry group. Note that each realization of $\{4, 5 : 5\}$ is also one of $\{4, 5 \mid 4\}$; the final pure realization of the latter is its collapse onto the segment $\{2\}$.

16B A Permutation Representation

Our starting point here is actually neither of the polytopes of types $\{3,5,3\}$ or $\{5,3,5\}$; instead, for several reasons we first treat the 4-polytope in the family of type $\{5,5,3\}$. The generatrix $(\boldsymbol{r}_0,\ldots,\boldsymbol{r}_3)$ of $\mathrm{L}_2(16).2$ is given by permutations on $0,1,\ldots,8,\bar{8},\ldots,\bar{1}$, where we write $\bar{k} := 17-k$ for $k = 1,\ldots,8$:

16B1
$$\begin{aligned} \boldsymbol{r}_0 &= (0\ \bar{2})(1\ 3)(\bar{1}\ \bar{6})(2\ 7)(\bar{4}\ 5)(\bar{5}\ 6), \\ \boldsymbol{r}_1 &= (1\ \bar{5})(\bar{1}\ 5)(2\ \bar{2})(3\ 4)(\bar{3}\ \bar{4})(7\ \bar{7}), \\ \boldsymbol{r}_2 &= (0\ 3)(1\ \bar{2})(\bar{1}\ 7)(2\ \bar{6})(\bar{3}\ 4)(\bar{7}\ 8), \\ \boldsymbol{r}_3 &= (1\ 5)(\bar{1}\ \bar{5})(3\ \bar{4})(\bar{3}\ 4)(6\ \bar{6})(8\ \bar{8}); \end{aligned}$$

see the notes at the end of the section. It is straightforward to show

16B2 Theorem *The generators $\boldsymbol{r}_0,\ldots,\boldsymbol{r}_3$ satisfy the natural relations for the Coxeter group $[5,5,3]$, together with*

$$(\boldsymbol{r}_0\boldsymbol{r}_1\boldsymbol{r}_2)^4 = \boldsymbol{e}.$$

The group satisfies the intersection property with respect to $(\boldsymbol{r}_0,\ldots,\boldsymbol{r}_3)$; thus it is a string C-group, and hence the automorphism group of an abstract regular 4-polytope.

16B3 Remark In fact, MAGMA [5] shows that the additional relation on $[5,5,3]$ determines the group $\mathrm{L}_2(16).2$. Hence, in the notation of [99, Section 2F], the polytope is denoted $\{5,5,3\}/\langle\langle(012)^4\rangle\rangle = \{\{5,5:4\},\{5,3\}\}$; the latter designation implies that it is the universal regular polytope with facet $\{5,5:4\}$ and vertex-figure the dodecahedron $\{5,3\}$. The group $\mathrm{L}_2(16).2$ can similarly be denoted $[5,5,3]/\langle\langle(012)^4\rangle\rangle$. We choose this polytope as our starting point precisely because this is the one in the family whose notation is the most concise.

16B4 Remark Recall that we described the polyhedron $\{5,5:4\}$ in detail in the previous Section 16A; we shall not repeat any of that description here.

16B5 Remark Since $\{\{5,5:4\},\{5,3\}\}$ is orientable, its rotation subgroup is $\mathrm{L}_2(16)$ itself; this coincides with the commutator subgroup, of course.

The generators have been chosen so that the Petrie polygon, with generatrix $(\boldsymbol{r}_0\boldsymbol{r}_2,\boldsymbol{r}_1\boldsymbol{r}_3)$, is given by

$$\begin{aligned} \boldsymbol{r}_0\boldsymbol{r}_2 &= (0\ 1)(\bar{1}\ 2)(\bar{2}\ 3)(\bar{3}\ 4)(\bar{4}\ 5)(\bar{5}\ 6)(\bar{6}\ 7)(\bar{7}\ 8), \\ \boldsymbol{r}_1\boldsymbol{r}_3 &= (1\ \bar{1})(2\ \bar{2})(3\ \bar{3})(4\ \bar{4})(5\ \bar{5})(6\ \bar{6})(7\ \bar{7})(8\ \bar{8}), \end{aligned}$$

so that

$$\boldsymbol{r}_1\boldsymbol{r}_3 \cdot \boldsymbol{r}_0\boldsymbol{r}_2 = (0\ 1\ \cdots\ 8\ \bar{8}\cdots\bar{1})$$

has period 17.

First observe that $\langle\boldsymbol{r}_1,\boldsymbol{r}_2,\boldsymbol{r}_3\rangle$ partitions the 17 points into $\{\bar{4},\bar{3},0,3,4\}$ and the complementary subset of twelve, so that we may label the initial vertex

$\bar{4}\bar{3}034$ (we denote a general vertex by concatenation). Thus the vertices of $\mathcal{S}$ correspond to certain quintuples of $0, 1, \ldots, 8, \bar{8}, \ldots, \bar{1}$; we shall look into the combinatorics later. The 68 vertices of $\mathcal{S}$ are obtained from $\bar{4}\bar{3}034$ and $\bar{8}\bar{6}068$, $\bar{5}\bar{1}015$, $\bar{7}\bar{2}027$, under the cyclic permutation $j \mapsto j{+}1 \pmod{17}$ of the Petrie polygon.

Notes to Section 16B

1. The permutation representation of $L_2(16).2$ was provided by Marston Conder using the computer algebra program MAGMA, and through this we can easily verify various subsequent claims. We relabelled what Conder sent us, to bring to the fore several features that we wished to emphasize.

16C The Polytope $\{\{5, 5 : 4\}, \{5, 3\}\}$

If we identify the points $0, 1, \ldots, 8, \bar{8}, \ldots, \bar{1}$ with the vertices of the 16-simplex, then from the permutation representation of $\boldsymbol{G} = [5, 5, 3]/\langle\langle (012)^4 \rangle\rangle$ we obtain the following data:

- the layer vector of $\mathcal{S} := \{\{5, 5 : 4\}, \{5, 3\}\}$ is $\Lambda = (1, 20, 20, 15, 12)$;
- $\mathcal{S}$ has a 16-dimensional realization S_1, whose vertices are the 68 centres of certain 4-faces of the 16-simplex;
- the cosine vector of S_1 is

$$\Gamma_1 = (1, \tfrac{3}{20}, \tfrac{3}{20}, -\tfrac{2}{15}, -\tfrac{5}{12});$$

- all the diagonal classes of $\mathcal{S}$ are symmetric, so that each pure realization of $\mathcal{S}$ has a 1-dimensional Wythoff space;
- $\mathcal{S}$ has 51 facets $\{5, 5 : 4\}$;
- the induced cosine vector of the facet $\{5, 5 : 4\}$ is $\Gamma^f = (1, \gamma_1, \gamma_4, \gamma_2)$.

The expression

$$\tfrac{1}{d} = \|\Gamma\|_\Lambda^2 = \tfrac{1}{68}\big(1 + 20\gamma_1^2 + 20\gamma_2^2 + 15\gamma_3^2 + 12\gamma_4^2\big)$$

of Theorem 3F5 for the dimension d from a general pure cosine vector $\Gamma = (1, \gamma_1, \ldots, \gamma_4)$ can be verified to give $d(\Gamma_1) = 16$.

Given these data, we can quickly find the remaining pure realizations. It should be clear that the entries γ_1 and γ_2 of a general cosine vector Γ can be interchanged; this will emerge from the discussion. We begin with an appeal to the centred facet criterion Corollary 4D10. Since the dual of $\mathcal{S}$ has only 51 vertices, there must be at least one pure realization of $\mathcal{S}$ with its facets centred. Moreover, S_1 is not one of them. For such a cosine vector, we can thus use the layer equation, Λ-orthogonality with respect to Γ_1 and the criterion, and solve

$$\begin{aligned}
1 + 20\gamma_1 + 20\gamma_2 + 15\gamma_3 + 12\gamma_4 &= 0,\\
1 + 3\gamma_1 + 3\gamma_2 - 2\gamma_3 - 5\gamma_4 &= 0,\\
1 + 5\gamma_1 + 5\gamma_2 \phantom{{}- 2\gamma_3} + 5\gamma_4 &= 0;
\end{aligned}$$

we obtain

$$\gamma_1+\gamma_2=-\tfrac{1}{5},\quad \gamma_3=\tfrac{1}{5},\quad \gamma_4=0.$$

Observe that we cannot separate $\gamma_1+\gamma_2$, so that we must have two solutions with the same dimension d, in which we can interchange γ_1 and γ_2. Taking half their sum and using Proposition 3F8, d is given by

$$\tfrac{1}{2d}=\|(1,-\tfrac{1}{10},-\tfrac{1}{10},0,\tfrac{1}{5})\|_\Lambda^2=\tfrac{1}{34};$$

that is, $d=17$.

We can now proceed in several ways. We only lack one further component, for which Theorem 3C11 yields

$$\Gamma_2=(1,0,0,-\tfrac{1}{3},\tfrac{1}{3}),$$

with $d(\Gamma_2)=17$. For Γ_3 and Γ_4 (with $\Gamma_3+\Gamma_4=2(1,-\frac{1}{10},-\frac{1}{10},0,\frac{1}{5})$), we can either appeal to

$$d(\Gamma_3)=17 \implies 1+20(\gamma_1^2+\gamma_2^2)+15(\tfrac{1}{5})^2=68\cdot\tfrac{1}{17}=4 \implies \gamma_1^2+\gamma_2^2=\tfrac{3}{25},$$

or use

$$\langle\Gamma_3,\Gamma_4\rangle_\Lambda=0 \implies 1+2\cdot 20\gamma_1\gamma_2+15(\tfrac{1}{5})^2=0 \implies \gamma_1\gamma_2=-\tfrac{1}{25}.$$

Both approaches give $\{\gamma_1,\gamma_2\}=\{\frac{1}{5}\tau^{-1},-\frac{1}{5}\tau\}$.

In conclusion, then, we have proved

16C1 Theorem *The cosine matrix of* $\{5,5,3\}/\langle\langle(012)^4\rangle\rangle$ *with* 68 *vertices and group* $\mathrm{O}(4,2^2,-1)$ *of order* 8160 *is*

$$\begin{bmatrix} 1 & 1 & 1 & 1 & 1 \\ 1 & \frac{3}{20} & \frac{3}{20} & -\frac{2}{15} & -\frac{5}{12} \\ 1 & 0 & 0 & -\frac{1}{3} & \frac{1}{3} \\ 1 & \frac{1}{5}\tau^{-1} & -\frac{1}{5}\tau & \frac{1}{5} & 0 \\ 1 & -\frac{1}{5}\tau & \frac{1}{5}\tau^{-1} & \frac{1}{5} & 0 \end{bmatrix},$$

with layer and dimension vectors

$$\Lambda=(1,20,20,15,12),\quad D=(1,16,17,17,17).$$

Let us begin to draw the parallels that we mentioned at the beginning of the chapter. For the present purposes, we rewrite the layer vector of $\{3,3,5\}/2=\{3,3,5:15\}$ as $\Lambda=(1,12,12,15,20)$. Reading off the data from Theorem 7E10, we can then rewrite its cosine matrix as

$$\begin{bmatrix} 1 & 1 & 1 & 1 & 1 \\ 1 & -\frac{1}{4} & -\frac{1}{4} & 0 & \frac{1}{4} \\ 1 & 0 & 0 & \frac{1}{5} & -\frac{1}{5} \\ 1 & \frac{1}{3}\tau & -\frac{1}{3}\tau^{-1} & -\frac{1}{3} & 0 \\ 1 & -\frac{1}{3}\tau^{-1} & \frac{1}{3}\tau & -\frac{1}{3} & 0 \end{bmatrix},$$

so that the dimension vector is now $D = (1, 16, 25, 9, 9)$.

Bear in mind as well that we have several polytopes with the same vertices; the definition here that is closest in spirit to what we just had will be

$$\{\{5, 5 \mid 3\}, \{5, 3\}\}/2 = \{5, 5, 3\}/\langle\langle (0121)^3, (0123)^{10} \rangle\rangle;$$

however, we have alternatively (and more simply) $\{3, 3, 5\}/2 = \{3, 3, 5 : 15\}$, with no analogue in the other family.

Notes to Section 16C

1. A more geometric way of looking at the construction of S_1 is to take as vertices the centres of the complementary 11-simplices, which correspond to icosahedral facets of the dual polytope.
2. There is an orthogonal projection of the 16-dimensional realization of $\mathcal{S}$ with 17-fold symmetry. The (unscaled) radii of the four circles on which the projected vertices lie are $\rho_k := 1 + 2\cos\frac{6k\pi}{17} + 2\cos\frac{8k\pi}{17}$ for $k = 1, 2, 4, 8$. Unfortunately, three of the four radii are nearly the same, forcing 17 pairs of the points to be too close; the resulting picture is thus impossible to draw clearly.

16D Layers and Strata of $\{\{5, 5 : 4\}, \{5, 3\}\}$

As the Schläfli symbol indicates, the 20 vertices of layers $\mathcal{L}_1$ and $\mathcal{L}_2$ are those of dodecahedra, the 15 vertices of $\mathcal{L}_3$ are those of a hemi-icosidodecahedron, while the 12 vertices of $\mathcal{L}_4$ are those of an icosahedron. We shall look into these in greater depth later.

The strata of $\{\{5, 5 : 4\}, \{5, 3\}\}$ help us to determine what the dual polytope looks like. From the permutation representation of Section 16B, we can deduce that the corresponding stratum vector is $C = (16, 16, 16, 20)$. The three sets of 16 vertices each give a facet. However, only the initial one has generatrix $(\boldsymbol{r}_0, \boldsymbol{r}_1, \boldsymbol{r}_2)$ as the obvious subset of (16B1); instead, the other two are related to it as in Proposition 16A8. This tells us that we have an automorphism of period 3 that permutes these facets cyclically.

A natural thought is that the remaining 20 vertices are those of $\{5, 4 : 5\}$, since we must have the same group $[5, 5 : 4]$ acting on them. However, this is not the case; what we actually have is the result of identifying the 40 vertices of $\{5, 4 \mid 4\}$ in pairs, as at the end of Section 16A.

16E The Dual Polytope $\{\{3,5\},\{5,5:4\}\}$

The dual polytope $\mathcal{S}^\delta = \{\{3,5\},\{5,5:4\}\}$ has 51 vertices. However, in the realization given by the permutation representation, the initial vertex is $\bar{8}$, and hence the vertex set is just $\{0,1,\ldots,8,\bar{8},\ldots,\bar{1}\}$. In other words, the vertices of the 16-dimensional realization coincide in threes.

From this and what we said in the previous section, we can see that the layer vector of $\{\{3,5\},\{5,5:4\}\}$ must be $\Lambda = (1,16,16,16,2^*)$; the asymmetry of the last layer $\mathcal{L}_4$ is a consequence of the aforementioned cyclic automorphisms of period 3 being the only ones that permute the three facets in the strata.

The 16-dimensional realization has cosine vector $\Gamma_1 = (1,-\frac{1}{16},-\frac{1}{16},-\frac{1}{16},1)$. The layer equation and Λ-orthogonality then easily lead to the remaining cosine vectors being of the form $\Gamma = (1,\alpha,\beta,\gamma,-\frac{1}{2})$, with $\alpha+\beta+\gamma=0$. The existence of the Petrie polygon of period 17 and the fact that we have an asymmetric diagonal class implies that the corresponding realizations must be 17-dimensional. It follows that

$$1+16(\alpha^2+\beta^2+\gamma^2)+2(-\tfrac{1}{2})^2 = 51\cdot\tfrac{1}{17} \implies \alpha^2+\beta^2+\gamma^2=\tfrac{3}{32}.$$

16F The Extended Family

We prove here that there is an extensive family of regular 4-polytopes related to $\{5,5,3\}/\langle\langle(012)^4\rangle\rangle$. However, to show how the connexions work, we first give in Table 16F1 half the corresponding family derived from the 600-cell $\{3,3,5\}$; this comes from Table 7G8, with the connecting operations given.

16F1

$$\begin{array}{ccc}
\{3,3,5\} & & \\
\varphi\updownarrow & & \\
\{\{3,5\},\{5,5\mid 3\}\} & \stackrel{\delta}{\longleftrightarrow} & \{\{5,5\mid 3\},\{5,3\}\} \\
\lambda\updownarrow & & \\
\{\{5,5\mid 3\},\{5,5\mid 3\}\} & & \\
\varphi\updownarrow & & \\
\{5,3,5\mid 3\} & &
\end{array}$$

Each of these operations is an involutory vertex-figure replacement, modulo an inner automorphism; indeed, φ is actually an edge-figure replacement. Recall from Section 5B that $\boldsymbol{\lambda}$ replaces a vertex-figure by an allomorph. In the present case, applied to $\{5,5\mid 3\}$, it is

$$(r_0,r_1,r_2)\mapsto(r_0r_1r_0r_1r_0,r_1,r_2r_1r_2)$$

(as a vertex-figure, the indices are raised by 1). It is convenient to present changes of generators in abbreviated form, so that here we have $\boldsymbol{\lambda}\colon (0,1,2,3) \mapsto (01010, 1, 212)$. Again, we write $\varphi := \varphi_2$, because this is the only faceting operation employed here.

In Table 16F2, we draw up a similar set of involutory mixing operations which connect the members of the new family; $\boldsymbol{\delta}$ is, of course, duality. This shows, incidentally, that all have the same automorphism group.

16F2

$$\begin{array}{ccc}
\{3,5,3 :: 4\} & & \\
\lambda \updownarrow & & \\
\{\{5,5:4\},\{5,3\}\} & \overset{\delta}{\longleftrightarrow} & \{\{3,5\},\{5,5:4\}\} \\
 & & \psi \updownarrow \\
\{\{5,5:4\},\{5,5 \mid 3\}\} & \overset{\delta}{\longleftrightarrow} & \{\{5,5 \mid 3\},\{5,5:4\}\} \\
\varphi \updownarrow & & \varphi \updownarrow \\
\{5,3,5 :: 4\} & & \{\{5,5:4\},\{5,5:4\}\}
\end{array}$$

There is a special notation for two polytopes of the family:

16F3 $$\{3,5,3 :: 4\} = \{3,5,3\}/\langle\langle\, (0121213)^4 \,\rangle\rangle,$$

16F4 $$\{5,3,5 :: 4\} = \{5,3,5\}/\langle\langle\, (01213)^4 \,\rangle\rangle.$$

Observe that we have here examples of branch contraction, as defined in (5B23); the notation is designed to suggest this. For instance, $(0,1,2,3) \mapsto (0,12121,3)$ on the generatrix of $\{3,5,3 :: 4\}$ gives that of $\{5,5:4\}$, while $(0,1,2,3) \mapsto (0,121,3)$ does the same for $\{5,3,5 :: 4\}$.

Once again, the operations in columns are vertex-figure replacements; the allomorphism is $\boldsymbol{\lambda}\colon (0,1,2,3) \mapsto (0,121232121,3,2)$, which interchanges the two regular dodecahedra with the same vertices. Like φ, quartering ψ is applied to the vertex-figure. Proposition 5B9 implies that $\{5,3,5 :: 4\}$ has deep hole $\{5\}$, while $\{\{5,5:4\},\{5,5:4\}\}$ has deep hole $\{3\}$. The polytopes in the first column have 68 vertices, while those in the second have 51.

In fact, we can say a little more. For the polytopes in the left column, layers $\mathcal{L}_1$ and $\mathcal{L}_2$ contain 20 vertices; each is the vertex-set of two dodecahedra. Layer $\mathcal{L}_4$ contains 12 vertices; these are the vertices of two icosahedra and two dodecahedra of the form $\{5,5 \mid 3\}$.

As with Table 16F1, we have actually only written down half the table. The polytopes in the right column have as vertex-figure the polyhedron $\{5,5:4\}$; as we saw in Section 16A, there are six such polyhedra with the same vertices. Applying the operations φ and ψ in turn to these polytopes yields a family of six, as depicted in Table 16F5. The pairs of polytopes in the same row are allomorphs. The operations on the horizontal arrows are φ and ψ, respectively, while those on the vertical arrows are ψ and φ, as in Table 16F2. That table then extends by the obvious symmetry, which also pairs allomorphic polytopes. Thus the total count gives $2 \cdot 2 + 4 + 6 = 14$ polytopes in the family, as the doubled Table 16F2 indicates.

16F5

$$\begin{array}{ccc}
\{\{3,5\},\{5,5:4\}\} & \overset{\varphi}{\longleftrightarrow} & \{\{3,5\},\{5,5:4\}\} \\
\psi\updownarrow & & \varphi\updownarrow \\
\{\{5,5\mid 3\},\{5,5:4\}\} & & \{\{5,5\mid 3\},\{5,5:4\}\} \\
\varphi\updownarrow & & \psi\updownarrow \\
\{\{5,5:4\},\{5,5:4\}\} & \overset{\psi}{\longleftrightarrow} & \{\{5,5:4\},\{5,5:4\}\}
\end{array}$$

An important observation follows from the permutation representation.

16F6 Theorem *The three polytopes of Table 16F2 with a component* $\{5,5:4\}$ *are universal.*

16F7 Remark Again, the parallel with the pentagonal polytopes of Section 7G is worth noting. All the star-polytopes of Table 16F1 are universal, with given facet and vertex-figure, except for $\{5,3,5 \mid 3\}$.

Notes to Section 16F

1. We guessed the defining relations for $\{3,5,3 :: 4\}$ and $\{5,3,5 :: 4\}$; these were verified by Marston Conder using MAGMA, which also showed – as we had surmised – that the second and third universal polytopes had the same group (of course, this also follows from the invertibility of the mixing operations).
2. All of these polytopes except $\{\{5,5:4\},\{5,5\mid 3\}\}$ were known to Asia Weiss by 1991, except that their polytopality had not been checked. The polytope $\{3,5,3\}$ occurs in [101] by Barry Monson and Egon Schulte. The others were variously rediscovered in 2012 by McMullen and Monson, who in particular verified that $\{\{5,5:4\},\{5,5:4\}\}$ is polytopal, this time using GAP. The size of its deep hole was guessed by McMullen and checked by Conder.
3. As we said, Table 16F2 should be doubled in size, as should Table 16F1; compare Table 7G8, where the geometric polytopes are listed. We saw that cycling the two operations φ and ψ on the right column gives isomorphic but distinct copies of the same polytopes. Similarly, suitable vertex-figure replacements applied to either of the last two polytopes in the left column again produce isomorphic but distinct polytopes; in particular, alllomorphism $\boldsymbol{\lambda}$: $(0,1,2,3) \mapsto (0,12321,2,323)$ interchanges two isomorphic but distinct copies of $\{5,3,5 :: 4\}$ with the same vertices. Observe, however, that we cannot (for example) jump between the two copies of $\{3,5,3 :: 4\}$ by simple vertex-figure replacement, because their vertex-figures fall into different layers.

16G $\{3,5,3::4\}$ and $\{5,3,5::4\}$

In view of previous remarks, we can say rather more about the abstract groups. The next result can be deduced from the permutation representation; further, see the notes at the end of the section.

16G1 Theorem *Let* p, q *stand for* $3, 5$ *in some order, so that* $\{3,5,3::4\}$ *and* $\{5,3,5::4\}$ *are of Schläfli type* $\{p,q,p\}$*. Then the edge replacement operation*

$$\varepsilon\colon (\boldsymbol{r}_0,\ldots,\boldsymbol{r}_3) \mapsto \big((\boldsymbol{r}_0\boldsymbol{r}_1\boldsymbol{r}_2)^5, \boldsymbol{r}_3, \boldsymbol{r}_2, \boldsymbol{r}_1\big) =: (\boldsymbol{s}_0,\ldots,\boldsymbol{s}_3)$$

on the generatrix is involutory, and interchanges $\{3,5,3::4\}$ *and* $\{5,3,5::4\}$*.*

It is therefore natural to ask if the group of $\{3,5,3::4\}$ or $\{5,3,5::4\}$ (as appropriate) is also determined by the extra relation $\big((\boldsymbol{r}_0\boldsymbol{r}_1\boldsymbol{r}_2)^5\boldsymbol{r}_3\big)^q = \boldsymbol{e}$ imposed on the Coxeter group $[p,q,p]$; it is possible that the dual relation would be needed as well.

Observe as well that applying the mixing operation, dualizing and then mixing again leads to

$$((\boldsymbol{r}_0\boldsymbol{r}_1\boldsymbol{r}_2)^5(\boldsymbol{r}_1\boldsymbol{r}_2\boldsymbol{r}_3)^5)^p = \boldsymbol{e},$$

for the appropriate p.

16G2 Remark Since q is odd, it would follow that $(\boldsymbol{r}_0\boldsymbol{r}_1\boldsymbol{r}_2)^5 \sim \boldsymbol{r}_3$, where $\sim$ denotes conjugacy in the group. In the hyperbolic honeycomb $\{p,q,p\}$, the corresponding geometric isometries $(R_0R_1R_2)^5$ and R_3 are reflexions in a point and hyperplane, respectively, and so are not conjugate. However, bear in mind that we are working in a quotient, which will mix up different kinds of reflexions.

The previous discussion and Theorem 16G1 show that we can make the following claims.

- The polytopes $\{3,5,3::4\}$ and $\{5,3,5::4\}$ have the same vertices and automorphism group $\mathrm{L}_2(16).2$.
- The edges of $\{5,3,5::4\}$ are the diameters (long diagonals) of the facets of $\{3,5,3::4\}$.
- The edges of $\{3,5,3::4\}$ are the diameters of the facets of $\{5,3,5::4\}$.
- There are two isomorphic copies of $\{3,5,3::4\}$ with the same vertices but different edges; their facets have the same vertices and are related as $\{3,5\}$ and $\{3,\frac{5}{2}\}$.
- There are two isomorphic copies of $\{5,3,5::4\}$ with the same vertices and edges; their facets lie in different layers.
- The facets of both copies of $\{3,5,3::4\}$ are inscribed in the vertex-figures of $\{5,3,5::4\}$.
- The facets of $\{5,3,5::4\}$ are inscribed in the vertex-figures of a copy of $\{3,5,3::4\}$ (depending on $\{5,3,5::4\}$), and are related to them as $\{5,3\}$ to $\{\frac{5}{2},3\}$.

We can augment this claim. Consider the hyperplane reflexions $R_0, \ldots, R_4$ in $\mathbb{H}^4$, whose mirrors have unit normals $u_0, \ldots, u_4$ given by

16G3

$$\begin{aligned} u_0 &:= \big(\tau^{1/2}, \tfrac{1}{2}\tau(1,1,1,1)\big), \\ u_1 &:= \big(0, \tfrac{1}{2}(0,-\tau^{-1},1,-\tau)\big), \\ u_2 &:= \big(0, \tfrac{1}{2}(0,1,-\tau,\tau^{-1})\big), \\ u_3 &:= \big(0, \tfrac{1}{2}(0,-\tau,\tau^{-1},1)\big), \\ u_4 &:= \big(\tau^{1/2}, \tfrac{1}{2}\tau(\tau^{-1},\tau,1,0)\big). \end{aligned}$$

Then $\langle R_0, \ldots, R_3\rangle = [3,5,3]$ and $\langle R_1, \ldots, R_4\rangle = [5,3,5]$. Moreover, with these generators, they are both subgroups of the Coxeter group $\langle R_0, \ldots, R_4\rangle$ with the diagram of Figure 16G4, which itself is a subgroup of infinite index in $[3,3,3,5]$; note that $\langle u_0, u_4\rangle = -\frac{1}{2}$. (In the discussion of $[5,3,5]$, we have – in effect – used the generators $R_1, \ldots, R_4$ in the reverse order.)

16G4

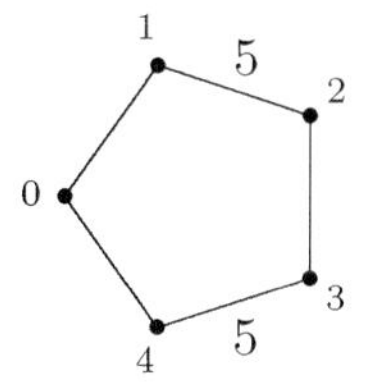

Notes to Section 16G

1. Another edge replacement is ε: $\{\{3,5\},\{5,5 : 4\}\} \longleftrightarrow \{\{5,5 \mid 3\},\{5,5 : 4\}\}$. Combinatorially, this has the same effect as ψ; however, geometrically it does not, since under ε the vertex-figures lie in different layers.

17

Two Families of 5-Polytopes

In this final chapter, we investigate two families of regular 5-polytopes, the second consisting of the double covers of the first. The starting point is the fact that the simple group $S_4(4)$ of order 9 79200 is a quotient of the Coxeter group $[3, 3, 3, 5]$ which satisfies the intersection property with respect to the naturally induced generators. This implies that there is a corresponding abstract regular polytope $\mathcal{P}$ of Schläfli type $\{3, 3, 3, 5\}$ with automorphism group $S_4(4)$.

Armed just with this information and a little reasonable guesswork, we show in Section 17A that the realization domain of $\mathcal{P}$ is very simple, in that $\mathcal{P}$ has only two non-trivial pure realizations of dimensions 50 and 85. In particular, the vertex-figure of $\mathcal{P}$ is the central quotient $\{3, 3, 5\}/2 = \{3, 3, 5 : 15\}$. In Section 17B we use two defining relations for the quotient to confirm this fact; subsequently, we show geometrically that one of the two relations is redundant.

Among the regular 5-polytopes with the same 136 vertices is one of Schläfli type $\{3, 5, 3, 5\}$, whose facet is the regular 4-polytope $\{3, 5, 3 :: 4\}$ which we met in Chapter 16; in Section 17C we look into its geometry. The dual polytope of type $\{5, 3, 5, 3\}$ has 120 vertices, whose realization domain we describe in Section 17D.

The relatives of the initial quotient of $\{3, 3, 3, 5\}$ are non-orientable. We describe the family of their double covers in Section 17E; these have members that are universal of their kind, particularly a certain quotient of $\{3, 5, 3, 5\}$ with 272 vertices and 240 facets. The processes of vertex-figure replacement and duality then lead in Section 17F to two extended families of regular polytopes with the same automorphism groups $S_4(4)$ and $S_4(4) \times 2$.

These families of polytopes correspond to actions of their automorphism groups on two of their maximal subgroups. The groups have another maximal subgroup, of index 85 in $S_4(4)$, and though there appear to be no nice related polytopes, nevertheless this gives rise to interesting symmetric sets of 85 and 255 points that we treat in Section 17G. The symmetric group $\boldsymbol{S}_6$ is also a maximal subgroup, which plays a rather different part.

Finally, in Section 17H we look briefly at other possible amalgamations. For instance, there is another quotient of type $\{3, 5, 3, 5\}$ and a close relative which one might initially think belong to one of our families; that they do not, with a completely unrelated group, is perhaps surprising.

17A An Intuitive Approach

In this section, we describe certain properties that the quotient $\mathcal{P}$ must have, based purely on the order of its automorphism group (see the notes at the end of the section). We shall confirm these ideas in Section 17B, but arguments such as these using numbers alone can often lead to deep results on realizations, as we shall see here and later.

There are only two possibilities for the vertex-figure $\mathcal{V}$ of the quotient $\mathcal{P}$: it must be either $\{3,3,5\}$ or its central quotient $\{3,3,5\}/2 = \{3,3,5:15\}$, since $\{3,3,5\}$ has no other regular quotients (see [99, Section 6C] and the further references quoted there). In the first case, $\mathcal{P}$ would have $979200/14400 = 68$ vertices. However, this cannot make sense, since $68 < 121 = 1 + 120$ (the initial vertex together with the vertex-figure). Thus the vertex-figure is $\mathcal{V} = \{3,3,5\}/2$, so that $\mathcal{P}$ has $979200/7200 = 136$ vertices.

Since $136 - 1 - 60 = 75$, this gives us 75 vertices other than the initial one and those of $\mathcal{V}$ to account for. Now the vertices of $\{3,3,3,5\}$ in layer 2 are those of a 120-cell $\{5,3,3\}$, and so are 600 in number. Since $600/75 = 8$, the only identifications which make sense (they must be compatible with the group $[3,3,5]/2$ of $\mathcal{V}$) are of vertex-sets of the 4-staurotopes (in the form $\{3,3^{1,1}\}$) inscribed in $\{5,3,3\}$, which exhibit the action of $[3,3,5]$ on its subgroup $[3^{1,1,1}]$.

This discussion has established some of the basic geometry of $\mathcal{P}$. Using just this information and a little intuition, we shall find its realization domain.

We have concluded that $\mathcal{P}$ must have layer vector $\Lambda = (1,60,75)$. The vertex-figure $\mathcal{V} = \{3,3,5\}/2$ has layer vector $\Lambda^{v} = (1,12,12,15,20)$, where the order of the layers follows the convention of Section 16F. In terms of the general cosine vector $\Gamma = (1,\alpha,\beta)$ of $\mathcal{P}$, we fairly easily see that the induced cosine vector of $\mathcal{V}$ is $\Gamma^{v} = (1,\alpha,\alpha,\beta,\beta)$. We shall revisit this point in more detail in Section 17B, but also see the notes at the end of the section.

Since $\mathcal{P}$ has only two non-trivial diagonal classes, both being symmetric (one corresponds to the vertex-figure and the other is odd), we can apply the vertex-figure criterion Theorem 4D4 to find the non-trivial pure realizations. Thus we have the two equations for α and β (the first is the layer equation of Theorem 3C7):

$$\begin{aligned} 0 &= 1 + 60\alpha + 75\beta, \\ 60\alpha^2 &= 1 + 12\alpha + 20\beta + 12\alpha + 15\beta = 1 + 24\alpha + 35\beta. \end{aligned}$$

Eliminating β yields

$$225\alpha^2 + 15\alpha - 2 = 0 \implies \alpha = -\tfrac{2}{15} \text{ or } \tfrac{1}{15}$$

(the order is chosen for future convenience), from which follows

17A1 Theorem *The quotient polytope $\mathcal{P}$ of type $\{3,3,3,5\}$ with automorphism group* $\mathrm{S}_4(4)$ *has cosine matrix*

$$\begin{bmatrix} 1 & 1 & 1 \\ 1 & -\frac{2}{15} & \frac{7}{75} \\ 1 & \frac{1}{15} & -\frac{1}{15} \end{bmatrix},$$

with layer and dimension vectors

$$\Lambda = (1, 60, 75), \quad D = (1, 50, 85).$$

Proof. For the dimensions, we appeal to [88, Theorem 4.5]. Thus we have

$$\|\Gamma_1\|_\Lambda^2 = \tfrac{1}{136}\big(1 + 60(-\tfrac{2}{15})^2 + 75(\tfrac{7}{75})^2\big) = \tfrac{1}{50},$$

and

$$\|\Gamma_2\|_\Lambda^2 = \tfrac{1}{136}\big(1 + 60(\tfrac{1}{15})^2 + 75(-\tfrac{1}{15})^2\big) = \tfrac{1}{85},$$

after minor calculations. □

17A2 Remark We may also verify that

$$\langle \Gamma_1, \Gamma_2 \rangle_\Lambda = \tfrac{1}{136}\big(1 + 60(-\tfrac{2}{15})(\tfrac{1}{15}) + 75(\tfrac{7}{75})(-\tfrac{1}{15})\big) = 0,$$

so that Γ_1 and Γ_2 are Λ-orthogonal, as they must be by Theorem 3F5.

Notes to Section 17A

1. In this section, we have only used the bare information (in the preamble to the chapter) sent to us by Marston Conder [12] in the first of two e-mail messages.
2. There are many other ways in which the group $\mathrm{S}_4(4)$ arises, so that alternative designations are $\mathrm{PSp}(4,4)$ or $\mathrm{Sp}_4(4)$); see $\mathbb{ATLAS}$.
3. According to $\mathbb{ATLAS}$, $\mathrm{S}_4(4)$ has no subgroup of order 14400; this confirms what we have seen, that $(\mathrm{A}_5 \times \mathrm{A}_5) \rtimes \mathrm{C}_2$ of order 7200 is indeed a maximal subgroup.
4. Though we never appeal to the degrees of the irreducible representations of $\mathrm{S}_4(4)$ as given in $\mathbb{ATLAS}$, nevertheless they give a useful check on our workings.
5. The polytope $\mathcal{P}$ is a quotient of the universal $\{\{3,3,3\},\{3,3,5:15\}\}$ and, since its automorphism group $\boldsymbol{G} \cong \mathrm{S}_4(4)$ is simple, it is (in a sense) a *minimal* quotient. This universal polytope is infinite, as shown in [61].
6. The locally toroidal regular polytope $\{\{4,4 \mid 3\},\{4,4 \mid 5\}\}$ has as automorphism group an extension of $\mathrm{S}_4(4)$ by $\mathrm{C}_2 \times \mathrm{C}_2$ (see [99, Section 10C] for further details). One wonders if there is any reasonably straightforward connexion between this toroidal polytope and $\mathcal{P}$.
7. Since $\mathcal{L}_3$ has 720 vertices, and 75 does not divide 720, we see that we cannot have $\mathcal{L}_3 \sim \mathcal{L}_2$, so that the induced cosine vector of $\mathcal{V}$ is of the form $\Lambda^{v} = (1, \alpha, \alpha, \beta, \gamma)$ for $\gamma = \alpha$ or β. With $\gamma = \alpha$ one ends up with a quadratic irrational α, and this would lead to two realizations with the same dimension; this is, of course, impossible. Indeed, because the two non-trivial pure realizations will necessarily have different dimensions, general considerations (compare Remark 3L4) imply that the cosine matrix must have rational entries.

17B Group and Geometry

In the subsequent discussion, we shall want to appeal to various properties of the geometric group $[3,3,3,5]$. We work in the hyperbolic space

$$\mathbb{H}^4 := \{x = (\eta, z) \in \mathbb{R}^5 \mid \eta^2 - \|z\|^2 = 1\};$$

the norm here is in $\mathbb{E}^4$. The (hyperbolic) distance μ between two points $x_1, x_2 \in \mathbb{H}^4$ is given by

$$\cosh\lambda = \langle x_1, x_2\rangle := \eta_1\eta_2 - \langle z_1, z_2\rangle,$$

where again the latter inner product is the usual one in $\mathbb{E}^4$. The hyperbolic cosine $\cosh\mu$ is the exact analogue of the cosine in the finite case; we often refer to it as a *height*.

As we shall see shortly, it is usually convenient to write 2λ for an edge-length. If $v \in \mathbb{E}^4$ is a unit vector, then the hyperbolic unit normal to the mirror of the hyperplane reflexion R that interchanges $(1, o)$ and $(\cosh 2\lambda, \sinh 2\lambda v)$ is $(\sinh\lambda, \cosh\lambda v)$, and R itself is given by

$$(\eta, z)R = \big(\eta\cosh 2\lambda - \langle z, v\rangle \sinh 2\lambda, (\eta\sinh 2\lambda - \langle z, v\rangle\cosh 2\lambda)v + z - \langle z, v\rangle v\big).$$

Generatrix

For the initial honeycomb $\{3,3,3,5\}$, the generatrix $(R_0, \ldots, R_4)$ consists of reflexions in hyperplanes, whose unit normals u_j are

17B1

$$\begin{aligned} u_0 &:= (\tau^{1/2}, \tau, 0, 0, 0),\\ u_1 &:= \tfrac{1}{2}(0, -\tau^{-1}, \tau, 1, 0),\\ u_2 &:= (0, 0, 0, -1, 0),\\ u_3 &:= \tfrac{1}{2}(0, 0, -\tau^{-1}, 1, \tau),\\ u_4 &:= (0, 0, 0, 0, -1). \end{aligned}$$

Note that a unit normal vector u in $\mathbb{H}^4$ has square norm $\|u\|^2 = -1$. For the most part, we need only know that $R_1, \ldots, R_4$ act on the component $z \in \mathbb{E}^4$ of a point $x = (\eta, z) \in \mathbb{H}^4$ alone, generating $[3,3,5]$ in a standard way. Our main concern is with R_0, for which we have

$$xR_0 = (\tau^3\eta - 2\tau^{3/2}\zeta_1, 2\tau^{3/2}\eta - \tau^3\zeta_1, \zeta_2, \zeta_3, \zeta_4).$$

Later, however, we shall find it convenient to replace R_0 by suitable conjugates under $\langle R_1, \ldots, R_4\rangle$.

In one context, quaternions prove useful (see the notes at the end of the section). We can identify the components z as vectors in $\mathbb{E}^4$, and further think of z as a quaternion $\zeta_1 + \zeta_2\mathbf{i} + \zeta_3\mathbf{j} + \zeta_4\mathbf{k}$. Then symmetries of $\mathbb{H}^4$ in the subgroup of $[3,3,3,5]$ that fix the initial vertex $v = (1, o)$ can be represented by quaternions, as in Chapter 7.

Point-reflexions are occasionally useful. It is easy to see that, if $c \in \mathbb{H}^4$, then

17B2 $$x \mapsto 2\langle x, c\rangle c - x$$

is the point-reflexion in c (it is an involutory isometry whose only fixed point is c – note that the same expression gives a point-reflexion in spherical space).

Heights and Vertices

As in (7E2), we write

$$G := \{(2,0,0,0)^s, (1,1,1,1)^s, (\tau, 1, \tau^{-1}, 0)^s\};$$

this is the vertex-set of $\{3,3,5\}$. Recall that the notation $(\cdots)^s$ means that we take all even permutations with arbitrary changes of sign of the coordinates within the brackets $(\cdots)$. Further, we denote by $\mathbb{Z}G$ the set of integer linear combinations of vectors of G. While we make little use of it, we shall establish the following

17B3 Proposition *The vertices of* $\{3,3,3,5\}$ *are precisely those of the form* $(r+s\sqrt{5}, \tau^{3/2}g) \in \mathbb{H}^4$ *such that* $g \in \mathbb{Z}G$ *and* $s = 0, 1, \ldots$, *where the non-negative integer* r *satisfies* $r+s$ *odd with* $|r - s\sqrt{5}| \leqslant 1$.

Proof. First, if we write $\kappa_s := r + s\sqrt{5}$ as above, then a point of the form $x = (\kappa_s, \tau^{3/2}g)$ is clearly taken into another by $R_1, \ldots, R_4$. For R_0, if $g = (\gamma_1, \ldots, \gamma_4) \in \mathbb{Z}G$, then $xR_0 = (\eta, \tau^{3/2}z)$ is given by

$$\eta = \tau^3(\kappa_s - 2\gamma_1),$$

which is of the right form since $2\gamma_1 = a + b\sqrt{5}$ with $a+b$ even. Further,

$$\begin{aligned} z &= (\tau^3(2\kappa_s - \gamma_1), \gamma_2, \gamma_3, \gamma_4) \\ &= 2(\tau^3\kappa_s - \tau^2\gamma_1)e_1 + g \in \mathbb{Z}G \end{aligned}$$

for the same reason.

For the next assertion, we employ a trick. We consider instead the non-discrete honeycomb $\{3,3,3,\frac{5}{2}\}$ in the unit sphere in $\mathbb{E}^5$. We obtain this just by applying the change of sign ‡ of $\sqrt{5}$ to $\{3,3,3,5\}$. Terms involving $\tau^{1/2}$ produce a factor $i = \sqrt{-1}$, but this just turns $\mathbb{H}^4$ into $\mathbb{S}^4$. We deduce that $\{3,3,3,\frac{5}{2}\}$ has vertices of the form $x = ({\kappa_s}^\ddagger, \tau^{-3/2}z)$ with $z \in \mathbb{Z}G^\ddagger$. But these lie on the unit sphere, from which we see that

$$|r - s\sqrt{5}| = |{\kappa_s}^\ddagger| \leqslant 1$$

for each $s \geqslant 0$, as claimed; observe that r is uniquely determined by s.

Last, write a general point of $\mathbb{H}^4$ simply as $x = (\eta, z)$. If $z \in \tau^{3/2}\mathbb{Z}G$, then, if need be, we may apply a symmetry in $[3,3,5]$, and replace z by an equivalent point in $\tau^{3/2}\mathbb{Z}G$ satisfying $\zeta_1 \geqslant (\tau^2/\sqrt{8})\|z\|$; compare [27, Table V(iv)] and see the notes at the end of the section. Using the easily proved inequality

$$\|z\| = \sqrt{\eta^2 - 1} \geqslant \eta - \eta^{-1}$$

for $\eta \geqslant 1$, we see that $xR_0 = (\eta', z')$ satisfies

$$\begin{aligned}\eta' &= \tau^3\eta - 2\tau^{3/2}\zeta_1 \leqslant \tau^3\eta - \tau^{7/2}2^{-1/2}\|z\| \\ &\leqslant \tau^3\left(1 - \sqrt{\frac{\tau}{2}}\right)\eta + \frac{\tau^{7/2}}{\sqrt{2}\eta} < \eta\end{aligned}$$

whenever $\eta \geqslant \kappa_2$, since

$$\tau^3\left(1 - \sqrt{\frac{\tau}{2}}\right) < \frac{1}{2}.$$

In other words, if $\eta = \kappa_s$ for some s, then we see that xR_0 is of the required form, and is nearer the initial vertex $(1, o)$ than x. We now appeal to the obvious inductive assumption: since xR_0 lies at height κ_t for some $t < s$, it is a vertex of $\{3,3,3,5\}$. Thus $x \in \text{vert}\{3,3,3,5\}$ also, and this completes the proof. □

What is missing from Proposition 17B3 is

17B4 Conjecture *For every $s \geqslant 0$, there are vertices of $\{3,3,3,5\}$ of the form $(\kappa_s, \tau^{3/2}g)$ with $g \in \mathbb{Z}G$.*

17B5 Remark Note that two (or more) layers may share the same height, so that we must distinguish them by $\mathcal{L}_{sa}, \mathcal{L}_{sb}, \ldots$, and so on.

Initial Layers

For our purposes, we have to know the first few layers $\mathcal{L}_j$ of $\{3,3,3,5\}$, and what configurations of vertices they form (we have actually described $\mathcal{L}_1$ and $\mathcal{L}_2$ already, and mentioned $\mathcal{L}_3$). Since certain sets of points occur frequently in what follows, it is convenient to use a shorthand for them. We already have the set G. Similarly, the vertex set H of the dual 120-cell $\{5,3,3\}$ in $\mathbb{E}^4$ is as in (7F3):

$$\begin{aligned}&\{(2,2,0,0)^s,\ (\sqrt{5},1,1,1)^s,\ (\tau^2,\tau^{-1},\tau^{-1},\tau^{-1})^s,\ (\tau^{-2},\tau,\tau,\tau)^s,\\ &(\sqrt{5},\tau^{-1},\tau,0)^s,\ (\tau^2,\tau^{-2},1,0)^s,\ (2,\tau,\tau^{-1},1)^s\}.\end{aligned}$$

In fact, we do not need this detail. All we must remember is that $v \in H$ just when $v = x + y$ for some orthogonal $x, y \in G$ (that is, $\langle x, y\rangle = 0$), and that this can happen in three ways. (We shall not need the alternative form of the vertex-set as given in Proposition 7F7.)

This example illustrates a general principle. If we have such a representation $v = x + y$, then $\langle v, x - y\rangle = 0$. The three representations of v then yield (with v) an orthogonal basis of $\mathbb{E}^4$ in H. Together with their opposites, they form an image of the vertex-set of a staurotope whose group (in the form $\mathbf{B}_4 = [3^{1,1,1}]$ – that is, half its full group) is a subgroup of $[3,3,5]$. Thus, for instance, we have

$$(2,2,0,0) = \begin{cases} (2,0,0,0)+(0,2,0,0) & \mapsto (2,-2,0,0), \\ (1,1,1,1)+(1,1,-1,-1) & \mapsto (0,0,2,2), \\ (1,1,1,-1)+(1,1,-1,1) & \mapsto (0,0,2,-2). \end{cases}$$

17B6 Remark A staurotope S with symmetry group $\mathbf{B}_4$ in $\{5,3,3\}$ is quasi-regular, in that the group preserves a 2-colouring of the facets. We shall call S a *q-staurotope*; recall that S is uniquely determined by any one of its vertices. As we saw in Section 7J there are other regular staurotopes with vertices in H, but with less symmetry. The alternative form of vertices of the 120-cell in Proposition 7F7 clearly illustrates the difference between the two kinds of staurotope.

Then there are three vertex-sets of polyhedra; these naturally embed in G, here taken as certain coordinates $(\zeta_2, \zeta_3, \zeta_4)$ of $(\zeta_1, \ldots, \zeta_4)$. For the icosahedron, dodecahedron and icosidodecahedron, respectively,

$$\begin{aligned} I &:= \{(\tau, 1, 0)^s\}, \\ D &:= \{(1,1,1)^s, (\tau, 0, \tau^{-1})^s\}, \\ T &:= \{(2,0,0)^s, (\tau, \tau^{-1}, 1)^s\}. \end{aligned}$$

Of course, with three coordinates an even permutation is cyclic. For instance, the vertex-figure of the 600-cell $\{3,3,5\}$ at its vertex $(2,0,0,0)$ is $(\tau, \tau^{-1}I) := \{(\tau, \tau^{-1}v) \mid v \in I\}$; this exemplifies a convention used throughout.

The corresponding diagonals can be found from the vertex-figure $\{3,3,5\}$; the first four subtend angles from the initial vertex $\vartheta = \pi/5, \pi/3, 2\pi/5, \pi/2$, the first being the original edge (we shall not need more than these). With λ_j the length of diagonal class j, and $\kappa_j := \cosh \lambda_j$, we have

$$\kappa_j = \begin{cases} \tau^3, & \text{if } j = 1, \\ \tau^3\sqrt{5}, & \text{if } j = 2, \\ 3\tau^3, & \text{if } j = 3, \\ \tau^6, & \text{if } j = 4. \end{cases}$$

The 120 vertices in $\mathcal{L}_1$ are those of the (broad) vertex-figure of $\{3,3,3,5\}$ at the initial vertex $(1, o)$; we can thus write $\mathcal{L}_1 = (\tau^3, \tau^{3/2}G)$.

To get to a vertex in $\mathcal{L}_2$, one passes through an initial cell, traversing a facet of the (broad) vertex-figure $\{3,3,5\}$, and going to the opposite vertex of the adjacent cell. We thus have the 600 vertices of a 120-cell $\{5,3,3\}$, which we can express as $(\tau^3\sqrt{5}, \tau^{5/2}H)$.

To get to $\mathcal{L}_3$, one stays in a planar section through a diametral decagon of the vertex-figure, going across the edge of an initial trigon to the opposite vertex of the adjacent trigon. We thus obtain the 720 points of the vertex-set J of a truncated 600-cell, with

$$J := \{a + b \mid a, b \in G,\ \langle a, b\rangle = 2\tau\};$$

the condition $\langle a, b\rangle = 2\tau$ just expresses the fact that a, b are adjacent vertices of $\{3, 3, 5\}$. As we have observed, the planar section is a hyperbolic tiling $\{3, 10\}$.

The 3600 vertices in $\mathcal{L}_4$ fall into icosidodecahedra, centred at the vertices in $\mathcal{L}_1$; these are central sections of the vertex-figures at these adjacent vertices. A vertex in $\mathcal{L}_4$ is of the form $(\tau^6, \tau^{9/2}u + \tau^{3/2}w)$, with $u, w \in G$ orthogonal vertices, so that $\langle u, w\rangle = 0$.

We shall not go into what happens in other layers in general terms; when it comes to identifications, we shall treat each case individually.

Identifications

We now move on to the identifications, which we treat here in more formal terms than before. In their original form (as produced by the computer algebra program MAGMA [5] – see the notes at the end of the section), the extra relations that have to be imposed on the Coxeter group [3, 3, 3, 5] to obtain the group $S_4(4)$ as the automorphism group of $\mathcal{P}$ looked rather formidable and conveyed no intuition about the geometry. In fact, they could both be simplified; in terms of the generatrix $(\boldsymbol{r}_0, \ldots, \boldsymbol{r}_4)$ of $\boldsymbol{G}$ they are equivalent to

17B7 $$(\boldsymbol{r}_1\boldsymbol{r}_2\boldsymbol{r}_3\boldsymbol{r}_4)^{15} = \boldsymbol{e},$$

17B8 $$(\boldsymbol{r}_0(\boldsymbol{r}_1\boldsymbol{r}_2\boldsymbol{r}_3\boldsymbol{r}_4)^3)^5 = \boldsymbol{e};$$

these relations were verified by MAGMA. Observe that (17B7) confirms the first of our claims in Section 17A, since $(\boldsymbol{r}_1\boldsymbol{r}_2\boldsymbol{r}_3\boldsymbol{r}_4)^{15}$ is the central symmetry of the vertex-figure $\{3, 3, 5\}$. We shall show later that this relation is actually redundant, though this is not obvious from the algebra. However, we have already noted from ATLAS that $S_4(4)$ does not have a subgroup [3, 3, 5], but only its central quotient [3, 3, 5 : 15].

17B9 Remark In either relation (17B7) or (17B8), the order of $\boldsymbol{r}_1, \ldots, \boldsymbol{r}_4$ within the brackets is immaterial. Moreover, in (17B8), we can clearly replace $(\boldsymbol{r}_1\boldsymbol{r}_2\boldsymbol{r}_3\boldsymbol{r}_4)^3$ by any conjugate under the subgroup $\boldsymbol{G}^e := \langle \boldsymbol{r}_2, \boldsymbol{r}_3, \boldsymbol{r}_4\rangle$.

For the geometry of (17B8), $(R_1R_2R_3R_4)^3$ is a Clifford displacement through a tenth of a turn about the initial vertex of $\{3, 3, 3, 5\}$. Recall that $\{3, 3, 3, 5\}$ has planar sections $\{3, 10\}$, such as those through the initial vertex and diametral decagons of its vertex-figure. The Clifford displacement then rotates each of twelve of these decagons by a tenth of a turn, and combines with R_0 to give the basic translation $R_0(R_1R_2R_3R_4)^3$ of a Petrie apeirogon of $\{3, 10\}$.

We have the following picture of what $Z := R_0(R_1R_2R_3R_4)^3$ does. Let $v = (1, o)$ be the intial vertex of the geometric honeycomb $\{3, 3, 3, 5\}$, let $a_0, \ldots, a_9$ be the vertices in order of the vertex-figure $\{10\}$ of the section $\{3, 10\}$, and let the next layer of its vertices be $b_0, \ldots, b_9$, with $b_j a_{2+j} a_{3+j}$ forming a trigon of the tiling for each $j \pmod{10}$. Then $a_0, a_1, v, a_2, a_3, b_0$ are six successive vertices of a Petrie apeirogon of $\{3, 10\}$, as depicted in Figure 17B10. In this, a_0v is an edge of the tiling, while vb_0 crosses the edge a_2a_3, making a_0vb_0 a right-angle. Since a_0 lies in layer 1 while b_0 lies in layer 3, this tells us that layers 3 and 1 are identified. More specifically, we write this as $a_0 \sim b_0$.

17B10

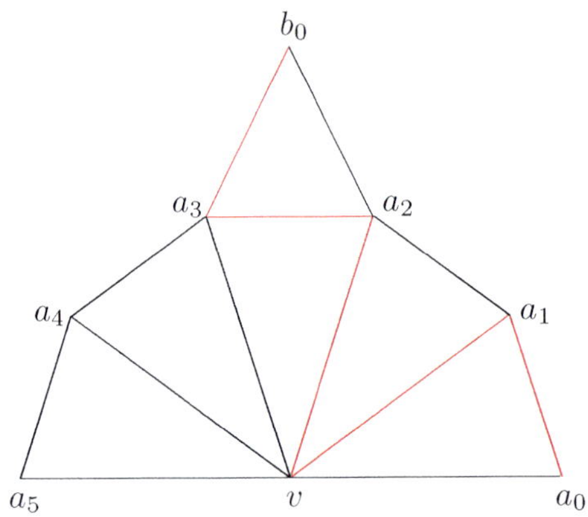

We choose a planar section $\{3, 10\}$ of $\{3, 3, 3, 5\}$ through the initial vertex $v = (1, o)$ in which a_0, b_0 are given by

17B11 $$a_0 = (\tau^3, 2\tau^{3/2}e_1), \quad b_0 = \big(3\tau^3, 2\tau^{5/2}(0, \tau, 1, 0)\big),$$

so that $a_0 \in \mathcal{L}_1$ and $b_0 \in \mathcal{L}_3$. With this choice of a_0, b_0, a suitable conjugate of $(R_1R_2R_3R_4)^3$ respecting Remark 17B9 is (left or right) multiplication by the quaternion $\mathbf{p} := \frac{1}{2}(\tau + \mathbf{i} + \tau^{-1}\mathbf{j})$, which acts as the cyclic permutation $a_j \mapsto a_{j+1}$. Note that $\langle a_0, b_0 \rangle = \tau^3 \cdot 3\tau^3 = 3\tau^6 =: \kappa_{12}$; in fact, $\{a_0, b_0\} \in \mathcal{D}_{12a}$, which consists of symmetric diagonals. The symmetries of $\{3, 3, 3, 5\}$ fixing $v = (1, o)$ and a_0 show that there are actually 12 such vertices in $\mathcal{L}_3$ associated with a_0, comprising the vertex-set of an icosahedron which is regular under symmetries in $\mathbf{G}$; we shall call it an *r-icosahedron*. Indeed, its vertex-set is $\big(3\tau^3, 2\tau^{5/2}(0, (\tau, 1, 0)^s)\big)$, with the usual convention.

17B12 Remark The vertices $\big(3\tau^3, 2\tau^{5/2}(0, (1, \tau, 0)^s)\big)$, which lie in the same hyperplane, form the vertex-set of an *n-icosahedron*, where 'n' stands for 'non-regular' (under the group $[3, 3, 3, 5]$). Indeed, the mirror of the hyperplane reflexion that swaps $\big(3\tau^3, 2\tau^{5/2}(0, 1, \tau, 0)\big)$ and $\big(3\tau^3, 2\tau^{5/2}(0, 0, 1, \tau)\big)$ has normal $\frac{1}{2}(0, 0, 1, \tau^{-1}, -\tau)$; however, $(0, 1, \tau^{-1}, -\tau) \notin G$. The diagonals from a_0 to these points lie in the asymmetric diagonal class $\mathcal{D}_{12b}$. The r-icosahedron containing $\big(3\tau^3, 2\tau^{5/2}(0, 1, \tau, 0)\big)$ actually has vertex-set $\big(3\tau^3, 0, 2\tau^{5/2}((0, 1, \tau)^s, 0)\big)$.

We now determine some identifications, writing $x \sim y$ if the vertices x, y are identified. First, we have already seen that $\mathcal{L}_1 \sim \mathcal{L}_3$. Moreover, from Figure 17B10 we obviously have $b_0 \sim a_5 = (\tau^3, -2\tau^{3/2}e_1)$, whence $a_0 \sim a_5$; that is, opposite vertices of the vertex-figure $\{3, 3, 5\}$ are identified. (More generally, if $a, b \in G$ satisfy $\langle a, b\rangle = 2\tau$, then the vertex $(3\tau^3, \tau^{5/2}(a+b)) \in \mathcal{L}_3$ is joined by diagonals in $\mathcal{D}_{12a}$ to the two vertices $(\tau^3, \pm\tau^{5/2}(a-b)) \in \mathcal{L}_1$.) This follows as well from the effect on the diametral $\{3, 10\}$, which collapses to $\{3, 10 : 5\}$. However, at this stage we may not conclude that this forces the collapse $\{3, 3, 5\} \searrow \{3, 3, 5 : 15\} = \{3, 3, 5\}/2$ of the whole vertex-figure.

Next, if T is the conjugate of R_0 with e_1 replaced by e_2, then

$$a_0T = \big(\tau^6, 2\tau^{3/2}(1, \tau^3, 0, 0)\big), \quad b_0T = \big(\tau^3\sqrt{5}, 2\tau^{5/2}(0, 1, 1, 0)\big),$$

and since $a_0T \in \mathcal{L}_4$ and $b_0T \in \mathcal{L}_2$, this shows that $\mathcal{L}_4 \sim \mathcal{L}_2$. Finally, we can apply R_2, which fixes a_0, commutes with T and changes the sign of ζ_3, to see that

$$\big(3\tau^3, 2\tau^{5/2}(0, \tau, 1, 0)\big) \sim \big(3\tau^3, 2\tau^{5/2}(0, \tau, -1, 0)\big),$$
$$\big(\tau^3\sqrt{5}, 2\tau^{5/2}(0, 1, 1, 0)\big) \sim \big(\tau^3\sqrt{5}, 2\tau^{5/2}(0, 1, -1, 0)\big).$$

The former two are the vertices of a typical edge of an r-icosahedron in $\mathcal{L}_3$, while the latter two are the vertices of a typical edge of a q-staurotope in $\mathcal{L}_2$; in particular, we have

17B13 Lemma *The relation (17B8) forces $\mathcal{D}_{20a} \sim \mathcal{D}_0$, and hence the collapse $\{5, 3, 3\} \searrow \{5, 3, 3\}/8$ of layer $\mathcal{L}_2$.*

17B14 Remark The asymmetric diagonals of the n-staurotopes in $\mathcal{L}_2$ fall into diagonal class $\mathcal{D}_{20b}$. In a similar way, the diagonal joining $(\tau^3, 2\tau^{3/2}e_4)$ to b_0 is a typical member of the asymmetric diagonal class $\mathcal{D}_{12b}$.

We now make the following observation. The quotient of $\{3, 3, 3, 5\}$ by (17B8) is compatible with its automorphisms, in particular those that fix an initial vertex (forming layer $\mathcal{D}_0$). In other words, the identifications in $\mathcal{L}_2$ and $\mathcal{L}_3$ are not just of vertices, but of faces of all ranks. But now each vertex in $\mathcal{L}_1$ is determined by the 20 adjacent vertices of a dodecahedron in $\mathcal{L}_2$, and carries the scaffolding formed by that vertex and the edges and trigons to the vertices and edges of the dodecahedron. We can now fill in facets of $\{3, 3, 3, 5\}$ that are, in turn, determined by these edges and trigons, first obtaining the vertex-figure $\{3, 3, 5\}$, and finally the facets through the initial vertex. We conclude that all opposite faces of the vertex-figure are identified by (17B8), and so have proved

17B15 Theorem *The relation (17B7) is redundant in defining the quotient polytope $\mathcal{P}$ of $\{3, 3, 3, 5\}$. In other words, employing the notation of (2D2),*

$$\mathcal{P} = \{3, 3, 3, 5\}/\langle\!\langle\, (0(1234)^3)^5 \,\rangle\!\rangle.$$

17B16 Remark We have in Theorem 17B15 a variant of the circuit criterion of Theorem 2D4, in that there is an imposed relator in the sense of Section 2D on the group of the vertex-figure itself (usually this group is assumed already known). Thus we have now shown that $\mathcal{D}_{8a}$, with $\kappa_8 = 2\tau^6 - 1 = 17 + 8\sqrt{5}$, arising from diameters of the vertex-figure, is also such that $\mathcal{D}_{8a} \sim \mathcal{D}_0$. It is important to note two further things. First, we did not need to appeal to identifications in $\mathcal{L}_3$. Second, the proof only used the fact that $\mathcal{D}_{20a} \sim \mathcal{D}_0$, rather than the stronger (17B8); we shall return to this point in Section 17E.

It may be worth noting that identifications by $\mathcal{D}_{20a} \sim \mathcal{D}_0$ alone (without the later geometric implications) are by a subgroup of $\mathbf{G} = [3,3,3,5]$ of infinite index. This is illustrated by Figure 17B17, in which one can clearly see the q-staurotope and r-icosahedron as subdiagrams.

17B17

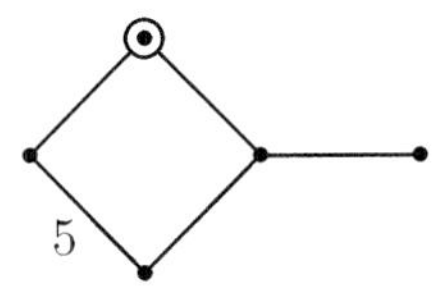

Let us be more specific about the way that the indentifications work (see the notes at the end of the section for more possibilities); we conjugate freely by elements of $\mathbf{G}^{\boldsymbol{v}} := \langle R_1, \ldots, R_4 \rangle$. As we have seen, $\mathcal{L}_1$ consists of the points $(\tau^3, \tau^{3/2}G)$, with identification $z \sim -z$ for $z \in G$. The identification in $\mathcal{L}_2$, consisting of $(\tau^3\sqrt{5}, \tau^{5/2}H)$, is as follows; again, we concern ourselves just with the $v \in H$. Then $v \sim w$ just when $\langle v, w \rangle = 0$ and $v \pm w \in 2G$. Thus, for example,

$$(\tau, \tau, -\tau, -\tau^{-2}) \sim (\tau, -\tau^{-2}, \tau, -\tau),$$
$$(\tau, -\tau, \tau, -\tau^{-2}) \not\sim (\tau, -\tau^{-2}, -\tau, \tau),$$

since $(0, 2, -2\tau, 2\tau^{-1}) \in 2G$ but $(0, -2\tau^{-1}, 2\tau, -2) \notin 2G$. (The symmetry in the first pair is deceptive; what is needed in the second pair is to change signs in the *same* coordinates of the first pair.) In fact, the three vertices obtained from the first set by cyclic permutation of the last three coordinates are all identified with a fourth vertex orthogonal to all three (the example was chosen with this end in mind):

17B18

$$\begin{aligned} z_1 &= (\tau, \tau, -\tau, -\tau^{-2}), \\ z_2 &= (\tau, -\tau^{-2}, \tau, -\tau), \\ z_3 &= (\tau, -\tau, -\tau^{-2}, \tau), \\ z_4 &= (\tau^{-2}, \tau, \tau, \tau). \end{aligned}$$

The vertices in $\mathcal{L}_4$ are of the form $\big(\tau^6, \tau^{3/2}(\tau^3 u + w)\big)$, with $u, w \in G$ such that $\langle u, w \rangle = 0$. Up to sign, there is a unique $v \in G$ with $\langle u, v \rangle = \langle v, w \rangle = 0$ such that $\frac{1}{2}(\tau u + v + \tau^{-1} w) \in G$ also. Then $u + v \in H$, and

17B19

$$(\tau^6, \tau^{9/2} u + \tau^{3/2} w) \sim \big(\tau^3\sqrt{5}, \tau^{5/2}(u + v)\big).$$

Once again, observe that this automatically identifies the vertices of $\mathcal{L}_2$ in eights, in view of the free choice of sign of v and conjugacy under $\boldsymbol{N}$; note that we already have the identification on changing the sign of w. Specifically, if $x \in G$ completes $\{u, v, w\}$ to an orthogonal set, then $\pm u \pm v \sim \pm w \pm x$.

In Section 17E, we shall need the details of layer $\mathcal{L}_{12a}$. We can move either a_0 or b_0 of (17B11) to $v = (1, o)$ and apply the symmetries fixing v. For the first, we apply R_0, obtaining

$$b_0 R_0 = \bigl(3\tau^6, 2\tau^{7/2}(3\tau, 1, \tau^{-1}, 0)\bigr).$$

Writing

$$(3\tau, 1, \tau^{-1}, 0) = \tau(2, 0, 0, 0) + (\tau, 1, \tau^{-1}, 0),$$

it is not hard to see that the general vertex in $\mathcal{L}_{12a}$ is $\bigl(3\tau^6, 2\tau^{7/2}(\tau g + h)\bigr)$, where $g, h \in G$ are such that $\langle g, h \rangle = 2\tau$ (that is, g and h are adjacent vertices of the 600-cell). For the second, we can apply the point-reflexion in the mid-point $c = \bigl(\tau^2, \tau^{1/2}(0, \tau, 1, 0)\bigr)$ of the edge $\{a_2, a_3\}$, under which

$$a_0, a_5 = (\tau^3, \pm 2\tau^{3/2} e_1) \mapsto \bigl(3\tau^6, 2\tau^{7/2}(\mp\tau^{-2}, \tau^3, \tau^2, 0)\bigr),$$

for which $g = (\mp\tau^{-1}, \tau, 1, 0)$, $h = (\pm\tau^{-1}, \tau, 1, 0)$.

In conclusion, we have proved

17B20 Proposition *The layer $\mathcal{L}_{12a}$ contains 1440 vertices, which are those of a truncated 600-cell*

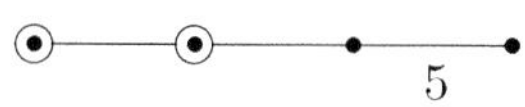

The icosahedral facets arise from the diametral r-icosahedra in $\mathcal{L}_3$, so that its edges belong to $\mathcal{L}_{20a}$; the remaining edges joining the icosahedra come from diameters of the vertex-figure in $\mathcal{L}_1$, and so lie in $\mathcal{L}_{8a}$. In particular, in view of Remark 17B16, if $\mathcal{D}_{20a} \sim \mathcal{D}_0$, then the vertices in $\mathcal{L}_{12a}$ are identified to a single point.

Petrie Polygon

The Petrie polygon of a regular polytope is often of interest; that of $\mathcal{P}$ is no different. We show here how to find it without having recourse to a computer algebra program. Working initially with $\{3, 3, 3, 5\}$ itself, we first use (17B1) to find its generatrix (S, T), with $S := R_0 R_2 R_4$ and $T := R_1 R_3$, namely,

$$S = \begin{bmatrix} \tau^3 & 2\tau^{3/2} & 0 & 0 & 0 \\ -2\tau^{3/2} & -\tau^3 & 0 & 0 & 0 \\ 0 & 0 & 1 & 0 & 0 \\ 0 & 0 & 0 & -1 & 0 \\ 0 & 0 & 0 & 0 & -1 \end{bmatrix}, \quad T = \tfrac{1}{2} \begin{bmatrix} 2 & 0 & 0 & 0 & 0 \\ 0 & \tau & 1 & \tau^{-1} & 0 \\ 0 & 1 & -1 & -1 & 1 \\ 0 & \tau^{-1} & -1 & 0 & -\tau \\ 0 & 0 & 1 & \tau & \tau^{-1} \end{bmatrix}.$$

We then apply S and T alternately, beginning with the initial vertex and using the identifications in $\mathcal{P}$ so as not to stray too far from it. We label the vertices $p_0 = (1, o), p_1 = p_0 S = \bigl(\tau^3, \tau^{3/2}(2, 0, 0, 0)\bigr), p_{-1} = p_1 T = \bigl(\tau^3, \tau^{3/2}(\tau, 1, \tau^{-1}, 0)\bigr),$

$p_2 = p_{-1}S = (\tau^3, \tau^{3/2}(\tau, 1, -\tau^{-1}, 0))$, and so on. The identifications employed are

$$p_7 = (3\tau^3, \tau^{5/2}(2\tau, \tau^{-1}, -1, -\tau)) \sim (\tau^3, \tau^{3/2}(0, 1, \tau, -\tau^{-1})),$$
$$p_9 = (\tau^6, \tau^{5/2}(3\tau, 2, -1, \tau^{-1})) \sim (\tau^3\sqrt{5}, \tau^{5/2}(\tau, 2, 1, -\tau^{-1})) = p_{-8}.$$

For the last identification, we have $(3\tau, 2, -1, \tau^{-1}) = \tau^2 u + \tau^{-1} w$, with

$$u = (\tau, 1, -\tau^{-1}, 0), \quad w = (1, -1, 1, 1).$$

The appropriate v for (17B19) is $v = (0, 1, \tau, -\tau^{-1})$, giving

$$u + v = (\tau, 1, -\tau^{-1}, 0) + (0, 1, \tau, -\tau^{-1}) = (\tau, 2, 1, -\tau^{-1}),$$

as required. In other words,

17B21 Proposition *The Petrie period of* $\{3, 3, 3, 5\}/\langle\langle (0(1234)^3)^5 \rangle\rangle$ *is* 17.

Notes to Section 17B

1. The honeycomb $\{3, 3, 3, 5\}$ is at the centre of [91]; there, quaternions play a key rôle in the investigation.
2. Further to the proof of Proposition 17B3, recall that a 600-cell with vertex-set $\frac{1}{2}G$ and dual 120-cell Q inscribed in the unit sphere in $\mathbb{E}^4$ have the same inradius $\tau^2/\sqrt{8}$. Thus, if $z \in \mathbb{E}^4$ satisfies $\langle z, u\rangle < \tau^2/\sqrt{8}$ for each $u \in \frac{1}{2}G$, then $z \in \operatorname{int}\mathsf{Q}$, so that $\|z\| < 1$. The claim in the proof then follows.
3. We are indebted to Marston Conder [12] for verifying the relations of (17B7) and (17B8); this information was conveyed in the second e-mail message. The proof that (17B8) implies (17B7) is our own.
4. Since we have $(\tau^3, 2\tau^{3/2}e_1) \sim (\tau^3, -2\tau^{3/2}e_1)$, we know that the pairs of diagonals from these two vertices to those in $\mathcal{L}_2$ or $\mathcal{L}_3$ are also equivalent. In turn, this leads to further identification of layers, if we need it.
5. As we noted, the relation (17B8) collapses the planar section $\{3, 10\}$ to $\{3, 10 : 5\}$. This has the six vertices of $\{3, 5 : 5\}$, but its edges and faces are doubly-covered. Hence, by themselves, the identifications in the section are insufficient to force the collapse of the vertex-figure to $\{3, 3, 5 : 15\}$.
6. As we saw in Section 7J, the symmetry group of an n-staurotope – as a subgroup of $[3, 3, 5]$, and hence also of $[3, 3, 3, 5]$ – has order 24.
7. An n-icosahedron in $\mathcal{L}_3$ has three pairs of opposite edges in $\mathcal{D}_{20a}$; the other 24 are in $\mathcal{D}_{20b}$.
8. On the other hand, general points of $\mathcal{L}_{12b}$ are of a similar form, with c replaced by $d \in G$ such that $\langle a, d\rangle = \langle b, d\rangle = 0$. They are obtained from those of $\mathcal{L}_{12a}$ by an odd permutation of the coordinates of z, and thus fall into two inequivalent half-layers. Similarly, the hyperplane reflexion that interchanges the points in $\mathcal{L}_{12b}$ with $z = 2(\tau^4, \tau^3, 0, \tau^{-1})$ and $2(3\tau^2, \tau, 0, -1)$ is not a symmetry of $\{3, 3, 3, 5\}$.
9. We had half-expected that the Petrie polygon of the (abstract) quotient $\mathcal{P}$ would be $\{17\}$; that the confirmation was relatively easy was a pleasant surprise.
10. The (suppressed) calculations actually tell us that the induced cosine vector of the Petrie polygon $\{17\}$ is $(1, \alpha, \alpha, \alpha, \alpha, \beta, \beta, \alpha, \beta)$ in terms of the cosine vector $(1, \alpha, \beta)$ of $\mathcal{P}$. We could employ this to construct planar projections of the realizations $\mathsf{P}_1, \mathsf{P}_2$ with 17-fold symmetry, but with the 136 vertices falling into 8 rings of 17 each they would not be very informative.

17C A Quotient of $\{3,5,3,5\}$

The process of vertex-figure replacement employed in Section 7G was introduced for finite regular polytopes in [71], where it was remarked that it applies equally to regular hyperbolic honeycombs; in other words, its systematic use would lead to a complete classification of the regular star-honeycombs. (See also [99, Section 7D], but the wider application is not mentioned there.) In the present case, it yields the family of Table 17C1. The convention here and elsewhere follows that of Section 7G, so that dual polytopes are on the same row, while those in a column – obtained by vertex-figure replacement – have the same vertices and edges; in all these cases, they also have the same 2-faces. In fact, $\{5,\frac{5}{2},5,3\}$ and $\{5,5,\frac{5}{2},3\}$ could be placed in column 2 as well, since the vertex-figures have the same vertices.

All the honeycombs here are of finite type (that is, with finite vertex-figures and facets), except for the last two. The facets $\{3,5,3\}$ of the first are (universal) hyperbolic honeycombs. Although honeycombs whose facets are themselves hyperbolic are usually excluded in the enumerations, we see no good reason why they should be. For the second, even its ridges are hyperbolic honeycombs $\{5,5\}$, but again there is no reason to exclude it. Note that any further attempts at vertex-figure replacement in $\mathbb{H}^4$ lead to 2-faces which are apeirogons; these we have good reason to ignore.

17C1

$\{5,3,3,3\}$	$\{3,3,3,5\}$		
	$\{3,3,5,\frac{5}{2}\}$	$\{\frac{5}{2},5,3,3\}$	
	$\{3,5,\frac{5}{2},5\}$		$\{5,\frac{5}{2},5,3\}$
	$\{3,5,3,\frac{5}{2}\}$		
			$\{5,5,\frac{5}{2},3\}$

However, what we can now do is pass to the quotients with automorphism group $S_4(4)$; note that $\{3,5,3,\frac{5}{2}\} \cong \{3,5,3,5\}/\langle\langle(123432)^3\rangle\rangle$. In fact, what we show has wider implications, as we shall see in Section 17E.

17C2 Theorem *The quotient* $\{3,5,3\} \searrow \{3,5,3 :: 4\}$ *on the facet* $\{3,5,3\}$ *of* $\{3,5,3,5\}/\langle\langle(123432)^3\rangle\rangle$ *is equivalent to identifying* $\mathcal{D}_{20a} \sim \mathcal{D}_0$ *in* $\{3,3,3,5\}$.

Proof. We choose a new generatrix $(R_0,\ldots,R_4)$ of $\{3,5,3,\frac{5}{2}\}$, rather than using the change of generatrix

$$(r_0,r_1,r_2,r_3,r_4) \mapsto (r_0,r_1,r_2r_3r_4r_3r_2,r_4r_3r_4,r_3),$$

which corresponds to the one that yields $\{5,3,5 \mid 3\}$ from $\{3,3,5\}$ (see [81] or [99, Section 7D]). The mirrors have normals $u_0, \ldots, u_4$, given by

17C3

$$\begin{aligned} u_0 &:= \big(\tau^{1/2}, \tfrac{1}{2}\tau(1,1,1,1)\big),\\ u_1 &:= \big(0, \tfrac{1}{2}(0,\tau,\tau^{-1},-1)\big),\\ u_2 &:= \big(0, \tfrac{1}{2}(0,\tau,-\tau^{-1},-1)\big),\\ u_3 &:= \big(0, \tfrac{1}{2}(0,1,-\tau.\tau^{-1})\big),\\ u_4 &:= \big(0, \tfrac{1}{2}(-\tau,\tau^{-1},0,1)\big). \end{aligned}$$

Now $\mathbf{G}^{v} = \langle R_1, \ldots, R_4\rangle$ preserves the hyperplane $L\colon \eta = 2\tau^{-1/2}\zeta_1$ which contains the initial facet $\{3,5,3\}$; indeed, $\operatorname{vert}\{3,5,3\} = L \cap \operatorname{vert}\{3,5,3,\frac{5}{2}\}$, since L captures the initial vertex of $\{3,5,3\}$ and its neighbours. If $S := R_0R_3$ and $T := R_1R_2R_1R_2R_1$ (our choice is such that the mirror of T has normal $(0,e_3)$), then starting with $a_0 = (1,o)$ and successively applying S and T, we obtain

$$a_0 = (1,o),\ a_{\pm1} = \big(\tau^3, \tau^{3/2}(1,1,\pm1,1)\big),\ a_{\pm2} = \big(\tau^3\sqrt{5}, \tau^{5/2}(\tau,\tau^{-1},\pm2,1)\big).$$

Since, by definition, $\{3,5,3 :: 4\} = \{3,5,3\}/\langle\langle(0121213)^4\rangle\rangle$ and $\{a_2, a_{-2}\} \in \mathcal{D}_{20a}$, the quotient $(ST)^4 = I$ is equivalent to the identification $\mathcal{D}_{20a} \sim \mathcal{D}_0$, as claimed. □

17C4 Corollary *The vertex-figure of the quotient of* $\{\{3,5,3\},\{5,3,5 \mid 3\}\}$ *with automorphism group* $S_4(4)$ *is the quotient* $\{5,3,5 \mid 3:6\}$.

For the time being, we shall say only a little about this quotient, instead leaving a more extended discussion to Section 17E (see the notes at the end of the section). Nonetheless, we do need a geometric picture. The detailed description in Section 17B will tell us how the quotient operates; in view of that description, our arguments can be largely synthetic. We first need to identify the facet $\mathcal{I}$ of $\mathcal{Q}$.

In $\mathbb{H}^4$, our choice of $\{3,5,3\}$ with initial vertex $(1,o)$ has vertex-figure in $\mathcal{L}_1$ consisting of the points $\big(\tau^3, \tau^{3/2}(1,D)\big)$, corresponding to a dodecahedral section $(1,D)$ of $\{3,3,5\}$. With L as in the proof of Theorem 17C2, this enables us easily to pin down the remaining vertices; of course, we keep the identifications in mind. Thus, in $\mathcal{L}_2$ we find the 60 vertices $(\tau^3\sqrt{5}, \tau^{5/2}z)$, with $z = (\tau, z')$ and

$$z' \in \big\{(\tau,\tau,\tau^{-2})^s, (\sqrt{5},\tau^{-1},0)^s, (2,1,\tau^{-1})^s\big\},$$

with our usual convention. In this layer, the identifications are in threes; the discussion in Section 17B leads to $(\tau^3\sqrt{5}, \tau^{1/2}, \tau^{7/2}D)$.

17C5 Remark We have already seen in (17B18) how the first 24 vertices are identified. A typical case for the others is

$$\{(\tau,\pm2,1,\tau^{-1}),(\tau,0,-\sqrt{5},-\tau^{-1})\} \mapsto (\tau^{-2},0,1,-\tau^2).$$

In $\mathcal{L}_3$, we find only the twelve vertices $(3\tau^3, \tau^{5/2}\sqrt{5}, \tau^{7/2}I)$, with I as before the vertex-set of the icosahedron; there are no identifications here.

In $\mathcal{L}_4$, we see that, if $(\tau^6, \tau^{9/2}u + \tau^{3/2}w) \in L$, with $u = (v_1, \dots, v_4)$ and $w = (\omega_1, \dots, \omega_4)$, then

$$\tau^3 v_1 + \omega_1 = 2\tau^2.$$

Now this can only happen if $v_1 = \omega_1 = 1$; recall that $u, w \in G$ with $\langle u, w\rangle = 0$. If we take a typical vertex, say with $u = (1,1,1,1)$, we see that $w = (1, \omega_2, \omega_3, \omega_4)$, with $(\omega_2, \omega_3, \omega_4)$ a cyclic permutation of $(1,-1,-1)$ or $(\tau^{-1}, -\tau, 0)$. Thus we count 120 vertices of $\{3,5,3\}$ in $\mathcal{L}_4$. Recalling how the identifications went earlier, and picking v of the form $v = (0, \dots)$, we see that each of the 120 vertices is equivalent to a unique one in $\mathcal{L}_2 \cap (\tau^3\sqrt{5}, 0, \tau^{5/2}T)$. However, these are further identified in opposite pairs, as we saw in Theorem 17B15, and thus the identifications in $\operatorname{vert}\{3,5,3\} \cap \mathcal{L}_4$ are in eights.

For the moment, we need go no further. What we have seen is that the layer vector of $\{3,5,3\}$ itself begins $(1, 20, 60, 12, 120, \dots)$, while that of the quotient $\mathcal{I}$ is $\Lambda^f = (1, 20, 20, 15, 12)$, if we switch $\mathcal{L}_3$ and $\mathcal{L}_4$. There is now no difficulty in recognizing that $\mathcal{I} = \{3,5,3 :: 4\}$ with automorphism group $\mathrm{L}_2(16).2$ of order 8160, as we saw in the previous chapter (see also [101]).

As a facet of the quotient $\mathcal{Q}$ of $\{3,5,3,5\}$, the induced cosine vector of its facet $\mathcal{I} := \{3,5,3 :: 4\}$ is therefore $\Gamma^f = (1, \alpha, \beta, \beta, \alpha)$. For the two non-trivial induced pure realizations $\mathsf{Q}_j \in \mathcal{Q}$, we therefore have

17C6
$$\eta_f(\mathsf{Q}_j) = \begin{cases} \frac{1}{68}\bigl(1 + 32(-\frac{2}{15}) + 35(\frac{7}{75})\bigr) = 0, & \text{for } j = 1, \\ \frac{1}{68}\bigl(1 + 32(\frac{1}{15}) + 35(-\frac{1}{15})\bigr) = \frac{1}{85}, & \text{for } j = 2. \end{cases}$$

where $\eta_f(\mathsf{Q})$ is the squared distance from o to the centre of a facet of Q; see Theorem 3C14.

17C7 Remark We know that $\eta_f(\mathsf{Q}_j) = 0$ for (at least) one j, since the dual of type $\{5,3,5,3\}$ has fewer vertices; see Section 17D which follows.

Since the facet $\mathcal{I} = \{3,5,3 :: 4\}$ of $\mathcal{Q}$ uses half its vertices, it follows that the other stratum of $\mathcal{Q}$ must also be a copy of $\mathcal{I}$. However, it cannot itself be a facet, and so it must have a different induced cosine vector Γ', say. But Theorem 3C14 must give the same squared distance η_f of the stratum from the origin o, and so the only possibility is to interchange γ_1 and γ_2 in the general cosine vector $(1, \gamma_1, \dots, \gamma_4)$. As a consequence,

17C8
$$\Gamma' = (1, \beta, \alpha, \beta, \alpha).$$

Petrie Polygon

We can find the Petrie polygon of $\mathcal{Q}$ in much the same way as we did for $\mathcal{P}$ in the previous section. Rather than using (17C3), we adopt a slight alteration of (17B1) which reverts to having the original initial vertex, namely,

17C9

$$\begin{aligned} v_0 &:= (\tau^{1/2}, \tau, 0, 0, 0), \\ v_1 &:= \tfrac{1}{2}(0, -\tau^{-1}, \tau, 1, 0), \\ v_2 &:= (0, 0, -1, 0, 0), \\ v_3 &:= \tfrac{1}{2}(0, 0, 1, -\tau, \tau^{-1}), \\ v_4 &:= (0, 0, 0, 0, -1). \end{aligned}$$

For the new generatrix (S, T) of the Petrie polygon, we have

$$S = \begin{bmatrix} \tau^3 & 2\tau^{3/2} & 0 & 0 & 0 \\ -2\tau^{3/2} & -\tau^3 & 0 & 0 & 0 \\ 0 & 0 & -1 & 0 & 0 \\ 0 & 0 & 0 & 1 & 0 \\ 0 & 0 & 0 & 0 & -1 \end{bmatrix}, \quad T = \frac{1}{2}\begin{bmatrix} 2 & 0 & 0 & 0 & 0 \\ 0 & \tau & 1 & \tau^{-1} & 0 \\ 0 & 1 & -\tau & 0 & -\tau^{-1} \\ 0 & \tau^{-1} & 0 & -\tau & 1 \\ 0 & 0 & -\tau^{-1} & 1 & \tau \end{bmatrix}.$$

Beginning with $q_0 = (1, o)$ and applying S and T alternately, we can calculate that $q_{-6} = q_6$, so that the Petrie period is 12 (we suppress the details – see the notes at the end of the section). The induced cosine vector of $\{12\}$ is $(1, \alpha, \alpha, \beta, \beta, \beta, \alpha)$.

Notes to Section 17C

1. When the symmmetry group is generated by hyperplane reflexions, vertex-figure replacement automatically leads from one polytope to another. However, while we have no right to expect this to carry over to more general situations, in the present context it is reasonable to assume that vertex-figure replacement in the original hyperbolic honeycombs will be respected by the quotient mapping. In any event, here vertex-figure replacement corresponds to invertible changes of generatrices; all that then remains to be verified is whether the intersection property is retained (we shall see in Section 17F an instance where it is not).
2. In layer $\mathcal{L}_4$, we find

$$q_4 = (\tau^6, \tau^{5/2}(3\tau, -\sqrt{5}, \tau^{-1}, 0)) \sim (\tau^3\sqrt{5}, \tau^{5/2}(\tau, -1, \tau^{-1}, 2)).$$

3. If we apply S directly to $q_{-4} = (\tau^3\sqrt{5}, \tau^{5/2}(1, 1, 1, \sqrt{5}))$, we end up in layer $\mathcal{L}_6$. Instead, a better way to proceed is to use

$$q_{-4} \sim (\tau^3\sqrt{5}, \tau^{5/2}(\sqrt{5}, -1, 1, -1)).$$

17D The Dual Polytope

We next consider the polytope $\mathcal{R}$ of type $\{5, 3, 5, 3\}$ dual to $\mathcal{Q}$; again, this will tell us things about the double cover. For ease of reference, we shall write

$\boldsymbol{G} := \mathrm{S}_4(4)$ (or later $\mathrm{S}_4(4) \times 2$), and $\boldsymbol{A} := [3,3,5]/2$ and $\boldsymbol{B} := \mathrm{L}_2(16).2$ for the two maximal subgroups of $\boldsymbol{G}$ under consideration. Polytopes with vertex-set $\boldsymbol{G}/\boldsymbol{A}$ will be said to belong to the $\boldsymbol{A}$-family; the $\boldsymbol{B}$-family is similarly defined.

Since the automorphism group of $\mathcal{I} = \{3,5,3 :: 4\}$ has order $68 \cdot 120 = 8160$, it follows that $\mathcal{R}$ has $979200/8160 = 120$ vertices. From (17C6) we know that the facets of the 85-dimensional realization Q_2 are not centred; this tells us that we can construct a geometric dual R_2 of Q_2, with vertices the normalized facet-centres. Moreover, R_2 must also be pure, since the corresponding symmetry group is irreducible.

As yet, we have very little idea about the structure of R_2; indeed, we do not know much about $\mathcal{R}$ itself, except that its layer vector is of the form $\Lambda = (1, 68, \ldots)$, where the remaining entries sum to $120 - 1 - 68 = 51$. (In view of Chapter 16, this strongly suggests that these 51 vertices form a single layer.) Nevertheless, since we have a pure realization of dimension 85, we do know that we only have a complementary dimension $120 - 1 - 85 = 34$ to account for. Observe that the centred facet criterion Corollary 4D10 forces the facets of the 50-dimensional realization Q_1 to be centred, because it cannot give rise to a geometric dual. If we write R_1 for the complementary component in $\mathcal{R}$ to $\{1\}$ and R_2, which may or may not be pure, then again Corollary 4D10 tells us that its facets too must be centred, otherwise we would obtain another pure realization of $\mathcal{Q}$ of dimension at most 34.

Suppose, if possible, that R_1 is blended. Each component of R_1 has a centred facet of type $\mathcal{F} = \mathcal{R}^{\boldsymbol{f}} = \{5,3,5 \mid 3:6\} \cong \{\frac{5}{2},3,5\}/2$ (recall the notation from (2D6) and (2D7) – further, see the notes at the end of the section). From Section 7E, we know that the non-trivial pure realizations of $\mathcal{F}$ have dimensions $9, 9, 16, 25$; if one of the 9-dimensional ones occurs, then so must the other, otherwise changing the sign of $\sqrt{5}$ in the cosine vector of the corresponding realization of $\mathcal{R}$ yields another of the same dimension. This narrows down the putative dimensions to 16 and $9+9=18$ (and excludes 25); since $16+18=34$, unless R_2 is pure its components must have dimensions 16 and 9 (twice).

Now $\mathcal{F}$ has layer vector $\Lambda^{\boldsymbol{f}} = (1,12,12,15,20)$ (as in Section 16F), and the cosine vector of its 16-dimensional realization is $(1, -\frac{1}{4}, -\frac{1}{4}, 0, \frac{1}{4})$, as we can see from Theorem 7E10. Feeding this (with unknown identifications of layers of $\mathcal{F}$ other than the first) into the layer equation for $\mathcal{R}$ itself would lead to an equation of the form

$$0 = 1 + 68(-\tfrac{1}{4}) + \lambda(\tfrac{1}{4}) + \mu(-\tfrac{1}{4}) = -16 + \tfrac{1}{4}\lambda - \tfrac{1}{4}\mu,$$

with non-negative integers λ and μ such that $\lambda + \mu \leqslant 51$. Since such an equation clearly admits no solutions, we deduce that $\mathcal{R}$ has no 16-dimensional realization; that is, R_1 must be pure of dimension 34. In other words, the dimension vector of $\mathcal{R}$ is $D = (1, 34, 85)$.

Since $\mathcal{R}$ has just two non-trivial pure realizations of different dimensions, it follows that it similarly has only two layers apart from $\mathcal{L}_0$. Since we already know that $\ell_1 = 68$, it follows that $\ell_2 = 51$.

We do not have a geometric model for $\{5,3,5,3\}$ to appeal to, and so we must proceed in a somewhat different way from that of Section 17A. Taking the cosine matrix of $\mathcal{R}$ to be (γ_{ij}) (with $\gamma_{i0} = \gamma_{0j} = 1$ for each $i, j = 0, 1, 2$), we apply to it the component and dimension equations for $\ell_2 = 51$ from Theorems 3C11 and 3F5. Thus we have

$$\begin{aligned} 0 &= 1 + 34\gamma_{12} + 85\gamma_{22}, \\ \tfrac{120}{51} &= 1 + 34\gamma_{12}^2 + 85\gamma_{22}^2. \end{aligned}$$

Writing $\xi = 17\gamma_{12}$ and $\eta = 17\gamma_{22}$ leads to

$$\begin{aligned} 0 &= 1 + 2\xi + 5\eta, \\ 40 &= 17 + 2\xi^2 + 5\eta^2. \end{aligned}$$

Eliminating η yields (after simplification)

$$7\xi^2 + 2\xi - 57 = 0,$$

with solutions $\xi = -3, \frac{19}{7}$.

Naturally, we guess that $\xi = -3$ is the correct solution, and so we have to exclude $\xi = \frac{19}{7}$, that is, $\gamma_{12} = \frac{19}{119}$, from which we find $\gamma_{11} = -\frac{16}{119}$ from the layer equation Theorem 3C7. Now $\mathcal{F}$ has layer vector Λ^f as before, and since its realization F_2 in the 34-dimensional R_2 is centred, if $\xi = \frac{19}{7}$, then we must have an equation holding of the form

$$1 + \tfrac{19}{119}\lambda - \tfrac{16}{119}\mu = 0 \implies 119 + 19\lambda - 16\mu = 0,$$

with λ, μ complementary sums from $12, 12, 15, 20$. By parity, λ has to have the contribution from 15; only a little work is then needed to show that, in fact, the equation has no solution.

From $\xi = -3$, simple calculations prove

17D1 Theorem *The cosine matrix of the regular polytope $\mathcal{R}$ of type $\{5, 3, 5, 3\}$ with automorphism group $\mathrm{S}_4(4)$ is*

$$\begin{bmatrix} 1 & 1 & 1 \\ 1 & \frac{2}{17} & -\frac{3}{17} \\ 1 & -\frac{1}{17} & \frac{1}{17} \end{bmatrix},$$

with layer and dimension vectors

$$\Lambda = (1, 68, 51), \quad D = (1, 34, 85).$$

We can confirm the dimension calculations using Theorem 3F5, which also shows that Γ_1 and Γ_2 are Λ-orthogonal.

We can use the argument that eliminated $\xi = \frac{19}{7}$ to find how identifications of layers of $\mathcal{F}$ contribute to the centred facet F_1 in R_1. Now we must solve an equation

$$1 + \tfrac{2}{17}\lambda - \tfrac{3}{17}\mu = 0,$$

with λ, μ as just before. Here, we quickly find that $\lambda = 12 + 20$, $\mu = 12 + 15$ is the only solution; the first 12 comes from the vertex-figure of $\mathcal{F}$. Moreover, the cosine vectors of the pure realizations of $\mathcal{F}$ of dimensions 9, 9, 16 are

$$(1, \tfrac{1}{3}\tau, -\tfrac{1}{3}\tau^{-1}, -\tfrac{1}{3}, 0), \quad (1, -\tfrac{1}{3}\tau^{-1}, \tfrac{1}{3}\tau, -\tfrac{1}{3}, 0), \quad (1, -\tfrac{1}{4}, -\tfrac{1}{4}, 0, \tfrac{1}{4});$$

with multiples $\frac{3}{17}\tau^2, \frac{3}{17}\tau^{-2}, \frac{8}{17}$, respectively, which sum to 1, we obtain the requisite cosine vector $(1, \frac{2}{17}, -\frac{3}{17}, -\frac{3}{17}, \frac{2}{17})$ of F_1. What this has shown is

17D2 Proposition *The induced cosine vector of the facet of the polytope $\mathcal{R}$ of type $\{5, 3, 5, 3\}$ is*

$$\Gamma^f = (1, \alpha, \beta, \beta, \alpha).$$

Now that we know how $\mathcal{F}$ fits into $\mathcal{R}$, we can calculate

$$\eta_f(\mathsf{R}_2) = \tfrac{1}{60}\big(1 + (12 + 20)(-\tfrac{1}{17}) + (12 + 15)\tfrac{1}{17}\big) = \tfrac{1}{85} = \eta_f(\mathsf{Q}_2).$$

This is as we should expect from Theorem 4D6, since $w(\mathsf{P}_2) = w(\mathsf{Q}_2) = 1$.

17D3 Remark Exactly as for the dual, there is one stratum other than the facet $\mathcal{F}$. This is a different copy of $\mathcal{F}$, and has induced cosine vector

$$\Gamma' = (1, \beta, \alpha, \beta, \alpha).$$

In a similar way, we can find the induced cosine vector Γ^v of the vertex-figure. The vertex-figure criterion Theorem 4D4 yields the equation

$$68\alpha^2 = 1 + \lambda\alpha + \mu\beta, \quad \text{for } (\alpha, \beta) = (\tfrac{2}{17}, -\tfrac{3}{17}) \text{ or } (-\tfrac{1}{17}, \tfrac{1}{17}),$$

with solution $\lambda = 40 = 20 + 20$, $\mu = 27 = 15 + 12$. In other words,

17D4 Proposition *The induced cosine vector of the vertex-figure $\{3, 5, 3 :: 4\}$ of the polytope $\mathcal{R}$ of type $\{5, 3, 5, 3\}$ is*

$$\Gamma^f = (1, \alpha, \alpha, \beta, \beta).$$

Notes to Section 17D

1. It is again worth noting how much information we have obtained from so little input. Observe also that we did not appeal to the layer equation for $\mathcal{R}^v$ in proving Proposition 17D4; this merely confirms our calculations.
2. The regular polytope $\{5, 3, 5 : 6\}$ is finite; from [11, 26], it has 30720 vertices and facets, and automorphism group of order 36 86400.

17E Double Covers

By Proposition 2D16, the initial quotient of $\{3,3,3,5\}$ with 136 vertices is non-orientable; it therefore has an orientable double cover. It is easy to see what its group must be: if $(r_0,\ldots,r_4)$ is the generatrix of $\{3,3,3,5\}/\langle\langle\,(0(1234)^3)^5\,\rangle\rangle$ and z is an involution that commutes with each r_j, then $(r_0z,\ldots,r_4z)$ is the generatrix of the double cover. In other words, the automorphism group of the double cover is $S_4(4)\times 2$ in the notation of $\mathbb{ATLAS}$ (see the notes at the end of the section).

Notation

At this stage, it is helpful to introduce concise notation for the polytopes that we consider; either way of describing them – by concatenating their facets and vertex-figures or by extra relations on the Schläfli type – will be rather clumsy. Most of these polytopes are universal of their kind, and so will be specified by their 3- and 4-infaces. Those of type $\{3,5,3\}$ must be isomorphic to $\{3,5,3::4\}$, and so do not need specification. Those of type $\{5,5\}$ are either $\{5,5\mid 3\}$ or $\{5,5:4\}$, while those of type $\{5,3,5\}$ are $\{5,3,5\mid 3\}$ or $\{5,3,5::4\}$.

We therefore mark the types of the infaces by suffixes 3 or 4, in the order that they occur. The usage is best illustrated by examples:

17E1
$$\begin{aligned}\{3,5,3,5\}_3 &= \{\{3,5,3::4\},\{5,3,5\mid 3\}\},\\ \{3,5,3,5\}_4 &= \{\{3,5,3::4\},\{5,3,5::4\}\},\\ \{3,5,5,5\}_{3,4} &= \{3,5,5,5\}/\langle\langle\,(1232)^3,(234)^4\,\rangle\rangle,\\ \{5,3,5,5\}_{4,3} &= \{5,3,5,5\}/\langle\langle\,(01213)^4,(2343)^3\,\rangle\rangle;\end{aligned}$$

these should adequately show how the notation is employed.

Universality

Our first step characterizes the polytope that we shall take as the starting point in working out the realizations. We shall prove

17E2 Theorem *The double cover of the polytope of Schläfli type* $\{3,5,3,5\}$ *with group* $S_4(4)$ *is the universal polytope* $\{3,5,3,5\}_3$ *in the notation of (17E1).*

Proof. We first observe that the double cover must be a quotient of $\{3,5,3,5\}_3$, since its group certainly satisfies the defining relations. Since Theorem 17C2 implies the identification $\mathcal{D}_{20a}\sim\mathcal{D}_0$, and then Remark 17B16 implies that $\mathcal{D}_{8a}\sim\mathcal{D}_0$ also, as a consequence the vertex-figure of $\{3,5,3,5\}_3$ undergoes the collapse $\{5,3,5\mid 3\}\searrow\{5,3,5\mid 3\}/2=\{5,3,5\mid 3:6\}$. Thus there is no harm in working with the corresponding quotient of $\{3,3,3,5\}$ instead.

Consider the element $z:=(r_0s)^5$, with $s:=(r_1r_2r_3r_4)^3$. As we saw in Section 17B, the corresponding symmetry $Z\in[3,3,3,5]$ takes the initial vertex v into $\mathcal{L}_{12a}$. By Proposition 17B20, the 1440 vertices of the truncated 600-cell in

$\mathcal{L}_{12a}$ are identified to a single point. Two things follow at once. First, $\boldsymbol{z}^{-1} = \boldsymbol{z}$ since Z^{-1} also takes v into $\mathcal{L}_{12a}$, so that $\boldsymbol{z}$ is involutory. Second, $\boldsymbol{z} \rightleftharpoons \boldsymbol{g}$ for every $\boldsymbol{g} \in \boldsymbol{G}^v = \langle \boldsymbol{r}_1, \ldots, \boldsymbol{r}_4 \rangle$ (recall that $\rightleftharpoons$ means 'commutes with'). In particular,

$$\boldsymbol{z} = \boldsymbol{s}\boldsymbol{z}\boldsymbol{s}^{-1} = \boldsymbol{s}(\boldsymbol{r}_0\boldsymbol{s})^3\boldsymbol{s}^{-1} = (\boldsymbol{s}\boldsymbol{r}_0)^3,$$

and then finally

$$\boldsymbol{z}\boldsymbol{r}_0 = (\boldsymbol{r}_0\boldsymbol{s})^3\boldsymbol{r}_0 = \boldsymbol{r}_0(\boldsymbol{s}\boldsymbol{r}_0)^3 = \boldsymbol{r}_0\boldsymbol{z},$$

so that $\boldsymbol{z} \rightleftharpoons \boldsymbol{r}_0$ as well. In other words, $\boldsymbol{z}$ is a central involution, and then we see at once that the mixing operation

$$(\boldsymbol{r}_0, \ldots, \boldsymbol{r}_4, \boldsymbol{z}) \mapsto (\boldsymbol{r}_0\boldsymbol{z}, \ldots, \boldsymbol{r}_4\boldsymbol{z})$$

yields the generatrix of $\{3,3,3,5\}/\langle\langle\,(0(1234)^3)^5\,\rangle\rangle$. The claim of the theorem is thus established. □

Realizations

We next look at the realization spaces; here, we find it convenient to work with the universal $\mathcal{P} := \{3,5,3,5\}_3$ and its dual, rather than with the double cover of $\{3,3,3,5\}/\langle\langle\,(0(1234)^3)^5\,\rangle\rangle$. Observe that $\mathcal{P}$ is centrally symmetric, with layer vector

$$\Lambda = (1, 60, 75, 75, 60, 1);$$

all its diagonals are symmetric, and so $w(\mathsf{P}) = 1$ for each pure $\mathsf{P} \in \mathcal{P}$. Since we have already treated $\mathcal{P}/2$, we need only find the components of the staurotope realization X; the diagonal count tells us that we have three pure ones. Since $\mathcal{P}$ has more vertices than its dual $\mathcal{Q}$ (say), it follows from Corollary 4D10 that at least one (pure) $\mathsf{P} \in \mathcal{P}$ has centred facets. The layer vectors and induced cosine vectors of the facet $\mathcal{F}$ and vertex-figure $\mathcal{V}$ are

$$\begin{aligned} \Lambda^f &= (1, 20, 20, 15, 12), \quad & \Gamma^f &= (1, \alpha, \beta, -\beta, -\alpha), \\ \Lambda^v &= (1, 12, 12, 15, 20), \quad & \Gamma^v &= (1, \alpha, -\alpha, -\beta, \beta), \end{aligned}$$

in terms of the general cosine vector $\Gamma = (1, \alpha, \beta, -\beta, -\alpha, -1)$ of $\mathcal{P}$, as we can see directly from the geometry of $\{3, 5, 3, \frac{5}{2}\}$. The fact that the facet is centred yields

$$0 = 1 + 20\alpha + 20\beta + 12(-\alpha) + 15(-\beta) = 1 + 8\alpha + 5\beta.$$

The vertex-figure criterion Theorem 4D4 (for $w = 1$) then gives

$$68\alpha^2 = 1 + 12\alpha + 20\beta + 15(-\beta) + 12(-\alpha) = 1 + 5\beta = -8\alpha,$$

from which we obtain

$$(\alpha, \beta) = (0, -\tfrac{1}{5}) \text{ or } (-\tfrac{2}{15}, \tfrac{1}{75}).$$

For the latter, we get dimension d given by

$$\tfrac{1}{d} = \tfrac{1}{136}\big(1 + 60(-\tfrac{2}{15})^2 + 75(\tfrac{1}{75})^2\big) = \tfrac{156}{136\cdot 75} = \tfrac{13}{850};$$

this fractional value must be discarded. For the former, we have

$$\tfrac{1}{d} = \tfrac{1}{136}\bigl(1 + 60 \cdot 0^2 + 75(-\tfrac{1}{5})^2\bigr) = \tfrac{4}{136} = \tfrac{1}{34};$$

that is, $d = 34$. Thus we have the cosine vector

$$\Gamma_3 = (1, 0, -\tfrac{1}{5}, \tfrac{1}{5}, 0, -1).$$

For the remaining two pure components, Λ-orthogonality with respect to Γ_3 implies that

$$0 = 1 + 75(-\tfrac{1}{5})\beta = 1 - 15\beta \implies \beta = \tfrac{1}{15}.$$

The vertex-figure criterion Theorem 4D4 now yields

$$60\alpha^2 = 1 + 5\beta = \tfrac{4}{3} \implies \alpha = \pm\tfrac{\sqrt{5}}{15}.$$

It is easy to check that the two cosine vectors

$$\Gamma_4 = \bigl(1, \tfrac{\sqrt{5}}{15}, \tfrac{1}{15}, -\tfrac{1}{15}, -\tfrac{\sqrt{5}}{15}, -1\bigr),$$
$$\Gamma_5 = \bigl(1, -\tfrac{\sqrt{5}}{15}, \tfrac{1}{15}, -\tfrac{1}{15}, \tfrac{\sqrt{5}}{15}, -1\bigr),$$

are Λ-orthogonal; moreover, we see that the common dimension is $d = 51$, as it must be. Finally, we find that

$$\eta_f = \tfrac{1}{68}(1 + 8\alpha + 5\beta) = \tfrac{1}{68}\bigl(1 \pm \tfrac{8\sqrt{5}}{15} + \tfrac{5}{15}\bigr) = \tfrac{1}{51\sqrt{5}}\tau^{\pm 3}.$$

Summarizing what we have done so far, we have

17E3 Theorem *The cosine matrix of* $\{3,5,3,5\}_3 = \{\{3,5,3 :: 4\}, \{5,3,5 \mid 3\}\}$ *is*

$$\begin{bmatrix} 1 & 1 & 1 & 1 & 1 & 1 \\ 1 & -\frac{2}{15} & \frac{7}{75} & \frac{7}{75} & -\frac{2}{15} & 1 \\ 1 & \frac{1}{15} & -\frac{1}{15} & -\frac{1}{15} & \frac{1}{15} & 1 \\ 1 & 0 & -\frac{1}{5} & \frac{1}{5} & 0 & -1 \\ 1 & \frac{\sqrt{5}}{15} & \frac{1}{15} & -\frac{1}{15} & -\frac{\sqrt{5}}{15} & -1 \\ 1 & -\frac{\sqrt{5}}{15} & \frac{1}{15} & -\frac{1}{15} & \frac{\sqrt{5}}{15} & -1 \end{bmatrix},$$

with layer and dimension vectors

$$\Lambda = (1, 60, 75, 75, 60, 1), \qquad D = (1, 50, 85, 34, 51, 51).$$

We end the present discussion by looking at the other strata. Their induced cosine vector $(1, \gamma_1, \ldots, \gamma_4)$ is obtained by lifting $\Gamma' = (1, \beta, \alpha, \beta, \alpha)$ of (17C8) in some way. In dimension 34, the stratum is also centred, resulting in an equation

$$1 - \tfrac{1}{5}\lambda + \tfrac{1}{5}\mu = 0$$

for some λ, μ, whose solution can only be $\lambda = 20$, $\mu = 15$; in other words, $\gamma_1 = \beta$, $\gamma_3 = -\beta$. Now the distance of the stratum from o cannot be the same as that for the facet. Indeed, since the two realizations in $\mathbb{E}^{51}$ have the same symmetry group $\mathbf{G}$, say, with their facets also having the same symmetry group $\mathbf{G}^f$, it must be the case that the cuts of facets and strata are swapped. The induced cosine vector is now forced on us: it must be $(1, \beta, -\alpha, -\beta, \alpha)$.

We now move on to the dual $\mathcal{Q} = \{5, 3, 5, 3\}_3 = \{\{5, 3, 5 \mid 3\}, \{3, 5, 3 :: 4\}\}$, observing as before that the relator $(0123)^6$ is imposed. Confining attention to components of X, we remark that we already know quite a lot about these realizations. There are two components of dimension 51, and hence a third (using the diagonal count) of dimension $120 - 2 \cdot 51 = 18$. By Corollary 4D10, the last must have centred facets, while the other two do not.

The layer vector is $\Lambda = (1, 68, 51, 51, 68, 1)$, and the typical cosine vector (in the present case) is of the form $\Gamma = (1, \alpha, \beta, -\beta, -\alpha, -1)$. As with $\mathcal{P}$, the diagonals of $\mathcal{Q}$ are all symmetric. The layer vectors of the facet and vertex-figure and induced cosine vectors are

$$\Lambda^f = (1, 12, 12, 15, 20), \quad \Gamma^f = (1, \alpha, \beta, -\beta, -\alpha),$$
$$\Lambda^v = (1, 20, 20, 15, 12), \quad \Gamma^v = (1, \alpha, -\alpha, -\beta, \beta),$$

which we obtain by lifting those of the central quotient with group $S_4(4)$. On the face of it, there is ambiguity about the sign of β in Γ^v; however, our analysis will resolve this.

For dimension 18, the centred facet criterion Corollary 4D10 gives

$$0 = 1 + 12\alpha + 12\beta + 15(-\beta) + 20(-\alpha) = 1 - 8\alpha - 3\beta.$$

For the dimension, Theorem 3F5 implies that

$$\tfrac{1}{18} = \tfrac{1}{120}(1 + 68\alpha^2 + 51\beta^2) \implies 12\alpha^2 + 9\beta^2 = 1,$$

after a little calculation. Eliminating β between these two equations leads to $\alpha = 0$ or $\frac{4}{19}$; clearly we must discard the latter. Thus we have $(\alpha, \beta) = (0, \frac{1}{3})$. As a check, the vertex-figure criterion Theorem 4D4 gives

$$68\alpha^2 = 1 + 20\alpha + 20(-\alpha) + 15(-\beta) + 12\beta = 1 - 3\beta = 8\alpha,$$

which holds for $(\alpha, \beta) = (0, \frac{1}{3})$. Thus the resulting cosine vector is

$$\Gamma_3 = \left(1, 0, \tfrac{1}{3}, -\tfrac{1}{3}, 0, -1\right).$$

For the other two, Λ-orthogonality with respect to Γ_3 tells us that

$$0 = 1 + 51 \cdot \tfrac{1}{3} \cdot \beta \implies \beta = -\tfrac{1}{17}.$$

The vertex-figure criterion then yields

$$60\alpha^2 = 1 - 3\beta = \tfrac{20}{17} \implies \alpha = \mp\tfrac{\sqrt{5}}{17},$$

leading to the cosine vectors

$$\Gamma_4 = \left(1, -\tfrac{\sqrt{5}}{17}, -\tfrac{1}{17}, \tfrac{1}{17}, \tfrac{\sqrt{5}}{17}, -1\right),$$
$$\Gamma_5 = \left(1, \tfrac{\sqrt{5}}{17}, -\tfrac{1}{17}, \tfrac{1}{17}, -\tfrac{\sqrt{5}}{17}, -1\right);$$

the reason for this order will appear in a moment.

We can easily check Λ-orthogonality and dimension. For instance,

$$\tfrac{1}{d} = \tfrac{1}{120}\left(1 + 68(\mp\tfrac{\sqrt{5}}{17})^2 + 51(-\tfrac{1}{17})^2\right) = \tfrac{40}{120\cdot 17} = \tfrac{1}{51},$$

as required. As a last check,

$$\eta_{\boldsymbol{f}} = \tfrac{1}{60}\left(1 - 8(\mp\tfrac{\sqrt{5}}{17}) - 3(-\tfrac{1}{17})\right) = \tfrac{1}{51\sqrt{5}}\tau^{\pm 3}.$$

The order of Γ_4 and Γ_5 was chosen so that $\mathsf{Q}_j = \mathsf{P}_j{}^{\delta}$ for $j = 4, 5$ (that is, they are geometric duals – we already have $\mathsf{Q}_2 = \mathsf{P}_2{}^{\delta}$).

In conclusion, we see that we have shown

17E4 Theorem *The cosine matrix of* $\{5,3,5,3\}/\langle\langle\,(012321)^3, (1232324)^4\,\rangle\rangle$ *is*

$$\begin{bmatrix} 1 & 1 & 1 & 1 & 1 & 1 \\ 1 & \frac{2}{17} & -\frac{3}{17} & -\frac{3}{17} & \frac{2}{17} & 1 \\ 1 & -\frac{1}{17} & \frac{1}{17} & \frac{1}{17} & -\frac{1}{17} & 1 \\ 1 & 0 & \frac{1}{3} & -\frac{1}{3} & 0 & -1 \\ 1 & -\frac{\sqrt{5}}{17} & -\frac{1}{17} & \frac{1}{17} & \frac{\sqrt{5}}{17} & -1 \\ 1 & \frac{\sqrt{5}}{17} & -\frac{1}{17} & \frac{1}{17} & -\frac{\sqrt{5}}{17} & -1 \end{bmatrix},$$

with layer and dimension vectors

$$\Lambda = (1, 68, 51, 51, 68, 1), \quad D = (1, 34, 85, 18, 51, 51).$$

Let us finally remark that exactly the same behaviour is exhibited by the strata as for the dual polytope.

Notes to Section 17E

1. Using MAGMA, Conder has confirmed Theorem 17E2, that $\{3,5,3,5\}_3$ is indeed universal with group $S_4(4) \times 2$.
2. ATLAS gives $[5,5,3,3]/\langle\langle\,(012)^4\,\rangle\rangle$ as a presentation of $S_4(4) \times 2$; the corresponding polytope $\{5,5,3,3\}/\langle\langle\,(012)^4\,\rangle\rangle$ belongs to the extended family to be discussed in the next Section 17F.

17F An Extended Family

Just as for the group $L_2(16).2$, we have an extended family of 5-polytopes with automorphism group $S_4(4) \times 2$. In fact, this family is so extensive that we shall not write it out with all the inter-connexions. (The reader may wish to construct such a table, but a very large sheet of paper is recommended.) However, what we can do is list all the polytopes that do occur, and point out that duality and vertex-figure replacement using Table 16F2 will ultimately connect them all.

There are polytopes in this extended family of types hitherto unmentioned; let us see where they arise. It is known from [91] that the geometric $\{5,3,3,5\}$ can be inscribed in $\{3,3,3,5\}$, meaning that $\mathrm{vert}\{5,3,3,5\} \subset \mathrm{vert}\{3,3,3,5\}$. Indeed, it is shown that there are self-dual compounds

$$k\{3,3,3,5\}[\mathit{26}k\{5,3,3,5\}]k\{5,3,3,3\},$$

for each $k = 1, 6, 12$.

While we shall not attempt to describe any details of these compounds here, what happens to these inscribed copies is of interest. Taking $(1, o)$ as a common vertex of both honeycombs, the vertex-figure of $\{5,3,3,5\}$ sits in layer $\mathcal{L}_3$ of $\{3,3,3,5\}$. Under the central quotient, this retracts to $\mathcal{L}_1$; it follows that the corresponding quotients have the same 136 vertices, and hence the same group $S_4(4)$. Moreover, this says that all the copies of $\{5,3,3,5\}$ inscribed in $\{3,3,3,5\}$ coincide in the quotient. The facet of the dual copy of $\{5,3,3,5\}$ sits in $\mathcal{L}_4$, which the identification forces into $\mathcal{L}_2$; thus the typical facet in the quotient collapses onto $\{5,3,3\}/8$.

If we do not pass to the central quotient, then (as we have seen) layer $\mathcal{L}_3$ of $\{3,3,3,5\}$ actually corresponds to layer $\mathcal{L}_4$ of the quotient. This obviously implies that there is an allomorphic quotient with vertex-figure in $\mathcal{L}_1$. Now Corollary 7G18 tells us how to obtain the allomorph of $\{3,3,5\}$; applying this operation to the vertex-figure of the (centrally symmetric) quotient of $\{3,3,3,5\}$ yields

$$(0,1,2,3,4) \mapsto (0, 1(234)^5 1, 23432, 434, 3)$$

in our abbreviated notation. Since

$$0 \cdot 1(234)^5 1 \sim 101 \cdot (234)^5 = 010 \cdot (234)^5 \sim 1 \cdot (234)^5,$$

which has period 5, it is clear that the vertex-figure replacement must result in $\{5,3,3,5\}/\langle\langle (0123214)^4 \rangle\rangle$.

17F1 Remark We can use this vertex-figure replacement to find a presentation of the quotient of $\{3,3,3,5\}$ with symmetry group $S_4(4) \times 2$, but even after some simplification it is far from nice.

We can apply the same idea of vertex-figure replacement to $\{5,3,3,5\}$, to obtain the family listed in Table 17F2. Strictly speaking, the two duals here are illegal, since their vertex-figures are infinite. However, once we pass to the quotient, everything is rectified. Note that there $\{5, \frac{5}{2}\}$ and $\{\frac{5}{2}, 5\}$ stand for

$\{5, 5 \mid 3\}$, while $\{5, 5\}$ stands for $\{5, 5 : 4\}$; in particular, we have three different polytopes of Schläfli type $\{5, 5, 5, 5\}$ (assuming polytopality, of course).

17F2

$$\{5, 3, 3, 5\}$$

$$\{5, 3, 5, \tfrac{5}{2}\} \quad \{\tfrac{5}{2}, 5, 3, 5\}$$

$$\{5, 5, \tfrac{5}{2}, 5\} \qquad\qquad \{5, \tfrac{5}{2}, 5, 5\}$$

$$\{5, 5, 3, \tfrac{5}{2}\}$$

$$\{\tfrac{5}{2}, 5, 5, \tfrac{5}{2}\}$$

The universal $\{5, 3, 3, 5\}/\langle\langle (0123214)^4 \rangle\rangle$ with this group is non-polytopal. Once again, we originally owed this information to MAGMA (see the notes at the end of the section); but it follows from the fact that its facet $\{5, 3, 3\}/8$ is non-polytopal. Hence its central quotient must also be non-polytopal, as an appeal to the quotient criterion of [99, Theorem 2E17] shows. This accounts for its omission from the listing below; however, observe that

$$(\{5, 3, 3, 5\}/\langle\langle (0123214)^4 \rangle\rangle)^{\varphi} = \{5, 3, 5, 5\}/\langle\langle (01213)^4, (2343)^3 \rangle\rangle,$$

which is polytopal. Conversely, the remaining cases are central quotients of various universal polytopes of the same type with this same group $S_4(4) \times 2$, and in these cases polytopality is preserved by the quotient.

At this point, recall the simplified notation of (17E1); we emphasize that it will only be employed in the present context. Since we refer to it elsewhere, we also write briefly

$$\{5, 3, 3, 5\}_4 := \{5, 3, 3, 5\}/\langle\langle (0123214)^4 \rangle\rangle.$$

The part played by $\mathcal{H}_8 := \{5, 3, 3\}/8$ in $\mathcal{L}_2$ or $\mathcal{L}_3$ is of particular interest. Of course, $\mathcal{H}_8$ is not itself polytopal, as we saw in Section 7F, and so (as we have already remarked) cannot be the facet of a polytope. As a vertex-figure it appears as the 60-cell $\mathcal{H}_2 = \{5, 3, 3\}/2$; what happens is that the 300 edges emanating from a vertex coincide in fours. There are three allomorphic pairs of polytopes with this vertex-figure, whose 3-faces are $\{3, 5\}$, $\{5, 5 \mid 3\}$ or $\{5, 5 : 4\}$. If we write $\mathcal{D}_j^v$ for the jth diagonal class of $\{5, 3, 3\}/8$, with its layer vector taken in the form $\Lambda^v = (1, 16, 16, 16, 24, 2^*)$, then we must determine to which diagonal class of the whole polytope it corresponds.

For this, it is easiest to use the known cosine matrix, and appeal to the vertex-figure criterion Theorem 4D4. We begin with the initial quotient with 136 vertices. Taking the realization P_2 in $\mathbb{E}^{85}$, we solve the vertex-figure equation

$$(-\tfrac{1}{15})^2 = \tfrac{1}{136}(1 + \mu \cdot \tfrac{1}{15} + \nu(-\tfrac{1}{15}))$$

for $\mu + \nu = 74$, yielding $\mu = 32$, $\nu = 42$. (We can also check that the equation holds for P_1 as well.) Now we know that we can interchange $\mathcal{D}_1^v$ and $\mathcal{D}_2^v$, because we can do this back in $\{5,3,3\}$ itself (giving the two allomorphic copies). Thus we must have a split $\mu = 16{+}16$, so that $\nu = 16{+}24{+}2$. Solving the vertex-figure equation for P_3 similarly leads to

$$(-\tfrac{1}{5})^2 = \tfrac{1}{136}\big(1 + \mu(-\tfrac{1}{5}) + \nu \cdot \tfrac{1}{5}\big)$$

with $\mu + \nu = 42$, and this gives $\mu = 16$, $\nu = 26$. Indeed, referring to the table of Lemma 7K7 shows that the induced cosine vector of $\{5,3,3\}/8$ as vertex-figure of the type $\{5,5,3,3\}/\langle\langle (0121)^3 \rangle\rangle = \{3,3,3,5\}^{\varphi\delta}$ must be

$$\Gamma^v = (1, \gamma_1, \gamma_2, \gamma_4, \gamma_3, \gamma_3).$$

Now we have seen that we have complete symmetry among the three diagonal classes $\mathcal{D}_1^v$, $\mathcal{D}_2^v$ and $\mathcal{D}_3^v$ in $\mathcal{H}_8 = \{5,3,3\}/8$. We first note that allomorphism of the 5-polytopes is induced by $\gamma_1 \leftrightarrow \gamma_4$, while the allomorphism $(0,\ldots,4) \mapsto (0,(123)^5,4,3,2)$ of the vertex-figure as $\mathcal{H}_2$ itself induces $\gamma_1 \leftrightarrow \gamma_2$.

If we now write the cosine vector (in the centrally symmetric case with 272 vertices) as $(1,\alpha,\beta,-\beta,-\alpha,-1)$, we conclude that the 3-faces are

- $\{5,5 \mid 3\}$, with cosine vector $(1,\pm\alpha,\beta,\mp\alpha,-\beta,-\beta)$,
- $\{3,5\}$, with cosine vector $(1,\beta,\pm\alpha,\mp\alpha,-\beta,-\beta)$,
- $\{5,5:4\}$, with cosine vector $(1,\pm\alpha,\mp\alpha,\beta,-\beta,-\beta)$.

Observe that the allomorphism of the vertex-figure $\mathcal{H}_2$ in the last case induces allorphism of the whole polytope, and so provides a bridge between allomorphs of the larger family. Therefore, while these types are obtainable by vertex-figure replacement from the others, it is not of any standard form. Instead, we must approach them from the dual $\{3,3,5,5\}/\langle\langle (234)^4 \rangle\rangle$ by vertex-figure replacement as in the right column of Table 16F2. Among their duals is $\{5,5,5,3\}_{4,3}$, which vertex-figure replacement – using Table 16F2 again – ties in with $\{3,5,3,5\}_3$.

17F3 Remark The foregoing analysis shows that there is symmetry between diagonal classes $\mathcal{D}_1$ and $\mathcal{D}_4$; however, there is no symmetry between $\mathcal{D}_2$ and $\mathcal{D}_3$, so that $\mathcal{L}_3$ cannot carry vertex-figures. Of course, all these polytopes have central quotients with group $\mathrm{S}_4(4)$.

Exactly the same phenomenon occurs in the dual class with 240 vertices; layers $\mathcal{L}_1, \mathcal{L}_2, \mathcal{L}_4$ can carry vertex-figures, while $\mathcal{L}_3$ cannot.

Because the family is so extensive, we list the polytopes in the following way. First, the members of the subfamily $\mathcal{A}$ have the 272 vertices of $\{3,5,3,5\}_4$, and can be further subdivided into $\mathcal{A}_k$ for $k = 1,2,4$, with $\mathcal{A}_k$ consisting of those with vertex-figures in layer $\mathcal{L}_k$. In exactly the same way, the members of $\mathcal{B}$ have the 240 vertices of $\{5,3,5,3\}_4$, with $\mathcal{B}$ split into $\mathcal{B}_1, \mathcal{B}_2, \mathcal{B}_4$ according to where their vertex-figures lie.

Note that allomorphism switches layers $\mathcal{L}_1$ and $\mathcal{L}_4$ in the classes $\mathcal{A}_k$ and $\mathcal{B}_k$ for $k = 1,4$, while in $\mathcal{A}_2$ and $\mathcal{B}_2$ it is induced by allomorphism of the vertex-figures. In each table, polytopes in a given row have allomorphic vertex-figures.

We list the polytopes according to their subclasses.

$\mathcal{A}_1, \mathcal{A}_4$: Table 7G8 gives ten possibilities for the vertex-figure in each case; however, the two allomorphs of $\{5,3,3,5\}_4$ are excluded as non-polytopal, and so there are only $2 \cdot 9 = 18$ polytopes here.

17F4		
	$\{3,3,3,5\}$	
	$\{3,3,5,5\}_3$	$\{5,3,5,5\}_{4,3}$
	$\{3,5,5,5\}_{3,3}$	$\{5,5,5,5\}_{4,3,3}$
	$\{3,5,3,5\}_3$	$\{5,5,3,5\}_{4,3}$
	$\{5,5,5,3\}_{3,3}$	$\{5,5,5,3\}_{4,3}$

$\mathcal{A}_2$: the six allomorphs of the vertex-figure give six polytopes in this class.

17F5			
	$\{5,5,3,3\}_3$	$\{3,5,3,3\}$	$\{5,5,3,3\}_4$

$\mathcal{B}_1, \mathcal{B}_4$: here, Table 16F2 shows that there are eight possible vertex-figures for each; hence we have a total of $2 \cdot 8 = 16$ polytopes.

17F6		
	$\{3,3,5,3\}$	$\{5,3,5,3\}_3$
	$\{3,5,5,3\}_4$	$\{5,5,5,3\}_{4,4}$
	$\{5,5,3,5\}_{3,4}$	$\{5,5,3,5\}_{4,4}$
	$\{5,5,5,5\}_{3,4,3}$	$\{5,5,5,5\}_{4,4,3}$

$\mathcal{B}_2$: up to isomorphism there are three different vertex-figures; however, since the three diagonal classes are freely interchangeable, we obtain $6 \cdot 3 = 18$ polytopes in this class.

17F7			
	$\{3,3,5,5\}_4$	$\{5,3,5,5\}_{3,4}$	$\{5,3,5,5\}_{4,4}$
	$\{3,5,5,5\}_{3,4}$	$\{5,5,5,5\}_{3,3,4}$	$\{5,5,5,5\}_{4,3,4}$
	$\{3,5,5,5\}_{4,4}$	$\{5,5,5,5\}_{3,4,4}$	$\{5,5,5,5\}_{4,4,4}$

In other words, the extended family contains $18+6+16+18 = 58$ members, which explains why we have no intention of trying to put them and their interconnexions into a single table. As we have said, the entries stand for allomorphic pairs, which will not be congruent unless $\gamma_1 = \gamma_4$ in their cosine vectors.

17F8 Remark The notation for $\{3,3,3,5\}$, $\{3,3,5,5\}_3$ and $\{3,5,5,5\}_{3,3}$ and their duals suppresses mention of the extra relators needed to specify that the automorphism group is $\mathrm{S}_4(4) \times 2$.

The polytopes in Tables 17F4, 17F5, 17F6 and 17F7 are connected in various ways. We have already mentioned allomorphism of vertex-figures. We also have the other vertex-figure replacement operations applicable to the 2- or 3-cofaces: $\varphi = \varphi_2$ leaves the 3-face fixed, while ψ (which can only be used in Table 17F7) induces φ on the 3-face. Finally, we have duality $\boldsymbol{\delta}$.

17F9 Remark There are various bridges between the subfamilies, particularly linking $\mathcal{A}_1$ and $\mathcal{A}_4$, and $\mathcal{B}_1$ and $\mathcal{B}_4$. For instance, the duals in $\mathcal{A}_1$ of $\{5,3,5,5\}_{3,4}$ and its image under ψ have vertex-figures in different layers; the same applies to the duals in $\mathcal{B}_1$ of $\{5,3,5,5\}_{4,4}$ and its image under ψ.

Those Schläfli types with at least one suffix 4, and by implication those with an inface $\{3,5,3\}$, are universal, in that no relator involves $0,\ldots,4$. Two of them have been treated already (again by implication). The final exception is derived from a universal polytope:

$$\{3,5,5,5\}_{3,3} = \{3,5,3,5\}_3{}^{\varphi} = \{\{3,5,3 :: 4\},\{5,3,5 \mid 3\}\}^{\varphi}.$$

When it comes to realizations, each combinatorial type of polytope occurs twice, the connexion being (as usual) allomorphism (corresponding to the change of sign of $\sqrt{5}$ in cosine vectors). Observe that facets of allomorphic polytopes lie in different strata, but also that different polytopes can have isomorphic facets. This corresponds to a phenomenon that we might call *facet replacement*, though generally replacing a facet by an allomorph will not yield a polytope (the new facets will not meet up). We list examples where facet replacement is valid:

$$\{3,3,5,3\} \longleftrightarrow \{3,3,5,5\}_4,$$

$$\{3,5,3,3\} \longleftrightarrow \{3,5,3,5\}_3,$$

$$\{5,3,5,3\}_3 \longleftrightarrow \{5,3,5,5\}_{3,4}.$$

17F10 Remark Among the star-polytopes of Section 7G, the only instances of facet replacement are $\{5,\frac{5}{2}\} \leftrightarrow \{\frac{5}{2},5\}$ and $\{5,3,\frac{5}{2}\} \leftrightarrow \{\frac{5}{2},3,5\}$. Further, facet replacement interchanges the two allomorphs of $\{3,5,3 :: 4\}$.

The universal polytope $\{3,3,5,5\}_4$ with 240 vertices, whose vertex-figure is $\{\{3,5\},\{5,5 : 4\}\}$ with 51 vertices, has an 18-dimensional realization with a nice symmetric projection into $\mathbb{E}^3$; unfortunately, any picture of it would be too crowded with details. This has single points at the vertices $(\pm 2,\pm 2,\pm 2)$ of a cube, the vertices all permutations of $(\pm 1,\pm 1,0)$ of the cuboctahedron are covered 16 times, and the origin o is covered 40 times. (This gives the count $8+12\cdot 16+2\cdot 20=240$ – the central section $\xi_1+\xi_2+\xi_3=0$ corresponds to layers $\mathcal{L}_1$ and $\mathcal{L}_4$, which is why we write $40=2\cdot 20$.) The vertex $(2,2,2)$ (say) is joined by edges to the points in the plane $\xi_1+\xi_2+\xi_3=2$; also occurring are (multiply covered) edges of the cuboctahedron and those joining the vertices of the cuboctahedron to o.

This suggests that the vertices of the whole realization take the following form. First, $\mathbb{E}^{18}$ is decomposed into six mutually orthogonal copies of $\mathbb{E}^3$; each of these contains a copy of the cube of side 4, giving $6\cdot 8=48$ vertices. The other $240-48=192$ vertices are certain of the $\bigl((\pm 1,\pm 1,0)^6\bigr)$, with appropriate permutations and changes of sign within the inner brackets; however, so far we have not worked out how to make choices of these consistent with the necessary inner products $\pm 12,\pm 4,0$ among them.

The mirrors of the generating reflexions are 12-dimensional; this induces a plane reflexion in the projection, typically changing sign of one of the three coordinates.

Among what we have done above, we found two 34-dimensional realizations, one of (say) $\mathcal{P} := \{5,5,3,3\}_4$ and the other of the central quotient $\mathcal{Q} := \mathcal{P}^\delta/2$. It is natural to ask whether their symmetry groups are related: in particular, is the first obtained from the second by replacing each of its matrices A by $\pm A$, mimicking the way we constructed the double covers? In fact, the answer is positive, but not in the obvious way. The group $\mathrm{S}_4(4)$ has an involutory outer automorphism, which interchanges the conjugacy class containing the generatrix of $\mathcal{Q}$ with another class (see the notes at the end of the section). It is this second class which lifts into that containing the generatrix of $\mathcal{P}$.

Notes to Section 17F

1. The fact that $\{5,3,3,5\}_4$ can only have facet $\{5,3,3\}/8$ makes it less surprising that it is non-polytopal. MAGMA shows that $|\boldsymbol{G}^f \cap \boldsymbol{G}^v| = 96$, reflecting the four copies of the tetrahedron $\{3,3\}$ which constitute the vertex-figure of $\mathcal{H}_8$.
2. The information about the 34-dimensional realizations can be extracted from the character table of $\mathrm{S}_4(4)$ in $\mathbb{ATLAS}$.

17G Another Symmetric Set

The group $\boldsymbol{G} = \mathrm{S}_4(4)$ has maximal subgroups of two other kinds than $\boldsymbol{A} = [3,3,5]^+ = (\mathrm{A}_5 \times \mathrm{A}_5) \rtimes \mathrm{C}_2$ and $\boldsymbol{B} = \mathrm{L}_2(16).2$; these are $\boldsymbol{C} = \mathrm{C}_2^6 \rtimes (\mathrm{C}_3 \times \mathrm{A}_5)$ and S_6. The index $[\boldsymbol{G} : \mathrm{S}_6] = 1360$ of the latter is a little large, and so we do not look for possible realizations; however, we shall return to this case later in the section. On the other hand, $\boldsymbol{C}$ – of order 11520 and index 85 – is of interest, even if we do not know what it looks like geometrically. We shall find the cosine matrix of $\boldsymbol{G}/\boldsymbol{C}$. Furthermore, $\boldsymbol{C}$ has a subgroup $\boldsymbol{D} = \mathrm{C}_2^6 \rtimes \mathrm{A}_5$ of index 3; the realizations of $\boldsymbol{G}/\boldsymbol{D}$ are then easily found.

Realizations of $\boldsymbol{G}/\boldsymbol{C}$

For the actions of $\boldsymbol{G}$ on $\boldsymbol{A}$ and $\boldsymbol{B}$, we have found the dimension vectors $(1,50,85)$ and $(1,34,85)$, respectively. Since $85 = 1 + 34 + 50$, it is obviously a reasonable assumption that the dimension vector of $\boldsymbol{G}/\boldsymbol{C}$ will be $(1,34,50)$; the data on $\mathrm{S}_4(4)$ in $\mathbb{ATLAS}$ confirm this. Moreover, the $\boldsymbol{A}$-cuts will be centred when $d = 34$, while the $\boldsymbol{B}$-cuts will be centred for $d = 50$.

The only realistic distribution of the 85 points into $\boldsymbol{A}$-cuts is $85 = 25 + 60$; the first corresponds to $\{5,3,3\}/24$, and – while there is some ambiguity – the second corresponds to $\{3,3,5\}/2$. (As we saw in Section 7K, one realization of $\{5,3,3\}/10$ is actually the geometric dual of the 16-dimensional realization of $\{3,3,5\}/2$. However, from Proposition 7K15 we see that the cosine vectors are much the same.)

In a similar way, for $\boldsymbol{B}$-cuts the distribution is $85 = 17+68$; the case 17 must occur, and only once since the realization with 17 points (of the three polytopes with vertex-figure $\{5,5:4\}$) is unique. When $d = 50$, the cut is centred; thus we only have the (reduced) cosine vector $(1, -\frac{1}{16})$ to play with, and this therefore implies that our cosine matrix must have an entry $-\frac{1}{16}$. Looking at the corresponding column of the dimension equation of Theorem 3J6 yields

$$1 + 34\alpha + 50(-\tfrac{1}{16}) = 0 \implies \alpha = \tfrac{1}{16};$$

the corresponding entry of the layer vector is thus given by

$$\tfrac{1}{85}\big(1 + 34(\tfrac{1}{16})^2 + 50(-\tfrac{1}{16})^2\big) = \tfrac{1}{85} \cdot \tfrac{340}{256} = \tfrac{1}{64}.$$

Our conclusion, therefore, is that the layer vector must be $\Lambda = (1, 64, 20)$, and then very little work is required to show

17G1 Theorem *The cosine matrix of* $\boldsymbol{G}/\boldsymbol{C}$ *is*

$$\begin{bmatrix} 1 & 1 & 1 \\ 1 & \frac{1}{16} & -\frac{1}{4} \\ 1 & -\frac{1}{16} & \frac{3}{20} \end{bmatrix},$$

with layer and dimension vectors

$$\Lambda = (1, 64, 20), \quad D = (1, 34, 50).$$

We note that it is straightforward to calculate that the dimensions are indeed what we said they had to be, and that these cosine vectors are Λ-orthogonal.

Now recall from Section 7K that there is a further quotient $\{5,3,3\}/24$ of $\{5,3,3\}/8$; its intrinsic layer vector is $(1,16,8)$. As we saw in Proposition 7K9, Γ_1 is also the cosine vector of the 16-dimensional pure realization of $\{5,3,3\}/24$, and this we know to be centred. The $\boldsymbol{A}$-cut corresponding to $\{3,3,5\}/2$ must also be centred, so that we have an equation of the form

$$1 + \lambda \cdot \tfrac{1}{16} + \mu(-\tfrac{1}{4}) = 0$$

with λ, μ complementary sums from $12, 12, 15, 20$. It is clear that 15 has to contribute to μ, leading to $(1, \lambda, \mu) = (1, 44, 15)$. From Γ_2, we then obtain the corresponding values of $\eta_{\boldsymbol{f}}$, namely,

$$\begin{aligned} \eta_1 &= \tfrac{1}{25}\big(1 + 16(-\tfrac{1}{16}) + 8 \cdot \tfrac{3}{20}\big) = \tfrac{6}{125}, \\ \eta_2 &= \tfrac{1}{60}\big(1 + 44(-\tfrac{1}{16}) + 15 \cdot \tfrac{3}{20}\big) = \tfrac{1}{120}. \end{aligned}$$

As was to be expected, both values are positive.

The case for $\boldsymbol{B}$ is treated in the same way, with Γ_2 now yielding the centred cuts. (By the way, there is no significance in the fact that $\frac{3}{20}$ occurs in the cosine matrix of $\{5,3,5\}/\langle\langle(01213)^4\rangle\rangle$.) We must now solve

$$1 + \lambda(-\tfrac{1}{16}) + \mu \cdot \tfrac{3}{20} = 0,$$

with λ, μ complementary sums from $20, 20, 15, 12$. Since 15 must contribute to μ, we quickly find that $(1, \lambda, \mu) = (1, 52, 15)$. From Γ_1, the corresponding values of η_f are

$$\begin{aligned}\eta_1 &= \tfrac{1}{17}\bigl(1 + 16 \cdot \tfrac{1}{16}\bigr) = \tfrac{2}{17},\\ \eta_2 &= \tfrac{1}{68}\bigl(1 + 52 \cdot \tfrac{1}{16} + 15(-\tfrac{1}{4})\bigr) = \tfrac{1}{136};\end{aligned}$$

again, both are positive.

There are consequences for the previous realizations.

17G2 Theorem *There are realizations of cuts of the following types:*

(a) ***C**-cuts of **G**/**A** with* $d = 50$ *and* $\eta = \frac{6}{125}, \frac{1}{120}$;
(b) ***C**-cuts of **G**/**B** with* $d = 34$ *and* $\eta = \frac{2}{17}, \frac{1}{136}$.

While we cannot say much about these cuts we can at least find their sizes. If these are, say, n_1, n_2, then – bearing in mind that η is the square of the distance to the centroid of the cut, and that the centroid of all the points is o – we must have $n_1\sqrt{\eta_1} = n_2\sqrt{\eta_2}$. Since $n_1 + n_2 = n$, the total number of points, we then calculate

$$\begin{aligned}\boldsymbol{A}: &\qquad \sqrt{\eta_1} : \sqrt{\eta_2} = 12 : 5 \implies (n_1, n_2) = 136(\tfrac{5}{17}, \tfrac{12}{17}) = (40, 96),\\ \boldsymbol{B}: &\qquad \sqrt{\eta_1} : \sqrt{\eta_2} = 4 : 1 \implies (n_1, n_2) = 120(\tfrac{1}{5}, \tfrac{4}{5}) = (24, 96).\end{aligned}$$

We conclude that $\boldsymbol{C}$ has transitive actions on 24, 40 and 96 points; it is notable that the last turns up in both $\boldsymbol{G}/\boldsymbol{A}$ and $\boldsymbol{G}/\boldsymbol{B}$.

Realizations of $\boldsymbol{G}/\boldsymbol{D}$

The symmetric set $\boldsymbol{G}/\boldsymbol{C}$ has a treble cover with 255 points; this is $\boldsymbol{G}/\boldsymbol{D}$ with $\boldsymbol{D} = \mathrm{C}_2^6 \rtimes \mathrm{A}_5$. Note that $\boldsymbol{D}$ is the group of the regular 4-polytope

$$\mathbf{2}^{\{3,5:5\}} = \{4, 3, 5\}/\langle\!\langle\, (123)^5 \,\rangle\!\rangle,$$

and its Petrial

$$\mathbf{2}^{\{5,5:3\}} = \{4, 5, 5\}/\langle\!\langle\, (0121)^4, (123)^3 \,\rangle\!\rangle.$$

Both are universal, since the vertex-figures are neighbourly (see Theorem 5C6), and each has 64 vertices. It is straightforward to show that the new layer vector is $\Lambda = (1, 64, 64, 64, 60, 2^*)$, and that the pure cosine vectors of the Λ-orthogonal complement of $\boldsymbol{G}/\boldsymbol{C}$ in $\boldsymbol{G}/\boldsymbol{D}$ are of the form

$$(1, \alpha, \beta, \gamma, 0, -\tfrac{1}{2}), \quad \text{with } \alpha + \beta + \gamma = 0,\ \alpha^2 + \beta^2 + \gamma^2 = \tfrac{3}{128};$$

the common dimension is $d = 85$. We see a familiar pattern here.

The cut vectors of $\boldsymbol{G}/\boldsymbol{D}$ are $(60, 60, 60, 75)$ for $\boldsymbol{A}$-cuts and $(68, 68, 68, 51)$ for $\boldsymbol{B}$-cuts. Those corresponding to the last entries have the vertices of $\{5, 3, 3\}/8$ and $\{3, 5, 5\}/\langle\!\langle\, (123)^4 \,\rangle\!\rangle$, respectively, and are centred. The cuts with 60 and 68 points each have induced cosine vectors $(1, \alpha, \beta, 0, \gamma)$ with cyclic permutation of (α, β, γ) (the previous conventions for the layer vectors are assumed), from which we find corresponding values

$$\eta_{\boldsymbol{A}} = \tfrac{1}{60}\big(1 + 12\alpha + 12\beta + 20\gamma\big) = \tfrac{1}{60}(1 + 8\gamma),$$
$$\eta_{\boldsymbol{B}} = \tfrac{1}{68}\big(1 + 20\alpha + 20\beta + 12\gamma\big) = \tfrac{1}{68}(1 - 8\gamma);$$

since $|\gamma| \leqslant \frac{1}{8}$, these values are non-negative, as they must be. Moreover, they imply that both $\boldsymbol{G}/\boldsymbol{A}$ and $\boldsymbol{G}/\boldsymbol{B}$ have pure 85-dimensional realizations, and so we have come full circle.

Observe from Proposition 3A8 that we will have $\boldsymbol{D}$-cuts of $\boldsymbol{G}/\boldsymbol{A}$ with cut vector $(32, 32, 32, 40)$, and $\boldsymbol{D}$-cuts of $\boldsymbol{G}/\boldsymbol{B}$ with cut vector $(32, 32, 32, 24)$. In the double covers, with group $\mathrm{S}_4(4) \times 2$, the cut vectors are also doubled; the cuts do not give rise to duals, since they are centred.

The subgroup S_6

As we said earlier, we also look at the part played by S_6. What we find is

17G3 Proposition *The polytope* $\{3, 3, 3, 5\}/\langle\!\langle\,(0(1234)^3)^5\,\rangle\!\rangle$ *and its double cover have inscribed regular 5-simplices* $\{3^4\}$.

Proof. We begin by looking at the original honeycomb $\mathsf{P} = \{3, 3, 3, 5\}$. A typical vertex in $\mathcal{L}_2$ is $v = \big(\tau^3\sqrt{5}, 2\tau^{5/2}(1, 1, 0, 0)\big)$. The mirror R of the reflexion that swaps $(1, o)$ and v has normal $(\tau^{1/2}, 1, 1, 0, 0)$, and so meets $\mathcal{L}_2$ in the 12 vertices of the 120-cell in its second layer $\mathcal{L}_2(\{5, 3, 3\})$. (These vertices are $(\tau^3\sqrt{5}, \tau^{5/2}h)$, with $h = (\tau^2, 1, 0, \tau^{-2})$, $(2, \tau, \tau^{-1}, 1)$, and those obtained from them under the symmetries that preserve v.)

We now pass to the quotients in the $\boldsymbol{A}$-family. With complete symmetry among its layers $\mathcal{L}_1, \mathcal{L}_2, \mathcal{L}_3$, it follows that $\mathcal{H}_8 = \{5, 3, 3\}/8$ has inscribed regular 4-simplices with edges in each diagonal class $\mathcal{D}_1, \mathcal{D}_2, \mathcal{D}_3$. (It is worth remarking that, under the quotient map, the type B simplices of the compound of (7J19) through a given vertex are identified in pairs and become regular.) Taking one of these 4-simplices with edges in $\mathcal{D}_2$ together with $\mathcal{L}_0$ thus yields a regular 5-simplex inscribed in $\{3, 3, 3, 5\}/\langle\!\langle\,(0(1234)^3)^5\,\rangle\!\rangle$ or its double cover. □

17G4 Remark To find a 4-simplex inscribed in $\mathcal{H}_8$ with edges in $\mathcal{D}_2$, it is more perspicuous to work with the coordinates for vert$\{5, 3, 3\}$ in Proposition 7F7. We have $(4, 0, 0, 0)$, the three points obtained from $(\tau\sqrt{5}, -\tau, -\tau^{-2}, -\tau^{-2})$ by permuting the last three coordinates cyclically, and $(-\sigma^{\ddagger}, \tau, \tau, \tau)$. The diagonals between each of the first four points lie in $\mathcal{D}_2$, while that from the last point to the first is in $\mathcal{D}_5$, with the remaining three edges lying in $\mathcal{D}_9$. Since $\mathcal{D}_2 \sim \mathcal{D}_5 \sim \mathcal{D}_9$ under the quotient, we have the required regular 4-simplex.

17G5 Remark Notice that the inscribed regular 5-simplices share their edges and trigonal faces with $\{3, 5, 3, 3\}/\langle\!\langle\,(0121213)^4\,\rangle\!\rangle$ and its central quotient. The existence of these inscribed 5-simplices with edges in $\mathcal{D}_2$ immediately implies that $\gamma_2 \geqslant -\frac{1}{5}$ for any of their realizations. In fact, $\gamma_2 = -\frac{1}{5}$ only for the 34-dimensional realizations of the double covers; see Theorem 17E3.

17G6 Remark Clearly, there are compounds of 5-simplices in these polytopes; there are 1360 in those with group $S_4(4)$, and 2720 in the double covers. There are no subcompounds. Taking the first, $(136, 6) = 2$ implies that any compound must cover the vertices $3k$ times for some k, leading to at least 68 simplices; with common subgroup S_6 this would make the group of the compound have order at least $68 \cdot 720 = 6 \cdot 8160$. However, $S_4(4)$ has no such proper subgroup.

While there are S_6-cuts of polytopes in the $\boldsymbol{B}$-family, none of them is a 5-simplex. Indeed, generically there are not even inscribed 4-simplices with the same edge-lengths. We do have tetrahedra inscribed in dodecahedra and hence in layers $\mathcal{L}_1$ and $\mathcal{L}_4$ (in the double covers), but their diagonal class $\mathcal{D}_3$ does not allow them to form equilateral 4-simplices either in the layers or in the whole polytopes. Of course, the 3-faces of $\{3, 3, 5, 5\}/\langle\langle (234)^4 \rangle\rangle$ are regular tetrahedra, but these do not lie in any layer.

17H Other Combinations

Another possible amalgamation of polytopes from the family of Table 16F2 is

$$\begin{aligned}\{3, 5, 3, 5\}_4 &= \{\{3, 5, 3 :: 4\}, \{5, 3, 5 :: 4\}\} \\ &= \{3, 5, 3, 5\}/\langle\langle (0121213)^4, (12324)^4 \rangle\rangle,\end{aligned}$$

in the abbreviated notation introduced in Section 17F. Somewhat surprisingly, perhaps, this is finite and polytopal, but with a group unrelated to $S_4(4)$. In fact, the polytope has 256 vertices; equally surprising is that it is not centrally symmetric, as the number 256 might lead one to think (see the notes at the end of the section). Its cosine matrix is

$$\begin{bmatrix} 1 & 1 & 1 & 1 & 1 \\ 1 & -\frac{3}{17} & \frac{1}{17} & \frac{1}{17} & \frac{1}{17} \\ 1 & \frac{1}{17} & -\frac{3}{17} & \frac{1}{17} & \frac{1}{17} \\ 1 & \frac{1}{17} & \frac{1}{17} & -\frac{3}{17} & \frac{1}{17} \\ 1 & \frac{1}{17} & \frac{1}{17} & \frac{1}{17} & -\frac{13}{51} \end{bmatrix},$$

with layer and dimension vectors $\Lambda = D = (1, 68, 68, 68, 51)$; we shall not go into the details, but again see the notes at the end of the section.

We then have

$$\{3, 5, 3, 5\}_4{}^{\varphi} = \{3, 5, 5, 5\}_{4,3},$$

which accounts for another omission from the lists of Section 17F. If we apply the operation $\psi_2 = (0, 121, 2, 323, 4)$ (in our abbreviated notation) to it, we merely obtain its dual, and so there is no extended family.

17H1 Remark The possibility of other nice finite quotients of hyperbolic honeycombs could also be considered. The honeycomb $\{4, 3, 3, 5\}$ can also be inscribed in $\{3,3,3,5\}$, but the quotient of the latter is incompatible with the inscription. Branch contractions also lead to

$$\{4,3,3,5\} \mapsto \{4,3,5\} \mapsto \{4,5\},$$

imposing the quotient mapping $\{4,5\} \mapsto \{4,5:5\}$ takes these to corresponding quotients of the hyperbolic honeycombs. We thus ask whether these quotients of higher rank are polytopal and, if so, finite. However, Dimitri Leemans [69] has confirmed using MAGMA that we do not actually obtain polytopes; the groups are just $[3,5]$ and $[3,3,5]$, respectively, with the initial generator forced to be the identity (see the notes at the end of the section).

We have already seen that the quotient $\{5,3,3,5\}/\langle\langle (0123214)^4 \rangle\rangle$ is not polytopal. In this context, Michael Hartley [61] has shown that the universal polytope $\{\{5,3,3:15\},\{3,3,5:15\}\}$ is infinite.

Notes to Section 17H

1. Once again, we thank Conder's applications of MAGMA for additional information in this section.
2. For $\{3,5,3,5\}_4$, since $\ell_1 = 68$, the layer vector can only be $\Lambda = (1,68,68,68,51)$; moreover, $\mathcal{L}_4$ must be symmetric. In turn, $\mathcal{L}_4$ must then have induced cosine vector $\Gamma = (1,\gamma_1,\gamma_2,\gamma_3,\gamma_4)$ corresponding to its layer vector $(1,16,16,16,2^*)$. Applying Theorem 4D4 and eliminating $\gamma_1,\gamma_2,\gamma_3$ using the layer equation leads to $\gamma_4 = -\frac{13}{51}$ or $\frac{1}{17}$. Symmetry will force $\gamma_1 = \gamma_2 = \gamma_3$, and the two solutions lead to $d_4 = 51$ or 204, respectively. The latter is impossible (no such realization can have dimension greater than 51), and a little further work then yields the given cosine matrix.
3. Leemans sent us the information given above after seeing an earlier version of the monograph.

Afterword

What we have written about in this book is by no means the end of the story. With the exception of discussion of realization domains, we have not considered regular polytopes or apeirotopes of corank greater than 1. For corank 2, the first open classes consist of the 4-dimensional regular apeirohedra and the 5-dimensional regular polyhedra. We have briefly looked at the former problem; it seems somewhat hard, but possibly by no means entirely intractible. (One curiosity that we cannot forbear to mention is that two families are derived from affine unitary groups, with involutory unitary reflexions in one corresponding to automorphisms induced by complex conjugation in the other, and conversely.) We have not investigated the latter case, where the main stumbling block is that the symmetry groups need not be particularly closely related to hyperplane reflexion groups.

In a rather different direction, there are the regular honeycombs and their relatives in hyperbolic spaces. Those with finite or euclidean facets and vertex-figures are well-known. From a geometric view-point, however, there seems to be no good reason to exclude honeycombs whose facets (or even faces of lower rank) are themselves hyperbolic, so long as their vertices are finite or lie on the absolute; this is a subject which, perhaps, deserves further attention.

Compounds of regular hyperbolic honeycombs have also been considered in the past, although some things that have been claimed about them (fortunately mostly unpublished) are not true. While we have results of our own in this area (we have mentioned [91] in Section 17H), we felt that the present monograph was already long enough, and so we have resiled from presenting them here. However, another basic reason for not discussing this topic is that it does not fit into our context – there are no analogues of the notions which go into realization theory.

In conclusion, then, while we have attempted to give a full account of the topics that we do cover, it must be recognized that the wider theory of geometric regular polytopes remains largely unexplored.

Bibliography

[1] A. D. Aleksandrov, A theorem on convex polyhedra (Russian). *Trudy Fiz.-Mat. Inst. Im Steklova* **4** (1933), 87.

[2] J. L. Arocha, J. Bracho and L. Montejano, Regular projective polyhedra with planar faces, I. *Aequ. Math.* **59** (2000), 55–73.

[3] L. Bieberbach, Über die Bewegungsgruppen der euklidischen Räume: erste Abhandlung. *Math. Ann.* **70** (1910), 297–336.

[4] G. Birkhoff and S. Mac Lane, *A Survey of Modern Algebra* (4th edition). Macmillan (New York, 1977).

[5] W. Bosma, J. Cannon and C. Playoust, The Magma algebra system, I: the user language. *J. Symbolic Comput.* **24** (1997), 235–265.

[6] J. Bracho, Regular projective polyhedra with planar faces, II. *Aequ. Math.* **59** (2000), 160–176.

[7] R. Brauer and H. S. M. Coxeter, A generalization of theorems of Schönhardt and Mehmke on polytopes. *Trans. R. Soc. Can.* **34** (1940), 29–34.

[8] C. J. Brianchon, Théorème nouveau sur les polyèdres. *J. Ec. (R.) Polytech.* **15** (1837), 317–319.

[9] A.-L. Cauchy, Recherches sur les polyèdres. *J. Ec. Polytech.* **16** (1813), 68–86. (In *Ostwalds Klassiker der exakten Wissenschaften* **151** (ed. R. Haussner), Engelmann (Leipzig, 1906), 49–72.)

[10] A.-L. Cauchy, Sur les polygones et polyèdres. *J. Ec. Polytech.* **16** (1813), 87–99. (In *Oeuvres T. 1* (Paris, 1905), 26–38.)

[11] C. J. Colbourn and A. I. Weiss, A census of regular 3-polystroma arising from honeycombs. *Discrete Math.* **50** (1984), 29–36.

[12] M. D. E. Conder, Private communications (2012–14).

[13] M. D. E. Conder and C. Maclachlan, Compact hyperbolic 4-manifolds of small volume. *Proc. Am. Math. Soc.* **133** (2005), 2469–2476.

[14] R. Connelly, Rigidity. In *Handbook of Convex Geometry* (eds. P. M. Gruber and J. M. Wills), North-Holland ((Amsterdam–London–New York–Tokyo, 1993).

[15] R. Connelly, A. I. Weiss and W. Whiteley (editors), *Rigidity and Symmetry, Fields Institute Communications* **70**, Springer (New York, etc., 2014).

[16] J. H. Conway, R. T. Curtis, S. P. Norton, R. A. Parker and R. A. Wilson, *Atlas of Finite Groups* (cited in text as $\mathbb{ATLAS}$). Clarendon Press (Oxford, 1985).

[17] H. S. M. Coxeter, The densities of the regular polytopes. *Proc. Cambridge Philos. Soc.* **27** (1930–31), 201–211. (In *Kaleidoscopes: Selected Writings of H. S. M. Coxeter* (eds. F. A. Sherk, P. McMullen, A. C. Thompson and A. I. Weiss), Wiley-Interscience (New York,etc., 1995), 35–45.)

[18] H. S. M. Coxeter, Groups whose fundamental regions are simplexes. *J. London Math. Soc.* **6** (1931), 132–136. (In *Kaleidoscopes: Selected Writings of H. S. M. Coxeter* (eds. F. A. Sherk, P. McMullen, A. C. Thompson and A. I. Weiss), Wiley-Interscience (New York, etc., 1995), 133–137.)

[19] H. S. M. Coxeter, The densities of the regular polytopes, II. *Proc. Cambridge Philos. Soc.* **28** (1931–32), 509–521. (In *Kaleidoscopes: Selected Writings of H. S. M. Coxeter* (eds. F. A. Sherk, P. McMullen, A. C. Thompson and A. I. Weiss), Wiley-Interscience (New York, etc., 1995), 47–59.)

[20] H. S. M. Coxeter, The densities of the regular polytopes, III. *Proc. Cambridge Philos. Soc.* **29** (1932–33), 1–22. (In *Kaleidoscopes: Selected Writings of H. S. M. Coxeter* (eds. F. A. Sherk, P. McMullen, A. C. Thompson and A. I. Weiss), Wiley-Interscience (New York, etc., 1995), 61–82.)

[21] H. S. M. Coxeter, Discrete groups generated by reflections. *Annals Math.* **35** (1934), 588–621. (In *Kaleidoscopes: Selected Writings of H. S. M. Coxeter* (eds. F. A. Sherk, P. McMullen, A. C. Thompson and A. I. Weiss), Wiley-Interscience (New York, etc., 1995), 145–178.)

[22] H. S. M. Coxeter, The complete enumeration of finite groups of the form $R_i^2 = (R_iR_j)^{k_{ij}} = 1$. *J. London Math. Soc.* **10** (1935), 21–35. (In *Kaleidoscopes: Selected Writings of H. S. M. Coxeter* (eds. F. A. Sherk, P. McMullen, A. C. Thompson and A. I. Weiss), Wiley-Interscience (New York, etc., 1995), 139–143.)

[23] H. S. M. Coxeter, Wythoff's construction for uniform polytopes. *Proc. London Math. Soc.* (2) **38** (1935), 327–339.

[24] H. S. M. Coxeter, Regular skew polyhedra in 3 and 4 dimensions and their topological analogues. *Proc. London Math. Soc.* (2) **43** (1937), 33–62. (Reprinted with amendments in *Twelve Geometric Essays*, Southern Illinois University Press (Carbondale, 1968), 76–105.)

[25] H. S. M. Coxeter, The abstract groups $G^{m,n,p}$. *Trans. Am. Math. Soc.* **45** (1939), 73–150.

[26] H. S. M. Coxeter, Twisted honeycombs. *Regional Conferences in Mathematics* **4**, *Am. Math. Soc.* (Providence, R.I., 1970).

[27] H. S. M. Coxeter, *Regular Polytopes* (3rd edition). Dover (New York, 1973).

[28] H. S. M. Coxeter, Regular and semi-regular polytopes, III. *Math. Z.* **200** (1988), 3–45. (Reprinted in *Kaleidoscopes: Selected Writings of H. S. M. Coxeter*, eds. F.A. Sherk, P. McMullen, A.C. Thompson and A.I. Weiss, Wiley-Interscience (New York, etc., 1995), 313–355.)

[29] H. S. M. Coxeter, Star polytopes and the Schläfli function $f(\alpha, \beta, \gamma)$. *Elem. Math.* **44** (2) (1989), 25–36. (In *Kaleidoscopes: Selected Writings of H. S. M. Coxeter* (eds. F. A. Sherk, P. McMullen, A. C. Thompson and A. I. Weiss), Wiley-Interscience (New York, etc., 1995), 121–132.)

[30] H. S. M. Coxeter, *Regular Complex Polytopes* (2nd edition). Cambridge University Press (Cambridge, 1991).

[31] H. S. M. Coxeter, The evolution of Coxeter-Dynkin diagrams. *Nieuw Archief voor Wiskunde* **9** (1991), 223–248. Reprinted in *Kaleidoscopes* (eds. F. A. Sherk, P. McMullen, A. C. Thompson and A. I. Weiss) Wiley-Interscience (New York, etc., 1995), and in substantially revised form in *Polytopes: Abstract, Convex and Computational* (eds. T. Bisztriczky, P. McMullen, R. Schneider and A. Ivić Weiss), NATO ASI Series C **440**, Kluwer (Dordrecht etc., 1994), 21–42.

[32] H. S. M. Coxeter, M. S. Longuet-Higgins and J. C. P. Miller, Uniform polyhedra. *Phil. Trans. R. Soc.* A **246** (1954), 401–450.

[33] H. S. M. Coxeter and W. O. J. Moser, *Generators and Relations for Discrete Groups* (4th edition). Springer (1980).

[34] C. W. Curtis and I. Reiner, *Representation Theory of Finite Groups and Associative Algebras.* Wiley-Interscience (New York-London, 1962).

[35] A. M. Cutler, Regular polyhedra of index two, II. *Beitr. Algebra Geom.* **52** (2011), 357–387.

[36] A. M. Cutler and E. Schulte, Regular polyhedra of index two, I. *Beitr. Algebra Geom.* **52** (2011), 133–161.

[37] L. Danzer, B. Grünbaum and V. L. Klee, Helly's theorem and its relatives. In *Convexity, Proc. Symposia Pure Math.*, vol VII (Am. Math. Soc., Providence, RI, 1963), 101–180.

[38] L. Danzer and E. Schulte, Reguläre Inzidenzkomplexe, I. *Geom. Dedicata* **13** (1982), 295–308.

[39] M. Davis, A hyperbolic 4-manifold. *Proc. Am. Math. Soc.* **93** (1985), 325–328.

[40] H. E. Debrunner, Dissecting orthoschemes into orthoschemes. *Geom. Dedicata* **33** (1990), 123–152.

[41] A. W. M. Dress, A combinatorial theory of Grünbaum's new regular polyhedra, I: Grünbaum's new regular polyhedra and their automorphism group. *Aequ. Math.* **23** (1981), 252–265.

[42] A. W. M. Dress, A combinatorial theory of Grünbaum's new regular polyhedra, II: complete enumeration. *Aequ. Math.* **29** (1985), 222–243.

[43] P. Du Val, *Homographies, Quaternions and Rotations.* Oxford University Press (Oxford, 1964).

[44] E. B. Dynkin, Classification of simple Lie groups (Russian). *Mat. Sbornik* **18** (60) (1940), 347–352.

[45] V. A. Efremovič and Ju. S. Il'jašenko, Regular polygons in E^n (Russian). *Vest. Mosk. Univ. Mat. Mekh.* **5** (1962), 18–24.

[46] E. L. Elte, *The Semiregular Polytopes of the Hyperspaces.* Hoitsema (Groningen), 1912.

[47] L. Fejes Tóth, *Reguläre Figuren.* Akadémiai Kiadó (Budapest, 1965). (English translation: *Regular Figures*, Pergamon Press (Oxford etc., 1964).)

[48] R. Frucht, J. E. Graver and M. E. Watkins, The groups of the generalized Petersen graphs. *Proc. Cambridge Philos. Soc.* **70** (1971), 211–218.

[49] W. Fulton and J. Harris, *Representation Theory* (*Graduate Texts in Mathematics*). Springer (New York, 1991).

[50] The GAP Group, *GAP – Groups, Algorithms and Programming, Version 4.4* (2011). `http://www.gap-system.org`.

[51] C. W. L. Garner, Compound honeycombs in hyperbolic space. *Proc. R. Soc. London A* **316** (1970), 441–448.

[52] P. A. Gordan, Ueber endliche Gruppen linearer Transformationen einer Veränderlichen. *Math. Ann.* **12** (1877), 23–46.

[53] T. Gosset, On the regular and semi-regular figures in space of n dimensions. *Messenger of Mathematics* **29** (1900), 43–48.

[54] E. Goursat, Sur les substitutions orthogonales et les divisions régulières de l'espace. *Ann. Sci. Ec. Norm. Super.* (3) **6** (1889), 9–102.

[55] J. P. Gram, Om Rumvinklerne i et Polyeder. *Tidsskr. Math. (Copenhagen)* (3) **4** (1874), 161–163.

[56] B. Grünbaum, *Convex Polytopes.* Wiley-Interscience (New York, 1967). (2nd edition prepared by V. Kaibel, V. L. Klee and G. M. Ziegler, Springer (2003).)

[57] B. Grünbaum, Regularity of graphs, complexes and designs. Colloque CNRS *Problèmes Combinatoire et Théorie des Graphes* (Orsay, 1976).

[58] B. Grünbaum, Regular polyhedra – old and new. *Aequ. Math.* **16** (1977), 1–20.

[59] H. Hadwiger, Über ausgezeichnete Vektorsterne und reguläre Polytope. *Comm. Math. Helvet.* **13** (1940), 90–107.

[60] M. I. Hartley, Polytopes of finite type. *Discrete Math.* **218** (2000), 97–108.

[61] M. I. Hartley, An exploration of locally projective polytopes. *Combinatorica* **28** (2008), 299–314.

[62] E. Hess, Über die regulären Polytope höherer Art. *Sitz. Ges. Bef. Ges. Naturwiss. Marburg* (1885), 31–57.

[63] J. E. Humphreys, *Reflection Groups and Coxeter Groups* (*Cambridge Studies in Advanced Mathematics* **29**). Cambridge University Press (Cambridge, 1990).

[64] N. Jacobson, *Structure of Rings* (revised edition). *Am. Math. Soc. Colloquium Publications* **37**, Am. Math. Soc. (Providence, 1964).

[65] N. W. Johnson, *Uniform Polytopes* (unpublished manuscript, 1991).

[66] N. W. Johnson, *Geometry and Transformations.* Cambridge University Press (Cambridge etc., 2018).

[67] F. Ladisch, Realizations of abstract regular polytopes from a representation theoretic view. *Aequ. Math.* **90** (2016), 1169–1193.

[68] F. Ladisch, Cosine vectors of the 120-cell (private communication, 2017).

[69] D. Leemans, Private communication (2018).

[70] P. McMullen, Combinatorially regular polytopes. *Mathematika* **14** (1967), 142–150.

[71] P. McMullen, Regular star-polytopes, and a theorem of Hess. *Proc. London Math. Soc. (3)* **18** (1968), 577–596.

[72] P. McMullen, Polytopes with centrally symmetric facets. *Israel J. Math* **23** (1976), 337–338.

[73] P. McMullen, Transforms, diagrams and representations. *Contributions to Geometry* (ed. J. Tölke and J.M. Wills), Birkhäuser (Basel, 1979), 92–130.

[74] P. McMullen, Angle-sum relations for polyhedral sets. *Mathematika* **33** (1986), 173–188.

[75] P. McMullen, Realizations of regular polytopes. *Aequ. Math.* **37** (1989), 38–56.

[76] P. McMullen, Regular polyhedra related to projective linear groups. *Discrete Math.* **91** (1991), 161–170.

[77] P. McMullen, The order of a finite Coxeter group. *Elem. Math.* **46** (1991), 121–130.

[78] P. McMullen, The regular polyhedra of type $\{p, 3\}$ with $2p$ vertices. *Geom. Ded.* **43** (1992), 285–289.

[79] P. McMullen, Locally projective regular polytopes. *J. Combin. Theory, Ser. A* **65** (1994), 1–10.

[80] P. McMullen, Realizations of regular apeirotopes. *Aequ. Math.* **47** (1994), 223–239.

[81] P. McMullen, The groups of the regular star-polytopes. *Can. J. Math.* (2) **50** (1998), 426–448.

[82] P. McMullen, Regular polytopes of full rank. *Discrete Comput. Geom.* **32** (2004), 1–35.

[83] P. McMullen, Four-dimensional regular polyhedra. *Discrete Comput. Geom.* **38** (2007), 355–387.

[84] P. McMullen, Regular apeirotopes of dimension and rank 4. *Discrete Comput. Geom.* **42** (2009), 224–260.

[85] P. McMullen, Realizations of regular polytopes, III. *Aequ. Math.* **82** (2011), 35–63.

[86] P. McMullen, Regular polytopes of nearly full rank. *Discrete Comput. Geom.* **46** (2011), 660–703.

[87] P. McMullen, Regular polytopes of nearly full rank: addendum. *Discrete Comput. Geom.* **49** (2013), 703–705.

[88] P. McMullen, Realizations of regular polytopes, IV. *Aequ. Math.* **87** (2014), 1–30.

[89] P. McMullen, Rigidity of regular polytopes. In *Rigidity and Symmetry* (ed. R. Connelly, A. I. Weiss and W. Whiteley), *Fields Institute Communications* **70**, Springer (New York, etc., 2014), 253–278.

[90] P. McMullen, New regular compounds of 4-polytopes. In *New Trends in Intuitive Geometry* (*Bolyai Soc. Math. Studies* **27**, ed. G. Ambrus et al.), 307–320.

[91] P. McMullen, Compounds of regular honeycombs in $\mathbb{H}^3$ and $\mathbb{H}^4$. *Adv. Math.* (to appear).

[92] P. McMullen, Realizations of the 120-cell (submitted for publication).

[93] P. McMullen and B. R. Monson, Realizations of regular polytopes, II. *Aequ. Math.* **65** (2003), 102–112.

[94] P. McMullen and E. Schulte, Constructions for regular polytopes. *J. Combinat. Theory* A **53** (1990), 1–28.

[95] P. McMullen and E. Schulte, Regular polytopes from twisted Coxeter groups and unitary reflexion groups. *Adv. Math.* **82** (1990), 35–87.

[96] P. McMullen and E. Schulte, Locally toroidal regular polytopes of rank 4. *Comment. Math. Helv.* **67** (1992), 77–118.

[97] P. McMullen and E. Schulte, Higher toroidal regular polytopes. *Adv. Math.* **117** (1996), 17–51.

[98] P. McMullen and E. Schulte, Regular polytopes in ordinary space. *Discrete Comput. Geom.* **17** (1997), 449–478.

[99] P. McMullen and E. Schulte, *Abstract Regular Polytopes* (*Encyclopedia of Mathematics and Its Applications* **92**). Cambridge University Press (Cambridge, 2002).

[100] P. McMullen and E. Schulte, Regular and chiral polytopes in low dimensions. *The Coxeter Legacy – Reflections and Projections* (ed. C. Davis and E. W. Ellers), Fields Institute Communications **46** (2006), 87–106.

[101] B. R. Monson and E. Schulte, Modular reduction in abstract polytopes. *Can. Math. Bull.* **52** (2009), 435–450.

[102] B. R. Monson and E. Schulte, Finite polytopes have finite regular covers. *J. Algebraic Combin.* **40** (2014), 75–82.

[103] B. R. Monson and A. I. Weiss, Realizations of regular toroidal maps. *Can. J. Math.* (6) **51** (1999), 1240–1257.

[104] J. Mowbray, Personal communication (2018).

[105] T. Muir, *A Treatise on the Theory of Determinants.* Dover (New York, 1960).

[106] D. Pellicer and I. Hubard, Personal communication (2011).

[107] D. Pellicer and E. Schulte, Regular polygonal complexes in space, I. *Trans. Am. Math. Soc.* **362** (2010), 6679–6714.

[108] D. Pellicer and E. Schulte, Regular polygonal complexes in space, II. *Trans. Am. Math. Soc.* **365** (2013), 2031–2061.

[109] Plato, *Timaeus* (and *Critias*) (trans. D. Lee). Penguin Books (Harmondsworth, Middlesex, England, 1971).

[110] J. G. Ratcliffe, *Foundations of Hyperbolic Manifolds. Graduate Texts in Mathematics*, Springer-Verlag (New York-Berlin-Heidelberg, 1994).

[111] G. de B. Robinson, On the fundamental region of a group, and the family of configurations which arise therefrom. *J. London Math. Soc.* **6** (1931), 70–75.

[112] L. Schläfli, On the multiple integral $\int^n dx\, dy \ldots dz$, whose limits are $p_1 = a_1x + b_1y + \cdots + h_1z > 0, p_2 > 0, \ldots, p_n > 0$ and $x^2 + y^2 + \cdots + z^2 < 1$. *Quart. J. Math.* **2** (1858), 269–301; **3** (1860), 54–68, 97-108. (In *Ges. Math. Abhandl.* Band II (Birkhäuser Verlag, 1953), 219–270.)

[113] L. Schläfli, *Theorie der vielfachen Kontinuität. Ges. Math. Abhandl.* Band I (Birkhäuser Verlag, 1950), 167–387.

[114] R. Schneider, *Convex Bodies: the Brunn-Minkowski Theory* (2nd expanded edition) (*Encyclopedia of Mathematics and Its Applications* **151**). Cambridge University Press (Cambridge, 2014).

[115] E. Schulte and A.I. Weiss, Chiral polytopes. In *Applied Geometry and Discrete Mathematics (The Victor Klee Festschrift)* (DIMACS Series in Discrete Mathematics and Theoretical Computer Science **4**), eds. P. Gritzmann and B. Sturmfels, Am. Math. Soc. and Assoc. Computing Machinery (1991), 493–516.

[116] G. C. Shephard, Regular complex polytopes. *Proc. London Math. Soc.* (3) **2** (1952), 82–97.

[117] G. C. Shephard, Unitary groups generated by reflections. *Can. J. Math.* **5** (1953), 364–383.

[118] G. C. Shephard, Polytopes with centrally symmetric faces. *Can. J. Math.* **19** (1967), 1206–1213.

[119] G. C. Shephard, An elementary proof of Gram's theorem for convex polytopes. *Can. J. Math.* **19** (1967), 1214–1217.

[120] G. C. Shephard, Diagrams for positive bases. *J. London Math. Soc.* (2) **4** (1971), 165–175.

[121] G. C. Shephard and J. A. Todd, Finite unitary reflection groups. *Can. J. Math.* **6** (1954), 274–304.

[122] B. Simon, *Representations of Finite and Compact Groups* (*Graduate Studies in Mathematics* **10**). American Mathematical Society (Providence, 1996).

[123] D. M. Y. Sommerville, The relations connecting the angle-sum and volume of a polytope in space of n dimensions. *Proc. R. Soc. London, Ser. A* **115** (1927), 103–119.

[124] S. L. van Oss, Die regelmässigen vierdimensionalen Polytope höherer Art. *Verh. Konink. Akad. Weten. Amsterdam* (Eerste sectie) **12.1** (1915).

[125] E. B. Vinberg, *Linear Representations of Groups.* Birkhäuser (Basel, 1989).

[126] J. M. Wills, Combinatorially regular polyhedra of index 2. *Aequ. Math.* **34** (1987), 206–220.

[127] J. M. Wills, On regular polyhedra with hidden symmetries. *Results Math.* **12** (1987), 450–458.

[128] J. M. Wills, A combinatorially regular dodecahedron of genus 3. *Discrete Math.* **67** (1987), 199–204.

[129] S. E. Wilson, Operators over regular maps. *Pacific J. Math.* **81** (1979), 559–568.

[130] E. Witt, Spiegelungsgruppen und Aufzählung halbeinfacher Liescher Ringe. *Abh. Math. Sem. Hamburg* **14** (1941), 289–322.

[131] W. A. Wythoff, A relation between the polytopes of the C_{600}-family. *K. Ned. Akad. Wet. Amsterdam, Proc.* **20** (1918), 966–970.

[132] G. M. Ziegler, *Lectures on Polytopes. Graduate Texts in Mathematics* **152**, Springer (New York, 1995).

Notation Index

Author Index

Subject Index

For the convenience of the reader, several items are – in effect – listed twice.